Recent Advances in the Synthesis, Functionalization and Applications of Pyrazole-Type Compounds I

Recent Advances in the Synthesis, Functionalization and Applications of Pyrazole-Type Compounds I

Editors

Vera L. M. Silva
Artur M. S. Silva

MDPI • Basel • Beijing • Wuhan • Barcelona • Belgrade • Manchester • Tokyo • Cluj • Tianjin

Editors
Vera L. M. Silva
Chemistry
University of Aveiro
Aveiro
Portugal

Artur M. S. Silva
Chemistry
University of Aveiro
Aveiro
Portugal

Editorial Office
MDPI
St. Alban-Anlage 66
4052 Basel, Switzerland

This is a reprint of articles from the Special Issue published online in the open access journal *Molecules* (ISSN 1420-3049) (available at: www.mdpi.com/journal/molecules/special_issues/pyrazole-type_compounds).

For citation purposes, cite each article independently as indicated on the article page online and as indicated below:

LastName, A.A.; LastName, B.B.; LastName, C.C. Article Title. *Journal Name* **Year**, *Volume Number*, Page Range.

ISBN 978-3-0365-5422-8 (Hbk)
ISBN 978-3-0365-5421-1 (PDF)

© 2022 by the authors. Articles in this book are Open Access and distributed under the Creative Commons Attribution (CC BY) license, which allows users to download, copy and build upon published articles, as long as the author and publisher are properly credited, which ensures maximum dissemination and a wider impact of our publications.

The book as a whole is distributed by MDPI under the terms and conditions of the Creative Commons license CC BY-NC-ND.

Contents

About the Editors vii

Vera L. M. Silva and Artur M. S. Silva
Special Issue "Recent Advances in the Synthesis, Functionalization and Applications of Pyrazole-Type Compounds"
Reprinted from: *Molecules* **2021**, *26*, 4989, doi:10.3390/molecules26164989 1

José Elguero and Ibon Alkorta
A Computational Study of Metallacycles Formed by Pyrazolate Ligands and the Coinage Metals M = Cu(I), Ag(I) and Au(I): $(pzM)_n$ for n = 2, 3, 4, 5 and 6. Comparison with Structures Reported in the Cambridge Crystallographic Data Center (CCDC)
Reprinted from: *Molecules* **2020**, *25*, 5108, doi:10.3390/molecules25215108 5

Kiyoshi Fujisawa, Takuya Nemoto, Yui Morishima and Daniel B. Leznoff
Synthesis and Structural Characterization of a Silver(I) Pyrazolato Coordination Polymer [†]
Reprinted from: *Molecules* **2021**, *26*, 1015, doi:10.3390/molecules26041015 35

Saied M. Soliman, Hessa H. Al-Rasheed, Jörg H. Albering and Ayman El-Faham
Fe(III) Complexes Based on Mono- and Bis-pyrazolyl-*s*-triazine Ligands: Synthesis, Molecular Structure, Hirshfeld, and Antimicrobial Evaluations
Reprinted from: *Molecules* **2020**, *25*, 5750, doi:10.3390/molecules25235750 47

Lyudmila Kayukova, Anna Vologzhanina, Kaldybai Praliyev, Gulnur Dyusembaeva, Gulnur Baitursynova and Asem Uzakova et al.
Boulton-Katritzky Rearrangement of 5-Substituted Phenyl-3-[2-(morpholin-1-yl)ethyl]-1,2,4-oxadiazoles as a Synthetic Path to Spiropyrazoline Benzoates and Chloride with Antitubercular Properties
Reprinted from: *Molecules* **2021**, *26*, 967, doi:10.3390/molecules26040967 61

Nikita P. Burlutskiy and Andrei S. Potapov
Approaches to the Synthesis of Dicarboxylic Derivatives of Bis(pyrazol-1-yl)alkanes
Reprinted from: *Molecules* **2021**, *26*, 413, doi:10.3390/molecules26020413 77

Urša Štanfel, Dejan Slapšak, Uroš Grošelj, Franc Požgan, Bogdan Štefane and Jurij Svete
Synthesis of 6,7-Dihydro-1*H*,5*H*-pyrazolo[1,2-*a*]pyrazoles by Azomethine Imine-Alkyne Cycloadditions Using Immobilized Cu(II)-Catalysts
Reprinted from: *Molecules* **2021**, *26*, 400, doi:10.3390/molecules26020400 89

Yoshihide Usami, Yuya Tatsui, Hiroki Yoneyama and Shinya Harusawa
C4-Alkylamination of C4-Halo-1*H*-1-tritylpyrazoles Using Pd(dba)$_2$ or CuI
Reprinted from: *Molecules* **2020**, *25*, 4634, doi:10.3390/molecules25204634 107

Rawan Alnufaie, Hansa Raj KC, Nickolas Alsup, Jedidiah Whitt, Steven Andrew Chambers and David Gilmore et al.
Synthesis and Antimicrobial Studies of Coumarin-Substituted Pyrazole Derivativesas Potent Anti-*Staphylococcus aureus* Agents
Reprinted from: *Molecules* **2020**, *25*, 2758, doi:10.3390/molecules25122758 119

Ching-Chun Tseng, Cheng-Yen Chung, Shuo-En Tsai, Hiroyuki Takayama, Naoto Uramaru and Chin-Yu Lin et al.
Selective Synthesis and Photoluminescence Study of Pyrazolopyridopyridazine Diones and *N*-Aminopyrazolopyrrolopyridine Diones
Reprinted from: *Molecules* **2020**, *25*, 2409, doi:10.3390/molecules25102409 135

Andres Arias-Gómez, Andrés Godoy and Jaime Portilla
Functional Pyrazolo[1,5-*a*]pyrimidines: Current Approaches in Synthetic Transformations and Uses As an Antitumor Scaffold
Reprinted from: *Molecules* **2021**, *26*, 2708, doi:10.3390/molecules26092708 . . . . **151**

Xuefei Li, Yanbo Yu and Zhude Tu
Pyrazole Scaffold Synthesis, Functionalization, and Applications in Alzheimer's Disease and Parkinson's Disease Treatment (2011–2020)
Reprinted from: *Molecules* **2021**, *26*, 1202, doi:10.3390/molecules26051202 . . . . **187**

Chiara Brullo, Federica Rapetti and Olga Bruno
Pyrazolyl-Ureas as Interesting Scaffold in Medicinal Chemistry
Reprinted from: *Molecules* **2020**, *25*, 3457, doi:10.3390/molecules25153457 . . . . **225**

Pedro M. O. Gomes, Pedro M. S. Ouro, Artur M. S. Silva and Vera L. M. Silva
Styrylpyrazoles: Properties, Synthesis and Transformations
Reprinted from: *Molecules* **2020**, *25*, 5886, doi:10.3390/molecules25245886 . . . . **261**

Steffen B. Mogensen, Mercedes K. Taylor and Ji-Woong Lee
Homocoupling Reactions of Azoles and Their Applications in Coordination Chemistry
Reprinted from: *Molecules* **2020**, *25*, 5950, doi:10.3390/molecules25245950 . . . . **293**

Shijie Zhang, Zhenguo Gao, Di Lan, Qian Jia, Ning Liu and Jiaoqiang Zhang et al.
Recent Advances in Synthesis and Properties of Nitrated-Pyrazoles Based Energetic Compounds
Reprinted from: *Molecules* **2020**, *25*, 3475, doi:10.3390/molecules25153475 . . . . **323**

Boris V. Lyalin, Vera L. Sigacheva, Anastasia S. Kudinova, Sergey V. Neverov, Vladimir A. Kokorekin and Vladimir A. Petrosyan
Electrooxidation Is a Promising Approach to Functionalization of Pyrazole-Type Compounds
Reprinted from: *Molecules* **2021**, *26*, 4749, doi:10.3390/molecules26164749 . . . . **365**

About the Editors

Vera L. M. Silva

Vera L. M. Silva graduated in Analytical Chemistry (1998), Master in Chemistry of Natural Products and Food Stuffs (2002) and PhD in Chemistry (2006) by the University of Aveiro. She is currently Assistant Professor at the Chemistry Department of University of Aveiro and Researcher at REQUIMTE-Associated Laboratory for Green Chemistry - Clean Technologies and Processes. She is an Organic Chemist, and member of the Molecular Synthesis group of REQUIMTE (https://laqv.requimte.pt/people/1747-vera_l_m_silva). Her main research interests comprise the synthesis and structural characterization of novel compounds for medicinal applications and development of sustainable organic synthesis methodologies, based on more contemporary tools such as ohmic heating. As part of her research activity, she published 2 book chapters, more than 55 SCI papers and 2 papers in national scientific journals with peer review, and she is co-author of 1 patent. Her activities also include the supervision and co-supervision of 6 PhD (2 concluded) and 9 MSc students, more than 21 Bachelor students (3rd year graduation projects) and 5 Erasmus students. She has participated in several national and one international research projects, and she presented several oral and poster communications at national and international scientific meetings.

Artur M. S. Silva

Artur M. S. Silva is currently full professor at the Chemistry Department and Vice-rector for research and innovation of the University of Aveiro (UA). He is also the President of the Portuguese Chemical Society. He obtained both the BSc (1987) and the PhD in Organic Chemistry (1993) degrees at the UA. He joined the Department of Chemistry of the same university in 1987 as Assistant and was appointed to Auxiliary Professor in 1996, Associate Professor in 1999 and Full Professor in 2001. He published more than 780 SCI papers, 1 e-book, 50 book chapters, delivered more than 60 lectures in scientific meetings and is co-author of 4 patents. His publications have been cited more than 18000 times and his Hirsch-index is 60. He supervised 19 post-doctoral fellows, 37 PhD students and 43 MSc students; he has also participated in 29 financed Portuguese and European projects and in 8 bilateral financed projects with European Research Groups.

Editorial

Special Issue "Recent Advances in the Synthesis, Functionalization and Applications of Pyrazole-Type Compounds"

Vera L. M. Silva * and Artur M. S. Silva *

LAQV-REQUIMTE, Department of Chemistry, University of Aveiro, 3810-193 Aveiro, Portugal
* Correspondence: verasilva@ua.pt (V.L.M.S.); artur.silva@ua.pt (A.M.S.S.); Tel.: +351-234-370704 (V.L.M.S.); +351-234-370714 (A.M.S.S)

Citation: Silva, V.L.M.; Silva, A.M.S. Special Issue "Recent Advances in the Synthesis, Functionalization and Applications of Pyrazole-Type Compounds". *Molecules* **2021**, *26*, 4989. https://doi.org/10.3390/molecules26164989

Received: 9 August 2021
Accepted: 13 August 2021
Published: 18 August 2021

Publisher's Note: MDPI stays neutral with regard to jurisdictional claims in published maps and institutional affiliations.

Copyright: © 2021 by the authors. Licensee MDPI, Basel, Switzerland. This article is an open access article distributed under the terms and conditions of the Creative Commons Attribution (CC BY) license (https://creativecommons.org/licenses/by/4.0/).

Pyrazoles and their reduced form, pyrazolines, are considered privileged scaffolds in medicinal chemistry, owing to their remarkable biological activities, physicochemical properties and occurrence in many low-molecular-weight compounds present in several marketed drugs (e.g., Celebrex® and Viagra®). Pyrazoles are also found in a variety of agrochemicals (fungicides, insecticides, and herbicides) and are versatile compounds for synthetic manipulations. Their structural features (mainly tautomerism, with possible implications in their reactivity), and diverse applications, have stimulated the work of several research groups towards the synthesis and functionalization of pyrazole-type compounds and study of their properties, and has inspired this Special Issue that aims to provide a broad survey of the most recent advances in pyrazole's chemistry.

In this Special Issue, nine original research articles and seven reviews covering some of the most recent advances in the synthesis, transformation, properties and relevant applications of pyrazoles are reported. Two articles deal with the synthesis and computational study of metallacycles formed by pyrazolate ligands and the coinage metals M = Cu(I), Ag(I) and Au(I). Elguero and Alkorta reported a computational study of metallacycles formed by pyrazolate ligands and the coinage metals $(pzM)_n$ for n = 2, 3, 4, 5 and 6 and carried out a comparison with structures reported in the Cambridge Crystallographic Data Center (CCDC). They discussed the most common size of the metallacycles' ring depending on the metal and analyzed the stability of different ring sizes [1]. By using an unexpected synthetic method, Leznoff and Fusijawa et al. obtained and characterized a silver(I) pyrazolate complex [Ag(μ-L1Clpz)]$_n$ as a coordination polymer, thus broadening the family of silver(I) coordination polymers [2]. El-Faham and Soliman et al. described the synthesis, the molecular structure, using Hirshfeld calculations, and antimicrobial evaluations of three iron(III) complexes obtained by self-assembly of iron(III) chloride and mono- and bis-pyrazolyl-s-triazine ligands. The Fe(III) complexes have shown enhanced antibacterial and antifungal activities as compared to the free ligands [3]. Kayukova et al. described a Boulton–Katritzky rearrangement of 5-substituted phenyl-3-[2-(morpholin-1-yl)ethyl]-1,2,4-oxadiazoles as a synthetic strategy to prepare spiropyrazoline benzoates and chloride. The tested compounds showed moderate to high in vitro antitubercular activity with minimum inhibitory concentration (MIC) values of 1–100 μg/mL. The highest activity in 1 μg/mL and 2 μg/mL on drug sensitive (DS) and multidrug-resistant (MDR) *Mycobacterium tuberculosis* strains, equal to the activity of the basic antitubercular drug rifampicin, was recorded for 2-amino-8-oxa-1,5-diazaspiro[4.5]dec-1-en-5-ium chloride [4]. Potapov et al. described different approaches to the synthesis of dicarboxylic derivatives of bis(pyrazol-1-yl)alkanes which are interesting building blocks for metal-organic frameworks [5]. Svete et al. reported on the synthesis of a series of 12 silica gel-bound enaminones and their Cu(II) complexes, which were used as suitable heterogeneous catalysts in [3+2] cycloadditions of 3-pyrazolidinone-derived azomethine imines to terminal ynones, for the regioselective synthesis of 6,7-dihydro-1*H*,5*H*-pyrazolo[1,2-*a*]pyrazoles [6]. Usami et al. reported their studies on alkylamino coupling reactions at C-4 of 4-halo-1*H*-1-tritylpyrazoles using readily accessible palladium or copper catalysts. They concluded that Pd(dba)$_2$-catalyzed reaction of 4-bromo-1*H*-1-tritylpyrazole is suitable for aromatic or bulky amines lacking β-hydrogen atoms, but not for cyclic amines or alkylamines possessing β-hydrogen atoms. On the other hand, the Cu(I)-catalyzed amination using 4-iodo-1*H*-1-tritylpyrazole was found to be favorable for alkylamines possessing β-hydrogen atoms, and not suitable for aromatic amines and bulky amines lacking β-hydrogens, showing the complementarity of the two catalysts [7]. Alam et al. reported the synthesis and antimicrobial studies of 31 novel coumarin-substituted pyrazole derivatives. Some of these compounds have shown potent activity against methicillin-resistant *Staphylococcus aureus* (MRSA) with MIC as low as 3.125 μg/mL and were potent at inhibiting the development of MRSA biofilm and the destruction of preformed biofilm. These are very significant results because MRSA strains have emerged as one of the most alarming pathogens of humans, bypassing human immunodeficiency viruses (HIV) in terms of fatality rate [8]. Wong et al. described the synthesis of novel pyrazolopyridopyridazine diones and *N*-aminopyrazolopyrrolopyridine diones and studied their photoluminescence, solvatofluorism, and quantum yield (Φf). They demonstrated that pyrazolopyridopyridazine diones generally exhibited the stronger fluorescence intensity and possessed a significant substituent effect, particularly for a *m*-chloro substituent, and that a more planar skeletal conformation led to a higher Φf, similar to that of the standard luminol. They also concluded that the introduction of the pyrazole and pyridine chromophores led to an increase in the conjugation and aromaticity of the synthetized compounds when compared with luminol [9]. Current approaches on the synthesis and post-functionalization of pyrazolo[1,5-*a*]pyrimidines were reviewed by Portilla et al., who have also highlighted the anticancer potential and enzyme inhibitory activity of these compounds, thus providing important insights for the rational and efficient design of new drugs bearing the pyrazolo[1,5-*a*]pyrimidine core [10]. Tu et al. described the advances in pyrazole scaffold's synthesis and functionalization of the last decade (2011–2020) and laid emphasis on the strategies applied in the development of therapeutics for neurodegenerative diseases, particularly for Alzheimer's disease (AD) and Parkinson's disease (PD), two of the most serious chronic neurodegenerative diseases. The various pyrazoles described can be regarded as candidate agents for the development of novel neurodegenerative drugs [11]. Brullo et al. highlighted the pharmacological relevance of pyrazolyl-ureas, which are known to possess a wide range of biological activities, such as antipathogenic, antimalarial, anti-inflammatory, antitumoral, among others, and summarized the biological data reported for these compounds from 2000 to date, including patented compounds [12]. Although there are studies on the synthesis of styrylpyrazoles dating back to the 1970s and even earlier, this type of compounds has rarely been studied. Silva et al. focused their attention in this particular type of pyrazoles, containing a styryl (2-arylvinyl) group linked in different positions of the pyrazole backbone, and described their properties, biological activity, methods of synthesis and transformation, highlighting the interest and huge potential for application of this kind of compounds [13]. Lee et al. summarized the recent advances in homocoupling reactions of pyrazoles and other azoles (imidazoles, triazoles and tetrazoles) to highlight the utility of homocoupling reactions in the synthesis of symmetric bi-heteroaryl systems compared with traditional synthesis. Metal-free reactions and transition-metal catalyzed homocoupling reactions were discussed with reaction mechanisms in detail [14]. In the field of energetic materials (EMs), the nitrated pyrazoles have attracted wide attention due to their high heat of formation, high density, tailored thermal stability, and detonation performance, having good applications in explosives, propellants, and pyrotechnics. Zhang and Kou et al. described the recent processes in the synthesis and the physical and explosive performances of the nitrated-pyrazole-based EMs, and analyzed the development trend of these compounds and the concepts for designing prominent high-performance nitropyrazole-based EMs [15]. Kokorekin et al. summarized, for the first time, the poorly studied electrooxidative functionalization of pyrazole derivatives towards the efficient synthesis of C–Cl, C–Br, C–I,

C–S and N–N coupling products with applied properties. Important aspects of aromatic hydrogen substitution were discussed, as well as the advantages of electrooxidative functionalization, mainly the predominantly galvanostatic electrolysis mode, the reusability of commercially available electrodes, salts and solvents, and gram-scalability of the processes. The increasing role of cyclic voltammetry as a tool to study mechanisms and to predict the efficiency of the synthesis was highlighted [16].

In conclusion, pyrazoles' chemistry is undoubtedly a current and pertinent topic of investigation. Several research groups have contributed to the advancement of this topic by designing novel pyrazole-type compounds and developing synthetic strategies for the preparation and post-functionalization of pyrazoles, or by studying their structures, properties and potential applications. Herein, important advances in pyrazoles' chemistry have been disclosed. We thank all of the authors for their valuable contributions to this Special Issue, all the peer reviewers for their valuable comments, criticisms, and suggestions and the staff members of MDPI for the editorial support.

Conflicts of Interest: The authors declare no conflict of interest.

References

1. Elguero, J.; Alkorta, I. A Computational Study of Metallacycles Formed by Pyrazolate Ligands and the Coinage Metals M = Cu(I), Ag(I) and Au(I): (pzM)$_n$ for n = 2, 3, 4, 5 and 6. Comparison with Structures Reported in the Cambridge Crystallographic Data Center (CCDC). *Molecules* **2020**, *25*, 5108. [CrossRef] [PubMed]
2. Fujisawa, K.; Nemoto, T.; Morishima, Y.; Leznoff, D.B. Synthesis and Structural Characterization of a Silver(I) Pyrazolato Coordination Polymer. *Molecules* **2021**, *26*, 1015. [CrossRef] [PubMed]
3. Soliman, S.M.; Al-Rasheed, H.H.; Albering, J.H.; El-Faham, A. Fe(III) Complexes Based on Mono- and Bis-pyrazolyl-s-triazine Ligands: Synthesis, Molecular Structure, Hirshfeld, and Antimicrobial Evaluations. *Molecules* **2020**, *25*, 5750. [CrossRef] [PubMed]
4. Kayukova, L.; Vologzhanina, A.; Praliyev, K.; Dyusembaeva, G.; Baitursynova, G.; Uzakova, A.; Bismilda, V.; Chingissova, L.; Akatan, K. Boulton-Katritzky Rearrangement of 5-Substituted Phenyl-3-[2-(morpholin-1-yl)ethyl]-1,2,4-oxadiazoles as a Synthetic Path to Spiropyrazoline Benzoates and Chloride with Antitubercular Properties. *Molecules* **2021**, *26*, 967. [CrossRef] [PubMed]
5. Burlutskiy, N.P.; Potapov, A.S. Approaches to the Synthesis of Dicarboxylic Derivatives of Bis(pyrazol-1-yl)alkanes. *Molecules* **2021**, *26*, 413. [CrossRef] [PubMed]
6. Štanfel, U.; Slapšak, D.; Grošelj, U.; Požgan, F.; Štefane, B.; Svete, J. Synthesis of 6,7-Dihydro-1H,5H-pyrazolo[1,2-a]pyrazoles by Azomethine Imine-Alkyne Cycloadditions Using Immobilized Cu(II)-Catalysts. *Molecules* **2021**, *26*, 400. [CrossRef] [PubMed]
7. Usami, Y.; Tatsui, Y.; Yoneyama, H.; Harusawa, S. C4-Alkylamination of C4-Halo-1H-1-tritylpyrazoles Using Pd(dba)$_2$ or CuI. *Molecules* **2020**, *25*, 4634. [CrossRef] [PubMed]
8. Alnufaie, R.; Raj, K.C.H.; Alsup, N.; Whitt, J.; Chambers, S.A.; Gilmore, D.; Alam, M.A. Synthesis and Antimicrobial Studies of Coumarin-Substituted Pyrazole Derivatives as Potent Anti-*Staphylococcus aureus* Agents. *Molecules* **2020**, *25*, 2758. [CrossRef]
9. Tseng, C.-C.; Chung, C.-Y.; Tsai, S.-E.; Takayama, H.; Uramaru, N.; Lin, C.-Y.; Wong, F.F. Selective Synthesis and Photoluminescence Study of Pyrazolopyridopyridazine Diones and N-Aminopyrazolopyrrolopyridine Diones. *Molecules* **2020**, *25*, 2409. [CrossRef]
10. Arias-Gómez, A.; Godoy, A.; Portilla, J. Functional Pyrazolo[1,5-a]pyrimidines: Current Approaches in Synthetic Transformations and Uses As an Antitumor Scaffold. *Molecules* **2021**, *26*, 2708. [CrossRef] [PubMed]
11. Li, X.; Yu, Y.; Tu, Z. Pyrazole Scaffold Synthesis, Functionalization, and Applications in Alzheimer's Disease and Parkinson's Disease Treatment (2011–2020). *Molecules* **2021**, *26*, 1202. [CrossRef] [PubMed]
12. Brullo, C.; Rapetti, F.; Bruno, O. Pyrazolyl-Ureas as Interesting Scaffold in Medicinal Chemistry. *Molecules* **2020**, *25*, 3457. [CrossRef] [PubMed]
13. Gomes, P.M.O.; Ouro, P.M.S.; Silva, A.M.S.; Silva, V.L.M. Styrylpyrazoles: Properties, Synthesis and Transformations. *Molecules* **2020**, *25*, 5886. [CrossRef] [PubMed]
14. Mogensen, S.B.; Taylor, M.K.; Lee, J.-W. Homocoupling Reactions of Azoles and Their Applications in Coordination Chemistry. *Molecules* **2020**, *25*, 5950. [CrossRef]
15. Zhang, S.; Gao, Z.; Lan, D.; Jia, Q.; Liu, N.; Zhang, J.; Kou, K. Recent Advances in Synthesis and Properties of Nitrated-Pyrazoles Based Energetic Compounds. *Molecules* **2020**, *25*, 3475. [CrossRef] [PubMed]
16. Lyalin, B.V.; Sigacheva, V.L.; Kudinova, A.S.; Neverov, S.V.; Kokorekin, V.A.; Petrosyan, V.A. Electrooxidation Is a Promising Approach to Functionalization of Pyrazole-Type Compounds. *Molecules* **2021**, *26*, 4749. [CrossRef]

Article

A Computational Study of Metallacycles Formed by Pyrazolate Ligands and the Coinage Metals M = Cu(I), Ag(I) and Au(I): (pzM)$_n$ for n = 2, 3, 4, 5 and 6. Comparison with Structures Reported in the Cambridge Crystallographic Data Center (CCDC)

José Elguero and Ibon Alkorta *

Instituto de Química Médica (IQM-CSIC), Juan de la Cierva, 3, E-28006 Madrid, Spain; iqmbe17@iqm.csic.es
* Correspondence: ibon@iqm.csic.es; Tel.: +34-915622900

Academic Editors: Vera L. M. Silva and Artur M. S. Silva
Received: 30 September 2020; Accepted: 21 October 2020; Published: 3 November 2020

Abstract: The structures reported in the Cambridge Structural Database (CSD) for neutral metallacycles formed by coinage metals in their valence (I) (cations) and pyrazolate anions were examined. Depending on the metal, dimers and trimers are the most common but some larger rings have also been reported, although some of the larger structures are not devoid of ambiguity. M06-2x calculations were carried out on simplified structures (without C-substituents on the pyrazolate rings) in order to facilitate a comparison with the reported X-ray structures (geometries and energies). The problems of stability of the different ring sizes were also analyzed.

Keywords: X-ray; pyrazolate; coinage metals; metallacycles; M06-2x

1. Introduction

There is a field in organometallic chemistry that has rightly demanded a great deal of attention, namely, the cyclic complexes between coinage metal cations and anionic pyrazolate ligands [1]. These metallacycles frequently have high symmetry and contain several nuclei with spin I = 1/2, which makes them ideally suited for NMR studies: ^{1}H, ^{13}C, ^{15}N, ^{19}F (from the much studied 3,5-bis-trifluoromethylpyrazole) and $^{107/109}$Ag. In contrast, $^{63/65}$Cu and ^{197}Au are quadrupolar (I = 3/2) and have therefore been explored to a much lesser extent.

It is common in publications in the field of coordination compounds that single-crystal X-ray structures are reported. In such cases the data are in the Results and Supporting Information sections. Note that there are some publications that concern other aspects of these compounds such as their use in sensors, optical properties, and theoretical calculations where crystal structures are not reported. Earlier papers should also be mentioned here because, although they do not contain crystal structures, they were key in generating interest in these metallacycles [2–4]. Lintang et al. reported that trinuclear group 11 metal pyrazolate complexes are phosphorescent chemosensors for the detection of benzene [5] (for two recent papers on photoluminescence of the Dias and Fujisawa groups see [6,7]). The Serrano group published several papers that describe compounds related to those discussed in the present paper but with interesting mesogen properties when the pyrazolate ligands have long chains in positions three and five and the metal is Cu, Ag or Au [8–11]. Cano's group published similar results for complexes with gold(I) [12].

Of particular relevance is a theoretical paper concerning the study of group 11 pyrazolate complexes. In this case, Caramori, Frenking et al. [13] discussed the trinuclear (pzM)$_3$ complexes (M = Cu(I), Ag(I) and Au(I)) in terms of different approaches including energy decomposition analysis

(EDA), natural bond orbital (NBO) and anisotropy of the induced current density (ACID). The main conclusions were that the pz-M bond has an elevated covalent character, especially when M = Au(I), and that the pyrazole ligands are strongly aromatic, although they are insulated because there is no through-bond metal-ligand conjugation.

Our group published a paper, in collaboration with Rasika Dias and another with Kiyoshi Fujisawa, on the NMR study of the organometallic nine-membered rings corresponding to trinuclear silver(I) complexes of pyrazolate ligands [14,15], and another on regium bonds between dinuclear silver(I) pyrazolates complexes and Lewis bases [16], two on regium bonds formed by Au_2 [17] and Ag_2, Cu_2 and mixed binary regium molecules [18], and finally, one on the comparison of acidity of Au(I) and Au(III) [19].

2. Results and Discussion

The present publication is divided into three sections. The first section concerns an exploration of the Cambridge Structural Database (CSD) [20] in a search for the structures of pyrazolates with coinage metals of valence (I): i.e., Cu(I), Ag(I) and Au(I); they will be reported using their refcodes. The second section covers a theoretical study of the stability of these metallacycles as a function of the ring size (dimers, trimers, tetramers, pentamers and hexamers) using the pyrazole itself as a model, i.e., without C-substituents and without supplementary ligands on the metals. The final section concerns the analysis of some metallacycles by Bader's quantum theory of atoms in molecules methodology (QTAIM) [21–24].

We will start with the exploration of the CSD [20]; this search was similar to one carried out by us on the cyclamers formed by NH-pyrazoles based on hydrogen bonds (HBs). NH-pyrazoles crystallize as catemers (chains) and cyclamers (rings with n pyrazoles), with examples reported for n = 2, 3, 4, and 6 (rare) but none for a pentamer [25,26] (Figure 1).

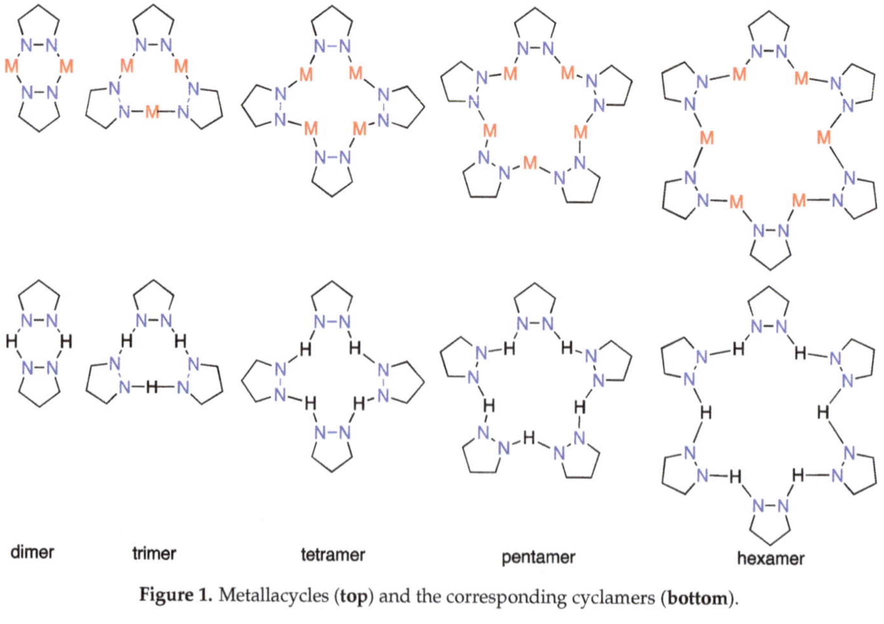

Figure 1. Metallacycles (**top**) and the corresponding cyclamers (**bottom**).

2.1. Analysis of the Reported CSD Structures (Hits) and Their Refcodes

Before discussing the most relevant hits, it is worth noting that there is only one compound that contains two different metals (Ag_2Au), namely a gold(I)imidazolate-silver(I)pyrazolate complex, MAMQIB [27] and MAMQIB01 [28] (Figure 2), but this only has two pyrazolates, with the third ligand being an carbeniate (C,N ligand). In any case, this example shows that [pzM(M′)]$_n$ compounds should be possible with different metals, although examples have not been reported to date.

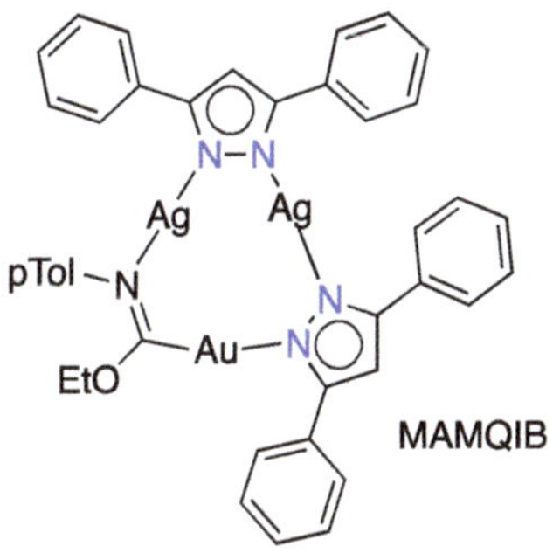

Figure 2. Structure of CSD Refcode MAMQIB.

2.1.1. Copper, Only Cu(I) Structures (Cu(II) Structures Were Excluded)

Dimers and trimers, (pzCu)$_2$ and (pzCu)$_3$, are common. As far as dimers are concerned, many of these are Cu(II) derivatives and all examples that contain Cu(I) have complex pyrazolates with arms that are able to coordinate the copper or they have supplementary ligands; dimers in which the Cu(I) atoms are "nude", i.e., linked only to the pyrazolate anion, have not been found. Amongst the supplementary ligands are CO (COCZAY [29], COCZOM [29]), C≡N–R (GITJUO [30], HEDFIF [31], JEMCAF [32]), PPh$_3$ (GITKEZ [30], PIRDAY [33], SATKAC [34]), pyridines (IPIGET [35], KUKLOR [36]) and N-heterocyclic carbenes NHCs (NETLUW [37], NETMAD [37]). The central metallacycle can adopt planar or folded conformations, either boat-type or chair-type (Figure 3), and both of these conformations are common. An examination of the Cu···Cu distances in (pzCu)$_2$ structures gave 320 values, with a mean value of 3.719 Å, a minimum of 3.013 Å (UTEWUM [38]) and a maximum of 4.074 Å (YADVEG [39]) (Figure 3).

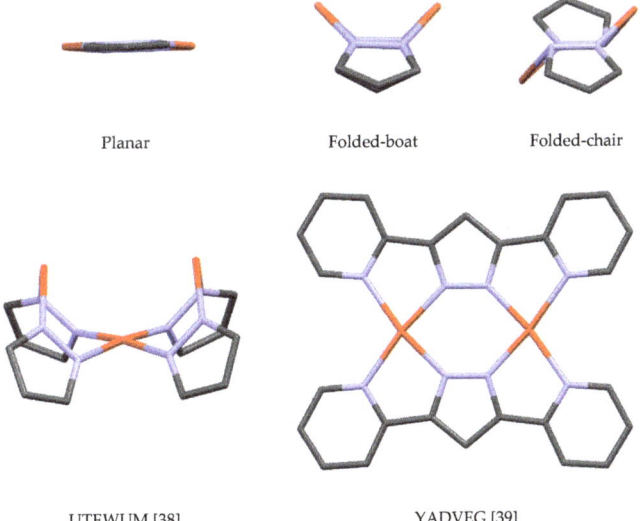

Figure 3. Planar and folded conformations of the metallacycle in (pzCu)$_2$ (simplified representations of UTEWUM and YADVEG).

As far as trimers are concerned, the two main classes are isolated trimers and double trimers and we proposed to denote the latter as "3 + 3". The simplest derivatives of isolated trimers include CODBAB (4-Cl-pyrazolate) [40], VIJMUW (3,5-diMe-pyrazolate) [41], BELTOC (3,5-di-*i*-Pr-pyrazolate) [42] and XELXAN (3,5-di-CF$_3$-pyrazolate) [43].

Simple examples of "3 + 3" double trimers (Figure 4) are IDUYOW (3-Ph-pyrazolate) [44], TANRUW (3-CF$_3$-pyrazolate) [32] and GITJIC (3,5-diMe-4-NO$_2$-pyrazolate) [30] with near one bond between the triangles formed by Cu atoms and XOGJOU (4-NO$_2$-pyrazolate) [45] with near three Cu-Cu bonds. The superimposed structures of double trimers can be a perfect fit, e.g., XOGJOU [45] or rotated (an example will be discussed later for another metal).

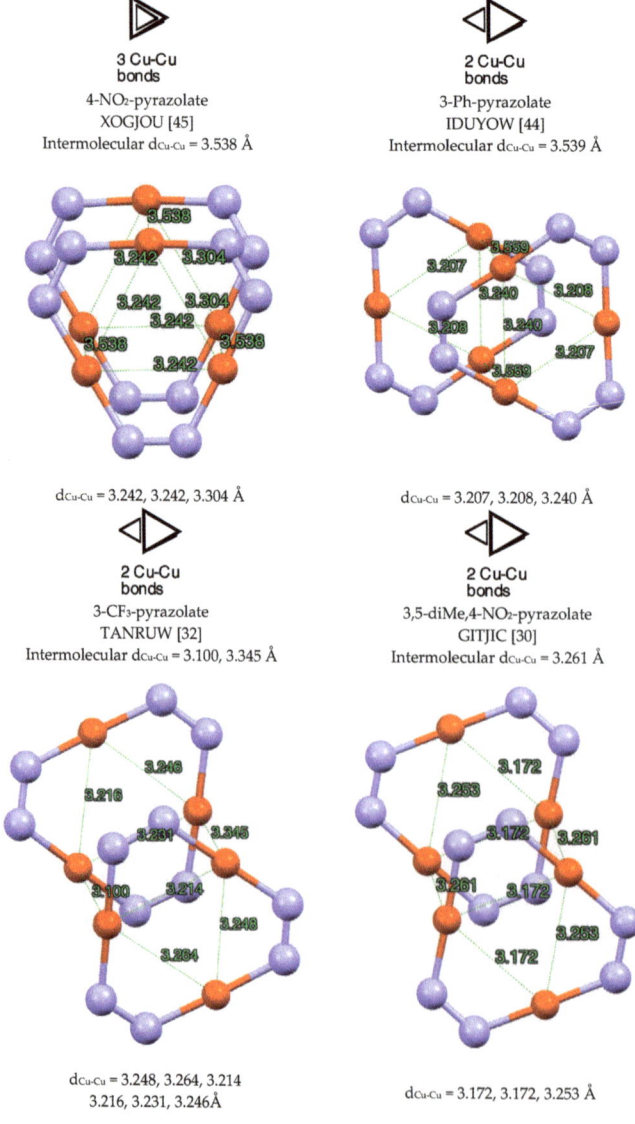

Figure 4. Different modes present in the "3 + 3" (pzCu)$_3$ complexes. Only the copper and nitrogen atoms are represented.

Tetramers (pzCu)$_4$ are frequently observed and the simplest derivatives include BELTUI [42] (and BELTUI01 [46]), HEDFEB [31], (and OMIPOP01 [42]), OMIPOP [47] and REWWOI [48] (Figure 5). Most Cu$_4$ rings are planar but in REWWOI [48] this ring is not planar. The structure of OMIPOP [42] was represented with two Cu atoms bonded to both N atoms of the corresponding pyrazole but this is only the result of a CSD convention that bonds are depicted when they are shorter than the sum of the van der Waals radii. In the original article [48] it is highlighted that the four Cu atoms form a rhombus with a Cu···Cu non-bonding interaction for the shortest distance corresponding to d^{10}-d^{10} contacts. Fujisawa re-examined this interesting structure (OMIPOO01 [42] and OMIPOO02 [46]) and noted the diamond-like disposition of the four Cu atoms with a short (3.40 Å) and a long (4.85 Å) structure.

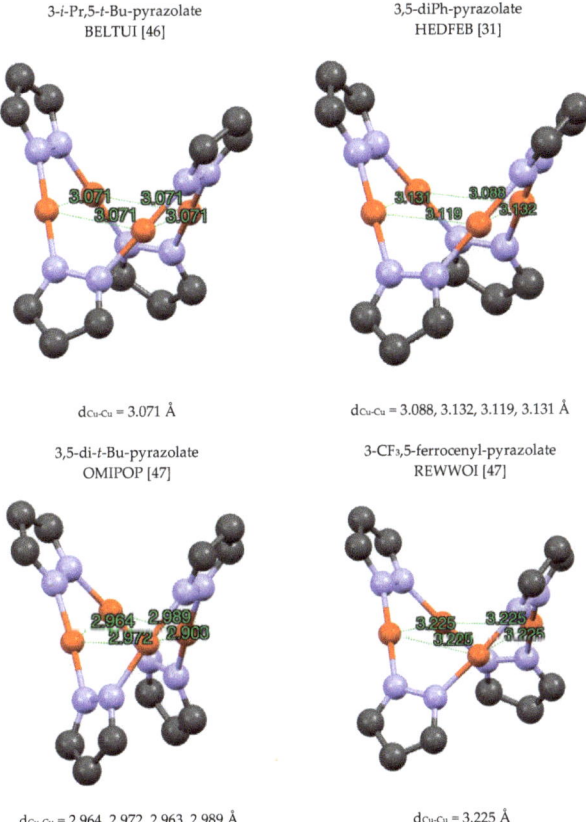

Figure 5. Different modes of (pzCu)$_4$ complexes. Only the copper atoms and the pyrazolate rings are represented. The sequential Cu-Cu distances are listed.

At this point it is worth commenting on hexamers, (pzCu)$_6$; these structures always contain six central OH bridges to form a star of alternated Cu atom and O atoms, which means that these are Cu(II) derivatives (QIMWOA and QIMWUG (Coronado et al. [49]) and SASXIW (Galassi, Martins et al. [34])). In summary, examples of Cu(I) pentamers or hexamers are not known.

2.1.2. Silver, Only Ag(I) Derivatives

Compounds with (pzAg)$_2$ and (pzAg)$_3$ structures are common. As in the case of (pzCu)$_2$, in (pzAg)$_2$ the hexagonal metallacycle adopts planar and folded conformations, with the latter being either boat-likes (the most common) or chair-like (as in cyclohexane), with the silver atoms located at the tips. The mean value of the Ag···Ag distance is 3.755 Å (shortest 3.425 Å, longest 4.305 Å).

Three topological dispositions of double trimers "3 + 3" were found in the CSD (Figure 6), namely three common sides, one common side and one common vertex. These structures are schematically represented in Figure 6 with triangles. These dispositions are illustrated with one or two examples for each situation: XOGJUA [45], DAZGIV [50], FISDIV01 [51] and DOJCUC [52] (EWEHAP [53] is similar with an intermolecular distance between silver metal centers of 3.179 Å). The intermolecular Ag···Ag distance decreases with the number of bonds (3.509 Å (three), 3.205 Å (two) and 2.986 Å (one)) and this is probably due to angular strain.

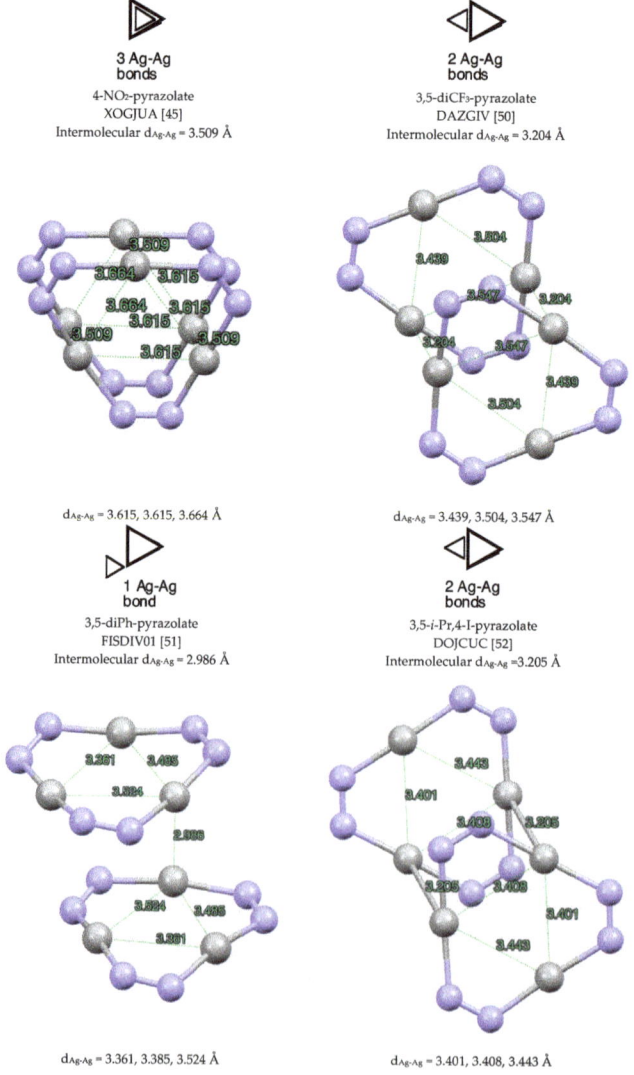

Figure 6. Different modes of "3 + 3" (pzAg)$_3$ complexes. Only the silver and nitrogen atoms are represented.

Tetramers (pzAg)$_4$ are frequent and, in this case, the simplest structures correspond to GAFJII [46], QEJJOH [54], QEZHEJ [55] and VUPGET [56] (Figure 7).

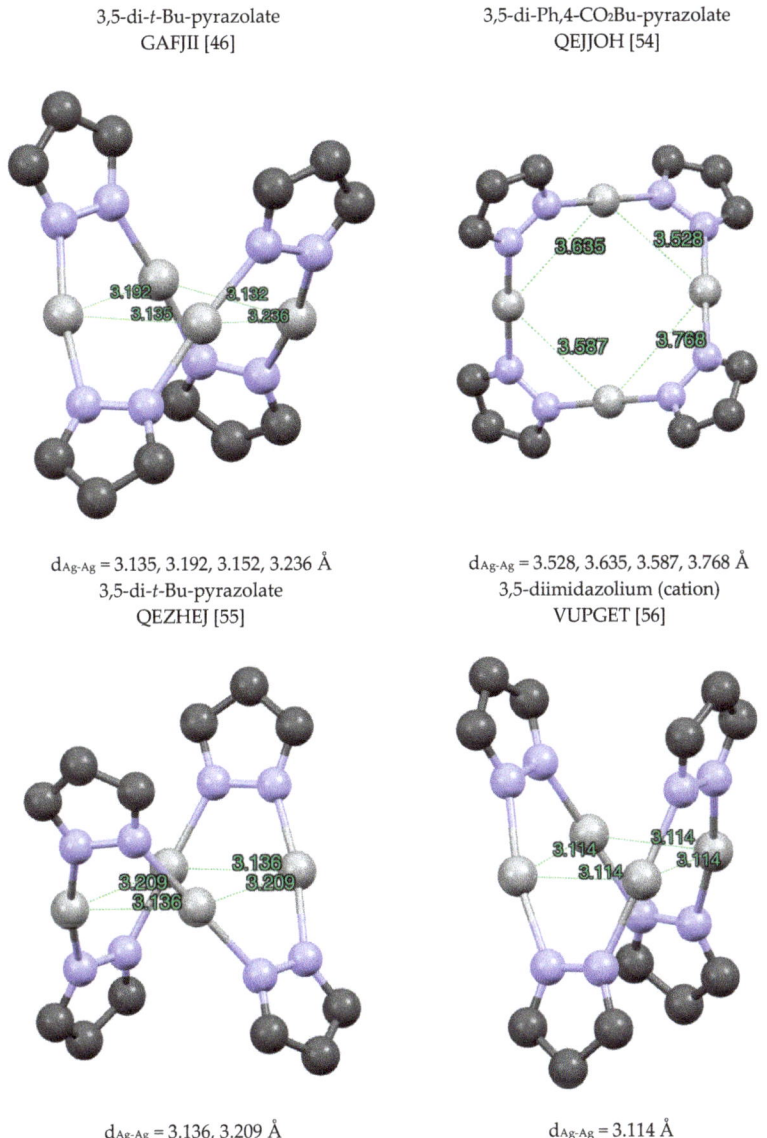

Figure 7. Different structures of (pzAg)₄ complexes. Only the silver atoms and the pyrazolate rings are represented. The four examples are very similar but different views have been used.

The structure of SOJCUS [57] is rather complex (Figure 8) in that the silver atoms are double-, triple- and quadruple-coordinated, thus allowing a structure that contains six pyrazolates and five silver atoms. Depending on the itinerary four, (pzAg)₄, and five, (pzAg)₅, complexes are present (Figure 8), but a true tetramer or pentamer does not exist.

3,5-di-CF$_3$-pyrazolate
SOJCUS [57]

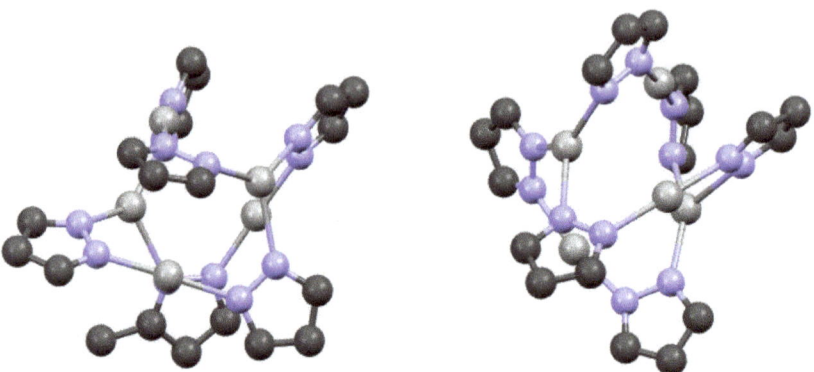

Figure 8. A view of the two independent molecules of SOJCUS. Only the silver atoms and the pyrazolate rings are represented.

The only examples of hexamers are QEJJEX [54] and QEJJIB [54] (almost identical), which are beautiful structures that form a loop with short Ag···Ag contacts (QEJJEX [54], Figure 9).

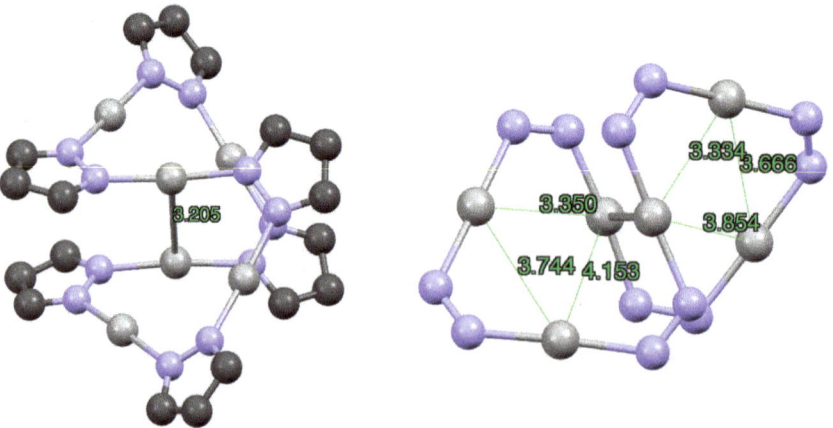

Figure 9. Two views of QEJJEX. Only the silver atoms and the pyrazolate rings are represented. Left Ag···Ag distance between loops; right Ag···Ag distance in the quasi-triangles.

In the case of silver, the numbers of structures found in the CSD search were 72 trimers, 16 dimers, 6 tetramers and 2 hexamers (2.1%) (pentamers were not found).

2.1.3. Gold, Only Au(I) Structures (Au(III) Structures Were Excluded)

Dimers (pzAu)$_2$ with Au(I) are not known and all of the examples found in the CSD are Au(III) derivatives. Trimers (pzAu)$_3$ are common, with some compounds isolated as trimers, e.g., the bis-3,5-CF$_3$-pyrazolate, COHFIO01 [50] (d_{Au-Au} = 3.341, 3.350, 3.360 Å) and bis-3,5-Ph-pyrazolate FUWXOK01 [58] (d_{Au-Au} = 3.361 Å, a regular triangle) while others (Figure 10) are "3 + 3" double trimers.

3 Au-Au
rotated
bonds
3,5-di-*i*-Pr-pyrazolate
GAFJOO [46]
Intermolecular d_{Au-Au} = 3.336, 3.403, 3.417 Å

3 Au-Au
bonds
3-Me,4-Ph-pyrazolate
JOSNOW01 [58]
Intermolecular d_{Au-Au} = 3.678, 3.704, 3.855 Å

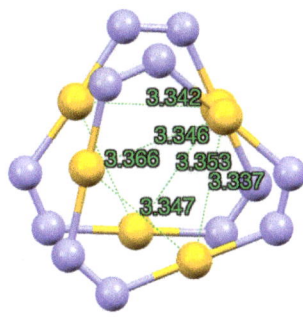

d_{Au-Au} = 3.337, 3.346, 3.347 and
3.342, 3.353, 3.366 Å

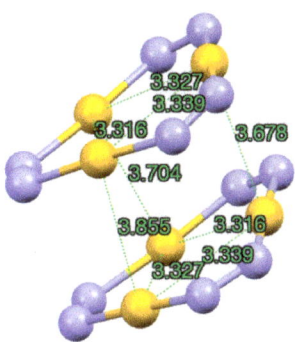

d_{Au-Au} = 3.316, 3.327, 3.339 Å

3 Au-Au
bonds
4-Me-pyrazolate
MUTLAO [59]
Intermolecular d_{Au-Au} = 3.673, 3.784, 3.786 Å

chain 2 Au⋯Au bonds
Unsubstituted pyrazolate
MUTKUH01 [60]
Intermolecular d_{Au-Au} = 3.108 Å

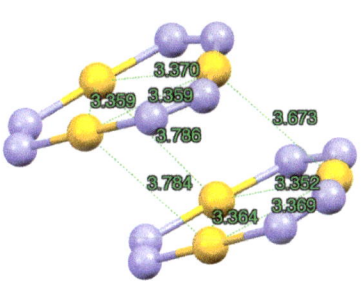

d_{Au-Au} = 3.359, 3.359, 3.370
3.352, 3.364, 3.369 Å

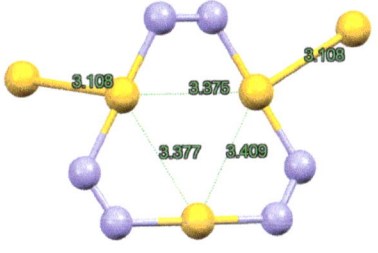

d_{Au-Au} = 3.375, 3.377, 3.409 Å

Figure 10. Different modes of (pzAu)$_3$ "3 + 3" complexes. Only the gold and nitrogen atoms are represented [59,60].

Au(I) tetramers, (pzAu)$_4$, are known and the simplest examples are GAFJUU [46], GAFKAB [46], OKALER [61] and OKALIV [61] (Figure 11). The gold centers can arrange in a regular square (GAFJUU [46]), in a rectangle (OKALER [61]), in a quadrilateral (GAFKAB [46]) or in a folded quadrilateral (OKALIV [61]).

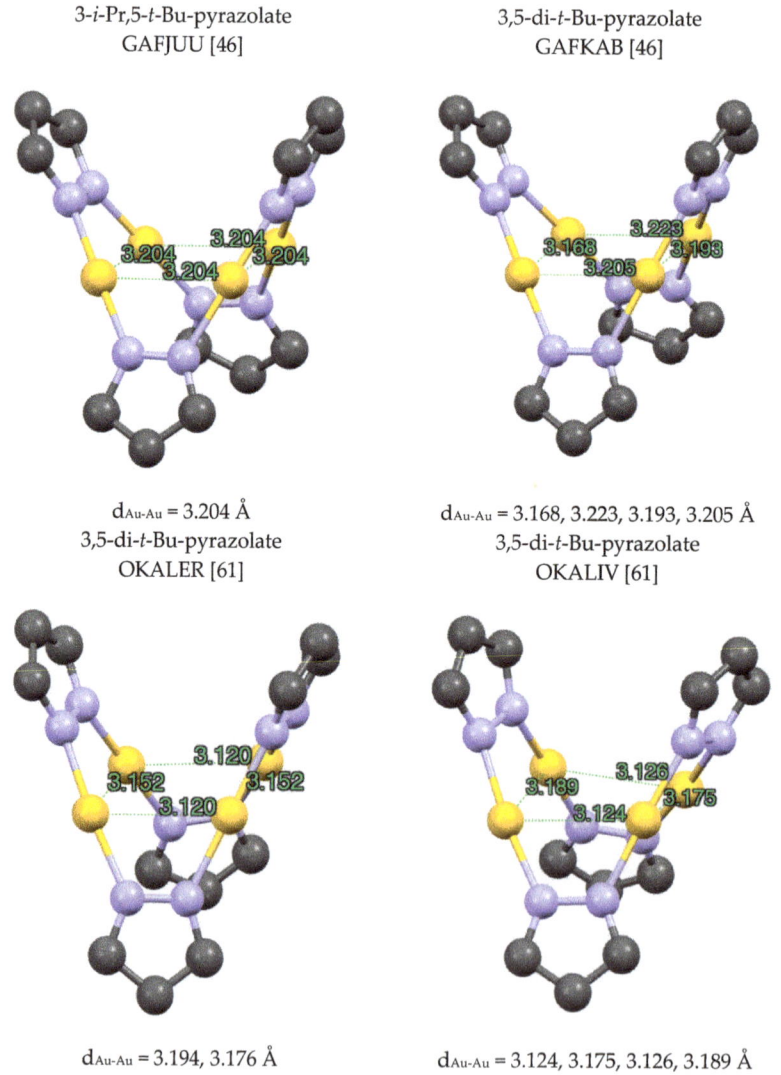

Figure 11. Different modes of (pzAu)$_4$ complexes. Only the gold atoms and the pyrazolate rings are represented.

Finally (pzAu)$_6$ structures are very rare: in 1988 Raptis reported FEJJAF10 [62], see Figure 12.

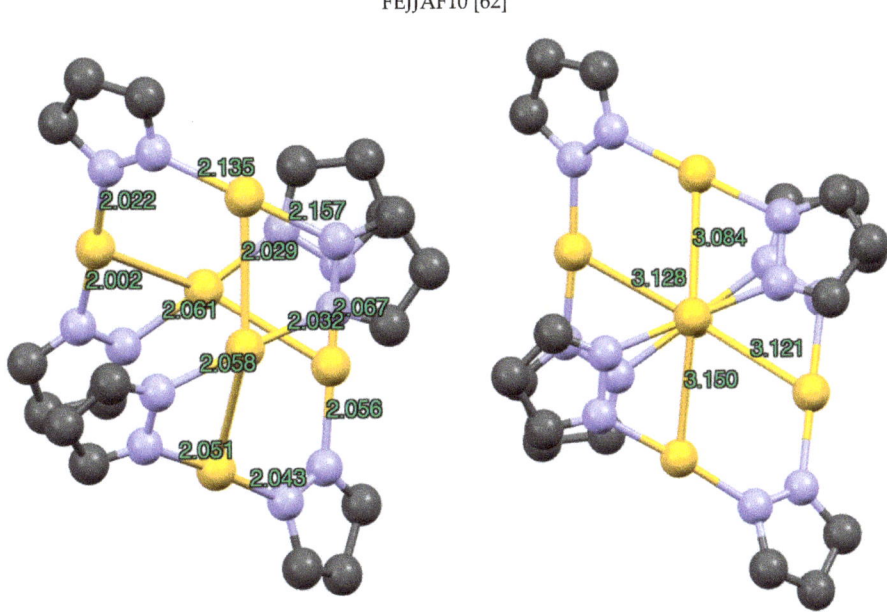

Figure 12. Two views of one of the two independent molecules of FEJJAF10 with Au-N distances (**left**) and Au-Au distances in Å (**right**). Only the gold atoms and the pyrazolate rings are represented.

On examining the structures of gold(I) reported in the CSD we found 28 trimers, 8 tetramers and 1 hexamer (dimers and pentamers were not found).

The X-ray structures previously discussed are summarized in Table 1 together with NH-pyrazole cyclamers.

Table 1. Structures found in the Cambridge Structural Database (CSD) for metallacycles formed by pyrazolate ligands and the coinage metals M = Cu(I), Ag(I) and Au(I): $(pzM)_n$ for n = 2, 3, 4, 5 and 6. The percentages are in brackets. For comparative purposes, the results for NH-pyrazoles (cyclamers) are also provided. The relative order from frequent to zero (pentamers) is in bold.

Metal/H	Total	Dimers	Trimers	Tetramers	Pentamers	Hexamers (Refcodes)
Cu(I)	81	**2** 22 [27.2]	**1** 41 [50.6]	**3** 18 [22.2]	**5** 0 [0.0]	**4** 0 [0.0]
Ag(I)	96	**2** 6 [6.2]	**1** 72 [75.0]	**3** 6 [6.2]	**5** 0 [0.0]	**4** 2 (QEJJEX [54], QEJJIB [54]) [2.1]
Au(I)	37	**4** 0 [0.0]	**1** 28 [75.7]	**2** 8 [21.6]	**5** 0 [0.0]	**3** 1 (FEJJAF10 [62]) [2.7]
H [26]	38	**1** 16 [42.1]	**3** 8 [21.1]	**2** 13 [34.2]	**5** 0 [0.0]	**4** 1 [2.6]

Thus, the situation has some similarities for H and for M in the sense that NH pyrazoles cyclamers with n = 2, 3, 4, and 6 (rare) have been reported but a pentamer (**5**) has not been reported [25,26]. However, while metallacycles trimers are the most common (**1**), in NH-pyrazoles they occupy only the third position (**3**).

2.2. Calculated Structures (Minima in All Cases)

2.2.1. Geometries

We calculated different dispositions of the metallacycles using the parent pyrazolate ligand as a model, i.e., without any C-substituent. In the case of dimers, all adopt the planar conformation and never the folded conformation found in the CSD (Figure 13 and Table 2).

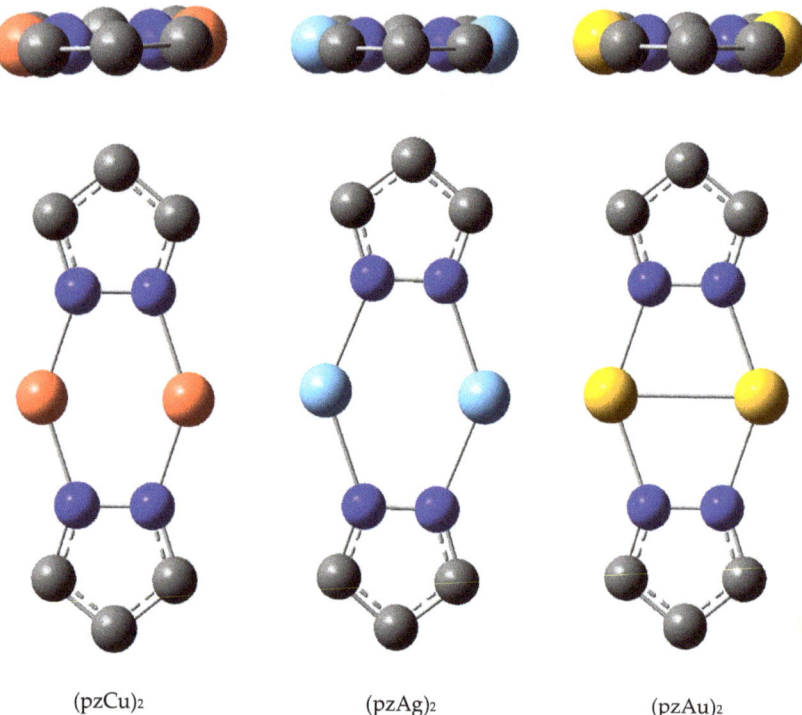

Figure 13. Two views of the calculated Cu, Ag and Au dimers; hydrogen atoms have been omitted.

Table 2. Calculated mean metal⋯metal and metal⋯N atom (Å) (averaged and parent pyrazoles).

Metal	Dimer		Trimer		Tetramer		Hexamer	
	M-M	N-M	M-M	N-M	M-M	N-M	M-M	N-M
Cu(I)	2.656	1.963	3.306	1.921	3.352	1.920	3.447	1.932
Ag(I)	2.953	2.192	3.507	2.124	3.448	2.122	3.631	2.145
Au(I)	2.808	2.124	3.440	2.041	3.478	2.037	3.605	2.042

The trimers lead to triangles and it is interesting to estimate how far they are from the equilateral case that results from a D_{3h} symmetry in the examples reported in the CSD. The tetramers will lead to squares (D_{4h}), planar deformed squares (rectangles, rhombs) and non-planar structures (folded about the M1–M3 edge). The situation increases in complexity as the number of metals increases; for hexamers there are the planar regular hexagon (D_{6h}) and several distorted hexagons, including the *ududud* structure (*u* or *d* refers to the *up* or *down* position of the pyrazole ring, as shown schematically in Figure 14).

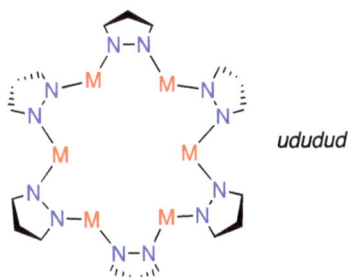

Figure 14. Schematic representation of the *ududud* structure.

We also calculated double dimers (2 + 2) and double trimers (3 + 3) in two orientations. The structures and some distances are represented in the following images for all metallacycles except dimers. The distances that were analyzed are M(I)···M(I) and N···M(I).

The studied Cu(I) derivatives, beyond monomers and dimers, are represented in Figure 15. The mean distances in the dimers are Cu···Cu = 2.656 Å and N···Cu = 1.963 Å.

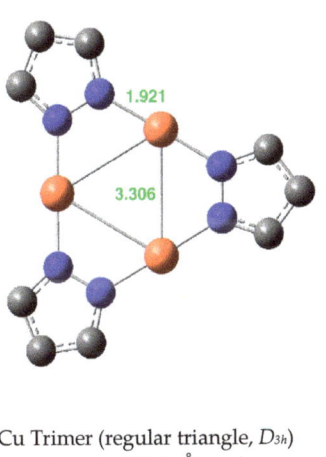

Cu Trimer (regular triangle, D_{3h})
Cu···Cu (3.306 Å) and
Cu···N distances (1.921 Å)

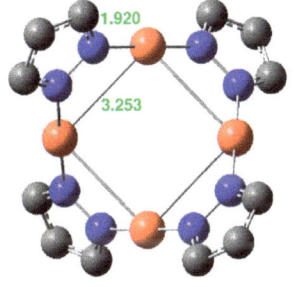

Cu Tetramer (planar square, D_{2d})
only one Cu···Cu (3.253 Å) and
one Cu···N distance (1.920 Å)

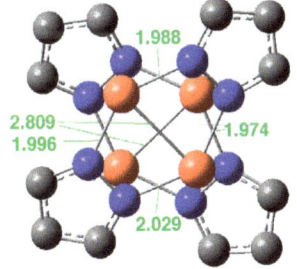

Cu$_4$ structure: 2 + 2
Cu···Cu (2.809 Å) and Cu···N
distances (1.974, 1.987, 1.996, 2029 Å)

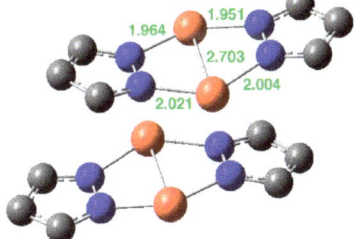

Cu$_4$, new structure 2 + 2 (twisted)
Cu···Cu (2.703 Å) and Cu···N
distances (1.964, 1.951, 2.004, 2.021 Å)

Figure 15. *Cont.*

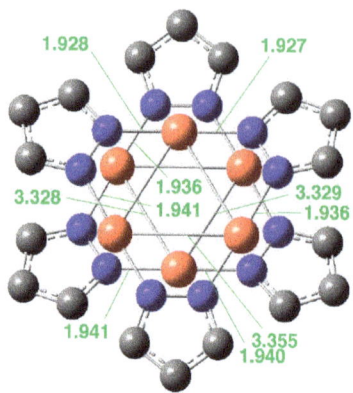

Cu₆ structure: 3 + 3
Cu···Cu (3.328, 3.329, 3.355 Å) and
Cu···N distances (1.928. 1.928. 1.936, 1.940, 1.940, 1936 Å)

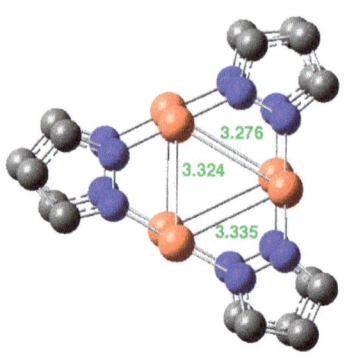

Cu₆, new structure 3 + 3 (twisted)
Cu···Cu (3.276, 3.324, 3.335 Å) and
Cu···N distances (1.924, 1.925, 1.929, 1.928, 1.937, 1.932Å)

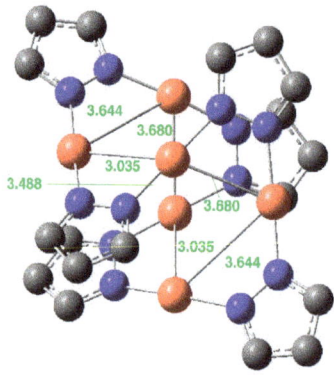

Cu analog of FEJJAF10 hexamer
Cu···Cu distances (3.644, 3.680, 3.035, 3.488 and 3.035, 3.644 Å)
Cu···N distances (1.916, 1.936, 1.944 Å)

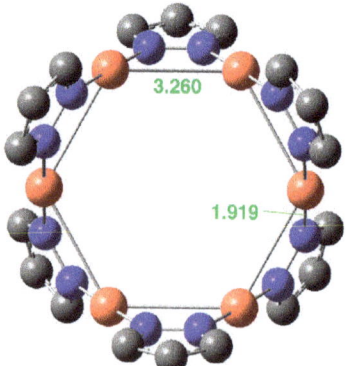

Cu *udadud*
Cu···Cu distance 3.260 Å
Cu···N distance 1.919 Å

Figure 15. Optimized structures of Cu(I) pyrazolates.

The Ag(I) structures are reported in Figure 16. The average distances for the dimer are Ag···Ag = 2.953 Å and N···Ag = 2.192 Å.

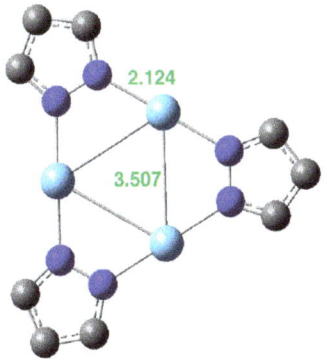

Ag Trimer (regular triangle, D_{3h})
Ag⋯Ag (3.507 Å) and
Ag⋯N distances (2.124 Å)

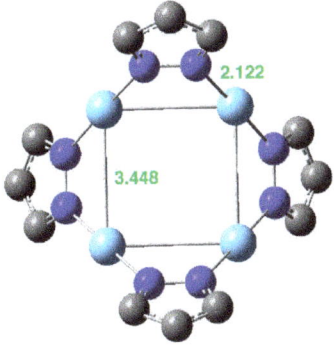

Ag Tetramer (planar square, D_{2d})
only one Ag⋯Ag (3.448 Å) and
one Ag⋯N distance (2.122 Å)

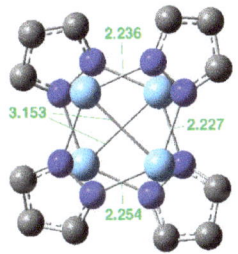

Ag$_4$ structure: 2 + 2
Ag⋯Ag (3.153 Å) and
Ag⋯N distances (2.227, 2.236, 2.254 Å)

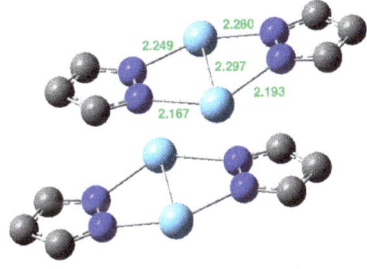

New 2 + 2 (twisted)
Ag⋯Ag (2.997 Å) and
Ag⋯N distances (2.167, 2.193, 2.249, 2.280 Å)

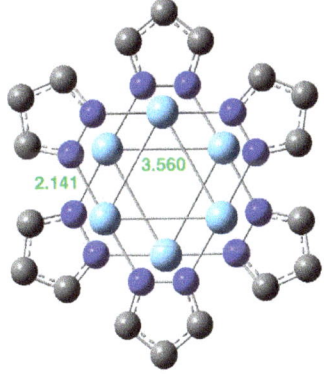

Ag$_6$ structure: 3 + 3
Ag⋯Ag (3.560 Å) and
Ag⋯N distances (2.141 Å)

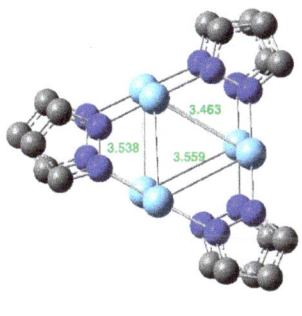

Ag$_6$, new structure 3 + 3 (twisted)
Ag⋯Ag (3.463, 3.538, 3.559 Å) and
Ag⋯N distances (2.126, 2.129, 2.133,
2.134, 2.153, 2.140 Å)

Figure 16. *Cont.*

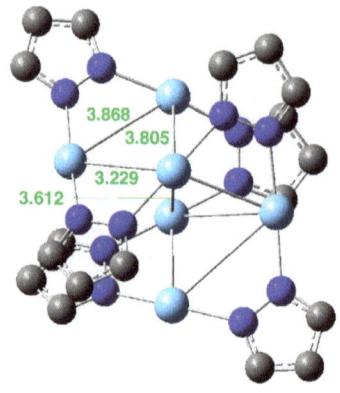 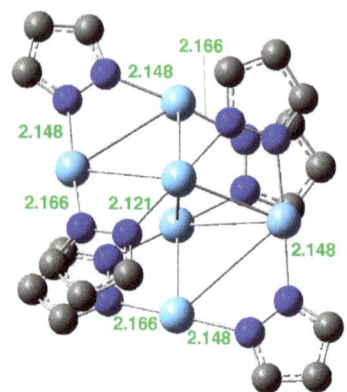

Ag analog of FEJJAF10 hexamer
Ag···Ag distances (3.229, 3.805, 3.868 and 3.612 Å)

Ag analog of FEJJAF10 hexamer (same structure than on the left) Ag···N distances (2.121, 2.148, 2.166 Å)

Figure 16. Optimized structures of Ag(I) pyrazolates.

Finally, the studied Au(I) derivatives are represented in Figure 17. For the dimer: Au···Au = 2.808 Å and N···Au = 2.124 Å.

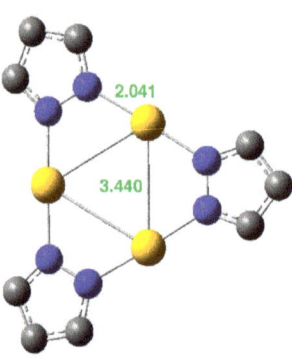 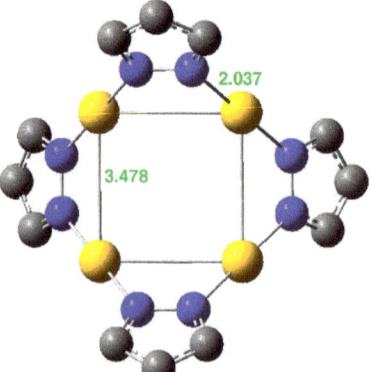

Au Trimer (regular triangle, D_{3h})
Au···Au (3.440 Å) and
Au···N distances (2.041 Å)

Au Tetramer (planar square, D_{2d})
only one Au···Au (3.478 Å) and
one Au···N distance (2.037 Å)

Figure 17. *Cont.*

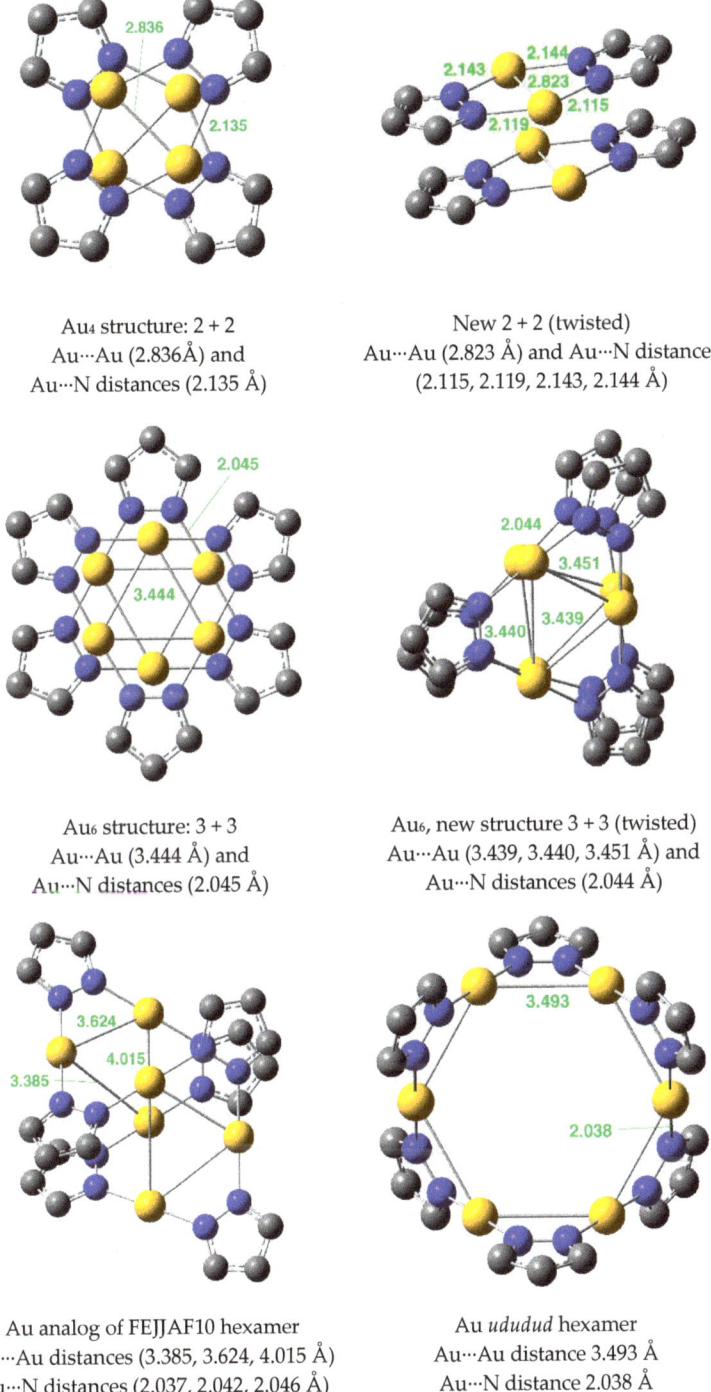

Au₄ structure: 2 + 2
Au···Au (2.836Å) and
Au···N distances (2.135 Å)

New 2 + 2 (twisted)
Au···Au (2.823 Å) and Au···N distances
(2.115, 2.119, 2.143, 2.144 Å)

Au₆ structure: 3 + 3
Au···Au (3.444 Å) and
Au···N distances (2.045 Å)

Au₆, new structure 3 + 3 (twisted)
Au···Au (3.439, 3.440, 3.451 Å) and
Au···N distances (2.044 Å)

Au analog of FEJJAF10 hexamer
Au···Au distances (3.385, 3.624, 4.015 Å)
Au···N distances (2.037, 2.042, 2.046 Å)

Au *ududud* hexamer
Au···Au distance 3.493 Å
Au···N distance 2.038 Å

Figure 17. Optimized structures of Au(I) pyrazolates.

2.2.2. Comparison of Calculated and Measured Geometries (Only Metal···Metal and Metal···Nitrogen Bond Lengths)

The average metal-metal and metal-nitrogen bond lengths in our calculations and in the structures found in the CSD search are gathered in Tables 2–4.

Table 3. Measured mean metal···metal and metal···N atom (Å) (averaged and parent pyrazoles).

Metal	Dimer (With Ligands)		Trimer		Tetramer		Hexamer	
	M-M	N-M	M-M	N-M	M-M	N-M	M-M	N-M
Cu(I)	3.424 [a]	1.974 [a]	3.228 [c]	1.858 [c]	3.095 (3)	1.849 (3)	-	-
Ag(I)	3.824 [b]	2.284 [b]	3.520 [d]	2.090 [d]	3.274 (5)	2.073 (5)	3.615 (5)	2.084 (5)
Au(I)	-	-	3.356 [e]	1.994 [e]	3.185 (9)	2.006 (5)	3.121 (10)	2.059 (10)

Experimental (averaged values) (the compounds used corresponds to those represented in the corresponding Figures in blue) or to compounds (refcodes) cited in the discussion. Some have been selected to build up this Table. [a] BITSAB [63], IPIGET [35], JEMCAF [32], NETMAD [37]; [b] FINWIL [64], KIRXIV [65], ZIGROZ [66], ZIGRUF [67]; [c] BELTOC [42], CODBAB [40], VIJMUW [41], XELXAN [43]; [d] AWAWUS [67], CENFIM [68], HICHIL [69], XOGJUA [45]; [e] COHFIO01 [50], FUWXOK01 [58].

Table 4. Experimental metal···metal and metal···N atom (Å) (averaged and parent pyrazoles).

Metal	Dimer (With Ligands)		Trimer		Tetramer		Hexamer	
	M-M	N-M	M-M	N-M	M-M	N-M	M-M	N-M
Cu(I)	3.726 [a]	2.010 [a]	3.251 [c]	1.861 [c]	3.394 [f]	1.962 [f]	-	-
Ag(I)	3.788 [b]	2.245 [b]	3.426 [d]	2.200 [d]	None	None	None	None
Au(I)	-	-	3.382 [e]	2.004 [e]	None	None	None	None

Experimental (unsubstituted pyrazoles or, at least, only 4-substituted pyrazoles). When neither HHH or HRH pyrazoles were found, none are written in the Table. [a] KIRXOB [65], NETLUW [37]; [b] RATFAT [70], RATFEX [70]; [c] No example with the parent pyrazole, instead the 4-chloro derivative (CODBAB [40]) was used; [d] HESBUC [4]; [e] MUTKUH [59]; [f] No example with the parent pyrazole, instead the 4-n-butyl derivative (FORGIE [71]) was used.

A statistical analysis of the results in Tables 2–4 provided the following three equations:

$$\text{Averaged} = (0.98 \pm 0.01) \text{ Parent}, n = 12, R^2 = 0.998, \text{RMS residual} = 0.12 \text{ Å} \quad (1)$$

$$\text{Averaged} = (0.96 \pm 0.01) + (0.93 \pm 0.09) \text{ dimer}, n = 20, R^2 = 0.998, \text{RMS residual} = 0.12 \text{ Å} \quad (2)$$

$$\text{Parent} = (1.00 \pm 0.01) + (0.96 \pm 0.06) \text{ dimer}, n = 12, R^2 = 0.999, \text{RMS residual} = 0.08 \text{ Å} \quad (3)$$

Equation (1) shows that averaged and parent pyrazole values are roughly proportional with a slope of 0.98 indicating that the averaged values are slightly smaller than the parent ones.

Equations (2) and (3) are similar, while (2) is better than (3), with a slope = 1.00 indicating that our calculated geometries that correspond to pyrazole itself are closer to a model of "parent" pyrazoles. It was found in a previous study [16] that the Ag···Ag distances of (pzAg)$_2$ are very sensitive to the ancillary ligands. If we assume that the situation is the same for the Cu(I) ligands (there are no examples of Au(I) dimers) it is sufficient to add a term (a dummy variable, one if dimers, zero if other metallacycles). The result is 0.93–0.96 Å and this indicates that the contraction of the Ag···Ag distance due to ancillary ligands is very important.

2.3. Energies

We start with a very simple premise that the more abundant a metallacycle of a given size found in the CSD the more stable the structure. A step further is to consider the percentages as a quantitative measurement of the stability in a sort of Maxwell-Boltzmann distribution. This implies two things: that the number of examples is very large and that the structures are in equilibrium (thermodynamic control). Clearly these conditions are not fulfilled, but it remains interesting to explore the possibility of partial agreement. In this work we explored the ring size, in NH-pyrazole cyclamers we successfully studied the effect of the C-substituents [25,26] and, finally, in the case of Ag(I) pyrazolate dimers we studied the effect of ancillary ligands, which can have a marked effect on the Ag···Ag distance with a concomitant decrease in stability that was compensated for by the ligands [16]. Consequently, the problem is of great complexity and it is useful to remember that the mechanism of crystal growth is also complex and is not fully understood [72,73].

In an effort to compare the stabilities of the different metallacycles we calculated their relative free energies, ΔG_{rel} in kJ mol^{-1}, per metallacycle and per monomer. The results are provided in Table 5. $\delta\Delta G_{rel} = [\Delta G_{rel} - \Delta G_{rel}$ (minimum)$] \times n$. The values corresponding to "true" dimers, trimers, tetramers, pentamers and hexamers are marked in bold for comparison with the percentages in Table 1. The more negative the ΔG_{rel}, the more stable the metallacycle (the monomer is not a metallacycle) while the higher the $\delta\Delta G_{rel}$ the less stable the metallacycle for any given n.

Table 5. Relative free energies per monomer, ΔG_{rel} and $\delta\Delta G_{rel}$, in kJ·mol^{-1}, of the metallacycles formed by the parent pyrazolate ligand and the coinage metals M = Cu(I), Ag(I) and Au(I): (pzM)$_n$ for n = 2, 3, 4, 5 and 6.

	Cu(I)		Ag(I)		Au(I)		H	
n-mer	ΔG_{rel}	$\delta\Delta G_{rel}$	ΔG_{rel}	$\delta\Delta G_{rel}$	ΔG_{rel}	$\delta\Delta G_{rel}$	ΔG_{rel}	$\delta\Delta G_{rel}$
Monomer	0	-	0	-	0	-	0	-
Dimer	**−211.4**	**0.0**	**−171.6**	**0.0**	**−165.5**	**0.0**	**−3.5**	**-**
1 + 1	−149.2	124.3	−126.4	90.4	−108.0	115.0	-	-
Trimer	**−254.4**	**-**	**−213.0**	**-**	**−256.0**	**-**	**−4.5**	**-**
Tetramer	**−256.0**	**0.0**	**−214.2**	**0.0**	**−257.8**	**0.0**	**−5.6**	**-**
2 + 2	−235.0	84.3	−202.8	45.4	−182.4	301.6	-	-
2 + 2 twisted	−229.4	106.7	−193.2	83.6	−182.1	302.8	-	-
Pentamer	**−259.4**	**-**	**−218.5**	**-**	**−253.9**	**-**	**-**	**-**
Hexamer	**−269.3**	**31.3**	**−230.0**	**33.5**	**−269.3**	**21.6**	**−2.0**	**10.6**
Hexamer ududud	−255.4	114.6	−213.6	131.9	−257.3	93.6	−3.7	0.0
3 + 3	−274.5	0.0	−235.6	0.0	−272.9	0.0	-	-
3 + 3 twisted	−271.3	19.4	−232.2	20.0	−272.6	1.8	-	-

To compare the data in Table 1 (crystal structures) and Table 5 (free energies) it is necessary to remember that in Table 1 the "2 + 2" and "3 + 3" structures are classified as dimers and trimers not as tetramers and hexamers, thus even if there are "3 + 3" structures that are more stable than hexamers, this does not affect the order of the values in bold.

Several main conclusions can be drawn from the values reported in Table 6:

1. Experimental metallacycles: mainly trimers, then dimers and tetramers, some hexamers, no pentamers.

2. Experimental cyclamers (NH-pyrazoles): dimers, tetramers and trimers, are common; hexamers are very rare and there are no pentamers. This is not identical but reasonably similar to the trend in experimental metallacycles. Note that the differences in cyclamers are insignificant (less than 6 kJ mol^{-1}) compared with metallacycles (Cu: −211.4/−269.3; Ag: −171.6/−232.3; Au: −165.5/−269.3 kJ mol^{-1}); this explains that steric effects of the substituents in cyclamers are sufficient to explain the size of the cycle (see point 6).

3. The absence of pentamers in the CSD can be due to the fact that pentameric species are crystallographically prohibited by "normal" rotational symmetry. Thus, perhaps more of these species exist but have not been crystallized for this reason.

4. Calculated metals: the order (in bold, Table 4) for Cu and Ag is hexamers, pentamers, tetramers, trimers and dimers; for Au the order is hexamers, tetramers, trimers, pentamers and dimers.

5. Metals: comparison of experimental vs. calculated values shows that there is no relationship between these, which means that the ring size is not the determining factor. Other factors such as steric effects of the substituents in the 3- and 5-positions, the roles of ancillary ligands, solvates and co-crystals as well as the kinetics of crystal growth could all play a determining role.

6. NH: there is a weak relationship for experimental vs. calculated values. Remember that in the experimental case the main factor is the steric effect of the substituents at the 3- and 5-positions [26].

7. There is no relationship between the order of calculated metals vs. that of calculated H.

Table 6. Comparison of the populations of Tables 1 and 5 (absolute values).

Metal/H	% Dimers	% Trimers	% Tetramers	% Pentamers	% Hexamers
Cu(I)	2 [27.2]	1 [50.6]	3 [22.2]	5 [0.0]	4 [0.0]
Ag(I)	2 [6.2]	1 [75.0]	3 [6.2]	5 [0.0]	4 [2.1]
Au(I)	4 [0.0]	1 [75.7]	2 [21.6]	5 [0.0]	3 [2.7]
H	1 [42.1]	3 [21.1]	2 [34.2]	5 [0.0]	4 [2.6]
	ΔG_{rel} dimers	ΔG_{rel} trimers	ΔG_{rel} tetramers	ΔG_{rel} pentamers	ΔG_{rel} hexamers
Cu(I)	5 211.4	4 254.4	3 256.0	2 259.4	1 269.3
Ag(I)	5 171.6	4 213.0	3 214.2	2 218.5	1 232.2
Au(I)	5 165.5	3 256.0	2 257.8	4 253.9	1 269.3
H	4 3.5	2 4.5	1 5.6	-	3 3.7 5 2.0

2.4. QTAIM Analysis

This study was limited to Au(I) because, as explained in the introduction, it is the most interesting metal and we have published two significant papers on this topic [17,19]. The AIM analysis was also employed successfully in a related work [16].

Analysis of the electron density within the QTAIM shows the presence of bond critical points (BCPs) that link the gold atoms with the nitrogen atoms, other gold atoms, and in one case, with a carbon atom (1 + 1). The molecular graph of all of the systems is provided in Figure 20 with an indication of the position of the electron density critical points and the bond paths that link the BCPs with the nuclei. The topological description is very simple for the monomer, trimer, tetramer, and hexamer (*ududud*), in which only sequential Au-N BCPs are found linking the different systems. In the rest of the cases, an Au-Au BCP and additional Au-N BCPs are found. The Au-N BCPs (17 unique contacts) are found for interatomic distances between 1.99 and 3.68 Å. The electron density at the BCPs ranges between 0.145 and 0.006 au, thus showing in all cases positive Laplacian values (between 0.423 and 0.018 au). These results are characteristics of BCPs between atoms with very different electronegativities. The negative value of the total energy at the BCP for those contacts with interatomic distances shorter than 2.2 Å is an indication of the partial covalent character of these interactions.

The Au-Au BCPs (seven unique cases) are present for interatomic distances between 2.81 and 4.02 Å. The electron density values range between 0.042 and 0.005 au with positive Laplacian values. As observed previously, some of the BCPs present negative values for the total energy density (interatomic distances shorter than 3.4 Å).

In the two types of BCPs analyzed in this research, excellent exponential relationships ($R^2 > 0.99$) were found between the electron density or the Laplacian at the BCP vs. the interatomic distance, a finding that it is consistent with previous reports in the literature for other contacts [16,74–78].

(pzAu) monomer

(pzAu)₂

(pzAu) + (pzAu) [1 + 1]

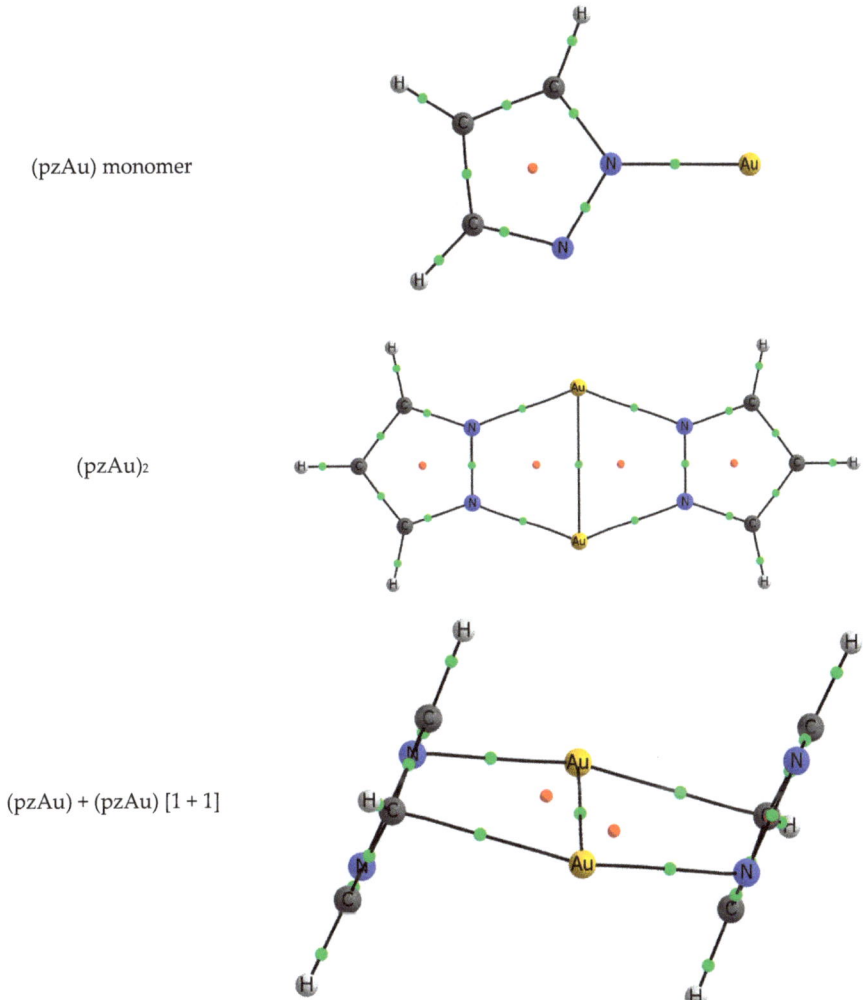

Figure 18. *Cont.*

(pzAu)₃

(pzAu)₄

(pzAu)₂ + (pzAu)₂ [2 + 2]

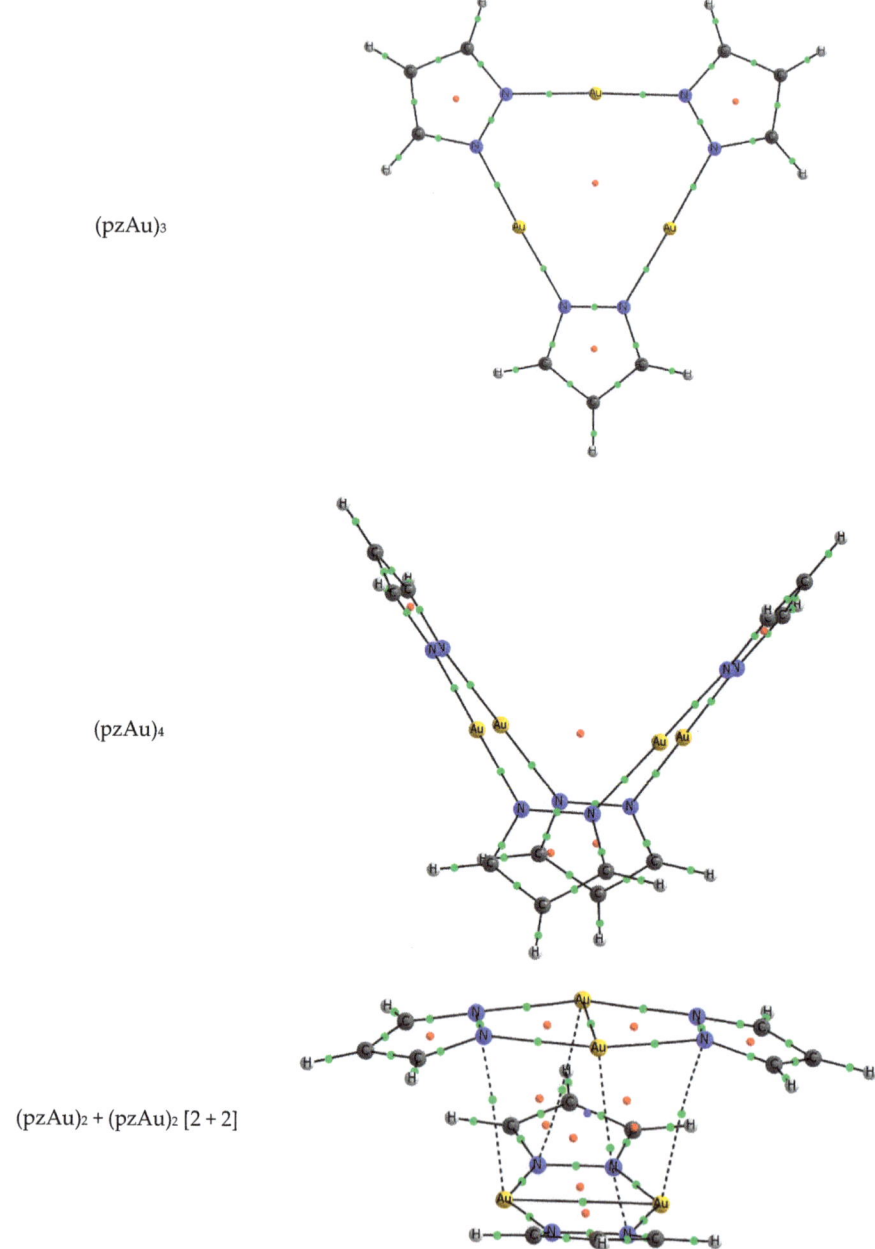

Figure 19. *Cont.*

Molecules **2020**, *25*, 5108

(pzAu)₅ [JALKIT]

(pzAu)₃ + (pzAu)₃ [3 + 3]

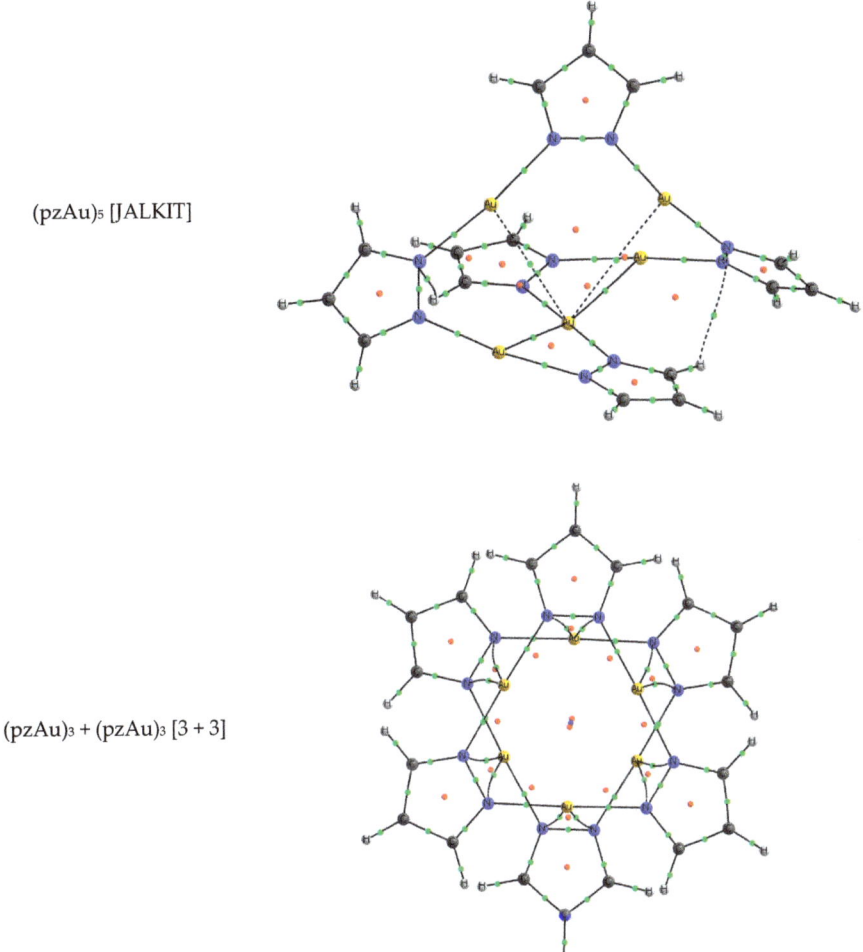

Figure 20. *Cont.*

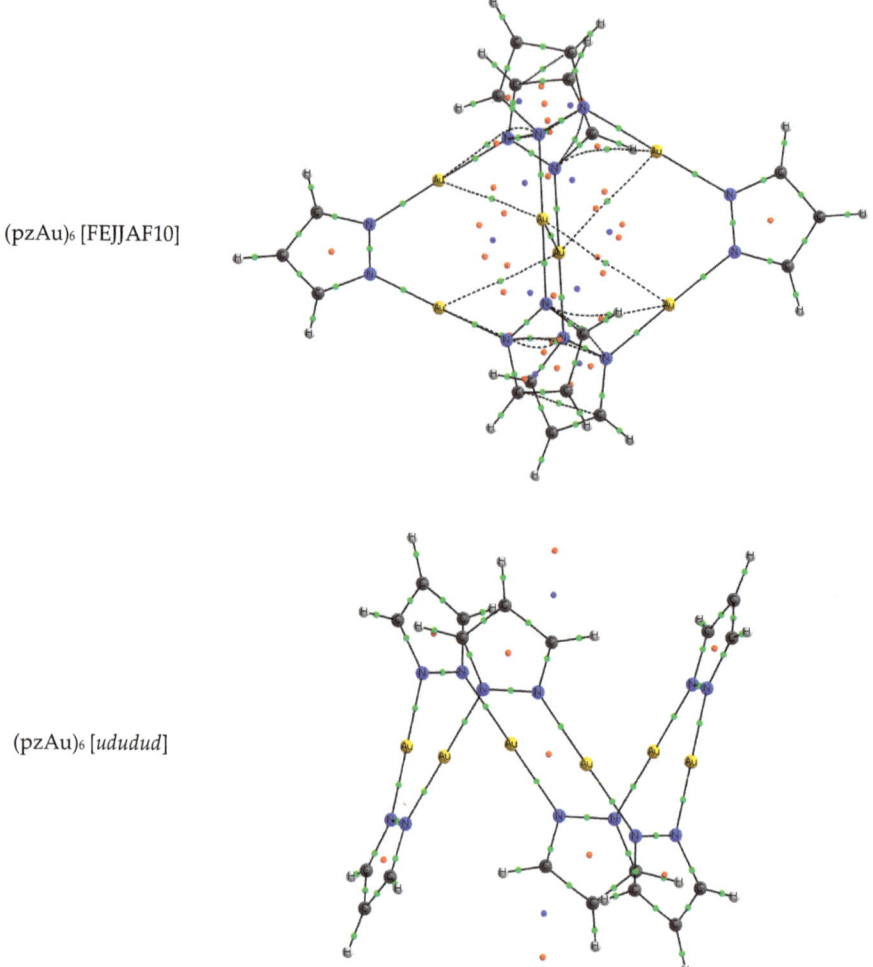

(pzAu)₆ [FEJJAF10]

(pzAu)₆ [ududud]

Figure 20. Molecular graph of the (pzAu)$_n$ complexes studied. The location of the bond, ring, and cage critical points are indicated with green red and blue dots.

3. Methods

The crystal structures with (pzM)$_n$ systems were searched in the CSD database 5.41 (November 2019) [20]. The M06-2x DFT functional [79] in combination with the jul-cc-pVDZ basis set [80,81] for the light atoms (C, N and H) and the aug-cc-pVDZ-PP effective core potential basis set [82] for the Cu, Ag and Au atoms were used for the theoretical calculations, all of them for isolated molecules in gas phase. The geometry optimization and frequency calculations were carried out with the Gaussian-16 package [83]. In all cases, the geometries obtained correspond to energetic minima (no imaginary frequencies).

The electron density of the systems was analyzed within the quantum theory of the atoms in molecules (QTAIM) [21,23] theory with the AIMAll program [84]. This program allows location and characterization of the critical points of the electron density (nuclear attractor, bond, ring and cage critical points).

4. Conclusions

The main conclusions of this work concerning the structure in the solid state of metallacycles of pyrazolates and coinage metals are:

1. The exploration of the CSD yielded a considerable number of crystal structures and this allowed a statistical analysis of the abundance of different cycles.

2. All examples contain only a single metal although it should not present any difficulties to prepare metallacycles with two or three metals.

3. Dimers and trimers are common in the case of Cu(I). Dimers in all cases contain other ligands. Double trimers (3 + 3) should not be confused with hexamers, which are not known. Pentamers are also not known. There is no reason why hexamers could not be prepared, but the main difficulty is that a method does not exist that allows selection a priori of the size of the ring.

4. Dimers and trimers are also common in the case of Ag(I). There are examples in which the hexagonal ring of dimers is planar, folded (boat-type) and folded (chain-type). In this case there are examples of "true" hexamers (no "3 + 3" double trimers), but otherwise Ag(I) and Cu(I) are similar.

5. In the case of Au(I) dimers are not known "all dimers are Au(III) derivatives". The double trimers form different patterns that can be classified according to the triangles formed by the three Au atoms. Tetramers are frequently found.

6. Calculations on simplified models (i.e., without C-substituents or other ligands) reproduce well the geometries but not the energies found experimentally, with stability increasing with ring size.

7. AIM analysis of the gold derivatives shows the presence of several Au-N and Au-Au BCPs and in one case an Au-C BCP.

Author Contributions: I.A. performed the calculations and the survey of the CSD. J.E. and I.A. contributed equally to the writing of this paper. Both authors have read and agreed to the published version of the manuscript.

Funding: We are grateful to the Spanish MICINN for financial support through project CTQ2018-094644-B-C22 and Comunidad de Madrid (P2018/EMT-4329 AIRTEC-CM).

Acknowledgments: Thanks are given to the CTI (CSIC) and the Irish Centre for High-End Computing (ICHEC) for their continued computational support. The authors would like to thank one of the reviewers for the idea that pentamers are crystallographically prohibited by "normal" rotational symmetry.

Conflicts of Interest: The authors declare no conflict of interest.

References

1. Zheng, J.; Lu, Z.; Wu, K.; Ning, G.-H.; Li, D. Coinage-Metal-Based Cyclic Trinuclear Complexes with Metal-Metal Interactions: Theories to Experiments and Structures to Functions. *Chem. Rev.* **2020**, *120*, 9675–9742. [CrossRef]
2. Okkersen, H.; Groenveld, W.L.; Reedijk, J. Pyrazoles and imidazoles as ligands. Part XVIII. Neutral and anionic pyrazole coordinated to Cu(I) and Ag(I). *Rec. Trav. Chim. Pays Bas.* **1973**, *92*, 945–953. [CrossRef]
3. Minghetti, G.; Banditelli, G.; Bonati, F. Metal derivatives of azoles. 3. The pyrazolato anion (and homologues) as a mono- or bidentate ligand: Preparation and reactivity of tri-. bi, and mononuclear gold(I) derivatives. *Inorg. Chem.* **1979**, *18*, 658–663. [CrossRef]
4. Masciocchi, N.; Moret, M.; Cairati, P.; Sironi, A.; Ardizzoia, G.A.; La Monica, G. The Multiphase Nature of the Cu(pz) and Ag(pz) (Hpz = Pyrazole) Systems: Selective Syntheses and Ab-Initio X-ray Powder Diffraction Structural Characterization of Copper(I) and Silver(I) Pyrazolates. *J. Am. Chem. Soc.* **1994**, *116*, 7668–7676. [CrossRef]
5. Ghazalli, N.F.; Yuliati, L.; Lintang, H.O. Molecular Self-Assembly of Group 11 Pyrazolate Complexes as Phosphorescent Chemosensors for Detection of Benzene. *IOP Conf. Ser. Mater. Sci. Eng.* **2018**, *299*, 12029. [CrossRef]
6. Lakhi, J.S.; Patterson, M.R.; Dias, H.V.R. Coinage metal metallacycles involving a fluorinated 3,5-diarylpyrazolate. *New J. Chem.* **2020**, *44*, 14814–14822. [CrossRef]
7. Fujisawa, K.; Saotome, M.; Takeda, S.; Young, D.J. Structures and Photoluminiscence of Coinage Metal(I) Phenylpyrazolato Trinuclear Complexes [M(3,5-Et2-4-Ph-pz)]3 and Arene Sandwich Complexes {[Ag(3,5-Et2-4-Ph-pz)]3}2(Ar) (Ar = Mesitylene and Toluene). *Chem. Lett.* **2020**, *49*, 670–673. [CrossRef]

8. Barberá, J.; Elduque, A.; Giménez, R.; Oro, L.A.; Serrano, J.L. Pyrazolate "Golden" Rings: Trinuclear Complexes That Form Columnar Mesophases at Room Temperature. *Angew. Chem. Int. Ed.* **1996**, *35*, 2832–2835. [CrossRef]
9. Barberá, J.; Lantero, I.; Moyano, S.; Serrano, J.L.; Elduque, A.; Giménez, R. Silver pyrazolates as coordination-polymer luminiscent metallomesogens. *Chem. Eur. J.* **2010**, *16*, 14545–14553. [CrossRef] [PubMed]
10. Beltrán, E.; Barberá, J.; Serrano, J.L.; Elduque, A.; Giménez, R. Chiral Cyclic Trinuclear Gold(I) Complexes with a Helical Columnar Phase. *Eur. J. Inorg. Chem.* **2014**, *2014*, 1165–1173. [CrossRef]
11. Cored, J.; Crespo, O.; Serrano, J.L.; Elduque, A.; Giménez, R. Decisive Influence of the Metal in Multifunctional Gold, Silver, and Copper Metallacycles: High Quantum Yield Phosphorescence, Color Switching, and Liquid Crystalline Behavior. *Inorg. Chem.* **2018**, *57*, 12632–12640. [CrossRef] [PubMed]
12. Ovejero, P.; Mayoral, M.J.; Cano, M.; Lagunas, M.C. Luminescence of neutral and ionic gold(I) complexes containing pyrazole or pyrazolate-type ligands. *J. Organomet. Chem.* **2007**, *692*, 1690–1697. [CrossRef]
13. Caramori, G.F.; Piccoli, R.M.; Segala, M.; Muñoz-Castro, A.; Guajardo-Maturana, R.; Andrada, D.M.; Frenking, G. Cyclic trinuclear copper(i), silver(i), and gold(i) complexes: A theoretical insight. *Dalton Trans.* **2015**, *44*, 377–385. [CrossRef] [PubMed]
14. Alkorta, I.; Elguero, J.; Dias, H.V.R.; Parasar, D.; Martín-Pastor, M. An experimental and computational NMR study of organometallic nine-membered rings: Trinuclear silver(I) complexes of pyrazolate ligands. *Magn. Reson. Chem.* **2020**, *58*, 319–328. [CrossRef] [PubMed]
15. Fujisawa, K.; Okano, M.; Martín-Pastor, M.; López-Sánchez, R.; Elguero, J.; Alkorta, I. Multinuclear magnetic resonance studies of five silver(I) trinuclear pyrazolate complexes. *Struct. Chem.* **2020**, in press. [CrossRef]
16. Alkorta, I.; Trujillo, C.; Sánchez-Sanz, G.; Elguero, J. Regium Bonds between Silver(I) Pyrazolates Dinuclear Complexes and Lewis Bases (N_2, OH_2, NCH, SH_2, NH_3, PH_3, CO and CNH). *Crystals* **2020**, *10*, 137. [CrossRef]
17. Sánchez-Sanz, G.; Trujillo, C.; Alkorta, I.; Elguero, J. Understanding Regium Bonds and their Competition with Hydrogen Bonds in Au_2:HX Complexes. *ChemPhysChem* **2019**, *20*, 1572–1580. [CrossRef] [PubMed]
18. Sánchez-Sanz, G.; Trujillo, C.; Alkorta, I.; Elguero, J. Rivalry between regium and hydrogen bonds established within diatomic coinage molecules and Lewis acids/bases. *ChemPhysChem* **2020**. accepted. [CrossRef] [PubMed]
19. Trujillo, C.; Sánchez-Sanz, G.; Elguero, J.; Alkorta, I. The Lewis acidities of gold(I) and gold(III) derivatives: A theoretical study of complexes of AuCl and AuCl3. *Struct. Chem.* **2020**, *31*, 1909–1918. [CrossRef]
20. Groom, C.R.; Bruno, I.J.; Lightfoot, M.P.; Ward, S.C. The Cambridge Structural Database. *Acta Crystallogr. Sect. B Struct. Sci. Cryst. Eng. Mater.* **2016**, *72*, 171–179. [CrossRef]
21. Bader, R.F.W. *Atoms in Molecules: A Quantum Theory*; (International Series of Monographs on Chemistry, 22); Clarendon Press: Oxford, UK, 1990.
22. Bader, R.F.W. A quantum theory of molecular structure and its applications. *Chem. Rev.* **1991**, *91*, 893–928. [CrossRef]
23. Popelier, P.L.A. *Atoms in Molecules: An Introduction*, 1st ed.; Prentice Hall: Harlow, UK, 2000.
24. Matta, C.F.; Boyd, R.J. *An Introduction to the Quantum Theory of Atoms in Molecules*; Wiley: Hoboken, NJ, USA, 2007; pp. 1–34.
25. Foces-Foces, C.; Alkorta, I.; Elguero, J. Supramolecular structure of 1H-pyrazoles in the solid state: A crystallographic and ab initio study. *Acta Crystallogr. Sect. B Struct. Sci.* **2000**, *56*, 1018–1028. [CrossRef]
26. Alkorta, I.; Elguero, J.; Foces-Foces, C.; Infantes, L. Classification of hydrogen-bond motives in crystals of NH-pyrazoles: A mixed empirical and theoretical approach. *Arkivoc* **2006**, *ii*, 15–30. [CrossRef]
27. Mohamed, A.A.; Burini, A.; Fackler, J.P. Mixed-Metal Triangular Trinuclear Complexes: Dimers of Gold–Silver Mixed-Metal Complexes from Gold(I) Carbeniates and Silver(I) 3,5-Diphenylpyrazolates. *J. Am. Chem. Soc.* **2005**, *127*, 5012–5013. [CrossRef]
28. Mohamed, A.A.; Galassi, R.; Papa, F.; Burini, A.; Fackler, J.P. Gold(I) and Silver(I) Mixed-Metal Trinuclear Complexes: Dimeric Products from the Reaction of Gold(I) Carbeniates or Benzylimidazolates with Silver(I) 3,5-Diphenylpyrazolate. *Inorg. Chem.* **2006**, *45*, 7770–7776. [CrossRef] [PubMed]
29. Parasar, D.; Jayaratna, N.B.; Muñoz-Castro, A.; Conway, A.E.; Mykhaailiuk, P.K.; Dias, H.V.R. Carbonyl complexes of copper(I) stabilized by bridging fluorinated pyrazolates and halide ions. *Dalton Trans.* **2019**, *48*, 6358–6371. [CrossRef]

30. Ardizzoia, G.A.; Cenini, S.; La Monica, G.; Masciocchi, N.; Maspero, A.; Moret, M. Syntheses, Structures, and Reactivity of Polynuclear Pyrazolato Copper(I) Complexes, Including an ab-Initio XRPD Study of [Cu(dmnpz)](3) (Hdmnpz = 3,5-Dimethyl-4-nitropyrazole). *Inorg. Chem.* **1998**, *37*, 4284–4292. [CrossRef]
31. Ardizzoia, G.A.; Cenini, S.; La Monica, G.; Masciocchi, N.; Moret, M. Synthesis, X-ray Structure, and Catalytic Properties of the Unprecedented Tetranuclear Copper(I) Species [Cu(dppz)]4 (Hdppz = 3,4-Diphenylpyrazole). *Inorg. Chem.* **1994**, *33*, 1458–1463. [CrossRef]
32. Dias, H.V.R.; Diyabalanage, H.V.K.; Eldabaja, M.G.; Elbjeirami, O.; Rawashdeh-Omary, M.A.; Omary, M.A. Brightly Phosphorescent Trinuclear Copper(I) Complexes of Pyrazolates: Substituent Effects on the Supramolecular Structure and Photophysics. *J. Am. Chem. Soc.* **2005**, *127*, 7489–7501. [CrossRef]
33. Titov, A.A.; Filippov, O.A.; Smol'yakov, A.F.; Baranova, K.F.; Titova, E.M.; Averin, A.A.; Shubina, E.S. Dinuclear CuI and AgI Pyrazolates Supported with Tertiary Phosphines: Synthesis, Structures, and Photophysical Properties. *Eur. J. Inorg. Chem.* **2019**, *2019*, 821–827. [CrossRef]
34. Galassi, R.; Simon, O.C.; Burini, A.; Tosi, G.; Conti, C.; Graiff, C.; Martins, N.M.R.; Da Silva, M.F.C.G.; Pombeiro, A.J.L.; Martins, L.M. Copper(I) and copper(II) metallacycles as catalysts for microwave assisted selective oxidation of cyclohexane. *Polyhedron* **2017**, *134*, 143–152. [CrossRef]
35. Omary, M.A.; Rawashdeh-Omary, M.A.; Diyabalanage, H.V.K.; Dias, H.V.R. Blue Phosphors of Dinuclear and Mononuclear Copper(I) and Silver(I) Complexes of 3,5-Bis(trifluoromethyl)pyrazolate and the Related Bis(pyrazolyl)borate. *Inorg. Chem.* **2003**, *42*, 8612–8614. [CrossRef] [PubMed]
36. Ardizzoia, G.A.; Beccalli, E.M.; La Monica, G.; Masciocchi, N.; Moret, M. Synthesis of poly(3,5-dicarbometho xypyrazolato)copper and its reactions with carbon monoxide and cyclohexyl isocyanide. Crystal structures of [Cu$_2$(dcmpz)$_2$(py)$_2$(CO)] and [Cu(dcmpz)(RNC)]$_2$ (Hdcmpz = 3,5-dicarbomethoxypyrazole, R = cyclohexyl). *Inorg. Chem.* **1992**, *31*, 2706–2711. [CrossRef]
37. Trose, M.; Nahra, F.; Poater, A.; Cordes, D.B.; Slawin, A.M.Z.; Cavallo, L.; Cazin, C.S. Investigating the Structure and Reactivity of Azolyl-Based Copper(I)-NHC Complexes: The Role of the Anionic Ligand. *ACS Catal.* **2017**, *7*, 8176–8183. [CrossRef]
38. Colombo, V.; Galli, S.; Choi, H.J.; Han, G.D.; Maspero, A.; Palmisano, G.; Masciocchi, N.; Long, J.R. High thermal and chemical stability in pyrazolate-bridged metal-organic frameworks with exposed metal sites. *Chem. Sci.* **2011**, *2*, 1311–1319. [CrossRef]
39. Mishima, A.; Fuyuhiro, A.; Kumagai, H.; Kawata, S. Bis[μ-3,5-bis-(2-pyrid-yl)pyrazolato]bis-(hydrogensulfato) -dicopper(II) methanol disolvate. *Acta Crystallogr. Sect. E Struct. Rep. Online* **2011**, *67*, m1523–m1524. [CrossRef]
40. Yang, H.; Zheng, J.; Peng, S.-K.; Zhu, X.-W.; Wan, M.-Y.; Lu, W.; Li, D. A chemopalette strategy for white light by modulating monomeric and excimeric phosphorescence of a simple Cu(I) cyclic trinuclear unit. *Chem. Commun.* **2019**, *55*, 4635–4638. [CrossRef]
41. Ehlert, M.K.; Rettig, S.J.; Storr, A.; Thompson, R.C.; Trotter, J. Synthesis and X-ray crystal structure of the 3,5-dimethylpyrazolato copper(I) trimer, [Cu(pz")]$_3$. *Can. J. Chem.* **1990**, *68*, 1444–1449. [CrossRef]
42. Fujisawa, K.; Ishikawa, Y.; Miyashita, Y.; Okamoto, K.-I. Crystal Structure of Pyrazolato-bridged Copper(I) Polynuclear Complexes. *Chem. Lett.* **2004**, *33*, 66–67. [CrossRef]
43. Dias, H.V.R.; A Polach, S.; Wang, Z. Coinage metal complexes of 3,5-bis(trifluoromethyl)pyrazolate ligand. *J. Fluor. Chem.* **2000**, *103*, 163–169. [CrossRef]
44. Morawitz, T.; Lemer, H.W.; Bolte, M. cyclo-Tri(μ2-3-phenyl-1H-pyrazole) tricopper(I). *Acta Crystallogr. Sect. E* **2006**, *62*, 1474–1476. [CrossRef]
45. Bertolotti, F.; Maspero, A.; Cervellino, A.; Guagliardi, A.; Masciocchi, N. Bending by Faulting: A Multiple Scale Study of Copper and Silver Nitropyrazolates. *Cryst. Growth Des.* **2014**, *14*, 2913–2922. [CrossRef]
46. Fujisawa, K.; Ishikawa, Y.; Miyashita, Y.; Okamoto, K.-I. Pyrazolate-bridged group 11 metal(I) complexes: Substituent effects on the supramolecular structures and physicochemical properties. *Inorganica Chim. Acta* **2010**, *363*, 2977–2989. [CrossRef]
47. Maspero, A.; Brenna, S.; Galli, S.; Penoni, A. Synthesis and characterization of new polynuclear copper(I) pyrazolate complexes and their catalytic activity in the cyclopropanation of olefins. *J. Organomet. Chem.* **2003**, *672*, 123–129. [CrossRef]
48. Titov, A.A.; Smol'yakov, A.F.; Rodionov, A.N.; Kosenko, I.D.; Guseva, E.A.; Zubavichus, Y.V.; Dorovatovskii, P.V.; Filippov, O.A.; Shubina, E.S. Ferrocene-containing tri- and tetranuclear cyclic copper(i) and silver(i) pyrazolates. *Russ. Chem. Bull.* **2017**, *66*, 1563–1568. [CrossRef]

49. Cañon-Mancisidor, W.; Gómez-García, C.J.; Espallargas, M.G.; Vega, A.; Spodine, E.; Venegas-Yazigi, D.; Coronado, E. Structural re-arrangement in two hexanuclear Cu(II) complexes: From a spin frustrated trigonal prism to a strongly coupled antiferromagnetic soluble ring complex with a porous tubular structure. *Chem. Sci.* **2014**, *5*, 324–332. [CrossRef]
50. Omary, M.A.; Rawashdeh-Omary, M.A.; Gonser, M.W.A.; Elbjeirami, O.; Grimes, T.; Cundari, T.R.; Diyabalanage, H.V.K.; Gamage, C.S.P.; Dias, H.V.R. Metal Effect on the Supramolecular Structure, Photophysics, and Acid−Base Character of Trinuclear Pyrazolato Coinage Metal Complexes†. *Inorg. Chem.* **2005**, *44*, 8200–8210. [CrossRef]
51. Fujisawa, K. CSD Private Communication, 2016.
52. Morishima, Y.; Young, D.J.; Fujisawa, K. Structure and photoluminescence of silver(i) trinuclear halopyrazolato complexes. *Dalton Trans.* **2014**, *43*, 15915–15928. [CrossRef] [PubMed]
53. Yamada, S.; Ishida, T.; Nogami, T. Supramolecular triangular and linear arrays of metal-radical solids using pyrazolato-silver(i) motifs. *Dalton Trans.* **2004**, 898–903. [CrossRef]
54. Lv, X.-P.; Wei, D.; Yang, G. Hexameric Silver(I) Pyrazolate: Synthesis, Structure, and Isomerization. *Inorg. Chem.* **2017**, *56*, 11310–11316. [CrossRef]
55. Yang, G.; Raptis, R.G. Synthesis and crystal structure of tetrameric silver(I) 3,5-di-tert-butyl-pyrazolate. *Inorganica Chim. Acta* **2007**, *360*, 2503–2506. [CrossRef]
56. Georgiou, M.; Wöckel, S.; Konstanzer, V.; Dechert, S.; John, M.; Meyer, F. Structural Variations in Tetrasilver(I) Complexes of Pyrazolate-bridged Compartmental N-Heterocyclic Carbene Ligands. *Z. Naturforsch. B* **2009**, *64*, s1542–s1554. [CrossRef]
57. Titov, A.A.; Filippov, O.A.; Smol'yakov, A.F.; Averin, A.A.; Shubina, E.S. Synthesis, structures and luminescence of multinuclear silver(i) pyrazolate adducts with 1,10-phenanthroline derivatives. *Dalton Trans.* **2019**, *48*, 8410–8417. [CrossRef]
58. Woodall, C.H.; Fuertes, S.; Beavers, C.M.; Hatcher, L.E.; Parlett, A.; Shepherd, H.J.; Christensen, J.; Teat, S.J.; Intissar, M.; Rodrigue-Witchel, A.; et al. Tunable Trimers: Using Temperature and Pressure to Control Luminescent Emission in Gold(I) Pyrazolate-Based Trimers. *Chem. Eur. J.* **2014**, *20*, 16933–16942. [CrossRef]
59. Yang, G.; Raptis, R.G. Supramolecular assembly of trimeric gold(I) pyrazolates through aurophilic attractions. *Inorg. Chem.* **2003**, *42*, 261–263. [CrossRef]
60. Fronczek, F.R. CSD Private Communication, 2014.
61. Yang, G.; Raptis, R.G. Synthesis, structure and properties of tetrameric gold(I) 3,5-di-tert-butyl-pyrazolate. *Inorg. Chim. Acta* **2003**, *352*, 98–104. [CrossRef]
62. Murray, H.H.; Raptis, R.G.; Fackler, J.P., Jr. Syntheses and X-ray structures of group 11 pyrazole and pyrazolate complexes. X-ray crystal structures of bis(3,5-diphenyl-pyrazole)copper(II) dibromide, tris(μ-3,5-diphenylpyrazolato -N,N') trisilver(I)-2-tetrahydro-furan, tris(μ-3,5-diphenylpyrazolato-N,N')trigold(I), and hexakis(μ-3,5-diphenyl -pyrazolato-N,N')hexagold(I). *Inorg. Chem.* **1988**, *27*, 26–33. [CrossRef]
63. Parasar, D.; Almotawa, R.M.; Jayaratna, N.B.; Ceylan, Y.S.; Cundari, T.R.; Omary, M.A.; Dias, H.V.R. Synthesis, Photophysical Properties, and Computational Analysis of Di- and Tetranuclear Alkyne Complexes of Copper(I) Supported by a Highly Fluorinated Pyrazolate. *Organometallics* **2018**, *37*, 4105–4118. [CrossRef]
64. Titov, A.A.; Smol'yakov, A.F.; Baranova, K.F.; Filippov, O.A.; Shubina, E.S. Synthesis, structures and photophysical properties of phosphorus-containing silver 3,5-bis(trifluoromethyl)pyrazolates. *Mendeleev Commun.* **2018**, *28*, 387–389. [CrossRef]
65. Yang, C. CSD Private Communication, 2019.
66. Kandel, S.; Stenger-Smith, J.; Chakraborty, I.; Raptis, R.G. Syntheses and X-ray crystal structures of a family of dinuclear silver(I)pyrazolates: Assessment of their antibacterial efficacy against P. aeruginosa with a soft tissue and skin infection model. *Polyhedron* **2018**, *154*, 390–397. [CrossRef]
67. Jayaratna, N.B.; Olmstead, M.M.; Kharisov, B.I.; Dias, H.V.R. Coinage Metal Pyrazolates [(3,5-$(CF_3)_2$Pz)M]$_3$ (M = Au, Ag, Cu) as Buckycatchers. *Inorg. Chem.* **2016**, *55*, 8277–8280. [CrossRef]
68. Yang, G.; Baran, P.; Martínez, A.R.; Raptis, R.G. Substituent Effects on the Supramolecular Aggregation of AgI-Pyrazolato Trimers. *Cryst. Growth Des.* **2012**, *13*, 264–269. [CrossRef]
69. Dias, H.V.R.; Gamage, C.S.P.; Keltner, J.; Diyabalanage, H.V.K.; Omari, I.; Eyobo, Y.; Dias, N.R.; Roehr, N.; McKinney, L.; Poth, T.; et al. Trinuclear Silver(I) Complexes of Fluorinated Pyrazolates. *Inorg. Chem.* **2007**, *46*, 2979–2987. [CrossRef]

70. Ardizzoia, G.A.; La Monica, G.; Maspero, A.; Moret, M.; Masciocchi, N. Silver(I) Pyrazolates. Synthesis and X-ray and (31)P-NMR Characterization of Triphenylphosphine Complexes and Their Reactivity toward Heterocumulenes. *Inorg. Chem.* **1997**, *36*, 2321–2328. [CrossRef]
71. Tian, A.-X.; Ning, Y.-L.; Ying, J.; Hou, X.; Li, T.-J.; Wang, X.-L. Three multi-nuclear clusters and one infinite chain induced by a pendant 4-butyl-1H-pyrazole ligand for modification of Keggin anions. *Dalton Trans.* **2015**, *44*, 386–394. [CrossRef]
72. Martins, M.A.P.; Salbego, P.R.S.; De Moraes, G.A.; Bender, C.R.; Zambiazi, P.J.; Orlando, T.; Pagliari, A.B.; Frizzo, C.P.; Hörner, M. Understanding the crystalline formation of triazene N-oxides and the role of halogen⋯π interactions. *CrystEngComm* **2018**, *20*, 96–112. [CrossRef]
73. Copetti, J.P.P.; Salbego, P.R.S.; Orlando, T.; Rosa, J.M.L.; Fiss, G.F.; de Oliveira, J.P.G.; Vasconcellos, M.L.A.A.; Zanatta, N.; Bonacorso, H.G.; Martins, M.A.P. Substituent effects of 7-chloro-4-substituted quinolines. *CrystEngComm* **2020**, *22*, 4094–4107. [CrossRef]
74. Alkorta, I.; Solimannejad, M.; Provasi, P.; Elguero, J. Theoretical Study of Complexes and Fluoride Cation Transfer Between N_2F^+ and Electron Donors. *J. Phys. Chem. A* **2007**, *111*, 7154–7161. [CrossRef]
75. Hugas, D.; Simon, S.; Duran, M. Electron Density Topological Properties are Useful to Assess the Difference Between Hydrogen and Dihydrogen Complexes. *J. Phys. Chem. A* **2007**, *111*, 4506–4512. [CrossRef]
76. Mata, I.; Alkorta, I.; Molins, E.; Espinosa, E. Universal Features of the Electron Density Distribution in Hydrogen-Bonding Regions: A Comprehensive Study Involving H⋯X (X=H, C, N, O, F, S, Cl, π) Interactions. *Chem. Eur. J.* **2010**, *16*, 2442–2452. [CrossRef] [PubMed]
77. Sánchez-Sanz, G.; Alkorta, I.; Elguero, J. Theoretical Study of Intramolecular Interactions in Peri-Substituted Naphthalenes: Chalcogen and Hydrogen Bonds. *Molecules* **2017**, *22*, 227. [CrossRef] [PubMed]
78. Alkorta, I.; Elguero, J.; Del Bene, J.E.; Mó, O.; Montero-Campillo, M.M.; Yáñez, M.; Romero, O.M. Mutual Influence of Pnicogen Bonds and Beryllium Bonds: Energies and Structures in the Spotlight. *J. Phys. Chem. A* **2020**, *124*, 5871–5878. [CrossRef]
79. Zhao, Y.; Truhlar, D.G. The M06 suite of density functionals for main group thermochemistry, thermochemical kinetics, noncovalent interactions, excited states, and transition elements: Two new functionals and systematic testing of four M06 functionals and 12 other functionals. *Theor. Chem. Accounts* **2008**, *120*, 215–241. [CrossRef]
80. Dunning, T.H. Gaussian basis sets for use in correlated molecular calculations. I. The atoms boron through neon and hydrogen. *J. Chem. Phys.* **1989**, *90*, 1007–1023. [CrossRef]
81. Papajak, E.; Zheng, J.; Xu, X.; Leverentz, H.R.; Truhlar, D.G. Perspectives on Basis Sets Beautiful: Seasonal Plantings of Diffuse Basis Functions. *J. Chem. Theory Comput.* **2011**, *7*, 3027–3034. [CrossRef]
82. Peterson, K.A.; Puzzarini, C. Systematically convergent basis sets for transition metals. II. Pseudopotential-based correlation consistent basis sets for the group 11 (Cu, Ag, Au) and 12 (Zn, Cd, Hg) elements. *Theor. Chem. Accounts* **2005**, *114*, 283–296. [CrossRef]
83. Frisch, M.J.; Trucks, G.W.; Schlegel, H.B.; Scuseria, G.E.; Robb, M.A.; Cheeseman, J.R.; Scalmani, G.; Barone, V.; Petersson, G.A.; Nakatsuji, H.; et al. *Gaussian 16 Rev. A.03*; Gaussian, Inc.: Wallingford, CT, USA, 2016.
84. Keith, T.A. *AIMAll, Version 19.10.12*; TK Gristmill Software: Overland Park, KS, USA, 2019.

Sample Availability: Not available.

Publisher's Note: MDPI stays neutral with regard to jurisdictional claims in published maps and institutional affiliations.

© 2020 by the authors. Licensee MDPI, Basel, Switzerland. This article is an open access article distributed under the terms and conditions of the Creative Commons Attribution (CC BY) license (http://creativecommons.org/licenses/by/4.0/).

Article

Synthesis and Structural Characterization of a Silver(I) Pyrazolato Coordination Polymer [†]

Kiyoshi Fujisawa [1,*], Takuya Nemoto [1], Yui Morishima [1] and Daniel B. Leznoff [2,*]

[1] Department of Chemistry, Ibaraki University, Mito, Ibaraki 310-8512, Japan; cu_peroxo@yahoo.co.jp (T.N.); fujisawa0608@gmail.com (Y.M.)
[2] Department of Chemistry, Simon Fraser University, 8888 University Drive, Burnaby, BC V5A1S6, Canada
* Correspondence: kiyoshi.fujisawa.sci@vc.ibaraki.ac.jp (K.F.); dleznoff@sfu.ca (D.B.L.); Tel.: +81-29-853-8373 (K.F.); +1-778-782-4887 (D.B.L.)
[†] Dedication: Dedicated to Professor Edward R.T. Tiekink on the occasion of his 60th birthday.

Citation: Fujisawa, K.; Nemoto, T.; Morishima, Y.; Leznoff, D.B. Synthesis and Structural Characterization of a Silver(I) Pyrazolato Coordination Polymer. *Molecules* 2021, 26, 1015. https://doi.org/10.3390/molecules26041015

Academic Editors: Vera L. M. Silva and Artur M. S. Silva

Received: 11 January 2021
Accepted: 12 February 2021
Published: 15 February 2021

Publisher's Note: MDPI stays neutral with regard to jurisdictional claims in published maps and institutional affiliations.

Copyright: © 2021 by the authors. Licensee MDPI, Basel, Switzerland. This article is an open access article distributed under the terms and conditions of the Creative Commons Attribution (CC BY) license (https://creativecommons.org/licenses/by/4.0/).

Abstract: Coinage metal(I)···metal(I) interactions are widely of interest in fields such as supramolecular assembly and unique luminescent properties, etc. Only two types of polynuclear silver(I) pyrazolato complexes have been reported, however, and no detailed spectroscopic characterizations have been reported. An unexpected synthetic method yielded a polynuclear silver(I) complex **[Ag(μ-L1Clpz)]$_n$** (L1Clpz$^-$ = 4-chloride-3,5-diisopropyl-1-pyrazolate anion) by the reaction of {[Ag(μ-L1Clpz)]$_3$}$_2$ with (^{n}Bu$_4$N)[Ag(CN)$_2$]. The obtained structure was compared with the known hexanuclear silver(I) complex {[Ag(μ-L1Clpz)]$_3$}$_2$. The Ag···Ag distances in **[Ag(μ-L1Clpz)]$_n$** are slightly shorter than twice Bondi's van der Waals radius, indicating some Ag···Ag argentophilic interactions. Two Ag–N distances in **[Ag(μ-L1Clpz)]$_n$** were found: 2.0760(13) and 2.0716(13) Å, and their N–Ag–N bond angles of 180.00(7)° and 179.83(5)° indicate that each silver(I) ion is coordinated by two pyrazolyl nitrogen atoms with an almost linear coordination. Every five pyrazoles point in the same direction to form a 1-D zig-zag structure. Some spectroscopic properties of **[Ag(μ-L1Clpz)]$_n$** in the solid-state are different from those of {[Ag(μ-L1Clpz)]$_3$}$_2$ (especially in the absorption and emission spectra), presumably attributable to this zig-zag structure having longer but differently arranged intramolecular Ag···Ag interactions of 3.39171(17) Å. This result clearly demonstrates the different physicochemical properties in the solid-state between 1-D coordination polymer and metalacyclic trinuclear (hexanuclear) or tetranuclear silver(I) pyrazolate complexes.

Keywords: polynuclear; silver; crystal structure; pyrazolate ligand; coordination polymer

1. Introduction

Cyclic trinuclear complexes with monovalent coinage metal ions have been of interest to coordination chemists for three decades [1–3]. One of the ligands known to form these cyclic trinuclear complexes is pyrazolate [4–6], which is known to act as an A-frame-like bridging ligand with some metal ions, with an Npz–M–Npz linear coordination mode (pz = pyrazolate anion, C$_3$H$_3$N$_2$$^-$) [1–4,6–9]. Early studies in 1970 suggested that silver(I) pyrazolato complexes existed as a polymeric 1-D chain [Ag(pz)]$_n$ (Figure 1, left) [10]. The structure of [Ag(pz)]$_n$ was discussed in that the deprotonated pyrazolato complexes are "at least trimeric but polymeric forms cannot be excluded", based on far-IR spectroscopy data [11]. The first structural characterization of pyrazolato complexes coordinated by coinage metal(I) ions was determined by Fackler and co-workers in 1988: their reported silver(I) pyrazolato complex had three intramolecular Ag···Ag distances of 3.306(2), 3.362(2), and 3.496(2) Å, forming a trinuclear complex [Ag(3,5-Ph$_2$pz)$_3$] (3,5-Ph$_2$pz = 3,5-diphenyl-1-pyrazolate anion) [12]. After this report, the hexanuclear complex {[Ag(3,5-Ph$_2$pz)$_3$]}$_2$ was also reported, having one intermolecular Ag···Ag interaction of 2.9712(14) Å without any crystalline solvents [13]. This dimerization to build the hexanuclear structure can be caused by an additional stabilization of the silver(I) ions provided by argentophilic interactions (Ag···Ag interactions) [14]. Subsequent ab initio powder X-ray diffraction (XRD) evidence also indicated two possibilities: silver(I) pyrazolato complexes could exist as either a coordination polymer [Ag(pz)]$_n$ with an intramolecular Ag···Ag distance of 3.40 Å or as a dimeric trinuclear {[Ag(pz)]$_3$}$_2$ structure with an intermolecular Ag···Ag distance of 3.431(2) Å, depending on the method of synthesis [15]. After that, the single-crystal characterization of the product made by the same synthetic method with aqueous NH$_3$ was reported as [Ag(pz)]$_n$ with an intramolecular Ag···Ag (argentophilic) interaction of 3.3718(7) Å, an intermolecular Ag···Ag interaction of 3.2547(6) Å, and an N–Ag–N angle of 169.98° (14) (Figures S1 and S2) [16]. However, its physicochemical properties such as solid-state photoluminescence have not been reported [15,16].

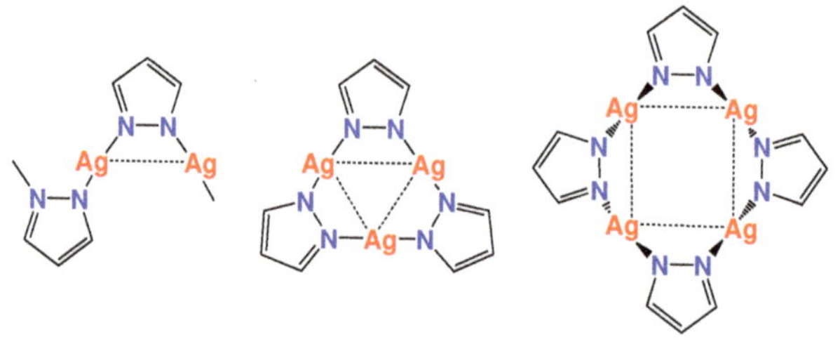

Figure 1. Schematic drawing of silver(I) pyrazolato complexes; left: polynuclear, center: trinuclear, right: tetranuclear. Dotted lines show intramolecular Ag···Ag interactions.

We have previously reported silver(I) pyrazolato complexes synthesized by alkyl- and aryl-substituted pyrazoles to form trinuclear and tetranuclear structures, depending on the method of synthesis and the nature of the substituents on the pyrazolate ring (Figure 1, center and right) [17–22]. Some Au···Ag (metallophilic) interactions in cyclic trinuclear coinage metal(I) complexes were found [1,2]. Au-Im complexes can be easily interacted with silver(I) ions to form Au···Ag interactions, since the order of π-acidity is Au < Cu < Ag for a given ligand, and Im$^-$ (imidazolate) < Pyridine < Cb$^-$ (carbeniate) < Pz$^-$ for a given metal [1–3]. As an example, the mixed metal(I) complex [(Ag{([Au(C^2,N^3-bzim)]$_3$)$_2$}](BF$_4$) with Au···Ag interactions was reported (bzim$^-$ = 1-benzylimidazole anion) [23].

With this electronic preference in mind, to construct Ag···Ag interactions by silver(I) pyrazolato complexes, it should be valuable to explore the different strategy of using electro-withdrawing substituents on the pyrazole ligand and [Ag(CN)$_2$]$^-$ as a silver(I) ion source. The dicyanoargentate(I) anion is very useful to build coordination polymers [24–28] and performs well as a building block in general [29]. In this article, we report the detailed structure and characterizations as well as the unexpected synthetic procedure that leads to the new polynuclear silver(I) pyrazolato complex **[Ag(μ-L1Clpz)]$_n$** (L1Clpz$^-$ = 4-chloride-3,5-diisopropyl-1-pyrazolate anion) (Figure 2).

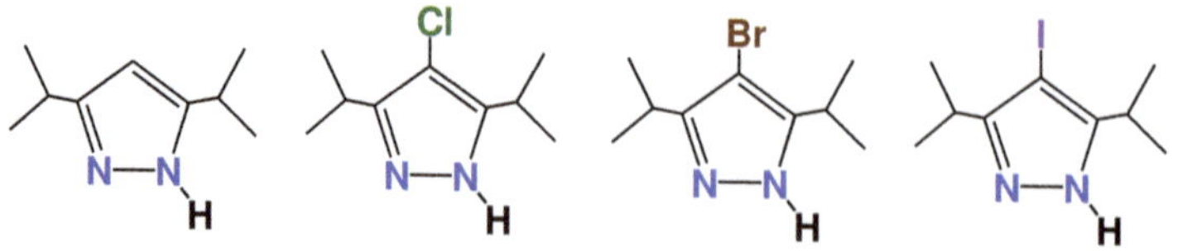

L1pz-H **L1Clpz-H** **L1Brpz-H** **L1Ipz-H**

Figure 2. Pyrazoles used in this research.

2. Results and Discussion

2.1. Synthesis

The reaction of {[Ag(μ-L1Clpz)]$_3$}$_2$ with 1.5 equivalents of (^{n}Bu$_4$N)[Ag(CN)$_2$] in ethyl acetate was carried out at room temperature as shown in Scheme 1. After one week of reaction time, a white powder was gradually generated. Its IR spectroscopic measurement revealed that the ν(C≡N) stretching peak of the starting material (^{n}Bu$_4$N)[Ag(CN)$_2$] at 2141 cm^{-1} disappeared and a new complex that did not contain any cyanide was generated. From a slow liquid-liquid diffusion method, we obtained a few single-crystals to reveal the structure (see Section 2.2). The above white powders prepared in the bulk synthetic method were the same as that obtained in single-crystal form and structurally characterized (below), as confirmed by the comparison of the powder X-ray diffractogram of the powder and the powder pattern generated from the single-crystal structure, as shown in Figure S3. As mentioned in the Introduction, although a [Ag(μ-L1Clpz)]$_3$/[Ag(CN)$_2$]$^-$ based polynuclear structure with intermolecular Ag···Ag interactions was initially targeted, only the polynuclear silver(I) complex **[Ag(μ-L1Clpz)]$_n$**, identified as a coordination polymer was obtained (Scheme 1); the detailed mechanism leading to this product was not explored. This coordination polymer assembly was only successfully performed by {[Ag(μ-L1Clpz)]$_3$}$_2$ but was not observed either by the non-halogenated derivative {[Ag(μ-L1pz)]$_3$}$_2$ or the other halogenated derivatives {[Ag(μ-L1Brpz)]$_3$}$_2$ and {[Ag(μ-L1Ipz)]$_3$}$_2$ (Figure 2). The reactions of these other hexanuclear complexes with (^{n}Bu$_4$N)[Ag(CN)$_2$] did not produce any powder precipitate. Indeed, after all of the solvent had slowly evaporated from the reaction mixture, powder XRD measurements were carried out on the resulting residues, which indicated that no new complexes were generated: only a physical mixture of the initial starting materials, i.e., the hexanuclear complexes {[Ag(L1Xpz)]$_3$}$_2$ (X = H, Br, and I) and (^{n}Bu$_4$N)[Ag(CN)$_2$] were observed. We also tried to make polynuclear [Ag(μ-L1Xpz)]$_n$ (X = H, Cl, Br, and I) by the literature method by using aqueous NH$_3$ [15,16], but only the same hexanuclear complexes {[Ag(L1Xpz)]$_3$}$_2$ (X = H, Cl, Br, and I) were obtained.

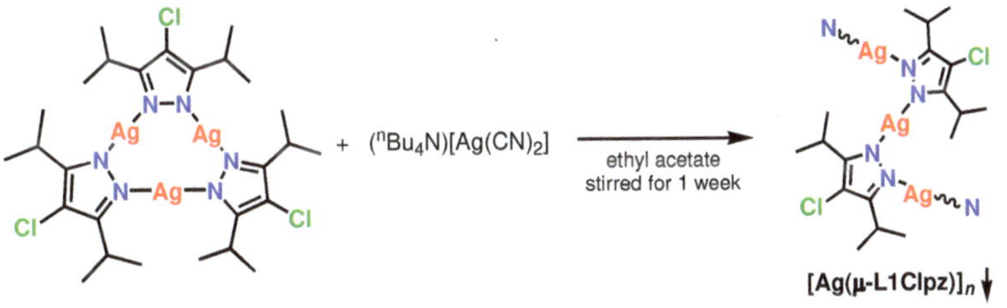

Scheme 1. Synthetic scheme of silver(I) coordination polymer **[Ag(μ-L1Clpz)]$_n$**.

2.2. Structure

Single-crystal X-ray structural analysis was performed on the polynuclear silver(I) material **[Ag(μ-L1Clpz)]$_n$**, the perspective drawing of which is shown in Figure 3. The relevant bond lengths (Å) and angles (°) are noted in the caption. The 1-D coordination polymer **[Ag(μ-L1Clpz)]$_n$** and packing diagram are drawn in Figure 4 and Figure S4, respectively.

[Ag(μ-L1Clpz)]$_n$ exists as a coordination polymer with an intramolecular argentophilic distance of Ag1···Ag2, 3.39171(17) Å, which is slightly shorter than twice Bondi's van der Waals radius (3.44 Å = 1.72 Å × 2) [30], indicating the presence of Ag···Ag argentophilic interactions [14]. Two Ag–N distances were found: Ag1–N1, 2.0760(13) Å and Ag2–N2, 2.0716(13) Å. The N–Ag–N bond angles of 180.00(7)° (N1–Ag1–N1′) and 179.83(5)° (N2′–Ag2′–N2″) indicate that each silver(I) ion has two pyrazolyl nitrogen atoms coordinated in an almost linear fashion. Two of the pyrazolate aromatic rings are completely co-planar with a dihedral angle of 0° (between pyrazole ring 1 and 2), and another pair is twisted relative to each other, with a dihedral angle of 79.24(7)° (between pyrazole ring 2 and 3) as shown in Figure 3. Every five pyrazole units point in the same direction, thereby forming a 1-D zig-zag structure. Each 1-D zig-zag structure is isolated and there are no significant interactions with the neighboring zig-zag chain. Within each 1-D zig-zag structure, two additional Ag···Ag distances are noteworthy: Ag1···Ag1#, 4.73335(10) Å and Ag2···Ag2′, 6.7834(3) Å. Moreover, this zig-zag structure has two Ag-based different angles with Ag2···Ag1···Ag2′, 180.000(4)° and Ag1···Ag2′···Ag1#, 88.499(6)° (Figure 3).

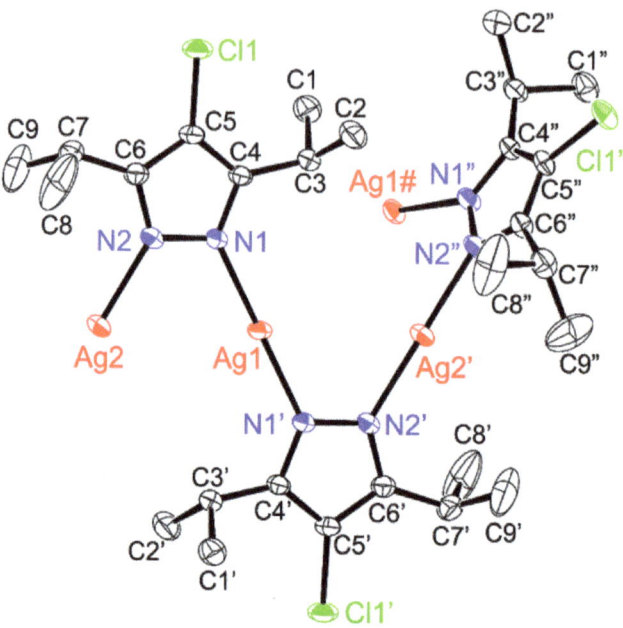

Figure 3. Crystal structure of **[Ag(μ-L1Clpz)]$_n$** showing 50% displacement ellipsoids and the atom labeling scheme. Hydrogen atoms were omitted for clarity. Relevant bond lengths (Å) and angles (°): Ag1–N1, 2.0760(13); Ag2–N2, 2.0716(13); Ag1–N1′, 2.0760(13); Ag2′–N2′, 2.0716(13); Ag2′–N2″, 2.0716(13); Ag1#–N1″, 2.0760(13); N1–N2, 1.3827(18); N1–Ag1–N1′, 180.00(7); N2′–Ag2′–N2″, 179.83(5); Ag1···Ag2, 3.39171(17); Ag1···Ag2′, 3.39171(17); Ag1#···Ag2′, 3.39171(17); Ag1···Ag1#, 4.73335(10); Ag2···Ag2′, 6.7834(3); Ag2···Ag1···Ag2′, 180.000(4); Ag1···Ag2′···Ag1#, 88.499(6). Symmetry codes: ′, −x+2, −y+1, −z+2; ″, x+1, −y+1, z; #, −x+1/2+2, y, −z+2.

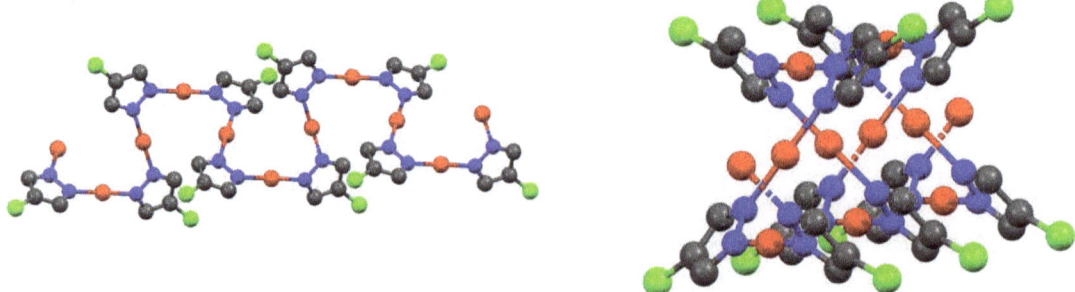

Figure 4. 1-D polynuclear structure of **[Ag(μ-L1Clpz)]$_n$** (**left**) top view and (**right**) side view. Carbon atoms of isopropyl group and hydrogen atoms were omitted for clarity. Color: red, silver; blue, nitrogen; green, chlorine; grey, carbon.

This **[Ag(μ-L1Clpz)]$_n$** structure is completely different from the reported one for [Ag(pz)]$_n$ which exhibits an intramolecular Ag···Ag distance of 3.3718(7) Å. The zig-zag chain in [Ag(pz)]$_n$ also interacts with neighboring zig-zag chains via inter-chain Ag···Ag distances of 3.2547(6) Å (Figures S1 and S2). The zig-zag structure in **[Ag(μ-L1Clpz)]$_n$** is also different from the original hexanuclear silver(I) complex {[Ag(μ-L1Clpz)]$_3$}$_2$, which features a dimeric trinuclear structure with three intermolecular argentophilic interactions of 3.1003(17), 3.1298(15), and 3.1051 Å to form an overall hexanuclear structure (Figures S5 and S6) [20]. These distances are significantly shorter than twice Bondi's van der Waals radius (3.44 Å) [30], indicating a strong Ag···Ag argentophilic interaction [14] (Figure S4). The layered structure of {[Ag(μ-L1Clpz)]$_3$}$_2$ did not interact with the neighboring layered one (Figure S6). The layered structure was formed with longer Ag···Ag distances between two {[Ag(μ-L1Clpz)]$_3$}$_2$ complexes of 7.3531(18), 7.3881(17), and 7.5258(18) Å.

2.3. Solution-State Properties

The ^{1}H-NMR spectrum of the obtained white powder **[Ag(μ-L1Clpz)]$_n$** in CDCl$_3$ was measured (Figure S7) and the observed chemical shifts are identical to those of {[Ag(μ-L1Clpz)]$_3$}$_2$ (Table S1) [20], indicating that the supramolecular solid-state 1-D zig-zag structure of **[Ag(μ-L1Clpz)]$_n$** is not stable upon dissolution, converting to form the known hexanuclear silver(I) complex {[Ag(μ-L1Clpz)]$_3$}$_2$. This observation is also supported by its solution-state UV-Vis spectrum in cyclohexane (Figure S8) and photoluminescence spectrum in cyclohexane (Figure S9). The maximum peak position in UV-Vis spectrum is 226 nm which is the same position as that of {[Ag(μ-L1Clpz)]$_3$}$_2$ (226 nm) [20]. Moreover, the emissive maximum is 307 nm (280 nm excitation), which is also the same position as that of {[Ag(μ-L1Clpz)]$_3$}$_2$. Therefore, detailed comparisons between **[Ag(μ-L1Clpz)]$_n$** and {[Ag(μ-L1Clpz)]$_3$}$_2$ were carried out by solid-state spectroscopies.

2.4. Solid-State Properties

IR and Raman spectra of the **[Ag(μ-L1Clpz)]$_n$** and {[Ag(μ-L1Clpz)]$_3$}$_2$ complexes are reproduced in Figures S10 and S11, respectively. The C=N stretching vibrations of 1506 cm^{-1} (IR) and of 1507 and 1495 cm^{-1} (Raman) in **[Ag(μ-L1Clpz)]$_n$** are almost the same as the C=N stretching vibrations of 1505 cm^{-1} (IR) and 1495 cm^{-1} (Raman) in {[Ag(μ-L1Clpz)]$_3$}$_2$. The C–Cl stretching vibrations could be observed in the far-IR region of 587 cm^{-1} (IR) and 575 cm^{-1} (Raman) in **[Ag(μ-L1Clpz)]$_n$**, which are slightly shifted from 579 cm^{-1} (IR) and 573 cm^{-1} (Raman) in {[Ag(μ-L1Clpz)]$_3$}$_2$ [20], respectively. The Ag–N stretching vibration should be observed at around 510 cm^{-1} [20,21,31] and was observed at 511 cm^{-1} (Raman) in **[Ag(μ-L1Clpz)]$_n$**, which was the same energy at 511 cm^{-1} (Raman) in {[Ag(μ-L1Clpz)]$_3$}$_2$ [20]. These vibrational spectroscopy comparisons indicate that each stretching vibration energy is almost the same as the other, although there is a clearly measurable difference in the C-Cl stretches, consistent with the presence of intercluster Cl···Cl interactions in the hexanuclear system (3.852 Å) that are absent in the 1-D zig-zag chain structure.

The solid-state UV-Vis absorption spectrum of **[Ag(μ-L1Clpz)]$_n$** acquired as a Nujol suspension is shown in Figure 5, along with that of {[Ag(μ-L1Clpz)]$_3$}$_2$ [20] for comparison. The characteristic absorption band at 225 nm in {[Ag(μ-L1Clpz)]$_3$}$_2$ [20] was obviously shifted to lower energy at 248 nm with a shoulder peak around 280 nm for the 1-D zig-zag structure. This is clearly different from the behavior of solution-state UV-Vis absorption spectra as shown in Figure S8. This shift in **[Ag(μ-L1Clpz)]$_n$** may be caused by polynuclear formation with intramolecular argentophilic interactions of Ag1···Ag2, 3.39171(17) Å as shown in Figure 3. For this detailed assignment, density functional theory calculations are required but are beyond the scope of this article. Nevertheless, the absorption band can be assigned to a silver(I) to pyrazolate change transfer (MLCT) based on the other reported hexanuclear coinage metal(I) complexes [18–20,32].

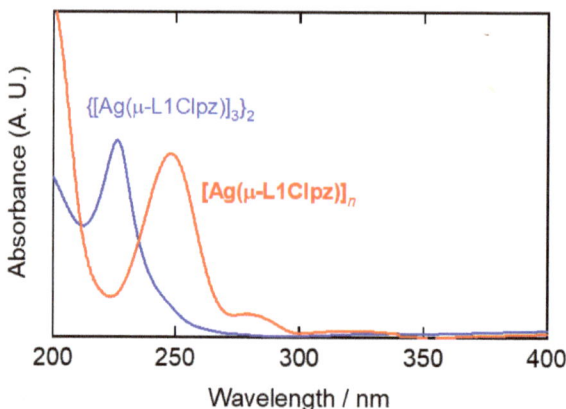

Figure 5. Solid-state UV-Vis absorption spectra of **[Ag(μ-L1Clpz)]**$_n$ (red line) and {[Ag(μ-L1Clpz)]$_3$}$_2$ (blue line) [20] recorded in a Nujol mull.

The emission spectrum of **[Ag(μ-L1Clpz)]**$_n$ in the solid-state was also somewhat different from that of the hexanuclear silver(I) analogue {[Ag(μ-L1Clpz)]$_3$}$_2$ [20] as shown in Figure 6 at room temperature, recorded using a 280 nm excitation wavelength. In {[Ag(μ-L1Clpz)]$_3$}$_2$, the main emissive band at 374 nm and an additional small one at 312 nm were observed. However, the emission of the coordination polymer **[Ag(μ-L1Clpz)]**$_n$ was clearly shifted to higher energy around 314 nm, and a new broad peak around 490 nm was observed.

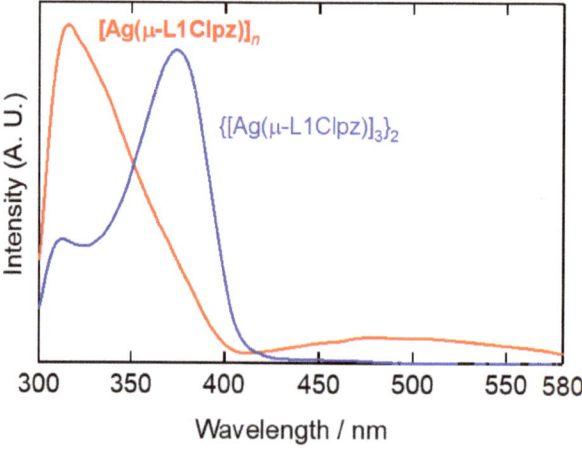

Figure 6. Solid-state photoluminescence spectra of **[Ag(μ-L1Clpz)]**$_n$ (red line) and {[Ag(μ-L1Clpz)]$_3$}$_2$ (blue line) [20] at room temperature recorded at 280 nm excitation wavelength.

The temperature-dependent photoluminescence spectra in **[Ag(μ-L1Clpz)]**$_n$ were recorded (Figure 7). The corresponding variable temperature emission spectra for {[Ag(μ-L1Clpz)]$_3$}$_2$ are also shown in Figure S12 [20]. The more intense 473 nm emission band of **[Ag(μ-L1Clpz)]**$_n$ at 83 K exhibited an additional vibrational fine structure around 346 nm, which was observed only at 83 K. From this vibrational behavior, this higher energy emission may be from ligand-based phosphorescence. On the other hand, the lower energy emission band was attributed to metal-based phosphorescence arising from closed shell d^{10}–d^{10} intramolecular Ag⋯Ag interactions [18–21,32–34]. For the original hexanuclear complex {[Ag(μ-L1Clpz)]$_3$}$_2$, the intensity of both bands at 374 and 312 nm gradually increased as the measurement temperature decreased. No bands around 470 nm were observed, even at low temperature. From these observations, the lower energy strong emission band at 473 nm in [Ag(μ-L1Clpz)]$_n$ is very unique due to its polynuclear supramolecular structure. We are now in the process of probing the origin of this stimulating behavior by theoretical and more detailed physicochemical researches.

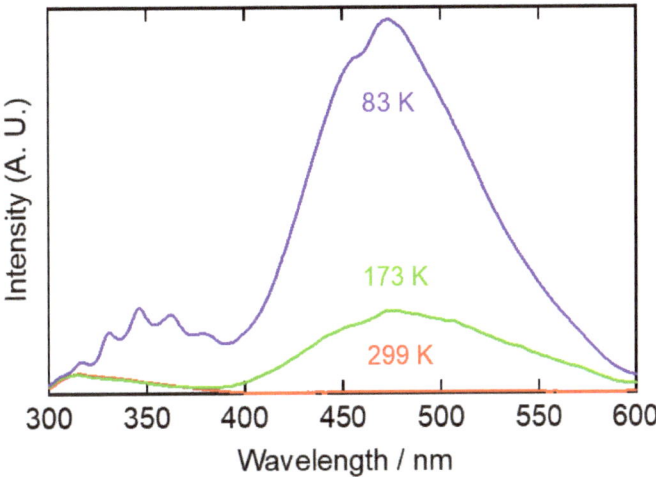

Figure 7. Solid-state temperature-dependent photoluminescence spectra at 83 K (violet line), 173 K (green line), and 299 K (red line) in [Ag(μ-L1Clpz)]$_n$ at 280 nm excitation.

3. Materials and Methods

3.1. Material and General Techniques

The preparation and handling of all complexes was performed under an argon atmosphere using standard Schlenk tube techniques. Ultra-dry ethyl acetate was purchased from Wako Pure Chemical Ind. Ltd. and deoxygenated by purging with argon gas. Deuteriochloroform was obtained from Cambridge Isotope Laboratories, Inc. Other reagents were commercially available and were used without further purification. The 3,5-diisopropyl-1-pyrazole (L1pz-H) [35] and its halogenated pyrazoles (L1Clpz-H, L1Brpz-H, and L1Ipz-H) [20] were prepared by published methods. (^{n}Bu$_4$N)[Ag(CN)$_2$] was obtained by the reaction of KAg(CN)$_2$ (1.016 g, 5.10 mmol) and ^{n}Bu$_4$NBr (1.7029 g, 5.28 mmol) in 20 mL of H$_2$O at room temperature for 2 hours to form a white powder of (^{n}Bu$_4$N)[Ag(CN)$_2$] (1.801 g, 4.48 mmol, 88% yield). Hexanuclear silver(I) complexes ({[Ag(μ-L1Clpz)]$_3$}$_2$, {[Ag(μ-L1Brpz)]$_3$}$_2$, and {[Ag(μ-L1Ipz)]$_3$}$_2$) were obtained by the published methods [20].

3.2. Instrumentation

IR spectra (4000–400 cm^{-1}) and far-IR spectra (680–150 cm^{-1}) were recorded as KBr pellets using a JASCO FT/IR-6300 spectrophotometer and as CsI pellets using a JASCO FT/IR 6700 spectrophotometer (JASCO, Tokyo, Japan), respectively. Raman spectra (4000–200 cm^{-1}) were measured as powders on a JASCO RFT600 spectrophotometer with a YAG laser 600 mW (JASCO, Tokyo, Japan). Abbreviations used in the description of vibration data are as follows: s, strong; m, medium; w, weak. ^{1}H-NMR (500 MHz) spectra were obtained on a Bruker AVANCE III-500 NMR spectrometer at room temperature (298 K) in CDCl$_3$ (Bruker, Yokohama, Japan). ^{1}H chemical shifts were reported as δ values relative to residual solvent peaks. UV-Vis spectra (solution and solid, 800–200 nm) were recorded on a JASCO V-570 spectrophotometer (JASCO, Tokyo, Japan). The values of ε were calculated per metal(I) ion. Solid samples (mulls) for spectroscopy were prepared by finely grinding microcrystalline material into powders with a mortar and pestle and then adding mulling agents (Nujol, poly(dimethylsiloxane), viscosity 10,000 (Aldrich)) before uniformly spreading it between quartz plates. Powder X-ray diffraction (XRD) measurements were conducted on a Rigaku SmartLab-SP/IUA X-ray diffractometer (Rigaku, Tokyo, Japan) with a Cu Kα radiation (λ = 1.54 Å) source (40 kV, 30 mA) and a high-speed one-dimensional detector D/teX Ultra 250. The 2θ was measured in the range of 5–90° with a scan step of 0.02° and scan speed of 10° $\min^{-1}$. Solid samples for XRD were prepared by finely grinding microcrystalline materials into powders with a mortar and pestle and then placing them on an aluminum dish (0.2 mm thickness). Luminescence spectra were recorded on a JASCO FP-6500 (solid, 700–300 nm) spectrofluorometer (JASCO, Tokyo, Japan). Low-temperature electronic absorption and luminescence spectra were recorded using solid samples cooled with a liquid nitrogen cryostat (CoolSpeK USP-203) from Unisoku Scientific Instruments (Osaka, Japan). The elemental analyses (C, H, and N) were performed by the Chemical Analysis Center of Ibaraki University.

3.3. Preparation of Complexes

[Ag(µ-L1Clpz)]$_n$

{[Ag(µ-L1Clpz)]$_3$}$_2$ (54.5 mg, 0.0309 mmol) and (^{n}Bu$_4$N)[Ag(CN)$_2$] (19.1 mg, 0.0475 mmol) were dissolved in ethyl acetate (6 cm^3) and the solution was stirred at room temperature for one week. A white solid was precipitated during this time. After this, the white solid was filtered and washed with a small amount of ethyl acetate. The obtained white powder as **[Ag(µ-L1Clpz)]$_n$** was dried by vacuum pump.

Yield: 22.7 mg, 0.077 mmol, 42%.

Calcd for C$_9$H$_{14}$AgClN$_2$·1/4(H$_2$O): C, 36.27; H, 4.90; N, 9.40. Found: C, 36.21; H, 4.57; N, 9.20. IR (KBr, cm^{-1}): 2971s(C-H), 2925s(C-H), 2862s(C-H), 1506m(C≡N), 1461m, 1400m, 1367m, 1361m, 1283m, 1145s, 1118s, 1092s, 1051s, 1035s, 804w. Far-IR(CsI, cm^{-1}): 685w, 587m(C-Cl), 561w, 540w, 501,476w, 437w, 411w, 389w, 371w, 349w, 276w, 237br. Raman (solid, cm^{-1}): 2974s(C-H), 2913s(C-H), 2865s(C-H), 1507m(C≡N), 1495m(C≡N), 1461m, 1448m, 1430m, 1382m, 1369s, 1300m, 1283m, 1176m, 1145w, 1107m, 957w, 880m, 706w, 657m, 575m(C-Cl), 511m(Ag-N), 438w, 385w, 341w, 268m. ^{1}H-NMR (CDCl$_3$, 500 MHz): δ 1.41 (d, J = 7 Hz, 36H, CH*Me*$_2$), δ 3.13 (sept, J = 7 Hz, 6H, C*H*Me$_2$). UV-Vis (solution, cyclohexane, λ$_{max}$/nm(ε/cm^{-1} mol^{-1} dm^3)) 226 (39200). UV-Vis (solid, Nujol, nm): 248, 280 (sh). Emission at 280 nm excitation wavelength (solution, cyclohexane, λ$_{max}$/nm): 307. Emission at 280 nm excitation wavelength (solid, λ$_{max}$/nm): 299 K, 315, 490; 173 K, 314, 474, 614sh; 83 K, 317, 331, 346, 362, 379, 456sh, 473.

Liquid-liquid diffusion was applied to obtain crystals at room temperature. The {[Ag(µ-L1Clpz)]$_3$}$_2$ (0.05 mmol) in 10mL of ethyl acetate was transferred to a 30 mL Erlenmeyer flask. On this solution, the solution containing (^{n}Bu$_4$N)[Ag(CN)$_2$] (0.1 mmol) dissolved in 10 mL of ethyl acetate was carefully layered and then the top was covered by parafilm. After a few weeks, some crystals were formed, which were suitable for single-crystal X-ray analysis.

3.4. X-ray Crystal Structure Determination

Crystal data and refinement parameters for the silver(I) pyrazolato coordination polymer **[Ag(µ-L1Clpz)]$_n$** are given in Table 1. All crystallographic data have been deposited at the Cambridge crystallographic data center (CCDC), 12 Union Road, Cambridge CB2 1EZ, UK and copies can be obtained on request, free of charge, by quoting the publication citation and the deposition number. CCDC number: 2053891.

The diffraction data were measured on a Rigaku/MSC Mercury CCD system (Rigaku, Tokyo, Japan) with graphite monochromated Mo Kα (λ = 0.71070 Å) radiation at −80 °C. The unit cell parameters of each crystal were determined using CrystalClear [36] from 6 images. The crystal to detector distance was 44.74 mm. Data were collected using 0.5° intervals in 65φ and ω to a maximum 2θ value of 55.0°. A total of 744 oscillation images were collected. The highly redundant data sets were reduced using CrysAlisPro [37]. An empirical absorption correction was applied for each complex. Structures were solved by direct methods (SIR2008) [38]. The position of the silver ions and their first coordination sphere were located from a direct method E-map; other non-hydrogen atoms were found in alternating difference Fourier syntheses and least squares refinement cycles. During the final refinement cycles the temperature factors were refined anisotropically. Refinement was carried out by a full matrix least-squares method on F^2. All calculations were performed with the CrystalStructure [39] crystallographic software package except for refinement, which was performed using SHELXL 2013 [40]. Hydrogen atoms were placed in calculated positions. Crystallographic data and structure refinement parameters including the final discrepancies (R and Rw) are listed in Table 1.

Table 1. Crystal data and structure refinement of [Ag(μ-L1Clpz)]$_n$.

Complex	[Ag(μ-L1Clpz)]$_n$
CCDC number	2053891
Empirical Formula	C$_9$H$_{14}$AgClN$_2$
Formula Weight	293.54
Crystal System	Monoclinic
Space Group	I2/a (#15)
a/Å	9.4667(11)
b/Å	11.0962(3)
c/Å	21.5508(7)
β/°	94.859(2)
V/Å^3	2255.66(11)
Z	8
D_{calc}/g cm^{-3}	1.729
μ(MoKα)/cm^{-1}	19.771
Temperature/°C	−80
2θ range, °	6–55
Reflections collected	8801
Unique reflections	2589
R_{int}	0.0111
Number of Variables	120
Refls./Para ratio	21.57
Residuals: R1 ($I > 2\sigma(I)$)	0.0155
Residuals: R (All reflections)	0.0167
Residuals: wR2 (All reflections)	0.0406
Goodness of fit indicator	1.046
Max/min peak,/e Å^{-3}	0.43/−0.40

$R1 = \Sigma||Fo| - |Fc||/\Sigma|Fo|$, $wR2 = [\Sigma(w(Fo^2 - Fc^2)^2)/\Sigma w(Fo^2)^2]^{1/2}$.

4. Conclusions

By using an unexpected synthetic method, we obtained a silver(I) pyrazolato complex [**Ag(μ-L1Clpz)**]$_n$ as a coordination polymer by the reaction of {[Ag(μ-L1Clpz)]$_3$}$_2$ with (^{n}Bu$_4$N)[Ag(CN)$_2$]. This polynuclear silver(I) structure was compared with the known hexanuclear silver(I) structure {[Ag(μ-L1Clpz)]$_3$}$_2$. Two N–Ag–N bond angles of 180.00(7)° and 179.83(5)° in the silver(I) coordination polymer [**Ag(μ-L1Clpz)**]$_n$ indicate that each silver(I) ion is coordinated by two pyrazolyl nitrogen atoms with an almost linear coordination. Every five pyrazoles point in the same direction to form a 1-D zig-zag structure. This zig-zag structure is not stable in solution, but it converts to the original hexanuclear silver(I) complex {[Ag(μ-L1Clpz)]$_3$}$_2$. In the solid-state photoluminescence spectrum, a lower energy strong emission band at 473 nm at lower temperature is very unique and attributable to the differences in polynuclear structure between the two systems. Silver(I) complexes are not generally so emissive, even at lower temperature and thus this 1-D zig-zag polynuclear structure is particularly noteworthy in coinage metal(I) pyrazolate research. This [**Ag(μ-L1Clpz)**]$_n$ broadens a family of silver(I) coordination polymers [24,25,41–43]. Moreover, some high antibacterial activity research using silver(I) coordination polymers are also reported [41–44]. Further efforts to probe how the structure of coinage metal(I) pyrazolates is affected by ligand and coordination environment are in progress.

Supplementary Materials: The following are available online, CIF and checkCIF reportr. Figure S1: 1-D polynuclear structure of $[Ag(\mu\text{-pz})]_n$, Figure S2: Packing diagram of $[Ag(\mu\text{-pz})]_n$, Figure S3: Powder X-ray diffraction spectra of $[Ag(\mu\text{-L1Clpz})]_n$ and calcd. pattern, Figure S4: Packing diagram of $[Ag(\mu\text{-L1Clpz})]_n$, Figure S5: Crystal structure of $\{[Ag(\mu\text{-L1Clpz})]_3\}_2$, Figure S6: Packing diagram of $\{[Ag(\mu\text{-L1Clpz})]_3\}_2$, Figure S7: ^{1}H-NMR spectrum of $[Ag(\mu\text{-L1Clpz})]_n$, Table S1: Comparisons of the ^{1}H-NMR chemical shifts, Figure S8: UV-Vis spectra of $[Ag(\mu\text{-L1Clpz})]_n$ and $\{[Ag(\mu\text{-L1Clpz})]_3\}_2$ in cyclohexane, Figure S9: Luminescence spectra of $[Ag(\mu\text{-L1Clpz})]_n$ and $\{[Ag(\mu\text{-L1Clpz})]_3\}_2$ in cyclohexane, Figure S10: IR spectra of $[Ag(\mu\text{-L1Clpz})]_n$ and $\{[Ag(\mu\text{-L1Clpz})]_3\}_2$, Figure S11: Raman spectra of $[Ag(\mu\text{-L1Clpz})]_n$ and $\{[Ag(\mu\text{-L1Clpz})]_3\}_2$, Figure S12: Temperature-dependent photoluminescence spectra in $\{[Ag(\mu\text{-L1Clpz})]_3\}_2$.

Author Contributions: K.F. and D.B.L. conceived and designed the project. T.N. and Y.M. performed the experiments. T.N. and K.F. analyzed the data. K.F. and D.B.L. wrote the paper. All authors have read and agreed to the published version of the manuscript.

Funding: This research received no directed/sponsored external funding.

Data Availability Statement: Data is contained within the article and supplementary material.

Acknowledgments: K.F. is grateful for the support from an Ibaraki University Priority Research Grant and the joint usage/research program "Artificial Photosynthesis" based at Osaka City University. DBL is grateful to NSERC of Canada for ongoing research support via the Discovery Grant program.

Conflicts of Interest: The authors declare no conflict of interest.

Sample Availability: Samples of the compounds $[Ag(\mu\text{-L1Clpz})]_n$ and $\{[Ag(\mu\text{-L1Clpz})]_3\}_2$ are available from the authors.

References

1. Zheng, J.; Lu, Z.; Wu, K.; Ning, G.-H.; Li, D. Coinage-metal-based cyclic trinuclear complexes with metal−metal interactions: Theories to experiments and structures to functions. *Chem. Rev.* **2020**, *120*, 9675–9742. [CrossRef] [PubMed]
2. Zheng, J.; Yang, H.; Xie, M.; Li, D. The π-acidity/basicity of cyclic trinuclear units (CTUs): From a theoretical perspective to potential applications. *Chem. Commun.* **2019**, *55*, 7134–7146. [CrossRef] [PubMed]
3. Galassi, R.; Rawashdeh-Omary, M.A.; Dias, H.V.R.; Omary, M.A. Homoleptic cyclic trinuclear d^{10} complexes: From self-association via metallophilic and excimeric bonding to the breakage thereof via oxidative addition, dative bonding, quadrupolar, and heterometal bonding interactions. *Comments Inorg. Chem.* **2019**, *39*, 287–348. [CrossRef]
4. Elguero, J.; Alkorta, I. A computational study of metallacycles formed by pyrazolate ligands and the coinage metals M = Cu(I), Ag(I) and Au(I): $(pzM)_n$ for $n = 2, 3, 4, 5$ and 6. comparison with structures reported in the Cambridge crystallographic data center (CCDC). *Molecules* **2020**, *25*, 5108. [CrossRef] [PubMed]
5. Omary, M.A.; Mohamed, A.A.; Rawashdeh-Omary, M.A.; Fackler, J.P., Jr. Photophysics of supramolecular binary stacks consisting of electron-rich trinuclear Au(I) complexes and organicelectrophiles. *Coord. Chem. Rev.* **2005**, *249*, 1372–1381. [CrossRef]
6. Zhang, J.-P.; Zhang, Y.-B.; Lin, J.-B.; Chen, X.-M. Metal azolate frameworks: From crystal engineering to functional materials. *Chem. Rev.* **2012**, *112*, 1001–1033. [CrossRef]
7. Halcrow, M.A. Pyrazoles and pyrazolides—Flexible synthons in self-assembly. *Dalton Trans.* **2009**, 2059–2073. [CrossRef]
8. Trofimenko, S. The coordination chemistry of pyrazole-derived ligands. *Chem. Rev.* **1972**, *72*, 497–509. [CrossRef]
9. Trofimenko, S. Recent advances in poly(pyrazolyl)borate (scorpionate) chemistry. *Chem. Rev.* **1993**, *93*, 943–982. [CrossRef]
10. Reimlinger, H.; Noels, A.; Jabot, J.; van Overstraeten, A. Synthesis with silver or sodium pyrazoles, I preparation of mono- and polypyrazoles. *Chem. Ber.* **1970**, *103*, 1942–1948. [CrossRef]
11. Okkersen, H.; Groeneveld, W.L.; Reedijk, J. Pyrazoles and imidazoles as ligands. Part XVIII: Neutral and anionic pyrazole coordinated to Cu(I) and Ag(I). *Recl. Trav. Chim. Pays-Bas* **1973**, *92*, 945–953. [CrossRef]
12. Murray, H.H.; Raptis, R.G.; Fackler, J.P., Jr. Syntheses and X-ray structures of group 11 pyrazole and pyrazolate complexes. X-ray crystal structures of bis(3,5-diphenylpyrazole)copper(II) dibromide, tris(μ-3,5-diphenylpyrazolato-N,N')trisilver(I)-2-tetrahydrofuran, tris(μ-3,5-diphenylpyrazolato-N,N')trigold(I), and hexakis(μ-3,5-diphenylpyrazolato- N,N')hexagold(I). *Inorg. Chem.* **1988**, *27*, 26–33.
13. Mohamed, A.A.; Pérez, L.M.; Fackler, J.P. Jr. Unsupported intermolecular argentophilic interaction in the dimer of trinuclear silver(I) 3,5-diphenylpyrazolates. *Inorg. Chim. Acta* **2005**, *358*, 1657–1662. [CrossRef]
14. Schmidbaur, H.; Schier, A. Argentophilic interactions. *Angew. Chem. Int. Ed.* **2015**, *54*, 746–784. [CrossRef] [PubMed]
15. Masciocchi, N.; Moret, M.; Cairati, P.; Sironi, A.; Ardizzoia, G.A.; Monica, G.L. The multiphase nature of the Cu(pz) and Ag(pz) (Hpz = pyrazole) systems: Selective syntheses and *ab-initio* X-ray powder diffraction structural characterization of copper(I) and silver(I) pyrazolates. *J. Am. Chem. Soc.* **1994**, *116*, 7668–7676. [CrossRef]

16. Zhang, C.-Y.; Feng, J.-B.; Gao, Q.; Xie, Y.-B. catena-Poly[silver(I)-µ-pyrazolato-$\kappa^2 N$:N']. *Acta Cryst.* **2008**, *64*, m352. [CrossRef] [PubMed]
17. Fujisawa, K.; Okano, M.; Martín-Pastor, M.; López-Sánchez, R.; Elguero, J.; Alkorta, I. Multinuclear magnetic resonance studies of five silver(I) trinuclear pyrazolate complexes. *Struct. Chem.* **2021**, *32*, 215–224. [CrossRef]
18. Fujisawa, K.; Saotome, M.; Takeda, S.; Young, D.J. Structures and photoluminescence of coinage metal(I) phenylpyrazolato trinuclear complexes [M(3,5-Et$_2$-4-Ph-pz)]$_3$ and arene sandwich complexes {[Ag(3,5-Et$_2$-4-Ph-pz)]$_3$}$_2$(Ar) (Ar = mesitylene and toluene). *Chem. Lett.* **2020**, *49*, 670–673. [CrossRef]
19. Saotome, M.; Shimizu, D.; Itagaki, A.; Young, D.J.; Fujisawa, K. Structures and photoluminescence of silver(I) and gold(I) cyclic trinuclear complexes with aryl substituted pyrazolates. *Chem. Lett.* **2019**, *48*, 533–536. [CrossRef]
20. Morishima, Y.; Young, D.J.; Fujisawa, K. Structure and photoluminescence of silver(I) trinuclear halopyrazolato complexes. *Dalton Trans.* **2014**, *43*, 15915–15928. [PubMed]
21. Fujisawa, K.; Ishikawa, Y.; Miyashita, Y.; Okamoto, K. Pyrazolate-bridged group 11 metal(I) complexes: Substituent effects on the supramolecular structures and physicochemical properties. *Inorg. Chim. Acta* **2010**, *363*, 2977–2989. [CrossRef]
22. Fujisawa, K.; Ishikawa, Y.; Miyashita, Y.; Okamoto, K. Crystal structure of pyrazolato-bridged copper(I) polynuclear complexes. *Chem. Lett.* **2004**, *33*, 66–67. [CrossRef]
23. Burini, A.; Bravi, R.; Fackler, J.P., Jr.; Galassi, R.; Grant, T.A.; Omary, M.A.; Pietroni, B.R.; Staples, R.J. Luminescent chains formed from neutral, triangular gold complexes sandwiching Tl^I and Ag^I. structures of {Ag([Au(μ-C^2,N^3-bzim)]$_3$)$_2$}BF$_4$·CH$_2$Cl$_2$, {Tl([Au(μ-C^2,N^3-bzim)]$_3$)$_2$}PF$_6$·0.5THF (bzim = 1-Benzylimidazolate), and {Tl([Au(μ-C(OEt)=NC$_6$H$_4$CH$_3$)]$_3$)$_2$}PF$_6$·THF, with MAu$_6$ (M = Ag$^+$, Tl$^+$) cluster cores. *Inorg. Chem.* **2000**, *39*, 3158–3165.
24. Baril-Robert, F.; Li, X.; Katz, M.J.; Geisheimer, A.R.; Leznoff, D.B.; Patterson, H. Changes in electronic properties of polymeric one-dimensional {[M(CN)$_2$]$^-$}$_n$ (M = Au, Ag) chains due to neighboring closed-shell Zn(II) or open-shell Cu(II) ions. *Inorg. Chem.* **2011**, *50*, 231–237. [CrossRef] [PubMed]
25. Shorrock, C.J.; Xue, B.-Y.; Kim, P.B.; Batchelor, R.J.; Patrick, B.O.; Leznoff, D.B. Heterobimetallic coordination polymers incorporating [M(CN)$_2$]$^-$ (M = Cu, Ag) and [Ag$_2$(CN)$_3$]$^-$ units: Increasing structural dimensionality via M–M' and M···NC interactions. *Inorg. Chem.* **2002**, *41*, 6743–6753. [CrossRef]
26. Rawashdeh-Omary, M.A.; Omary, M.A.; Patterson, H.H.; Fackler, J.P., Jr. Excited-state interactions for [Au(CN)$_2$$^-$]$_n$ and [Ag(CN)$_2$$^-$]$_n$ oligomers in solution. Formation of luminescent gold–gold bonded excimers and exciplexes. *J. Am. Chem. Soc.* **2001**, *123*, 11237–11247. [CrossRef]
27. Omary, M.A.; Webb, T.R.; Assefa, Z.; Shankle, G.E.; Patterson, H.H. Crystal structure, electronic structure, and temperature-dependent Raman spectra of Tl[Ag(CN)$_2$]: Evidence for ligand-unsupported argentophilic interactions. *Inorg. Chem.* **1998**, *37*, 1380–1386. [CrossRef]
28. Omary, M.A.; Patterson, H.H. Temperature-dependent photoluminescence properties of Tl[Ag(CN)$_2$]: Formation of luminescent metal-metal-bonded inorganic exciplexes in the solid state. *Inorg. Chem.* **1998**, *37*, 1060–1066. [CrossRef]
29. Dragulescu-Andrasi, A.; Hietsoi, O.; Üngör, Ö.; Dunk, P.W.; Stubbs, V.; Arroyave, A.; Kovnir, K.; Shatruk, M. Dicyanometalates as building blocks for multinuclear iron(II) spin-crossover complexes. *Inorg. Chem.* **2019**, *58*, 11920–11926. [CrossRef]
30. Bondi, A. van der Waals volume and radii. *J. Phys. Chem.* **1964**, *68*, 441–451. [CrossRef]
31. Nakamoto, K. *Infrared and Raman spectra of inorganic and coordination compounds*, 6th ed.; John Wiley and Sons, Inc.: New York, NY, USA, 2009.
32. Omary, M.A.; Rawashdeh-Omary, M.A.; Gonser, M.W.A.; Elbjeirami, O.; Grimes, T.; Cundari, T.R.; Diyabalanage, H.V.K.; Gamage, C.S.P.; Dias, H.V.R. Metal Effect on the supramolecular structure, photophysics, and acid-base character of trinuclear pyrazolato coinage metal complexes. *Inorg. Chem.* **2005**, *44*, 8200–8210. [CrossRef] [PubMed]
33. Grimes, T.; Omary, M.A.; Dias, H.V.R.; Cundari, T.R. Intertrimer and intratrimer metallophilic and excimeric bonding in the ground and phosphorescent states of trinuclear coinage metal pyrazolates: A computational study. *J. Phys. Chem. A* **2006**, *110*, 5823–5830. [CrossRef] [PubMed]
34. Hettiarachchi, C.V.; Rawashdeh-Omary, M.A.; Korir, D.; Kohistani, J.; Yousufuddin, M.; Dias, H.V.R. Trinuclear copper(I) and silver(I) adducts of 4-chloro-3,5-bis(trifluoromethyl)pyrazolate and 4-bromo-3,5-bis(trifluoromethyl)pyrazolate. *Inorg. Chem.* **2013**, *52*, 13576–13583. [CrossRef]
35. Kitajima, N.; Fujisawa, K.; Fujimoto, C.; Moro-oka, Y.; Hashimoto, S.; Kitagawa, T.; Toriumi, T.; Tatsumi, K.; Nakamura, A. A new model for dioxygen binding in hemocyanin. Synthesis, characterization, and molecular structure of the μ-η^2:η^2 peroxo dinuclear copper(II) complexes, [Cu(HB(3,5-R$_2$pz)$_3$)]$_2$(O$_2$) (R = i-Pr and Ph). *J. Am. Chem. Soc.* **1992**, *114*, 1277–1291. [CrossRef]
36. *CrystalClear: Data Collection and Processing Software*; Rigaku Corporation: Tokyo, Akishima, Japan, 1999.
37. *CrysAlisPro: Data Collection and Processing Software*; Rigaku Corporation: Tokyo, Akishima, Japan, 2015.
38. Burla, M.C.; Caliandro, R.; Camalli, M.; Carrozzini, B.; Cascarano, G.L.; De Caro, L.; Giacovazzo, C.; Polidori, G.; Siliqi, D.; Spagna, R. IL MILIONE: A suite of computer programs for crystal structure solution of proteins. *J. Appl. Cryst.* **2007**, *40*, 609–613. [CrossRef]
39. *Crystal Structure 4.3: Crystal Structure Analysis Package*; Rigaku Corporation: Tokyo, Akishima, Japan, 2019.
40. Sheldrick, G.M. A short history of SHELX. *Acta Cryst.* **2008**, *A64*, 112–122. [CrossRef]
41. Jaros, S.W.; da Silva, M.F.C.G.; Florek, M.; Smoleński, P.; Pombeiro, A.J.L.; Kirillov, A.M. Silver(I) 1,3,5-triaza-7-phosphaadamantane coordination polymers driven by substituted glutarate and malonate building blocks: Self-assembly synthesis, structural features, and antimicrobial properties. *Inorg. Chem.* **2016**, *55*, 5886–5894. [CrossRef]

42. Lu, X.; Ye, J.; Zhang, D.; Xie, R.; Bogale, R.F.; Sun, Y.; Zhao, L.; Zhao, Q.; Ning, G. Silver carboxylate metal–organic frameworks with highly antibacterial activity and biocompatibility. *J. Inorg. Biochem.* **2014**, *138*, 114–121. [CrossRef] [PubMed]
43. Jaros, S.W.; Smoleński, P.; da Silva, M.F.C.G.; Florek, M.; Król, J.; Staroniewicz, Z.; Pombeiro, A.J.L.; Kirillov, A.M. New silver BioMOFs driven by 1,3,5-triaza-7-phosphaadamantane-7-sulfide (PTA=S): Synthesis, topological analysis and antimicrobial activity. *CrystEngComm* **2013**, *15*, 8060–8064. [CrossRef]
44. Domb, A.J.; Kunduru, K.R.; Farah, S. *Antimicrobial Materials for Biomedical Applications*; The Royal Society of Chemistry: Croydon, UK, 2019.

Article

Fe(III) Complexes Based on Mono- and Bis-pyrazolyl-s-triazine Ligands: Synthesis, Molecular Structure, Hirshfeld, and Antimicrobial Evaluations

Saied M. Soliman [1,*], Hessa H. Al-Rasheed [2], Jörg H. Albering [3] and Ayman El-Faham [1,2,*]

[1] Department of Chemistry, Faculty of Science, Alexandria University, P.O. Box 426, Ibrahimia, Alexandria 21321, Egypt
[2] Department of Chemistry, College of Science, King Saud University, P.O. Box 2455, Riyadh 11451, Saudi Arabia; halbahli@ksu.edu.sa
[3] Graz University of Technology, Mandellstr. 11/III, A-8010 Graz, Austria; joerg.albering@tugraz.at
* Correspondence: saied1soliman@yahoo.com (S.M.S.); aelfaham@ksu.edu.sa (A.E.-F.)

Academic Editors: Vera L. M. Silva and Artur M. S. Silva
Received: 15 November 2020; Accepted: 3 December 2020; Published: 5 December 2020

Abstract: The self-assembly of iron(III) chloride with three pyrazolyl-s-triazine ligands, namely 2,4-bis(3,5-dimethyl-1H-pyrazol-1-yl)-6-(piperidin-1-yl)-1,3,5-triazine (^{Pip}BPT), 4-(4,6-bis(3,5-dimethyl-1H-pyrazol-1-yl)-1,3,5-triazin-2-yl)morpholine (^{Morph}BPT), and 4,4′-(6-(3,5-dimethyl-1H-pyrazol-1-yl)-1,3,5-triazine-2,4-diyl)dimorpholine (^{bisMorph}PT) afforded [Fe(^{Pip}BPT)Cl$_2$][FeCl$_4$] (**1**), [Fe(^{Morph}BPT)Cl$_2$][FeCl$_4$] (**2**), and [H(^{bisMorph}PT)][FeCl$_4$]. ^{bisMorph}PT.2H$_2$O (**3**), respectively, in good yield. In complexes **1** and **2**, the Fe(III) is pentacoordinated with three Fe-N interactions from the pincer ligand and two coordinated chloride anions in the inner sphere, and FeCl$_4^-$ in the outer sphere. Complex **3** is comprised of one protonated ligand as cationic part, one FeCl$_4^-$ anion, and one neutral ^{bisMorph}PT molecule in addition to two crystallized water molecules. Analysis of molecular packing using Hirshfeld calculations indicated that H . . . H and Cl . . . H are the most important in the molecular packing. They comprised 40.1% and 37.4%, respectively in **1** and 32.4% and 37.8%, respectively in **2**. Complex **1** exhibited the most bioactivity against the tested microbes while **3** had the lowest bioactivity. The ^{bisMorph}PT and ^{Morph}BPT were inactive towards the tested microbes while ^{Pip}BPT was active. As a whole, the Fe(III) complexes have enhanced antibacterial and antifungal activities as compared to the free ligands.

Keywords: pyrazolyl-s-triazine; Fe(III); self-assembly; Hirshfeld; antimicrobial activity

1. Introduction

Iron is a readily available element, as it is considered to be one of the most abundant. It is cheap and has almost-negligible hazardous effects on the environment as it has low toxicity [1–3]. Iron compounds play a crucial role in ammonia production by the Haber–Bosch process. On other hand, iron and its compounds have a key role in homogenous molecular catalysis [4–7].

Bis-pyrazolyl-s-triazine (**BPT**) ligands are a class of chelators which have been utilized in the synthesis of several divalent metal ion complexes with interesting molecular and supramolecular structures [8–13]. These s-triazine pincer-type complexes can be easily synthesized using self-assembly in a water-alcohol mixture. Additionally, they have extra-stability due to the chelate effect. Although iron has low toxicity, there are many problems due to high iron overload because it plays a major role in the generation of free radicals [14,15]. **BPT** ligands have key characteristics to act as a solution for this problem because they are powerful chelators.

On other hand, several organic-based antibacterial and antifungal drugs were discovered over the last few years [16]. Many of these antibiotics cannot overcome the problem of multidrug-resistant microbes [17–19]. Therefore, the replacement of these traditional antibiotics by other medications that can solve the problem of antibiotic-resistant pathogens has become an urgent need [17–19]. In this regard, some Fe(III) complexes have good antibacterial activity against a broad range of bacteria, but not fungi [20]. Others were found to have good antibacterial and moderate antifungal activities [20].

In the present work, we self-assembled three Fe(III) complexes by the direct reaction of $FeCl_3$ with the mono- and bis-pyrazolyl-s-triazine ligands shown in Figure 1. Their structure aspects were studied using single-crystal X-ray diffraction in combination with Hirshfeld analysis. The antibacterial and antifungal activities of these Fe(III) complexes are also presented.

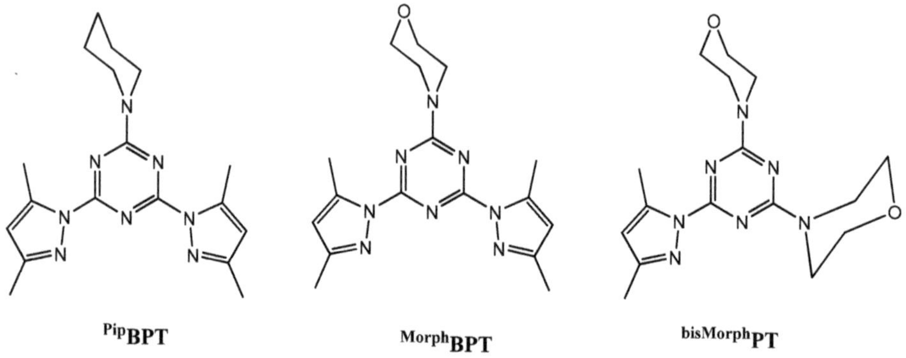

Figure 1. Structure of the *mono-* and bis-pyrazolyl-s-triazine ligands [21,22]. Ligands shown are: 2,4-bis(3,5-dimethyl-1H-pyrazol-1-yl)-6-(piperidin-1-yl)-1,3,5-triazine (^{Pip}BPT), 4-(4,6-bis(3,5-dimethyl-1H-pyrazol-1-yl)-1,3,5-triazin-2-yl)morpholine (^{Morph}BPT), and 4,4′-(6-(3,5-dimethyl-1H-pyrazol-1-yl)-1,3,5-triazine-2,4-diyl)dimorpholine (^{bisMorph}PT).

2. Results and Discussion

2.1. Structure Description

The crystals of the synthesized complexes were simply obtained from the direct reaction of the Fe(III) salt with the functional ligand in water-ethanol solvent mixture at room temperature using self-assembly. The X-ray single-crystal structure of the Fe(III) complexes are presented for the first time and the crystal data are listed in Table 1.

2.1.1. Crystal Structure Description of [Fe(^{Pip}BPT)Cl$_2$][FeCl$_4$] (1)

Complex **1** crystallizes in the monoclinic crystal system with the space group $P2_1/c$ and $Z = 4$; the asymmetric unit comprises one [Fe(^{Pip}BPT)Cl$_2$][FeCl$_4$] unit. The structure of the inner sphere complex in **1** consists of one cationic complex unit in which the Fe(III) ion is coordinated by ^{Pip}BPT in a tridentate pincer fashion and two chloride ions. The outer sphere is an anion: a tetrahedral FeCl$_4^-$ unit (Figure 2). The Fe-N distances are significantly shorter for the Fe-N(s-triazine) than the Fe-N(pyrazole), where the two Fe-N(pyrazole) bonds are only slightly different (Table 2). The two Fe1-Cl1 and Fe1-Cl2 bonds have very close bond distances of 2.1699(6) Å and 2.1766(6) Å, respectively. The coordination geometry of the five-coordinated Fe(III) ion is described using Addison criteria [23]. The coordination geometry, as shown in Figure 2, lies between the square pyramid and the trigonal bipyramid with a N3-Fe1-N2 angle (β) of 146.46(6)° and N1-Fe1-Cl2 angle (α) of 133.34(5)°, giving a $\tau = ((β − α)/60)$ value of 0.22. As a result, the coordination geometry around Fe(III) could be described as a distorted square pyramid.

Table 1. Crystal data and structure refinement for the studied complexes.

Compound	1	2	3
Empirical formula	$C_{18}H_{24}Cl_6Fe_2N_8$	$C_{17}H_{22}Cl_6Fe_2N_8O$	$C_{32}H_{51}Cl_4FeN_{14}O_6$
Formula weight (g/mol)	676.85	678.82	925.51
Temperature (K)	119(2)	124(2)	293(2)
λ (Å)	0.71073	0.71073	0.71073
Crystal system	Monoclinic	Orthorhombic	Triclinic
Space group	$P2_1/c$	Pbcm	P-1
Unit cell dimensions			
a (Å)	8.9549(3)	8.5201(3)	12.4352(15)
b (Å)	15.7871(6)	13.7094(5)	12.8632(16)
c (Å)	19.9063(7)	23.2383(9)	15.6509(19)
α (°)	90	90	76.955(3)
β (°)	99.457(2)	90	89.531(3)
γ (°)	90	90	66.926(3)
Volume (Å^3)	2775.9(2)	2714.4(2)	2234.7(5)
Z	4	4	2
Density (calculated, g/cm^3)	1.620	1.661	1.370
Absorption coefficient (mm^{-1})	1.647	1.687	0.633
F(000)	1368	1368	958
Crystal size (mm^3)	0.29 × 0.16 × 0.09	0.04 × 0.12 × 0.15	0.26 × 0.20 × 0.08
θ range (°)	2.31 to 25.49	2.81 to 24.99	2.33 to 25.09
Index ranges	−10 ≤ h ≤ 10, −19 ≤ k ≤ 19, −24 ≤ l ≤ 24	−10 ≤ h ≤ 10, −16 ≤ k ≤ 16, −27 ≤ l ≤ 27	−14 ≤ h ≤ 14, −15 ≤ k ≤ 15, −18 ≤ l ≤ 18
Reflections collected	38,064	21,560	64,462
Independent reflections	5139 [R(int) = 0.0427]	2453 [R(int) = 0.0294]	7915 [R(int) = 0.0769]
Completeness to theta (%)	99.8	99.90	99.5
Refinement method		Full-matrix least-squares on F^2	
Data/restraints/parameters	5139/0/311	2453/0/207	7909/0/528
Goodness-of-fit on F^2	1.045	1.083	1.008
Final R indices [I > 2sigma(I)]	R1 = 0.0238, wR2 = 0.0530	R1 = 0.0397, wR2 = 0.0988	R_1 = 0.0932, wR_2 = 0.2007
R indices (all data)	R1 = 0.0334, wR2 = 0.0571	R1 = 0.0440, wR2 = 0.1024	R_1 = 0.1654, wR_2 = 0.2464
Largest diff. peak and hole	0.291 and −0.306	0.621 and −1.035	0.87 and −0.64
CCDC No.	2044018	2044016	2044017

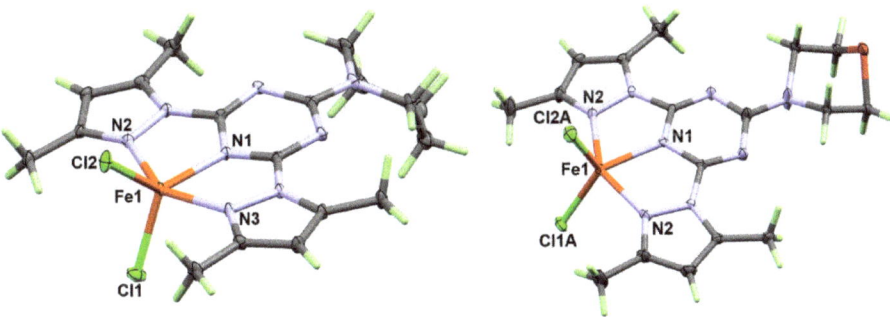

Figure 2. X-ray structure of complexes **1** (left) and **2** (right), the atoms have been drawn at a 30% probability level. The [FeCl$_4$]$^-$ anions in the outer sphere were omitted for better clarity.

The molecules of **1** are packed mainly by Cl ... H hydrogen bonds as shown in Figure 3 (upper part) and listed in Table 3. The donor–acceptor distances are 3.428(2) Å, 3.622(2) Å, and 3.723(2) Å for C8-H8 ... Cl1, C3-H3 ... Cl3, and C5-H5B ... Cl6 hydrogen bonding interactions, respectively. The packing of complex molecules is shown in Figure 4 (upper part). The network connected via Cl ... H bridge interactions shows a 3D connectivity.

Table 2. Bond distances and angles in **1**.

Atoms	Distance (Å)	Atoms	Distance (Å)
Fe1-N1	2.0295(15)	Fe2-Cl3	2.1840(6)
Fe1-N3	2.0940(16)	Fe2-Cl4	2.1836(6)
Fe1-N2	2.1092(16)	Fe2-Cl5	2.1791(6)
Fe1-Cl1	2.1699(6)	Fe2-Cl6	2.1873(6)
Fe1-Cl2	2.1766(6)		
Atoms	**Angle (°)**	**Atoms**	**Angle (°)**
N1-Fe1-N3	73.24(6)	N2-Fe1-Cl1	100.76(5)
N1-Fe1-N2	73.65(6)	N1-Fe1-Cl2	133.34(5)
N3-Fe1-N2	146.46(6)	N3-Fe1-Cl2	99.11(5)
N1-Fe1-Cl1	117.06(5)	N2-Fe1-Cl2	99.58(5)
N3-Fe1-Cl1	98.84(5)	Cl1-Fe1-Cl2	109.58(3)

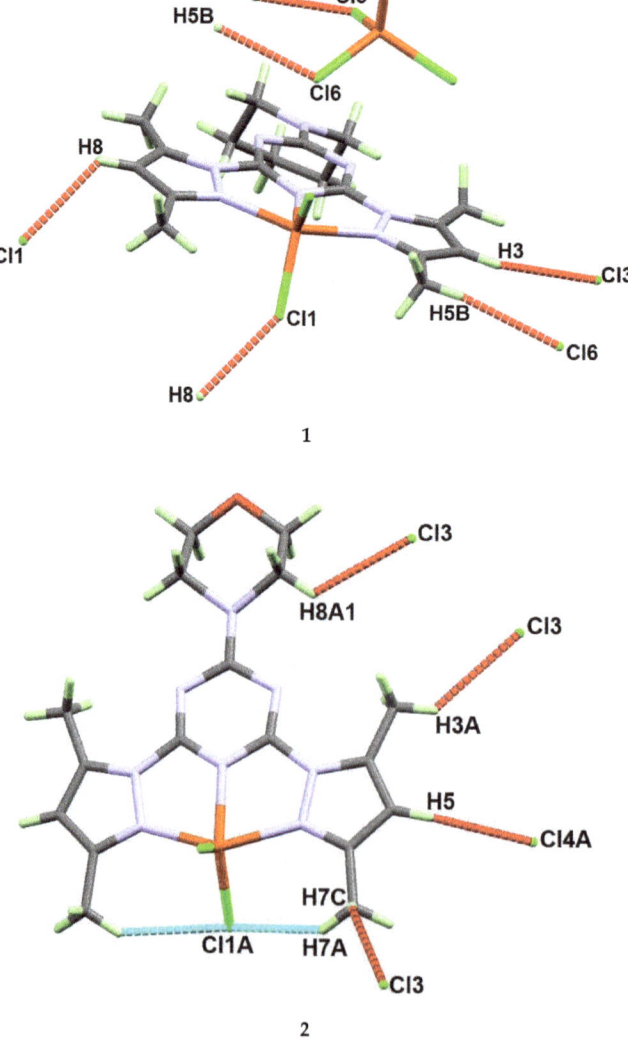

Figure 3. The hydrogen bond contacts in complexes **1** (upper) and **2** (lower).

Table 3. Hydrogen bond parameters of complexes 1 and 2.

Atoms	D-H (Å)	H...A (Å)	D...A (Å)	D-H...A (°)
Complex 1				
C3-H3...Cl3 [i]	0.95	2.77	3.622(2)	149
C5-H5B...Cl6 [i]	0.98	2.8	3.723(2)	158
C8-H8...Cl1 [ii]	0.95	2.78	3.428(2)	126
[i] 1 + x,y,z [i] 1 + x,y,z [ii] 1-x,-y,1-z and				
Complex 2				
C5-H5...Cl4A [i]	0.95	2.79	3.651(5)	152
C9A-H9A1...Cl1A [ii]	0.99	2.55	2.916(7)	101
[i] x,-1 + y,z and [ii] 1-x,1/2 + y,3/2-z				

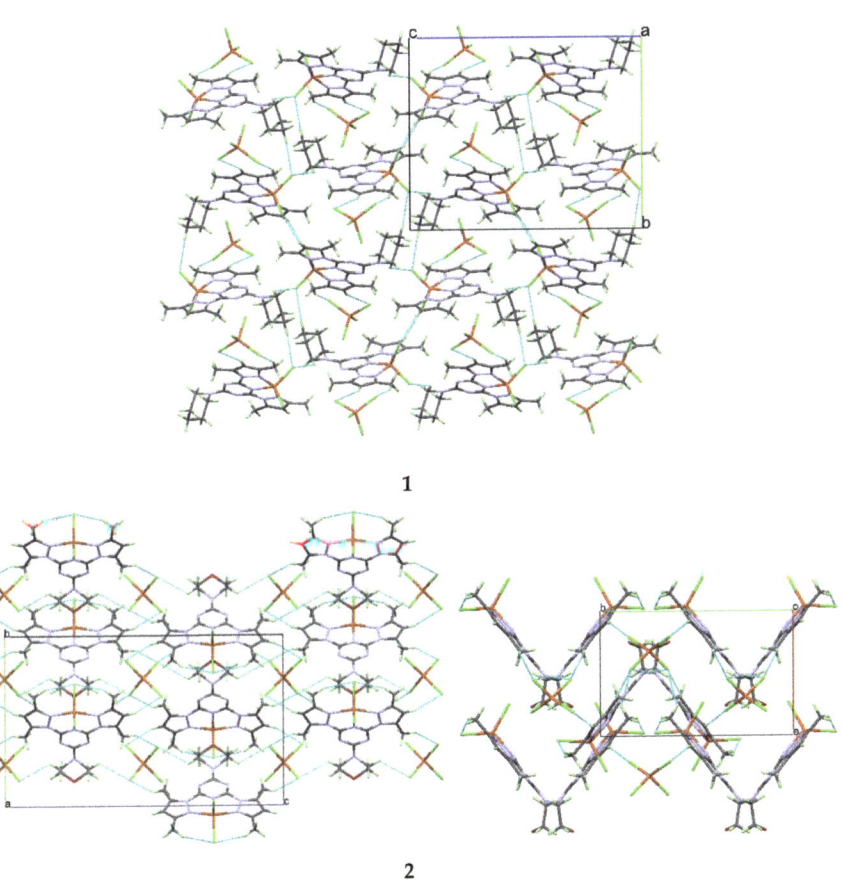

Figure 4. The hydrogen bond polymers in the crystal structures of complexes **1** (upper, view along the *a*-axis) and **2** (lower, views along the *a*- and *c*-axes). The hydrogen bridges are shown as light-blue dotted lines.

2.1.2. Crystal Structure Description of [Fe(^{Morph}BPT)Cl$_2$][FeCl$_4$] (**2**)

Complex **2** has a very close structure to **1**, with one [Fe(^{Morph}**BPT**)Cl$_2$]$^+$ as an inner sphere complex and [FeCl$_4$]$^−$ as a counter ion. The major difference is that complex **2** crystallizes in the more symmetric orthorhombic crystal system and space group Pbcm with half molecular formula as asymmetric unit.

The molecule comprises a symmetrical plane passing vertically through the molecule, intersecting the Fe(III) center and the two chloride anions, and splitting the organic ligand into two halves. In this regard, there are two equidistant Fe-N(pyrazole) bonds with iron to nitrogen distance of 2.099(2) Å and one shorter Fe-N(s-triazine) bond (2.036(3) Å). List of the most important bond distances are given in Table 4. The coordination sphere is completed by the two coordinated chloride anions with iron to chlorine distances ranging from 2.090(6)–2.262(5) Å for the two disordered parts (Figure S4, Supplementary Materials). The Addison criteria τ for the two complex parts are 0.26 and 0.05 for the disordered parts A and B, respectively. These calculations indicate that the two complex parts have different coordination geometries: part **B** is closer to being a more perfect square pyramid than part **A**. Regarding the scale factors for the two domains A and B, both are close 0.5. Thus, if the Cl1A and Cl2B atoms, as well as Cl1B and Cl2A, are assumed to belong to the same polyhedron, the τ values are calculated to be same (τ = 0.26 and 0.05), which confirms our conclusion.

The different hydrogen bridge contacts controlling the molecular packing of complex **2** are listed in Table 3 and shown in Figure 3, while the molecular packing showing the different molecular units packed via C-H … Cl interactions is shown in the lower part of Figure 4. Complex **2** also shows a 3D network connected via Cl … H interactions. Although the coordination modes of both complexes **1** and **2** look very similar, the hydrogen bridge networks turn out to be quite different. Small changes of the ligand molecule lead to significant differences in the packing.

Table 4. Bond distances and angles in **2**.

Atoms	Distance (Å)	Atoms	Distance (Å)
Fe1-N1	2.038(3)	Fe2-Cl4B	2.085(3)
Fe1-N2 [1]	2.099(3)	Fe2-Cl3 [2]	2.1667(10)
Fe1-N2	2.099(3)	Fe2-Cl3	2.1668(10)
Fe1-Cl1A	2.250(3)	Fe2-Cl4A	2.329(3)
Fe1-Cl2A	2.262(5)		
Fe1-Cl1B	2.151(2)		
Fe1-Cl2B	2.090(6)		

Atoms	Angle (°)	Atoms	Angle (°)
N1-Fe1-N2 [1]	73.66(7)	Cl2B-Fe1-N2	101.82(8)
N1-Fe1-N2	73.66(7)	N1-Fe1-Cl1B	149.41(12)
N2 [1]-Fe1-N2	146.19(13)	Cl2B-Fe1-Cl1B	90.49(17)
N1-Fe1-Cl1A	130.50(12)	N2 [1]-Fe1-Cl1B	101.81(7)
N2[1]-Fe1-Cl1A	97.28(7)	N2-Fe1-Cl1B	101.81(7)
N2-Fe1-Cl1A	97.28(7)	Cl1A-Fe1-Cl2A	122.17(15)
N1-Fe1-Cl2A	107.33(16)	Cl4B-Fe2-Cl3 [2]	106.87(8)
N2 [1]-Fe1-Cl2A	98.86(8)	Cl4B-Fe2-Cl3	120.70(11)
N2-Fe1-Cl2A	98.86(8)	Cl3[2]-Fe2-Cl3	109.97(6)
N1-Fe1-Cl2B	120.10(17)	Cl3[2]-Fe2-Cl4A	105.43(8)
Cl2B-Fe1-N2 [1]	101.82(8)	Cl3-Fe2-Cl4A	102.31(10)

[1] X,Y,3/2-Z and [2] +X,3/2-Y,1-Z.

2.1.3. Crystal Structure Description of [H(bisMorphPT)][FeCl4] bisMorphPT.2H$_2$O (3)

Attempts to synthesize a coordination complex compound of the bisMorphPT ligand with FeCl$_3$ have failed so far. The only crystalline compound that was found was a [H(bisMorphPT)][FeCl4] salt with one co-crystallized bisMorphPT ligand and two crystal water molecules in the asymmetric unit (Z = 2) of the triclinic unit cell with the symmetry P-1.

Compound **3** comprises four parts: the protonated organic ligand [H(bisMorphPT)]$^+$ as a cationic part, one electrically neutral bisMorphPT molecule, a negatively charged [FeCl4]$^-$ ion, and two crystal water molecules in a void of the packing (Figure 5). The crystal quality of this compound was not very good. We found some disorder in the organic part of the crystal structure, and the protons of the crystal water molecules were not detectable. For these reasons, we only give the crystallographic data in this publication and do not further describe its molecular and supramolecular aspects in detail. Although our attempts to synthesize a complex containing the ^{bisMorph}**PT** ligand and FeCl$_3$ were not successful, it does not necessarily mean that such a compound does not exist. In any case, it seemed to be useful to publish the data of compound **3** found in this context in order to create a reference for subsequent work. One possible reason for not obtaining a coordination complex of Fe(III) with the bidentate ^{bisMorph}PT ligand is its lower denticity compared to the tridentate ^{Pip}BPT and ^{Morph}BPT pincer chelates. Another possible reason is the steric effect resulting from the replacement of one pyrazole moiety by the morpholine one. The latter has no coordinating ability and a more bulky character that prevents the ^{bisMorph}**PT** from coordinating to the Fe(III) ion.

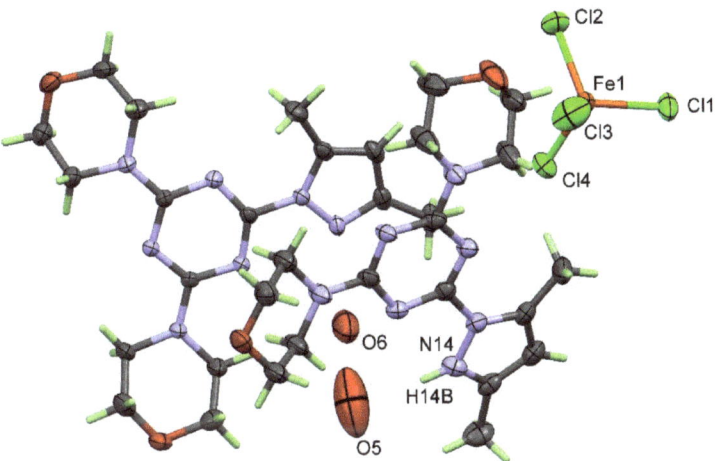

Figure 5. X-ray structure of compound **3**. All atoms have been drawn at a 50% probability level.

2.2. Analysis of Molecular Packing

Hirshfeld surfaces mapped over d_{norm}, shape index (SI), and curvedness for complexes **1** and **2** are shown in Figure S5 (Supplementary Materials). A summary of the most important contacts and their percentages are shown in Figure 6, while the decomposed d_{norm} maps of the short and most significant contacts in the studied complexes are collected in Figure 7. The decomposed fingerprint plots indicate the same common contacts in both complexes, which are H...H and Cl...H interactions, the most abundant intermolecular interactions in the studied complexes. The percentages of these contacts are 40.1% and 37.4% in complex **1**, respectively while they are 32.4% and 37.8% in complex **2**, respectively. The Cl...H hydrogen bonds appear as red regions in the Hirshfeld d_{norm} maps in both complexes and indicate shorter contact distances than the van der Waals (vdW) radii sum of H and Cl atoms. The anion (FeCl$_4^-$)-π stacking interactions are significant in both complexes. Complexes **1** and **2** show significantly short C...Cl and N...Cl contacts, with interaction distances also found to be shorter than the van der Waals radii sum of the two elements sharing this contact (Figure 7). The contact distances of the N...Cl interactions are 3.213 Å and 3.226 Å for complexes **1** and **2**, respectively, while the C...Cl contact distances are 3.257 Å and 3.381 Å for complexes **1** and **2**, respectively. In complex **1**, there is one short Fe1...Cl3 interaction (3.708 Å) between the complex cation and one of the Cl atoms from the complex anion (FeCl$_4^-$).

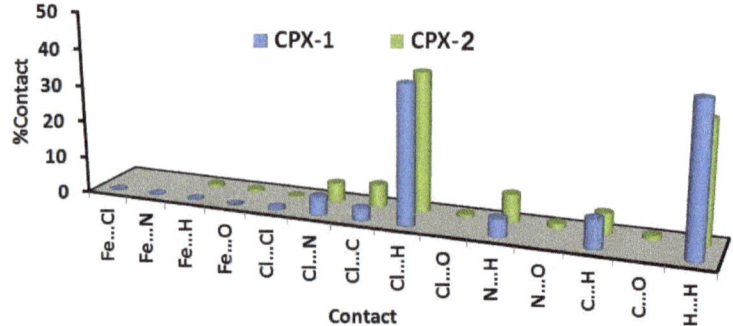

Figure 6. All intermolecular interactions in complexes (CPXs) **1** and **2**.

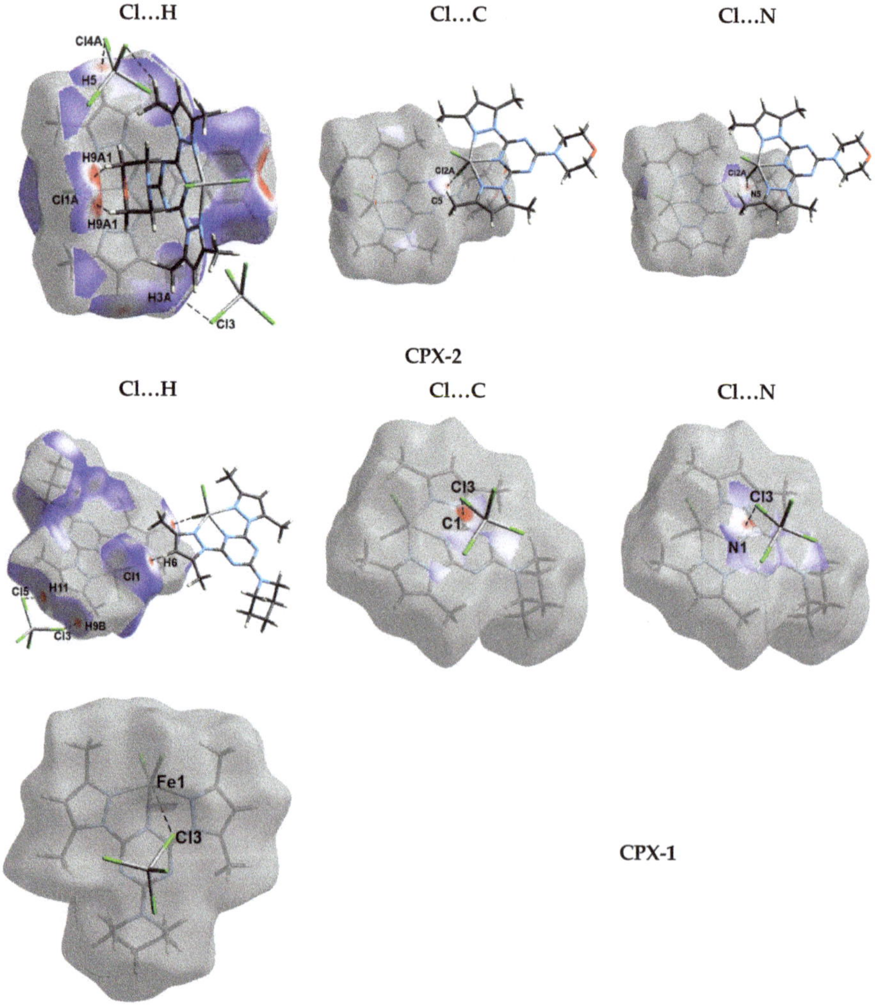

Figure 7. The decomposed d_{norm} maps and fingerprint plots of the most important contacts in **1** and **2**.

2.3. Antimicrobial Activity of the Studied Compounds

2.3.1. Inhibition Zones

In the current study, the antibacterial activity of the free ligands as well as compounds 1–3 were tested against Gram-positive bacteria, namely, *Staphylococcus aureus* (ATCC 29213) and *Staphylococcus epidermidis* (ATCC 12228); and Gram-negative bacteria, namely, *Escherichia coli* (ATCC 25922) and *Pseudomonas aeruginosa* (ATCC 27853) [24–27]. The free ^{Morph}BPT and ^{bisMorph}PT ligands were inactive against the target microbes at the applicable concentration. On other hand, ^{Pip}BPT showed good activity against *S. aureus*, *S. epidermidis*, *P. aeruginosa*, and *Candida albicans* (ATCC 60193) and it was completely inactive against *E. coli* (Table 5). In contrast, Fe(III) compounds 1–3 showed more potent activities against the target pathogenic microbes than did the corresponding free ligands, as illustrated from the values of the inhibition zones (mm) in Table 5. The values are considered an indicator for the bioactivity of the tested compounds at a concentration of 200 µg/mL. Compounds 1–3 appeared to have a more potent bioactivity against the target Gram-positive pathogenic bacteria than against the Gram-negative ones, and showed potent activity against the tested fungus (*C. albicans*).

Table 5. Anti-microbiological activities of the studied compounds against some tested microbes at 200 µg by the agar well diffusion method.

Test Compounds	Microbes				
	Staphylococcus aureus	*Staphylococcus epidermidis*	*Escherichia coli*	*Pseudomonas aeruginosa*	*Candida albicans*
^{Pip}BPT	11	17	-	13	12
^{Morph}BPT	-	-	-	-	-
^{bisMorph}PT	-	-	-	-	-
1	25	23	19	22	18
2	18	19	16	17	14
3	17	16	14	15	12
Fluconazole	-	-	-	-	14
Gentamycin	28	22	21	19	-

Complex 1 showed the most potency as an antibacterial and antifungal agent against all the target microbes, while 3 showed the lowest bioactivity. Additionally, complex 1 (18 mm) had better antifungal activity than the standard fluconazole (14 mm). Complex 1 had better antibacterial action against *S. epidermidis* (23 mm) and *P. aeruginosa* (22 mm) than the standard drug gentamycin (22 mm and 19 mm, respectively).

The antimicrobial activities of the studied compounds were also evaluated at different concentrations of 100 µg/mL, 200 µg/mL, and 300 µg/mL per disc, as shown in Table 6. Compounds 1–3 at these concentrations showed moderate to strong activity against all tested microbes, even at the lowest concentration of 100 µg/mL, where the best results were obtained for complex 1. On the other hand, as the concentration of the tested compound increased, the inhibition zone also increased. This result reveals that the presence of the piperidine/bis-pyrazolo combination with Fe(III) in one compound is the key for the bioactivity. These data agree with the literature, where the presence of piperidine enhanced the activity compared to the analogous morpholine derivatives [27].

2.3.2. Minimum Inhibitory Concentration (MIC) and Minimum Bactericidal Concentration (MBC)

The minimum inhibitory concentrations (MIC) of compounds 1–3 are given in Table 7. All tested compounds were active against *S. aureus*, *S. epidermidis*, *E. coli*, *P. aeruginosa*, and *C. albicans*. Again, complex 1 had the lowest MIC and MBC values, indicating its higher potency against all tested microbes as compared to 2 and 3 (Table 7).

Table 6. Antimicrobial activities of ^{Pip}BPT and **1–3** at different concentrations.

Compounds	Organism	Concentration		
		100	200	300
1	E. coli	16	19	21
	P. aeruginosa	20	22	23
	S. aureus	19	25	26
	S. epidermidis	18	23	25
	C. albicans	14	18	20
^{Pip}BPT	E. coli	-	-	-
	P. aeruginosa	13	15	18
	S. aureus	17	18	19
	S. epidermidis	11	13	15
	C. albicans	10	11	13
2	E. coli	14	16	18
	P. aeruginosa	14	17	18
	S. aureus	14	18	19
	S. epidermidis	15	19	20
	C. albicans	12	14	16
3	E. coli	12	15	17
	P. aeruginosa	12	15	17
	S. aureus	14	17	19
	S. epidermidis	13	16	17
	C. albicans	11	12	15

Table 7. Minimum inhibitory concentrations (MIC) (µg/mL) and minimum bactericidal concentrations (MBC) (µg/mL) of **1–3** against the growth of target microbes.

Microbes	[Fe(^{Pip}BPT)Cl$_2$][FeCl$_4$] (1)		[Fe(^{Morph}BPT)Cl$_2$][FeCl$_4$] (2)		[H(^{bisMorph}PT)][FeCl$_4$]. ^{bisMorph}PT (3)	
	MIC	MBC	MIC	MBC	MIC	MBC
S. epidermidis	8.3	16.6	9.7	19.4	18.8	37.5
S. aureus	8.7	17.5	9.8	19.6	18.8	37.5
E. coli	8.7	17.5	9.8	19.6	18.8	37.5
P. aeruginosa	8.2	16.5	9.8	19.6	18.8	37.5
C. albicans	18.8	100.0	37.5	150.0	37.5	150.0

3. Experimental

3.1. Materials and Physical Measurements

Chemicals were purchased from Sigma-Aldrich Company. CHN analyses were performed using a Perkin Elmer 2400 Elemental Analyzer.

3.2. Syntheses

3.2.1. Synthesis of s-Triazine-Based Ligands

The ligands ^{Pip}BPT, ^{Morph}BPT, and ^{bisMorph}PT were prepared following the method reported by our research group [21,22] (Supplementary Materials, Method S1 and Method S2, Figures S1–S3).

3.2.2. Synthesis of Fe(III) Complexes

All of the studied complexes were synthesized using a self-assembly technique, by mixing the aqueous solution of FeCl$_3$ (1 mmol, 162 mg) with the ethanolic solution of the functional ligand. The resulting clear solutions were left for slow evaporation until plate-like brown crystals of the target complexes were formed. The resulting crystals were collected by filtration and were found suitable for single-crystal X-ray diffraction measurements.

Yield: $C_{18}H_{24}N_8Fe_2Cl_6$ (**1**) 73% with respect to the ligand. Anal. Calc. C, 31.94; H, 3.57; N, 16.56%. Found: C, 31.80; H, 3.51; N, 16.43%.

Yield: $C_{17}H_{22}N_8OFe_2Cl_6$ (**2**) 76% with respect to the ligand. Anal. Calc. C, 30.08; H, 3.27; N, 16.51%. Found: C, 29.90; H, 3.21; N, 16.38%.

Yield: $C_{32}H_{51}N_{14}O_6$ Fe Cl_4 (**3**) 70% with respect to the ligand. Anal. Calc. C, 41.53; H, 5.55; N, 21.19%. Found: C, 41.35; H, 5.49; N, 21.01%.

3.3. Crystal Structure Determination

The crystal structures of complexes **1–3** were determined using a Bruker D8 Quest diffractometer employing SHELXTL and SADABS programs [28–30]. Hirshfeld calculations were performed using the default parameters of the Crystal Explorer 17.5 program [31–35].

3.4. Antimicrobial Studies

We determined the antimicrobial activities of the free ligands, as well as those of the corresponding Fe(III) complexes, against different microbes [22]. More details regarding the antimicrobial assay are found in Supplementary Materials (Methods S3–S6).

4. Conclusions

Three self-assembled Fe(III) complexes were synthesized by direct reaction of iron(III) chloride and the functional ligand in a water-alcohol medium. All complexes were obtained in good yield and their structures were confirmed by single-crystal X-ray diffraction. The supramolecular structures of complexes **1** and **2** were analyzed using Hirshfeld calculations with the aid of the CIF data. The Fe(III) complexes were bioactive against the target microbes and generally more active than the functional ligands. It was found that the combination of piperidine and bispyrazolo moieties with Fe(III) in one compound (**1**) had the best bioactivity in comparison with the corresponding complexes (**2** and **3**) comprised of morpholine group(s).

Supplementary Materials: The following are available online. Figure S1: ^{1}H-NMR and ^{13}C-NMR of ^{Morph}BPT, Figure S2: ^{1}H-NMR and ^{13}C-NMR of ^{Pip}BPT, Figure S3: ^{1}H- and ^{13}C-NMR for ^{bisMorph}PT, Figure S4: Structure showing the disordered parts in **2**. Figure S5: Hirshfeld surfaces mapped over d_{norm}, shape index, and curvedness. In addition, the detailed synthesis of the ligands and biological experiments are given in the Supplementary data file. Methods S1–S6.

Author Contributions: The work was designed and supervised by S.M.S. X-ray structure analyses were performed by J.H.A. and S.M.S. Computational calculations as well as the synthesis of complexes **1–3** were carried out by S.M.S. and H.H.A.-R., and A.E.-F. carried out the preparation of the organic ligands, analysis, and the biological evaluation. All authors contributed in the first draft and the final version. All authors have read and agreed to the published version of the manuscript.

Funding: The Deanship of Scientific Research at King Saud University funded this research, grant number RGP-1441-234, Saudi Arabia.

Acknowledgments: The authors extend their thanks to the Deanship of Scientific Research at King Saud University for funding this work through research group number RGP-1441-234, Saudi Arabia.

Conflicts of Interest: The authors declare no conflict of interest.

References

1. Egorova, K.S.; Ananikov, V.P. Toxicity of Metal Compounds: Knowledge and Myths. *Organometallics* **2017**, *36*, 4071–4090. [CrossRef]
2. Egorova, K.S.; Ananikov, V.P. Which Metals are Green for Catalysis? Comparison of the Toxicities of Ni, Cu, Fe, Pd, Pt, Rh, and Au Salts. *Angew. Chem. Int. Ed.* **2016**, *55*, 12150–12162. [CrossRef] [PubMed]
3. Bauer, E.B. Iron Catalysis II. *Top. Organomet. Chem.* **2015**, *50*, 1–18.
4. Bolm, C.; Legros, J.; Le Paih, J.; Zani, L. Iron-Catalyzed Reactions in Organic Synthesis. *Chem. Rev.* **2004**, *104*, 6217–6254. [CrossRef] [PubMed]
5. Bauer, I.; Knölker, H.-J. Iron Catalysis in Organic Synthesis. *Chem. Rev.* **2015**, *115*, 3170–3387. [CrossRef]

6. Wei, D.; Darcel, C. Iron Catalysis in Reduction and Hydrometalation Reactions. *Chem. Rev.* **2019**, *119*, 2550–2610. [CrossRef]
7. Ludwig, J.R.; Schindler, C.S. Catalyst: Sustainable Catalysis. *Chemistry* **2017**, *2*, 313–316. [CrossRef]
8. Soliman, S.M.; El-Faham, A. One pot synthesis of two Mn(II) perchlorate complexes with s-triazine NNN-pincer ligand; molecular structure, Hirshfeld analysis and DFT studies. *J. Mol. Struct.* **2018**, *1164*, 344–353. [CrossRef]
9. Soliman, S.M.; El-Faham, A. Synthesis, characterization, and structural studies of two heteroleptic Mn(II) complexes with tridentate N,N,N-pincer type ligand. *J. Coord. Chem.* **2018**, *71*, 2373–2388. [CrossRef]
10. Soliman, S.M.; El-Faham, A. Synthesis, molecular structure and DFT studies of two heteroleptic nickel(II) s-triazine pincer type complexes. *J. Mol. Struct.* **2019**, *1185*, 461–468. [CrossRef]
11. Soliman, S.M.; El-Faham, A. Synthesis, X-ray structure, and DFT studies of five and eight-coordinated Cd(II) complexes with s-triazine N-pincer chelate. *J. Coord. Chem.* **2019**, *72*, 1621–1636. [CrossRef]
12. Soliman, S.M.; Almarhoon, Z.; El-Faham, A. Synthesis, Molecular and Supramolecular Structures of New Cd(II) Pincer-Type Complexes with s-Triazine Core Ligand. *Crystals* **2019**, *9*, 226. [CrossRef]
13. Soliman, S.M.; Almarhoon, Z.; El-Faham, A. Bis-pyrazolyl-s-triazine Ni(II) pincer complexes as selective Gram positive antibacterial agents; synthesis, structural and antimicrobial studies. *J. Mol. Struct.* **2019**, *1195*, 315–322. [CrossRef]
14. Crisponi, G.; Remelli, M. Iron chelating agents for the treatment of iron overload. *Coord. Chem. Rev.* **2008**, *252*, 1225–1240. [CrossRef]
15. Brittenham, G.M. *Hematology: Basic Principles and Practice*; Hoffman, R., Benz, E., Shattil, S., Furie, B., Cohen, H., Eds.; Churchill Livingstone: New York, NY, USA, 1991; p. 327.
16. Al-Resayes, S.I.; Shakir, M.; Shahid, N.; Azam, M.; Khan, A.U. Synthesis, spectroscopic characterization and in vitro antimicrobial studies of Schiff base ligand, H2L derived from glyoxalic acid and 1,8-diaminonaphthalene and its Co(II), Ni(II), Cu(II) and Zn(II) complexes. *Arab. J. Chem.* **2016**, *9*, 335–343. [CrossRef]
17. Wenzel, M.; Patra, M.; Senges, C.H.R.; Ott, I.; Stepanek, J.J.; Pinto, A.; Prochnow, P.; Vuong, C.; Langklotz, S.; Metzler-Nolte, N.; et al. Analysis of the Mechanism of Action of Potent Antibacterial Hetero-tri-organometallic Compounds: A Structurally New Class of Antibiotics. *ACS Chem. Biol.* **2013**, *8*, 1442–1450. [CrossRef]
18. Patra, M.; Gasser, G.; Metzler-Nolte, N. Small organometallic compounds as antibacterial agents. *Dalton Trans.* **2012**, *41*, 6350–6358. [CrossRef]
19. Albada, H.B.; Prochnow, P.; Bobersky, S.; Bandow, J.E.; Metzler-Nolte, N. Highly active antibacterial ferrocenoylated or ruthenocenoylated Arg-Trp peptides can be discovered by an L-to-D substitution scan. *Chem. Sci.* **2014**, *5*, 4453–4459. [CrossRef]
20. Pansuriya, P.B.; Patel, M.N. Iron(III) complexes: Preparation, characterization, antibacterial activity and DNA-binding. *J. Enzyme Inhib. Med. Chem.* **2008**, *23*, 230–239. [CrossRef]
21. Farooq, M.; Sharma, A.; Almarhoon, Z.; Al-Dhfyan, A.; El-Faham, A.; Abu Taha, N.; Wadaan, M.A.M.; de laTorre, B.G.; Albericio, F. Design and synthesis of mono-and di-pyrazolyl-s-triazine derivatives, their anticancer profile in human cancer cell lines, and in vivo toxicity in zebrafish embryos. *Bioorg. Chem.* **2019**, *87*, 457–464. [CrossRef]
22. Sharma, A.; Ghabbour, H.; Khan, S.T.; de la Torre, B.G.; Albericio, F.; El-Faham, A. Novel Pyrazolyl-s-Triazine Derivatives, Molecular Structure and Antimicrobial Activity. *J. Mol. Struct.* **2017**, *1145*, 244–253. [CrossRef]
23. Addison, A.W.; Rao, T.N.; Reedijk, J.; Rijn, J.V.; Verschoor, G.C. Synthesis, structure, and spectroscopic properties of copper(II) compounds containing nitrogen–sulphur donor ligands; the crystal and molecular structure of aqua[1,7-bis(N-methylbenzimidazol-2′-yl)-2,6-dithiaheptane]copper(II) perchlorate. *J. Chem. Soc. Dalton Trans.* **1984**, 1349–1356. [CrossRef]
24. De Souza, S.M.; Delle Monache, F.; Smânia, A. Antibacterial activity of coumarins. *Z. Nat. C* **2005**, *60*, 693–700. [CrossRef] [PubMed]
25. Arshad, M.; Bhat, A.R.; Hoi, K.K.; Choi, I.; Athar, F. Synthesis, characterization and antibacterial screening of some novel 1, 2, 4-triazine derivatives. *Chin. Chem. Lett.* **2017**, *28*, 1559–1565. [CrossRef]
26. Hassan, M.T.; Sareh, Z.J. Synthesis, Characterization and in Vitro Antimicrobial Screening of the Xanthate Derivatives and their Iron(II) Complexes. *Iran. J. Chem. Chem. Eng.* **2017**, *36*, 43–54.
27. Dhokale, N.T.; Karale, B.K.; Nagawadw, A.V. Synthesis, Characterization and Antibacterial Studies on Mn(II) and Fe(II) Complexes of N, O-Donor Salicyloyl Pyrazole Oxime Schiff Bases. *Orient. J. Chem.* **2017**, *33*, 165–172. [CrossRef]

28. Sheldrick, G.M.; SADABS. *Program for Empirical Absorption Correction of Area Detector Data*; University of Göttingen: Göttingen, Germany, 1996.
29. Sheldrick, G.M. SHELXT—Integrated space-group and crystal-structure determination. *Acta Crystallogr. Sect. A* **2015**, *71*, 3–8. [CrossRef]
30. Spek, A.L. Structure validation in chemical crystallography. *Acta Crystallogr. Sect. D* **2009**, *65*, 148–155. [CrossRef]
31. Turner, M.J.; McKinnon, J.J.; Wolff, S.K.; Grimwood, D.J.; Spackman, P.R.; Jayatilaka, D.; Spackman, M.A. Crystal Explorer 17. 2017. University of Western Australia. Available online: http://hirshfeldsurface.net (accessed on 12 June 2017).
32. Spackman, M.A.; Jayatilaka, D. Hirshfeld surface analysis. *CrystEngComm* **2009**, *11*, 19–32. [CrossRef]
33. Spackman, M.A.; McKinnon, J.J. Fingerprinting intermolecular interactions in molecular crystals. *CrystEngComm* **2002**, *4*, 378–392. [CrossRef]
34. Bernstein, J.; Davis, R.E.; Shimoni, L.; Chang, N.-L. Patterns in hydrogen bonding: Functionality and graph set analysis in crystals. *Angew. Chem. Int. Ed. Engl.* **1995**, *34*, 1555–1573. [CrossRef]
35. McKinnon, J.J.; Jayatilaka, D.; Spackman, M.A. Towards quantitative analysis of intermolecular interactions with Hirshfeld surfaces. *Chem. Commun.* **2007**, 3814–3816. [CrossRef] [PubMed]

Sample Availability: Samples of the compounds are available from the authors.

Publisher's Note: MDPI stays neutral with regard to jurisdictional claims in published maps and institutional affiliations.

© 2020 by the authors. Licensee MDPI, Basel, Switzerland. This article is an open access article distributed under the terms and conditions of the Creative Commons Attribution (CC BY) license (http://creativecommons.org/licenses/by/4.0/).

Article

Boulton-Katritzky Rearrangement of 5-Substituted Phenyl-3-[2-(morpholin-1-yl)ethyl]-1,2,4-oxadiazoles as a Synthetic Path to Spiropyrazoline Benzoates and Chloride with Antitubercular Properties

Lyudmila Kayukova [1,*], Anna Vologzhanina [2,*], Kaldybai Praliyev [1], Gulnur Dyusembaeva [1], Gulnur Baitursynova [1], Asem Uzakova [1], Venera Bismilda [3], Lyailya Chingissova [3] and Kydyrmolla Akatan [4]

1. JSC A. B. Bekturov Institute of Chemical Sciences, 106 Shokan Ualikhanov St., Almaty 050010, Kazakhstan; praliyevkd@mail.ru (K.P.); g_gazinovna@mail.ru (G.D.); guni-27@mail.ru (G.B.); a7_uzakova@mail.ru (A.U.)
2. A. N. Nesmeyanov Institute of Organoelement Compounds, Russian Academy of Sciences, 28 Vavilov St., B-334, 119991 Moscow, Russia
3. National Scientific Center of Phthisiopulmonology of Ministry of Health of the Republic of Kazakhstan, 5 Bekkhozhin St., Almaty 050010, Kazakhstan; venerabismilda@mail.ru (V.B.); lchingissova@mail.ru (L.C.)
4. National scientific laboratory, S. Amanzholov East Kazakhstan State University, 18/1 Amurskaya St., Ust-Kamenogorsk 070002, Kazakhstan; ahnur.hj@mail.ru
* Correspondence: lkayukova@mail.ru (L.K.); vologzhanina@mail.ru (A.V.); Tel.: +7-(705)-168-81-53 (L.K.)

Abstract: The analysis of stability of biologically active compounds requires an accurate determination of their structure. We have found that 5-aryl-3-(2-aminoethyl)-1,2,4-oxadiazoles are generally unstable in the presence of acids and bases and are rearranged into the salts of spiropyrazolinium compounds. Hence, there is a significant probability that it is the rearranged products that should be attributed to biological activity and not the primarily screened 5-aryl-3-(2-aminoethyl)-1,2,4-oxadiazoles. A series of the 2-amino-8-oxa-1,5-diazaspiro[4.5]dec-1-en-5-ium (spiropyrazoline) benzoates and chloride was synthesized by Boulton–Katritzky rearrangement of 5-substituted phenyl-3-[2-(morpholin-1-yl)ethyl]-1,2,4-oxadiazoles and characterized using FT-IR and NMR spectroscopy and X-ray diffraction. Spiropyrazolylammonium chloride demonstrates in vitro antitubercular activity on DS (drug-sensitive) and MDR (multidrug-resistant) of MTB (*M. tuberculosis*) strains (1 and 2 μg/mL, accordingly) equal to the activity of the basic antitubercular drug rifampicin; spiropyrazoline benzoates exhibit an average antitubercular activity of 10–100 μg/mL on MTB strains. Molecular docking studies revealed a series of *M. tuberculosis* receptors with the energies of ligand–receptor complexes (−35.8––42.8 kcal/mol) close to the value of intermolecular pairwise interactions of the same cation in the crystal of spiropyrazolylammonium chloride (−35.3 kcal/mol). However, only in complex with transcriptional repressor EthR2, both stereoisomers of the cation realize similar intermolecular interactions.

Keywords: 1,2,4-oxadiazoles; spiropyrazolinium compounds; in vitro antitubercular screening; X-ray diffraction; molecular docking

1. Introduction

There are pyrazoline-containing compounds that act as active pharmaceutical ingredients of such commercially available drugs as *aminopyrine* (*aminophenazone*; analgesic and antipyretic), *dipyrone* (*metamizole*, *noramidopyrine*; analgesic), *antipyrine* (*benzocaine*; non-narcotic analgesic, an antipyretic and antirheumatic), *zaleplon*(hypnotic and sedative), *celecoxib* (*Aclarex*, *Celebrex*; anti-inflammatory and antirheumatic drug), *allopurinol* (uricostatic agent, xanthine oxidase inhibitor) [1]. Therefore, there is always a demand for new molecules, methodologies and improved synthetic approaches to novel pyrazoline derivatives.

Pyrazolines, as noticeable, practically meaningful nitrogen-containing heterocyclic compounds, can be synthesized by a variety of methods. However, one of the most popular methods is the Fischer and Knoevenagel synthesis based on the reaction of α,β-unsaturated ketones with phenylhydrazine in acetic acid under refluxing conditions. However, depending on the reactivity of molecules and the need of the chemist, they had synthesized the pyrazolines under different solvent media and acidic or basic conditions [2–4].

Information on pyrazolinium structures with a quaternary nitrogen atom is limited. Thus, two examples of biologically active pyrazolinium salts were found: 3-amino-1-ethyl-1-phenyl-4,5-dihydro-1H-pyrazolinium iodide (PubChemCID: 13585073 structure, [5]) and 3-amino-1,4-dimethyl-1-phenyl-2-pyrazolinium iodide (PubChemCID: 13064197 structure, [6]) (Scheme 1):

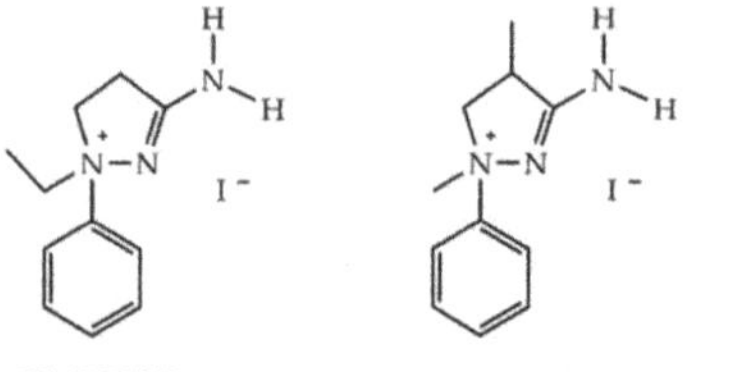

Scheme 1. Amino-1-phenyl-4,5-dihydro-1H-pyrazolinium iodides.

In some works, we found spiropyrazolinium compounds with a quaternary nitrogen atom, which is common for two heterocycles. When studying the stability of 3-(2-aminoethyl)-5-aryl-1,2,4-oxadiazoles having six-membered cyclic tertiary 2-amino groups towards hydrolysis, we found that they were capable of rearranging to spiropyrazoline benzoates or chlorides [7–9].

Particularly, upon keeping 3-[2-(4-phenylpiperazin-1-yl)ethyl]-5-phenyl-1,2,4-oxadiazole recrystallized in 2-PrOH under conditions of air moisture access for nine months for growing single crystals for X-ray structural analysis or by exposure of 3-[2-thiomorpholin-1-yl)ethyl]-5-aryl-1,2,4-oxadiazoles in ethanol with ethereal HCl solution the above-mentioned 1,2,4-oxadiazoles underwent the rearrangement to spiropyrazolinium benzoates or chlorides (Scheme 2) [7,8]:

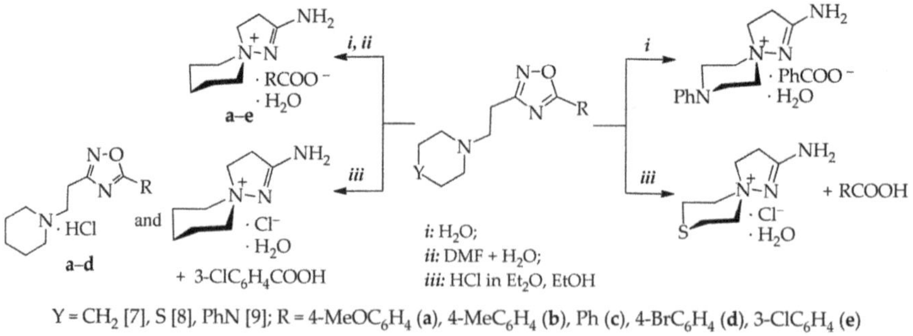

Y = CH$_2$ [7], S [8], PhN [9]; R = 4-MeOC$_6$H$_4$ (**a**), 4-MeC$_6$H$_4$ (**b**), Ph (**c**), 4-BrC$_6$H$_4$ (**d**), 3-ClC$_6$H$_4$ (**e**)

Scheme 2. 3-(2-Aminoethyl)-5-aryl-1,2,4-oxadiazoles in the conditions of exposure to H$_2$O and HCl.

Furthermore, at targeted exposure on 3-[2-(piperidin-1-yl)ethyl]-5-aryl-1,2,4-oxadiazoles with: (*i*) water, (*ii*) water in DMF or (*iii*) ethereal HCl they underwent rearrangement with the formation 2-amino-1,5-diazaspiro[4.5]dec-1-en-5-ium benzoates or chlorides (Scheme 2). In the latter case, along with the formation of spiropyrazolinium chloride hydrate from 3-[2-(piperidin-1-yl)ethyl]-5-(3-chloro-phenyl)-1,2,4-oxadiazole hydrochlorides of starting 1,2,4-oxadiazoles were obtained as secondary products [9].

These facts are consistent with the known for 3,5-substituted 1,2,4-oxadiazoles with a saturated side chain spontaneous thermally induced monomolecular Boulton–Katritzky rearrangement and provided the first examples of spirocompound formation through such reaction. In general, a variety of Boulton–Katritzky rearrangements could be represented as the following scheme (Scheme 3) [10,11]:

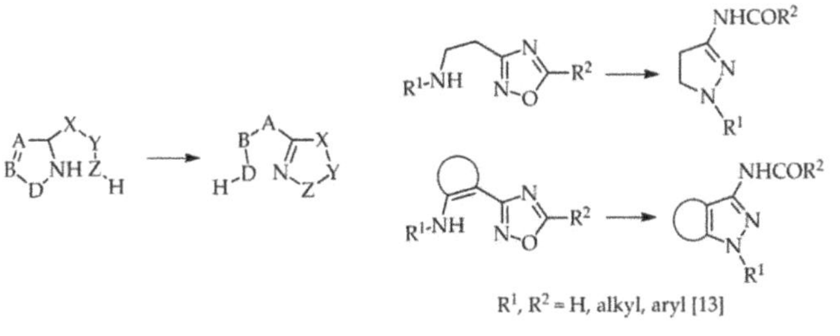

Scheme 3. Mononuclear heterocyclic Boulton-Katritzky rearrangement of 3,5-substituted 1,2,4-oxadiazoles.

In addition, spiropyrazolinium structures–2-amino-8-oxa-1,5-diazaspiro[4.5]dec-1-ene-5-ammonium arylsulfonates are formed at arylsulfochlorination of β-aminopropioamidoximes (Scheme 4) [12].

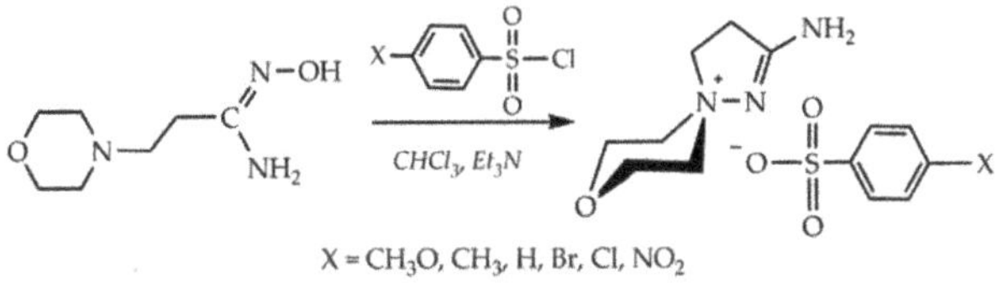

Scheme 4. Amino-4,5-dihydrospiropyrazolylammonium formation at arylsulfochlorination of β-aminopropioamidoximes.

The practical interest in the class of β-aminopropioamidoxime derivatives is supported by their pronounced local anesthetic, antitubercular, and antidiabetic activities [13–16].

Herein we report on the stability of 5-aryl-3-[β-(morpholin-1-yl)ethyl]-1,2,4-oxadiazoles towards hydrolysis at: (*i*) DMF with the two equivalent amount of water when heated to 60–70 °C; (*ii*) alcohol/ethereal HCl mixture. A number of previously unknown spiropyrazolinium salts were obtained and characterized using FT-IR and NMR spectroscopy and X-ray diffraction. In vitro antitubercular screening of spiropyrazoline benzoates and chloride was carried out, and their molecular docking was performed. It was shown that the hydrolysis of 1,2,4-oxadiazoles with a 3-morpholinoethyl substituent leads to spiropyrazoline compounds within 25–40 h. Acid hydrolysis of 1,2,4-oxadiazoles occurs immediately after reagents adding. In vitro antitubercular screening of benzoates and chloride of spiropyrazoline drug-susceptible and multidrug-resistant strains, *M. tuberculosis* revealed compounds with significant activity, and the results are in accordance with molecular docking studies.

2. Results and Discussion

2.1. Synthesis and Spectra

The synthesis of the starting compounds **1**, **2a–e**, **3a–e** and **4a–e** was described earlier [17]. 5-Substituted phenyl-3-[2-(morpholin-1-yl)ethyl]-1,2,4-oxadiazoles (**4a–e**) were obtained by heating of -aroyl-(β-morpholin-1-yl)propioamidoximes (**3a–e**) in DMF at 70 °C for several hours, evaporating off the solvent in an oil pump vacuum and treating of the residue with acetone. 1,2,4-Oxadiazoles (**4a–e**) are obtained as crystalline precipitates from acetone (Scheme 5).

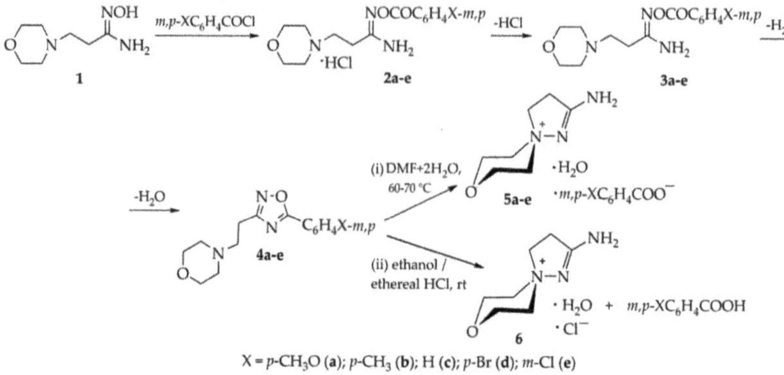

Scheme 5. Obtaining of 5-substituted phenyl-3-[2-(morpholin-1-yl)ethyl]-1,2,4-oxadiazoles (**4a–e**), and 2-amino-8-oxa-1,5-diazaspiro[4.5]dec-1-en-5-ium benzoates and hloride (**5a–e, 6**).

Physicochemical, FT-IR and NMR spectral characteristics of representatives of 1,2,4-oxadiazoles of the morpholine series (**4a–e**) were recorded immediately after isolation from DMF, and they correspond to the structure of 1,2,4-oxadiazoles. However, the X-ray diffraction analysis of single crystals grown for nine months from 1,2,4-oxadiazoles **4c–e** recrystallized from 2-PrOH showed a complete transition of 1,2,4-oxadiazoles into rearranged spiropyrazolinium compounds **5c–e** (Section 2.3.). This indicates the hydrolysis of 1,2,4-oxadiazoles **4a–e** by way of Boulton–Katritsky rearrangement to spiropyrazolinium compounds **5a–e** under the influence of air moisture. The rearrangement products **5a–e** have increased values of the mobility index R_f and m.p. in comparison with the initial compounds **4a–e** (Table 1).

Table 1. Physicochemical data of 5-substituted phenyl-3-[β-(morpholin-1-yl)ethyl]-1,2,4-oxadiazoles (**4a–e**) and spiropyrazolinium compounds (**5a–e**) fixed in the conditions for obtaining single crystals for XRD analysis and in condition (*i*).

Compd	X	Yield,%	m.p., °C	R_f *	Compd	t, h **	Yield,%	m.p., °C	R_f *
4a	p-CH$_3$O	62	230	0.71	5a	40	63	248	0.80
4b	p-CH$_3$	70	220	0.62	5b	40	63	235	0.75
4c	H	92	216	0.66	5c	25	77	220	0.75
4d	p-Br	83	224	0.67	5d	25	80	240	0.77
4e	m-Cl	56	190	0.62	5e	25	43	200	0.70

* R_f determined in the system ethanol:benzene, 3:1; ** the heating time of 1,2,4-oxadiazoles **4a–e** in DMF + 2H$_2$O.

Further in this work, we investigated the conditions for the Boulton–Katritsky rearrangement of 5-substituted phenyl-3-[2-(morpholin-1-yl)ethyl]-1,2,4-oxadiazoles (**4a–e**) under the deliberate establishment of hydrolysis conditions: (*i*) DMF with the two equivalent amount of water when heated to 60–70 °C; (*ii*) alcohol/ethereal HCl mixture in the presence of air moisture (Scheme 5).

As can be seen from Table 1, the heating time has an increased value (40 h) for electron-donor substituents in the phenyl ring of 1,2,4-oxadiazoles **4a, 4b** in comparison with 1,2,4-oxadiazoles with an unsubstituted phenyl ring and with a phenyl ring having electron-withdrawing substituents—**4c–e** (25 h). Spiropyrazolinium compounds **5a–e** were obtained after evaporation of DMF in an oil pump vacuum, treatment of the residue with acetone with the isolation of rearranged products **5a–e** and their recrystallization from 2-PrOH. In the case of the action of ethereal HCl solution on alcohol solutions of 1,2,4-oxadiazoles **4a–e** in all cases, 2-amino-8-oxa-1,5-diazaspiro[4.5]dec-1-en-5-ium chloride (**6**) and the corresponding benzoic acids were isolated.

IR spectra view of spirocompounds **5a–e** and **6** differ from the IR spectra of 1,2,4-oxadiazoles **4a–e**. First, the former compounds have symmetric and asymmetric ν(N–H) stretching bands at 3152–3485 and 3158–3457 cm^{-1}, respectively; second, there are pronounced bands of asymmetric and symmetric stretching vibrations of strong intensity ν(C$^-$) at 1545–1557 cm^{-1} and 1420–1442 cm^{-1}, respectively for **5a–e** and no stretching bonds of aromatic protons for salt **6**.

The ^{1}H-NMR spectra of 1,2,4-oxadiazole **4a–e** were recorded immediately with isolation; if they were recorded after 1–2 weeks, then the emergence and increase in the intensity of the NH$_2$ group signal of the rearranged spiropyrazolinium products **5a–e** in the region of δ 7.51–7.7.57 ppm was observed. It indicates a transition of 1,2,4-oxadiazole to the spiropyrazolinium compounds **4a–e**→**5a–e** in the presence of air moisture.

Comparison of the NMR spectra (^{1}H and ^{13}C) of compounds **4a–e** and **5a–e** shows almost no differences in the regional characteristics of the groups of protons and carbon atoms of the structures of 5-substituted phenyl-3-[2-(morpholin-1-yl)ethyl]-1,2,4-oxadiazoles (**4a–e**) and of 2-amino-8-oxa-1,5-diazaspiro[4.5]dec-1-en-5-ium benzoates (**5a–e**). Hence, protons of α-CH$_2$ and β-CH$_2$ groups in the first case give signals in the range δ 3.15–3.17 ppm and δ 3.65–3.66 ppm and in the region δ 3.15–3.19 ppm and δ 3.64–3.66 ppm—in the second. The signals of the aromatic protons of *para*-substituted 1,2,4-oxadiazoles **4a, 4b, 4d** and spiropyrazolinium benzoates **5a, 5b, 5d** have the form of two symmetric doublets with the spin–spin coupling constant *J* equal 7.5 and 8.0 Hz. Aromatic protons of 1,2,4-oxadiazoles and spiropyrazolinium benzoates with unsubstituted and *meta*-substituted phenyl rings have signals at δ 7.23–7.83 ppm (**4c**) and 7.22–7.80 ppm (**4e**) and 7.25–7.78 ppm (**5c, 5e**). Proton-containing substituents CH$_3$O and CH$_3$ have singlet signals with an intensity of 3 protons at δ 3.73 and 2.27 ppm for compounds **4a, 4b** and **5a, 5b**.

A distinctive feature of ^{1}H-NMR spectra of benzoates **5a–e** from the spectra of 1,2,4-oxadiazoles **4a–e** is the presence of NH$_2$ proton signal with the integral intensity of 2H at δ 7.51–7.57 ppm.

The ^{1}H-NMR spectrum of spirocompound **6** contained the triplet proton signals of α- and β-CH$_2$ groups at δ 3.16 and 3.68 ppm and signals of N(+)(CH$_2$)$_2$(CH$_2$)$_2$ and N(+)(CH$_2$)$_2$(CH$_2$)$_2$ groups at δ 3.40 and 3.92 ppm, respectively; no aromatic proton signals were observed.

Of the remarkable features of the ^{1}H-NMR spectra of compounds **4a–e**, **5a–e**, and **6**, is that the axial and equatorial protons of the methylene groups located at the nitrogen atom of the morpholine ring give independent multiplet signals. In one case, these signals are superimposed with the common signal of two groups methylene protons located at the oxygen atom of the morpholine ring at δ 3.91–3.93 ppm with a total intensity of six protons and in the other case have a multiplet signal at ~δ 3.40 ppm intensity of two protons. The diastereotopicity of discussed geminal protons of compounds **4a–e**, **5a–e**, and **6** is associated with a dynamic cause due to slow rotation of the morpholine heterocycle. The effect of hindered inversion of six-membered heterocycles, with a chair-like conformer with fixed positions of the axial and equatorial protons being predominant, in the ^{1}H-NMR spectra is a known fact reported in reference data [18]. In addition, the diastereotopicity of these geminal protons of compounds **5a–e** and **6** is associated with asymmetry due to the presence of the spirocyclic system.

In the ^{13}C spectra of the compounds **4a–e**, **5a–e** and **6**, all signals of aliphatic and aromatic protons were recorded in the expected regions.

So, in the ^{13}C NMR spectra the characteristic groups of compounds **4a–e**, **5a–e** and **6** include the signals of: α-methylene groups at δ 31.4 ppm; β-methylene groups at δ 62.1 ppm; signals of carbon atoms of methylene groups with intensity 2C located at nitrogen and oxygen atoms of the heterocycle are in the regions δ 62.4–62.5 ppm and δ 63.2–63.3 ppm. Two signals of carbon atoms of C=N bonds of 1,2,4-oxadiazoles **4a–e** are in the regions δ 167.0–168.9 ppm and δ 169.2–169.5 ppm. The carbon atoms of the C=N bond of the pyrazoline ring of compounds **5a–e** have two signals at δ 167.0–168.8 ppm and δ 169.2 ppm, which may be due to the existence of enantiomers A and B. The carbon atom of the C=N bond of compound **6** has a chemical shift at δ 169.1 ppm. The signals of aromatic carbon atoms for compounds **4a–e** and **5a–e** are in the range δ 112.5–161.0 ppm; *para*-C$_3$ and *para*-CH$_3$ groups of the compounds **4a**, **5a** and **4b**, **5b**, respectively, give signals at δ 55.4 ppm and δ 21.3 ppm.

As we have proved in this article, 1,2,4-oxadiazoles **4a–e** are recorded spectroscopically (FT-IR and NMR spectral data), and they are the initial ones during hydrolysis to pyrazolinium compounds. The Boulton–Katritsky rearrangement mechanism **4a–e→5a–e**, and **4a–e→6** can be represented as a sequence of protonation, proton transfer and nucleophilic attack steps, representing hydrolysis during the reaction of 1,2,4-oxadiazoles **4a–e** with water and wet HCl in the same way as we indicated for piperidine derivatives [9].

All these data demonstrate that the biological activity of compounds under discussion should be associated with their spiropyrazolinium form.

2.2. In Vitro Antitubercular Screening of 2-amino-8-oxa-1,5-diazaspiro[4.5]dec-1-en-5-ium Benzoates and Chloride (5a–e, 6)

In vitro antitubercular bacteriostatic activity of spiropyrazolinium compounds **5a–e** and **6** on drug-sensitive (DS) and multidrug-resistant (MDR) of *M. tuberculosis* (MTB) strains was studied using the method of serial dilution on the liquid Shkolnikova medium (Table 2).

Table 2. In vitro antitubercular activity (MIC) of 2-amino-8-oxa-1,5-diazaspiro[4.5]dec-1-en-5-ium benzoates and chloride (**5a–e, 6**), μg/mL * on DS (H37Rv) and MDR (I) strains of MTB.

Compd		5a	5b	5c	5d	5e	6	Rifampicin
MIC, μg/mL	H37Rv	50	10	20	100	100	1	1
	I	50	50	50	100	100	2	2

* In the upper line of MIC values, the activity on the DS of MTB strains (H37Rv) is shown; in the lower–on the wild MDR strains (I), isolated from the patient, resistant to rifampicin and isoniazid.

A number of compounds **5a–e** on DS and MDR MTB strains exhibits an average antitubercular activity of 10–100 μg/mL. Moreover, an improvement in activity is observed with a decrease in MIC to 10 μg/mL (**5b**); 20 μg/mL (**5c**) and 50 μg/mL (**5a**) on the DS MTB strains and up to 50 μg/mL on the MDR MTB strains for the compounds **5a–c** containing donor substituents in the phenyl ring or with an unsubstituted phenyl ring. Spiropyrazolylammonium chloride **6** demonstrates high in vitro antitubercular activity equal to the activity of the basic antitubercular drug of the first-row rifampicin: on the DS strain as low as 1 μg/mL; on the wild MDR strain—2 μg/mL.

To rationalize the results of the antitubercular activity of 2-amino-8-oxa-1,5-diazaspiro dec-1-en-5-ium salts, X-ray diffraction studies and molecular docking studies were carried out.

2.3. X-ray Diffraction

Molecular structures of compounds **5c–e** and **6** are given in Figure 1. Asymmetric units of **5c**, **5e** and **6** contain one water molecule besides the target ions, and the asymmetric unit of **5d** contains two cations and two symmetrically independent anions. All hydrogen atoms could be located on residual density maps; thus, it was confirmed that the structures are salts with deprotonated carboxylic acids. The six-membered oxo-containing cycles adopt the chair conformation, and the five-membered aza-containing cycles realize the envelope conformation. C(1) carbon atom deviates from the meanplane of N1=N2-C3-C2 atoms at 0.40(1)–0.49(1) Å. Although the cation is rather rigid, it can realize two conformational isomers depending on the shift of the C(1) atom from the meanplane of the rest atoms of the five-membered ring. In crystals of **5c**, **5e** and **6**, these isomers are related with each other by an inversion center, and in acentric crystal **5d**, two symmetrically independent cations realize different conformations (Figure 2a). Nevertheless, for cations in **5c–e** and **6** with similar conformations, the mean atomic deviation doesn't exceed 0.5Å (Figure 2b). The positive charge on quaternary ammonium atom causes elongation of N(1)-N(2) and

N(1)-C(1) bonds as it was previously demonstrated for 5-aryl-3-[2-(piperidin-1-yl)ethyl]-1,2,4-oxadiazoles [9], 2-amino-8-thia-1-aza-5-azoniaspiro[4.5]dec-1-ene [7], 1-(*tert*-butyl)-4,5-dihydro-1*H*-pyrazol-1-ium [19] and 1,1,3-trimethyl-Δ^2-pyrazolinium [20] analogs.

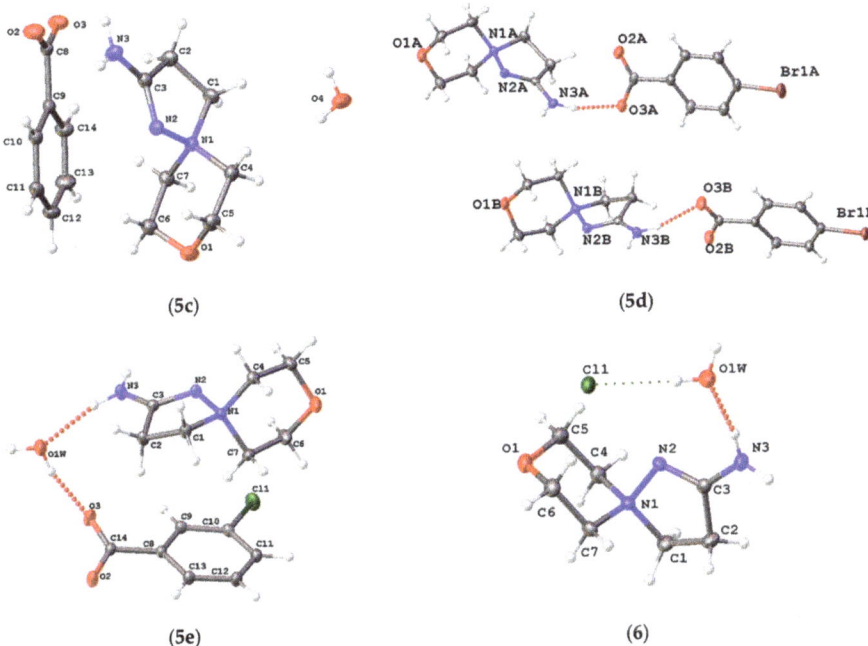

Figure 1. Molecular view of **5c–e** and **6** depicted in thermal ellipsoids ($p = 50\%$).

Figure 2. Molecular conformations of (**a**) two independent cations in **5d** (blue and red; H atoms are omitted); (**b**) similar conformations of cations in **5c** (red), **5d** (blue), **5e** (green) and **6** (magenta).

Crystal packing of a molecule can give valuable information about the most abundant intermolecular interactions of a polytopic molecule. The discussed cations can act as donors of two hydrogen bonds and, theoretically, acceptors of three H-bonds (through O(1), N(2) and N(3) atoms). However, the Full Interaction Maps tool [21,22] implemented within the Mercury 2020.1 package [23] undoubtedly indicates the "inertness" of this cation as an acceptor of H-bonds (Figure 3). A monocarboxylate can only be an acceptor of 2–4 H-bonds. Thus, in the absence of water molecules, the cations and anions form simple H-bonded chains (Figure 4) [9]. In **5d**, these chains are further connected through halogen C-Br..O interactions (Figure 4a). In crystals of **5c**, **5e**, and **6**, water molecules act as linkers between the anion and cation and additionally bind two chains through O-H...O(anion) or O-H...Cl⁻ interactions (Figure 5) so that topologically identical H-bonded chains of the ladder-type are formed.

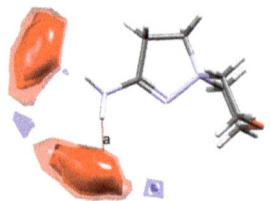

Figure 3. Interaction map in the crystal of **6** probed with H-bond donors (blue) and H-bond acceptors (red).

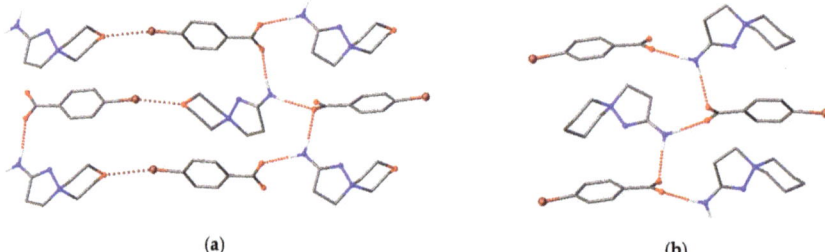

Figure 4. Fragment of H-bonded chains in the crystal of (**a**) **5d** and (**b**) 2-amino-1,5-diazaspiro[4.5]dec-1-en-5-ium 4-bromobenzoate (deposit CCDC 1496456). Color code: Br—brown, C—grey, H—white, N—blue, O—red. The H(C) atoms are omitted. Intemolecular interactions are shown with dotted lines.

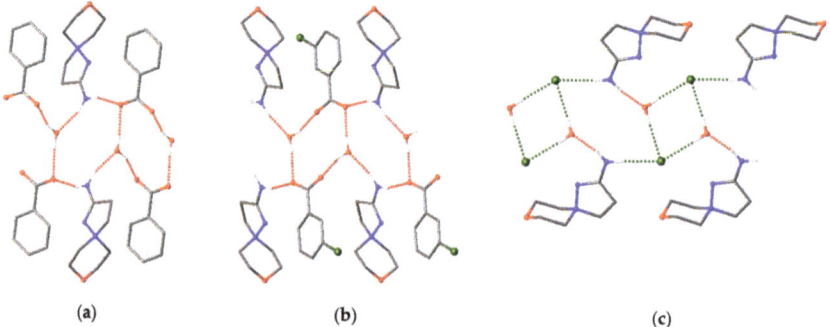

Figure 5. Fragment of H-bonded chains in the crystal of (**a**) **5c**, (**b**) **5e**, (**c**) **6**. Color code: C—grey, Cl—green, H—white, N—blue, O—red. The H(C) atoms are omitted. Intemolecular interactions are shown with dotted lines.

To sum up, X-ray diffraction data demonstrated that molecular docking studies should be carried out for two molecular conformations and that for stable ligand–receptor complexes, the most likely hydrogen bond occurs through the NH_2 group, while acceptor atoms of heterocycles less readily take part in the hydrogen bonds.

2.4. Molecular Docking Studies

Our results indicate that compounds **4a–e** at neutral pH in aqueous solutions undergo transformation to **5a–e**, and antitubercular activity demonstrated by the latter compounds can be referred to either anion of benzoic acids, or the cation, or a mixture of an anion and cation. Some benzoic acids and their anions previously showed activity against *M. tuberculosis* in vitro [24–26]. However, it was assumed that benzoic acids do not have a specific cellular target for *M. tuberculosis* besides their general effect on disrupting the membrane function [25], which is supported by X-ray data of some benzoic acid derivatives with receptors, where ligands are situated on the protein surface [see, for example,

PDB id 5TJZ, 5TJY, 2HRG]. Moreover, the activity of **5a–e** decreased in comparison with monocarboxylate-free salt **6** allows suggesting that the antitubercular activity of these salts should be referred to as the cation mainly.

As we mentioned above, the cation has a rigid conformation with spirocyclic motifs known to be present in some drugs [27,28]. Thus, the mutual disposition of donor, acceptor and hydrophobic fragments of this molecule can be compared with previously X-rayed drugs and ligand–receptor complexes. We used CSD-Crossminer 2020.1 package to search in the Protein Data Bank [29] ligand–receptor complexes where a ligand contains an (*i*) planar heterocycle with (*ii*) donor amine group involved in two H-bonds and (*iii*) neighboring ring, and a receptor was obtained from *M. tuberculosis*. The search gave three hits with PDB id codes 1NBU, 4FOG, and 4XT4, and R.M.S.D. of pharmacophore model from experimental ligand–receptor complexes of 0.398–0.717 Å. For all these hits, 2-amino-8-oxa-1,5-diazaspiro[4.5]dec-1-ene-5-ammonium emulates the pterin moiety, known as a precursor of tetrahydrofolate synthesis, a coenzyme that acts as a carrier for one-carbon units in the biosynthesis of thymidylate, purine nucleotides, and some amino acids [30]. Receptors for these complexes include dihydroneopterin aldolase, thymidylate synthase or oxidoreductase Rv2671. The second search was carried out for structural analogs of spiro[4.5]decane in complexes with receptors obtained from *M. tuberculosis*. The second search gave two hits (PDB codes 5ICJ and 5N7O) of *M. tuberculosis* regulatory protein with 1-oxa-2,8-diazaspiro[4.5]dec-2-en-8-yl derivatives. These receptors were taken as targets for molecular docking calculations together with more typical for theoretical antitubercular screening UDP-galactopyranose mutase.

For all cases, both stereoisomers of the cation were docked independently. Docking without constraints does not give any ligand: receptor complexes; thus, docking was performed with H-bond constraints similar to that in the initial ligand–receptor complexes. Sterical clashes in the binding pocket of thymidylate synthase and UDP-galactopyranose mutase (PRB id codes, respectively, 4FOG and 4RPJ) do not allow the cation to realize appropriate H-bonding; thus, these receptors were excluded from consideration. For 1NBU and 4XT4, we succeeded in overcoming some steric clashes when flexible residues were allowed to rotate freely along with single bonds. Energies of interactions between the cations and three receptors are listed in Table 3; the closest environment of the cation is schematically represented in Figure 6.

Table 3. Energies of interactions for the 2-amino-8-oxa-1,5-diazaspiro[4.5]dec-1-ene-5-ammonium complexes with receptors [kcal/mol].

| Receptor | 1NBU | | 4XT4 | | 5ICJ | |
Ligand	Type A	Type B	Type A	Type B	Type A	Type B
H-bonds	−3.00	−2.00	−3.12	−2.00	−2.79	−3.00
Electrostatic	−4.37	−1.68	−6.34	−5.80	−0.51	−2.12
Nonpolar	−35.72	−32.51	−31.93	−34.80	−38.43	−35.43
Repulsive	0.25	0.35	0.99	0.36	0.04	0.01
Total	−42.84	−35.84	−40.40	−42.24	−41.69	−40.54

Overall, energies of ligand–receptor interactions vary from −35.84 to −42.84 kcal/mol, with the most prominent contribution from nonpolar C-H...π interactions. These values are close to the value of intermolecular pairwise interactions of the same cation in the crystal of **6** (−35.3 kcal/mol) estimated using UNI potentials [31,32]. The cation within binding pockets is involved in two or three hydrogen bonds, and for dihydroneopterin aldolase and thymidylate synthase complexes, the best solution for molecules type A and type B differs in H-bonding patterns. For 1NBU, the best docking solution for molecule type A includes three hydrogen bonds (two N-H...O interactions of amine and oxygen atoms of Tyr52 and Glu74 residues, additionally supported with N-H...N bonds between amide of Asn44 and heterocycle), while the solution for type B molecules includes two H-bonds (N-H...O between amine and Tyr52 and N-H...O between ammonium of Lys99 and the oxygen atom of the cation). The latter solution with the poorest system of H-bonds also has the highest energy of intermolecular interactions among the six solutions under discussion. For 4XT4, both isomers of cation form similar H-bonds with amide group of Asn44 and carboxylate group of Asp67, and interact not only with receptor but also cofactor. The nature of ligand-cofactor interactions differs for two isomers, as well as the system of H-bonds. Only in 5ICJ, both isomers realize nearly similar ligand–receptor interactions through two H-bonds between hydrogen atoms of NH$_2$ group, and OH group of Ser134 and COO group of Asp168 and numerous hydrophobic interactions with indolyl fragment of Trp100 additionally supported by hydrophobic interactions with Thr138 and Leu167. To summarize, based on molecular docking calculations, the cation is able to bind with *M. tuberculosis* dihydroneopterin aldolase, thymidylate synthase and regulatory proteins of transcription. Only in the latter case binding energy, the system of H-bonds and contributions of various types of interactions are independent on stereoisomer of the cation. However, other studies are needed to reveal the mechanism of antitubercular activity of compounds under discussion.

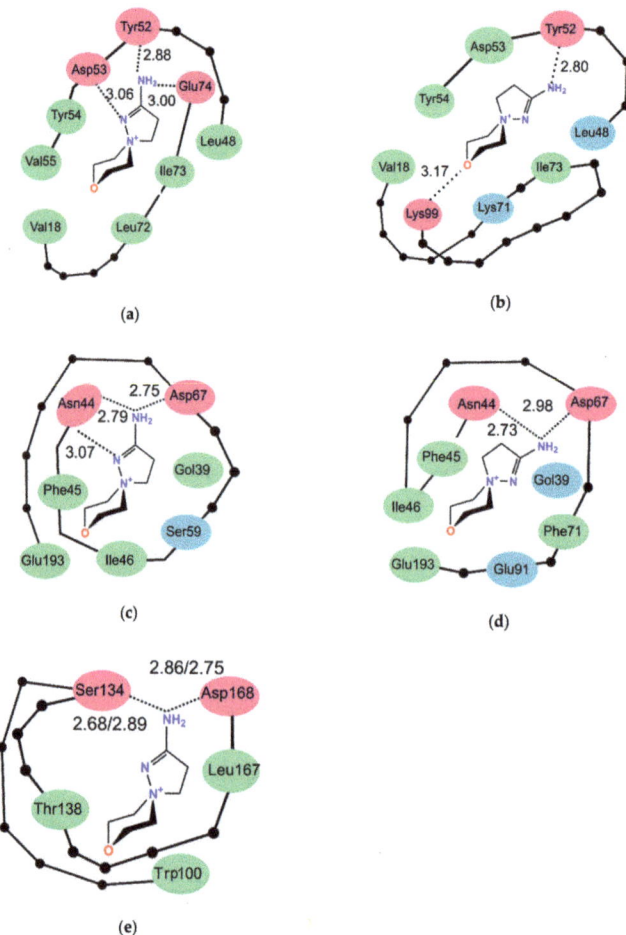

Figure 6. Schematic representation of the interactions between 2-amino-8-oxa-1,5-diazaspiro[4.5]dec-1-ene-5-ammonium and the residues within binding pockets of 1NBU (type A (**a**), type B (**b**)), 4XT4 (type A (**c**) and type B (**d**)), and 5ICJ (**e**). The colors of the residues indicate the character of bonding: pink correspond to H-bonds, green—hydrophobic interactions, blue—electrostatic interactions. H-bonds and their distances are depicted with dotted lines. **Left**—column corresponds to type A cations, and the **right**—to type B molecules.

3. Materials and Methods

3.1. Synthesis

The reagents were purchased from different chemical suppliers and were purified before use. FT-IR spectra were obtained on a Thermo Scientific Nicolet 5700 FTIR instrument (Waltham, MA USA) in KBr pellets. ^{1}H- and ^{13}C-NMR spectra were acquired on a Bruker Avance III 500 MHz NMR spectrometer (500 and 126 MHz, respectively) (Bruker, BioSpin GMBH, Rheinstetten, Germany). The signals of DMSO-$d6$ were used as the internal reference for ^{1}H-NMR (2.50 ppm) and ^{13}C-NMR (39.5 ppm) spectra. Elemental analysis was carried out on a CE440 elemental analyzer (Exeter Analytical, Inc., Shanghai, China). Melting points were determined in glass capillaries on a PTP(M) apparatus (Khimlabpribor, Klin, Russia). The reaction progress and purity of the obtained products were controlled using Sorbfil (Sorbpolymer, Krasnodar, Russia) TLC plates coated with CTX-1A silica gel, grain size 5–17 μm, containing UV-254 indicator. The eluent for TLC analysis was a mixture of benzene–EtOH, 1:3. The solvents for synthesis, recrystallization, and TLC analysis (ethanol, 2-PrOH, benzene, DMF, acetone) were purified according to the standard techniques.

3.1.1. A General Procedure for the Synthesis of 5-Aryl-3-[2-(morpholin-1-yl)ethyl]-1,2,4-oxadiazoles (4a–e)

A solution of -aroyl-(β-morpholin-1-yl)propioamidoximes (3a–e) in dry DMF in a ratio of 1 g:5 mL, respectively, was heated on an oil bath at 70 °C for different times characteristic for compounds 3a–e: 3.5 h (3a, 3b), 2.5 h (3c), 2 h (3d), 1.5 h (3e) with TLC control (eluent was mixture benzene—EtOH, 1:3). The reaction mixture was evaporated to dryness under an oil pump vacuum at 50 °C/1 mm Hg. The organic residue was treated with dry acetone at a ratio of 1 g of parent compounds (3a–e): 5 mL of dry acetone. The obtained crude precipitates of the compounds 4a–e were recrystallized from 2-PrOH.

5-(4-Methoxyphenyl)-3-[2-(morpholin-1-yl)ethyl]-1,2,4-oxadiazole (**4a**). Starting from (4.93 g, 0.016 mol) of **3a** in 25 mL of dry DMF to the resulting 2.89 g (62%), colorless solid **4a**, m.p. 230 °C, R_f 0.71. IR (KBr, cm^{-1}): 1670 (C=N), 1597 (C=N), 1553 (C=C), 1362 (C–O). ^{1}H-NMR (500 MHz, DMSO-d_6): 3.17 (t, J = 7.0 Hz, 2H, CH$_2$CH$_2$N), 3.40 [m, 2H, N(CH eq)$_2$] and 3.93 [m, 6H (2H, N(CH ax)$_2$ and 4H, O(CH$_2$)$_2$)], 3.65 (t, J = 7.0 Hz, 2H, CH$_2$CH$_2$N), 3.73 (s, 3H, p-CH$_3$O), 6.75 (d, J = 8.0 Hz, 2H, o-H Ar), 7.75 (d, J = 8.0 Hz, 2H, m-H Ar). ^{13}C-NMR (126 MHz, DMSO-d_6): 31.4, 55.4, 62.0, 62.4, 112.5, 130.9, 136.0, 160.0, 168.9, 169.3. Anal. Calcd for C$_{15}$H$_{19}$N$_3$O$_3$ (289,33): C, 62.27; H, 6.62. Found: C, 62.60; H, 7.02.

5-(4-Tolyl)-3-[2-(morpholin-1-yl)ethyl]-1,2,4-oxadiazole (**4b**). Starting from (2.81 g, 0.0096 mol) of **3a** in 15 mL of dry DMF to the resulting 1.84 g (70%), colorless solid **4b**, m.p. 220 °C, R_f 0.62. IR (KBr, cm^{-1}): 1648 (C=N), 1593 (C=N), 1553 (C=C), 1376 (C–O). ^{1}H-NMR (500 MHz, DMSO-d_6):2.27 (s, 3H, p-CH$_3$), 3.15 (t, J = 7.0 Hz, 2H, CH$_2$CH$_2$N), 3.40 [m, 2H, N(CHeq)$_2$] and 3.91 [6H, m: 2H, N(CH ax)$_2$ and 4H, O(CH$_2$)$_2$], 3.66 (t, J = 7.0 Hz, 2H, CH$_2$CH$_2$N), 7.01 (d, J = 8.0 Hz, 2H, o-H Ar), 7.70 (d, J = 8.0 Hz, 2H, m-H Ar). ^{13}C-NMR (126 MHz, DMSO-d_6): 21.3, 31.4, 62.1, 62.4, 63.2, 112.5, 130.9, 136.0, 160.0, 169.0, 169.3. Anal. Calcd for C$_{15}$H$_{19}$N$_3$O$_2$ (273,33): C, 65.91; H, 7.01. Found: C, 65.50; H 7.22.

5-Phenyl-3-[2-(morpholin-1-yl)ethyl]-1,2,4-oxadiazole (**4c**). Starting from (1.00 g, 0.0036 mol) of **3** in 5 mL of dry DMF to the resulting 0.86 g (92%) colorless solid **4c**, m.p. 216 °C, R_f 0.66. IR (KBr, cm^{-1}): 1650 (C=N), 1596 (C=N), 1559 (C=C), 1381 (C–O). ^{1}H-NMR (500 MHz, DMSO-d_6): 3.18 (t, J = 7.0 Hz, 2H, CH$_2$CH$_2$N), 3.40 [2H, m, N(CHeq)$_2$] and 3.93 [6H, m: 2H, N(CHax)$_2$ and 4H, O(CH$_2$)$_2$], 3.65 (2H, t, J = 7.0 Hz, CH$_2$CH$_2$N), 7.23–7.83 (m, 5H, C$_6$H$_5$). ^{13}C-NMR (126 MHz, DMSO-d_6): 31.4, 62.1, 62.4, 63.3, 127.5, 127.8, 128.9, 129.1, 129.4, 167.0, 169.2. Anal. Calcd for C$_{14}$H$_{17}$N$_3$O$_2$ (259,30):C, 64.85; H, 6.61. Found: C, 65.30; H, 7.02.

5-(4-Bromophenyl)-3-[2-(morpholin-1-yl)ethyl]-1,2,4-oxadiazole (**4d**). Starting from (1.44 g, 0.004 mol) of **3d** in 8 mL of dry DMF to the resulting 1.12 g (83%) colorless solid **4d**, m.p. 224 °C, R_f 0.67. IR (KBr, cm^{-1}): 1650 (C=N), 1596 (C=N), 1559 (C=C), 1381 (C–O). ^{1}H-NMR (500 MHz, DMSO-d_6): 3.17 (t, J = 7.0 Hz, 2H, CH$_2$CH$_2$N), 3.39 [m, 2H, N(CHeq)$_2$] and 3.93 [m, 6H: 2H, N(CH ax)$_2$ and 4H, O(CH$_2$)$_2$], 3.65 (2H, t, J = 7.0 Hz, CH$_2$CH$_2$N), 7.40 (d,

J = 7.5 Hz, 2H, o-H Ar), 7.74 (d, J = 7.5 Hz, 2H, m-H Ar). ^{13}C-NMR (126 MHz, DMSO-d_6): 31.4, 62.1, 62.4, 63.3, 122.7, 130.2, 131.6, 141.5, 168.3, 169.5. Anal. Calcd for $C_{14}H_{16}BrN_3O_2$ (338.20): C, 49.72; H, 4.77. Found,%: C, 49.29; H, 4.97.

5-(3-Chlorophenyl)-3-[2-(morpholin-1-yl)ethyl]-1,2,4-oxadiazole (**4e**). Starting from (2.70 g, 0.0087 mol) of **3e** in 15 mL of dry DMF to the resulting 1.43 g (56%) colorless solid **4e**, m.p. 190 °C, R_f 0.62. IR (KBr, cm^{-1}): 1680 (C=N), 1596 (C=N), 1557 (C=C), 1377 (C–O). ^{1}H-NMR (500 MHz, DMSO-d_6): 3.15 (t, J = 7.0 Hz, 2H, CH$_2$CH$_2$N), 3.40 [m, 2H, N(CHeq)$_2$] and 3.91 [m, 6H: 2H, N(CHax)$_2$ and 4H, O(CH$_2$)$_2$], 3.65 (t, J = 7.0 Hz, 2H, CH$_2$CH$_2$N), 7.25–7.78 (m, 4H, 3-Cl-C$_6$H$_4$). ^{13}C-NMR (126 MHz, DMSO-d_6): 31.4, 62.1, 62.4, 63.2, 127.7, 127.8, 129.2, 132.4, 144.5, 162.1, 169.2. Anal. Calcd for $C_{14}H_{16}ClN_3O_2$ (293.75): C, 57.24; H, 5.49. Found,%: C, 57.55; H, 5.92.

3.1.2. A General Procedure of the Formation of 2-Amino-8-oxa-1,5-diazaspiro[4.5]dec-1-ene-5-ammonium benzoates (**5a–e**)

To the solutions of 5-(3,4-substituted phenyl)-3-[2-(morpholin-1-yl)ethyl]-1,2,4-oxadiazoles (**4a–e**) in DMF in a ratio of 0.5 g: 10 mL, respectively, two equivalents of H$_2$O was added dropwise. The reaction solutions were heated at 70 °C with TLC control during the time typical for groups of compounds **4a–e**: **4a, b** (40 h), **4c–e** (25 h). After the disappearance of the 1,2,4-oxadiazoles **4a–e** spots on the silufol plate and the appearance of products **5a–e** spots, the solvent was evaporated to dryness in an oil vacuum pump. Acetone was added to the viscous residues of the reaction mixtures **5a–e** at the rate of 0.5 g of the starting compounds **4a–e**: 15 mL of acetone. The formed precipitates of 2-amino-8-oxa-1,5-diazaspiro[4.5]dec-1-ene-5-ammonium 3,4-substituted benzoates (**5a–e**) were filtered off and dried at room temperature.

2-Amino-8-oxa-1,5-diazaspiro[4.5]dec-1-ene-5-ammonium 4-methoxybenzoate hydrate (**5a**). Starting from 0.5 g (0.0017 mol) of compound **4a** in 10 mL of DMF and 0.062 mL (0.0034 mol) H$_2$O to the resulting 0.35 g (63%) colorless solid **5a**, m.p. 235 °C, R_f 0.75. IR (KBr, cm^{-1}): 1657 (C=N), 1594 (C=C), 1548 (νCOO$^-$ as), and 1420 (νCOO$^-$ sy), 3309, 3414, 3459 [νN(-H)$_2$]. ^{1}H-NMR (500 MHz, DMSO-d_6): 2.27 (s, 3H, p-CH$_3$O), 3.17 [t, J = 7.0 Hz, 2H, CH$_2$CH$_2$N(+)], 3.41 [m, 2H, N(CHeq)$_2$] and 3.92 [m, 6H: 2H, N(CHax)$_2$ and 4H, O(CH$_2$)$_2$], 3.65 [t, J = 7.0 Hz, 2H, CH$_2$CH$_2$N(+)], 6.78 (d, J = 8.0 Hz, 2H, o-H Ar), 7.55 and 7.60 (s, 2H, NH$_2$), 7.78 (d, J = 8.0 Hz, 2H, m-H Ar). ^{13}C-NMR (126 MHz, DMSO-d_6): 31.4, 62.4, 63.2, 112.5, 130.9, 134.8, 159.9, 144.5, 168.8, 169.2. Anal. Calcd for $C_{15}H_{23}N_3O_5$ (325,36): C, 55.37; H, 7.13. Found,%: C, 55.79; H, 7.57.

2-Amino-8-oxa-1,5-diazaspiro[4.5]dec-1-ene-5-ammonium 4-methylbenzoate hydrate (**5b**). Starting from 0.5 g (0.0018 mol) of the compound **4b** in 10 mL of DMF and 0.066 mL (0.0036 mol) H$_2$O to the resulting 0.35 g (63%) colorless solid **5b**, m.p. 248 °C, R_f 0.80. IR (KBr, cm^{-1}): 1655 (C=N), 1597 (C=C), 1550 (COO$^-$as), and 1442 (COO$^-$ sy), 3157, 3312, 3457 [N(-H)$_2$]. ^{1}H-NMR (500 MHz, DMSO-d_6): 3.15 [t, J = 7.0 Hz, 2H, CH$_2$CH$_2$N(+)], 3.41 [m, 2H, N(CHeq)$_2$] and 3.93 [m, 6H: 2H, N(CHax)$_2$ and 4H, O(CH$_2$)$_2$], 3.64 [t, J = 7.0 Hz, 2H, CH$_2$CH$_2$N(+)], 3.73 (s, 3H, p-CH$_3$), 7.02 (d, J = 8.0 Hz, 2H, o-H Ar), 7.54 (s, 2H, NH$_2$), 7.69 (d, J = 8.0 Hz, 2H, m-H Ar). ^{13}C-NMR (126 MHz, DMSO-d_6): 21.3, 31.38, 62.1, 62.4, 63.2, 127.9, 128.3, 129.4, 139.7, 168.8, 169.2. Anal. Calcd for $C_{15}H_{23}N_3O_4$. (309,36): C, 58.24; H, 7.49. Found: C, 58.29; H, 7.59.

2-Amino-8-oxa-1,5-diazaspiro[4.5]dec-1-ene-5-ammonium benzoate hydrate (**5c**). Starting from 0.5 g (0.0019 mol) of **4c** in 10 mL of DMF and 0.07 mL (0.0038 mol) H$_2$O to the resulting 0.43 g (77%) colorless solid **5c**, m.p. 220 °C, R_f 0.75. IR (KBr, cm^{-1}):1658 (C=N), 1596 (C=C), 1556 (COO$^-$ as), and 1426 (COO$^-$ sy), 3152, 3311, 3456 [N(-H)$_2$]. ^{1}H-NMR (500 MHz, DMSO-d_6): 3.19 [t, J = 7.0 Hz, 2H, CH$_2$CH$_2$N(+)], 3.39 [m, 2H, N(CHeq)$_2$] and 3.93 [m, 6H: 2H, N(CHax)$_2$ and 4H, O(CH$_2$)$_2$], 3.65 [t, J = 7.0 Hz, 2H, CH$_2$CH$_2$N(+)], 7.22–7.80 (m, 5H, C$_6$H$_5$), 7.55 and 7.64 (s, 2H, NH$_2$). ^{13}C-NMR (126 MHz, DMSO-d_6): 31.4, 61.3, 62.4, 63.2, 127.26, 128.3, 129.4, 142.4, 168.7, 169.2. Anal. Calcd for $C_{14}H_{21}N_3O_4$ (295,33): C, 56.94; H, 7.17. Found: C, 56.79; H, 7.50.

2-Amino-8-oxa-1,5-diazaspiro[4.5]dec-1-ene-5-ammonium 4-bromobenzoate (**5d**). Starting from 0.5 g (0.0015 mol) of **4d** in 10 mL of DMF and 0.05 mL (0.0028 mol) H$_2$O to the resulting 0.43 g (80%) colorless solid **5d**, m.p. 240 °C, R_f 0.77. IR (KBr, cm^{-1}): 1657 (C=N), 1591 (C=C), 1545 (COO$^-$ as), and 1423 (COO$^-$ sy), 3302, 3411, 3485 [N(-H)$_2$]. ^{1}H-NMR (500 MHz, DMSO-d_6): 3.15 [t, J = 7.0 Hz, 2H, CH$_2$CH$_2$N(+)], 3.40 [m, 2H, N(CHeq)$_2$] and 3.92 [m, 6H: 2H, N(CHax)$_2$ and 4H, O(CH$_2$)$_2$], 3.65 [t, J = 7.0 Hz, 2H, CH$_2$CH$_2$N(+)], 7.39 (d, J = 7.5 Hz, 2H, o-H Ar), 7.55 (s, 2H, NH$_2$), 7.73 (d, J = 7.5 Hz, 2H, m-H Ar). ^{13}C-NMR (126 MHz, DMSO-d_6): 31.4, 62.1, 62.4, 63.2, 121.9, 130.1, 131.6, 141.8, 167.4, 169.2. Anal. Calcd for C$_{14}$H$_{18}$BrN$_3$O$_3$ (356,22): C, 47.20; H, 5.09. Found: C, 47.49; H, 4.97.

2-Amino-8-oxa-1,5-diazaspiro[4.5]dec-1-ene-5-ammonium 3-chlorobenzoate hydrate (**5e**). Starting from 0.5 g (0.0017 mol) **4e** in 10 mL of DMF and 0.06 mL (0.034 mol) H$_2$O to the resulting 0.24 g (43%) colorless solid **5e**, m.p. 200 °C, R_f 0.70. IR (KBr, cm^{-1}): 1657 (C=N), 1594 (C=C), 1557 (COO$^-$ as), and 1420 (COO$^-$ sy), 3155, 3309, 3457 [N(-H)$_2$]. ^{1}H-NMR (500 MHz, DMSO-d_6): 3.16 [t, J = 7.0 Hz, 2H, CH$_2$CH$_2$N(+)], 3.40 [m, 2H, N(CH eq)$_2$] and 3.93 [6H, m: 2H, N(CHax)$_2$ and 4H, O(CH$_2$)$_2$], 3.66 [t, J = 7.0 Hz, 2H, CH$_2$CH$_2$N(+)], 7.25–7.78 (m, 4H, 3-Cl-C$_6$H$_4$), 7.57 and 7.61 (s, 2H, NH$_2$). ^{13}C-NMR (126 MHz, DMSO-d_6): 31.4, 62.1, 62.5, 63.2, 127.9, 128.2, 129.2, 129.3, 132.4, 144.7, 167.0, 169.2. Anal. Calcd for C$_{14}$H$_{20}$ClN$_3$O$_4$ (329,78): C, 50.99; H, 6.11. Found: C, 50.52; H, 6.47.

3.1.3. A General Method of the Formation of 2-Amino-8-oxa-1,5-diazaspiro[4.5]dec-1-ene-5-ammonium chloride (6) and Substituted Benzoic Acids

To a solution of 0.2 g (0.0007 mol) of 5-(3,4-substituted phenyl)-3-[2-(morpholin-1-yl)ethyl]- 1,2,4-oxadiazoles (**4a–e**) in 10 mL ethanol a HCl solution in ether was added dropwise to pH 2. A white precipitate of 2-amino-1,5-diazaspiro[4.5]dec-1-en-5-ium chloride hydrate (**6**) in all cases was formed at once in 75–82% yields and collected by filtration. 3,4-Substituted benzoic acids were precipitated during the evaporation of mother liquors obtained after the filtration of chloride monohydrate **6**. All characteristics of substituted benzoic acids corresponded to the reference data.

2-Amino-8-oxa-1,5-diazaspiro[4.5]dec-1-ene-5-ammonium chloride hydrate (**6**)—white opaque powder, m.p. 257–260 °C, R_f 0.21. IR (KBr, cm^{-1}): 1639 (νC=N), [1615, δN(–H)$_2$], 1362 (νC–O), 3158, 3310, 3457 [νN(-H)$_2$]. ^{1}H-NMR (500 MHz, DMSO-d_6): 3.16 [t, J = 7.0 Hz, 2H, CH$_2$CH$_2$N(+), 3.40 [m, 2H, N(CH eq)$_2$] and 3.92 [m, 6H: 2H, N(CH ax)$_2$ and 4H, O(CH$_2$)$_2$], 3.68 [t, J = 7.0 Hz, 2H, CH$_2$CH$_2$N(+)], 7.51 (s, 2H, NH$_2$). ^{13}C-NMR spectrum, δ, ppm: 31.4, 31,5, 62.1, 62.4, 63.3, 169.2. Anal. Calcd for C$_7$H$_{16}$N$_3$O$_2$ (209.67): C, 40.10, H 7.69. Found: C, 40.29; H, 7.97.

3.2. Single-Crystal X-ray Diffraction

The X-ray diffraction data of **5c–e, 6** were collected at 120 K on a Bruker Apex II diffractometer (Bruker AXS, Inc., Madison, WI, USA) equipped with an Oxford Cryostream cooling unit and a graphite monochromated Mo anode (λ = 0.71073 Å). Crystal structures were solved using SHELXT [33] program and refined with SHELXL [34] using OLEX2 software [35]. The structures were refined by full-matrix least-squares procedure against F^2. Non-hydrogen atoms were refined anisotropically. The H(C) positions were calculated, the H(N) and H(O) atoms were located on difference Fourier maps and refined using the riding model. Experimental details and crystal parameters are given in Table S1 (Electronic Supporting Information).

3.3. Molecular Docking Studies

The molecular docking studies were performed following the previously described protocol [36,37] and processed with GOLD software (version 2020.2.0, Cambridge Crystallographic Data Center, Cambridge, UK).

The three-dimensional (3D) crystal structures of *M. tuberculosis* regulatory protein (PDB ID: 5ICJ), 7,8-dihydroneopterin aldolase complexed with 2-amino-6-(hydroxymethyl) pteridin-4(3H)-one (PH2; PDB ID: 1NBO), thymidylate synthase in complex with 5-fluoro DUMP (UFP; PDB ID: 4FOG), UDP-galactopyranose mutase from Mtb docked with UDP (UGM; PDB ID: 4RPJ), and Rv2671 protein in complex with dihydropteridine (44W;

PDB ID: 4XT4) were obtained from the RCSB Protein Data Bank [29], while the 3D-structure of cation was taken from X-ray diffraction data of **6** (two stereoisomers were analyzed independently).

To validate the molecular docking outcomes, 4,4,4-trifluoro-1-(3-phenyl-1-oxa-2,8-diazaspiro[4.5]dec-2-en-8-yl)butan-1-one, PH2, UFP, UGM, and 44W were removed from their receptors and re-docked back into their receptors. The docking results were expressed as binding energy values (kcal/mol) of ligand–receptor complexes; these are based on hydrogen bond, hydrophobic, and electrostatic interactions. All required docking settings, including the preparation of mol2 files for the receptors and ligands, determination of binding sites, the protonation state, calculations, and the overall charges, were established as hitherto described [38].

4. Conclusions

5-Substituted phenyl-3-(2-aminoethyl)-1,2,4-oxadiazoles in the presence of moisture and acids undergo Boulton–Katritsky rearrangement to the salts of spiropyrazolinium compounds—2-amino-8-oxa-1,5-diazaspiro[4.5]dec-1-en-5-ium benzoates and chloride. Hence, biological screening results should be associated with rearranged products and not with the original taken on trials 5-substituted phenyl-3-(2-aminoethyl)-1,2,4-oxadiazoles. A small library of the newly 2-amino-8-oxa-1,5-diazaspiro[4.5]dec-1-en-5-ium benzoates and chloride has been used in the drug design of antitubercular drugs. The tested compounds show moderate to high in vitro antitubercular activity with MIC values of 1–100 µg/mL. The highest activity in 1 µg/mL and 2 µg/mL on DS and MDR of *M. tuberculosis* strains, equal to the activity of the basic antitubercular drug rifampicin, was recorded for 2-amino-8-oxa-1,5-diazaspiro[4.5]dec-1-en-5-ium chloride. 3D molecular structure of the cation extracted from crystal structure was used for molecular docking studies with various *M. tuberculosis* receptors. It was demonstrated that two stereoisomers of the rigid cation form different sets of hydrogen bonds in complexes with dihydroneopterin aldolase or oxidoreductase Rv2671 and similar H-bonds in complex with thymidylate synthase. However, energies of all ligand–receptor complexes vary from -35.8 to -42.8 kcal/mol.

Supplementary Materials: The following are available online, supplementary data including the NMR spectra of 5-substituted phenyl-3-[2-(morpholin-1-yl)ethyl]-1,2,4-oxadiazole and 2-amino-8-oxa-1,5-diazaspiro [4.5]dec-1-en-5-ium benzoate and chloride associated with this article are available online. CCDC 2049798-2049801 contains crystallographic information for **5c–e, 6**. Crystallographic information files are available from the Cambridge Crystallographic Data Center upon request (http://www.ccdc.cam.ac.uk/structures).

Author Contributions: Conceptualization, L.K.; data curation, K.P.; formal analysis, K.P., V.B. and K.A.; funding acquisition, L.K.; investigation, A.V., G.D., G.B., A.U., V.B. and L.C.; methodology, L.K., G.D., G.B., A.U. and L.C.; resources, K.P., G.D., G.B., A.U., V.B., L.C. and K.A.; validation, A.V., K.P., G.D., G.B., A.U., V.B., L.C. and K.A.; visualization, L.K., A.V. and K.A.; writing—original draft, L.K. and A.V.; writing—review and editing, L.K. and A.V. All authors have read and agreed to the published version of the manuscript.

Funding: This study was supported by the Committee of Science of the Ministry of Education and Science of the Republic of Kazakhstan (grant AP08856440). A. Vologzhanina acknowledges the support of the Russian Science Foundation for molecular docking studies (grant 20-13-00241).

Data Availability Statement: The data presented in this study are available in this article.

Acknowledgments: Single-crystal X-ray diffraction data were measured using the equipment of the Center for Molecular Studies of INEOS RAS.

Conflicts of Interest: The authors declare no conflict of interest.

Sample Availability: The samples may be available.

References

1. List of Drugs–Wikipedia. Available online: https://en.wikipedia.org/wiki/List_of_drugs (accessed on 12 December 2020).
2. Fischer, E.; Knovenagel, O. Ueber die Verbindungen des Phenylhydrazinsmit Acroleïn, Mesityloxyd und Allylbromid. *Ann. Chem.* **1887**, *239*, 194–206. [CrossRef]
3. Santos, C.M.M.; Silva, V.L.M.; Silva, A.M.S. Synthesis of Chromone-Related Pyrazole Compounds. *Molecules* **2017**, *22*, 1665. [CrossRef] [PubMed]
4. Sharma, S.; Kaur, S.; Bansal, T.; Gaba, J. Review on Synthesis of Bioactive Pyrazoline Derivatives. *Chem. Sci. Trans.* **2014**, *3*, 861–875. [CrossRef]
5. 1-H Pyrazolium at National Library of Medicine. Available online: https://pubchem.ncbi.nlm.nih.gov/compound/13585073 (accessed on 12 December 2020).
6. 3-Amino-1,4-Dimethyl-1-Phenyl-2-Pyrazolinium Iodide at National Library of Medicine. Available online: https://pubchem.ncbi.nlm.nih.gov/compound/13064197 (accessed on 12 December 2020).
7. Kayukova, L.A.; Orazbaeva, M.A.; Gapparova, G.I.; Beketov, K.M.; Espenbetov, A.A.; Faskhutdinov, M.F.; Tashkhodjaev, B.T. Rapid acid hydrolysis of 5-aryl-3-(β-thiomorpholinoethyl)-1,2,4-oxadiazoles. *Chem. Heterocycl. Compd.* **2010**, *46*, 879–886. [CrossRef]
8. Kayukova, L.A.; Beketov, K.M.; Baitursynova, G.P. 5-Aryl-3-(β-aminoethyl)-1,2,4-oxadiazoles in Boulton-Katritzky Rearrangement. In Proceedings of the International Conference Catalysis in Organic Synthesis ICCOS-2012 Book of Abstracts, Moscow, Russia, 15–20 September 2012; p. 214; Available online: http://www.spsl.nsc.ru/FullText/konfe/ICCOS2012.pdf (accessed on 26 January 2021).
9. Kayukova, L.A.; Uzakova, A.B.; Vologzhanina, A.V.; Akatan, K.; Shaymardan, E.; Kabdrakhmanova, S.K. Rapid Boulton–Katritzky rearrangement of 5-aryl-3-[2-(piperidin-1-yl)ethyl]-1,2,4-oxadiazoles upon exposure to water and HCl. *Chem. Heterocycl. Compd.* **2018**, *54*, 643–649. [CrossRef]
10. Li, J.J. Boulton–Katritzky rearrangement. In *Name Reactions*; Springer: Cham, Switzerland, 2014. [CrossRef]
11. Korbonits, D.; Kanzel-Szvoboda, I.; Horváth, K. Ring transformation of 3-(2-aminoaryl)-1,2,4-oxadiazoles into 3-acylaminoindazoles; extension of the Boulton–Katritzky scheme. *J. Chem. Soc. Perkin Trans.* **1982**, *1*, 759–766. [CrossRef]
12. Kayukova, L.A.; Praliyev, K.D.; Myrzabek, A.B.; Kainarbayeva, Z.N. Arylsulfochlorination of β-aminopropioamidoximes giving 2-aminospiropyrazolylammonium arylsulfonates. *Rus. Chem. Bull.* **2020**, *69*, 496–503. [CrossRef]
13. Kayukova, L.A.; Praliev, K.D.; Akhelova, A.L.; Kemel'bekov, U.S.; Pichkhadze, G.M.; Mukhamedzhanova, G.S.; Kadyrova, D.M.; Nasyrova, S. Local anesthetic activity of new amidoxime derivatives. *Pharm. Chem. J.* **2011**, *45*, 468–471. [CrossRef]
14. Kayukova, L.A.; Jussipbekov, U.; Praliyev, K. Amidoxime Derivatives with Local Anesthetic, Antitubercular, and Antidiabetic Activity. In *Heterocycles-Synthesis and Biological Activities*; Nandeshwarappa, B.P., Sadashiv, S.O., Eds.; IntechOpen: London, UK, 17 December 2019. [CrossRef]
15. Kayukova, L.A.; Uzakova, A.B.; Baitursynova, G.P.; Dyusembaeva, G.T.; Shul'gau, Z.T.; Gulyaev, A.E.; Sergazy, S.D. Inhibition of α-Amylase and α-Glucosidase by New β-Aminopropionamidoxime Derivatives. *Pharm. Chem. J.* **2019**, *53*, 129–133. [CrossRef]
16. The Use of a Derivative of Beta-Morpholinopropioamidoxime as an Antidiabetic Agent. Patent Owner: JSC «A.B. Institute of Chemical Sciences». Russia Patent No. 2684779, 15 April 2019.
17. Ashenhurst, J. Homotopic, Enantiotopic, and Diastereotopic Groups: What Does It Mean? Last updated: October 16th, 2020. Available online: https://www.masterorganicchemistry.com/homotopic-enantiotopic-diastereotopic/ (accessed on 26 January 2021).
18. Kayukova, L.A. Conditions for the Heterocyclization of O-Aroyl-β-morpholinopropioamide Oximes to 5-Aryl-3-(β-morpholino)ethyl-1,2,4-oxadiazoles. *Chem. Heterocycl. Compd.* **2003**, *39*, 223–227. [CrossRef]
19. Usachev, S.V.; Nikiforov, G.A.; Lyssenko, K.A.; Nelyubina, Y.V.; Levkin, P.A.; Kostyanovsky, R.G. Sterically hindered and completely arrested nitrogen inversion in pyrazolidines. *Tetrahedron Asymmetry* **2007**, *18*, 1540–1547. [CrossRef]
20. Yap, G.P.A.; Alkorta, I.; Elguero, J.; Jagerovic, N. The structure of 1,1,3-trimethyl-Δ^2-pyrazolinium perchlorate: An X-ray crystallographic and GIAO/DFT multinuclear NMR study. *Spectrosc. Int. J.* **2004**, *18*, 605–611. [CrossRef]
21. Wood, P.A.; Olsson, T.S.G.; Cole, J.C.; Cottrell, S.J.; Feeder, N.; Galek, P.T.A.; Groom, C.R.; Pidcock, E. Evaluation of molecular crystal structures using Full Interaction Maps. *Cryst. Eng. Comm.* **2013**, *15*, 65–72. [CrossRef]
22. Vologzhanina, A.V. Intermolecular Interactions in Functional Crystalline Materials: From Data to Knowledge. *Crystals* **2019**, *9*, 478. [CrossRef]
23. Macrae, C.F.; Sovago, I.; Cottrell, S.J.; Galek, P.T.A.; McCabe, P.; Pidcock, E.; Platings, M.; Shields, G.P.; Stevens, J.S.; Towler, M.; et al. Mercury 4.0: From visualization to analysis, design and prediction. *J. Appl. Cryst.* **2020**, *53*, 226–235. [CrossRef]
24. French, F.A.; Freedlander, B.L. Derivatives of benzoic acid and simple phenols in the chemotherapy of tuberculosis. *J. Am. Pharm. Assoc.* **1949**, *38*, 343–346. [CrossRef]
25. Zhang, Y.; Zhang, H.; Sun, Z. Susceptibility of *Mycobacterium tuberculosis* to weak acids. *J. Antimicrob. Chemother.* **2003**, *52*, 56–60. [CrossRef]
26. Maresca, A.; Vullo, D.; Scozzafava, A.; Manole, G.; Supuran, C.T. Inhibition of the β-class carbonic anhydrases from Mycobacterium tuberculosis with carboxylic acids. *J. Enz. Inhib. Med. Chem.* **2013**, *28*, 392–396. [CrossRef]
27. Taylor, F.F.; Faloon, W.W. The role of potassium in the natriuretic response to a steroidal lactone (SC-9420). *J. Clin. Endocrinol. Metab.* **1959**, *19*, 1683–1687. [CrossRef]

28. Brunner, H.R. The New Angiotensin II Receptor Antagonist, Irbesartan: Pharmacokinetic and Pharmacodynamic Considerations. *Am. J. Hypertens.* **1997**, *10*, 311S–317S. [CrossRef]
29. Berman, H.M.; Westbrook, J.; Feng, Z.; Gilliland, G.; Bhat, T.N.; Weissig, H.; Shindyalov, I.N.; Bourne, P.E. The Protein Data Bank. *Nucleic Acids Res.* **2000**, *28*, 235–242. [CrossRef] [PubMed]
30. Hanson, A.D.; Gregory, J.F., III. Folate Biosynthesis, Turnover, and Transport in Plants. *Annu. Rev. Plant Biol.* **2011**, *62*, 105–125. [CrossRef] [PubMed]
31. Gavezzotti, A. Are crystal structures predictable? *Acc. Chem. Res.* **1994**, *27*, 309–314. [CrossRef]
32. Gavezzotti, A.; Filippini, G. Geometry of the Intermolecular X-H . . . Y (X, Y = N, O) Hydrogen Bond and the Calibration of Empirical Hydrogen-Bond Potentials. *J. Phys. Chem.* **1994**, *98*, 4831–4837. [CrossRef]
33. Sheldrick, G.M. SHELXT–Integrated space-group and crystal-structure determination. *ActaCryst.* **2015**, *A71*, 3–8. [CrossRef]
34. Sheldrick, G.M. Crystal structure refinement with SHELXL. *Acta Cryst.* **2015**, *C71*, 3–8. [CrossRef]
35. Dolomanov, O.V.; Bourhis, L.J.; Gildea, R.J.; Howard, J.A.K.; Puschmann, H. OLEX2: A complete structure solution, refinement and analysis program. *J. Appl. Cryst.* **2009**, *42*, 339–341. [CrossRef]
36. Verdonk, M.L.; Cole, J.C.; Hartshorn, M.J.; Murray, C.W.; Taylor, R.D. Improved protein-ligand docking using GOLD. *Proteins* **2003**, *52*, 609–623. [CrossRef]
37. Liebeschuetz, J.W.; Cole, J.C.; Korb, O. Pose prediction and virtual screening performance of GOLD scoring functions in a standardized test. *J. Comput. Mol. Des.* **2012**, *26*, 737–748. [CrossRef]
38. GOLD User Guide. Available online: https://www.ccdc.cam.ac.uk/support-and-resources/ccdcresources/GOLD_User_Guide_2020_1.pdf (accessed on 12 December 2020).

Article

Approaches to the Synthesis of Dicarboxylic Derivatives of Bis(pyrazol-1-yl)alkanes

Nikita P. Burlutskiy [1] and Andrei S. Potapov [2,*]

1 Kizhner Research Center, National Research Tomsk Polytechnic University, 30 Lenin Ave., 634050 Tomsk, Russia; npb1@tpu.ru
2 Nikolaev Institute of Inorganic Chemistry, Siberian Branch of the Russian Academy of Sciences, 3 Lavrentiev Ave., 630090 Novosibirsk, Russia
* Correspondence: potapov@niic.nsc.ru; Tel.: +7-383-330-94-90

Abstract: Carboxylation of bis(pyrazol-1-yl)alkanes by oxalyl chloride was studied. It was found that 4,4′-dicarboxylic derivatives of substrates with electron-donating methyl groups and short linkers (from one to three methylene groups) can be prepared using this method. Longer linkers lead to significantly lower product yields, which is probably due to instability of the intermediate acid chlorides that are initially formed in the reaction with oxalyl chloride. Thus, bis(pyrazol-1-yl)methane gave only monocarboxylic derivative even with a large excess of oxalyl chloride and prolonged reaction duration. An alternative approach involves the reaction of ethyl 4-pyrazolecarboxylates with dibromoalkanes in a superbasic medium (potassium hydroxide–dimethyl sulfoxide) and is suitable for the preparation of bis(4-carboxypyrazol-1-yl)alkanes with both short and long linkers independent of substitution in positions 3 and 5 of pyrazole rings. The obtained dicarboxylic acids are interesting as potential building blocks for metal-organic frameworks.

Keywords: pyrazole; bis(pyrazol-1-yl)alkanes; carboxylation; oxalyl chloride; dicarboxylic acids; alkylation; superbasic medium

1. Introduction

Bis(pyrazol-1-yl)alkanes are bidentate ligands widely used for the synthesis of the coordination compounds [1], some of which were shown to exhibit catalytic [2–6], anticancer [7,8], antibacterial [9] and SOD-like activity [10,11], and electroluminescent properties [12,13]. The properties of the ligands can be varied by introducing the functional groups into the pyrazole rings [14–17], besides, carboxylic groups themselves can act as donor groups for the formation of coordination bond with the metal ions. Carboxy-substituted pyrazoles were used for the construction of highly porous metal-organic frameworks [18–21]. Carboxylic acids based on bis(pyrazol-1-yl)alkanes are much less explored and only a few research papers were published so far [22–25]. In addition, a series of works devoted to the synthesis of metal-organic frameworks based on structurally related bis(4-carboxyphenylpyrazol-1-yl)methane was carried out by Sumby et al., the presence of the free chelating units in the structure of the framework allowed them to prepare catalysts with single metal sites [26,27].

In this contribution we report a facile synthesis of a series of 4,4′-dicarboxy-substituted bis(pyrazol-1-yl)alkanes with varied linker length and substitution in positions 3 and 5 of the pyrazole rings.

2. Results and Discussion

To introduce carboxyl groups into bis(pyrazol-1-yl)alkanes, a reaction with oxalyl chloride was used, in which it was both a reagent and a solvent. Previously oxalyl chloride was successfully used for the carboxylation of 1-phenyl- and 1-alkylpyrazoles [28]. In this reaction, a pyrazole-containing derivative of oxalic acid chloride is initially formed, which is converted to a carboxylic acid chloride with the release of carbon monoxide. The acid chloride is hydrolyzed without isolation to form a carboxylic acid. Other methods for the introduction of carboxylic groups into pyrazole rings include oxidation of alkyl [29,30] or formyl [31] groups, substitution of halogens via intermediate organolithium derivatives [32], and hydrolysis of trichloromethyl derivatives [33].

Carboxylation of bis(3,5-dimethylpyrazol-1-yl)methane **1a** with subsequent hydrolysis gave the corresponding dicarboxylic acid **1b** in high yield (Scheme 1). Similarly, 1,2-bis(3,5-dimethylpyrazol-1-yl)ethane **2a** and 1,3-bis(3,5-dimethylpyrazol-1-yl)propane **3a** gave dicarboxylic acids **2b** and **3b**, albeit in lower yield (Scheme 1). When bis(3,5-dimethylpyrazol-1-yl)alkanes with longer spacers (tetramethylene, $(CH_2)_4$ and longer) were introduced into the reaction, full conversion of the starting materials was achieved (GC–MS control), but no dicarboxylic acids were isolated. Instead, mixtures of monosubstituted carboxy-derivatives (not isolated) and starting substrates were obtained, which is probably a result of instability of the intermediate acid chlorides leading to their decarboxylation during the hydrolysis.

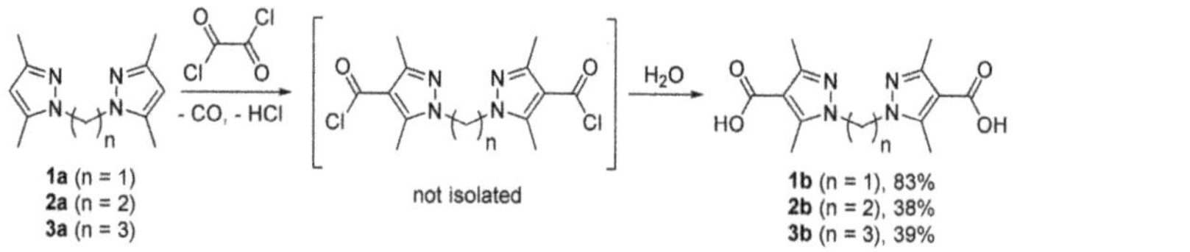

Scheme 1. Carboxylation of bis(3,5-dimethylpyrazol-1-yl)alkanes with short linkers.

To overcome the instability of acid chlorides their methanolysis instead of hydrolysis was attempted and indeed led to the dimethyl esters **4c–6c**, although in low yields (Scheme 2). Alkaline hydrolysis of the esters proceeded smoothly with the formation of dicarboxylic acids **4b–6b** (Scheme 2).

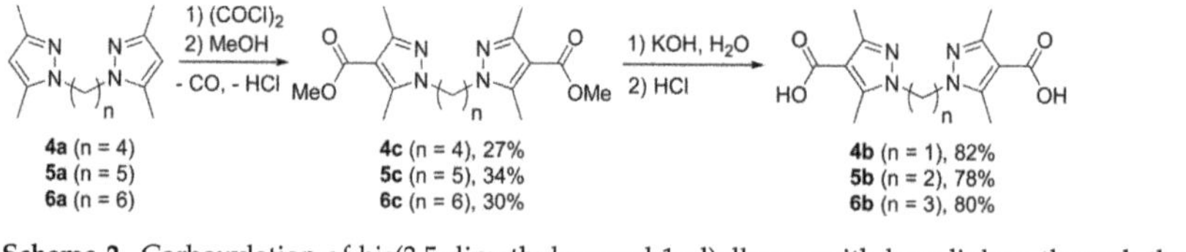

Scheme 2. Carboxylation of bis(3,5-dimethylpyrazol-1-yl)alkanes with long linkers through the methanolysis step.

Carboxylation by oxalyl chloride proved to be poorly suitable for bis(pyrazol-1-yl)alkanes without electron-donating methyl groups. Thus, only in the case of 1,3-bis(pyrazol-1-yl)propane **3d** a dicarboxylation product could be isolated, while bis(pyrazol-1-yl)methane **1d** gave solely a monocarboxylated product **7** (Scheme 3). Even a large excess of oxalyl chloride and prolonged heating did not lead to the formation of the dicarboxylic acid. For derivatives with longer spacers (four to six methylene groups) only traces of carboxylation products (after conversion to methyl esters) were detected (for substrate **4c**) but were not isolated (Scheme 3).

In order to gain insight into the reasons for low reactivity of bis(pyrazol-1-yl)methane **1d**, DFT calculations of charge distribution in the starting compound **1d** and compounds **1a**, **3a**, and **3d** for comparison and the corresponding monosubstituted acid chlorides formed during the first electrophilic substitution step were carried out. The structures were optimized at the B3LYP 6-31G(d) level of theory and absence of imaginary frequencies confirmed that the found stationary points correspond to minima on the potential energy surface. In order to get a more precise electron density distribution, MP2 6-31G(d) single point calculations were carried out for the found structures. Next, charges on carbon atoms (q_C) in position 4 of the pyrazole ring (since an electrophilic attack is directed at this position) and the hydrogen atom bound to it (q_{H_1}) were calculated using the Bader's theory of atoms in molecules [34]. The calculation results are presented in Table 1.

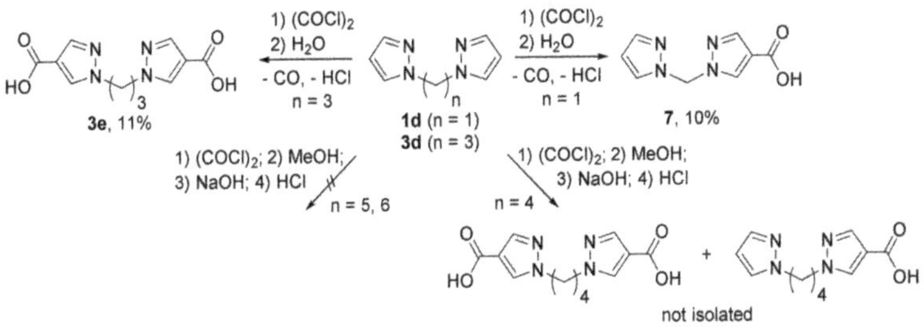

Scheme 3. Carboxylation of bis(pyrazol-1-yl)alkanes.

Table 1. Selected atomic charges in some bis(pyrazol-1-yl)alkanes and their monoacid chloride derivatives.

Structure	q_C	q_H	q_{CH} [1]	Δq_C [2]	Δq_{CH} [2]
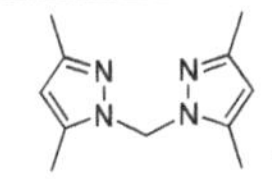 1a	−0.09691	+0.07896	−0.01795	0.00493	0.01012
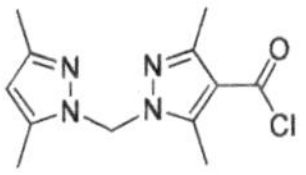 1a′	−0.09199	+0.08416	−0.00783		
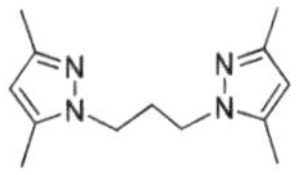 3a	−0.10241	+0.07636	−0.02606	0.00349	0.00682
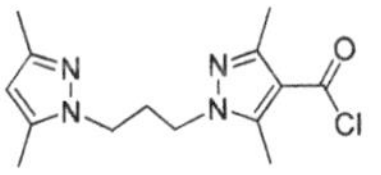 3a′	−0.09901	+0.07978	−0.01924		
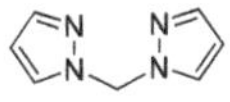 1d	−0.07525	+0.09219	+0.01694	0.00609	0.01195
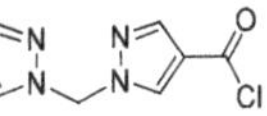 1d′	−0.06916	+0.09805	+0.02889		
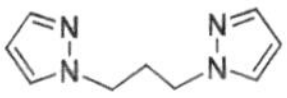 3d	−0.08297	+0.08930	+0.00633	0.01016	0.00904
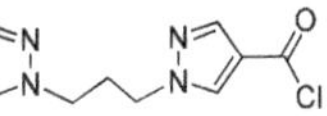 3d′	−0.07775	+0.09313	+0.01538		

[1] $q_{CH} = q_C + q_H$; [2] Δq_C and Δq_{CH} are calculated as the difference of the corresponding charges in the monoacid chloride and the starting substrate.

The following conclusions can be drawn from the obtained charge distribution:

(1) The introduction of methyl groups into pyrazole rings (in pairs of compounds **1a–1d** and **3a–3d**) noticeably increases the negative charge at position 4 of the heterocycle, i.e., makes it more active in the electrophilic substitution reaction, and the effect of electron-donor groups is best manifested when comparing the sum of charges on carbon and hydrogen atoms;

(2) An increase in the length of the linker from one to three methylene groups also increases the excess negative charge at position 4, which is apparently associated with the negative inductive effect of the pyrazole ring;
(3) The introduction of an electron-withdrawing acid chloride group into one of the pyrazole rings deactivates the other cycle in the electrophilic substitution reaction, and to a greater extent, deactivation manifests itself in pyrazole derivatives without methyl substituents, and with a short methylene linker.

Based on the charge distribution, bis(pyrazol-1-yl)methane **1d** is the least active in electrophilic substitution reactions, the charge on the 4-CH group in which is close to the charge on this group in the acid chloride of the propane derivative **3d′**. Experimental data show that monochloro anhydride **3d′** can undergo carboxylation with the formation of dicarboxylic acid **3e**. At the same time, bis(pyrazol-1-yl)methane **1d** at the first step of electrophilic substitution gives acid chloride **1d′**, in which position 4 of the other pyrazole ring is so deactivated by the electron-withdrawing effect of the already substituted heterocycle that the reaction halts on monochloroanhydride **1d′**, the hydrolysis of which gives monocarboxylic acid **7**. Therefore, the pyrazole ring with an electron-withdrawing functional group (i.e., chlorocarbonyl) is a substituent with a strong negative inductive effect, which affects the reactivity of the neighboring pyrazole ring located even across two aliphatic bonds.

Taking into account the limitations of direct carboxylation of bis(pyrazol-1-yl)alkanes, an alternative approach involving the double alkylation of 4-pyrazolecarboxylic acid and its 3,5-dimethylderivative by α,ω-dibromoalkanes was evaluated. Ethyl esters of 4-pyrazolecarboxylic acid and 3,5-dimethyl-4-pyrazole carboxylates were smoothly alkylated by α,ω-dibromoalkanes in a superbasic KOH-DMSO system and gave the corresponding diethyl dicarboxylates in high yields (Scheme 4). Alkaline hydrolysis of the diesters and subsequent neutralization gave the target dicarboxylic acids, including the ones with longer spacers, unavailable by direct carboxylation (Scheme 4). Unfortunately, derivatives of 1,2-bis(pyrazol-1-yl)ethane cannot be prepared by this route, since, as it is known, the reaction of pyrazole and its derivatives with 1,2-dibromoethane leads to the formation of 1-vinylpyrazoles, which undergo hydrolysis to the starting materials upon isolation [35].

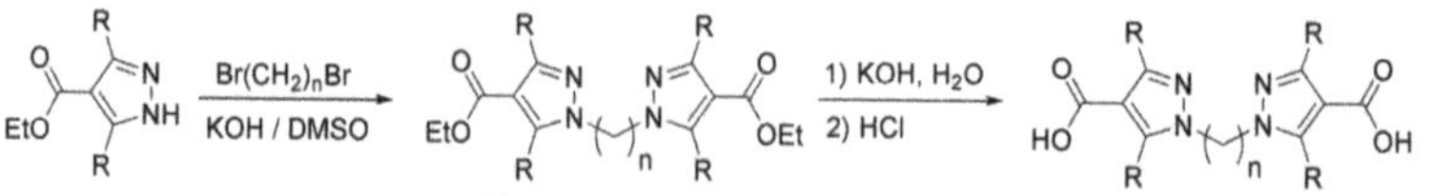

Scheme 4. Synthesis of bis(pyrazol-1-yl)alkane-4,4′-dicarboxylic acids from ethyl 4-pyrazolecarboxylates.

1f (n = 1, R = H), 89%
3f (n = 3, R = H), 80%
4f (n = 4, R = H), 86%
5f (n = 5, R = H), 95%
6f (n = 6, R = H), 73%
1g (n = 1, R = Me), 87%
3g (n = 3, R = Me), 92%
4g (n = 4, R = Me), 86%
5g (n = 5, R = Me), 79%
6g (n = 6, R = Me), 81%

1e (n = 1, R = H), 90%
3e (n = 3, R = H), 76%
4e (n = 4, R = H), 86%
5e (n = 5, R = H), 67%
6e (n = 6, R = H), 73%
1b (n = 1, R = Me), 88%
3b (n = 3, R = Me), 84%
4b (n = 4, R = Me), 80%
5b (n = 5, R = Me), 70%
6b (n = 6, R = Me), 67%

Since both alkylation and ester hydrolysis take place under basic catalysis, we explored the possibility to carry out the synthesis of diacid **1e** in a one-pot process. Indeed, after the alkylation was complete (TLC control), two additional equivalents of KOH and excess of water were added to the reaction mixture, which gave the hydrolysis product in several minutes in 92% overall yield.

3. Materials and Methods

Gas chromatography-mass spectrometry analysis was performed using Agilent 7890A gas chromatograph (Santa Clara, CA, USA) equipped with Agilent MSD 5975C mass-selective detector with quadrupole mass-analyzer (electron impact ionization energy 70 eV).

NMR spectra were recorded on Bruker DRX400 and Bruker Advance 500 instruments (Billerica, MA, USA), solvent residual peaks were used as internal standards. Elemental analyses were carried out on a Vario Micro-Cube analyzer (Elementar Analysensysteme GmbH, Langenselbold, Germany). IR-spectra of solid samples were recorded on Agilent Cary 630 FT IR (Santa Clara, CA, USA) spectrophotometer equipped with diamond ATR accessory. Melting points of compounds **1b**, **3b**, and **1d** (with melting points higher than 300 °C) were determined in helium atmosphere on NETZSCH TG 209 F1 thermoanalyzer (NETZSCH TAURUS Instruments GmbH, Weimar, Germany) with the heating rate of 10°/min.

The calculations were performed using the Gaussian 09 package, revision D.01 [36]. Atomic charges were calculated using the AIMAll Professional 10.05.04 package.

Ethyl 4-pyrazolecarboxylate [37], ethyl 3,5-dimethyl-4-pyrazolecarboxylate [25], bis(pyrazol-1-yl)alkanes **1a** and **1d** [38], **2a** [35], **3d** and **3a** [39], **4a–6a**, and **4d–6d** [40] were prepared following the literature procedures. Dimethyl sulfoxide was distilled from KOH under vacuum (10 Torr) prior to use. Other commercially reagents and solvents were of reagent grade and used as received. FT-IR and NMR plots of the synthesized compounds can be found in the Supplementary Material.

Bis(3,5-dimethylpyrazol-1-yl)methane-4,4′-dicarboxylic acid (**1b**). Method A (carboxylation by oxalyl chloride). Oxalyl chloride (1.91 g, 1.29 mL, 15 mmol) was added dropwise to 0.51 g (2.5 mmol) of compound **1a**. The mixture was refluxed during 3 h, after which the excess of oxalyl chloride was removed in vacuum of water aspiration pump. The residue was treated by 10 mL of water, the precipitate was filtered and washed with water. Yield 0.606 g, 83%, colorless power, m.p. 322–324 °C (EtOH, dec.). $C_{13}H_{16}N_4O_4$ (292.12): calcd. C 53.42; H 5.52; N 19.17; found C 53.4; H 5.7; N 19.0. ^{1}H-NMR (400 MHz, D_2O): δ 2.21 (s, 6H, 3-CH_3), 2.51 (s, 6H, 5-CH_3), 6.16 (s, 2H, CH_2), ppm. ^{13}C-NMR (100 MHz, D_2O): δ 10.1, 12.2, 58.6, 113.9, 148.0, 152.6, 189.5, ppm. FT-IR (cm^{-1}): ν = 2598 (w), 2532 (w), 1668 (m), 1554 (m), 1472 (w), 1435 (m), 1370 (w), 1316 (m), 1251 (m), 1164 (m), 1119 (m), 1040 (w), 1005 (w), 928 (w), 868 (w), 790 (m), 770 (m), 732 (w), 691 (m). Compounds **2b** and **3b** can be prepared similarly.

Method B (hydrolysis of diethyl diesters). Compound **1f** (0.696 g, 2 mmol) was suspended in 10 mL of the aqueous solution of KOH (0.448 g, 8 mmol) and heated at 80 °C until complete dissolution of a starting diester. After cooling, the obtained solution of potassium salts was acidified by concentrated HCl solution, the resulting precipitate was filtered, washed with water and dried. Yield 0.514 g, 88%.

1,2-Bis(3,5-dimethylpyrazol-1-yl)ethane-4,4′-dicarboxylic acid (**2b**) was prepared similarly to compound **1b** by Method A. Yield 38%, colorless crystals, m.p. 224–225 °C (H_2O). $C_{14}H_{18}N_4O_4$ (306.32): calcd. C 54.89; H 5.92; N 18.29; found C 54.7; H 6.0; N 18.1. ^{1}H-NMR (400 MHz, $(CD_3)_2SO$): δ 2.09 (s, 6 H, 3-CH_3), 2.25 (s, 6 H, 5-CH_3), 4.43 (s, 4 H, CH_2), ppm. ^{13}C-NMR (100 MHz, $(CD_3)_2SO$): δ 9.6 (5-CH_3), 13.1 (3-CH_3), 47.4 (CH_2), 113.1 (C^4(Pz)), 145.8 (C^5(Pz)), 149.9 (C^3(Pz)), 184.0 (COOH), ppm. FT-IR (cm^{-1}): 3500 (br., COOH), 1664 (C=O), 1535 (Pz), 1007 (Pz breathing).

1,3-Bis(3,5-dimethylpyrazol-1-yl)propane-4,4′-dicarboxylic acid (**3b**) was prepared similarly to compound **1b** by Methods A and B. Yield 39% (Method A) and 84% (Method B), colorless crystals, m.p. 290–292 °C (EtOH). $C_{15}H_{20}N_4O_4$ (320.35): calcd. C 56.24; H 6.29; N 17.49; found C 56.4, H 6.1, N 17.1. ^{1}H-NMR (400 MHz, $(CD_3)_2SO$): δ 2.07 (p, 2H, β-CH_2), 2.24 (s, 6H, 3-CH_3), 2.42 (s, 6H, 5-CH_3), 4.07 (t, 4H, CH_2), ppm ^{13}C-NMR (100 MHz, $(CD_3)_2SO$): δ 10.4 (5-CH_3), 13.2 (3-CH_3), 28.5 (PzCH$_2$$\underline{C}H_2$), 45.3 (Pz$\underline{C}H_2CH_2$), 112.9 ($C^4$(Pz)), 144.9 ($C^5$(Pz)), 149.3 ($C^3$(Pz)), 185.0 (COOH) ppm. FT-IR (cm^{-1}): 3430 (br., COOH), 1651 (C=O), 1536 (Pz), 1005 (Pz breathing).

1,4-Bis(3,5-dimethylpyrazol-1-yl)butane-4,4′-dicarboxylic acid (**4b**) was prepared similarly to compound **1b** by Method B from diesters **4c** or **4g**. Yield 80%, colorless crystals, m.p. 273–275 °C (reprecipitation; dec.). $C_{16}H_{22}N_4O_4$ (334.38): calcd. C 57.47; H 6.63; N 16.76; found C 57.1; H 6.5; N 16.9. ^{1}H-NMR (400 MHz, $(CD_3)_2SO$): δ 1.66 (p, 4 H, β-CH_2), 2.25 (s, 6 H, 3-CH_3), 2.43 (s, 6 H, 5-CH_3), 3.98 (t, 4 H, α-CH_2), 12.12 (s, 2 H, COOH), ppm. ^{13}C-NMR

(100 MHz, (CD$_3$)$_2$SO): δ 11.1, 14.5, 26.8, 47.9, 109.4, 143.7, 149.4, 165.8 ppm. FT-IR (cm^{-1}): ν = 3464 (w), 2932 (w), 2649 (w), 1674 (m), 1556 (w), 1506 (w), 1458 (w), 1431 (m), 1394 (w), 1366 (m), 1302 (m), 1264 (m), 1189 (m), 1118 (m), 1089 (m), 1038 (w), 1005 (w), 887 (w), 863 (w), 787 (m), 757 (s).

1,5-Bis(3,5-dimethylpyrazol-1-yl)pentane-4,4'-dicarboxylic acid (**5b**) was prepared similarly to compound **1b** by Method B from diesters **5c** or **5g**. Yield 70%, colorless crystals, m.p. 243–244 °C (reprecipitation; dec.). C$_{17}$H$_{24}$N$_4$O$_4$ (348.40): calcd. C 58.61; H 6.94; N 16.08; found C 58.3; H 6.7; N 15.9. ^{1}H-NMR (400 MHz, (CD$_3$)$_2$SO): δ 1.20 (p, 2 H, γ-CH$_2$), 1.69 (p, 4 H, β-CH$_2$), 2.25 (s, 6 H, 3-CH$_3$), 2.42 (s, 6 H, 5-CH$_3$), 3.95 (t, 4 H, α-CH$_2$), ppm. ^{13}C-NMR (100 MHz, (CD$_3$)$_2$SO): δ 11.1, 14.5, 23.6, 29.3, 48.3, 109.4, 143.6, 149.3, 165.8, ppm. FT-IR (cm^{-1}): ν = 2933 (w), 2866 (w), 1689 (m), 1669 (m), 1545 (m), 1499 (w), 1435 (w), 1384 (w), 1301 (w), 1268 (m), 1237 (w), 1216 (m), 1165 (m), 1114 (m), 1043 (w), 1012 (w), 824 (w), 787 (w), 752 (m).

1,6-Bis(3,5-dimethylpyrazol-1-yl)hexane-4,4'-dicarboxylic acid (**6b**) was prepared similarly to compound **1b** by Method B from diesters **6c** or **6g**. Yield 67%, colorless crystals, m.p. 193–195 °C (reprecipitation; dec.). ^{1}H-NMR (400 MHz, (CD$_3$)$_2$SO): δ 1.23 (p, 4 H, γ-CH$_2$), 1.66 (p, 4 H, β-CH$_2$), 2.26 (s, 6 H, 3-CH$_3$), 2.42 (s, 6 H, 5-CH$_3$), 3.94 (t, 4 H, α-CH$_2$), 12.13 (s, 2 H, COOH), ppm. ^{13}C-NMR (100 MHz, (CD$_3$)$_2$SO): δ 11.1, 14.5, 26.0, 29.6, 48.3, 109.3, 143.60, 149.3, 165.8, ppm. FT-IR (cm^{-1}): ν = 2933 (w), 2861 (w), 1697 (s), 1543 (m), 1497 (m), 1470 (m), 1430 (m), 1380 (m), 1277 (m), 1235 (m), 1178 (m), 1108 (s), 1044 (w), 1002 (m), 977 (w), 862 (m), 787 (m), 746 (s), 730 (m), 664 (w).

1,4-Bis(4-methoxycarbonyl-3,5-dimethylpyrazol-1-yl)butane (**4c**). Oxalyl chloride (1.91 g, 1.29 mL, 15 mmol) was added dropwise to 0.51 g (2.5 mmol) of compound **4a**. The mixture was heated in a screw-cup vial for 33 h, after which the reaction mixture was transferred to 50 mL of chloroform and the solvent was evaporated in vacuum of water aspiration pump to remove the excess of oxalyl chloride. Methanol (5 mL) was added dropwise to the residue, 50 mL of water was added and the precipitate was filtered and washed with water. Yield 27%, colorless crystals, m.p. 129–132 °C (EtOAc/hexane, 1:1). C$_{18}$H$_{26}$N$_4$O$_4$ (362.42): calcd. C 59.65; H 7.23; N 15.46; found C 59.8; H 7.5; N 15.2. ^{1}H-NMR (400 MHz, CDCl$_3$): δ =1.80 (p, 4 H, β-CH$_2$), 2.40 (s, 6 H, 3-CH$_3$), 2.47 (s, 6 H, 5-CH$_3$), 3.82 (s, 6 H, OCH$_3$), 4.01 (t, 4 H, α-CH$_2$), ppm. ^{13}C-NMR (100 MHz, CDCl$_3$): δ 11.2, 14.2, 26.9, 48.1, 50.9, 109.4, 143.7, 150.4, 164.9, ppm. FT-IR (cm^{-1}): ν = 2988 (w), 2930 (w), 2861 (w), 1685 (s), 1544 (s), 1493 (m), 1442 (s), 1430 (s), 1303 (m), 1261 (m), 1223 (m), 1191(m), 1175 (m), 1104 (s), 1082 (s), 1041 (m), 1005 (m), 960 (m), 842 (w), 803 (m), 785 (m), 768 (m), 657 (w). m/z: 362 (35%), 347 (12%), 331 (16%), 315 (4%), 303 (4%), 209 (29%), 195 (31%), 181 (57%), 167 (100%).

1,5-Bis(4-methoxycarbonyl-3,5-dimethylpyrazol-1-yl)pentane (**5c**) was prepared similarly to compound **4c**. Yield 34%, colorless crystals, m.p. 86–87 °C (EtOAc/hexane, 1:1). C$_{19}$H$_{28}$N$_4$O$_4$ (376.46): calcd. C 60.62; H 7.50; N 14.88; found C 60.3; H 7.4; N 14.6. ^{1}H-NMR (400 MHz, CDCl$_3$): δ 1.30 (p, 2 H, γ-CH$_2$), 1.81 (p, 4 H, β-CH$_2$), 2.40 (s, 6 H, 3-CH$_3$), 2.48 (s, 6 H, 5-CH$_3$), 3.82 (s, 6 H, OCH$_3$), 3.99 (t, 4 H, α-CH$_2$), ppm. ^{13}C-NMR (100 MHz, CDCl$_3$): δ 11.2, 14.2, 23.7, 29.5, 48.5, 50.9, 109.3, 143.6, 150.3, 164.9, ppm. FT-IR (cm^{-1}): ν = 2934 (w), 2863 (w), 1690 (s), 1546 (s), 1495 (w), 1439 (s), 1377 (m), 1324 (w), 1297 (s), 1271 (m), 1236 (m), 1213 (s), 1192 (m), 1160 (m), 1102 (s), 1040 (w), 1001 (w), 959 (m), 856 (w), 825 (w), 805 (w), 785 (s), 736 (w), 713 (w). m/z: 376 (27%), 361 (11%), 345 (14%), 329 (5%), 317 (4%), 223 (24%), 209 (39%), 195 (25%), 181 (37%), 167 (100%).

1,6-Bis(4-methoxycarbonyl-3,5-dimethylpyrazol-1-yl)hexane (**6c**) was prepared similarly to compound **4c**. Yield 30%, colorless crystals, m.p. 112–114 °C (EtOAc/hexane, 1:1). ^{1}H-NMR (400 MHz, CDCl$_3$): δ 1.23 (p, 2 H, γ-CH$_2$), 1.65 (p, 4 H, β-CH$_2$), 2.26 (s, 6 H, 3-CH$_3$), 2.43 (s, 6 H, 5-CH$_3$), 3.72 (s, 6 H, OCH$_3$), 3.95 (t, 4 H, α-CH$_2$), ppm. ^{13}C-NMR (100 MHz, CDCl$_3$): δ 11.2, 14.5, 25.9, 29.5, 48.4, 51.1, 108.6, 143.7, 149.1, 164.6, ppm. FT-IR (cm^{-1}): ν = 2925 (m), 2862 (w), 1701 (s), 1540 (s), 1488 (s), 1485 (m), 1431 (s), 1375 (m), 1293 (s), 1241 (s), 1205 (s), 1191 (s), 1161 (m), 1147 (m), 1101 (s), 1053 (s), 998 (m), 965 (m), 917 (m), 876 (w), 845 (w), 802 (m), 792 (m), 780 (s), 738 (m). m/z: 390 (10%), 375 (4%), 359 (13%), 343 (4%), 331 (2%), 237 (21%), 223 (88%), 209 (17%), 195 (25%), 181 (19%), 167 (100%).

Bis(pyrazol-1-yl)methane-4,4'-dicarboxylic acid (**1e**) was prepared similarly to compound **1b** by Method B from diester **1f**. Yield 90%, colorless crystals, m.p. 335–337 °C (reprecipitation; dec.). $C_9H_8N_4O_4$ (236.19): calcd. C 45.77; H 3.41; N 23.72; found C 45.9; H 3.6; N 23.5. ^{1}H-NMR (500 MHz, $(CD_3)_2SO$): δ 6.49 (s, 2 H, CH_2), 7.87 (s, 2 H, 3-H), 8.53 (s, 2 H, 5-H), 12.52 (s, 2 H, COOH), ppm. ^{13}C-NMR (125 MHz, $(CD_3)_2SO$): δ 64.6, 115.9, 134.6, 141.7, 163.4, ppm. FT-IR (cm^{-1}): ν = 3108 (m), 1669 (s), 1560 (s), 1419 (m), 1400 (m), 1366 (m), 1345 (m), 1294 (m), 1241 (s), 1133 (m), 1003 (m), 995 (s), 944 (m), 915 (s), 863 (m), 784 (m), 769 (s), 729 (m).

One-pot procedure. A suspension of finely powdered KOH (1.68 g, 30 mmol) and ethyl 4-pyrazole carboxylate (1.40 g, 10 mmol) in 5 mL of DMSO was vigorously stirred for 30 min at 80 °C. After that, the reaction mixture was cooled to room temperature and solution of dibromomethane (0.87 g, 0.35 mL, 5 mmol) in 5 mL of DMSO was added dropwise with stirring and cooling by water over 30 min. Stirring and heating at 80 °C was continued for additional 4 h (TLC control), then 20 mL of water and 1.12 g (20 mmol) of KOH were added to the reaction flask. Heating was continued until complete dissolution of the initially formed precipitate (1 h), after that, additional 80 mL of water were added and the resulting solution was acidified by concentrated HCl. The precipitate was filtered, washed with water and dried. Yield 1.34 g, 92%.

1,3-Bis(pyrazol-1-yl)propane-4,4'-dicarboxylic acid (**3e**) was prepared similarly to compound **1b** by Method A (yield 11%) or by Method B from diester **3f**. Yield 76%, colorless crystals, m.p. 225–227 °C (EtOH). $C_{11}H_{12}N_4O_4$ (264.24): calcd. C 50.00; H 4.58; N 21.20; found C 50,4, H 4,8, N 21,1. ^{1}H-NMR (400 MHz, $(CD_3)_2SO$): δ 2.35 (p, 2 H, β-CH_2), 4.22 (t, 4 H, α-CH_2), 8.07 (s, 2 H, 5-H), 8.60 (s, 2 H, 3-H), ppm. ^{13}C-NMR (100 MHz, $(CD_3)_2SO$): δ 29.9, 49.0, 118.8, 137.0, 141.4, 179.4, ppm. FT-IR (cm^{-1}): ν = 3127 (w), 3093 (w), 2948 (w), 1695 (s), 1556 (s), 1473 (w), 1369 (w), 1407 (m), 1356 (m), 1343 (m), 1298 (m), 1224 (s), 1172 (s), 1149 (m), 1083 (m), 1051 (m), 982 (s), 889 (s), 863 (s), 822 (m), 765 (s), 743 (s).

1,4-Bis(pyrazol-1-yl)butane-4,4'-dicarboxylic acid (**4e**) was prepared similarly to compound **1b** by Method B from diester **4f**. Yield 86%, colorless crystals, m.p. 300 °C (dec.). $C_{12}H_{14}N_4O_4$ (278.27): calcd. C 51.80; H 5.07; N 20.13; found C 51.6; H 5.1; N 20.3. ^{1}H-NMR (500 MHz, $(CD_3)_2SO$): δ 1.71 (p, 4 H, β-CH_2), 4.15 (t, 4 H, α-CH_2), 7.79 (s, 2 H, 5-H), 8.25 (s, 2 H, 3-H), 12.30 (s, 2 H, COOH), ppm. ^{13}C-NMR (125 MHz, $(CD_3)_2SO$): δ 26.5, 50.8, 114.6, 133.5, 140.4, 163.8, ppm. FT-IR (cm^{-1}): ν = 3122 (w), 3088 (w), 2953 (w), 1685 (s), 1558 (s), 1438 (m), 1404 (m), 1379 (m), 1349 (m), 1304 (m), 1237 (s), 1221 (s), 1188 (s), 1149 (m), 1122 (m), 1018 (s), 996 (m), 986 (s), 947 (m), 896 (m), 867 (m), 805 (s), 775 (s), 748 (s).

1,5-Bis(pyrazol-1-yl)pentane-4,4'-dicarboxylic acid (**5e**) was prepared similarly to compound **1b** by Method B from diester **5f**. Yield 67%, colorless powder, m.p. 216–219 °C (reprecipitation; dec.). $C_{13}H_{16}N_4O_4$ (292.30): calcd. C 53.42; H 5.52; N 19.17; found C 53.6; H 5.7; N 19.0. ^{1}H-NMR (500 MHz, $(CD_3)_2SO$): δ 1.15 (p, 2 H, γ-CH_2), 1.80 (p, 4 H, β-CH_2), 4.11 (t, 4 H, α-CH_2), 7.78 (s, 2 H, 5-H), 8.24 (s, 2 H, 3-H), 12.27 (s, 2 H, COOH), ppm. ^{13}C-NMR (125 MHz, $(CD_3)_2SO$): δ 22.7, 28.9, 51.2, 114.5, 133.5, 140.3, 163.8, ppm. FT-IR (cm^{-1}): ν = 3128 (w), 2974 (w), 1696 (s), 1554 (s), 1472 (w), 1438 (w), 1403 (m), 1376 (w), 1343 (w), 1333 (w), 1298 (w), 1269 (w), 1217 (s), 1098 (m), 1041 (w), 998 (m), 987 (s), 915 (m), 885 (s), 769 (s), 745 (s), 672 (m).

1,6-Bis(pyrazol-1-yl)hexane-4,4'-dicarboxylic acid (**6e**) was prepared similarly to compound **1b** by Method B from diester **6f**. Yield 73%, colorless powder, m.p. 252–254 °C (reprecipitation; dec.). $C_{14}H_{18}N_4O_4$ (306.32): calcd. C 54.89; H 5.92; N 18.29; found C 55.0; H 6.1; N 18.0. ^{1}H-NMR (500 MHz, $(CD_3)_2SO$): δ 1.22 (p, 4 H, γ-CH_2), 1.75 (p, 4 H, β-CH_2), 4.10 (t, 4 H, α-CH_2), 7.78 (s, 2 H, 5-H), 8.24 (s, 2 H, 3-H), 12.27 (s, 2 H, COOH), ppm ^{13}C-NMR (125 MHz, $(CD_3)_2SO$): δ 25.3, 29.3, 51.4, 114.5, 133.4, 140.3, 163.8, ppm. FT-IR (cm^{-1}): ν = 3114 (w), 2940 (w), 2861 (w), 1652 (m), 1549 (m), 1497 (w), 1469 (w), 1438 (w), 1384 (w), 1352 (w), 1258 (w), 1228 (w), 1129 (w), 1053 (w), 995 (w), 982 (w), 931 (w), 860 (w), 778 (m), 736 (w).

Bis(4-ethoxycarbonylpyrazol-1-yl)methane (**1f**). A suspension of finely powdered KOH (1.68 g, 30 mmol) and ethyl 4-pyrazole carboxylate (1.40 g, 10 mmol) in 5 mL of DMSO was vigorously stirred for 30 min at 80 °C. After that, the reaction mixture was cooled to room temperature and solution of dibromomethane (0.87 g, 0.35 mL, 5 mmol) in 5 mL of DMSO was added dropwise with stirring and cooling by water over 30 min. Stirring and heating at 80 °C was continued for additional 4 h (TLC control), and the reaction mixture was poured into 100 mL of water, the precipitate was filtered and washed by water. Yield 1.30 g, 89%, colorless crystals, m.p. 144–145 °C (EtOAc/hexane, 1:1). $C_{13}H_{16}N_4O_4$ (292.30): calcd. C 53.42; H 5.52; N 19.17; found C 53.6; H 5.7; N 18.9. ^{1}H-NMR (400 MHz, CDCl$_3$): δ 1.34 (t, 6 H, CH$_3$), 4.29 (q, 4 H, OCH$_2$), 6.31 (s, 2 H, α-CH$_2$), 7.96 (s, 2 H, 3-H), 8.18 (s, 2 H, 5-H), ppm. ^{13}C-NMR (100 MHz, CDCl$_3$): δ 14.3, 60.5, 65.7, 117.0, 133.3, 142.5, 162.3, ppm. FT-IR (cm^{-1}): ν = 3149 (w), 3099 (w), 3069 (w), 2987 (w), 1724 (s), 1705 (s), 1558 (s), 1478 (w), 1445 (m), 1412 (w), 1383 (m), 1345 (w), 1290 (m), 1192 (s), 1230 (s), 1161 (m), 1133 (s), 1109 (m), 1029 (s), 1015 (s), 993 (s), 979 (m), 953 (m), 908 (m), 886 (m), 827 (m), 764 (s), 739 (s), 652 (m). m/z: 292 (23%), 247 (68%), 219 (13%), 153 (100%).

1,3-Bis(4-ethoxycarbonylpyrazol-1-yl)propane (**3f**) was prepared similarly to compound **1f**. Yield 80%, colorless crystals, m.p. 92–94 °C (EtOAc/hexane, 1:1). $C_{15}H_{20}N_4O_4$ (320.35): calcd. C 56.24; H 6.29; N 17.49; found C 56.5; H 6.0; N 17.2. ^{1}H-NMR (400 MHz, CDCl$_3$): δ 1.36 (t, 6 H, CH$_3$), 2.48 (p, 2 H, β-CH$_2$), 4.16 (t, 4 H, α-CH$_2$), 4.30 (q, 4 H, OCH$_2$), 7.94 (s, 2 H, 5-H), 7.95 (s, 2 H, 3-H), ppm. ^{13}C-NMR (100 MHz, CDCl$_3$): δ 14.4, 30.4, 49.0, 60.3, 115.3, 133.1, 141.3, 162.8, ppm. FT-IR (cm^{-1}): ν = 3126 (w), 3091 (w), 2984 (w), 2940 (w), 2875 (w), 1691 (s), 1556 (s), 1472 (w), 1448 (w), 1405 (m), 1373 (m), 1350 (m), 1131 (m), 1286 (m), 1238 (s), 1124 (s), 1218 (s), 1111 (m), 1023 (s), 1084 (s), 976 (s), 874 (s), 830 (m), 769 (s), 708 (w), 653 (m). m/z: 320 (3%), 291 (1%), 275 (24%), 181 (26%), 167 (100%), 153 (76%), 139 (13%).

1,4-Bis(4-ethoxycarbonylpyrazol-1-yl)butane (**4f**) was prepared similarly to compound **1f**. Yield 86%, colorless crystals, m.p. 121 °C (EtOAc/hexane, 1:1). $C_{16}H_{22}N_4O_4$ (334.38): calcd. C 57.47; H 6.63; N 16.76; found C 57.5; H 6.5; N 16.9. ^{1}H-NMR (400 MHz, CDCl$_3$): δ 1.27 (t, 6 H, CH$_3$), 1.81 (p, 4 H, β-CH$_2$), 4.06 (t, 4 H, α-CH$_2$), 4.22 (q, 4 H, OCH$_2$), 7.78 (s, 2 H, 5-H), 7.83 (s, 2 H, 3-H), ppm. ^{13}C-NMR (100 MHz, CDCl$_3$): δ 14.4, 27.0, 51.8, 60.2, 115.2, 132.5, 141.2, 162.9, ppm. FT-IR (cm^{-1}): ν = 3122 (w), 3089 (w), 2980 (w), 2948 (w), 2869 (w), 1691 (s), 1554 (s), 1464 (m), 1432 (w), 1405 (m), 1394 (m), 1373 (s), 1355 (w), 1338 (w), 1298 (w), 1276 (m), 1223 (s), 1194 (s), 1172 (s), 1113 (s), 1092 (m), 1018 (s), 984 (s), 884 (s), 832 (m), 770 (s), 740 (m), 657 (m). m/z: 334 (2%), 305 (19%), 289 (50%), 195 (44%), 181 (19%), 167 (59%), 153 (100%), 139 (10%).

1,5-Bis(4-ethoxycarbonylpyrazol-1-yl)pentane (**5f**) was prepared similarly to compound **1f**. Yield 95%, colorless crystals, m.p. 82–83 °C (EtOAc/hexane, 1:1). $C_{17}H_{24}N_4O_4$ (348.40): calcd. C 58.61; H 6.94; N 16.08; found C 58.6; H 6.7; N 16.0. ^{1}H-NMR (400 MHz, CDCl$_3$): δ 1.21 (p, 2 H, γ-CH$_2$, partially overlapped with 1.28), 1.28 (t, 6 H, CH$_3$), 1.84 (p, 4 H, β-CH$_2$), 4.05 (t, 4 H, α-CH$_2$), 4.22 (q, 4 H, OCH$_2$), 7.78 (s, 2 H, 5-H), 7.83 (s, 2 H, 3-H), ppm. ^{13}C-NMR (100 MHz, CDCl$_3$): δ 14.4, 23.4, 29.5, 52.2, 60.2, 115.0, 132.5, 141.0, 163.0, ppm. FT-IR (cm^{-1}): ν = 3129 (w), 2970 (w), 2935 (w), 2874 (w), 1699 (s), 1554 (s), 1459 (m), 1441 (m), 1404 (s), 1372 (m), 1352 (m), 1333 (m), 1304 (m), 1263 (w), 1224 (s), 1209 (s), 1190 (s), 1113 (s), 1096 (s), 1021 (s), 980 (s), 877 (m), 826 (w), 762 (s). m/z: 348 (11%), 319 (24%), 303 (47%), 275 (4%), 209 (14%), 195 (47%), 181 (22%), 167 (33%), 153 (100%), 139 (4%).

1,6-Bis(4-ethoxycarbonylpyrazol-1-yl)hexane (**6f**) was prepared similarly to compound **1f**. Yield 73%, colorless crystals, m.p. 85–86 °C (EtOAc/hexane, 1:1). $C_{18}H_{26}N_4O_4$ (362.43): calcd. C 59.65; H 7.23; N 15.46; found C 59.8; H 7.4; N 15.2. ^{1}H-NMR (400 MHz, CDCl$_3$): δ 1.24 (p, 4 H, γ-CH$_2$, partially overlapped with 1.27), 1.27 (t, 6 H, CH$_3$), 1.80 (p, 4 H, β-CH$_2$), 4.04 (t, 4 H, α-CH$_2$), 4.22 (q, 4 H, OCH$_2$), 7.79 (s, 2 H, 5-H), 7.83 (s, 2 H, 3-H), ppm. ^{13}C-NMR (100 MHz, CDCl$_3$): δ 14.4, 25.9, 29.8, 52.4, 60.2, 115.0, 132.4, 140.9, 163.0, ppm. FT-IR (cm^{-1}): ν = 3130 (w), 2983 (w), 2940 (w), 2866 (w), 1693 (s), 1546 (s), 1466 (w), 1442 (w), 1408 (w), 1373 (m), 1352 (w), 1256 (s), 1219 (s), 1201 (s), 1123 (s), 1110 (m), 1052 (m), 1028 (s), 995 (m), 978 (s), 899 (m), 869 (w), 830 (m), 775 (m), 739 (m), 654 (m). m/z: 362 (10%), 333 (11%), 317 (35%), 289 (4%), 223 (7%), 209 (44%), 195 (29%), 181 (19%), 167 (22%), 153 (100%), 139 (3%).

Bis(4-ethoxycarbonyl-3,5-dimethylpyrazol-1-yl)methane (**1g**). A suspension of finely powdered KOH (1.68 g, 30 mmol) and ethyl 3,5-dimethyl-4-pyrazolecarboxylate (1.68 g, 10 mmol) in 5 DMSO was vigorously stirred for 30 min at 80 °C. After that, the reaction mixture was cooled to room temperature and solution of dibromomethane (0.87 g, 0.35 mL, 5 mmol) in 5 mL of DMSO was added dropwise under stirring and cooling by water over 30 min. Stirring and heating at 80 °C was continued for additional 4 h (TLC control), and the reaction mixture was poured into 100 mL of water, the precipitate was filtered and washed by water. Yield 1.51 g, 87%, colorless crystals, m.p. 152–153 °C (EtOAc/hexane, 1:2). $C_{17}H_{24}N_4O_4$ (348.18): calcd. C 58.61; H 6.94; N 16.08; found C 58.4; H 7.1; N 16.2. ^{1}H-NMR (400 MHz, CDCl$_3$): δ 1.35 (t, 6 H, OCH$_2$CH$_3$), 2.39 (s, 6 H, 3-CH$_3$), 2.75 (s, 6 H, 5-CH$_3$), 4.29 (q, 4 H, OCH$_2$), 6.12 (s, 2 H, α-CH$_2$), ppm. ^{13}C-NMR (100 MHz, CDCl$_3$): δ 11.4, 14.3, 14.4, 59.8, 59.9, 110.9, 145.8, 151.4, 164.2, ppm. FT-IR (cm^{-1}): ν = 2990 (w), 2937 (w), 1701 (s), 1557 (s), 1476 (m), 1426 (m), 1394 (m), 1352 (m), 1309 (s), 1299 (m), 1241 (s), 1212 (m), 1166 (s), 1121 (m), 1101 (s), 1042 (m), 1002 (m), 853 (m), 815 (m), 782 (s), 735 (m), 691 (s). *m/z*: 348 (50%), 333 (5%), 303 (33%), 275 (2%), 181 (100%).

1,3-Bis(4-ethoxycarbonyl-3,5-dimethylpyrazol-1-yl)propane (**3g**) was prepared similarly to compound **1g**. Yield 92%, colorless crystals, m.p. 112–113 °C (EtOAc/hexane, 1:1). $C_{19}H_{28}N_4O_4$ (376.46): calcd. C 60.62; H 7.50; N 14.88; found C 60.3; H 7.7; N 14.9.^{1}H-NMR (400 MHz, CDCl$_3$): δ 1.36 (t, 6 H, OCH$_2$CH$_3$), 2.38 (p, 2 H, β-CH$_2$, partially overlapped with 2.41), 2.41 (s, 6 H, 3-CH$_3$), 2.46 (s, 6 H, 5-CH$_3$), 4.04 (t, 4 H, α-CH$_2$), 4.29 (q, 4 H, OCH$_2$), ppm. ^{13}C-NMR (100 MHz, CDCl$_3$): δ 11.0, 14.3, 14.4, 29.3, 45.5, 59.6, 109.7, 144.0, 150.6, 164.4, ppm. FT-IR (cm^{-1}): ν = 2980 (w), 2929 (w), 1712 (m), 1681 (m), 1545 (m), 1489 (m), 1425 (m), 1385 (m), 1370 (m), 1323 (w), 1293 (s), 1249 (w), 1231 (m), 1200 (w), 1138 (s), 1120 (m), 1100 (s), 1043 (m), 1013 (m), 1000 (m), 984 (w), 853 (w), 815 (w), 782 (s), 716 (w). *m/z*: 376 (11%), 331 (10%), 209 (18%), 195 (100%), 181 (51%), 167 (6%).

1,4-Bis(4-ethoxycarbonyl-3,5-dimethylpyrazol-1-yl)butane (**4g**) was prepared similarly to compound **1g**. Yield 86%, colorless crystals, m.p. 122–123 °C (EtOAc/hexane, 1:1). $C_{20}H_{30}N_4O_4$ (390.48): calcd. C 61.52; H 7.74; N 14.35; found C 61.7; H 8.0; 14.4. ^{1}H-NMR (400 MHz, CDCl$_3$): δ 1.36 (t, 6 H, OCH$_2$CH$_3$), 1.82 (p, 4 H, β-CH$_2$), 2.42 (s, 6 H, 3-CH$_3$), 2.49 (s, 6 H, 5-CH$_3$), 4.03 (t, 4 H, α-CH$_2$), 4.29 (q, 4 H, OCH$_2$), ppm. ^{13}C-NMR (100 MHz, CDCl$_3$): δ 11.1, 14.3, 14.4, 26.9, 48.0, 59.6, 109.7, 143.7, 150.4, 164.4, ppm. FT-IR (cm^{-1}): ν = 2985 (w), 2933 (w), 2875 (w), 1695 (s), 1547 (s), 1491 (m), 1472 (m), 1436 (m), 1400 (m), 1383 (m), 1364 (m), 1298 (s), 1248 (s), 1188 (s), 1122 (m), 1102 (s), 1076 (m), 1044 (m), 1015 (m), 993 (m), 910 (w), 888 (w), 849 (m), 813 (w), 784 (s), 742 (w), 711 (m). *m/z*: 390 (36%), 375 (12%), 361 (12%), 345 (29%), 329 (1%), 317 (4%), 223 (25%), 209 (29%), 195 (68%), 181 (100%).

1,5-Bis(4-ethoxycarbonyl-3,5-dimethylpyrazol-1-yl)pentane (**5g**) was prepared similarly to compound **1g**. Yield 79%, colorless crystals, m.p. 100–101 °C (EtOAc/hexane, 1:1). $C_{21}H_{32}N_4O_4$ (404.51): calcd. C 62.35; H 7.97; N 13.85; found C 62.0; H 8.1; N 13.9. ^{1}H-NMR (400 MHz, CDCl$_3$): δ 1.32 (p, 2 H, γ-CH$_2$, partially overlapped with 1.36), 1.36 (t, 6 H, OCH$_2$CH$_3$), 1.81 (p, 4 H, β-CH$_2$), 2.42 (s, 6 H, 3-CH$_3$), 2.49 (s, 6 H, 5-CH$_3$), 3.99 (t, 4 H, α-CH$_2$), 4.29 (q, 4 H, OCH$_2$), ppm. ^{13}C-NMR (100 MHz, CDCl$_3$): δ 11.2, 14.3, 14.4, 23.7, 29.5, 48.5, 59.6, 109.5, 143.5, 150.3, 164.5, ppm. FT-IR (cm^{-1}): ν = 2981 (w), 2930 (w), 2866 (w), 1686 (s), 1548 (s), 1496 (m), 1473 (m), 1436 (m), 1399 (m), 1371 (m), 1296 (s), 1281 (m), 1246 (s), 1219 (m), 1190 (s), 1166 (m), 1122 (s), 1101 (s), 1042 (m), 997 (m), 889 (m), 852 (m), 812 (w), 783 (s), 746 (w), 718 (w), 698 (w). *m/z*: 404 (29%), 389 (8%), 375 (10%), 359 (30%), 343 (6%), 331 (4%), 237 (28%), 223 (47%), 209 (35%), 195 (47%), 181 (100%).

1,6-Bis(4-ethoxycarbonyl-3,5-dimethylpyrazol-1-yl)hexane (**6g**) was prepared similarly to compound **1g**. Yield 81%, colorless crystals, m.p. 107–108 °C (EtOAc/hexane, 1:1). $C_{22}H_{34}N_4O_4$ (418.54): calcd. C 63.13; H 8.19; N 13.39; found C 63.2; H 8.3; N 13.4. ^{1}H-NMR (400 MHz, CDCl$_3$): δ 1.33 (p, 4 H, γ-CH$_2$, partially overlapped with 1.36), 1.36 (t, 6 H, OCH$_2$CH$_3$), 1.79 (p, 4 H, β-CH$_2$), 2.42 (s, 6 H, 3-CH$_3$), 2.49 (s, 6 H, 5-CH$_3$), 3.98 (t, 4 H, α-CH$_2$), 4.29 (q, 4 H, OCH$_2$), ppm. ^{13}C-NMR (100 MHz, CDCl$_3$): δ 11.2, 14.3, 14.4, 26.2, 29.8, 48.6, 59.6, 109.4, 143.5, 150.2, 164.6, ppm. FT-IR (cm^{-1}): ν = 2983 (w), 2929 (w), 2863 (w), 1686 (s), 1548 (s), 1481 (m), 1469 (m), 1436 (m), 1374 (m), 1361 (m), 1320 (m), 1311 (m),

1291 (s), 1238 (s), 1185 (s), 1121 (s), 1102 (s), 1041 (m), 1018 (m), 997 (m), 851 (m), 819 (w), 783 (s), 744 (w), 716 (m). m/z: 418 (21%), 403 (2%), 389 (6%), 373 (29%), 357 (8%), 345 (4%), 251 (25%), 237 (100%), 223 (22%), 209 (26%), 195 (20%), 181 (87%).

Bis(pyrazol-1-yl)methane-4-carboxylic acid (7) was prepared following the procedure for compound 1b (Method A). Yield 10%, colorless crystals, m.p. 205–207 °C (EtOH). $C_8H_8N_4O_2$ (192.18): calcd. C 50.00; H 4.20; N 29.15; found C 50.4; H 4.5; N 28.8. ^{1}H-NMR (400 MHz, $(CD_3)_2SO$): δ 6.32 (t, 1 H, 4-H-Pz), 6.50 (s, 2 H, CH_2), 7.53 (d, 1 H, 3-H-Pz), 8.01 (d, 1 H, 5-H-Pz), 8.08 (s, 1 H, 3-H-PzCOOH), 8.12 (s, 1 H, 5-H-PzCOOH), ppm. ^{13}C-NMR (100 MHz, $(CD_3)_2SO$): δ 64.5, 106.7, 119.7, 131.4, 136.8, 140.9, 142.2, 179.6, ppm. FT-IR (cm^{-1}): ν = 3430 (br., COOH), 1682 (C=O), 1544 ($ν_{Pz}$), 1002 (Pz breathing).

4. Conclusions

In summary, approaches to the synthesis of 4,4′-dicarboxy-substituted bis(pyrazol-1-yl)alkanes were evaluated. It was found that direct carboxylation by oxalyl chloride is feasible only for the preparation of bis(3,5-dimethylpyrazol-1-yl)methane derivates due to electron-donating methyl groups and short methylene linker. Longer linkers lead to significantly lower product yields, which is probably due to the instability of the intermediate acid chlorides that are formed in the reaction with oxalyl chloride. A more universal method is based on the reaction of ethyl 4-pyrazolecarboxylates with dibromoalkanes in a superbasic medium and is applicable for the preparation of bis(4-carboxypyrazol-1-yl)alkanes with both short and long linkers. The obtained dicarboxylic acids are interesting as potential building blocks for metal-organic frameworks.

Supplementary Materials: The following are available online: FT-IR and NMR plots of the synthesized compounds.

Author Contributions: Conceptualization, A.S.P.; methodology, N.P.B.; investigation, N.P.B.; writing—original draft preparation, N.P.B.; writing—review and editing, A.S.P.; supervision, A.S.P.; funding acquisition, A.S.P. All authors have read and agreed to the published version of the manuscript.

Funding: This research was funded by the Ministry of Science and Higher Education of the Russian Federation.

Institutional Review Board Statement: Not applicable.

Informed Consent Statement: Not applicable.

Data Availability Statement: The data presented in this study are available on request from the corresponding author.

Acknowledgments: The Siberian Branch of the Russian Academy of Sciences (SB RAS) Siberian Supercomputer Center is gratefully acknowledged for providing supercomputer facilities. The authors thank the Analytical laboratory and thermal analysis facilities of NIIC SB RAS for carrying out the corresponding analyses.

Conflicts of Interest: The authors declare no conflict of interest. The funders had no role in the design of the study; in the collection, analyses, or interpretation of data; in the writing of the manuscript, or in the decision to publish the results.

Sample Availability: Samples of the dicarboxylic acids reported in this article are available from the authors.

References

1. Pettinari, C.; Pettinari, R. Metal Derivatives of Poly(pyrazolyl)alkanes: II. Bis(pyrazolyl)alkanes and Related Systems. *Coord. Chem. Rev.* **2005**, *249*, 663–691. [CrossRef]
2. Dehury, N.; Tripathy, S.K.; Sahoo, A.; Maity, N.; Patra, S. Facile Tandem Suzuki Coupling/Transfer Hydrogenation Reaction with a Bis-Heteroscorpionate Pd-Ru Complex. *Dalton Trans.* **2014**, *43*, 16597–16600. [CrossRef] [PubMed]
3. Martínez, J.; Otero, A.; Lara-Sánchez, A.; Castro-Osma, J.A.; Fernández-Baeza, J.; Sánchez-Barba, L.F.; Rodríguez, A.M. Heteroscorpionate Rare-Earth Catalysts for the Hydroalkoxylation/Cyclization of Alkynyl Alcohols. *Organometallics* **2016**, *35*, 1802–1812. [CrossRef]

4. Otero, A.; Fernández-Baeza, J.; Lara-Sánchez, A.; Sánchez-Barba, L.F. Metal Complexes with Heteroscorpionate Ligands Based on the Bis(pyrazol-1-yl)methane Moiety: Catalytic Chemistry. *Coord. Chem. Rev.* **2013**, *257*, 1806–1868. [CrossRef]
5. Zubkevich, S.V.; Tuskaev, V.A.; Gagieva, S.C.; Pavlov, A.A.; Khrustalev, V.N.; Polyakova, O.V.; Zarubin, D.N.; Kurmaev, D.A.; Kolosov, N.A.; Bulychev, B.M. Catalytic Systems Based on Nickel(ii) Complexes with Bis(3,5-dimethylpyrazol-1-yl)methane—Impact of PPh$_3$ on the Formation of Precatalysts and Selective Dimerization of Ethylene. *New J. Chem.* **2020**, *44*, 981–993. [CrossRef]
6. Moegling, J.; Hoffmann, A.; Herres-Pawlis, S. Insights into Copper-Poly(pyrazolyl)methane-Catalyzed Reactions for Organic Transformations. *Synthesis* **2017**, *49*, 225–236.
7. Montani, M.; Pazmay, G.V.B.; Hysi, A.; Lupidi, G.; Pettinari, R.; Gambini, V.; Tilio, M.; Marchetti, F.; Pettinari, C.; Ferraro, S.; et al. The Water Soluble Ruthenium(II) Organometallic Compound [Ru(p-cymene)(bis(3,5 dimethylpyrazol-1-yl)methane)Cl]Cl Suppresses Triple Negative Breast Cancer Growth by Inhibiting Tumor Infiltration of Regulatory T Cells. *Pharmacol. Res.* **2016**, *107*, 282–290. [CrossRef] [PubMed]
8. Silvestri, S.; Cirilli, I.; Marcheggiani, F.; Dludla, P.; Lupidi, G.; Pettinari, R.; Marchetti, F.; Di Nicola, C.; Falcioni, G.; Marchini, C.; et al. Evaluation of Anticancer Role of a Novel Ruthenium(II)-Based Compound Compared with NAMI-A and Cisplatin in Impairing Mitochondrial Functionality and Promoting Oxidative Stress in Triple Negative Breast Cancer Models. *Mitochondrion* **2021**, *56*, 25–34. [CrossRef]
9. Fonseca, D.; Páez, C.; Ibarra, L.; García-Huertas, P.; Macías, M.A.; Triana-Chávez, O.; Hurtado, J.J. Metal Complex Derivatives of Bis(pyrazol-1-yl)methane Ligands: Synthesis, Characterization and Anti-Trypanosoma Cruzi Activity. *Transit. Met. Chem.* **2019**, *44*, 135–144. [CrossRef]
10. Schepetkin, I.; Potapov, A.; Khlebnikov, A.; Korotkova, E.; Lukina, A.; Malovichko, G.; Kirpotina, L.; Quinn, M.T. Decomposition of Reactive Oxygen Species by Copper(II) bis(1-pyrazolyl) Methane Complexes. *J. Biol. Inorg. Chem.* **2006**, *11*, 499–513. [CrossRef]
11. Potapov, A.S.; Nudnova, E.A.; Domina, G.A.; Kirpotina, L.N.; Quinn, M.T.; Khlebnikov, A.I.; Schepetkin, I.A. Synthesis, Characterization and Potent Superoxide Dismutase-Like Activity of Novel Bis(pyrazole)-2,2′-bipyridyl Mixed Ligand Copper(II) Complexes. *Dalton Trans.* **2009**, 4488–4498. [CrossRef] [PubMed]
12. Zhang, F.; Li, W.; Wei, D.; Li, C.; Pan, C.; Dong, X.; Li, Z.; Li, S.; Wei, B.; Zhang, F.; et al. Synthesis, Characterization, Photo- and Electro-Luminescent Properties of Blue Cationic Iridium Complexes with Nonconjugated Bis(pyrazole-1-yl)methane as the Ancillary Ligand. *Dye. Pigment.* **2016**, *134*, 19–26. [CrossRef]
13. Meng, S.; Jung, I.; Feng, J.; Scopelliti, R.; Di Censo, D.; Grätzel, M.; Nazeeruddin, M.K.; Baranoff, E. Bis(pyrazol-1-yl)methane as Non-Chromophoric Ancillary Ligand for Charged Bis-Cyclometalated Iridium(III) Complexes. *Eur. J. Inorg. Chem.* **2012**, *2012*, 3209–3215. [CrossRef]
14. Alkorta, I.; Claramunt, R.M.; Díez-Barra, E.; Elguero, J.; de la Hoz, A.; López, C. The Organic Chemistry of Poly(1H-pyrazol-1-yl)methanes. *Coord. Chem. Rev.* **2017**, *339*, 153–182. [CrossRef]
15. Potapov, A.S.; Khlebnikov, A.I.; Vasilevskii, S.F. Synthesis of Monomeric and Oligomeric 1,1′-methylenebis-(1H-pyrazoles) Contaning Ethynyl Fragments. *Russ. J. Org. Chem.* **2006**, *42*, 1368–1373. [CrossRef]
16. Zhang, S.; Gao, Z.; Lan, D.; Jia, Q.; Liu, N.; Zhang, J.; Kou, K. Recent Advances in Synthesis and Properties of Nitrated-Pyrazoles Based Energetic Compounds. *Molecules* **2020**, *25*, 3475. [CrossRef] [PubMed]
17. Usami, Y.; Tatsui, Y.; Yoneyama, H.; Harusawa, S. C4-Alkylamination of C4-Halo-1H-1-tritylpyrazoles Using Pd(dba)2 or CuI. *Molecules* **2020**, *25*, 4634. [CrossRef]
18. Liu, Q.; Song, Y.; Ma, Y.; Zhou, Y.; Cong, H.; Wang, C.; Wu, J.; Hu, G.; O'Keeffe, M.; Deng, H. Mesoporous Cages in Chemically Robust MOFs Created by a Large Number of Vertices with Reduced Connectivity. *J. Am. Chem. Soc.* **2019**, *141*, 488–496. [CrossRef]
19. Jia, Y.-Y.; Ren, G.-J.; Li, A.-L.; Zhang, L.-Z.; Feng, R.; Zhang, Y.-H.; Bu, X.-H. Temperature-Related Synthesis of Two Anionic Metal–Organic Frameworks with Distinct Performance in Organic Dye Adsorption. *Cryst. Growth Des.* **2016**, *16*, 5593–5597. [CrossRef]
20. Wang, L.-D.; Tao, F.; Cheng, M.-L.; Liu, Q.; Han, W.; Wu, Y.-J.; Yang, D.-D.; Wang, L.-J. Syntheses, Crystal Structures, and Luminescence of Two Main-Group Metal Complexes Based on 3,4-pyrazoledicarboxylic acid. *J. Coord. Chem.* **2012**, *65*, 923–933. [CrossRef]
21. Chen, Y.; Liu, C.-B.; Gong, Y.-N.; Zhong, J.-M.; Wen, H.-L. Syntheses, Crystal Structures and Antibacterial Activities of Six Cobalt(II) Pyrazole Carboxylate Complexes with Helical Character. *Polyhedron* **2012**, *36*, 6–14. [CrossRef]
22. Li, F.-L.; Chen, Q.; Song, H.-B.; Dai, B.; Tang, L.-F. Synthesis of Organotin bis(pyrazol-1-yl)methane-tetracarboxylates and Tris(pyrazol-1-yl)methane-hexacarboxylates. *Polyhedron* **2014**, *83*, 102–107. [CrossRef]
23. Cheng, M.; Wang, Q.; Bao, J.; Wu, Y.; Sun, L.; Yang, B.; Liu, Q. Synthesis and Structural Diversity of d^{10} Metal Coordination Polymers Constructed from New Semi-Rigid Bis(3-methyl-1H-pyrazole-4-carboxylic acid)alkane Ligands. *New J. Chem.* **2017**, *41*, 5151–5160. [CrossRef]
24. Radi, S.; El-Massaoudi, M.; Benaissa, H.; Adarsh, N.N.; Ferbinteanu, M.; Devlin, E.; Sanakis, Y.; Garcia, Y. Crystal Engineering of a Series of Complexes and Coordination Polymers Based on Pyrazole-carboxylic Acid Ligands. *New J. Chem.* **2017**, *41*, 8232–8241. [CrossRef]
25. Kivi, C.E.; Gelfand, B.S.; Dureckova, H.; Ho, H.T.K.; Ma, C.; Shimizu, G.K.H.; Woo, T.K.; Song, D. 3D Porous Metal-Organic Framework for Selective Adsorption of Methane over Dinitrogen under Ambient Pressure. *Chem. Commun.* **2018**, *54*, 14104–14107. [CrossRef]

26. Bloch, W.M.; Burgun, A.; Coghlan, C.J.; Lee, R.; Coote, M.L.; Doonan, C.J.; Sumby, C.J. Capturing Snapshots of Post-Synthetic Metallation Chemistry in Metal-Organic Frameworks. *Nat. Chem.* **2014**, *6*, 906. [CrossRef]
27. Burgun, A.; Coghlan, C.J.; Huang, D.M.; Chen, W.; Horike, S.; Kitagawa, S.; Alvino, J.F.; Metha, G.F.; Sumby, C.J.; Doonan, C.J. Mapping-Out Catalytic Processes in a Metal-Organic Framework with Single-Crystal X-ray Crystallography. *Angew. Chem. Int. Ed.* **2017**, *56*, 8412–8416. [CrossRef]
28. Chiriac, C.I. The Direct Carboxylation of Pyrazoles. *Synthesis* **1986**, *1986*, 753–755. [CrossRef]
29. Padial, N.M.; Quartapelle Procopio, E.; Montoro, C.; López, E.; Oltra, J.E.; Colombo, V.; Maspero, A.; Masciocchi, N.; Galli, S.; Senkovska, I.; et al. Highly Hydrophobic Isoreticular Porous Metal-Organic Frameworks for the Capture of Harmful Volatile Organic Compounds. *Angew. Chem. Int. Ed.* **2013**, *52*, 8290–8294. [CrossRef]
30. Zhao, B.; Liang, Q.; Ren, H.; Zhang, X.; Wu, Y.; Zhang, K.; Ma, L.-Y.; Zheng, Y.-C.; Liu, H.-M. Discovery of Pyrazole Derivatives as Cellular Active Inhibitors of Histone Lysine Specific Demethylase 5B (KDM5B/JARID1B). *Eur. J. Med. Chem.* **2020**, *192*, 112161. [CrossRef]
31. Batista, D.C.; Silva, D.P.B.; Florentino, I.F.; Cardoso, C.S.; Gonçalves, M.P.; Valadares, M.C.; Lião, L.M.; Sanz, G.; Vaz, B.G.; Costa, E.A.; et al. Anti-Inflammatory Effect of a New Piperazine Derivative: (4-methylpiperazin-1-yl)(1-phenyl-1H-pyrazol-4-yl)methanone. *Inflammopharmacology* **2018**, *26*, 217–226. [CrossRef] [PubMed]
32. Hüttel, R.; Schön, M.E. Über Pyrazolyl-lithium-Verbindungen. *Justus Liebigs Ann. Chem.* **1959**, *625*, 55–65. [CrossRef]
33. Janin, Y.L. Synthetic Accesses to 3/5-pyrazole Carboxylic Acids. *Mini Rev. Org. Chem.* **2010**, *7*, 314–323. [CrossRef]
34. Bader, R.F.W. *Atoms in Molecules. A Quantum Theory*; Clarendon Press: Oxford, UK, 1994.
35. Torres, J.; Lavandera, J.L.; Cabildo, P.; Claramunt, R.M.; Elguero, J. Synthesis and Physicochemical Studies on 1,2-bisazolylethanes. *J. Heterocycl. Chem.* **1988**, *25*, 771–782. [CrossRef]
36. Frisch, M.J.; Trucks, G.W.; Schlegel, H.B.; Scuseria, G.E.; Robb, M.A.; Cheeseman, J.R.; Scalmani, G.; Barone, V.; Mennucci, B.; Petersson, G.A.; et al. *Gaussian 09, Revision D.01*; Gaussian, Inc.: Wallingford, UK, 2013.
37. Skidmore, J.; Heer, J.; Johnson, C.N.; Norton, D.; Redshaw, S.; Sweeting, J.; Hurst, D.; Cridland, A.; Vesey, D.; Wall, I.; et al. Optimization of Sphingosine-1-phosphate-1 Receptor Agonists: Effects of Acidic, Basic, and Zwitterionic Chemotypes on Pharmacokinetic and Pharmacodynamic Profiles. *J. Med. Chem.* **2014**, *57*, 10424–10442. [CrossRef]
38. Potapov, A.S.; Khlebnikov, A.I. Synthesis of Mixed-Ligand Copper(II) Complexes Containing Bis(pyrazol-1-yl)methane Ligands. *Polyhedron* **2006**, *25*, 2683–2690. [CrossRef]
39. Potapov, A.S.; Domina, G.A.; Khlebnikov, A.I.; Ogorodnikov, V.D. Facile Synthesis of Flexible bis(pyrazol-1-yl)alkane and Related Ligands in a Superbasic Medium. *Eur. J. Org. Chem.* **2007**, 5112–5116. [CrossRef]
40. Zatonskaya, L.V.; Schepetkin, I.A.; Petrenko, T.V.; Ogorodnikov, V.D.; Khlebnikov, A.I.; Potapov, A.S. Synthesis and Cytotoxicity of Bis(pyrazol-1-yl)-Alkane Derivatives with Polymethylene Linkers and Related Mono- and Dipyrazolium Salts. *Chem. Heterocycl. Compd.* **2016**, *52*, 388–401. [CrossRef]

Article

Synthesis of 6,7-Dihydro-1*H*,5*H*-pyrazolo[1,2-*a*]pyrazoles by Azomethine Imine-Alkyne Cycloadditions Using Immobilized Cu(II)-Catalysts

Urša Štanfel, Dejan Slapšak, Uroš Grošelj ⬤, Franc Požgan ⬤, Bogdan Štefane ⬤ and Jurij Svete *⬤

Faculty of Chemistry and Chemical Technology, University of Ljubljana, Večna pot 113, 1000 Ljubljana, Slovenia; ursa.stanfel@gmail.com (U.Š.); dejan.slapsak@gmail.com (D.S.); Uros.Groselj@fkkt.uni-lj.si (U.G.); Franc.Pozgan@fkkt.uni-lj.si (F.P.); Bogdan.Stefane@fkkt.uni-lj.si (B.Š.)
* Correspondence: jurij.svete@fkkt.uni-lj.si; Tel.: +386-1-479-8562

Abstract: A series of 12 silica gel-bound enaminones and their Cu(II) complexes were prepared and tested for their suitability as heterogeneous catalysts in azomethine imine-alkyne cycloadditions (CuAIAC). Immobilized Cu(II)–enaminone complexes showed promising catalytic activity in the CuAIAC reaction, but these new catalysts suffered from poor reusability. This was not due to the decoordination of copper ions, as the use of enaminone ligands with additional complexation sites resulted in negligible improvement. On the other hand, reusability was improved by the use of 4-aminobenzoic acid linker, attached to 3-aminopropyl silica gel via an amide bond to the enaminone over the more hydrolytically stable *N*-arylenamine C-N bond. The study showed that silica gel-bound Cu(II)–enaminone complexes are readily available and suitable heterogeneous catalysts for the synthesis of 6,7-dihydro-1*H*,5*H*-pyrazolo[1,2-*a*]pyrazoles.

Keywords: 1,3-dipolar cycloadditions; 2,3-dihydropyrazolo[1,2-*a*]pyrazoles; copper-catalyzed azomethine imine-alkyne cycloaddition (CuAIAC); azomethine imines; ynones

Citation: Štanfel, U.; Slapšak, D.; Grošelj, U.; Požgan, F.; Štefane, B.; Svete, J. Synthesis of 6,7-Dihydro-1*H*,5*H*-pyrazolo[1,2-*a*]pyrazoles by Azomethine Imine-Alkyne Cycloadditions Using Immobilized Cu(II)-Catalysts. *Molecules* **2021**, *26*, 400. https://doi.org/10.3390/molecules26020400

Academic Editor: Vera L. M. Silva
Received: 9 December 2020
Accepted: 11 January 2021
Published: 13 January 2021

Publisher's Note: MDPI stays neutral with regard to jurisdictional claims in published maps and institutional affiliations.

Copyright: © 2021 by the authors. Licensee MDPI, Basel, Switzerland. This article is an open access article distributed under the terms and conditions of the Creative Commons Attribution (CC BY) license (https://creativecommons.org/licenses/by/4.0/).

1. Introduction

1,3-Dipolar cycloadditions of azomethine imines are important reactions to obtain pyrazoles with variable degree of saturation [1,2]. Since the end of the 20th century, this field has gained much attention; most azomethine imines have been recognized as stable compounds that are easy to prepare, store, and handle [1,2]. In this context, 1-alkylidene-3-oxopyrazolidin-1-ium-2-ides (3-oxopyrazolidin-1-azomethine imines), accessible by condensation of 1,2-unsubstituted pyrazolidin-3-ones with aldehydes or ketones, have been extensively used for regio- and stereoselective synthesis of pyrazolo[1,2-*a*]pyrazoles (bicyclic pyrazolidinones). Bicyclic pyrazolidinones exhibit antibiotic [3–5] and anti-Alzheimer activity [6], as well as inhibition of lymphocyte-specific protein tyrosine kinase [7,8] and *Plasmodium falciparum* dihydroorotate dehydrogenase (PfDHODH) [9]. The most prominent examples of bioactive bicyclic pyrazolidinones are Eli Lilly's γ-lactam antibiotics, which exhibit antibiotic activity similar to that of penicillins and cephalosporins (Figure 1) [3–5]. These antibiotics are based on 6,7-dihydro-1*H*,5*H*-pyrazolo[1,2-*a*]pyrazole scaffold, which is accessible by [3 + 2] cycloaddition of 3-oxopyrazolidin-1-ium-2-ides to acetylenes [1,2]. In this context, copper-catalyzed azomethine imine-alkyne cycloadditions (CuAIAC) [1,2,10–16] provide easy access to 6,7-dihydro-1*H*,5*H*-pyrazolo[1,2-*a*]pyrazoles in a regio- and stereoselective manner under mild conditions that are compliant with requirements of "click" chemistry (Figure 1) [17–23]. In contrast to the CuAAC reaction, which is catalyzed only by Cu(I), the azomethine imine analogue (CuAIAC) is also catalyzed by Cu(II) [10–12,24–27]. This is a major advantage in terms of catalyst scope and simplicity of workup as the use of reducing agent, such as sodium ascorbate, can be avoided when Cu(II) catalyst is used (Figure 1).

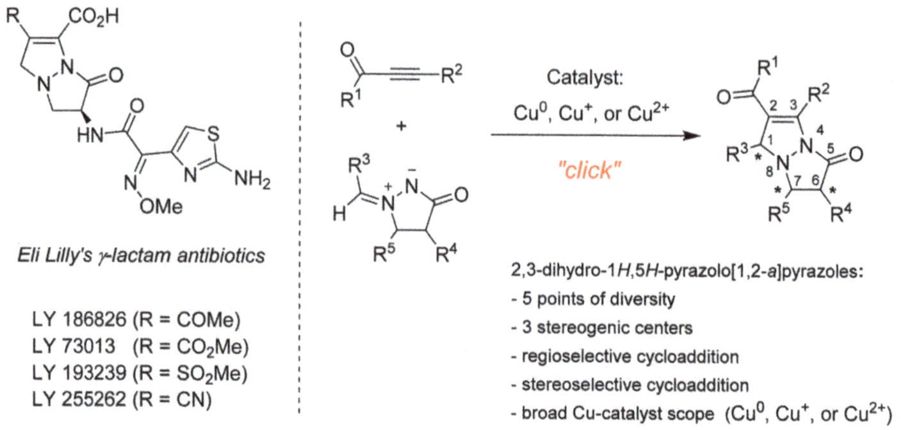

Figure 1. Examples of bioactive bicyclic pyrazolidinones (**left**) and CuAIAC reaction (**right**).

Alkyl 2-substituted-3-(dimethylamino)propenoates and related enaminones are readily available and stable enamino-masked β-keto aldehydes, which are useful 1,3-dielectrophilic reagents in synthetic organic chemistry. Acid-catalyzed reactions with *N*-, *C*-, and *O*-nucleophiles take place under mild conditions by substitution of the dimethylamino group to give β-functionalized propenoates. With ambident nucleophiles, enaminones undergo cyclization into different heterocyclic systems [28–33]. Enaminones are also used as alkenes in cycloaddition reactions [34–38] and as bidentate N, O ligands [39–49] and tetradentate acacen-type ligands [27,50–54] to coordinate metal ions.

In recent years, an important part of our ongoing research on the chemistry of 3-pyrazolidinones [55] has been focused on CuAIAC reactions catalyzed by Cu(0) [56,57], Cu(I) [58–61], and Cu(II) [27]. In extension, we were interested in the use of immobilized Cu(II) complexes with enaminone-type ligands attached to the solid support in CuAIAC reactions. In contrast to the rather extensive use of immobilized copper complexes in azide-alkyne cycloadditions (CuAAC) [62], their applications in CuAIAC reactions are almost unknown [26]. 3-Aminopropyl silica gel-immobilized Cu(II)-enaminone complexes would be easy to prepare via a transamination reaction [28–33,63,64], could serve as heterogeneous Cu(II) catalysts for the synthesis of pyrazolo[1,2-*a*]pyrazoles, and would complement well the known examples of heterogeneous Cu(0)- [41], Cu(I)- [65–69], and Cu(II)-catalysts [26] in the CuAIAC reaction. Herein, we report the results of this study confirming the suitability of these new enaminone-based heterogeneous copper catalysts in regioselective [3 + 2] cycloadditions of 1-benzylidene-5,5-dimethyl-3-oxopyrazolidin-1-ium-2-ides to methyl propiolate leading to methyl 1-aryl-7,7-dimethyl-5-oxo-6,7-dihydro-1*H*,5*H*-pyrazolo[1,2-*a*]pyrazole-2-carboxylates.

2. Results

2.1. Synthesis and Catalytic Activity of Silica Gel-Bound Cu–Enaminone Complexes **5a–g**

First, the starting enaminones **2a–g** were prepared from active methylene compounds **1a–g** by treatment with *N*,*N*-dimethylformamide dimethylacetal (DMFDMA) or *tert*-butoxy-bis(dimethylamino)methane (TBDMAM) at 20–110 °C following literature procedure [27]. Next, the enaminones **2a–g** were reacted with equimolar amount of 3-aminopropyl silica gel (**3**) in methanol for 48 h to give the immobilized enaminones **4a–g**. Subsequent treatment of **4a–g** with one equivalent of Cu(OAc)$_2$·H$_2$O in methanol at room temperature for 48 h then furnished the desired complexes **5a–g**. The complex **3-Cu** was prepared by treatment of **3** with Cu(OAc)$_2$·H$_2$O in methanol (Scheme 1). Absorption bands at around 1600 cm^{-1} (C = O/C = N) in the IR spectra of compounds **5a–h**, the results of combustion analyses for compounds **5a–g**, and the results of characterization of the catalyst **5f** by SEM and

EDX spectroscopy were in line with attachment of copper–enaminone complexes to **3** (For characterization details see the Supporting Information).

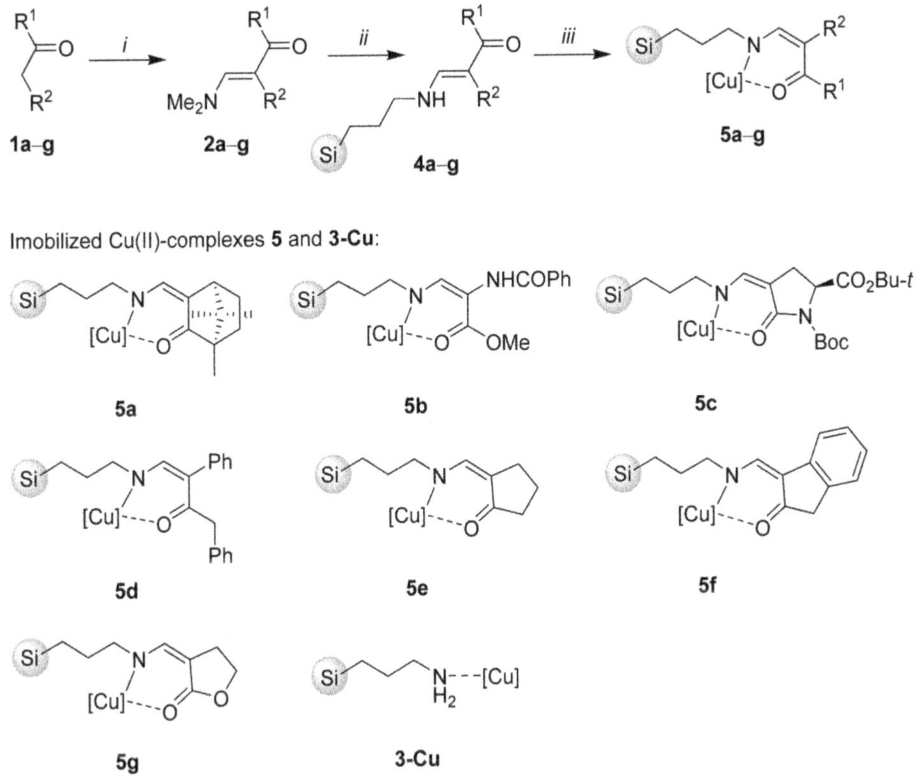

Scheme 1. Reaction conditions: (i) DMFDMA or TBDMAM, CH_2Cl_2 or toluene, 20–110 °C; (ii) 3-aminopropyl silica gel (**3**), MeOH, 20 °C, 48 h; (iii) $Cu(OAc)_2 \cdot H_2O$, MeOH, 20 °C, 48 h.

Compounds **5a–g** and **3-Cu** were then evaluated for their catalytic activity in [3 + 2] cycloaddition of (Z)-3,3-dimethyl-5-oxo-2-(3,4,5-trimethoxybenzylidene)pyrazolidin-2-ium-1-ide (**6a**) to methyl propiolate (**7**). The reaction was performed in CH_2Cl_2 at room temperature for 5 h with 30 mg (~20 mol%) catalyst loading (Table 1). Quantitative conversion was obtained only with 2-indanone-derived catalyst **5f** (Table 1, entry 6), while the conversion above 50% was also obtained from related enamino ketone-derived catalysts **5d** and **5e** (Table 1, entries 4 and 5). Catalysts **5a–c** and **5g** were less active and the respective conversions ranged from 33% to 47% (Table 1, entries 1–3 and 7). Moderate activity of **3-Cu** (Table 1, entry 8) was in line with complexation of $Cu(OAc)_2$ to 3-aminopropyl silica gel (**3**), which itself was found inactive (Table 1, entry 9).

Table 1. Evaluation of catalytic activity of 5a–g, 3-Cu, and 3 in model cycloaddition reaction [1].

Entry	Catalyst	Conversion (%) [2]	Entry	Catalyst	Conversion (%) [2]
1	5a	33	6	5f	quant.
2	5b	47	7	5g	39
3	5c	33	8	3-Cu	31
4	5d	64	9	3	0
5	5e	58			

[1] Reaction conditions: **6a** (37 mg, 0.125 mmol), **7a** (13 mg, 0.150 mmol), catalyst **3** or **5** (30 mg, ~20 mol%), CH_2Cl_2 (4 mL), 20 °C, 5 h. [2] Determined by ^1H NMR.

The most active catalyst **5f** was tested further. The model reaction was carried out varying reaction time (1–3 h) and catalyst loading (10–30 mg). The results are presented in Table 2. In the presence of 30 mg of the catalyst, the conversion was around 50% after one hour, around 90% after two hours, and 100% after three hours (Table 2, entries 1–3). Complete conversion was also achieved with 25 mg and 20 mg of the catalyst (Table 2, entries 4 and 5), while further lowering of the catalyst loading to 15 mg (89%) and to 10 mg (61%) gave incomplete conversions (Table 2, entries 6 and 7).

Table 2. Evaluation of catalyst **5f** in model cycloaddition reaction [1].

Entry	Loading (mg)	Time (h)	Conversion (%) [2]
1	30	1	51
2	30	2	89
3	30	3	quant.
4	25	3	quant.
5	20	3	97
6	15	3	89
7	10	3	61

[1] Reaction conditions: **6a** (37 mg, 0.125 mmol), **7a** (13 mg, 0.150 mmol), catalyst **5f** (10–30 mg, ~7–20 mol%), CH_2Cl_2 (4 mL), 20 °C, 1–5 h. [2] Determined by ^1H NMR.

Next, the substrate scope was investigated using 20 mg (~13 mol%) of catalyst **5f** in reactions with azomethine imines **6a–f** (Scheme 2). After 3 h, only dipoles **6a** and **6d** were transformed quantitatively into the corresponding cycloadducts **8a** and **8d**, while conversions of other dipoles ranged from 23% to 95%. The highest conversions (95–100%) were obtained with 3,4,5-trimethoxyphenyl- (**6a**), phenyl- (**6d**), and 4-nitrophenyl-substituted dipole (**6f**), whereas poor conversions (23–29%) were observed with 4-methoxy- (**6b**), 4-methyl- (**6c**), and 4-chloro-substituted dipole (**6e**). Since closely related Cu^0- and Cu^+-catalyzed cycloadditions did not show any significant substrate dependence [56,57], incomplete conversions may seem surprising, yet they are explainable by much shorter reaction time (i.e., 12–48 h [56,57] vs. 3 h in the present case). Quantitative conversion of dipole **6e** into cycloadduct **8e** after 48 h was in line with this rationale (Scheme 2).

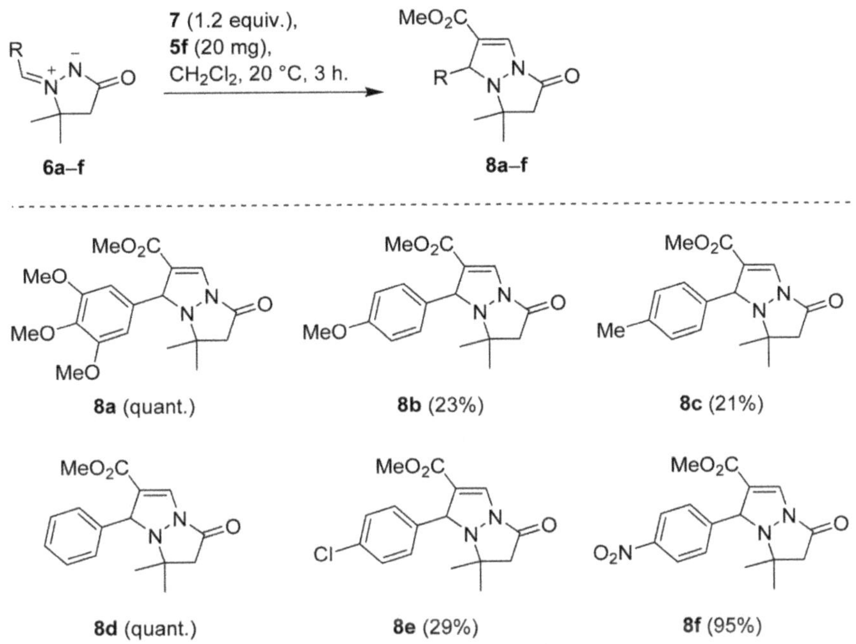

Scheme 2. The conversions in CuAIAC reactions of dipoles **6a–f** with methyl propiolate (**7**) catalyzed by 20 mg (~13 mol%) of **5f**. The conversions were determined by ^{1}H NMR of the crude reaction mixtures.

To further explore the reaction scope, azomethine imine **6a** was reacted also with nonpolar phenylacetylene in the presence of catalyst **5f** under the above standard reaction conditions. This reaction gave no conversion, even after prolonged treatment for 150 h. This result indicated a limitation of the reaction scope to polar electron-poor alkynes.

Reusability of the catalyst **5f** in the standard model reaction (**6a** + **7** → **8a**, 3 h, 30 mg of **5f**) was tested next. Much to our disappointment, the quantitative conversion in the first run dropped significantly in the second (29%) and the third run (5%) and the catalyst was inactive upon the third run (Figure 2). If poor reusability of catalyst **5f** is explainable by decomplexation of copper ions from the heterogeneous ligand **4f**, then reusability should be improved by stronger coordination of copper(II) to the ligand. Therefore, we decided to address the reusability issue by attaching stronger coordinating acacen ligands **9a,b** [27,50–54,70] and pyridine-enaminone ligands **9c** [71] to 3-aminopropyl silica gel (**3**).

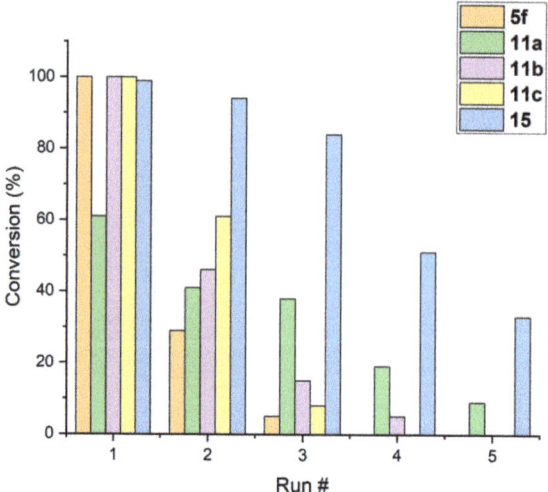

Figure 2. Reusability of catalysts **5f** (☐), **11a** (☐), **11b** (☐), **11c** (☐), and **15** (☐) in the model reaction **6a** + **7** → **8a**. The conversions were determined by ^{1}H NMR.

2.2. Synthesis and Catalytic Activity of Silica Gel-Bound Cu–Enaminone Complexes 11a–c and 15

Bis-enaminone compounds **9a** and **9b** [70] (Scheme 3) contain two terminal N,N-(dimethyl)enaminone groups that enable transaminative attachment to 3-aminopropyl silica gel (**3**). Thus, treatment of **9a** and **9b** with **3** in methanol at room temperature afforded the immobilized acacen ligands **10a** and **10b**, which were subsequently reacted with Cu(OAc)$_2$·H$_2$O in methanol to furnish the desired immobilized Cu–acacen complexes **11a** and **11b** (Scheme 3). To obtain pyridine-type catalyst **11c**, bis-enaminone ligand **9c** [71], was reacted with 3-aminopropyl silica gel (**3**) to give silica gel-bound ligand **10c**, followed by treatment with Cu(OAc)$_2$·H$_2$O in methanol to furnish the copper complex **11c** (Scheme 3). Absorption bands at around 1600 cm^{-1} (C = O/C = N) in the IR spectra of compounds **11a–c** and the combustion analyses for compounds **11a–c** were in line with attachment of copper–enaminone complexes to **3** (For characterization details see the Supporting Information).

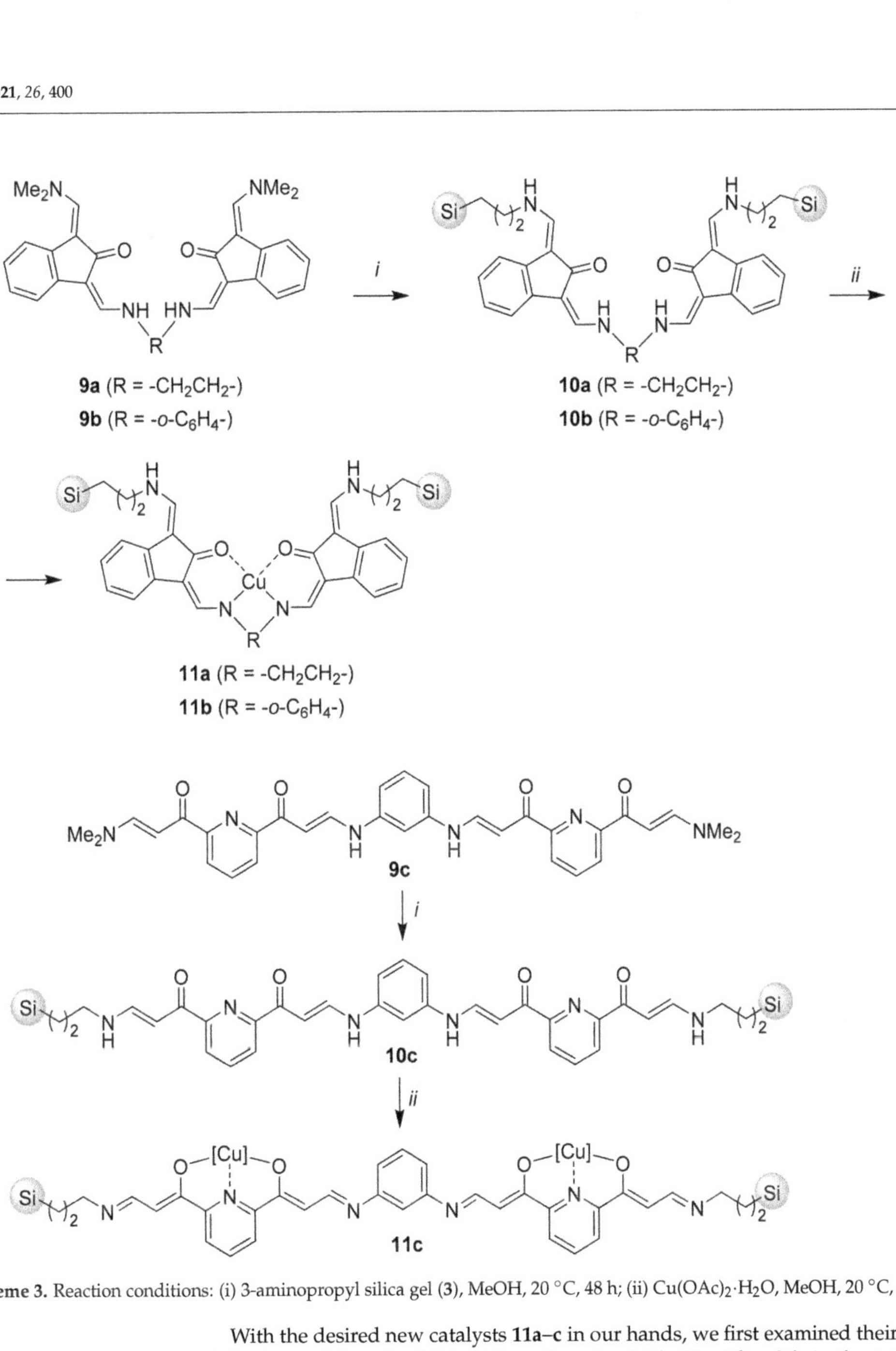

Scheme 3. Reaction conditions: (i) 3-aminopropyl silica gel (**3**), MeOH, 20 °C, 48 h; (ii) Cu(OAc)$_2$·H$_2$O, MeOH, 20 °C, 48 h.

With the desired new catalysts **11a–c** in our hands, we first examined their catalytic activity in model cycloaddition (**6a** + **7** → **8a**, Table 3). After 3 h in the presence of 30 mg (~20 mol%) of the catalyst **11**, 1,2-ethylenediamine-based catalyst **11a** showed only moderate performance (61% conversion, Table 3, entry 1), while activities of 1,2-phenylenediamine-based catalyst **11b** and pyridine-based catalyst **11c** (Table 3, entries 2 and 3) were similar to that of catalyst **5f** (cf. Table 2, entry 3). Further evaluation of catalysts **11b** and **11c** in terms of catalyst loading (Table 3, entries 4–7) and reaction time (Table 3, entries 8–11) confirmed the performance of **11b** and **11c**, which was similar to that of catalyst **5f** (cf. Table 2, entries 1–7).

Table 3. Catalytic activity of heterogeneous Cu(II) catalysts **11a–c** in model reaction [1].

Entry	Catalyst 11	Loading (mg)	Time (h)	Conversion (%) [2]
1	11a	30	3	61
2	11b	30	3	quant.
3	11c	30	3	quant.
4	11b	20	3	quant.
5	11c	10	3	73
6	11b	20	3	quant.
7	11c	10	3	50
8	11b	30	2	quant.
9	11b	30	1	69
10	11c	30	2	quant.
11	11c	30	1	82

[1] Reaction conditions: **6a** (37 mg, 0.125 mmol), **7** (13 mg, 0.150 mmol), catalysts **11a–c** (10–30 mg, ~7–20 mol%), CH_2Cl_2 (4 mL), 20 °C, 1–3 h. [2] Determined by ^1H NMR.

The substrate scope of catalysts **11a–c** was then checked by measuring conversions in the reactions of azomethine imines **6a–f** with methyl propiolate (**7**) in dichloromethane using ~13 mol% (20 mg) catalyst loading (Scheme 4). Quantitative conversions after 3 h were achieved only with dipole **6a** in the presence of catalysts **11b** and **11c**, and with electron-poor dipole **6f**, relatively good conversions above 80% were obtained with all three catalysts. The conversions after 3 h were low to moderate (15–69%) with dipoles **6b–e**. For the most part, these results were in line with those obtained with catalyst **5f**. Notably, also the less reactive dipole **6e** underwent full conversion within 48 h with catalyst **11c** (Scheme 4, cf. Scheme 3).

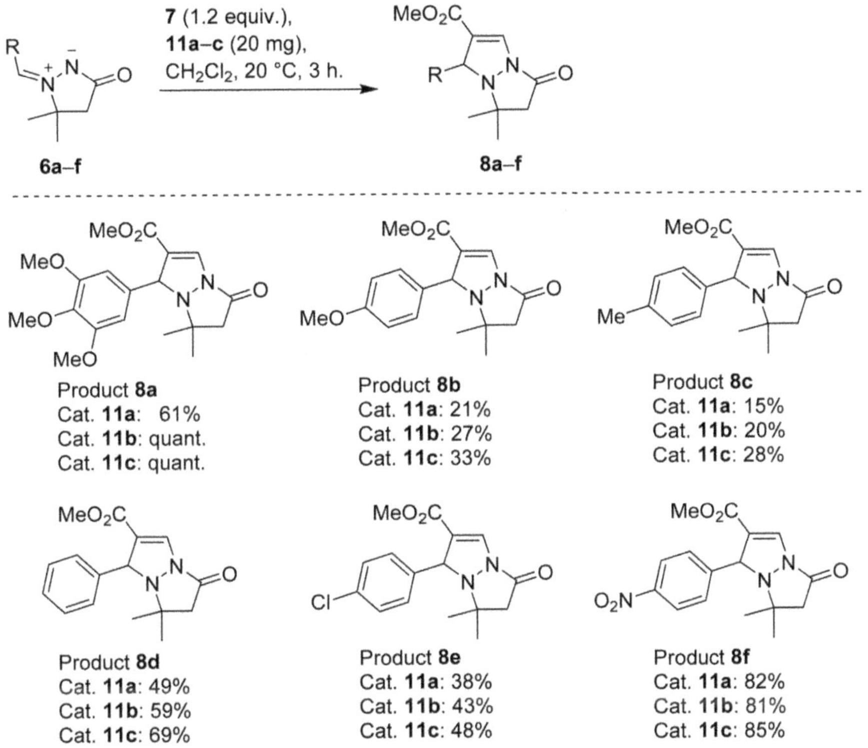

Scheme 4. The conversions in CuAIAC reactions of dipoles **6a–f** with methyl propiolate (**7**) catalyzed by ~13 mol% of **11a–c**. The conversions were determined by ^{1}H NMR of the crude reaction mixtures.

To our disappointment, reusability tests for catalysts **11a–c** in the standard model reaction (**6a** + **7** → **8a**, 3 h, 30 mg of **11**) revealed only minor improvement of reusability of catalysts **11a–c** in comparison to catalyst **5f**. Initially highly active catalysts **11b** and **11c** became inactive upon the third run (see Figure 2 at the end of Section 2.1). On the basis of these data, it became clear that decoordination of Cu(II) from the ligand was not the main reason for low reusability of **5f** and **11a–c**. We then considered that loss of catalytic activity could also be explainable by detachment of Cu(II)-enaminone complex from 3-aminopropyl silica gel (**3**), for example, through hydrolytic cleavage of the enamine C-N bond, as proposed in Scheme 5. Hydrolysis of enaminone complex **5** gives the complex **5′**, which can release Cu(II)-1,3-dicarbonyl complex **5″** in solution through decoordination from aminopropyl silica gel **3**.

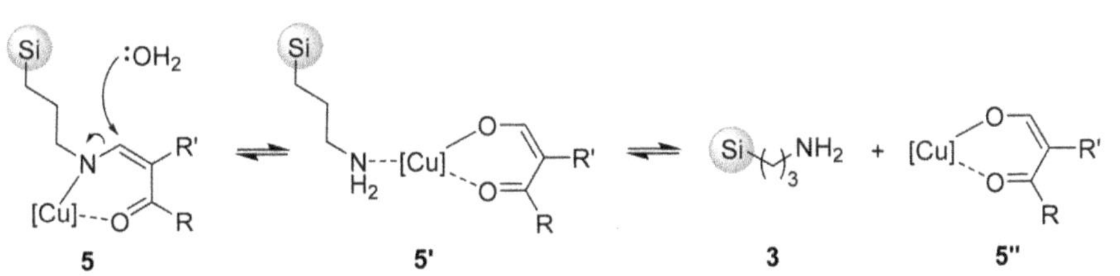

Scheme 5. A plausible mechanism for detachment of Cu(II)-enaminone complex from **3**.

According to the proposed mechanism, the use of hydrolytically more stable enamine C-N bond should reduce detachment of Cu(II)-enaminone complex from the solid support and, thus, improve reusability of the catalyst. To confirm this hypothesis, we prepared silica gel-bound enaminone **14** using 4-aminobenzoic acid (**12**) as a bifunctional linker, which was bound to 3-aminopropyl silica gel (**3**) via a robust amide bond and to the enaminone **2f** through a stronger N-arylenamine C-N bond (Scheme 6) [28–33,63,64]. Acid-catalyzed transamination of **2f** with 4-aminobenzoic acid (**12**) gave the carboxy-functionalized enaminone **13**, which was amidated with **3** using 1,1′-carbonyldiimidazole (CDI) as activating reagent. Subsequent treatment of the silica gel-bound enaminone **14** with copper(II) acetate in methanol then furnished the desired catalyst **15** (Scheme 6). Absorption bands at around 1600 cm^{-1} (C = O/C = N) in the IR spectra of compound **15** and the combustion analyses for **15** were in line with attachment of copper–enaminone complex to **3** (For characterization details see the Supporting Information).

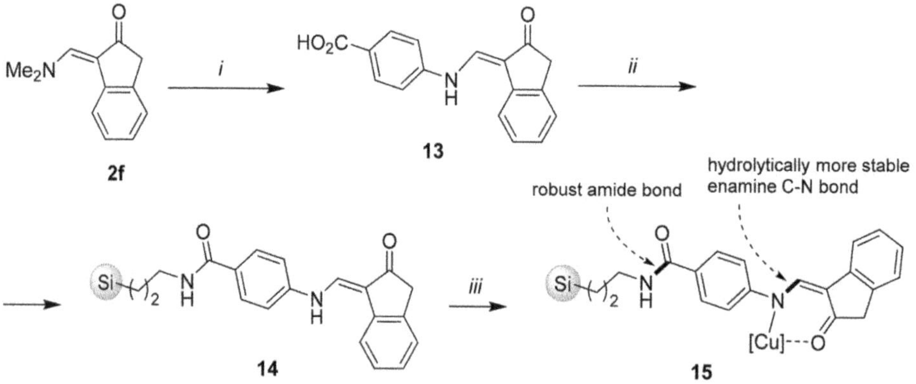

Scheme 6. Reaction conditions: (*i*) 4-aminobenzoic acid (**12**), 37% aq. HCl (1 equiv.), MeOH, 20 °C; (*ii*) CDI, MeCN, 20 °C, 1 h, then 3-aminopropyl silica gel (**3**), MeCN, 20 °C, 120 h; (*iii*) Cu(OAc)$_2$·H$_2$O, MeOH, 20 °C, 48 h.

Activity and reusability of catalysts **5f**, **11a–c**, and **15** were tested in the standard model reaction (**6a** + **7** → **8a**, CH$_2$Cl$_2$, 20 °C, 3 h, 30 mg catalyst loading). The results are summarized in Figure 2. In the first run, quantitative conversion was obtained with catalysts **5f**, **11b**, **11c**, and **15**, while catalyst **11a** gave only 61% conversion. The catalytic activity of **5f** and **11b,c** dropped significantly and they became practically inactive after the second run. Surprisingly, the initially least active catalyst **11a** lost catalytic activity more slowly than analogues **5f** and **11b,c** and remained only weakly active in the fifth run. On the other hand, catalyst **15** gave a near quantitative conversion in the second run (94%), followed by a gradual decrease of catalytic activity leading to 31% conversion in the fifth run. Thus, the reusability of N-arylenaminone catalyst **15** was significantly better than that of N-alkylenaminone analogues **5** and **11** (Figure 2). This result was consistent with the hypothesis that the decrease of catalytic activity was largely due to the detachment of the copper-enaminone complex from the solid support by hydrolysis of the C-N bond of the enamine (cf. Scheme 5).

3. Conclusions

Transamination of enaminones **2a–g** and bis-enaminones **9a–c** with 3-aminopropyl silica gel (**3**) in methanol gives the corresponding silica gel-bound enaminones **4a–g** and **10a–c**. Subsequent treatment of the immobilized enaminones **4a–g** and **10a–c** with copper(II) acetate in methanol gives the corresponding silica gel-bound Cu(II) complexes **5a–g** and **11a–c**. Both reactions are general and take place with different types of enaminones **2** and **9** under mild conditions. The obtained copper(II) complexes **5a–g** and **11a–c** exhibit catalytic activity in azomethine imine-alkyne cycloadditions (CuAIAC). The 2-indanone-derived catalyst **5f** and the bis-enaminone-derived catalysts **11b** and **11c** showed the most promising activity, unfortunately, with poor reusability. The main cause of the poor reusability appears to be hydrolytic cleavage of the Cu(II)-enaminone complex from the 3-aminopropyl silica gel (**3**), rather than decomplexation of the copper(II) ions from the ligand. This hypothesis was confirmed by the synthesis of a modified catalyst **15** with hydrolytically more stable enamine C-N bond of the enamine attached to 3-aminopropyl silica gel (**3**) via a robust amide bond. Catalyst **15** exhibited better reusability while still retaining the same catalytic activity as analogues **5** and **11**. In conclusion, silica gel-bound Cu(II)-enaminone complexes **5**, **11**, and **15** are easily available heterogeneous catalysts for the regioselective synthesis of pyrazolo[1,2-*a*]pyrazoles via [3 + 2] cycloaddition of 3-pyrazolidinone-derived azomethine imines to terminal ynones.

4. Materials and Methods

4.1. General Information

All solvents and reagents were used as received. Melting points were determined on SRS OptiMelt MPA100—Automated Melting Point System (Stanford Research Systems, Sunnyvale, CA, USA). The ^{1}H NMR, ^{13}C NMR, and 2D NMR spectra were recorded in CDCl$_3$ and DMSO-d_6 as solvents using Me$_4$Si as the internal standard on a Bruker Avance III UltraShield 500 plus instrument (Bruker, Billerica, MA, USA) at 500 MHz for ^{1}H and at 126 MHz for ^{13}C nucleus, respectively. IR spectra were recorded on a Bruker FTIR Alpha Platinum spectrophotometer (Bruker, Billerica, MA, USA). Microanalyses were performed by combustion analysis on a Perkin-Elmer CHN Analyzer 2400 II (PerkinElmer, Waltham, MA, USA). Mass spectra were recorded on an Agilent 6224 Accurate Mass TOF LC/MS (Agilent Technologies, Santa Clara, CA, USA). Parallel stirring was carried out on a Tehtnica Vibromix 313 EVT orbital shaker (400 rpm in all cases) (Domel, Železniki, Slovenia). Flash column chromatography was performed on silica gel (Silica gel 60, particle size: 0.035–0.070 mm, Sigma-Aldrich, St. Louis, MO, USA).

Active methylene compounds **1a–g**, 3-aminopropyl silica gel (**3**) (for preparative chromatography, 40–63 μm, 0.9 mmol/g amino groups, pore size ~9 nm), 4-aminobenzoic acid (**12**), Cu(OAc)$_2$·H$_2$O, *N*,*N*-dimethylformamide dimethylacetal (DMFDMA, for synthesis, ≥96%), *tert*-butoxy-bis(dimethylamino)methane (TBDMAM, technical grade), and 1,1′-carbonyldiimidazole (CDI) are commercially available (Sigma-Aldrich). Enaminones **2a** [72], **2b** [73], **2c** [74], **2d** [75], **2e** [76], **2f** [77], and **2g** [78], bis-enaminones **9a**, **9b** [70], and **9c** [71], and azomethine imines **6a,f** [79], **6b,e** [80], **6c** [81], and **6d** [82] were prepared following the literature procedures.

4.2. Synthesis of 3-Aminopropyl Silica Gel-Bound Copper(II)-Catalyst **3-Cu**

A mixture of 3-aminopropyl silica gel (**3**) (5.015 g, 4.5 mmol of amino group), Cu(OAc)$_2$·H$_2$O (903 mg, 4.5 mmol), and methanol (25 mL) was stirred at 20 °C for 48 h. The insoluble material was collected by filtration, washed carefully with methanol until the filtrate was colorless (around 10 × 5 mL), and air-dried to give the copper(II) catalyst **3-Cu**. Blue powder (5.163 g).

4.3. General Procedure for the Synthesis of 3-Aminopropyl Silica Gel-Bound Copper(II) Catalysts **5a–g**

Enaminone **2** (1.381 mmol) was added to a suspension of 3-aminopropyl silica gel (**3**) (1.534 g, 1.381 mmol of amino group) in methanol (4 mL) and the mixture was stirred at 20 °C for 48 h. The insoluble material was collected by filtration, washed with methanol until the filtrate was colorless (around 10 × 5 mL), and air-dried to give **4**. The immobilized enaminone **4** was resuspended in methanol (8 mL), Cu(OAc)$_2$·H$_2$O (275 mg, 1.381 mmol) was added, and the mixture was stirred at room temperature for 48 h. The insoluble material was collected by filtration, washed carefully with methanol until the filtrate was colorless (around 10 × 5 mL), and air-dried to give the copper(II) catalyst **5**. The following compounds were prepared in this manner:

4.3.1. Compound 5a

Prepared from **2a** (934 mg, 4.5 mmol) and **3** (5.015 g, 4.5 mmol of amino group) in MeOH (15 mL); then Cu(OAc)$_2$·H$_2$O (903 mg, 4.5 mmol), MeOH (25 mL). Blue powder (5.392 g), ν_{max} 1558 (C = O/C = N), 1418 cm^{-1}.

4.3.2. Compound 5b

Prepared from **2b** (1.119 g, 4.5 mmol) and **3** (5.015 g, 4.5 mmol of amino group) in MeOH (15 mL); then Cu(OAc)$_2$·H$_2$O (903 mg, 4.5 mmol), MeOH (25 mL). Blue powder (5.611 g), ν_{max} 1558 (C = O/C = N), 1419 cm^{-1}.

4.3.3. Compound 5c

Prepared from **2c** (1.119 g, 4.5 mmol) and **3** (5.015 g, 4.5 mmol of amino group) in MeOH (15 mL); then Cu(OAc)$_2$·H$_2$O (903 mg, 4.5 mmol), MeOH (25 mL). Blue powder (5.415 g), ν_{max} 1565 (C = O/C = N), 1418 cm^{-1}.

4.3.4. Compound 5d

Prepared from **2d** (366 mg, 1.4 mmol) and **3** (1.534 g, 1.4 mmol of amino group) in MeOH (4 mL); then Cu(OAc)$_2$·H$_2$O (275 mg, 1.4 mmol), MeOH (8 mL). Blue powder (1.643 g), ν_{max} 1567 (C = O/C = N), 1416 cm^{-1}.

4.3.5. Compound 5e

Prepared from **2e** (366 mg, 1.4 mmol) and **3** (1.534 g, 1.4 mmol of amino group) in MeOH (4 mL); then Cu(OAc)$_2$·H$_2$O (275 mg, 1.4 mmol), MeOH (8 mL). Light brown powder (1.598 g), ν_{max} 1565 (C = O/C = N), 1416 cm^{-1}.

4.3.6. Compound 5f

Prepared from **2f** (844 mg, 4.5 mmol) and **3** (5.015 g, 4.5 mmol of amino group) in MeOH (15 mL); then Cu(OAc)$_2$·H$_2$O (903 mg, 4.5 mmol), MeOH (25 mL). Dark brown powder (5.514 g), ν_{max} 1606 (C = O/C = N), 1436 cm^{-1}.

4.3.7. Compound 5g

Prepared from **2g** (195 mg, 1.4 mmol) and **3** (1.534 g, 1.4 mmol of amino group) in MeOH (4 mL); then Cu(OAc)$_2$·H$_2$O (275 mg, 1.4 mmol), MeOH (8 mL). Blue powder (1.607 g), ν_{max} 1565 (C = O/C = N), 1416 cm^{-1}.

4.4. General Procedure for the Synthesis of Silica Gel-Bound Copper(II) Catalysts **11a–c**

Bis-enaminone **9** (0.5 mmol) was added to a suspension of 3-aminopropyl silica gel (**3**) (1.111 g, 1 mmol of amino group) in methanol (4 mL) and the mixture was stirred at 20 °C for 48 h. The insoluble material was collected by filtration, washed with methanol until the filtrate was colorless (around 10 × 5 mL), and air-dried to give the silica gel-bound bis-enaminone **10**. The immobilized enaminone **10** was resuspended in methanol (8 mL), Cu(OAc)$_2$·H$_2$O (200 mg, 1 mmol) was added, and the mixture was stirred at room temperature for 48 h. The insoluble material was collected by filtration, washed carefully with methanol until the filtrate was colorless (around 10 × 5 mL), and air-dried to give the copper(II) catalyst **11**. The following compounds were prepared in this manner:

4.4.1. Compound 11a

Prepared from **9a** (100 mg, 0.2 mmol) and **3** (490 g, 0.4 mmol of amino group); then Cu(OAc)$_2$·H$_2$O (80 mg, 0.4 mmol). Brown powder (527 mg), ν_{max} 1569 (C = O/C = N), 1411 cm^{-1}.

4.4.2. Compound 11b

Prepared from **9b** (200 mg, 0.4 mmol) and **3** (884 g, 0.8 mmol of amino group); then Cu(OAc)$_2$·H$_2$O (160 mg, 0.8 mmol). Green-blue powder (906 mg), ν_{max} 1564 (C = O/C = N), 1417 cm^{-1}.

4.4.3. Compound 11c

Prepared from **9c** (322 mg, 0.56 mmol) and **3** (1.265 g, 1.12 mmol of amino group); then Cu(OAc)$_2$·H$_2$O (160 mg, 0.8 mmol). Brown powder (1.294 mg), ν_{max} 1552 (C = O/C = N), 1414 cm^{-1}.

4.5. Synthesis of 3-Aminopropyl Silica Gel-Bound Copper(II) Complex 15

4.5.1. (E)-4-{[(2-Oxo-2,3-dihydro-1H-inden-1-ylidene)methyl]amino}benzoic acid (13)

Enaminone **2f** (374 mg, 2 mmol) was added to a mixture of 4-aminobenzoic acid (**12**) (274 mg, 2 mmol), methanol (10 mL), and 37% aq. HCl (0.15 mL, 1.8 mmol) and the mixture was stirred at room temperature for 12 h. The precipitate was collected by filtration and washed with methanol (2 × 5 mL) and diethyl ether (2 × 5 mL) to give **13**. Beige solid (385 mg, 69%); E/Z = 85:15; mp 263–264 °C (with slow decomposition above 200 °C); ν_{max}/cm^{-1} (ATR) 3014, 2813, 1669 (C = O), 1594, 1566, 1424, 1261, 1180, 1199, 1092, 947, 848, 753, 713, 634; δ_H (500 MHz; DMSO-d_6; Me$_4$Si): major isomer 3.49 (2H, s), 7.09 (1H, td, J = 7.5, 1.1 Hz), 7.23–7.27 (2H, m), 7.52 (2H, d, J = 8.8 Hz), 7.67 (1H, d, J = 7.6 Hz), 7.92 (2H, d, J = 8.8 Hz), 8.39 (1H, d, J = 12.2 Hz), 11.03 (1H, d, J = 12.2 Hz), 12.74 (1H, s), minor isomer 3.46 (3H, s), 7.19 (1H, td, J = 7.5, 1.1 Hz), 7.23–7.27 (1H, m), 7.33 (2H, br d, J = 7.4 Hz), 7.48 (2H, br d, J = 8.8 Hz), 7.71 (1H, d, J = 13.4 Hz), 8.04 (1H, d, J = 7.4 Hz), 9.47 (1H, d, J = 13.3 Hz), 12.74 (1H, s); δ_C (126 MHz; DMSO-d_6; Me$_4$Si): major isomer 41.9, 111.5, 115.6, 117.6, 124.7, 124.8, 124.9, 126.9, 131.1, 134.3, 134.4, 140.0, 143.8, 166.8, 204.4, minor isomer 41.4, 112.8, 116.2, 121.9, 124.6, 124.7, 125.7, 126.8, 131.1, 131.8, 135.5, 138.2, 145.5, 166.9, 202.4; HRMS (ESI): MH$^+$, found 280.0968 (MH$^+$). [C$_{17}$H$_{14}$NO$_3$]$^+$ requires 280.0968; (found: C, 71.98; H, 4.24; N, 4.73. C$_{17}$H$_{13}$NO$_3$·$\frac{1}{4}$H$_2$O requires C, 71.95; H, 4.79; N, 4.94%).

4.5.2. Synthesis of Silica Gel-Bound Enaminone 14

1,1'-Carbonyldiimidazole (85 mg, 0.52 mmol) was added to a stirred suspension of carboxylic acid **13** (140 mg, 0.5 mmol) in acetonitrile (5 mL) and the mixture was stirred at room temperature for 1 h. Then, 3-aminopropyl silica gel (**3**) (500 mg, 0.45 mmol of NH$_2$ group) was added and the suspension was stirred at room temperature for 120 h. Ethanol (2 mL) was added and the insoluble material was collected by filtration using a short column with fritted bottom (d = 1.5 cm, l = 10 cm) and the functionalized silica gel **14** was washed with EtOH-MeCN (1:1, 3 × 5 mL), EtOH (2 × 5 mL), DMF (2 × 5 mL), EtOH (3 mL), and Et$_2$O (2 × 5mL) and air-dried. Brown powder (536 mg, 31%, loading ~0.3 mmol/g); FT-IR (ATR): ν_{max} 1603 (C = O/C = N) cm^{-1}.

4.5.3. Synthesis of Silica Gel-Bound Copper(II) Catalyst 15

3-Enaminopropyl silica gel **14** (300 mg, ~0.1 mmol of the enaminone) was added to a solution of Cu(OAc)$_2$·H$_2$O (50 mg, 0.25 mmol) in methanol (10 mL) and the mixture was stirred at 20 °C for 48 h. The insoluble material was collected by filtration, washed carefully with methanol until the filtrate was colorless (around 5 × 5 mL), and air-dried to give the copper(II) catalyst **15**. Brown powder (280 mg, 81%, loading ~0.3 mmol/g); FT-IR (ATR): ν_{max} 1603 (C = O/C = N) cm^{-1}.

4.6. Synthesis of Methyl 1-Aryl-7,7-dimethyl-5-oxo-6,7-dihydro-1H,5H-pyrazolo[1,2-a]pyrazole-2-carboxylates 8a–f by [3 + 2] Cycloadditions of Azomethine Imines 6a–f to Methyl Propiolate (7) in the Presence of Catalysts 3-Cu, 5, 11, and 15

4.6.1. Determination of Conversion. General Procedure A

Catalyst **3-Cu**, **5**, **11**, or **15** (10–30 mg) was added to a mixture of azomethine imine **6a–f** (25–37 mg [83], 0.125 mmol), methyl propiolate (**7**) (12.5 µL, 0.15 mmol), and CH_2Cl_2 (4 mL) and the mixture was stirred at room temperature for 1–5 h. The reaction mixture was filtered to remove the catalyst and the filtrate was evaporated in vacuo to give **8a–f** and ^1H NMR spectrum of the residue was measured in $CDCl_3$ to determine the conversion. ^1H NMR data of compounds **8a,f** [79], **8d** [56,84], and **8e** [56] were in agreement with the literature data.

4.6.2. Determination of Reusability of Catalysts. General Procedure B

Catalyst **5**, **11**, or **15** (30 mg) was added to a mixture of azomethine imine **6a** (37 mg, 0.125 mmol), methyl propiolate (**7**) (12.5 µL, 0.15 mmol), and CH_2Cl_2 (4 mL) and the mixture was stirred at room temperature for 3 h. Stirring was stopped, the catalyst was allowed to settle down for 2 min, and the supernatant was carefully decanted and filtered. Dichloromethane (4 mL) was added to the catalyst, the mixture was stirred for 2 min, the catalyst was allowed to settle down for 2 min, and the supernatant was carefully decanted and filtered. The catalyst was washed once more with dichloromethane (4 mL) as described above. The combined filtrate was evaporated in vacuo and ^1H NMR spectrum of the residue was measured to determine conversion, while the washed catalyst was used in the next run.

4.6.3. Synthesis of 1-aryl-7,7-dimethyl-5-oxo-6,7-dihydro-1H,5H-pyrazolo[1,2-a]pyrazole-2-carboxylates 8b and 8c. General Procedure C

Catalyst **5f** (120 mg) was added to a mixture of azomethine imine **6b** or **6c** (0.5 mmol), methyl propiolate (**7**) (50 µL, 0.6 mmol), and CH_2Cl_2 (5 mL) and the mixture was stirred at room temperature for 24 h. The catalyst was removed by filtration and washed with dichloromethane (3 mL). The combined filtrate was evaporated in vacuo and the residue was purified by flash column chromatography (Et_2O). Fractions containing the product **8** were combined and evaporated in vacuo to give **8b** and **8c**.

7,7-Dimethyl-1-(4-methoxyphenyl)-5-oxo-6,7-dihydro-1H,5H-pyrazolo[1,2-a]pyrazole-2-carboxylate (**8b**). Prepared from azomethine imine **6b** (116 mg, 0.5 mmol), methyl propiolate (**7**) (50 µL, 0.6 mmol), and CH_2Cl_2 (4 mL). Yellow oil (74 mg, 47%); v_{max}/cm^{-1} (ATR) 2955, 1696 (C = O), 1599, 1511, 1444, 1408, 1371, 1323, 1200, 1173, 1099, 1031, 959, 824, 727; δ_H (500 MHz; $CDCl_3$; Me_4Si): 1.14 (3H, s), 1.22 (3H, s), 2.38 (1H, d, J = 15.7 Hz), 2.86 (1H, d, J = 15.7 Hz), 3.62 (3H, s), 3.79 (3H, s), 5.43 (1H, d, J = 1.3 Hz), 6.87 (2H, d, J = 8.7 Hz), 7.35 (2H, d, J = 8.7 Hz), 7.48 (1H, d, J = 1.3 Hz); δ_C (126 MHz; DMSO-d_6; Me_4Si): 19.1, 25.1, 49.6, 51.6, 55.3, 64.1, 64.4, 113.9, 117.1, 129.0, 129.3, 134.2, 159.3, 164.3, 166.5; HRMS (ESI): MH$^+$, found 317.1493 (MH$^+$). $[C_{17}H_{21}N_2O_4]^+$ requires 317.1496.

7,7-Dimethyl-1-(4-methylphenyl)-5-oxo-6,7-dihydro-1H,5H-pyrazolo[1,2-a]pyrazole-2-carboxylate (**8c**). Prepared from azomethine imine **6c** (98 mg, 0.45 mmol), methyl propiolate (**7**) (50 µL, 0.6 mmol), and CH_2Cl_2 (4 mL). Yellow solid (83 mg, 61%); mp 152–155 °C; v_{max}/cm^{-1} (ATR) 3082, 2946, 1731 (C = O), 1687 (C = O), 1601, 1514, 1323, 1275, 1225, 1192, 1120, 1099, 1039, 1007, 947, 818, 733; δ_H (500 MHz; $CDCl_3$; Me_4Si): 1.12 (3H, s), 1.19 (3H, s), 2.33 (3H, s), 2.38 (1H, d, J = 14.7 Hz), 2.82 (1H, d, J = 14.7 Hz), 3.58 (3H, s), 5.40 (1H, d, J = 1.2 Hz), 7.12 (2H, d, J = 7.8 Hz), 7.29 (2H, d, J = 8.1 Hz), 7.46 (1H, d, J = 1.5 Hz); δ_C (126 MHz; DMSO-d_6; Me_4Si): 19.1, 21.3, 25.1, 49.5, 51.6, 64.4, 64.6, 117.0, 127.8, 129.2, 129.5, 137.6, 139.1, 164.3, 166.7; HRMS (ESI): MH$^+$, found 301.1546 (MH$^+$). $[C_{17}H_{21}N_2O_3]^+$ requires 301.1547.

Following the above Procedure C, also known cycloadducts **8a,d–f** were obtained in the following isolated yields: compound **8a** (92%), **8d** (89%), **8e** (84%), and **8f** (88%).

Spectral data for compounds **8a** [56,79], **8d** [56,84], **8e** [56], and **8f** [56,79] were in agreement with the literature data.

Supplementary Materials: The following are available online, copies of ^{1}H and ^{13}C NMR spectra of new compounds **8b**, **8c**, and **13**, copies of IR spectra of catalysts **5a–g**, **11a–c**, and **15**.

Author Contributions: Individual contributions of the authors are the following: conceptualization, U.Š., D.S., and J.S.; methodology, U.Š., D.S., U.G., F.P., B.Š., and J.S.; validation, U.Š., D.S, U.G., F.P., B.Š., and J.S.; formal analysis, U.Š., D.S., and J.S.; investigation, U.Š., D.S., and J.S.; resources, U.G., F.P., B.Š., and J.S.; data curation, U.Š., D.S., U.G., F.P., B.Š., and J.S.; writing—original draft preparation, U.G., F.P., B.Š., and J.S.; writing—review and editing, U.Š., D.S, U.G., F.P., B.Š., and J.S.; supervision, U.G., F.P., B.Š., and J.S.; project administration, U.G., F.P., B.Š., and J.S.; funding acquisition, U.G., F.P., B.Š., and J.S. All authors have read and agreed to the published version of the manuscript.

Funding: This research was funded by Slovenian Research Agency (ARRS), research core funding No. P1-0179.

Data Availability Statement: The data presented in this study are available in the Supporting Information.

Acknowledgments: We thank EN-FIST Centre of Excellence, Trg Osvobodilne fronte 13, 1000 Ljubljana, Slovenia, for using BX FTIR spectrophotometer. We thank Marjan Marinšek, University of Ljubljana, for characterization of the catalyst **5f** by SEM and EDX spectroscopy.

Conflicts of Interest: The authors declare no conflict of interest.

Sample Availability: Samples of the compounds **5a–g**, **8a–f**, **11a–c**, and **13–15** are available from the authors.

References and Notes

1. Grošelj, U.; Svete, J. [3+2] Cycloadditions of Azomethine Imines. *Org. React.* **2020**, *103*, 529–930. [CrossRef]
2. Nájera, C.; Sansano, J.M.; Yus, M. 1,3-Dipolar cycloadditions of azomethine imines. *Org. Biomol. Chem.* **2015**, *13*, 8596–8636. [CrossRef] [PubMed]
3. Jungheim, L.N.; Sigmund, S.K.; Fisher, J.W. Bicyclic pyrazolidinones, a new class of antibacterial agent based on the β-lactam model. *Tetrahedron Lett.* **1987**, *28*, 285–288. [CrossRef]
4. Ternansky, R.J.; Draheim, S.E. The chemistry of substituted pyrazolidinones; applications to the synthesis of bicyclic derivatives. *Tetrahedron* **1992**, *48*, 777–796. [CrossRef]
5. Ternansky, R.J.; Draheim, S.E.; Pike, A.J.; Counter, F.T.; Eudaly, J.A.; Kasher, J.S. Structure-activity relationship within a series of pyrazolidinone antibacterial agents. 2. Effect of side-chain modification on in vitro activity and pharmacokinetic parameters. *J. Med. Chem.* **1993**, *36*, 3224–3229. [CrossRef]
6. Kosower, E.M.; Hershkowitz, E. 1,5-Diazabicyclo [3,3,0—] octanediones and pharmaceutical compositions containing them. IL 94658 A, 1994. *Chem. Abstr.* **1995**, *122*, 214077.
7. Sabat, M.; Van Rens, J.C.; Brugel, T.A.; Maier, J.; Laufersweiler, M.J.; Golebiowski, A.; De, B.; Easwaran, V.; Hsieh, L.C.; Rosegen, J.; et al. The development of novel 1,2-dihydro-pyrimido[4,5-c]pyridazine-based inhibitors of lymphocyte specific kinase (Lck). *Bioorg. Med. Chem. Lett.* **2006**, *16*, 4257–4261. [CrossRef]
8. VanRens, J.C.; Sabat, M.; Brugel, T.A.; Maier, J.; Laufersweiler, M.J.; Golebiowski, A.; Berberich, S.; De, B. Synthesis of novel pyrimido[4,5-c]pyridazines and 1,2-dihydro-3a,7,9,9b-tetraaza-cyclopenta[a]naphthalen-3-ones as potent inhibitors of lymphocyte specific kinase (LCK). *Heterocycles* **2006**, *68*, 2037–2044. [CrossRef]
9. Strašek, N.; Lavrenčič, L.; Oštrek, A.; Slapšak, D.; Grošelj, U.; Klemenčič, M.; Brodnik Žugelj, H.; Wagger, J.; Novinec, M.; Svete, J. Tetrahydro-1H,5H-pyrazolo[1,2-a]pyrazole-1-carboxylates as inhibitors of Plasmodium falciparum dihydroorotate dehydrogenase. *Bioorg. Chem.* **2019**, *89*, 102982. [CrossRef]
10. Grošelj, U.; Požgan, F.; Štefane, B.; Svete, J. Copper-Catalyzed Azomethine Imine–Alkyne Cycloadditions (CuAIAC). *Synthesis* **2018**, *50*, 4501–4524. [CrossRef]
11. Požgan, F.; Al Mamari, H.; Grošelj, U.; Svete, J.; Štefane, B. Synthesis of Non-Racemic Pyrazolines and Pyrazolidines by [3+2] Cycloadditions of Azomethine Imines. *Molecules* **2018**, *23*, 3. [CrossRef]
12. Decuypère, E.; Plougastel, L.; Audisio, D.; Taran, F. Sydnone–alkyne cycloaddition: Applications in synthesis and bioconjugation. *Chem. Commun.* **2017**, *53*, 11515–11527. [CrossRef]
13. Nelina-Nemtseva, J.I.; Gulevskaya, A.V.; Suslonov, V.V.; Misharev, A.D. 1,3-Dipolar cycloaddition of azomethine imines to ethynyl hetarenes: A synthetic route to 2,3-dihydropyrazolo[1,2-a]pyrazol-1(5H)-one based heterobiaryls. *Tetrahedron* **2018**, *74*, 1101–1109. [CrossRef]

14. Decuypère, E.; Bernard, S.; Feng, M.; Porte, K.; Riomet, M.; Thuéry, P.; Audisio, D.; Taran, F. Copper-Catalyzed Aza-Iminosydnone-Alkyne Cycloaddition Reaction Discovered by Screening. *ACS Catal.* **2018**, *8*, 11882–11888. [CrossRef]
15. Zhang, M.; Wu, F.; Wang, F.; Wu, J.; Chen, W. Copper-Catalyzed Sequential Azomethine Imine-Alkyne Cycloaddition and Umpolung Thiolation Reactions. *Adv. Synth. Catal.* **2017**, *359*, 2768–2772. [CrossRef]
16. Koler, A.; Paljevac, M.; Cmager, N.; Iskra, J.; Kolar, M.; Krajnc, P. Poly(4-vinylpyridine) polyHIPEs as catalysts for cycloaddition click reaction. *Polymer* **2017**, *126*, 402–407. [CrossRef]
17. Kolb, H.C.; Finn, M.G.; Sharpless, K.B. Click Chemistry: Diverse Chemical Function from a Few Good Reactions. *Angew. Chem. Int. Ed.* **2001**, *40*, 2004–2021. [CrossRef]
18. Tornøe, C.W.; Christensen, C.; Meldal, M. Peptidotriazoles on Solid Phase: [1,2,3—]-Triazoles by Regiospecific Copper(I)-Catalyzed 1,3-Dipolar Cycloadditions of Terminal Alkynes to Azides. *J. Org. Chem.* **2002**, *67*, 3057–3064. [CrossRef]
19. Rostovtsev, V.V.; Green, L.G.; Fokin, V.V.; Sharpless, K.B. A Stepwise Huisgen Cycloaddition Process: Copper(I)-Catalyzed Regioselective "Ligation" of Azides and Terminal Alkynes. *Angew. Chem. Int. Ed.* **2002**, *41*, 2596–2599. [CrossRef]
20. Meldal, M.; Tornøe, C.W. Cu-Catalyzed Azide–Alkyne Cycloaddition. *Chem. Rev.* **2008**, *108*, 2952–3015. [CrossRef]
21. Fokin, V.V.; Matyjaszewski, K. CuAAC: The Quintessential Click Reaction. In *Organic Chemistry–Breakthroughs and Perspectives*, 1st ed.; Ding, K., Dai, L.-X., Eds.; Wiley-VCH: Weinheim, Germany, 2012; pp. 247–277.
22. Berg, R.; Straub, B.F. Advancements in the mechanistic understanding of the copper-catalyzed azide–alkyne cycloaddition. *Beilstein J. Org. Chem.* **2013**, *9*, 2715–2750. [CrossRef]
23. Haldón, E.; Nicasio, M.C.; Pérez, P.J. Copper-catalyzed azide–alkyne cycloadditions (CuAAC): An update. *Org. Biomol. Chem.* **2015**, *13*, 9528–9550. [CrossRef] [PubMed]
24. Safaei, S.; Mohammadpoor-Baltork, I.; Khosropour, A.R.; Moghadam, M.; Tangestaninejad, S.; Mirkhani, V. Copper(II) ionic liquid catalyzed cyclization–aromatization of hydrazones with dimethyl acetylenedicarboxylate: A green synthesis of fully substituted pyrazoles. *New J. Chem.* **2013**, *37*, 2037–2042. [CrossRef]
25. Hori, M.; Sakakura, A.; Ishihara, K. Enantioselective 1,3-Dipolar Cycloaddition of Azomethine Imines with Propioloylpyrazoles Induced by Chiral π–Cation Catalysts. *J. Am. Chem. Soc.* **2014**, *136*, 13198–13201. [CrossRef]
26. Comas-Barceló, J.; Blanco-Ania, D.; van den Broek, S.A.M.W.; Nieuwland, P.J.; Harrity, J.P.A.; Rutjes, F.P.J.T. Cu-catalyzed pyrazole synthesis in continuous Flow. *Catal. Sci. Technol.* **2016**, *6*, 4718–4723. [CrossRef]
27. Tomažin, U.; Grošelj, U.; Počkaj, M.; Požgan, F.; Štefane, B.; Svete, J. Combinatorial Synthesis of Acacen-Type Ligands and Their Coordination Compounds. *ACS Comb. Sci.* **2017**, *19*, 386–396. [CrossRef]
28. Huang, J.; Yu, F. Recent Advances in Organic Synthesis Based on N,N-Dimethyl Enaminones. *Synthesis* **2021**, *53*. in press. [CrossRef]
29. Fu, L.; Wan, J.-P. C3-Functionalized Chromones Synthesis by Tandem C-H Elaboration and Chromone Annulation of Enaminones. *Asian J. Org. Chem.* **2021**, in press. [CrossRef]
30. Stanovnik, B. Enaminone, enaminoesters, and related compounds in the metal-free synthesis of pyridines and fused pyridines. *Eur. J. Org. Chem.* **2019**, 5120–5132. [CrossRef]
31. Gaber, H.M.; Bagley, M.C.; Muhammad, Z.A.; Gomha, S.M. Recent developments in chemical reactivity of N,N-dimethylamino ketones as synthons for various heterocycles. *RSC Adv.* **2017**, *7*, 14562–14610. [CrossRef]
32. Chattopadhyay, A.K.; Hanessian, S. Cyclic enaminones. Part I: Stereocontrolled synthesis using diastereoselective and catalytic asymmetric methods. *Chem. Commun.* **2015**, *51*, 16437–16449. [CrossRef] [PubMed]
33. Chattopadhyay, A.K.; Hanessian, S. Cyclic enaminones. Part II: Applications as versatile intermediates in alkaloid synthesis. *Chem. Commun.* **2015**, *51*, 16450–16467. [CrossRef]
34. Efimov, I.V.; Matveeva, M.D.; Luque, R.; Bakulev, V.A.; Voskressensky, L.G. [3+2] Anionic Cycloaddition of Isocyanides to Acyclic Enamines and Enaminones: A New, Simple, and Convenient Method for the Synthesis of 2,4-Disubstituted Pyrroles. *Eur. J. Org. Chem.* **2020**, 1108–1113. [CrossRef]
35. Feng, J.; He, T.; Xie, Y.; Yu, Y.; Baell, J.B.; Huang, F. I2-Promoted [4 + 2] cycloaddition of in situ generated azoalkenes with enaminones: Facile andefficient synthesis of 1,4-dihydropyridazines andpyridazines. *Org. Biomol. Chem.* **2020**, *18*, 9483–9493. [CrossRef] [PubMed]
36. Zhou, S.; Yan, B.-W.; Fan, S.-X.; Tian, J.-S.; Loh, T.-P. Regioselective Formal [4 + 2] Cycloadditions of Enaminones with Diazocarbonyls through Rh^{III}-Catalyzed C–H Bond Functionalization. *Org. Lett.* **2018**, *20*, 3975–3979. [CrossRef] [PubMed]
37. Prek, B.; Bezenšek, J.; Stanovnik, B. Synthesis of pyridines with an amino acid residue by [2+2] cycloadditions of electron-poor acetylenes on enaminone systems derived from N-Boc protected amino acids. *Tetrahedron* **2017**, *73*, 5260–5267. [CrossRef]
38. Wan, J.-P.; Cao, S.; Liu, Y. Base-Promoted Synthesis of N-Substituted 1,2,3-Triazoles via Enaminone–Azide Cycloaddition Involving Regitz Diazo Transfer. *Org. Lett.* **2016**, *18*, 6034–6037. [CrossRef]
39. Fu, L.; Cao, X.; Wan, J.-P.; Liu, Y. Synthesis of Enaminone-Pd(II) Complexes and Their Application in Catalysing Aqueous Suzuki-Miyaura Cross Coupling Reaction. *Chin. J. Chem.* **2020**, *38*, 254–258. [CrossRef]
40. Huma, R.; Mahmud, T.; Mitu, L.; Ashraf, M.; Iqgal, A.; Iftikhar, K.; Hayat, A. Synthesis, Characterization, Molecular Docking and Enzyme InhibitionStudies of Some Novel Enaminone Derivatives and Their Complexes with Cu(II), Cd(II) and Co(II) Ions. *Rev. Chim.* **2019**, *70*, 3564–3569. [CrossRef]
41. Sasinska, A.; Leduc, J.; Frank, M.; Czympiel, L.; Fischer, T.; Christiansen, S.H.; Mathur, S. Competitive interplay of deposition and etching processes in atomic layer growth of cobalt and nickel metal films. *J. Mater. Res.* **2018**, *33*, 4241–4250. [CrossRef]

42. Chopin, N.; Novitchi, G.; Médebielle, M.; Pilet, G. A versatile ethanolamine-derived trifluoromethyl enaminone ligand for the elaboration of nickel(II) and copper(II)–dysprosium(III) multinuclear complexes with magnetic properties. *J. Fluorine Chem.* **2015**, *179*, 169–174. [CrossRef]
43. Jeragh, B.; Elassar, A.-Z. Enaminone Complexes: Synthesis, Characterization amd Bioactivity. *Chem. Sci. Trans.* **2015**, *4*, 113–120. [CrossRef]
44. Chopin, N.; Médebielle, M.; Pilet, G. 8-Aminoquinoline and 2-aminopyridine trifluoromethyl-derived enaminone ligands and their Co(II), Ni(II) and Zn(II) metal complexes: Structural arrangements due to F···H interactions. *J. Fluor. Chem.* **2013**, *155*, 89–96. [CrossRef]
45. Beckmann, U.; Hägele, G.; Frank, W. Square-Planar 2-Toluenido(triphenylphosphane)nickel(II) Complexes Containing Bidentate N,O Ligands: An Example of Planar Chirality. *Eur. J. Inorg. Chem.* **2010**, 1670–1678. [CrossRef]
46. Shi, Y.-C.; Hu, Y.-Y. Syntheses and structures of copper complexes of tetradentate enaminones derived from condensation of benzoylacetone and ferrocenoylacetone with 1,2-bis(2-aminophenoxy)ethane. *J. Coord. Chem.* **2009**, *62*, 1302–1312. [CrossRef]
47. Pilet, G.; Tommasino, J.-B.; Fenain, F.; Matrak, R.; Médebielle, M. Trifluoromethylated enaminones and their explorative coordination chemistry with Cu(II): Synthesis, redox properties and structural characterization of the complexes. *Dalton Trans.* **2008**, 5621–5626. [CrossRef]
48. Shi, Y.-C.; Yang, H.-M.; Song, H.-B.; Liu, Y.-H. Syntheses and crystal structures of a ferrocene-containing enaminone and its copper complex. *Polyhedron* **2004**, *23*, 1541–1546. [CrossRef]
49. Aumann, R.; Göttker-Schnetmann, I.; Fröhlich, R.; Saarenketo, P.; Holst, C. Enaminone Substituents Attached to Cyclopentadienes: 3E/3Z Stereochemistry of 1-Metalla-1,3,5-hexatriene Intermediates (M = Cr,W) as a Functional Criterion for the Formation of Cyclopentadienes and Six-Membered Heterocycles, Respectively. *Chem. Eur. J.* **2001**, *7*, 711–720. [CrossRef]
50. Weber, B.; Jäger, E.-G. Structure and Magnetic Properties of Iron(II/III) Complexes with $N_2O_2^{2-}$ Coordinating Schiff Base Like Ligands. *Eur. J. Inorg. Chem.* **2009**, 465–477. [CrossRef]
51. Schuhmann, K.; Jäger, E.-G. Equilibrium Studies of the Axial Addition of Chiral Bases to Chiral Metal Schiff Base Complexes. *Eur. J. Inorg. Chem.* **1998**, 2051–2054. [CrossRef]
52. Jäger, E.-G.; Keutel, H.; Rudolph, M.; Krebs, B.; Wiesemann, F. Activation of Molecular Oxygen by Biomimetic Schiff Base Complexes of Manganese, Iron, and Cobalt. *Chem. Ber.* **1996**, *129*, 503–514. [CrossRef]
53. Liao, J.-H.; Cheng, K.-Y.; Fang, J.-M.; Cheng, M.-C.; Wang, Y. Oxidation of Alkenes and Sulfides with Transition Metal Catalysts. *J. Chin. Chem. Soc.* **1995**, *42*, 847–860. [CrossRef]
54. Weissenfels, M.; Thust, U.; Mühlstädt, M. Über Umsetzungen von Hydroxymethylenecyclanonen. Die Synthese von 2,3-Benzo-5,6-cyclano-1,4-diazepinen. *J. Prakt. Chem.* **1963**, *20*, 117–124. [CrossRef]
55. Grošelj, U.; Svete, J. Recent advances in the synthesis of polysubstituted 3-pyrazolidinones. *Arkivoc 2015* **2015**, 175–205. [CrossRef]
56. Pušavec Kirar, E.; Grošelj, U.; Mirri, G.; Požgan, F.; Strle, G.; Štefane, B.; Jovanovski, V.; Svete, J. "Click" Chemistry: Application of Copper Metal in Cu-Catalyzed Azomethine Imine−Alkyne Cycloadditions. *J. Org. Chem.* **2016**, *81*, 5988–5997. [CrossRef]
57. Mirnik, J.; Pušavec Kirar, E.; Ričko, S.; Grošelj, U.; Golobič, A.; Požgan, F.; Štefane, B.; Svete, J. Cu^0-catalysed 1,3-dipolar cycloadditions of α-amino acid derived N,N-cyclic azomethine imines to ynones. *Tetrahedron* **2017**, *73*, 3329–3337. [CrossRef]
58. Pezdirc, L.; Stanovnik, B.; Svete, J. Copper(I) Iodide-Catalyzed Cycloadditions of (1Z,4R*,5R*)-4-Benzamido-5-phenylpyrazolidin-3-on-1-azomethine Imines to Ethyl Propiolate. *Aust. J. Chem.* **2009**, *62*, 1661–1666. [CrossRef]
59. Pušavec, E.; Mirnik, J.; Šenica, L.; Grošelj, U.; Stanovnik, B.; Svete, J.Z. Cu(I)-catalyzed [3+2] Cycloadditions of tert-Butyl (S)-(3-Oxopent-4-yn-2-yl)carbamate to 1-Benzylidenepyrazole-3-one-derived Azomethine Imines Naturforsch. *Z. Naturforsch. B.* **2014**, *69*, 615–626. [CrossRef]
60. Pušavec Kirar, E.; Drev, M.; Mirnik, J.; Grošelj, U.; Golobič, A.; Dahmann, G.; Požgan, F.; Štefane, B.; Svete, J. Synthesis of 3D-Rich Heterocycles: Hexahydropyrazolo[1,5-a]pyridin-2(1H)-ones and Octahydro-2H-2a,2a¹-diazacyclopenta[cd]inden-2-ones. *J. Org. Chem.* **2016**, *81*, 8920–8933. [CrossRef]
61. Pušavec Kirar, E.; Grošelj, U.; Golobič, A.; Požgan, F.; Pusch, S.; Weber, C.; Andernach, L.; Štefane, B.; Opatz, T.; Svete, J. Absolute Configuration Determination of 2,3-Dihydro-1H,5H;-pyrazolo[1,2-a]pyrazoles Using Chiroptical Methods at Different Wavelengths. *J. Org. Chem.* **2016**, *81*, 11802–11812. [CrossRef]
62. Shiri, P.; Aboonajmi, J. A systematic review on silica-, carbon-, and magnetic materials-supported copper species as efficient heterogeneous nanocatalysts in "click" reactions. *Beilstein J. Org. Chem.* **2020**, *16*, 551–586. [CrossRef] [PubMed]
63. Stanovnik, B.; Svete, J. Synthesis of Heterocycles from Alkyl 3-(Dimethylamino)propenoates and Related Enaminones. *Chem. Rev.* **2004**, *104*, 2433–2480. [CrossRef] [PubMed]
64. Stanovnik, B.; Svete, J. Alkyl 2-Substituted 3-(Dimethylamino)propenoates and Related Compounds-Versatile Reagents in Heterocyclic Chemistry. *Synlett* **2000**, 1077–1091. [CrossRef]
65. Keller, M.; Sido, A.S.S.; Pale, P.; Sommer, J. Copper(I) Zeolites as Heterogeneous and Ligand-Free Catalysts: [3+2] Cycloaddition of Azomethine Imines. *Chem. Eur. J.* **2009**, *15*, 2810–2817. [CrossRef]
66. Chassaing, S.; Alix, A.; Bonongari, T.; Sido, K.S.S.; Keller, M.; Kuhn, P.; Louis, B.; Sommer, J.; Pale, P. Copper(I)-Zeolites as New Heterogeneous and Green Catalysts for Organic Synthesis. *Synthesis* **2010**, 1557–1567. [CrossRef]
67. Mizuno, N.; Kamata, K.; Nakagawa, Y.; Oishi, T.; Yamaguchi, K. Scope and reaction mechanism of an aerobic oxidative alkyne homocoupling catalyzed by a di-copper-substituted silicotungstate. *Catal. Today* **2010**, *157*, 359–363. [CrossRef]

68. Oishi, T.; Yoshimura, K.; Yamaguchi, K.; Mizuno, N. An Efficient Copper-mediated 1,3-Dipolar Cycloaddition of Pyrazolidinone-based Dipoles to Terminal Alkynes to Produce N,N-Bicyclic Pyrazolidinone Derivatives. *Chem. Lett.* **2010**, *39*, 1086–1087. [CrossRef]
69. Yoshimura, K.; Oshi, T.; Yamaguchi, K.; Mizuno, N. An Efficient, Ligand-Free, Heterogeneous Supported Copper Hydroxide Catalyst for the Synthesis of N,N-Bicyclic Pyrazolidinone Derivatives. *Chem. Eur. J.* **2011**, *17*, 3827–3831. [CrossRef]
70. Tomažin, U.; Alič, B.; Kristl, A.; Ručigaj, A.; Grošelj, U.; Požgan, F.; Krajnc, M.; Štefane, B.; Šebenik, U.; Svete, J. Synthesis of polyenaminones by acid-catalyzed amino–enaminone 'click' polymerisation. *Eur. Polym. J.* **2018**, *108*, 603–616. [CrossRef]
71. Zupanc, A.; Kotnik, T.; Štanfel, U.; Brodnik Žugelj, H.; Kristl, A.; Ručigaj, A.; Matoh, L.; Pahovnik, D.; Grošelj, U.; Opatz, T.; et al. Chemical recycling of polyenaminones by transamination reaction via amino–enaminone polymerisation/depolymerisation. *Eur. Polym. J.* **2019**, *121*, 109282. [CrossRef]
72. Grošelj, U.; Rečnik, S.; Svete, J.; Meden, A.; Stanovnik, B. Stereoselective synthesis of (1R,3R,4R)-3-(1,2,4-triazolo[4,3-x]azin-3-yl)-1,7,7-trimethylbicyclo[2.2.1]heptan-2-ones. *Tetrahedron Asymmetry* **2002**, *13*, 821–833. [CrossRef]
73. Yamato, E.; Okumura, K. 2-Acylamino-3-(indol-3-yl)acrylic acid esters. JP 50058063 A, 1975. *Chem. Abstr.* **1975**, *83*, 193075y.
74. Malavašič, Č.; Brulc, B.; Čebašek, P.; Dahmann, G.; Heine, N.; Bevk, D.; Grošelj, U.; Meden, A.; Stanovnik, B.; Svete, J. Combinatorial Solution-Phase Synthesis of (2S,4S)-4-Acylamino-5-oxopyrrolidine-2-carboxamides. *J. Comb. Chem.* **2007**, *9*, 219–229. [CrossRef] [PubMed]
75. Abdelkhalik, M.M.; Eltoukhy, A.M.; Agamy, S.M.; Elnagdi, M.H. Enaminones as Building Blocks in Heterocyclic Synthesis: New Syntheses of Nicotinic Acid and Thienopyridine Derivatives. *J. Heterocycl. Chem.* **2004**, *41*, 431–434. [CrossRef]
76. Schuda, P.F.; Ebner, C.B.; Morgan, T.M. The Synthesis of Mannich Bases from Ketones and Esters via Enaminones. *Tetrahedron Lett.* **1986**, *23*, 2567–2570. [CrossRef]
77. Altomare, C.; Cellamare, S.; Summo, L.; Catto, M.; Carotti, A.; Thull, U.; Carrupt, P.-A.; Testa, B.; Stoeckli-Evans, H. Inhibition of Monoamine Oxidase-B by Condensed Pyridazines and Pyrimidines: Effects of Lipophilicity and Structure-Activity Relationships. *J. Med. Chem.* **1998**, *41*, 3812–3820. [CrossRef]
78. Wasserman, H.H.; Ives, J.L. Reaction of Singlet Oxygen with Enamino Lactones. Conversion of Lactones to α-Keto Lactones. *J. Org. Chem.* **1978**, *43*, 3238–3240. [CrossRef]
79. Turk, C.; Svete, J.; Stanovnik, B.; Golič, L.; Golič-Grdadolnik, S.; Golobič, A.; Selič, L. Regioselective 1,3-Dipolar Cycloadditions of (1Z)-1-(Arylmethylidene)-5,5-dimethyl-3-oxopyrazolidin-1-ium-2-ide Azomethine Imines to Acetylenic Dipolarophiles. *Helv. Chim. Acta* **2001**, *84*, 146–156. [CrossRef]
80. Sibi, M.P.; Rane, D.; Stanley, L.M.; Soeta, T. Copper(II)-Catalyzed Exo and Enantioselective Cycloadditions of Azomethine Imines. *Org. Lett.* **2008**, *10*, 2971–2974. [CrossRef]
81. Petek, N.; Grošelj, U.; Svete, J.; Požgan, F.; Kočar, D.; Štefane, B. Eosin Y-Catalyzed Visible-Light-Mediated Aerobic Transformation of Pyrazolidine-3-One Derivatives. *Catalysts* **2020**, *10*, 981. [CrossRef]
82. Shintani, R.; Fu, G.C. A New Copper-Catalyzed [3 + 2] Cycloaddition: Enantioselective Coupling of Terminal Alkynes with Azomethine Imines to Generate Five-Membered Nitrogen Heterocycles. *J. Am. Chem. Soc.* **2003**, *125*, 10778–10779. [CrossRef] [PubMed]
83. 0.125 Mmol of dipoles 6a–g correspond to 37 mg of dipole 6a, 29 mg of 6b, 27 mg of 6c, 25 mg of 6d, 30 mg of 6e, 28 mg of 6f, and 31 mg of 6g.
84. Arai, T.; Ogino, Y. Chiral Bis(Imidazolidine)Pyridine-Cu Complex-Catalyzed Enantioselective [3+2]-Cycloaddition of Azomethine Imines with Propiolates. *Molecules* **2012**, *17*, 6170–6178. [CrossRef] [PubMed]

Article

C4-Alkylamination of C4-Halo-1*H*-1-tritylpyrazoles Using Pd(dba)₂ or CuI

Yoshihide Usami *, Yuya Tatsui, Hiroki Yoneyama and Shinya Harusawa

Department of Pharmaceutical Organic Chemistry, Osaka University of Pharmaceutical Sciences, 4-20-1 Nasahara, Takatsuki, Osaka 569-1094, Japan; e18902@gap.oups.ac.jp (Y.T.); yoneyama@gly.oups.ac.jp (H.Y.); harusawa@gly.oups.ac.jp (S.H.)
* Correspondence: usami@gly.oups.ac.jp; Tel.: +81-726-90-1087

Academic Editors: Vera L. M. Silva and Artur M. S. Silva
Received: 10 September 2020; Accepted: 8 October 2020; Published: 12 October 2020

Abstract: Alkylamino coupling reactions at the C4 positions of 4-halo-1*H*-1-tritylpyrazoles were investigated using palladium or copper catalysts. The Pd(dba)₂ catalyzed C-N coupling reaction of aryl- or alkylamines, lacking a β-hydrogen atom, proceeded smoothly using ^{t}BuDavePhos as a ligand. As a substrate, 4-Bromo-1-tritylpyrazole was more effective than 4-iodo or chloro-1-tritylpyrazoles. Meanwhile, the CuI mediated C-N coupling reactions of 4-iodo-1*H*-1-tritylpyrazole were effective for alkylamines possessing a β-hydrogen atom.

Keywords: amination; 4-halopyrazole; Buchwald-Hartwig coupling; Pd(dba)₂; CuI mediated coupling; aliphatic amine

1. Introduction

Synthetic methodologies towards a range of substituted pyrazoles have been developed, as they commonly exhibit bioactivities such as antitumor, antiviral, and antifungal activities. Furthermore, the synthetic study of pyrazoles provides diverse building blocks for the discovery of new drugs, biological probes, herbicides, and other new useful materials [1–3]. Therefore, introduction of various functional groups at specific positions on a pyrazole ring is an important and attractive endeavor in synthetic organic chemistry. In particular, the synthesis of C4-aminated pyrazoles has become a prominent research topic, due to the important bioactivities exhibited by this compound class, as shown in Figure 1.

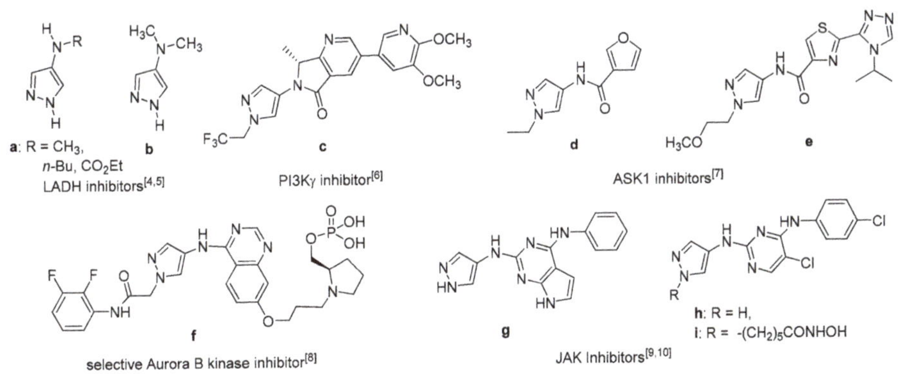

Figure 1. Examples of bioactive 4-aminopyrazoles (**a**–**i**).

Simple 4-alkylaminopyrazoles (**a** and **b**) have been reported to exhibit weak inhibitory activities against horse lever alcohol dehydrogenase (LADH) [4,5]. Azaisoindolinone derivative (**c**) exhibits potent lipid kinase phosphoinositide 3-kinase γ (PI3Kγ) inhibition, with the distinct advantage of being orally administered and central nervous system (CNS)-penetrant [6]. Two 4-heteroarylamidopyrazoles (**d** and **e**) have been presented as apoptosis signal-regulating kinase 1 (ASK 1) inhibitors [7]. 1-Acetoanilide-4-aminopyrazole-substituted quinazoles are selective Aurora B protein kinase inhibitors with potent anti-tumor activity, and structure **f** is the most potent among them. The 3-aminopyrazole analog of compound **f** is AZD1152, which was the first Aurora B selective inhibitor to enter clinical trials [8]. 7H-pyrrolo[2,3-d]pyrimidine-based 4-amino-(1H)-pyrazole derivative (**g**) and pyrimidine-based 4-amino-(1H)-pyrazole derivatives (**h** and **i**) are Janus kinase (JAK) inhibitors. Specifically, compound **i**, a dual inhibitor of JAK and histone deacetylase (HDAC), comprises a zinc-binding moiety (HONHCO) linked to the pyrazole N1, via a $(CH_2)_5$-aliphatic chain [9,10].

The well-known and widely utilized Buchwald-Hartwig coupling reaction is one of the most powerful methods for the amination of aromatic rings. Moreover, the applicability and efficiency of the reaction are continually being improved with the design and development of efficient palladium catalysts, precatalysts, and bulky ligands. Numerous combinations of catalysts and ligands exist that are suitable for specific coupling reactions [11–18].

In spite of such developments, there have been only a few reports of Buchwald-Hartwig coupling at the C4 position of pyrazoles. In 2011, the first example involving the C4 coupling of pyrazoles with aromatic amines was reported by Buchwald, as shown in Scheme 1, Equation (1) [19]. In the following year, the same group described the amidation of five-membered heterocycles with aromatic amides, wherein three examples using 1-benzyl-4-bromopyrazoles and one example using 4-bromo-1-methylpyrazole were reported (Equation (2)) [20]. In their subsequent study on the amination of unprotected five-membered bromoheterocycles, Pd-catalyzed coupling reactions of 4-bromopyrazole with eleven aromatic amines, as well as one benzylic amine, were disclosed (Equation (3)) [21]. Recently, Buchwald et al. described visible-light-mediated amination of aryl halides in the presence of nickel and photoredox catalysts, for which one example of the reaction between 1-benzyl-4-bromopyrazole and pyrrolidine was included (Equation (4)) [22].

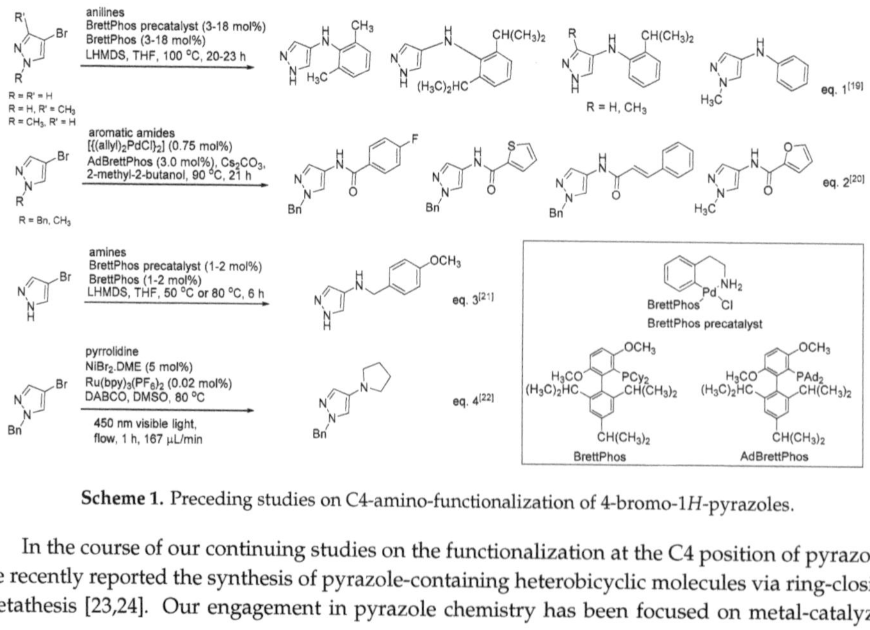

Scheme 1. Preceding studies on C4-amino-functionalization of 4-bromo-1H-pyrazoles.

In the course of our continuing studies on the functionalization at the C4 position of pyrazoles, we recently reported the synthesis of pyrazole-containing heterobicyclic molecules via ring-closing metathesis [23,24]. Our engagement in pyrazole chemistry has been focused on metal-catalyzed coupling reactions, such as Kumada-Tamao, Suzuki-Miyaura, and Sonogashira couplings, and the Heck-Mizoroki reaction [25–28]; while the Buchwald-Hartwig coupling reaction for the C4 amination of pyrazoles has remained unchallenged. Encouraged by the above-mentioned successful results, our interest has shifted to Buchwald coupling between 4-halo-1H-1-tritylpyrazoles and alkyl amines, which has not been investigated in detail, with readily accessible palladium or copper catalysts, such as bis(benzylideneacetone) palladium(0) (Pd(dba)$_2$), or copper (I) iodide (CuI). Herein, we report C4-alkylamino coupling reactions using Pd(dba)$_2$ or CuI with 4-halo-1H-1-tritylpyrazoles.

2. Results and Discussion

2.1. Pd(dba)$_2$-Catalyzed Buchwald-Hartwig Coupling for C4-Amination of 4-halo-1H-1-tritylpyrazoles

First, we investigated the Buchwald-Hartwig coupling between 4-halo-1H-1-tritylpyrazoles (**1**) and piperidine, as a representative secondary amine [16], in order to determine the optimum reaction conditions. The results are summarized in Table 1.

Table 1. Buchwald-Hartwig coupling between 4-halo-1H-1-tritylpyrazoles (**1**) and piperidine.

Entry [a]	Substrate	Pd Catalyst	Ligand [d]	Solvent	Temperature (°C)	Time	Yield 2a (%)
1	**1$_I$**: X = I	Pd(dba)$_2$	L1	xylene	160 (MW [b])	10 min	0
2	**1$_I$**	Pd(dba)$_2$	L2	xylene	160 (MW)	10 min	0
3	**1$_I$**	Pd(dba)$_2$	L3	xylene	160 (MW)	10 min	0
4	**1$_I$**	Pd(dba)$_2$	L4	xylene	160 (MW)	10 min	21
5	**1$_I$**	PdCl$_2$	L4	xylene	160 (MW)	10 min	9
6	**1$_I$**	Pd(OAc)$_2$	L4	xylene	160 (MW)	10 min	20
7	**1$_I$**	PEPPSI-IPr	L4	xylene	160 (MW)	10 min	13
8 [c]	**1$_I$**	Pd(dba)$_2$	L4	xylene	160 (MW)	10 min	52
9 [c]	**1$_I$**	Pd(dba)$_2$	L4	toluene	160 (MW)	10 min	30
10 [c]	**1$_I$**	Pd(dba)$_2$	L4	mesitylene	160 (MW)	10 min	49
11 [c]	**1$_I$**	Pd(dba)$_2$	L4	1,4-dioxane	160 (MW)	10 min	32
12 [c]	**1$_I$**	Pd(dba)$_2$	L4	THF	160 (MW)	10 min	0
13	**1$_I$**	Pd(dba)$_2$	L4	xylene	rt	24 h	7
14	**1$_I$**	Pd(dba)$_2$	L4	xylene	60	24 h	19
15	**1$_I$**	Pd(dba)$_2$	L4	xylene	90	24 h	48
16	**1$_{Br}$**: X = Br	Pd(dba)$_2$	L4	xylene	90	24 h	60
17	**1$_{Cl}$**: X = Cl	Pd(dba)$_2$	L4	xylene	90	24 h	40
18	**1$_{Br}$**	Pd(dba)$_2$	L4	xylene	70	24 h	43
19	**1$_{Br}$**	Pd(dba)$_2$	L4	xylene	140	24 h	23

[a]. general reaction conditions: substrate (50 mg, 0.13 mmol); solvent (2 mL), others are seen in the scheme in this table. [b]. MW: microwave, [c]. 40 mol% of **L4** was used. [d].

As the Buchwald–Hartwig coupling reaction for 4-halo-1H-pyrazoles requires high temperatures (>80 °C) as well as prolonged time [19–21], we utilized microwave (MW) apparatus to expedite the experimental process. Ligand screening was performed with the fixed conditions of 4-iodo-1*H*-1-tritylpyrazole (**1**$_I$, X = I), Pd(dba)$_2$, xylene, 160 °C, and 10 min under MW irradiation (entries 1–4). In the case of commonly used bidentate ligands, namely 1,1'-bis(diphenylphosphino)ferrocene (dppf, **L1**), 1,2-bis(diphenylphosphino)ethane (dppe, **L2**), and 2,2'-bis(diphenylphosphino)diphenyl ether (DPEPhos, **L3**), the reaction did not proceed (entries 1–3), while with the use of the bulky *t*BuDavePhos ligand (**L4**) the desired coupled product **2a** was obtained in 21% yield; hence **L4** was deemed a suitable ligand for this coupling reaction (entry 4). The use of **L4** with palladium(II) chloride (PdCl$_2$), palladium(II) acetate (Pd(OAc)$_2$), or pyridine-enhanced precatalyst preparation stabilization and initiation-isopropyl (PEPPSI-IPr) catalysts did not improve the yield of **2a** (entries 5–7). Although increasing the amount of **L4** to 40 mol% yielded 52% of **2a**, this created an additional problem for the purification of **2a** (entry 8). Solvent screening with the use of **L4** (40 mol%) did not improve results upon that of entry 8 (entries 9–12). Prolonged reaction time (24 h) with **L4** (20 mol%) at room temperature (rt) under MW irradiation gave **2a** in only 7% yield (entry 13). Conducting the reaction at 60 °C and 90 °C afforded **2a** in 19% and 48% yields, respectively (entries 14 and 15). Alternatively, when 4-bromo- and 4-chloropyrazoles (**1**$_{Br}$: X = Br and **1**$_{Cl}$: X = Cl) were used as substrates at 90 °C for 24 h (entries 16 and 17), the 4-bromo analogue delivered the highest yield of **2a** (60%) (entry 16). Reaction conditions using bromo compound **1**$_{Br}$ at lower or higher temperatures (70 or 140 °C in a sealed reaction vial) delivered inferior results compared to that of entry 16 (entries 18 and 19). Based on these results, further experiments were performed employing the reaction conditions listed in entry 16.

Next, optimized reaction conditions were applied to various amines, and the results are summarized in Table 2. Reactions of **1**$_{Br}$ (X=Br) with piperidine and morpholine afforded desired products **2a** and **2b** in 60% and 67% yields, respectively (entries 1 and 2), while the reactions with pyrrolidine and allylamine afforded **2c** (7%) and **2d** (6%) in low yields (entries 3 and 4). The coupling reactions of **1**$_{Br}$ with various primary amines produced the corresponding 4-alkylaminopyrazoles **2e–g**, **2k**, and **2l** in low yields (17–34%) (entries 5–8, 11, and 12). Meanwhile, in the cases of isopropylamine and benzylamine, the desired products **2i** and **2j** were not obtained (entries 9 and 10). The reactions of **1**$_{Br}$ with adamantylamine or *tert*-butylamine afforded the corresponding products **2m** and **2n** in 90% and 53% yields, respectively (entries 13 and 14). Furthermore, reactions with aromatic amines (anilines and 1-naphtylamine) gave the corresponding **2o** (94%), **2p** (91%), and **2q** (85%) in high yields (entries 15–17) as being analogous to Buchwald's findings [21]. As the reaction with diphenylamine afforded **2r** in 45% yield, we surmised that bulkiness at the reaction center depresses the chemical yield (entry 18).

Reactions of **1**$_{Br}$ with pyrrolidine, allylamine, or primary amines bearing a β-hydrogen atom resulted in low yields (entries 3–12), while amines lacking a β-hydrogen afforded good yields (entries 13–18). These contrasting results are likely due to β-elimination occurring in the palladium complex during the coupling process.

Table 2. Buchwald-Hartwig coupling of 4-bromo-1H-1-tritylpyrazole (**1**$_{Br}$) with various amines.

Pd(dba)$_2$ (10 mol%)
^{t}BuDavePhos (**L4**) (20 mol%)
amine (2.0 eq.), ^{t}BuOK (2.0 eq.)
xylene, 90 °C, 24 h

Entry	Amine	Product	Yield (%)
1	piperidine	**2a**: R = R' = -CH$_2$CH$_2$CH$_2$CH$_2$CH$_2$-	60
2	morpholine	**2b**: R = R' = -CH$_2$CH$_2$OCH$_2$CH$_2$-	67
3	pyrrolidine	**2c**: R = R' = -CH$_2$CH$_2$CH$_2$CH$_2$-	7
4	allylamine	**2d**: R = CH$_2$CH=CH$_2$, R' = H	6
5	n-propylamine	**2e**: R = CH$_2$CH$_2$CH$_3$, R' = H	24
6	n-butylamine	**2f**: R = CH$_2$CH$_2$CH$_2$CH$_3$, R' = H	17
7	isobutylamine	**2g**: R = CH$_2$CH(CH$_3$)$_2$, R' = H	28
8	isoamylamine	**2h**: R = CH$_2$CH$_2$CH(CH$_3$)$_2$, R' = H	20
9	isopropylamine	**2i**: R = CH(CH$_3$)$_2$, R' = H	0
10	PhCH$_2$NH$_2$	**2j**: R = CH$_2$Ph, R' = H	0
11	PhCH$_2$CH$_2$NH$_2$	**2k**: R = CH$_2$CH$_2$Ph, R' = H	30
12	PhCH$_2$CH$_2$CH$_2$NH$_2$	**2l**: R = CH$_2$CH$_2$CH$_2$Ph, R' = H	34
13 [a]	adamantylamine	**2m**: R = adamantyl, R' = H	90
14 [a]	tert-butylamine	**2n**: R = CH(CH$_3$)$_3$, R' = H	53
15 [a]	aniline	**2o**: R = Ph, R' = H	94
16 [a]	2-methoxyaniline	**2p**: R = 2-MeOPh, R' = H	91
17 [a]	1-naphthylamine	**2q**: R = naphth-1-yl, R' = H	85
18 [a]	N,N-diphenylamine	**2r**: R = R' = Ph	45

[a]. Entries 13–18 were performed with 1.1 equivalents of amine.

2.2. CuI-Catalyzed Coupling for C4-Amination of 4-Halo-1H-1-tritylpyrazoles

Copper-catalyzed C-N coupling reactions have been extensively studied [29], and Buchwald has reportedly implemented this type of reaction using bromo- or iodobenzenes as substrates progressively, but not with five-membered heterocyclic compounds such as pyrazoles [30–36]. As the C-N coupling reaction of 4-halopyrazoles **1** with allyl- or alkylamines bearing β-hydrogen atoms revealed low reactivities in the above investigation (Table 2, entries 4–12), the copper-catalyzed reaction of **1** was further studied.

For this purpose, the reaction of allylamine with 4-iodopyrazole **1**$_I$ (X = I), which could be got easier than 4-bromopyrazole, was investigated, as presented in Table 3. First, the reaction was performed using the conditions similar to those used in Buchwald's procedure [32]: CuI (5 mol%), 2-isobutyrylcyclohexanone (**L5**: 20 mol%) as the ligand, N,N-dimethylformamide (DMF), 100 °C, 24 h, and t-BuOK (2 eq). Although the desired 4-allylaminopyrazole **2d** was obtained in only 17% yield (entry 1), increasing the amount of CuI from 5 to 20 mol% improved the chemical yield of **2d** to 72% (entry 2). The use of 2-acetylcyclohexanone (**L6**) as an alternative ligand, which is nearly 10-fold cheaper than **L5**, afforded a good yield (68%, entry 3), while the use of 3,4,7,8-tetramethyl-1,10-phenanthroline (**L7**) resulted in a poor yield (12%, entry 4). Hence, **L6** was applied in the following experiments (entries 5–15 in Table 3). The reaction temperature was varied in entries 5–7, however 100 °C proved optimal (entry 3). Furthermore, various copper catalysts were investigated in entries 9–13, and it was found that the use of the high-cost (CuOTf)$_2$·C$_6$H$_6$ catalyst (entry 13) furnished a comparable yield (70%) to that of CuI (72%) (entry 2). In addition, while the use of 4-bromopyrazole **1** (X = Br) provided **2i** in 66% yield (entry 14), chloropyrazole **1**$_{Cl}$ (X = Cl) did not react (entry 15).

Table 3. CuI-catalyzed allylamination of 4-halo-1H-1-tritylpyrazoles **1**.

Cu catalyst (20 mol%), ligand (20 mol%), allylamine (2.0 eq.), ^{t}BuOK (2.0 eq.), DMF, 24 h

Entry [a]	Substrate	Cu Catalyst	Ligand [c]	Temperature (°C)	Yield 2d (%)
1 [b]	1_I: X = I	CuI	L5	100	17
2	1_I	CuI	L5	100	72
3	1_I	CuI	L6	100	68
4	1_I	CuI	L7	100	12
5	1_I	CuI	L6	rt	0
6	1_I	CuI	L6	70	41
7	1_I	CuI	L6	130	9
8	1_I	CuI	L6	100	52
9	1_I	CuI$_2$	L6	100	57
10	1_I	Cu(OAc)$_2$	L6	100	58
11	1_I	Cu$_2$O	L6	100	16
12	1_I	CuCT	L6	100	50
13	1_I	[CuOTf]$_2$·C$_6$H$_6$	L6	100	70
14	1_{Br}: X = Br	CuI	L6	100	66
15	1_{Cl}: X = Cl	CuI	L6	100	0

[a]. general reaction conditions: substrate (50 mg, 0.12 mmol); solvent (2 mL), others are seen in the scheme in this table. [b]. CuI (5 mol%), L5 (20 mol%), Cs$_2$CO$_3$ (2.0 Equation). [c].

L5, L6, L7, CuCT

Therefore, to evaluate the scope of this transformation, additional coupling reactions between iodopyrazole 1_I and various amines were performed, by applying the optimized reaction conditions (entry 3 of Table 3), as shown in Table 4. It should be noted that there were a number of distinct contrasts between the outcomes of the CuI-catalyzed (Table 4) and those of the Pd-catalyzed coupling reactions (Table 2). In the case of CuI coupling, reactions of **1i** with piperidine and morpholine afforded **2a** and **2b** (21% and 22%, respectively) in lower yields (Table 4, entries 1 and 2) than those obtained (60% and 67%, respectively) in the corresponding Pd-catalyzed reaction of 1_{Br} (entries 1 and 2 in Table 2). The CuI catalyst provided the pyrrolidine derivative **2c** in 43% yield (Table 4, entry 3), while the Pd catalyst yielded **2c** in only 7% yield (Table 2, entry 3). CuI-catalyzed reactions with primary alkylamines gave moderate to good yields of products **2d**–**2l** (entries 4–12), while reactions with adamantyl, tert-butyl, and aromatic amines did not afford the desired products (entries 13–17), and only aniline furnished a low yield of **2o** (15%) (entry 15); these trends were reversed in the case of Pd-catalyzed processes. These negative results may be ascribed to the increase in bulkiness as well as a decrease in the basicity of the amine sources.

Table 4. CuI-catalyzed coupling of **1**$_I$ with various amines.

CuI (20 mol%), **L6** (20 mol%), amine (2.0 eq.), ^{t}BuOK (2.0 eq.), DMF, 100 °C, 24 h

Entry	Amine	Product	Yield (%)
1	piperidine	2a	21
2	morpholine	2b	22
3	pyrrolidine	2c	43
4	allylamine	2d	68
5	n-propylamine	2e	75
6	n-butylamine	2f	62
7	isobutylamine	2g	70
8	isoamylamine	2h	62
9	isopropylamine	2i	57
10	PhCH$_2$NH$_2$	2j	55
11	Ph CH$_2$CH$_2$NH$_2$	2k	53
12	Ph CH$_2$ CH$_2$CH$_2$NH$_2$	2l	69
13	adamantylamine	2m	0
14	*tert*-butylamine	2n	0
15	aniline	2o	15
16	1-naphthylamine	2q	0
17	N,N-diphenylamine	2r	0

3. Conclusions

We have studied the C4 amination of pyrazole derivatives using readily accessible Pd(dba)$_2$ or CuI catalysts. The Pd(dba)$_2$-catalyzed reaction of 4-bromo-1H-1-tritylpyrazole proved to be suitable for aromatic or bulky amines lacking β-hydrogen atoms, but not for cyclic amines (piperidine and morpholine); additionally it was not suitable for alkylamines possessing β-hydrogen atoms. On the other hand, the CuI-catalyzed amination using 4-iodo-1H-1-tritylpyrazole was revealed to be favorable for alkylamines possessing β-hydrogen atoms, and not suitable for aromatic amines and bulky amines lacking β-hydrogens, indicating the complementarity of the two catalysts. Although further improvements are required for practical synthesis, such as the reduction of catalyst or ligand loading, the findings of the present study offer a useful synthetic method for the construction of 4-functionalized pyrazoles. Further application of the methodology developed in this study to the C-O coupling reaction of halopyrazoles with alkylated alcohols will be evaluated and reported in the near future.

4. Materials and Methods

General: Nuclear magnetic resonance (NMR) spectra were recorded at 27 °C on an Agilent 400-MR-DD2 spectrometer (Agilent Tech., Inc., Santa Clara, CA, USA) in CDCl$_3$ with tetramethylsilane (TMS) as an internal standard. Abbreviations for splitting patterns in ^{1}H-NMR spectra are noted as d = doublet; t = triplet; q = quartet; quin = quintet; sept = septet. Electron impact-high-resolution mass spectra (EI-HRMS) were measured with a JEOL JMS-700 (2) mass spectrometer (JEOL, Tokyo, Japan). Melting points were determined on a Yanagimoto micromelting point apparatus and were uncorrected. Liquid column chromatography was conducted with silica gel (FL-60D, Fuji Silysia Chemical Ltd., Kasugai, Aichi, Japan). Analytical thin layer chromatography (TLC) was performed on silica gel 70 F$_{254}$ plates (Wako Pure Chemical Industries, Tokyo, Japan), and compounds were detected by dipping the plates into an EtOH solution of phosphomolybdic acid followed by heating. MW-aided reactions were carried out in a Biotage Initiator® reactor (PartnerTech Atvidaberg AB for Biotage Sweden AB, Uppsala, Sweden). Pd(dba)$_2$, mesitylene, dppf (**L2**), copper (I) thiophene-2-carboxylate (CuCT), piperidine, pyrrolidine, allylamine, *n*-propylamine, isobutylamine, isoamylamine, isopropylamine, benzylamine, 2-phenylethylamine, 3-phenylpropylamine, adamantylamine, *tert*-butylamine, aniline, 2-methoxyaniline, 1-naphthylamine, and *N,N*-diphenylamine were purchased from Tokyo Chemical Industry (TCI) Co. (Tokyo, Japan). ^{t}BuOK, CuI, and 3,4,7,8-tetramethyl-1,10-phenanthroline (**L7**) were purchased from Nacalai Tesque, Inc. (Kyoto, Japan). Dry xylene, THF, 1,4-dioxane, and DMF were purchased from FUJIFILM Wako Pure Chemical Co. (Osaka, Japan). PEPSI-IPr, dppf (**L1**), DEPPhos (**L3**), ^{t}BuDavePhos (**L4**), morpholine, 2-isobutyrylcyclohexanone (**L5**), and 2-acetylcyclohexanone (**L6**) were purchased from Sigma-Aldrich Co. LLC (St. Louis, MI, USA).

Palladium-catalyzed coupling reaction with 1 and amines (Tables 1 and 2)

Typical procedure (Table 1, entry 16): To a solution of **1$_{Br}$** (50.0 mg, 1.28 × 10^{-1} mmol) in xylene (2 mL) in a MW vial were added ^{t}BuDavePhos (8.8 mg, 2.56 × 10^{-2} mmol, 20 mol%), Pd(dba)$_2$ (7.4 mg, 1.28 × 10^{-2} mmol, 10 mol%), potassium *t*-butoxide (^{t}BuOK) (28.8 mg, 2.57 × 10^{-1} mmol, 2.0 Equation) and piperidine (0.03 mL, 2.57 × 10^{-1} mmol, 2.0 Equation). The reaction vial was sealed and heated at 90 °C with stirring in an oil bath for 24 h. The reaction mixture was quenched by the addition of sat. aq. NH$_4$Cl (1 mL) and extracted with CH$_2$Cl$_2$ (1 mL × 3). The combined organic layers were dried over MgSO$_4$, filtered, and evaporated to give a crude residue, which was purified by silica gel column chromatography (eluent: Hexane/AcOEt = 4:1) to afford 1-(1-trityl-1*H*-pyrazol-4-yl)piperidine (**2a**) (30.9 mg, 60%) as a white powder.

CuI-catalyzed coupling reaction with 1 and amines (Tables 3 and 4)

Typical procedure (Table 3, entry 3); To a solution of **1$_I$** (50.0 mg, 1.15 × 10^{-1} mmol) in DMF (2 mL) in a MW vial, were added 2-acetylcyclohexanone (3.0 μL, 2.30 × 10^{-2} mmol, 20 mol%), CuI (4.4 mg, 2.30 × 10^{-2} mmol, 20 mol%), ^{t}BuOK (25.7 mg, 2.30 × 10^{-1} mmol, 2.0 Equation) and allylamine (0.03 mL, 2.30 × 10^{-1} mmol, 2.0 Equation). The reaction vial was sealed and heated at 100 °C with stirring in an oil bath for 24 h. The reaction mixture was quenched by the addition of sat. aq. NH$_4$Cl (1 mL) and extracted with CH$_2$Cl$_2$ (1 mL × 3). The combined organic layers were dried over MgSO$_4$, filtered, and evaporated to give a crude residue, which was purified by silica gel column chromatography (eluent: Hexane/AcOEt = 4:1) to afford 2d (28.6 mg, 68%).

1-(1-Trityl-1H-pyrazol-4-yl)piperidine (**2a**): white powder; mp 170–174 °C; ^{1}H-NMR (400 MHz, CDCl$_3$): δ 1.49 (2H, quin, *J* = 5.7 Hz, -CH$_2$C*H*$_2$CH$_2$-), 1.64 (4H, quin, *J* = 5.7 Hz, -C*H*$_2$CH$_2$C*H*$_2$-), 2.83 (4H, t, *J* = 5.7 Hz, -NC*H*$_2$CH$_2$) 6.88 (1H, d, *J* = 0.8 Hz, pyrazole-H), 7.13–7.18 (6H, m, Ph-H), 7.28–7.31 (9H, m, Ph-H), 7.39 (1H, d, *J* = 0.8 Hz, pyrazole-H); ^{13}C-NMR (100 MHz, CDCl$_3$): δ 23.9, 25.5, 52.4, 78.4, 118.9, 127.5, 127.6, 129.5, 130.1, 137.7, 143.4; EI-HRMS *m/z* calcd. for C$_{27}$H$_{27}$N$_3$ (M$^+$) 393.2205, found 393.2210.

4-(1-Trityl-1H-pyrazol-4-yl)morpholine (**2b**): white powder; mp 209–211 °C; ^{1}H-NMR (400 MHz, CDCl$_3$): δ 2.87 (4H, t, *J* = 4.7 Hz, -NC*H*$_2$CH$_2$), 3.78 (4H, t, *J* = 4.7 Hz, -OC*H*$_2$CH$_2$-), 6.90 (1H, s, pyrazole-H), 7.14–7.17 (6H, m, Ph-H), 7.27–7.30 (9H, m, Ph-H), 7.39 (1H, s, pyrazole-H); ^{13}C-NMR (100 MHz, CDCl$_3$): δ 51.4, 66.5, 78.5, 118.8, 127.7, 129.0, 130.1, 136.9, 143.3 (two signals are overlapping to give one signal); EI-HRMS *m/z* calcd. for C$_{25}$H$_{25}$N$_3$O (M$^+$) 395.1996, found 395.1997.

4-(Pyrrolidin-1-yl)-1-trityl-1H-pyrazole (**2c**): white powder; mp 189–190 °C; ^{1}H-NMR (400 MHz, CDCl$_3$): δ 1.90 (4H, br t, *J* = 6.5 Hz, -NCH$_2$C*H*$_2$-), 2.99 (4H, t, *J* = 6.5 Hz, -NC*H*$_2$CH$_2$), 6.74 (1H, d, *J* = 0.8 Hz, pyrazole-H), 7.16–7.18 (6H, m, Ph-H), 7.25–7.30 (10H, m, Ph-H and pyrazole-H); ^{13}C-NMR (100 MHz, CDCl$_3$): δ 24.7, 51.0, 78.3, 116.9, 127.5, 127.6, 128.2, 130.1, 135.1, 143.5; EI-HRMS *m/z* calcd. for C$_{26}$H$_{25}$N$_3$ (M$^+$) 379.2049, found 379.2048.

N-Allyl-1-trityl-1H-pyrazol-4-amine (**2d**): oil; ^{1}H-NMR (400 MHz, CDCl$_3$): δ 3.53 (2H, dt, *J* = 5.7, 1.6 Hz, -NHC*H*$_2$CH=CH$_2$), 5.09–5.12 (1H, dq, *J* = 10.1, 1.4 Hz, -NHCH$_2$CH=CH*H*), 5.16–5.21 (1H, dq, *J* = 17.1, 1.6 Hz, -NHCH$_2$CH=C*H*H), 5.86–5.96 (1H, ddt, *J* = 17.1, 10.1, 5.7 Hz, -NHCH$_2$C*H*=CH$_2$), 6.88 (1H, d, *J* = 0.8 Hz, pyrazole-H), 7.14–7.18 (6H, m, Ph-H), 7.25–7.30 (9H, m, Ph-H), 7.32 (1H, d, *J* = 0.8, pyrazole-H); ^{13}C-NMR (100 MHz, CDCl$_3$): δ 50.5, 78.3, 116.3, 119.2, 127.6, 129.9, 130.1, 132.3, 135.8, 143.4; EI-HRMS *m/z* calcd. for C$_{25}$H$_{23}$N$_3$ (M$^+$) 365.1892, found 365.1892.

N-Propyl-1-trityl-1H-pyrazol-4-amine (**2e**): white powder; mp 145–148 °C; ^{1}H-NMR (400 MHz, CDCl$_3$): δ 0.94 (3H, t, *J* = 7.4 Hz, -NHCH$_2$CH$_2$CH$_3$), 1.56 (2H, sext, *J* = 7.4 Hz, -NHCH$_2$CH$_2$CH$_3$), 2.86 (2H, t, *J* = 7.0 Hz, -NHC*H$_2$*CH$_2$CH$_3$), 6.86 (1H, d, *J* = 0.8 Hz, pyrazole-H), 7.14–7.20 (6H, m, Ph-H), 7.26–7.35 (10H, m, Ph-H and pyrazole-H); ^{13}C-NMR (100 MHz, CDCl$_3$): δ 11.6, 23.0, 49.7, 78.2, 118.7, 127.5, 127.6, 129.7, 130.1, 132.9, 143.5; EI-HRMS *m/z* calcd. for C$_{25}$H$_{25}$N$_3$ (M$^+$) 367.2048, found 367.2049.

N-Butyl-1-trityl-1H-pyrazol-4-amine (**2f**): white amorhous; mp 112–116 °C; ^{1}H-NMR (400 MHz, CDCl$_3$): δ 0.91 (3H, t, *J* = 7.4 Hz, -CH$_2$CH$_3$-), 1.36 (2H, br sext, *J* = 7.4 Hz, -CH$_2$CH$_2$CH$_3$), 1.52 (2H, br quint, *J* = 7.4 Hz, -CH$_2$CH$_2$CH$_2$-),2.89 (2H, t, *J* = 7.0 Hz, -NHC*H$_2$*CH$_2$-), 6.86 (1H, s, pyrazole-H), 7.14–7.19 (6H, m, Ph-H), 7.27–7.33 (10H, m, Ph-H and pyrazole-H); ^{13}C-NMR (100 MHz, CDCl$_3$): δ 14.0, 20.2, 32.0, 47.6, 78.2, 118.7, 127.5, 127.6, 129.7, 130.1, 132.9, 143.4 EI-HRMS *m/z* calcd. for C$_{26}$H$_{27}$N$_3$ (M$^+$) 381.2205, found 381.2215.

N-Isobutyl-1-trityl-1H-pyrazol-4-amine (**2g**): white powder; mp 135–136 °C; ^{1}H-NMR (400 MHz, CDCl$_3$): δ 0.93 (6H, d, *J* = 6.6 Hz, -NHCH$_2$CH(CH$_3$)$_2$), 1.78 (1H, nonet, *J* = 6.6 Hz, -NHCH$_2$C*H*(CH$_3$)$_2$), 2.70 (2H, d, *J* = 6.6 Hz, -NHC*H$_2$*CH(CH$_3$)$_2$), 6.85 (1H, d, pyrazole-H), 7.11–7.19 (6H, m, Ph-H), 7.25–7.32 (10H, m, Ph-H and pyrazole-H); ^{13}C-NMR (100 MHz, CDCl$_3$): δ 20.5, 28.4, 55.7, 78.2, 118.4, 127.5, 127.6, 129.6, 130.1, 133.1, 143.5; EI-HRMS *m/z* calcd. for C$_{26}$H$_{27}$N$_3$ (M$^+$) 381.2205, found 381.2210.

N-Isoamyl-1-trityl-1H-pyrazol-4-amine (**2h**): white amorphous; mp 110–113 °C; ^{1}H-NMR (400 MHz, CDCl$_3$): δ 0.89 (6H, d, *J* = 6.7 Hz, -CH(CH$_3$)$_2$), 1.48 (2H, q, *J* = 7.4 Hz, -CH$_2$CH$_2$CH-), 1.64 (1H, nonet, *J* = 6.6 Hz, -CH$_2$C*H*(CH$_3$)$_2$), 2.89 (2H, br t, *J* = 7.3 Hz, -NHC*H$_2$*CH$_2$-), 6.86 (1H, s, pyrazole-H), 7.15–7.18 (6H, m, Ph-H), 7.26–7.32 (10H, m, Ph-H and pyrazole-H); ^{13}C-NMR (100 MHz, CDCl$_3$): δ 22.6, 25.9, 38.9, 46.0, 78.3, 118.7, 127.5, 127.6, 129.7, 130.1, 132.9, 143.5; EI-HRMS *m/z* calcd. for C$_{27}$H$_{29}$N$_3$ (M$^+$) 395.2361, found 395.2359.

N-Isopropyl-1-trityl-1H-pyrazol-4-amine (**2i**): white powder; mp 130–133 °C; ^{1}H-NMR (400 MHz, CDCl$_3$): δ 1.11 (6H, d, *J* = 6.3 Hz, -NHCH(CH$_3$)$_2$), 3.18 (1H, sept, *J* = 6.3 Hz, -NHC*H*(CH$_3$)$_2$), 6.88 (1H, s, pyrazole-H), 7.15–7.19 (6H, m, Ph-H), 7.26–7.35 (10H, m, Ph-H and pyrazole-H); ^{13}C-NMR (100 MHz, CDCl$_3$): δ 23.0, 48.4, 78.2, 120.5, 127.5, 127.6, 130.1, 131.1, 143.4 (two carbon signals overlapped); EI-HRMS *m/z* calcd. for C$_{25}$H$_{25}$N$_3$ (M$^+$) 367.2048, found 367.2046

N-Benzyl-1-trityl-1H-pyrazol-4-amine (**2j**): white powder; mp 148–151 °C; ^{1}H-NMR (400 MHz, CDCl$_3$): δ 4.06 (2H, s, -CH$_2$Ph), 6.84 (1H, s, pyrazole-H), 7.13–7.16 (6H, m, Ph-H), 7.24–7.30 (14H, m, Ph-H), 7.32 (1H, s, pyrazole-H); ^{13}C-NMR (100 MHz, CDCl$_3$): δ52.2, 78.3, 119.2, 127.2, 127.5, 127.6, 127.9, 128.5, 129.9, 130.1, 132.4, 139.4, 143.4; EI-HRMS *m/z* calcd. for C$_{29}$H$_{25}$N$_3$ (M$^+$) 415.2048, found 415.2046.

N-Phenethyl-1-trityl-1H-pyrazol-4-amine (**2k**): white powder; mp 134–137 °C; ^{1}H-NMR (400 MHz, CDCl$_3$): δ 2.84 (2H, t, *J* = 6.9 Hz, -NHCH$_2$CH$_2$Ph), 3.16 (2H, t, *J* = 6.9 Hz, -NHC*H$_2$*CH$_2$Ph), 6.85 (1H, d, *J* = 0.9 Hz, pyrazole-H), 7.14–7.32 (21H, m, Ph-H and pyrazole-H); ^{13}C-NMR (100 MHz, CDCl$_3$) δ 35.8, 48.9, 78.3, 119.0, 126.4, 127.5, 127.6, 128.6, 128.8, 129.8, 130.1, 132.3, 139.3, 143.4; EI-HRMS *m/z* calcd. for C$_{30}$H$_{27}$N$_3$ (M$^+$) 429.2205, found 429.2200.

N-(3-Phenyl)propyl-1-trityl-1H-pyrazol-4-amine (**2l**): white powder; mp 114–117 °C; ^{1}H-NMR (400 MHz, CDCl$_3$): δ 1.86 (2H, br quint, *J* = 7.3 Hz, -CH$_2$CH$_2$ CH$_2$-), 2.67 (2H, br t, *J* = 7.5 Hz, -CH$_2$CH$_2$Ph), 2.94 (2H, t, *J* = 7.1 Hz, -NHC*H$_2$*CH$_2$-), 6.84 (1H, s, pyrazole-H), 7.14–19 (8H, m, Ph-H and pyrazole-H), 7.24–7.30 (13H, m, Ph-H, and pyrazole-H); ^{13}C-NMR (100 MHz, CDCl$_3$) δ 31.4, 33.3, 47.4, 78.3, 118.8, 125.9, 127.5, 127.6, 128.3, 128.4, 129.8, 130.1, 132.6, 141.8, 143.4; EI-HRMS *m/z* calcd. for C$_{31}$H$_{29}$N$_3$ (M$^+$) 443.2362, found 443.2365.

N-((3s,5s,7s)-Adamantan-1-yl)-1-trityl-1H-pyrazol-4-amine (**2m**): white powder; mp 204–205 °C; ^{1}H-NMR (400 MHz, CDCl$_3$): δ 1.59 (12H, m, Ad-H), 2.05 (4H, br n, Ad-H, and -NHAd), 7.00 (1H, s, pyrazole),

7.14–7.18 (6H, m, Ph-H), 7.28–7.30 (9H, m, Ph-H), 7.34 (1H, s, pyrazole-H); ^{13}C-NMR (100 MHz, CDCl$_3$): δ 29.6, 36.4, 43.2, 51.7, 78.3, 125.4, 127.2, 127.5, 127.6, 130.1, 136.8, 143.3; EI-HRMS m/z calcd. for C$_{32}$H$_{33}$N$_3$ (M$^+$) 459.2674, found 459.2673.

N-(tert-Butyl)-1-trityl-1H-pyrazol-4-amine (**2n**): white powder; mp 137–140 °C; ^{1}H-NMR (400 MHz, CDCl$_3$): δ 1.11 (9H, s, -C(CH$_3$)$_3$), 7.01 (1H, s, pyrazole-H), 7.15–7.18 (6H, m, Ph-H), 7.28–7.30 (9H, m, Ph-H), 7.36 (1H, s, pyrazole-H); ^{13}C-NMR (100 MHz, CDCl$_3$): δ 29.5, 51.9, 78.3, 126.7, 127.0, 127.5, 127.6, 130.1, 136.2, 143.3; EI-HRMS m/z calcd. for C$_{26}$H$_{27}$N$_3$ (M$^+$) 381.2205, found 381.2206.

N-Phenyl-1-trityl-1H-pyrazol-4-amine (**2o**): white powder; mp 191–192 °C; ^{1}H-NMR (400 MHz, CDCl$_3$): δ 5.05 (1H, br, -NHPh), 6.70–6.76 (3H, m, Ph-H and pyrazole-H), 7.14–7.20 (7H, m, Ph-H), 7.24–7.32 (11H, m,Ph-H), 7.61 (1H, s, pyrazole-H); ^{13}C-NMR (100 MHz, CDCl$_3$): δ 78.8, 113.4, 118.5, 123.5, 127.2, 127.8, 129.3, 130.0, 130.1, 136.0, 143.1, 146.6; EI-HRMS m/z calcd. for C$_{28}$H$_{23}$N$_3$ (M$^+$) 401.1892, found 401.1890.

N-(o-Methoxy)phenyl-1-trityl-1H-pyrazol-4-amine (**2p**): white powder; mp 133–136 °C; ^{1}H-NMR (400 MHz, CDCl$_3$): δ 3.87 (3H, s, -OCH$_3$), 5.70 (1H, br, -NHAr), 6.70–6.76 (1H, m, Ph-H), 6.82–6.84 (2H, m, Ph-H), 7.22–7.25 (8H, m, Ph-H), 7.32–7.68 (9H, m, Ph-H, pyrazole-H), 7.68 (1H, s, pyrazole-H); ^{13}C-NMR (100 MHz, CDCl$_3$): δ 55.4, 78.6, 109.8, 110.9, 117.6, 121.1, 123.3, 126.6, 127.6, 130.1, 135.7, 136.2, 143.1, 146.5; EI-HRMS m/z calcd. for C$_{29}$H$_{25}$N$_3$ (M$^+$) 431.1998, found 431.1998.

N-(Naphthalen-1-yl)-1-trityl-1H-pyrazol-4-amine (**2q**): white powder; mp 175–178 °C; ^{1}H-NMR (400 MHz, CDCl$_3$): δ 5.66 (1H, s, -NH-naphthyl), 6.82–6.84 (1H, m, naphtyl-H), 7.21–7.26 (8H, m, Ph-H and naphthyl-H), 7.28–7.36 (9H, m, Ph-H), 7.39 (1H, s, pyrazole-H), 7.43–7.48 (2H, m, naphthyl-H), 7.69 (1H, s, pyrazole-H), 7.79–7.87 (2H, m, naphthyl-H); ^{13}C-NMR (100 MHz, CDCl$_3$): δ 78.8, 106.9, 118.9, 119.8, 123.4, 123.6, 125.1, 125.9, 126.3, 127.6, 127.8, 128.7, 130.1, 130.4, 134.4, 136.4, 142.2, 143.1; EI-HRMS m/z calcd. for C$_{32}$H$_{25}$N$_3$ (M$^+$) 451.2049, found 451.2052.

N,N-Diphenyl-1-trityl-1H-pyrazol-4-amine (**2r**): white powder; mp 175–177 °C; ^{1}H-NMR (400 MHz, CDCl$_3$): δ6.92 (2H, t, J = 7.3 Hz, Ph-H), 7.04–7.06 (4H, m, Ph-H and pyrazole-H), 7.16–7.22 (10H, m, Ph-H), 7.29–7.33 (10H, m, Ph-H and pyrazole-H), 7.52 (1H, s, pyrazole-H); ^{13}C-NMR (100 MHz, CDCl$_3$): δ 78.9, 121.5, 121.9, 127.7, 127.9, 127.74, 127.78, 127.8, 128.6, 129.1, 129.2, 130.1, 130.2, 137.1, 143.0, 147.7; EI-HRMS m/z calcd. for C$_{34}$H$_{27}$N$_3$ (M$^+$) 477.2205, found 477.2197.

Author Contributions: Y.U. and Y.T. conceived, designed, and performed the synthetic experiments; Y.U., H.Y. and S.H. wrote the manuscript. All authors have read and agreed to the published version of the manuscript.

Funding: This research received no external funding.

Acknowledgments: M. Fujitake of our University is appreciated for performing MS measurements.

Conflicts of Interest: The authors declare no conflict of interest.

References

1. Li, J.J. Pyrazoles, pyrazolones, and indazoles. In *Heterocyclic Chemistry in Drug Discovery*; Li, J.J., Ed.; Wiley-VCH: Weinheim, Germany, 2013; pp. 198–229.
2. Brown, A.W. Recent developments in the chemistry of pyrazoles. *Adv. Heterocycl. Chem.* **2018**, *126*, 55–107.
3. Karrouchi, K.; Radi, S.; Ramli, Y.; Taoufik, J.; Mabkhot, Y.N.; Al-aizari, F.A.; M'hammed, A. Synthesis and pharmacological activities of pyrazole derivatives: A review. *Molecules* **2018**, *23*, 134. [CrossRef] [PubMed]
4. Tolf, B.-R.; Dahlbom, R.; Aakeson, A.; Theorell, H. Synthetic inhibitors of alcohol dehydrogenase. Pyrazoles containing polar groups directly attached to the pyrazole ring in the 4-position. *Acta Pharm. Suec.* **1985**, *22*, 147–156.
5. Tolf, B.-R.; Dahlbom, R.; Akeson, A.; Theorell, H. 4-Substituted pyrazoles as inhibitors of liver alcohol dehydrogenase. Structure-activity relationships. In *Biologically Active Principles of Natural Products*; Voelter, W., Daves, D.G., Eds.; Thieme Medical Publishers: Stuttgart, Germany, 1984; pp. 265–277.

6. Come, J.H.; Collier, P.N.; Henderson, J.A.; Pierce, A.C.; Davies, R.J.; Le, T.A.; O'Dowd, H.; Bandarage, U.K.; Cao, J.; Deininger, D.; et al. Design and Synthesis of a Novel Series of Orally Bioavailable, CNS-Penetrant, Isoform Selective Phosphoinositide 3-Kinase γ (PI3Kγ) Inhibitors with Potential for the Treatment of Multiple Sclerosis (MS). *J. Med. Chem.* **2018**, *61*, 5245–5256. [CrossRef] [PubMed]
7. Gibson, T.S.; Johnson, B.; Fanjul, A.; Halkowycz, P.; Dougan, D.R.; Cole, D.; Swann, S. Structure-based drug design of novel ASK1 inhibitors using an integrated lead optimization strategy. *Bioorg. Med. Chem. Lett.* **2017**, *27*, 1709–1713. [CrossRef]
8. Foote, K.M.; Mortlock, A.A.; Heron, N.M.; Jung, F.H.; Hill, G.B.; Pasquet, G.; Brady, M.C.; Green, S.; Heaton, S.P.; Kearney, S.; et al. Synthesis and SAR of 1-acetanilide-4-aminopyrazole-substituted quinazolines: Selective inhibitors of Aurora B kinase with potent antitumor activity. *Bioorg. Med. Chem. Lett.* **2008**, *18*, 1904–1909. [CrossRef]
9. Liang, X.; Zang, J.; Zhu, M.; Gao, Q.; Wang, B.; Xu, W.; Zhang, Y. Design, Synthesis, and Antitumor Evaluation of 4-Amino-(1H)-pyrazole Derivatives as JAKs Inhibitors. *ACS Med. Chem. Lett.* **2016**, *7*, 950–955. [CrossRef]
10. Liang, X.; Zang, J.; Li, X.; Tang, S.; Huang, M.; Geng, M.; Chou, C.J.; Li, C.; Cao, Y.; Xu, W.; et al. Discovery of novel Janus kinase (JAK) and histone deacetylase (HDAC) dual inhibitors for the treatment of hematological malignancies. *J. Med. Chem.* **2019**, *62*, 3898–3923. [CrossRef]
11. Jiang, L.; Buchwald, S.L. *Palladium-Catalyzed Aromatic Carbon-Nitrogen Bond Formation in Metal-Catalyzed Cross-Coupling Reactions*, 2nd ed.; De Meijere, A., Diederich, F., Eds.; Wiley-VCH Verlag GmbH & Co.: Weinheim, Germany, 2004; Volume 2, pp. 699–760.
12. Dorel, R.; Grugel, C.P.; Haydl, A.M. The Buchwald–Hartwig Amination after 25 Years. *Angew. Chem. Int. Ed.* **2019**, *58*, 17118–17129. [CrossRef]
13. Muci, A.R.; Buchwald, S.L. Practical palladium catalysts for C-N and C-O bond formation. *Top. Curr. Chem.* **2002**, *219*, 131–209.
14. Altman, R.A.; Fors, B.P.; Buchwald, S.L. Pd-catalyzed amination reactions of aryl halides using bulky biarylmonophosphine ligands. *Nat. Protoc.* **2007**, *2*, 2881–2887. [CrossRef] [PubMed]
15. Hartwig, J.F. Evolution of a fourth-generation catalyst for the amination and thioetherification of aryl halides. *Acc. Chem. Res.* **2008**, *41*, 1534–1544. [CrossRef] [PubMed]
16. Hartwig, J.F. "Palladium-catalyzed amination of aryl halides and related reactions". In *Handbook of Organopalladium Chemistry for Organic Synthesis*; Negishi, E., Ed.; John Wiley & Sons, Inc.: Hoboken, NJ, USA, 2002; Volume 1, pp. 1051–1096.
17. Louie, J.; Hartwig, J.F. Palladium-catalyzed synthesis of arylamines from aryl halides. Mechanistic studies lead to coupling in the absence of tin reagents. *Tetrahedron Lett.* **1995**, *36*, 3609–3612. [CrossRef]
18. Ingoglia, B.T.; Wagen, C.C.; Buchwald, S.L. Biaryl monophosphine ligands in palladium-catalyzed C–N coupling: An updated User's guide. *Tetrahedron* **2019**, *75*, 4199–4211. [CrossRef]
19. Maiti, D.; Fors, B.P.; Henderson, J.L.; Nakamura, Y.; Buchwald, S.L. Palladium-catalyzed coupling of functionalized primary and secondary amines with aryl and heteroaryl halides: Two ligands suffice in most cases. *Chem. Sci.* **2011**, *2*, 57–68. [CrossRef]
20. Su, M.; Buchwald, S.L. Bulky biaryl phosphine ligands for palladium-catalyzed amidation of five-membered heterocycles as electrophiles. *Angew. Chem. Int. Ed.* **2012**, *51*, 4710–4713. [CrossRef]
21. Su, M.; Hoshiya, N.; Buchwald, S.L. Palladium-catalyzed amidation of five-membered heterocyclic bromides. *Org. Lett.* **2014**, *16*, 832–835. [CrossRef]
22. Park, B.Y.; Pirnot, N.T.; Buchwald, S.L. Visible light-mediated (hetero)aryl amidation using Ni(II) salts and photoredox catalysis in flow: A synthesis of tetracaine. *J. Org. Chem.* **2020**, *85*, 3234–3244. [CrossRef]
23. Usami, Y.; Tsujiuchi, Y.; Machiya, Y.; Chiba, A.; Ikawa, T.; Yoneyama, H.; Harusawa, S. Synthetic challenges in the construction of 8- to 10-Membered Pyrazole-fused rings via ring-closing metathesis. *Heterocycles* **2020**, *101*, 496–511. [CrossRef]
24. Usami, Y.; Tatsui, Y.; Sumimoto, K.; Miyamoto, A.; Koito, N.; Yoneyama, H.; Harusawa, S. 3-Trifluoromethansulfonyloxy-4,7-dihidropyrazolo[1,5-a]pyridine via ring-closing metathesis: Synthesis and transformation to withasomnine homologs. *Heterocycles* **2020**, *103*. in press.
25. Ichikawa, H.; Ohno, Y.; Usami, Y.; Arimoto, M. Synthesis of 4-Arylpyrazoles via PdCl$_2$(dppf) catalyzed cross-coupling reaction with Grignard reagents. *Heterocycles* **2006**, *68*, 2247–2252. [CrossRef]

26. Ichikawa, H.; Nishioka, M.; Arimoto, M.; Usami, Y. Synthesis of 4-Aryl-1H-pyrazoles by Suzuki-Miyaura cross coupling reaction between 4-Bromo-1H-1-tritylpyrazole and arylboronic acids. *Heterocycles* **2010**, *81*, 1509–1516. [CrossRef]
27. Ichikawa, H.; Ohfune, H.; Usami, Y. Microwave-assisted selective synthesis of 2H-indazoles via double Sonogashira coupling of 3,4-diiodopyrazoles and Bergman-Masamune cycloaromatization. *Heterocycles* **2010**, *81*, 1651–1659. [CrossRef]
28. Usami, Y.; Ichikawa, H.; Harusawa, H.S. Heck-Mizoroki reaction of 4-Iodo-1H-pyrazoles. *Heterocycles* **2011**, *83*, 827–835. [CrossRef]
29. Bhunia, S.; Pawar, G.G.; Kumar, S.V.; Jiang, Y.; Ma, D. Selected copper-based reactions for C-N, C-O, C-S, and C-C bond formation. *Angew. Chem. Int. Ed.* **2017**, *56*, 16136–16179. [CrossRef] [PubMed]
30. Kwong, F.Y.; Klapars, A.; Buchwald, S.L. Copper-catalyzed coupling of alkylamines and aryl iodide: An efficient system even in an air atmosphere. *Org. Lett.* **2002**, *4*, 581–584. [CrossRef] [PubMed]
31. Kwong, F.Y.; Buchwald, S.L. Mild and efficient copper-catalyzed amination of aryl bromides with primary alkylamines. *Org. Lett.* **2003**, *5*, 793–796. [CrossRef]
32. Shafir, A.; Buchwald, S.L.J. Highly selective room-temperature copper-catalyzed C–N coupling reactions. *J. Am. Chem. Soc.* **2006**, *128*, 8742–8743. [CrossRef]
33. Jiang, S.; Dong, X.; Qiu, Y.; Xiaoxing, D.C.; Wu, X.; Jiang, S. A new ligand for copper-catalyzed amination of aryl halides to primary (hetero)aryl amines. *Tetrahedron Lett.* **2020**, *61*, 151683. [CrossRef]
34. Luo, T.; Wan, J.-P.; Liu, Y. Toward C2-nitrogenated chromones by copper-catalyzed β-C(sp2)–H N-heteroarylation of enaminones. *Org. Chem. Front.* **2020**, *7*, 1107. [CrossRef]
35. Liu, W.; Xu, J.; Chen, X.; Zhang, F.; Xu, Z.; Wang, D.; He, Y.; Xia, X.; Zhang, X.; Liang, Y. CuI/2-Aminopyridine 1-Oxide Catalyzed Amination of Aryl Chlorides with Aliphatic Amines. *Org. Lett.* **2020**, *22*, 7486–7490. [CrossRef] [PubMed]
36. Lyakhovich, M.S.; Averin, A.D.; Grigorova, O.K.; Roznyatovsky, V.A.; Maloshitskaya, O.A.; Beletskaya, I.P. Cu(I)- and Pd(0)-catalyzed arylation of oxadiamines with fluorinated halogenobenzenes: Comparison of efficiency. *Molecules* **2020**, *25*, 1084. [CrossRef] [PubMed]

Sample Availability: Not available.

 © 2020 by the authors. Licensee MDPI, Basel, Switzerland. This article is an open access article distributed under the terms and conditions of the Creative Commons Attribution (CC BY) license (http://creativecommons.org/licenses/by/4.0/).

Article

Synthesis and Antimicrobial Studies of Coumarin-Substituted Pyrazole Derivatives as Potent Anti-*Staphylococcus aureus* Agents

Rawan Alnufaie [1], Hansa Raj KC [1], Nickolas Alsup [1], Jedidiah Whitt [1], Steven Andrew Chambers [1], David Gilmore [2] and Mohammad A. Alam [1,*]

[1] Department of Chemistry and Physics, College of Science and Mathematics, Arkansas State University, Jonesboro, AR 72467, USA; rawan.alnufaie@smail.astate.edu (R.A.); hansa.kc@smail.astate.edu (H.R.K.C.); nickolas.alsup@smail.astate.edu (N.A); jedidiah.whitt@smail.astate.edu (J.W.); steven.chambers@smail.astate.edu (S.A.C.)

[2] Department of Biological Sciences, College of Science and Mathematics, Arkansas State University, Jonesboro, AR 72467, USA; dgilmore@astate.edu

* Correspondence: malam@astate.edu; Tel.: +1-870-972-3319

Received: 21 May 2020; Accepted: 12 June 2020; Published: 15 June 2020

Abstract: In this paper, synthesis and antimicrobial studies of 31 novel coumarin-substituted pyrazole derivatives are reported. Some of these compounds have shown potent activity against methicillin-resistant *Staphylococcus aureus* (MRSA) with minimum inhibitory concentration (MIC) as low as 3.125 µg/mL. These molecules are equally potent at inhibiting the development of MRSA biofilm and the destruction of preformed biofilm. These results are very significant as MRSA strains have emerged as one of the most menacing pathogens of humans and this bacterium is bypassing HIV in terms of fatality rate.

Keywords: pyrazole; coumarin; antimicrobial; hydrazone; biofilm; *Staphylococcus aureus*; *S. epidermidis*; MRSA

1. Introduction

Drug-resistant infections kill more than 700,000 people each year and the number could increase to 10 million per year by 2050 in the world. The menace of antimicrobial resistance could force up to 24 million people into extreme poverty [1]. According to a Centers for Disease Control and Prevention (CDC) report, each year more than 2.8 million antibiotic infections occur and more than 35,000 people die as a result of these infections in the United States alone. One of the four guidelines recommended by CDC to combat antibiotic resistance is promoting the development of new antibiotics and new diagnostic tests for dealing with drug-resistant bacteria [2]. *Staphylococcus aureus* and its drug-resistant variants are among the most common bacterial pathogens. Methicillin-resistant *S. aureus* (MRSA) causes ten-fold more infections than all multidrug resistant Gram-negative bacteria. A special pathogenic feature of *S aureus* is its ability to survive on abiotic and biotic surfaces in a slimy extracellular matrix called biofilm. The bacteria comprising biofilms are more resistant to antibiotics and the host immune system by a variety of mechanisms [3].

Pyrazole, a five-membered 1,2-diazole, is found in a number of widely used drugs such as celecoxib and lonazolac [4]. The pyrazole nucleus is less susceptible to oxidative metabolism than other five-membered heterocycles [5]. Derivatives of this diazole have been reported as antimicrobial agents in a number of publications by us [6–11] and others [12,13]. Similarly, natural and synthetic coumarin derivatives have attracted considerable interest because of their various pharmacological activities (Figure 1) [4]. Warfarin and novobiocin are examples of widely used coumarin-derived medications.

Coumarin derivatives have been reported as potent growth inhibitors of a wide variety of bacteria such as methicillin-resistant *S. aureus* (MRSA) [14], *Pseudomonas aeruginosa* [15], *Escherichia coli* [16], and several other bacterial species [17].

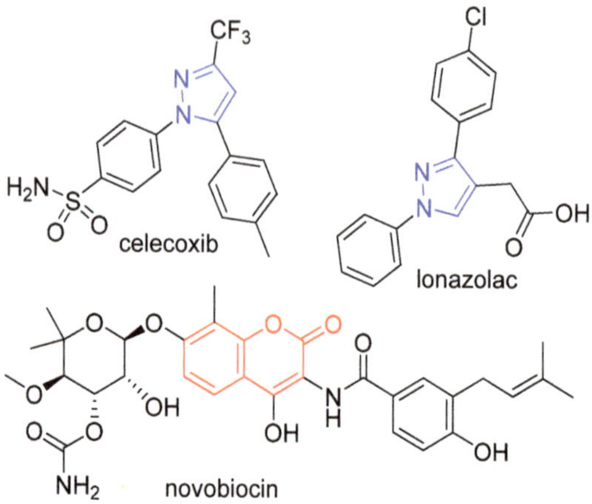

Figure 1. Representative examples of medicinally important pyrazole and coumarin derivatives.

In our efforts to develop potent antimicrobial agents, we have reported the synthesis and antimicrobial studies of phenyl [6,10], fluorophenyl [8,9], and naphthyl-substituted pyrazole derivatives [11]. We recently reported the synthesis, antimicrobial, and toxicological studies of coumarin-substituted pyrazole-derived hydrazones. Some of the molecules are potent growth inhibitors of *Acinetobacter baumannii* with an MIC value as low as 1.56 µg/mL. These potent molecules are non-toxic to human healthy and cancer cell lines and in vivo mouse model studies [7]. Encouraged with these results, we synthesized and studied the antimicrobial properties of 31 new fluoro and hydroxy-substituted coumarin-derived pyrazole compounds. These new coumarin-pyrazole-hydrazone derivatives are readily synthesized by using commercially available starting materials and reagents using benign reaction conditions.

2. Results and Discussion

2.1. Chemistry

The reaction of 4-hydrazinobenzoic acid (**1**) with fluoro (**2a**) and hydroxy (**2b**) substituted 3-acetylcoumarin formed the corresponding hydrazones (**3a–b**), which were subjected to further reaction with POCl$_3$/DMF to get the formyl-substituted pyrazole derivatives (**4a–b**) in overall excellent yields (Scheme 1). Multi-gram scale synthesis of pure pyrazole-derived aldehydes (**4a–b**) has helped to synthesize a series of hydrazone derivatives (Scheme 2).

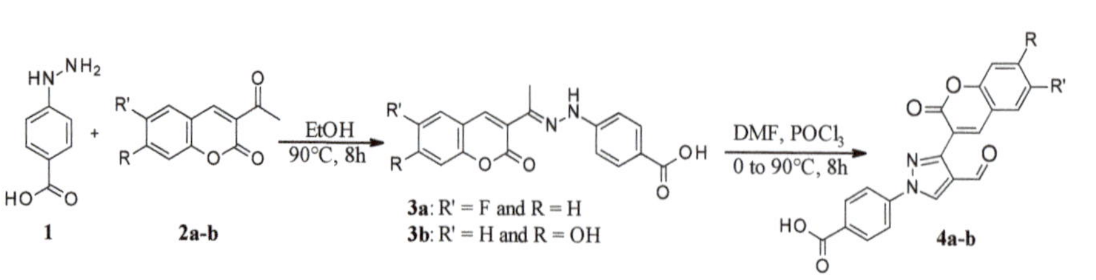

Scheme 1. Synthesis of coumarin-substituted pyrazole-derived aldehyde.

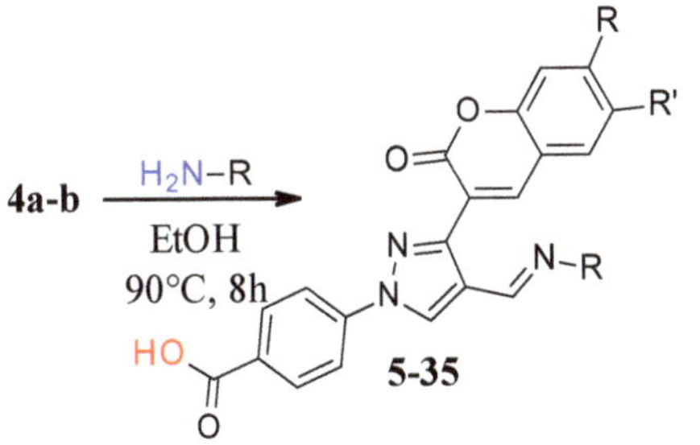

Scheme 2. Synthesis of coumarin-substituted pyrazole-derived hydrazones.

The reaction of phenylhydrazine with aldehyde derivatives (**4a**) formed the product (**5**) in good yield (Table 1). N,N-Disubstituted hydrazine derivative also reacted with the aldehyde derivative (**4**) to give the desired compounds (**6, 7,** and **8**) in 77%, 84%, and 82%, yields, respectively (Table 1). Products containing electron-donating groups (**9** and **10**) were obtained in 89% and 79% yields, respectively. Moderately electron-withdrawing groups: fluoro (**11**), chloro (**12**), and bromo (**13**) substituted coumarin-derived hydrazone products were formed in very good yield. Dihalogen substituted compounds formed accordingly and the pure products (**14, 15, 16, 17,** and **18**) were isolated in 75–95% yield. Strong electron-withdrawing groups such as trifluoromethyl (**19**), cyano (**20**), and nitro (**21**) formed the corresponding hydrazones. Reaction of N,N-dimethylhydrazine also formed the product (**22**) efficiently (Table 1).

Table 1. Fluoro-coumarin-substituted pyrazole-derived hydrazone derivatives. HRMS: high resolution mass spectrometry, Ph: phenyl, and Bn: benzyl.

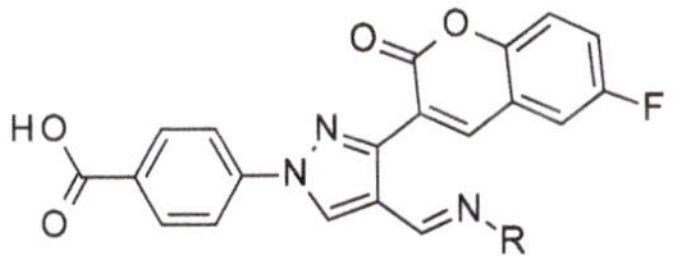

SN	R	HRMS (M + H)exp+	Yield (%)
5	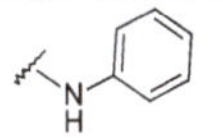 NH–Ph	469.1298	73
6	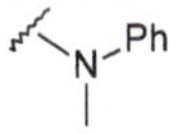 N(Ph)–N	483.1466	77
7	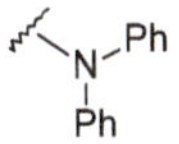 N(Ph)–N–Ph	545.1613	84
8	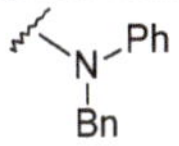 N(Ph)–N–Bn	559.1768	82
9	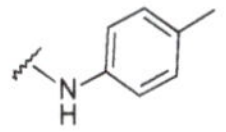 NH–C6H4–CH3	483.1455	89

Table 1. Cont.

SN	R	HRMS (M + H)exp$^+$	Yield (%)
10	2-Et-C$_6$H$_4$-NH-	497.1609	79
11	3-F-C$_6$H$_4$-NH-	487.1198	81
12	3-Cl-C$_6$H$_4$-NH-	503.0902, 505.0905	85
13	3-Br-C$_6$H$_4$-NH-	547.0400, 549.0394	78
14	2,4-F$_2$-C$_6$H$_3$-NH-	505.1110	75
15	3,4-F$_2$-C$_6$H$_3$-NH-	505.1106	84
16	2,4-Cl$_2$-C$_6$H$_3$-NH-	537.0512, 539.0482	77
17	3-Cl-2-F-C$_6$H$_3$-NH-	521.0818, 523.0793	80
18	3-Cl-4-F-C$_6$H$_3$-NH-	521.0806, 523.0783	95
19	4-CF$_3$-C$_6$H$_4$-NH-	537.1177	73
20	4-CN-C$_6$H$_4$-NH-	494.1252	88
21	4-NO$_2$-C$_6$H$_4$-NH-	514.115	89
22	-N(CH$_3$)$_2$	421.1298	79

Similarly, hydroxy-substituted coumarin derivative (**4b**) reacted with the hydrazines smoothly to form the corresponding hydrazones (**23–35**) efficiently (Table 2). The *N*-phenyl-derived product (**23**) formed in 79% yield. 3-Fluoro (**24**), 4-chloro (**25**), and 4-bromo (**26**) derivatives formed in 86%, 71%, and 80% yields, respectively. Difluoro (**27** and **28**) and dichloro (**29**) hydrazones were synthesized in 75% average yield. Mixed halo compounds (**30** and **31**) were formed efficiently. Strong electron-withdrawing groups such as cyano (**32**) and nitro (**33**) also formed in very good yield. Last but not least, *N*-carbamate product (**34**) formed in 72% yield.

Table 2. Hydroxy-coumarin-substituted pyrazole-derived hydrazone derivatives.

SN	R	HRMS (M + H)exp$^+$	Product Yield (%)
23	–NH–C$_6$H$_5$	449.1107	79
24	–NH–C$_6$H$_4$–F (3-F)	485.1257	86
25	–NH–C$_6$H$_4$–Cl (4-Cl)	501.0962, 503.0932	71
26	–NH–C$_6$H$_4$–Br (4-Br)	545.0456, 547.0435	80
27	–NH–C$_6$H$_3$–F$_2$ (2,4-F)	503.1158	91
28	–NH–C$_6$H$_3$–F$_2$ (3,4-F)	503.1162	75
29	–NH–C$_6$H$_3$–Cl$_2$ (3,4-Cl)	535.0570, 537.0545	90
30	–NH–C$_6$H$_3$–F,Cl (2-F,?-Cl)	519.0869, 521.0844	84
31	–NH–C$_6$H$_3$–F,Cl	519.0866, 521.0837	73
32	–NH–C$_6$H$_4$–CF$_3$ (4-CF$_3$)	535.1223	83
33	–NH–C$_6$H$_4$–CN (4-CN)	492.1300	81
34	–NH–C(O)–OCH$_3$	467.1352	72

2.2. Antimicrobial Studies

Synthesized molecules were tested against eight strains of Gram-positive bacteria and three strains of *A. baumannii*, a Gram-negative bacterium (Table 3). *N*-Phenyl and *N*-methyl-*N*-phenyl substituted compounds (**5** and **6**) did not show any significant activity up to 50 µg/mL against the tested bacterial strains. Excitingly, *N*,*N*-diphenyl substituted compound (**7**) showed potent activity against all the tested Gram-positive bacterial strains but no activity against *A. baumannii*. This molecule (**7**) inhibited the growth of methicillin-sensitive *S. aureus* (MSSA) with an MIC value as low as 3.125 µg/mL. Four methicillin-resistant strains were inhibited by this molecule with an MIC value as low as 1.56 µg/mL. It is worth mentioning that three of the MRSA strains: *S. aureus* ATCC 33591 (Sa91), *S. aureus* ATCC 700699 (Sa99), and *S. aureus* ATCC 33592 (Sa92) possess the SCC mec II or III genomic islands conferring multidrug resistance. This molecule also inhibited the growth of *S. epidermidis* 700296 (Se), and *B. subtilis* ATCC 6623 (Bs) with an MIC value of 6.25 µg/mL. Replacing one phenyl group with a benzyl group eliminated the activity of the resultant molecule (**8**). Methyl (**9**) and ethyl (**10**) substitution showed partial activity against the tested strains. Fluoro-phenyl substituted compound (**11**) did not show any growth inhibition of the tested bacteria. Chloro (**12**) and bromo (**13**) substitution showed moderate activity against Gram-positive bacterial strains. These two molecules (**12** and **13**) also showed similar activity against three *A. baumannii* strains with MIC value as low as 6.25 µg/mL. 2,5-Difluoro substitution (**14**) failed to show any activity against the tested bacteria. Nevertheless, 3,4-fluoro substitution (**15**) showed moderate activity against both the Gram-positive and Gram-negative strains with MIC value of 6.25 µg/mL. Other dihalogen-substituted compounds (**16**, **17**, and **18**) did not show any antimicrobial potency. Very strong electron withdrawing groups such as trifluoromethyl (**19**) showed good activity against Gram-positive strains with MIC value as low as 3.125 µg/mL although the other two compounds (**20** and **21**) with very strong electron withdrawing substitution did not show any activity. N,N-Dimethyl compound (**22**) also failed to show any activity against the tested strains. Surprisingly, hydroxy-substituted coumarin compounds (**23–34**) did not show any activity

Based on these results, we can deduce the following structure activity relationship. Fluoro-substitution in the coumarin moiety produced potent antimicrobial compounds, whereas the hydroxy-substitution almost eliminated the activity of the molecules. This finding could be due to the different nature of the hydroxy and fluoro substituents. Among the fluoro-substituted coumarins, N,N-diphenyl derivative showed potent activity but replacing one of the phenyl groups with methyl (**6**) or benzyl (**8**) eliminated the potency of the compounds. This observation is in agreement with our previous report [7]. Hydrophobic mono-substituted compounds showed better activity than the disubstituted compounds, 4-trifluoromethyl substituent (**19**) showed the most potent overall activity among all the compounds against the tested Gram-positive bacteria. Substituent with a very strong electron withdrawing nature such as cyano (**20**) and nitro (**21**) eliminated activity of the compounds.

After finding potent molecules against planktonic bacteria, we studied their ability to affect *S. aureus* biofilms (Figure 2). The most potent compound (**7**) inhibited biofilm by 85% at 0.5 × MIC. Higher concentrations were even more effective. Similarly, compound **12** inhibited the biofilm formation at 2× and 1 × MIC values but its potency decreased at 0.5 × MIC. Compounds (**13** and **19**) also inhibited biofilm formation similarly to compound (**7**). The positive control, vancomycin, also inhibited biofilm formation at 2× and 1×MICs but showed significantly less effectiveness at 0.5 × MIC. It should be noted that inhibition of bacterial growth would be expected to result in reduced biofilm formation. Therefore, it is very encouraging that significant inhibition of biofilm formation by our compounds occurred at a concentration of 0.5 × MIC. Three of these potent compounds (**7**, **12**, and **13**) were efficient eliminators of preformed biofilms of *S. aureus*. Compound **19** removed more than 90% of the biofilm at 2× and 1 × MICs, although 0.5 × MIC showed some reduced effectiveness against the preformed biofilm of *S. aureus*. These results are very significant as the positive control could only eliminate ~70% preformed biofilm at 2 × MIC.

Table 3. Antimicrobial activities of the synthesized compounds against various bacteria. Gram-positive: antibiotic susceptible *S. aureus* ATCC 25923 (Sa23); antibiotic-resistant *S. aureus* BAA-2312 (Sa12), *S. aureus* ATCC 33591 (Sa91), *S. aureus* ATCC 700699 (Sa99), *S. aureus* ATCC 33592 (Sa92); and antibiotic susceptible *S. epidermidis* 700296 (Se) and *B. subtilis* ATCC 6623 (Bs); Gram-negative: *A. baumannii* ATCC 19606 (type strain, AB06), *A. baumannii* ATCC BAA-1605 (Ab05), and *A. baumannii* ATCC 747 (Ab47). V = vancomycin (positive control), C = colistin (positive control), and NA = no activity up to 50 µg/mL.

SN	Sa23	Sa12	Sa91	Sa92	Sa99	Se	Bs	Ab05	Ab47	Ab06
5	NA	NA	NA	NA	NA	NA	NA	NA	NA	NA
6	NA	NA	NA	NA	NA	NA	NA	NA	NA	NA
7	3.125	3.125	6.25	1.56	3.125	3.125	6.25	NA	NA	NA
8	NA	NA	NA	NA	NA	NA	NA	NA	NA	NA
9	25	25	25	25	12.5	25	25	25	25	25
10	25	25	25	12.5	25	NA	25	NA	NA	NA
11	NA	NA	NA	NA	NA	NA	NA	NA	NA	NA
12	12.5	12.5	12.5	6.25	25	12.5	12.5	25	12.5	6.25
13	6.25	12.5	12.5	6.25	6.25	12.5	12.5	12.5	12.5	6.25
14	NA	NA	NA	NA	NA	NA	NA	NA	NA	NA
15	25	6.25	12.5	6.25	12.5	NA	25	25	6.25	25
16	NA	NA	NA	NA	NA	NA	NA	NA	NA	NA
17	NA	NA	NA	NA	NA	NA	NA	NA	NA	NA
18	NA	NA	NA	NA	NA	NA	NA	NA	NA	NA
19	3.125	12.5	6.25	12.5	3.125	12.5	6.25	NA	NA	NA
20	NA	NA	NA	NA	NA	NA	NA	NA	NA	NA
21	NA	NA	NA	NA	NA	NA	NA	NA	NA	NA
22	NA	NA	NA	NA	NA	NA	NA	NA	NA	NA
23	NA	NA	NA	NA	NA	NA	NA	25	25	25
24	NA	NA	NA	NA	NA	NA	NA	NA	NA	NA
25	NA	NA	NA	NA	NA	NA	NA	NA	NA	NA
26	NA	NA	NA	NA	NA	NA	NA	50	50	50
27	NA	NA	NA	NA	NA	NA	NA	NA	NA	NA
28	NA	NA	NA	NA	NA	NA	NA	50	NA	NA
29	NA	NA	NA	NA	NA	NA	NA	NA	NA	NA
30	NA	NA	NA	NA	NA	NA	NA	NA	NA	NA
31	NA	NA	NA	NA	NA	NA	NA	NA	NA	NA
32	NA	NA	NA	NA	NA	NA	NA	NA	NA	NA
33	NA	NA	NA	NA	NA	NA	NA	NA	NA	NA
34	NA	NA	NA	NA	NA	NA	NA	NA	NA	NA
V	0.78	0.78	1.56	1.56	3.125	3.125	0.195			
C								3.125	1.56	3.125

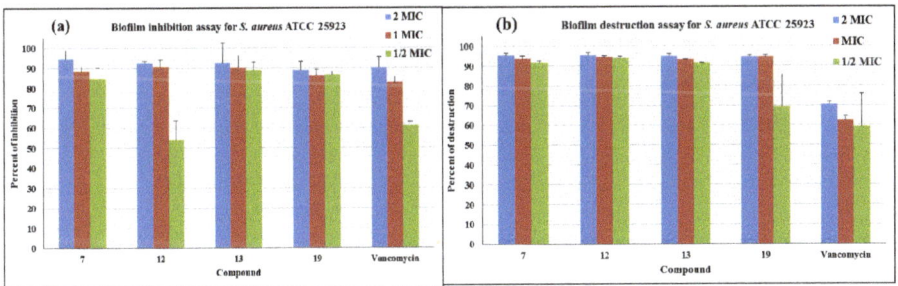

Figure 2. Representation of the biofilm inhibitory (**a**) and destructive capacity (**b**) of the active compounds against *S. aureus* ATCC 25923. Inhibition values are in percentage.

The two most effective compounds against *A. baumannii* (**12, 13**) and the two most effective against *S. aureus* (**7, 19**) were used in time kill assays over a 24-h period. At 4 × MIC, both (**12**) and (**13**) were bacteriostatic against *A. baumannii*. Compound (**19**) was likewise bacteriostatic against the MRSA strain tested, but compound (**7**) had marked bactericidal activity, reducing colony-forming units (CFU)/mL to undetectable by 4 h as shown in Figure 3.

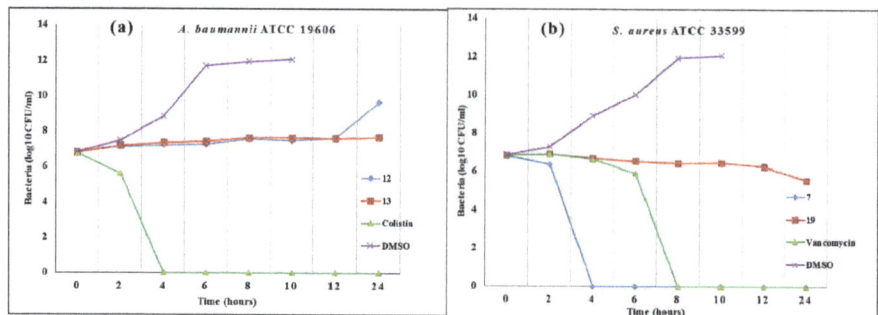

Figure 3. Time kill assay. Compounds were tested at 4 × MIC (except indicated) against (**a**) *A. baumannii* ATCC 19606 and (**b**) *S. aureus* ATCC 33599 (MRSA) over an incubation period of 24 h at 35 °C.

We tested the potent compounds for their possible toxicity for the human embryonic kidney cell line (HEK293). Potent compounds (**7**, **12**, and **13**) inhibited the growth of cell lines with an IC_{50} around 15 µg/mL but compound **19** showed much less toxicity. More than 90% of cells survived at 25 µg/mL (Figure 4).

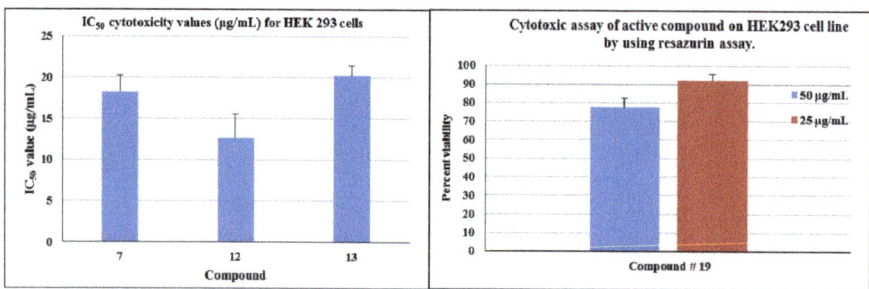

Figure 4. Cytotoxicity assay of potent compounds **7**, **12**, **13**, and **19** against HEK293 cells.

3. Materials and Methods

All of the reactions were carried out under an air atmosphere in round-bottom flasks. Reagents, substrate, and solvents for reactions, recrystallizations, and deuterated solvents for ^{1}H and ^{13}C-NMR spectroscopy were purchased from Fisher Scientific (Hanover Park, IL, USA) and Oakwood chemical (Estill, SC, USA). The more details of spectroscopy in Supplementary Materials.

A mixture of 4-hydrazinobenzoic acid (**1**, 10 mmol, 1.521 g) and 3-acetylcoumarin derivatives (**2a–b**, 10.5 mmol) in ethanol was refluxed for 8 h to obtain the hydrazone derivative (**3**) (Scheme 1). The solvent was evaporated under the reduced pressure at 60 °C, and the hydrazone derivative was further dried in vacuo. The dried product was used for further reactions without isolation or purification. The hydrazone derivative (**3**) was dissolved in *N*,*N*-dimethyl formamide (DMF, 30 mL) and the flask was sealed by a rubber septum. The mixture was stirred for 15 min to dissolve the solid material completely. The clear solution was stirred at 0 °C in an ice bath, and phosphorous oxychloride ($POCl_3$, 5.43 mL) was added dropwise to form the Vilsmeier reagent. After 30 min, the reaction mixture was heated for 8 h at 90 °C. After the completion of the reaction, the mixture was poured onto ice in a beaker and then stirred for 12 h to precipitate the product, which was filtered and washed with water repeatedly until the filtrate was clear. The final product was dried under vacuum.

Novel coumarin-derived hydrazones were synthesized by the reaction of the aldehyde product (**3a** and **3b**, 1 mmol) with commercially available substituted-hydrazines (1.1 mmol) in ethanol (Scheme 2). Sodium acetate (1.1 mmol, 0.088 g) and acetic acid were added in case of the hydrochloride salt of the hydrazine derivatives. The resulting product was filtered and washed with ethanol (~15 mL) followed by washing with water (~20 mL) to get the pure product.

3.1. Experimental Data

4-[3-(7-fluoro-2-oxo-3,8a-dihydrochromen-3-yl)-4-formyl-pyrazol-1-yl]benzoic acid (**4**). Brownish; (3.567 g, 93.7%) ^{1}H-NMR (300 MHz, DMSO-d_6): δ 9.91 (s, 1H), 9.35 (s, 1H), 8.68 (s, 1H), 8.09 (s, 4H), 7.72 (d, J = 8.0 Hz, 1H), 7.55–7.53 (m, 2H); ^{13}C-NMR (75 MHz, DMSO-d_6): δ 186.0, 166.9, 158.6 (d, ^1J = 239.6 Hz), 159.5, 150.3, 148.0, 142.6, 141.8, 133.6, 131.4, 130.1, 124,2, 121.1, 120.4 (d, ^2J = 24.5 Hz), 120.1 (d, ^3J = 9.7 Hz), 119.4, 118.7 (d, ^3J = 8.7 Hz), 114.6 (d, ^2J = 24.3 Hz). HRMS (ESI-FTMS) Mass (*m/z*): calcd for $C_{20}H_{11}FN_2O_5$ [M + H]$^+$ = 379.0725, found 379.0738.

4-[3-(7-fluoro-2-oxo-3,8a-dihydrochromen-3-yl)-4-[(E)-(phenylhydrazono)methyl]pyrazol-1-yl]benzoic acid (**5**). Yellow; (343 mg, 73%) ^{1}H-NMR (300 MHz, DMSO-d_6): δ 10.22 (s, 1H), 8.98 (s, 1H), 8.32 (s, 1H), 8.12–8.05 (m, 4H), 7.84 (s, 1H), 7.74 (d, J = 9.3 Hz, 1H), 7.60–7.58 (m, Hz, 2H), 6.99 (t, J = 8.1 Hz, 2H), 6.75–6.73 (m, 2H), 6.63 (t, J = 7.2 Hz, 1H); ^{13}C-NMR (75 MHz, DMSO-d_6): δ 167.0, 158.6 (d, ^1J = 239.3 Hz), 159.1, 150.2, 146.0, 145.6, 142.4, 141.9, 131.4, 129.2, 128.99, 128.96, 127.3, 123.3, 121.3, 120.3 (d, ^3J = 9.5 Hz), 120.0 (d, ^2J = 24.6 Hz), 118.8, 118.5 (d, ^3J = 8.5 Hz), 118.3, 114.4 (d, ^2J = 24.2 Hz), 112.0. HRMS (ESI-FTMS) Mass (*m/z*): calcd for $C_{26}H_{20}N_4O_5$ [M + H]$^+$ = 469.1506, found 469.1298.

4-[3-(6-fluoro-2-oxo-3,8a-dihydrochromen-3-yl)-4-[(E)-[methyl(phenyl)hydrazono]methyl]pyrazol-1-yl]benzoic acid (**6**). Yellow; (373 mg, 77%) ^{1}H-NMR (300 MHz, DMSO-d_6): 8.95 (s, 1H), 8.32 (s, 1H), 8.09 (s, 4H), 7.76–7.72 (m, 1H), 7.62–7.56 (m, 3H), 7.10–7.07 (m, 2H), 6.98 (t, J = 7.3 Hz, 2H), 6.75 (t, J = 7.2 Hz, 1H), 2.50–2.49 (m, 3H); ^{13}C-NMR (75 MHz, DMSO-d_6): δ 167.1, 158.6 (d, ^1J = 239.3 Hz), 159.2, 150.3, 147.5, 146.0, 142.4, 141.8, 131.4, 129.0, 128.9, 127.2, 125.4, 123.7, 122.2, 120.3 (d, ^3J = 9.5 Hz), 120.1, 119.7, 118.6 (d, ^3J = 8.5 Hz), 118.3, 114.5, 114.2, 32.7. HRMS (ESI-FTMS) Mass (*m/z*): calcd for $C_{27}H_{19}FN_4O_4$ [M + H]$^+$ = 483.1463, found 483.1466.

4-[4-[(E)-(diphenylhydrazono)methyl]-3-(6-fluoro-2-oxo-3,8a-dihydrochromen-3-yl)pyrazol-1-yl]benzoic acid (**7**). Yellow; (459 mg, 84%) ^{1}H-NMR (300 MHz, DMSO-d_6): δ 8.94 (s, 1H), 8.33 (s, 1H), 8.09–8.06 (m, 2H), 8.01–7.98 (m, 1H), 7.77–7.74 (m, 1H), 7.65–7.58 (m, 2H), 7.30–7.24 (m, 4H), 7.16–7.11 (m, 3H), 6.94–6.91 (m, 4H); ^{13}C-NMR (75 MHz, DMSO-d_6): δ 167.0, 158.6 (d, ^1J = 239.3 Hz), 159.0, 150.2, 145.8, 143.1, 142.3, 141.9, 131.4, 130.2, 129.0, 128.6, 128.0, 124.9, 123.7, 122.2, 121.0, 120.2 (d, ^3J = 9.7 Hz), 119.8, 118.6 (d, ^3J = 8.7 Hz), 118.3, 144.4 (d, ^2J = 24.3 Hz). HRMS (ESI-FTMS) Mass (*m/z*): calcd for $C_{32}H_{21}FN_4O_4$ [M + H]$^+$ = 545.1620, found 545.1613.

4-[4-[(E)-[benzyl(phenyl)hydrazono]methyl]-3-(6-fluoro-2-oxo-3,8a-dihydrochromen-3-yl)pyrazol-1-yl]benzoic acid (**8**). Yellow; (459 mg, 82%) ^{1}H-NMR (300 MHz, DMSO-d_6): δ 8.88 (s, 1H), 8.28 (s, 1H), 8.09–8.00 (m, 4H), 7.73–7.71 (m, 1H), 7.58 (s, 3H), 7.53–7.12 (m, 5H), 6.99–6.86 (m, 4H), 6.75–6.71 (m, 1H), 5.20 (s, 2H); ^{13}C-NMR (75 MHz, DMSO-d_6): δ 167.0, 158.6 (d, ^1J = 239.4 Hz), 159.0, 150.2, 147.2, 145.9, 142.4, 141.7, 136.2, 131.4, 129.1, 128.9, 127.8, 127.4, 126.6, 125.3, 123.9, 121.8, 120.3, 120.2, 119.9 (d, ^2J = 24.6 Hz), 118.6 (d, ^3J = 8.7 Hz), 118.3, 114.4 (d, ^2J = 24.1 Hz), 113.9. 48.2. HRMS (ESI-FTMS) Mass (*m/z*): calcd for $C_{33}H_{23}FN_4O_4$ [M + H]$^+$ = 559.1776, found 559.1768.

4-[3-(7-fluoro-2-oxo-3,8a-dihydrochromen-3-yl)-4-[(E)-(p-tolylhydrazono)methyl]pyrazol-1-yl]benzoic acid (**9**). Orange; (481 mg, 89%) ^{1}H-NMR (300 MHz, DMSO-d_6): δ 10.10 (s, 1H), 8.96 (s, 1H), 8.31 (s, 1H), 8.11–8.05 (m, 4H), 7.79–7.72 (m, 2H), 7.60–7.59 (m, 2H), 6.83–6.80 (m, 2H), 6.67–6.65 (m, 2H), 2.14 (s, 3H); ^{13}C-NMR (75 MHz, DMSO-d_6): δ 167.1, 158.6 (d, ^1J = 239.4 Hz), 159.1, 150.3, 145.9, 143.3, 142.4, 141.9, 131.4, 129.7, 128.9, 128.3, 127.3, 127.1, 123.3, 121.4, 120.3 (d, ^3J = 9.6 Hz), 120.0 (d, ^2J = 24.2 Hz), 118.5 (d, ^3J = 8.4 Hz), 118.3, 114.4 (d, ^2J = 24.0 Hz), 112.0, 20.6. HRMS (ESI-FTMS) Mass (*m/z*): calcd for $C_{27}H_{19}FN_4O_4$ [M + H]$^+$ = 483.1463, found 483.1455.

4-[4-[(E)-[(2-ethylphenyl)hydrazono]methyl]-3-(7-fluoro-2-oxo-3,8a-dihydrochromen-3-yl)pyrazol-1-yl]benzoic acid (**10**). Orange; (393 mg, 79%) ^{1}H-NMR (300 MHz, DMSO-d_6): 9.52 (s, 1H), 9.00 (s, 1H), 8.32 (s, 1H), 8.12–8.10 (m, 5H), 7.75–7,72 (m, 1H), 7.60–7.58 (m, 2H), 6.99–6.91 (m, 2H), 6.70–6.60 (m, 2H), 2.56–2.49 (m, 2H), 1.11 (t, J = 7.3 Hz, 3H); ^{13}C-NMR (75 MHz, DMSO-d_6): δ 167.1, 158.6 (d, J = 239.3 Hz), 159.1, 150.2, 146.1, 142.7, 142.4, 142.0, 131.4, 130.2, 129.9, 128.9, 128.6, 127.3, 126.5, 123.4,

121.5, 120.3 (d, ^3J = 9.6 Hz), 119.8, 119.0, 118.6 (d, ^3J = 8.8 Hz), 118.3, 114.4 (d, ^2J = 24.3 Hz), 112.3, 23.5, 14.3. HRMS (ESI-FTMS) Mass (m/z): calcd for $C_{28}H_{21}FN_4O_4$ [M + H]$^+$ = 497.1620, found 497.1609.

4-[3-(7-fluoro-2-oxo-3,8a-dihydrochromen-3-yl)-4-[(E)-[(3-fluorophenyl)hydrazono]methyl]pyrazol-1-yl] benzoic acid (**11**). Orange; (395 mg, 81%) ^{1}H-NMR (300 MHz, DMSO-d_6): 10.45 (s, 1H), 9.03 (s, 1H), 8.34 (s, 1H), 8.09–8.06 (m, 4H), 7.86 (s,1H), 7.73 (d, J = 7.1 Hz, 1H), 7.58–7.56 (m, 2H), 7.09–7.01 (m, 1H), 6.56–6.51 (m, 2H), 6.39 (t, J = 6.3 Hz, 1H); ^{13}C-NMR (75 MHz, DMSO-d_6): 167.0, 163.7 (d, ^1J = 239.0 Hz), 158.6 (d, ^1J = 239.4 Hz), 159.1, 150.2, 147.6 (d, ^3J = 11.0 Hz), 146.0, 142.4, 142.1, 131.4, 130.8 (d, ^3J = 9.9 Hz), 130.3, 129.0, 127.7, 123.1, 121.0, 120.2 (d, ^3J = 9.7 Hz), 119.8, 118.6 (d, ^3J = 8.5 Hz), 118.4, 114.4 (d, ^2J = 24.1 Hz), 108.2, 104.8 (d, ^2J = 21.3 Hz), 98.5 (d, ^2J = 26.1 Hz). HRMS (ESI-FTMS) Mass (m/z): calcd for $C_{26}H_{16}F_2N_4O_4$ [M + H]$^+$ = 487.1212, found 487.1198.

4-[4-[(E)-[(3-chlorophenyl)hydrazono]methyl]-3-(7-fluoro-2-oxo-3,8a-dihydrochromen-3-yl)pyrazol-1-yl] benzoic acid (**12**). Orange; (429 mg, 85%) ^{1}H-NMR (300 MHz, DMSO-d_6): 10.42 (s, 1H), 9.02 (s, 1H), 8.34 (s, 1H), 8.12–8.05 (m, 4H), 7.88 (s, 1H), 7.74–7.72 (m, 1H), 7.58–7.56 (m, 2H), 7.05 (t, J = 8.0 Hz, 1H), 6.77 (s, 1H), 6.66 (t, J = 8.1 Hz, 2H); ^{13}C-NMR (75 MHz, DMSO-d_6): δ 167.0, 158.6 (d, ^1J = 239.3 Hz), 159.1, 150.3, 147.0, 146.0, 142.4, 142.1, 134.3, 131.4, 130.7 (d, ^3J = 22.5 Hz), 129.0, 127.9, 123.1, 120.9, 120.3, 120.1, 119.8, 118.8 (d, ^3J = 8.7 Hz), 118.4, 118.2, 114.4 (d, ^2J = 24.3 Hz), 111.1, 110.8. HRMS (ESI-FTMS) Mass (m/z): calcd for $C_{26}H_{16}ClFN_4O_4$ [M + H]$^+$ = 503.0917, 505.0890, found 503.0902, 505.0905.

4-[4-[(E)-[(3-bromophenyl)hydrazono]methyl]-3-(7-fluoro-2-oxo-3,8a-dihydrochromen-3-yl)pyrazol-1-yl] benzoic acid (**13**). Orange; (428 mg, 78%) ^{1}H-NMR (300 MHz, DMSO-d_6): δ 10.40 (s, 1H), 9.02 (s, 1H), 8.33 (s, 1H), 8.12–8.05 (m, 4H), 7.88 (s, 1H), 7.74–7.71 (m, 1H), 7.58–7.52 (m, 2H), 7.01–7.93 (m, 2H), 6.79-6.70 (m, 2H); ^{13}C-NMR (75 MHz, DMSO-d_6): δ 167.0, 158.6 (d, ^1J = 239.3 Hz), 159.1, 150.3, 147.1, 146.0, 142.4, 142.1, 131.4, 131.2, 130.6, 129.0, 127.9, 123.1, 122.9, 121.1, 120.8, 120.2 (d, ^3J = 9.7 Hz), 119.8, 118.9 (d, ^3J = 8.4 Hz), 118.4, 114.4 (d, ^2J = 24.1 Hz), 113.9, 111.2. HRMS (ESI-FTMS) Mass (m/z): calcd for $C_{26}H_{16}BrFN_4O_4$ [M + H]$^+$ = 547.0412, 549.0393, found 547.0400, 549.0394.

4-[4-[(E)-[(2,5-difluorophenyl)hydrazono]methyl]-3-(6-fluoro-2-oxo-3,8a-dihydrochromen-3-yl) pyrazol-1-yl] benzoic acid (**14**). Orange; (379 mg, 75%) ^{1}H-NMR (300MHz, DMSO-d_6): 10.38 (s, 1H), 9.08 (s, 1H), 8.33 (s, 1H), 8.14–8.05 (m, 5H), 7.72 (d, J = 7.2 Hz, 1H), 7.60–7.50 (m, 2H), 7.13–7.05 (m, 1H), 6.75–6.68 (m, 1H), 6.45–6.38 (m, 1H); ^{13}C-NMR (75 MHz, DMSO-d_6): δ 167.1, 159.5 (d, ^1J = 236.5 Hz), 158.6 (d, ^1J = 239.5 Hz), 159.1, 150.2, 145.4 (d, ^1J = 233.5 Hz), 146.1, 142.3, 142.2, 135.1 (t, ^3J = 11.8 Hz), 133.1, 131.4, 129.2, 128.2, 122.9, 120.7, 120.2, 120.0 (d, ^3J = 13.1 Hz), 118.6 (d, ^3J = 8.6 Hz), 118.4, 116.4-116.0 (m), 114.4 (d, ^2J = 24.2 Hz), 103.8 (d, ^3J = 17.7 Hz), 100.2 (d, ^2J = 30.0 Hz). HRMS (ESI-FTMS) Mass (m/z): calcd for $C_{26}H_{15}F_3N_4O_4$ [M + H]$^+$ = 505.1118, found 505.1110.

4-[4-[(E)-[(3,4-difluorophenyl)hydrazono]methyl]-3-(6-fluoro-2-oxo-3,8a-dihydrochromen-3-yl)pyrazol-1-yl]benzoic acid (**15**). Orange; (425 mg, 84%) ^{1}H-NMR (300 MHz, DMSO-d_6): 10.40 (s, 1H), 9.02 (s, 1H), 8.33 (s, 1H), 8.11–8.04 (m, 4H), 7.84 (s, 1H), 7.75–7.72 (m, 1H), 7.57 (d, J = 5.8 Hz, 2H), 7.16–7.07 (m, 1H), 6.71–6.65 (m, 1H), 6.52–6.49 (m, 1H); ^{13}C-NMR (75 MHz, DMSO-d_6): δ 167.0, 158.6 (d, ^1J = 239.4 Hz), 159.2, 150.4 (dd, J = 13.0, 240.6 Hz), 150.6, 146.0, 142.8 (dd, J = 12.8, 233.6 Hz), 143.2 (d, ^3J = 8.4 Hz), 142.4, 142.1, 131.4, 130.4, 129.0, 127.7, 122.9, 120.9, 120.2 (d, ^3J = 9.5 Hz), 120.25, 119.9, 118.6 (d, ^3J = 8.5 Hz), 118.4, 118.1, 117.9, 114.4 (d, ^2J = 24.1 Hz). HRMS (ESI-FTMS) Mass (m/z): calcd for $C_{26}H_{15}F_3N_4O_4$ [M + H]$^+$ = 505.1118, found 505.1106.

4-[4-[(E)-[(2,4-dichlorophenyl)hydrazono]methyl]-3-(6-fluoro-2-oxo-3,8a-dihydrochromen-3-yl)pyrazol-1-yl]benzoic acid (**16**). Yellowish; (415 mg, 77%) ^{1}H-NMR (300 MHz, DMSO-d_6): 13.11 (br s, 1H), 9.98 (s, 1H), 9.08 (s, 1H), 8.34 (s, 1H), 8.21 (s, 1H), 8.10 (s, 4H), 7.76–7.72 (m, 1H), 7.62–7.59 (m, 2H), 7.418–7.410 (m, 1H), 7.12–7.09 (s, 1H), 6.96–6.92 (m, 1H); ^{13}C-NMR (75 MHz, DMSO-d_6): δ 167.0, 158.6 (d, ^1J = 239.7 Hz), 159.1, 150.2, 146.3, 142.3, 140.9, 133.8, 131.4, 129.1, 129.0, 127.9, 127.8, 122.7, 122.2, 120.8, 120.21 (d, ^2J = 24.6 Hz), 120.22 (d, ^3J = 9.7 Hz) 118.7, 118.54, 118.53, 116.8, 114.8, 114.5 (d, ^2J = 24.1 Hz). HRMS (ESI-FTMS) Mass (m/z): calcd for $C_{26}H_{15}Cl_2FN_4O_4$ [M + H]$^+$ = 537.0527, 539.0499, found 537.0512, 539.0482.

4-[4-[(E)-[(3-chloro-2-fluoro-phenyl)hydrazono]methyl]-3-(6-fluoro-2-oxo-3,8a-dihydrochromen-3-yl) pyrazol-1-yl]benzoic acid (**17**). Orange; (418 mg, 80%) ^{1}H-NMR (300 MHz, DMSO-d_6): 10.39 (s, 1H),

9.06 (s, 1H), 8.33 (s, 1H), 8.14–8.08 (m, 5H), 7.75–7.72 (m, 1H), 7.60–7.58 (m, 2H), 6.95 (t, J = 7.6 Hz, 1H), 6.81–6.70 (m, 2H); ^{13}C-NMR (75 MHz, DMSO-d_6): δ 167.0, 158.6 (d, ^1J = 239.7 Hz), 159.1, 150.2, 144.6 (d, ^1J = 240 Hz), 146.2, 142.3, 142.2, 135.2 (d, ^3J = 9.3 Hz), 133.1, 131.4, 129.1, 128.0, 125.4, 125.3, 123.0, 120.7, 120.3, 120.1 (d, ^3J = 8.5 Hz), 119.9 (d, ^3J = 7.3 Hz), 118.7 (d, ^3J = 13.7 Hz), 118.5, 114.5 (d, ^2J = 24.4 Hz), 112.5. HRMS (ESI-FTMS) Mass (m/z): calcd for $C_{26}H_{15}ClF_2N_4O_4$ [M + H]$^+$ = 521.0823, 523.0796, found 521.0806, 523.0783.

4-[4-[(E)-[(3-chloro-4-fluoro-phenyl)hydrazono]methyl]-3-(6-fluoro-2-oxo-3,8a-dihydrochromen-3-yl) pyrazol-1-yl]benzoic acid (**18**). Orange; (496 mg, 95%) ^{1}H-NMR (300 MHz, DMSO-d_6): 10.36 (s, 1H), 8.99 (s, 1H), 8.32 (s, 1H), 8.11–8.03 (m, 4H), 7.85 (s, 1H), 7.74–7.71 (m, 1H), 7.58–7.56 (m, 2H), 7.10 (t, J = 9.0 Hz, 1H), 6.83–6.80 (m, 1H), 6.70–6.65 (m, 1H); ^{13}C-NMR (75 MHz, DMSO-d_6): δ 167.1, 158.7 (d, ^1J = 239.3 Hz), 159.1, 150.8 (d, ^1J = 235.0 Hz), 150.2, 146.0, 143.0, 142.3, 142.1, 131.1, 130.5, 129.0, 127.9, 123.1, 120.8, 120.3 (d, ^3J = 8.6 Hz), 120.0 (d, ^2J = 24.4 Hz), 120.12, 118.8 (d, ^3J = 8.6 Hz), 118.4, 117.5 (d, ^2J = 21.9 Hz), 114.4 (d, ^2J = 24.1 Hz), 112.2, 111.9 (d, ^3J = 6.0 Hz). HRMS (ESI-FTMS) Mass (m/z): calcd for $C_{26}H_{15}ClF_2N_4O_4$ [M + H]$^+$ = 521.0823, 523.0796, found 521.0818, 523.0793.

4-[3-(7-fluoro-2-oxo-3,8a-dihydrochromen-3-yl)-4-[(E)-[[4-(trifluoromethyl)phenyl]hydrazono]methyl] pyrazol-1-yl]benzoic acid (**19**). Yellowish; (393 mg, 73%) ^{1}H-NMR (300 MHz, DMSO-d_6): 10.70 (s, 1H), 9.04 (s, 1H), 8.34 (s, 1H), 8.12–8.09 (m, 4H), 7.92 (s, 1H), 7.75–7.73 (m, 1H), 7.62–7.59 (m, 2H), 7.35–7.32 (m, 2H), 6.92–6.89 (m, 2H); ^{13}C-NMR (75 MHz, DMSO-d_6): δ 167.0, 158.6 (d, ^1J = 239.6 Hz), 159.2, 150.2, 148.5, 146.2, 142.3, 142.2, 131.8, 131.4, 129.2, 127.9, 126.6, 125.7 (q, J = 268.6 Hz) 122.9, 120.7, 120.2 (d, ^3J = 9.6 Hz), 118.6 (d, ^3J = 10.8 Hz), 118.3, 117.8, 114.5 (d, ^2J = 24.0 Hz), 111.7. HRMS (ESI-FTMS) Mass (m/z): calcd for $C_{27}H_{16}F_4N_4O_4$ [M + H]$^+$ = 537.1180, found 537.1177.

4-[4-[(E)-[(4-cyanophenyl)hydrazono]methyl]-3-(7-fluoro-2-oxo-3,8a-dihydrochromen-3-yl)pyrazol-1-yl] benzoic acid (**20**). Yellow; (436 mg, 88%) ^{1}H-NMR (300 MHz, DMSO-d_6): δ 10.86 (s, 1H), 9.08 (s, 1H), 8.36 (s, 1H), 8.13–8.06 (m, 4H), 7.95 (s, 1H), 7.76–7.73 (m, 1H), 7.65–7.55 (m, 2H), 7.46–7.43 (m, 2H), 6.90–6.87 (m, 2H); ^{13}C-NMR (75 MHz, DMSO-d_6): δ 167.1, 159.2, 158.6 (d, ^1J = 239.7 Hz), 150.2, 148.8, 146.2, 142.3, 133.8, 133.0, 131.4, 129.3, 128.1, 122.7, 120.54, 120.57, 120.2 (d, ^3J = 9.8 Hz), 120.0, 118.7, 118.59, 118.55, 114.5 (d, ^2J = 24.0 Hz), 112.1, 99.3. HRMS (ESI-FTMS) Mass (m/z): calcd for $C_{27}H_{16}FN_5O_4$ [M + H]$^+$ = 494.1259, found 494.1252.

4-[3-(7-fluoro-2-oxo-3,8a-dihydrochromen-3-yl)-4-[(E)-[(4-nitrophenyl)hydrazono]methyl]pyrazol-1-yl] benzoic acid (**21**). Orange; (458 mg, 89%) ^{1}H-NMR (30 0MHz, DMSO-d_6): δ 11.2 (s, 1H), 9.08 (s, 1H), 8.35 (s, 1H), 8.11–8.02 (m, 5H), 7.94 (d, J = 9.2 Hz, 2H), 7.75–7.72 (m, 1H), 7.61–7.55 (m, 2H), 6.90–6.87 (m, 2H); ^{13}C-NMR (75 MHz, DMSO-d_6): δ 167.0, 158.6 (d, ^1J = 239.4 Hz), 159.2, 150.8, 150.3, 146.4, 142.3, 142.2, 138.4, 134.9, 131.4, 129.3, 128.4, 126.3, 122.5, 120.37, 120.32, 120.2, 120.1 (d, ^3J = 10.6 Hz), 118.3, 114.5 (d, ^2J = 24.4 Hz), 111.2. HRMS (ESI-FTMS) Mass (m/z): calcd for $C_{26}H_{16}FN_5O_6$ [M + H]$^+$ = 514.1157, found 514.115.

4-[4-[(E)-(dimethylhydrazono)methyl]-3-(6-fluoro-2-oxo-3,8a-dihydrochromen-3-yl)pyrazol-1-yl]benzoic acid (**22**). Yellow; (333 mg, 79%) ^{1}H-NMR (300 MHz, DMSO-d_6): 8.75 (s, 1H), 8.27 (s, 1H), 8.12–8.06 (m, 4H), 7.73–7.70 (m, 1H), 7.54 (d, J = 6.4 Hz, 2H), 7.20 (s, 1H), 2.49 (s, 3H); ^{13}C-NMR (75 MHz, DMSO-d_6): δ 167.1, 158.5 (d, ^1J = 239.1 Hz), 159.1, 150.3, 146.1, 142.5, 141.8, 131.4, 128.8, 125.7, 125.5, 122.9, 122.3, 120.3 (d, ^3J = 9.6 Hz), 119.7 (d, ^2J = 24.8 Hz), 118.5 (d, ^3J = 8.7 Hz), 118.3, 114.3 (d, ^2J = 24.1 Hz), 42.9. HRMS (ESI-FTMS) Mass (m/z): calcd for $C_{22}H_{17}FN_4O_4$ [M + H]$^+$ = 421.1307, found 421.1298.

4-[3-(7-hydroxy-2-oxo-chromen-3-yl)-4-[(E)-(phenylhydrazono)methyl]pyrazol-1-yl]benzoic acid (**23**). Brown; (370 mg, 79%) ^{1}H-NMR (300 MHz, DMSO-d_6): δ 10.17 (s, 1H), 8.94 (s, 1H), 8.22 (s, 1H), 8.08 (s, 5H), 7.81 (s, 1H), 7.67 (d, J = 8.4 Hz, 1H), 7.04 (t, J = 7.9 Hz, 2H), 6.83 (t, J = 14.7 Hz, 4H), 6.65 (t, J = 7.0 Hz, 1H); ^{13}C-NMR (75 MHz, DMSO-d_6) δ 167.1, 162.2, 159.9, 155.9, 146.9, 145.7, 143.7, 142.5, 131.4, 130.7, 129.5, 129.3, 128.8, 126.8, 121.2, 118.7, 118.3, 117.0, 114.0, 112.1, 111.8, 102.4. HRMS (ESI-FTMS) Mass (m/z): calcd for C26H18N4O5 [M + H]$^+$ = 467.1350, found 467.1352.

4-[4-[(E)-[(3-fluorophenyl)hydrazono]methyl]-3-(7-hydroxy-2-oxo-chromen-3-yl)pyrazol-1-yl]benzoic acid (**24**). Brownish; (428 mg, 86%) ^{1}H-NMR (300 MHz, DMSO-d_6): δ 10.95 (br s, 1H), 10.54 (s, 1H), 9.00 (s, 1H), 8.22 (s, 1H), 8.08 (s, 5H), 7.87 (s, 1H), 7.66 (d, J = 8.1 Hz, 1H), 7.07–7.06 (m, 1H), 6.90–6.86

(m, 2H), 6.66–6.60 (m, 2H), 6.41 (t, J = 8.1 Hz, 1H); ^{13}C-NMR (75 MHz, DMSO-d_6) δ 167.0, 163.8 (d, ^1J = 238.9 Hz), 162.4, 159.9, 155.9, 147.7 (d, ^3J = 11.0 Hz), 147.0, 143.8, 142.5, 131.4, 130.8, 130.7, 130.6, 128.9, 127.2, 120.9, 118.3, 116.7, 114.1, 111.7, 108.2, 104.7 (d, ^2J = 21.7 Hz), 102.4, 98.5 (d, ^2J = 26.0 Hz). HRMS (ESI-FTMS) Mass (m/z): calcd for $C_{26}H_{17}FN_4O_5$ [M + H]$^+$ = 485.1256, found 485.1257.

4-[4-[(E)-[(4-chlorophenyl)hydrazono]methyl]-3-(7-hydroxy-2-oxo-chromen-3-yl)pyrazol-1-yl]benzoic acid (**25**). Brown; (357 mg, 71%) ^{1}H-NMR (300 MHz, DMSO-d_6): δ 10.82 (br s, 1H), 10.32 (s, 1H). 8.95 (s, 1H), 8.22 (s, 1H), 8.06 (s, 4H), 7.80 (s, 1H), 7.66 (d, J = 8.2 Hz, 1H), 7.06 (d, J = 8.5 Hz, 3H), 6.83–6.81(m, 4H); ^{13}C-NMR (75 MHz, DMSO-d_6) δ 167.1, 162.2, 159.9, 156.0, 147.0, 144.6, 143.8, 142.5, 131.4, 130.7, 130.4, 129.0, 128.9, 127.0, 121.9, 120.9, 118.3, 116.7, 114.0, 113.5, 111.8, 102.4. HRMS (ESI-FTMS) Mass (m/z): calcd for $C_{26}H_{17}ClN_4O_5$ [M + H]$^+$ = 501.0960, 503.0933, found 501.0962, 503.0932.

4-[4-[(E)-[(4-bromophenyl)hydrazono]methyl]-3-(7-hydroxy-2-oxo-chromen-3-yl)pyrazol-1-yl]benzoic acid (**26**). Brown; (437 mg, 80%) ^{1}H-NMR (300 MHz, DMSO-d_6): δ 10.88 (br s, 1H), 10.39 (s, 1H), 8.96 (s, 1H), 8.22 (s, 1H), 8.07 (s, 5H), 7.83 (s, 1H), 7.67 (d, J = 7.9 Hz, 1H), 7.18 (d, J = 7.8 Hz, 2H), 6.90–6.78 (m, 4H); ^{13}C-NMR (75 MHz, DMSO-d_6) δ 167.0, 162.3, 159.9, 155.9, 147.0, 145.0, 143.8, 142.5, 131.8, 131.4, 130.7, 130.4, 128.8, 127.0, 120.9, 118.3, 116.7, 114.1, 114.0, 111.8, 109.4, 102.4. HRMS (ESI-FTMS) Mass (m/z): calcd for $C_{26}H_{17}BrN_4O_5$ [M + H]$^+$ = 545.0455, 547.0436, found 545.0456, 547.0435.

4-[4-[(E)-[(2,5-difluorophenyl)hydrazono]methyl]-3-(7-hydroxy-2-oxo-chromen-3-yl)pyrazol-1-yl]benzoic acid (**27**). Brownish; (459 mg, 91%) ^{1}H-NMR (300 MHz, DMSO-d_6): δ 10.94 (br s, 1H), 10.33 (s, 1H), 9.03 (s, 1H), 8.20 (s, 1H), 8.13–8.07 (m, 5H), 7.65 (d, J = 8.3 Hz, 1H), 7.12–7.04 (m, 1H), 6.90–6.84 (m, 3H), 6.44–6.39 (m, 1H); ^{13}C-NMR (75 MHz, DMSO-d_6) δ 167.0, 162.4, 159.7 (d, ^1J = 235.9 Hz), 159.8, 155.9, 145.5 (d, ^1J = 241.0 Hz), 147.0, 142.4, 135.2 (m), 133.5, 131.4, 130.6, 128.9, 127.7, 120.6, 118.4, 116.6, 116.4, 116.2 (d, ^3J = 9.6 Hz), 114.1, 111.6, 103.7 (d, ^2J = 25.0 Hz), 102.4, 100.4 (d, ^2J = 29.2 Hz). HRMS (ESI-FTMS) Mass (m/z): calcd for $C_{26}H_{16}F_2N_4O_5$ [M + H]$^+$ = 503.1162, found 503.1158.

4-[4-[(E)-[(3,4-difluorophenyl)hydrazono]methyl]-3-(7-hydroxy-2-oxo-chromen-3-yl)pyrazol-1-yl]benzoic acid (**28**). Black; (378 mg, 75%) ^{1}H-NMR (300 MHz, DMSO-d_6): δ 10.41 (s, 1H), 8.99 (s, 1H), 8.22 (s, 1H), 8.07 (s, 4H), 7.83 (s, 1H), 7.66 (d, J = 8.2 Hz, 1H), 7.17–7.07 (m, 1H), 6.88–6.76 (m, 3H), 6.57 (s, 1H); ^{13}C-NMR (75 MHz, DMSO-d_6) δ 167.1, 162.3, 159.9, 155.9, 152.1 (dd, J = 13.0, 254.0 Hz), 146.9, 144.4 (dd, J = 12.9, 246.3 Hz), 143.8, 142.4, 141.2 (d, ^3J = 12.8 Hz), 131.4, 130.7, 128.8, 127.3, 120.8, 119.3, 118.3, 118.0 (d, ^3J = 17.5 Hz), 116.7, 114.1, 111.7, 107.7, 102.4, 100.2 (d, ^2J = 21.9 Hz). HRMS (ESI-FTMS) Mass (m/z): calcd for $C_{26}H_{16}F_2N_4O_5$ [M + H]$^+$ = 503.1162, found 503.1162.

4-[4-[(E)-[(2,4-dichlorophenyl)hydrazono]methyl]-3-(7-hydroxy-2-oxo-chromen-3-yl)pyrazol-1-yl]benzoic acid (**29**). Brownish; (483 mg, 90%) ^{1}H-NMR (300 MHz, DMSO-d_6): δ 10.94 (br s, 1H), 9.96 (s, 1H), 9.04 (s, 1H), 8.24 (d, J = 5.7 Hz, 2H), 8.09 (s, 5H), 7.67 (d, J = 8.1 Hz, 1H), 7.40 (s, 1H), 7.23 (d, J = 8.7 Hz, 1H), 7.00 (d, J = 8.9 Hz, 1H) 6.92–6.88 (m, 2H); ^{13}C-NMR (75 MHz, DMSO-d_6) δ 167.0, 162.4, 159.9, 155.9, 147.3, 144.0, 142.4, 141.0, 134.2, 131.4, 130.7, 128.9, 127.9, 127.3, 122.1, 120.7, 118.4, 116.7, 116.4, 115.0, 114.1, 111.7, 102.4. HRMS (ESI-FTMS) Mass (m/z): calcd for $C_{26}H_{16}Cl_2N_4O_5$ [M + H]$^+$ = 535.0571, 537.0543, found 535.0570, 537.0545.

4-[4-[(E)-[(3-chloro-2-fluoro-phenyl)hydrazono]methyl]-3-(7-hydroxy-2-oxo-chromen-3-yl)pyrazol-1-yl] benzoic acid (**30**). Brown; (437 mg, 84%) ^{1}H-NMR (300 MHz, DMSO-d_6): δ 10.33 (s, 1H), 9.00 (s, 1H), 8.32 (s, 1H), 8.11–8.07 (m, 5H), 7.66 (d, J = 8.1 Hz, 1H), 7.08 (s, 1H), 6.89–6.79 (m, 4H); ^{13}C-NMR (75 MHz, DMSO-d_6) δ 167.1, 162.3, 159.9, 155.9, 147.1, 144.7 (d, ^1J = 240.2 Hz), 143.8, 142.4, 135.3 (d, ^3J = 9.3 Hz), 133.5, 131.4, 130.7, 128.9, 127.5, 125.4, 120.6, 119.9 (d, ^3J = 14.2 Hz), 118.7, 118.4, 116.7, 114.1, 112.6, 111.7, 102.4. HRMS (ESI-FTMS) Mass (m/z): calcd for $C_{26}H_{16}ClFN_4O_5$ [M + H]$^+$ = 519.0866, 521.0839, found 519.0869, 521.0844.

4-[4-[(E)-[(3-chloro-4-fluoro-phenyl)hydrazono]methyl]-3-(7-hydroxy-2-oxo-chromen-3-yl)pyrazol-1-yl] benzoic acid (**31**). Brown; (380 mg, 73%) ^{1}H-NMR (300 MHz, DMSO-d_6): δ 10.42 (s, 1H), 8.99 (s, 1H), 8.22 (s, 1H), 8.10–8.07 (m, 5H), 7.84 (s, 1H), 7.66 (d, J = 7.8 Hz, 1H), 7.12 (t, J = 9.5 Hz, 1H), 6.94–6.85 (m, 3H), 6.75 (br s, 1H); ^{13}C-NMR (75 MHz, DMSO-d_6) δ 167.0, 162.3, 159.9, 155.9, 150.8 (d, ^1J = 234.6 Hz), 146.9, 143.7, 143.2, 142.5, 131.4, 130.7 (d, ^3J = 14.9 Hz), 128.9, 127.4, 120.7, 120.2 (d, ^2J = 18.1 Hz), 119.4,

118.3, 117.4 (d, ^2J = 21.6 Hz), 116.8, 114.0, 112.4, 111.9 (d, ^3J = 6.8 Hz), 111.7, 102.6. HRMS (ESI-FTMS) Mass (m/z): calcd for $C_{26}H_{16}ClFN_4O_5$ [M + H]$^+$ = 519.0866, 521.0839, found 519.0866, 521.0837.

4-[3-(7-hydroxy-2-oxo-3,8a-dihydrochromen-3-yl)-4-[(E)-[[4-(trifluoromethyl)phenyl]hydrazono]methyl] pyrazol-1-yl]benzoic acid (**32**). Yellow; (445 mg, 83%) ^{1}H-NMR (300 MHz, DMSO-d_6): δ 10.7 (s, 1H), 9.00 (s, 1H), 8.24 (s, 1H), 8.08 (s, 4H), 7.90 (s, 1H), 7.68 (d, J = 8.3 Hz, 1H), 7.36 (d, J = 8.5 Hz, 1H), 6.96–6.86 (m, 3H); ^{13}C-NMR (75 MHz, DMSO-d_6) δ 167.1, 162.3, 159.9, 156.0, 148.6, 147.1, 143.9, 142.4, 132.1, 131.4, 130.8, 129.0, 127.4, 127.2, 126.6, 123.6, 120.6, 118.6-118.1 (m), 116.6, 114.1, 111.8, 111.7, 102.3. HRMS (ESI-FTMS) Mass (m/z): calcd for $C_{27}H_{17}F_3N_4O_5$ [M + H]$^+$ = 535.1224, found 535.1223.

4-[4-[(E)-[(4-cyanophenyl)hydrazono]methyl]-3-(7-hydroxy-2-oxo-3,8a-dihydrochromen-3-yl)pyrazol-1-yl] benzoic acid (**33**). Yellow; (399 mg, 81%) ^{1}H-NMR (300 MHz, DMSO-d_6): δ 10.83 (s, 1H), 9.02 (s, 1H), 8.25 (s,1H), 8.08 (s, 4H), 7.92 (s, 1H), 7.67 (d, J = 8.3 Hz, 1H), 7.46 (d, J = 8.7 Hz, 2H), 6.95–6.85 (m, 4H); ^{13}C-NMR (75 MHz, DMSO-d_6) δ 167.1, 162.3, 159.9, 156.0, 148.9, 147.2, 143.9, 142.4, 133.8, 133.3, 131.4, 130.8, 129.0, 127.6, 120.6, 120.4, 118.4, 116.4, 114.1, 112.1, 111.8, 102.4, 99.2. HRMS (ESI-FTMS) Mass (m/z): calcd for $C_{27}H_{17}N_5O_5$ [M + H]$^+$ = 492.1302, found 492.1300.

4-[3-(7-hydroxy-2-oxo-3,8a-dihydrochromen-3-yl)-4-[(E)-(methoxycarbonylhydrazono)methyl]pyrazol-1-yl] benzoic acid (**34**). Brownish; (324 mg, 72%) ^{1}H-NMR (300MHz, DMSO-d_6): δ 10.97 (br s, 1H), 8.95 (s, 1H), 8.24 (s, 2H), 8.06 (s, 4H), 7.95 (s, 1H), 7.66 (d, J = 8.4 Hz, 1H), 6.87–6.81 (m, 2H); ^{13}C-NMR (75 MHz, DMSO-d_6) δ 167.4, 162.6, 160.0, 156.0, 154.2, 147.7, 144.3, 142.0, 131.3, 130.8, 130.3, 127.5, 119.6, 118.9, 118.6, 115.6, 114.1, 111.6, 102.4, 52.2. HRMS (ESI-FTMS) Mass (m/z): calcd for $C_{22}H_{16}N_4O_7$ [M + H]$^+$ = 449.1092, found 449.1107.

3.2. Culturing of Bacteria

Bacterial cultures were maintained on tryptic soy agar (TSA) slants. Bacteria prepared for MIC testing or plated for time kill assays were grown on blood agar (TSA with 5% sheep blood) plates. Bacteria grown in liquid culture including 96-well plates for MIC testing were grown in Mueller Hinton Broth (cation adjusted, CAMHB). Bacterial dilutions were made in normal saline or phosphate buffered saline (PBS)

3.3. Minimum Inhibitory Concentration (MIC)

The broth microdilution method was utilized to determine MIC values of coumarin-substituted pyrazole derivatives against different clinically important *S. aureus* strains, *B. subtilis* and *S. epidermidis* according to the guidelines outlined by the Clinical and Laboratory Standards Institute (CLSI) as reported in our recent papers [9]. The starting concentration of compounds for MIC determination was 50 µg/mL serially diluted down the wells and the MIC values were recorded in duplicates in three independent experiments on different days.

3.4. Time Kill Assay

Time kill assay was performed against two strains of bacteria (*A. baumannii* ATCC 19606 and *S. aureus* ATCC 33599) using our most potent compounds. Bacteria in the logarithmic growth phase (OD600 ~2.00) were diluted to 1.5×10^6 colony-forming units (CFU/mL) and exposed to concentrations equivalent to 4×MIC (in triplicates) of tested compounds, vancomycin and colistin as control drugs for *S. aureus* and *A. baumannii* respectively. Aliquots were collected from treatment after 0, 2, 4, 6, 8, 10, 12, and 24 h of incubation at 35 °C and subsequently serially diluted in PBS. The diluted aliquots were then transferred to blood agar plates and incubated at 35 °C for 18–20 h before viable CFU/mL was determined by using the 6 × 6 drop plate method.

3.5. Biofilm Inhibition Assay

The biofilm-forming strain *S. aureus* ATTC 25923 was grown in the presence of four different coumarin-substituted pyrazole derivatives with the lowest MIC values to determine if the compounds would interfere with biofilm formation. An overnight culture of bacteria was suspended in PBS solution to match the 0.5 McFarland standard and then diluted 1:100 into CAMHB with 1% glucose to give an approximate final concentration of 1×10^6 CFU/mL. Bacterial broth suspension (195 µL) was transferred to each well in 96-well polystyrene flat bottom plate. Compounds (5 µL) at 2 × MIC, MIC, and 0.5 × MIC were added to wells in triplicate along with broth only, and bacteria along with DMSO controls, and plates were incubated at 35 °C for 24 h. After incubation, the contents of the wells were removed and wells were washed with 1×PBS solution three times to remove any planktonic cells. The plate was dried in an oven at 60 °C for about 15 min and 0.1% (w/v) crystal violet (250 µL) was added to each well and left for 15 min for staining biofilms. Excess crystal violet was removed by draining and washing three times with deionized water, and the plate was again dried in oven for 10 min. After drying, 33% acetic acid (250 µL) was added to each well to dissolve the stained biofilm. The optical density of the solubilized crystal violet in each well was measured at 620 nm using a Bio Tek™ Cytation™ 5 plate reader.

3.6. Biofilm Destruction Assay

This assay was performed to test whether our four most effective compounds could destroy the preformed biofilm in vitro. Bacterial cultures were prepared as for the biofilm inhibition test (195 µL culture per well) but without addition of compounds. Bacteria and growth medium only controls were incubated overnight at 35 °C for 24 h to allow bacterial formation of enough biofilm. After incubation, soluble and suspended well contents were carefully removed, and wells were washed with sterile PBS solution to remove any unadhered cells. Next, 195 µL sterile CAMHB with 1% glucose was added to each well with 5 µL of 2 × MIC, MIC, and 0.5 × MIC concentrations of compounds or DMSO in triplicate and the plate was incubated at 35 °C for 24 h. This extra 24 h incubation was required in order to provide enough time for compounds to destroy the preformed biofilm in the plate. After incubation, washing, drying, staining, dissolving stained dye, and measuring optical density in a plate reader were performed as described above.

3.7. Processing of Data

As these biofilm assays were performed in triplicates mean and standard deviation of plate reading data were processed. Results were expressed as a percentage by using the formula:

$$\text{Percentage biofilm inhibition/destruction} = \left[1 - \frac{OD_{compound} - OD_{broth}}{OD_{dmso} - OD_{broth}}\right]$$

where, $OD_{compound}$ = optical density of well with compound, OD_{broth} = OD of well with broth only, OD_{dmso} = OD of well with bacteria broth + DMSO.

The data were processed and represented in graphical form in Microsoft® Excel® for Office 365 MSO.

4. Conclusions

Because of the ease of synthesis without column purification or work-up, we have reported the synthesis of new hydrazone derivatives of coumarin-derived pyrazoles. We synthesized 31 new pyrazole derivatives. These new molecules were tested against several bacterial strains, and we found several molecules, which showed promising results with MIC values as low as 1.56 µg/mL. We found that fluoro-substituted compounds are more potent than the hydroxy-substituted compounds. Potent molecules are growth inhibitor of MRSA biofilms and eliminated the preformed biofilm more efficiently than the positive control, vancomycin. One of the potent molecules (**19**) showed very mild toxicity when comparing the IC_{50} against HEK293 cells to the MIC against bacteria. Future direction will be the synthesis of molecules related to the potent compounds to target bacterial biofilms.

5. Patents

Alam, M. A. Antimicrobial agents and the method of synthesizing the antimicrobial agents. US Patent. 10,596,153, 2020.

Supplementary Materials: The following are available online at http://www.mdpi.com/1420-3049/25/12/2758/s1, ^{1}H and ^{13}C-NMR spectra.

Author Contributions: Conceptualization, M.A.A.; methodology, R.A., N.A., J.W., S.A.C.; writing—original draft preparation, M.A.A.; writing—review and editing, M.A.A., D.G., H.R.K.C.; supervision, M.A.A., D.G.; project administration, M.A.A.; funding acquisition, M.A.A., D.G. All authors have read and agreed to the published version of the manuscript.

Funding: This publication was made possible by the Arkansas INBRE program, supported by a grant from the National Institute of General Medical Sciences, (NIGMS), P20 GM103429 from the National Institutes of Health.

Acknowledgments: This publication was made possible by the Arkansas INBRE program, INBRE voucher program, Arkansas Statewide Mass Spectrometry Facility, University of Arkansas, Fayetteville, Arkansas Biosciences, and Department of Chemistry and Physics, Arkansas State University, Jonesboro.

Conflicts of Interest: The authors declare no conflict of interest.

References

1. WHO New Report Calls for Urgent Action to Avert Antimicrobial Resistance Crisis. Available online: https://www.who.int/news-room/detail/29-04-2019-new-report-calls-for-urgent-action-to-avert-antimicrobial-resistance-crisis (accessed on 28 February 2020).
2. CDC Antibiotic/Antimicrobial Resistance (AR/AMR). Available online: https://www.cdc.gov/drugresistance/biggest-threats.html (accessed on 28 February 2020).
3. Craft, K.M.; Nguyen, J.M.; Berg, L.J.; Townsend, S.D. Methicillin-resistant *Staphylococcus aureus* (MRSA): Antibiotic-resistance and the biofilm phenotype. *Med. Chem. Comm.* **2019**, *10*, 1231–1241. [CrossRef] [PubMed]
4. Alam, M.J.; Alam, O.; Khan, S.A.; Naim, M.J.; Islamuddin, M.; Deora, G.S. Synthesis, anti-inflammatory, analgesic, COX1/2-inhibitory activity, and molecular docking studies of hybrid pyrazole analogues. *Drug Des. Devel. Ther.* **2016**, *10*, 3529–3543. [CrossRef] [PubMed]
5. Driscoll, J.P.; Sadlowski, C.M.; Shah, N.R.; Feula, A. Metabolism and Bioactivation: It's Time to Expect the Unexpected. *J. Med. Chem.* **2020**. [CrossRef] [PubMed]
6. Brider, J.; Rowe, T.; Gibler, D.J.; Gottsponer, A.; Delancey, E.; Branscum, M.D.; Ontko, A.; Gilmore, D.; Alam, M.A. Synthesis and antimicrobial studies of azomethine and *N*-arylamine derivatives of 4-(4-formyl-3-phenyl-*1H*-pyrazol-1-yl)benzoic acid as potent anti-methicillin-resistant *Staphylococcus aureus* agents. *Med. Chem. Res.* **2016**, *25*, 2691–2697. [CrossRef]
7. Whitt, J.; Duke, C.; Sumlin, A.; Chambers, S.A.; Alnufaie, R.; Gilmore, D.; Fite, T.; Basnakian, A.G.; Alam, M.A. Synthesis of Hydrazone Derivatives of 4-[4-Formyl-3-(2-oxochromen-3-yl)pyrazol-1-yl]benzoic acid as Potent Growth Inhibitors of Antibiotic-resistant *Staphylococcus aureus* and *Acinetobacter baumannii*. *Molecules* **2019**, *24*, 2051. [CrossRef] [PubMed]
8. Whitt, J.; Duke, C.; Ali, M.A.; Chambers, S.A.; Khan, M.M.K.; Gilmore, D.; Alam, M.A. Synthesis and Antimicrobial Studies of 4-[3-(3-Fluorophenyl)-4-formyl-*1H*-pyrazol-1-yl]benzoic Acid and 4-[3-(4-Fluorophenyl)-4-formyl-*1H*-pyrazol-1-yl]benzoic Acid as Potent Growth Inhibitors of Drug-Resistant Bacteria. *ACS Omega* **2019**, *4*, 14284–14293. [CrossRef] [PubMed]
9. Zakeyah, A.A.; Whitt, J.; Duke, C.; Gilmore, D.F.; Meeker, D.G.; Smeltzer, M.S.; Alam, M.A. Synthesis and antimicrobial studies of hydrazone derivatives of 4-[3-(2,4-difluorophenyl)-4-formyl-*1H*-pyrazol-1-yl]benzoic acid and 4-[3-(3,4-difluorophenyl)-4-formyl-*1H*-pyrazol-1-yl]benzoic acid. *Bioorg. Med. Chem. Lett.* **2018**, *28*, 2914–2919. [CrossRef] [PubMed]
10. Allison, D.; Delancey, E.; Ramey, H.; Williams, C.; Alsharif, Z.A.; Al-khattabi, H.; Ontko, A.; Gilmore, D.; Alam, M.A. Synthesis and antimicrobial studies of novel derivatives of 4-(4-formyl-3-phenyl-*1H*-pyrazol-1-yl)benzoic acid as potent anti-*Acinetobacter baumannii* agents. *Bioorg Med. Chem. Lett.* **2017**, *27*, 387–392. [CrossRef] [PubMed]
11. Alam, M.A. Antimicrobial Agents and the Method of Synthesizing the Antimicrobial Agents. US Patent US10596153B2, 24 March 2020.
12. Ahn, M.; Gunasekaran, P.; Rajasekaran, G.; Kim, E.Y.; Lee, S.-J.; Bang, G.; Cho, K.; Hyun, J.-K.; Lee, H.-J.; Jeon, Y.H.; et al. Pyrazole derived ultra-short antimicrobial peptidomimetics with potent anti-biofilm activity. *Eur. J. Med. Chem.* **2017**, *125*, 551–564. [CrossRef] [PubMed]

13. Khan, M.F.; Alam, M.M.; Verma, G.; Akhtar, W.; Akhter, M.; Shaquiquzzaman, M. The therapeutic voyage of pyrazole and its analogs: A review. *Eur. J. Med. Chem.* **2016**, *120*, 170–201. [CrossRef] [PubMed]
14. Hu, C.-F.; Zhang, P.-L.; Sui, Y.-F.; Lv, J.-S.; Ansari, M.F.; Battini, N.; Li, S.; Zhou, C.-H.; Geng, R.-X. Ethylenic conjugated coumarin thiazolidinediones as new efficient antimicrobial modulators against clinical methicillin-resistant *Staphylococcus aureus*. *Bioorg. Chem.* **2020**, *94*, 103434. [CrossRef] [PubMed]
15. Widelski, J.; Luca, S.V.; Skiba, A.; Chinou, I.; Marcourt, L.; Wolfender, J.-L.; Skalicka-Wozniak, K. Isolation and Antimicrobial Activity of Coumarin Derivatives from Fruits of *Peucedanum luxurians* Tamamsch. *Molecules* **2018**, *23*, 1222. [CrossRef] [PubMed]
16. Kovvuri, J.; Nagaraju, B.; Ganesh Kumar, C.; Sirisha, K.; Chandrasekhar, C.; Alarifi, A.; Kamal, A. Catalyst-free synthesis of pyrazole-aniline linked coumarin derivatives and their antimicrobial evaluation. *J. Saudi Chem. Soc.* **2018**, *22*, 665–677. [CrossRef]
17. Anamika; Divya, U.; Ekta; Nisha, J.; Shivali, S. Advances in Synthesis and Potentially Bioactive of Coumarin Derivatives. *Curr. Org. Chem.* **2018**, *22*, 2509–2536. [CrossRef]

Sample Availability: Samples of all the compounds are available from the authors.

© 2020 by the authors. Licensee MDPI, Basel, Switzerland. This article is an open access article distributed under the terms and conditions of the Creative Commons Attribution (CC BY) license (http://creativecommons.org/licenses/by/4.0/).

Article

Selective Synthesis and Photoluminescence Study of Pyrazolopyridopyridazine Diones and N-Aminopyrazolopyrrolopyridine Diones

Ching-Chun Tseng [1,2], Cheng-Yen Chung [2], Shuo-En Tsai [1,2], Hiroyuki Takayama [3], Naoto Uramaru [4], Chin-Yu Lin [5] and Fung Fuh Wong [1,2,*]

1. The Ph.D. Program for Biotech Pharmaceutical Industry, China Medical University, No. 91, Hsueh-Shih Rd., Taichung 40402, Taiwan; u106308201@cmu.edu.tw (C.-C.T.); u100003044@cmu.edu.tw (S.-E.T.)
2. School of Pharmacy, China Medical University, No. 91 Hsueh-Shih Rd., Taichung 40402, Taiwan; u108071001@cmu.edu.tw
3. Department of Medico Pharmaceutical Science, Nihon Pharmaceutical University, Komuro Inamachi Kita-adachi-gun, Saitama-ken 10281, Japan; h.takayama@nichiyaku.ac.jp
4. Department of Environmental Science, Nihon Pharmaceutical University, Komuro Inamachi Kita-adachi-gun, Saitama-ken 10281, Japan; uramaru@nichiyaku.ac.jp
5. Institute of New Drug Development, College of Biopharmaceutical and Food Sciences, China Medical University, No. 91 Hsueh-Shih Rd., Taichung 40402, Taiwan; geant@mail.cmu.edu.tw
* Correspondence: wongfungfuh@yahoo.com.tw or ffwong@mail.cmu.edu.tw; Tel.: +886-422-053-366 (ext. 5603); Fax: +886-422-078-083

Academic Editors: Vera L. M. Silva, Artur M. S. Silva and Elena Cariati
Received: 28 March 2020; Accepted: 19 May 2020; Published: 21 May 2020

Abstract: The newly designed luminol structures of pyrazolopyridopyridazine diones and N-aminopyrazolopyrrolopyridine diones were synthesized from versatile 1,3-diaryfuropyrazolopyridine -6,8-diones, 1,3-diarylpyrazolopyrrolopyridine-6,8-diones, or 1,3-diaryl-7- methylpyrazolopyrrolopyridine -6,8-diones with hydrazine monohydrate. Photoluminescent and solvatofluorism properties containing UV–Vis absorption, emission spectra, and quantum yield (Φ_f) study of pyrazolopyridopyridazine diones and N-aminopyrazolopyrrolopyridine diones were also studied. Generally, most of pyrazolopyrrolopyridine-6, 8-diones **6** exhibited the significant fluorescence intensity and the substituent effect when compared with N-aminopyrazolopyrrolopyridine diones, particularly for **6c** and **6j** with a *m*-chloro group. Additionally, the fluorescence intensity of **6j** was significantly promoted due to the suitable conjugation conformation. Based on the quantum yield (Φ_f) study, the value of compound **6j** (0.140) with planar structural skeletal was similar to that of standard luminol (**1**, 0.175).

Keywords: pyrazolopyridopyridazine dione; N-aminopyrazolopyrrolopyridine dione; luminol; photoluminescence

1. Introduction

Sleep-disorders are one of the largest public health concerns in the whole world [1]. New functionalized pyrazolo [3,4-*b*]pyrrolo[3,4-*d*]pyridine derivatives were enthusiastically investigated to develop the increased potency and reduced side effects of novel sedative/hypnotic drug compounds for treatment of sleep-disorders [2,3]. On the other hand, pyrazolopyridopyridazine diones are well-known as the versatile precursors for synthesis of pyrazolopyridopyridazine phosphodiesterase type 5 (PDE5) inhibitors [4,5]. In recent years, chemiluminescent luminol derivatives have been an attractive detection technique in analytical applications such as presumptive test agents for latent blood detection [6–9], high-performance liquid chromatography (HPLC) [10,11], DNA, immunoassay, and cancer screening detection [12,13]. Since now, many newly designed luminol structures have been enthusiastically investigated to increase the chemiluminescence efficiency, intensity, sensitivity, quantum yield, or the recognition ability of the resulting chemiluminogens (Figure 1) [14–19].

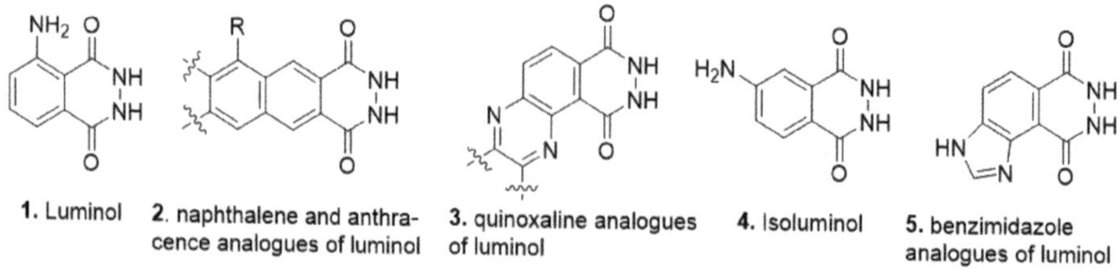

1. Luminol **2.** naphthalene and anthracence analogues of luminol **3.** quinoxaline analogues of luminol **4.** Isoluminol **5.** benzimidazole analogues of luminol

Figure 1. Luminol **1**, naphthalene and anthracene **2**, quinoxaline analogues of luminol **3**, isoluminol **4**, and benzimidazole analogues of luminol **5**.

Furthermore, N-aminophthalimides were considered as phthalazine 1,4-dione tautomeric pairs [20,21]. N-Amino maleimides with pyridine heterocycle series also presented as a very important privileged substructure in organic synthesis for preparing diverse biologically active molecules [22]. Typically, the most important pharmacological effects that have been reported are potential antimicrobial [22] and anticancer activities [23]. Herein, we judiciously explore the insertion of pyridazinedione and N-Amino maleimide units into the pyrazolopyridine core ring for construction of the new designed luminol structures **6a–j** and **7a–i** from versatile 1,3-diarylpyrazolopyrrolopyridine-6,8-diones **11**. Observably, we found that the series of pyridazinediones **6a–j** would not only provide conjugation systems but also allow to modify the fluorescence intensity and biological activity (Figure 2).

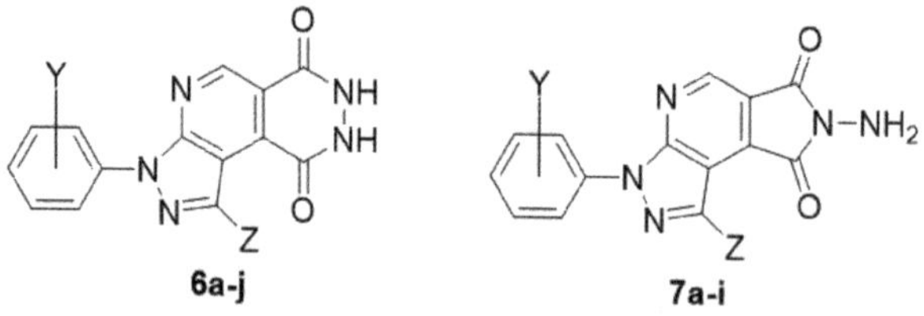

6a-j   **7a-i**

Figure 2. Pyrazolopyridopyridazine diones **6a–j** and N-Aminopyrazolopyrrolopyridine diones **7a–i** as luminol analogues.

2. Results and Discussion

Initially, dimethyl 1,3-diphenyl-1H-pyrazolo[3,4-b]pyridine-4,5-dicarboxylate **8** and diethyl 1,3-diphenyl-1H-pyrazolo[3,4-b]pyridine-4,5-dicarboxylate **9** were prepared by following our previously reported literature [24] from N,N-diisopropylamidinyl pyrazolylimine and chosen as the model substrate for this investigation on the construction of pyrazolopyridopyridazine diones **6a** (Scheme 1). Compounds **8** and **9** were reacted with hydrazine hydrate at reflux in methanol or ethanol solution under the basic condition for 24–36 h [25,26]. However, all the efforts for the predominant formation of **6a** were unsuccessful. We also attempted to perform the hydrolysis of ester groups of compounds **8** and **9** under basic conditions to obtain 1,3-diphenyl-1H-pyrazolopyridine-4,5-dicarboxylic acid **10** [27,28]. Subsequently, pyrazolopyridine-4,5-dicarboxylic acid **10** was refluxed with hydrazine in acetic acid to carry out the cyclization for 8 h, but without success (Scheme 1).

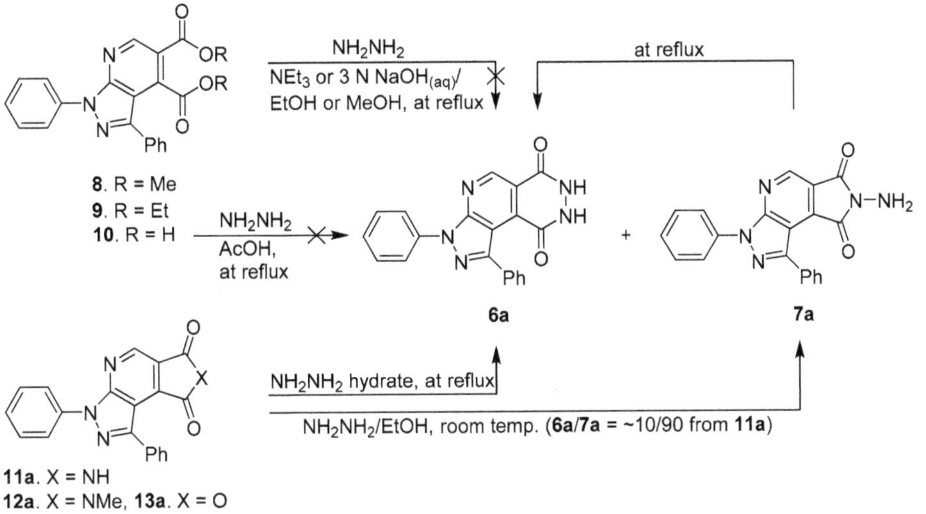

Scheme 1. Synthesis study of pyrazolopyridopyridazine dione **6a** as luminol analogue.

In other attempts, we preliminarily tried to synthesize 1,3-phenylpyrazolopyrrolopyridine-6, 8-dione **11a** from *N,N*-diisopropylamidinyl pyrazolylimine with maleimide via our published $InCl_3$/silica gel catalyzed hetero Diels-Alder reaction [29]. Subsequently, the resulting compound **11a** was reacted with an excess of hydrazine hydrate in $EtOH/H_2O$ co-solution at room temperature for ~7 h [29,30]. The formation of the *N*-aminopyrazolopyrrolopyridine dione **7a** was observed in 83% yield as the major product and accompanied with a trace amount of luminol-type pyrazolopyridopyridazine dione **6a** (<10%, Scheme 1). Fortunately, compounds **6a** and **7a** can be successfully and selectively prepared via kinetic and thermodynamic control reactions [31,32].

For further searching optimal conditions, we also prepared 7-methyl-1,3-phenylpyrazolopyrrolopyridine -6,8-dione **12a** [29] and 1,3-diphenylfuropyrazolopyridine-6,8-dione **13a** [33,34] as probes for monitoring cyclization tendency with hydrazine hydrate [29,30]. Most of the compounds **11a–13a** were refluxed in neat hydrazine hydrate solution for ~5 h (Scheme 1 and Table 1). The reactions were monitored until the consumption of starting materials **11a–13a** by TLC and produced the luminol-type pyrazolopyridopyridazine dione **6a**. Compound **11a** smoothly underwent the cyclization reaction to give luminol-type analogue **6a** in better yield (84%, Entry 1, Table 1). However, compounds **12a–13a** resulted in 32% and 18% low yields, respectively (Entries 2 and 3, Table 1). For further demonstration of reactivity efficiency, compounds **11b–c**, **12b–c**, and **13b–c** bearing various substituents including *o*- and *m*-Cl in *N*-1-phenyl ring and phenyl at *C*-3 position of pyrazole moiety were synthesized and refluxed under the same condition (Entries 4–9, Table 1). Based on the experimental data of Table 1, the better yields of pyrazolopyridopyridazine dione products **6b–c** were provided from 1,3-diarylpyrazolopyrrolopyridine-6,8-diones **11b–c** (74% and 71%, Entries 4 and 7, Table 1). Unfortunately, 1,3-diaryl-7-methylpyrazolopyrrolopyridine-6,8-diones **12b–c** and 1,3-diarylfuropyrazolopyridine-6,8-diones **13b–c** showed poor reactivity for the formation of pyrazolopyridopyridazine diones **6b–c** (Entries 5–6 and 8–9, Table 1).

Table 1. The results of pyrazolopyridopyridazine diones 6a–j from reactants 11a–j, 12a–c, or 13a–c with hydrazine hydrate.

11a-j, 12a-c, 13a-c   →   6a-j

Entry	S.M.	X	Y	Z	Reaction Time	Products	Yields (%)
1	11a	NH	Ph	Ph	5	6a	84
2	12a	NMe	Ph	Ph	5	6a	32
3	13a	O	Ph	Ph	5	6a	18
4	11b	NH	o-Cl-Ph	Ph	5	6b	74
5	12b	NMe	o-Cl-Ph	Ph	5	6b	24
6	13b	O	o-Cl-Ph	Ph	5	6b	13
7	11c	NH	m-Cl-Ph	Ph	5	6c	71
8	12c	NMe	m-Cl-Ph	Ph	5	6c	38
9	13c	O	m-Cl-Ph	Ph	5	6c	11
10	11d	NH	p-Cl-Ph	Ph	5	6d	81
11	11e	NH	p-Br-Ph	Ph	5	6e	77
12	11f	NH	p-Me-Ph	Ph	5	6f	84
13	11g	NH	p-OMe-Ph	Ph	5	6g	81
14	11h	NH	p-CN-Ph	Ph	5	6h	73
15	11i	NH	p-NO$_2$-Ph	Ph	5	6i	69
16	11j	NH	m-Cl-Ph	H	5	6j	71

Furthermore, we applied this reliable procedure to reactants 11d–j bearing p-Cl-Ph, p-Br-Ph, p-Me-Ph, p-OMe-Ph, p-CN-Ph, p-NO$_2$-Ph, and m-Cl-Ph at the N-1 position and phenyl and H at C-3 position of pyrazolic ring. Various substituted reactants 11d–j were demonstrated to proceed smoothly. Both electron-donating and electron-withdrawing substituents were all well-tolerated in good yields (69–84%, Entries 10–16, Table 1). All of 1,3-diarylpyrazolopyrrolopyridine-6,8-diones 6a–j were fully characterized by spectroscopic methods. For example, compound 6a presented one singlet at δ 9.41 ppm for pyrazolopyridine ring N=CH–C=C in ^{1}H-NMR and two peaks at δ 153.1 and 155.7 ppm for pyridazine dione carbon O=C–NH in ^{13}C-NMR spectrum. Its IR absorptions showed peaks at 3161 cm^{-1} for stretching of the –NH group and at 1014 cm^{-1} for stretching of the N–N group.

For the further controlled experiment for photoluminescence study, we also tried to prepare a series of N-aminopyrazolopyrrolopyridine diones 7a–i as the comparison cases (Scheme 2). Treatment of pyrazolopyrrolopyridine-6,8-diones 11a–i with 5.0 equivalents of hydrazine hydrate in EtOH/H$_2$O co-solution was performed in an ice-bath to room temperature for 48 h. The corresponding N-aminopyrazolopyrrolopyridine diones 7a–i were obtained in 71–87% yields and characterized by spectroscopic methods. For example, compound 7a presented one singlet peak at δ 8.83 ppm for pyrazolopyridine ring N=CH–C=C in ^{1}H-NMR and two peaks at δ 164.1 and 164.4 ppm for phthalimide moiety carbon O=C–NH in ^{13}C-NMR spectrum. Its IR absorptions showed peaks at 3172 and 3276 cm^{-1} for stretching of the –NH$_2$ group and at 1014 cm^{-1} for stretching of the N–N group. Due to the structural skeletons being very similar between 6 and 7, the identify method should be our next future evaluation.

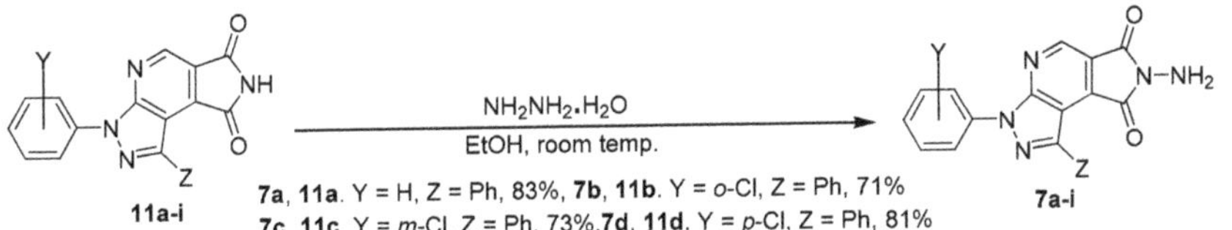

Scheme 2. The results of *N*-aminopyrazolopyrrolopyridine diones **7a–i** from pyrazolopyrrolopyridine-6, 8-diones **11a–i** with hydrazine hydrate.

7a, 11a. Y = H, Z = Ph, 83%, **7b, 11b.** Y = *o*-Cl, Z = Ph, 71%
7c, 11c. Y = *m*-Cl, Z = Ph, 73%, **7d, 11d.** Y = *p*-Cl, Z = Ph, 81%
7e, 11e. Y = *p*-Br, Z = Ph, 79%, **7f, 11f.** Y = *p*-Me, Z = Ph, 86%
7g, 11g. Y = *p*-OMe, Z = Ph, 87%, **7h, 11h.** Y = *p*-CN, Z = Ph, 73%
7i, 11i. Y = *p*-NO$_2$, Z = Ph, 74%

Luminol (**1**), compounds **6a** and **7a** were dissolved in DMSO to prepare a stock solution (1 × 10^{-3} M). Then the stock solutions of compounds **6a** and **7a** were individually diluted to a concentration of 10 μM in the presence of various solvents such as toluene, THF, ethyl acetate (EA), CH$_2$Cl$_2$, MeCN, acetone, and DMSO. The standard stock solution of luminol (10 μM) was diluted in DMSO solution as the standard sample. The UV–Vis absorption and fluorescence emission spectra of the pyrazolopyridopyridazine dione **6a** and *N*-aminopyrazolopyrrolopyridine dione **7a** compounds in the above-mentioned solution of varying polarities were reported in Table 2. The pyrazolopyridopyridazine dione **6a** has better solubility in polar organic solvents, such as DMSO > THF > acetone, but *N*-aminopyrazolopyrrolopyridine dione **7a** has the solubility only in highly polar solvents like DMSO. Luminol (**1**) was also measured and used as the standard sample. The UV–Vis absorption spectra of the compounds **6a** and **7a** in all the studied solvents were almost nearly the same; their absorption property is independent of the solvent polarity (Figure 3 and Table 2). All these compounds exhibit two highly intense absorption maxima peaks. Among these two, the first one was a high energy absorption between 253 nm and 286 nm for **6a** and **7a** probably due to the π–π* transition of the aryl core [35] while the low energy band between 329 nm and 366 nm is attributed to the intramolecular charge transfer transition (ICT). However, the rigidity in the structure of compounds **6a** and **7a** exhibited the stronger blue-shifted absorption (~15 nm) than luminol (**1**) in DMSO solution, as shown in Figure 3 and Table 2. In comparison with **6a** and **7a**, they demonstrated a similar absorption intensity, and compound **6a** has obvious red-shift ~20 nm with respect to **7a**.

Table 2. UV-Vis absorption maximum and fluorescence emission peak wavelength of luminol (**1**), pyrazolopyridopyridazine dione **6a** and *N*-aminopyrazolopyrrolopyridine dione **7a** in the different solvents.

Compound	Solvent	λmax/nm of UV-Vis	λmax/nm of PL
6a	Toluene	-[1], 366	469
6a	THF	271, 358	471
6a	Ethyl acetate	268, 356	473
6a	CH$_2$Cl$_2$	271, 350	483
6a	MeCN	268, 351	488
6a	Acetone	-[1], 353	477
6a	DMSO	264, 338	486
7a	Toluene	286,[1] ,348	452
7a	THF	264, 344	454
7a	Ethyl acetate	262, 343	459
7a	CH$_2$Cl$_2$	264, 344	471
7a	MeCN	261, 329	425, 461
7a	Acetone	-[1], 338	452
7a	DMSO	264, 335	429, 478
Luminol (1)	DMSO	350	392

[1] It was overlapped with solvent absorption band.

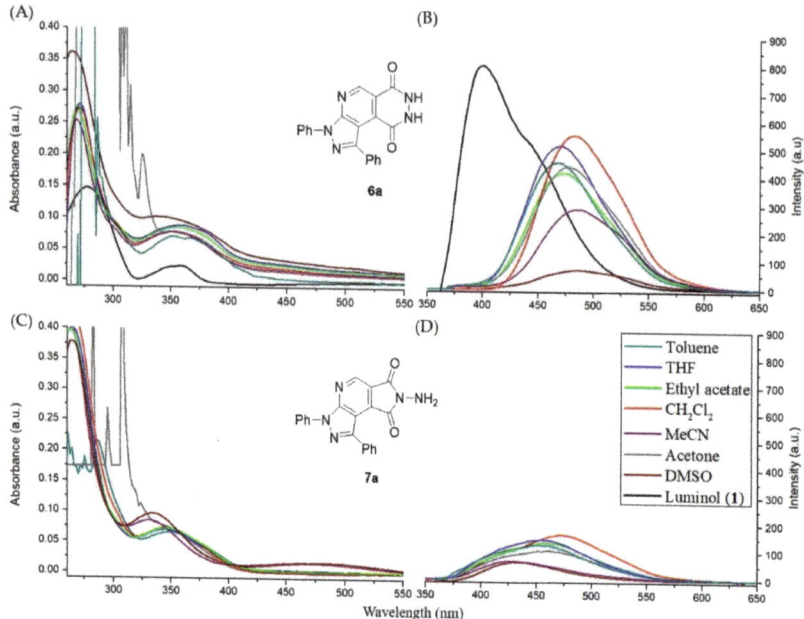

Figure 3. Photoluminescence spectra of luminol (**1**), pyrazolopyridopyridazine dione **6a** and N-aminopyrazolopyrrolopyridine dione **7a** in the different solvents. (**A**) Absorption and (**B**) emission spectra of luminol (**1**) and compound **6a**. (**C**) Absorption and (**D**) emission spectra of compound **7a**.

Consequently, we investigated the photoluminescence properties of the compounds **6a** and **7a** with luminol (**1**). For the fluorescence spectra, as shown in Figure 3 and Table 2, both the fluorescence intensity and the maximal position slightly varied depending on the solvent. Compound **6a** displayed a characteristic emission band of the excitation wavelengths between 400 and 600 nm, and the λ_{max}s of PL was ~480 nm with the intense greenish-blue fluorescence in Figures 3 and 4. For compound **7a**, it's emission spectrum was between 350 and 550 nm, and the λ_{max}s of PL was ~450 nm with the intense bluish-green fluorescence in Figures 3 and 4. Compounds **6a** and **7a** exhibited a red-shift ~80 nm or ~60 nm as compared to luminol (**1**). Therefore, new luminol analogues **6a** and **7a** were efficiently conjugate and connect two chromophores (pyrazole and pyridine) to lead to an increase of aromaticity and provide the greenish-blue or bluish-green fluorescent materials (Table 2 and Figure 4) [36]. Particularly, the best positive solvatofluorism phenomenon was presented in CH$_2$Cl$_2$ solution. It was also beneficial for the visibility of the naked eye due to the bathochromic (red-shift) phenomenon from blue color to green (Figure 4).

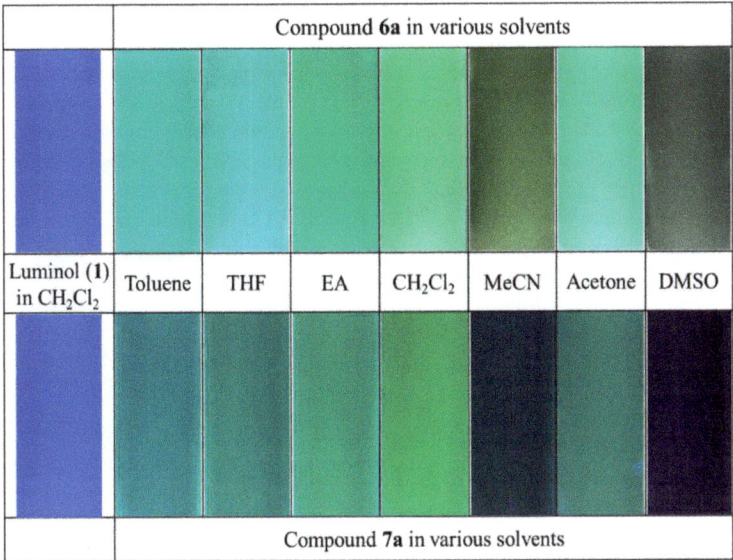

Figure 4. Color pictures of the fluorescence of compounds **6a** and **7a** in various solvents under excitation at 365 nm.

Moreover, the maximum of fluorescence wavelength and intensity, as shown in Figure 3, significantly vary with the diluted solvent. Further, we surprisingly observed the more significant solvent effect on compound **6a** when compared with compound **7a**. As shown in Figure 3, similar fluorescence spectra but a significant difference in intensity (~6 times) were observed in varying solvents. Of note, it was interesting that toluene, THF, EA, and CH_2Cl_2 had differences in their polarity (toluene: 0.099, THF: 0.207, EA: 0.228, CH_2Cl_2: 0.309, with respect to the reference polarity of DMSO: 0.444) [37,38]. However, for the above solvents, we observed a strong intensity, in comparison to that for protic or/and polar solvent (DMSO). The intensities of fluorescence bands were reversed in protic or/and polar solvents. Therefore, the solvent polarity modulation of fluorescence was quite interesting. It was well studied that amide tautomer of pyrazolopyridopyridazine dione **6a** was efficiently produced in toluene, THF, EA, and CH_2Cl_2 solvents [39–44]. In alcoholic (protic) and DMSO solvent, there exists competition between intermolecular bonding of the nearest hydrogen with the hydroxyimine tautomer of 6-hydroxypyrazolopyridopyridazin-9-one **6a**. Therefore, different intensities of behavior were observed in different polarity solutions. On the other hand, the different fluorescence intensity between structural isomers **6a** and **7a** was also observed [45]. The aromaticity of compound **6a** possessed the bathochromic shift of fluorescence maximum λ_{max} by 12 nm and ~4 times significant intensity in CH_2Cl_2 solution when compared with compound **7a** (Table 2 and Figure 3) [46]. However, the intramolecular and intermolecular hydrogen bondings between the amino and carbonyl groups of *N*-aminopyrazolopyrrolopyridine dione **7a** were formed to lead to the poor intensity in solution.

For further investigation of substituent efficiency of compounds **6** and **7** in photoluminescence properties, we synthesized a series of pyrazolopyridopyridazine diones **6a–j** and *N*-aminopyrazolo pyrrolopyridine diones **7a–i** bearing various substituents including *o*-, *m*- and *p*-Cl, *p*-Br, *p*-Me, *p*-OMe, *p*-CN, and *p*-NO$_2$ groups in *N*1-phenyl ring of pyrazole moiety. Generally, most of the substituents such as *o*-, *m*- and *p*-Cl, *p*-Br, *p*-Me, and *p*-CN in *N*1-phenyl of pyrazolic ring of compounds **6** possessed the blue-shift phenomenon range ~10 to 30 nm with significant fluorescence intensity when compared with compound **6a**, particularly for **6c** with *meta*-chloro group (Figure 5). For compounds **6g** and **6i** with the strong electron-donating (*p*-OMe) or electron-withdrawing groups (*p*-NO$_2$), they exhibited negative photoluminescence properties (Figure 5). While we modified the skeletal structure of pyrazolopyridopyridazine dione **6j**, in which Ph-group was replaced to H atom on C-3 position of pyrazolic ring, the blue-shift phenomenon was remarkably observed in photoluminescence spectra. Additionally, the fluorescence intensity of **6j** was significantly promoted about 2.3 times in comparison with compound **6a** (Figure 5). Based on the result of the substituent study, we conceived that compound **6j** was an effective substrate that possessed suitable conjugation conformation without the torsion effect to facilitate the photoluminescence properties [26]. For compounds **7a–i** bearing the above various substituents, they provided the weak fluorescence intensity [45] and possessed the blue-shift phenomenon when compared with **7a**, except for **7c** with *m*-chloro group and **7h** with *p*-CN group (Figure 6). Generally, compounds **7a–i** were the inappropriate photoluminescent substrates [45].

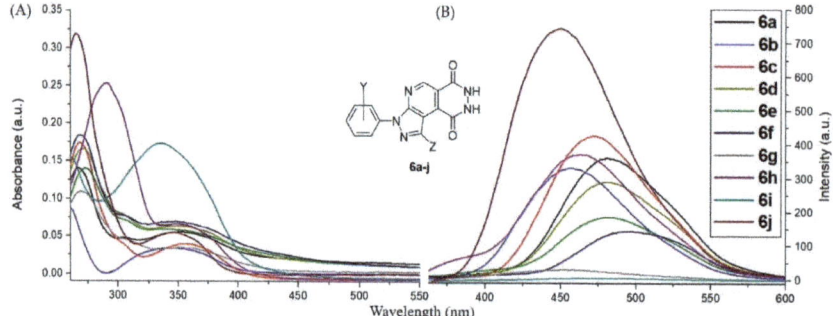

Figure 5. Photoluminescence spectra of pyrazolopyridopyridazine diones **6a–j** dissolved in DMSO to prepare a stock solution (1.0 mM). Then the stock solutions were diluted with CH_2Cl_2 to a concentration of 10 µM. (**A**) Absorption and (**B**) emission spectra of compounds **6a–j**.

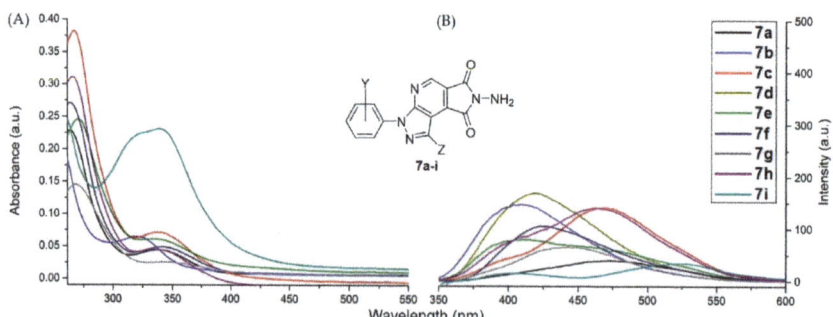

Figure 6. Photoluminescence spectra of *N*-aminopyrazolopyrrolopyridine diones **7a–i** dissolved in DMSO to prepare a stock solution (1.0 mM). Then the stock solutions were diluted with CH_2Cl_2 to a concentration of 10 µM. (**A**) Absorption and (**B**) emission spectra of compounds **7a–i**.

The quantum yields (Φ_f) of luminol (**1**) and pyrazolopyridopyridazine diones **6a**, **6c**, and **6j** were measured in the CH_2Cl_2 solution using quinine sulfate in 0.05M H_2SO_4 ($\Phi_f = 0.60$) as the standard (excitation wavelength 350 nm) [47,48]. The quantum yields (Φ_f) values of luminol (**1**) and pyrazolopyridopyridazine diones **6a**, **6c**, and **6j** were estimated as 0.175, 0.056, 0.067, and 0.140 in CH_2Cl_2 solution, respectively, indicating that the Φ_f value of **6j** was similar to that of luminol (**1**, Table 3). Moreover, we also investigated the quantum yields of **6j** in various solvents by using the same condition. The estimated values order trendy was as 0.218 (THF) > 0.209 (Toluene) > 0.140 (CH_2Cl_2) > 0.083 (acetone) > 0.049 (EA), indicating THF provided the largest Φ_f value among them (Table 3). On the other hand, most of the quantum yields (Φ_f) pyrazolopyridopyridazine diones **6a–i** in CH_2Cl_2 solution were predicted to be an almost identical value (ca. 0.05–0.06). Interestingly, the high

Φ_f value of **6j** was obtained and possibly caused by a particular improvement in the planar skeletal conformation (Table 3 and Figure 7).

Table 3. Quantum yields of fluorescence of luminol (**1**) and pyrazolopyridopyridazine diones **6a**, **6c**, and **6j**.

Compound	Solvent	λ_{fl} [1]/nm	Φ_f [2]
6a	CH$_2$Cl$_2$	481	0.056
6c	CH$_2$Cl$_2$	472	0.067
6j	CH$_2$Cl$_2$	450	0.140
6j	THF	435	0.218
6j	Toluene	438	0.209
6j	Acetone	437	0.083
6j	Ethyl acetate	437	0.049
Luminol (1)	CH$_2$Cl$_2$	399	0.175

[1] Fluorescence maximum wavelength (λ_{fl}). [2] Φ_f: Fluorescence quantum efficiency, relative to quinine sulfate ($\Phi_f = 0.60$).

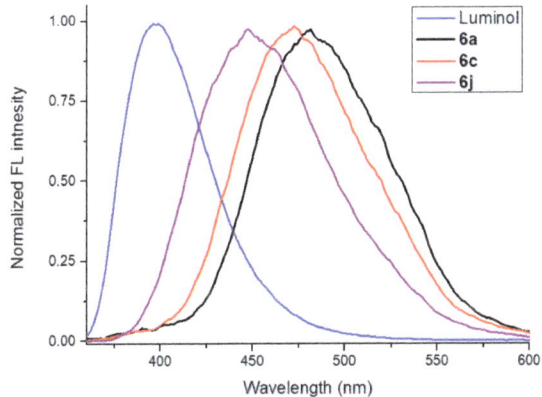

Figure 7. Normalized fluorescence spectra of luminol and pyrazolopyridopyridazine diones **6a**, **6c**, and **6j** in the CH$_2$Cl$_2$ solution (excitation wavelength 350 nm).

3. Experimental Section

3.1. General Information

All reagents were used as obtained commercially. All reactions were carried out under argon or nitrogen atmosphere and monitored by thin-layer chromatography (TLC). Flash column chromatography was carried out on silica gel (230–400 mesh). Analytical thin-layer chromatography was performed using pre-coated plates (silica gel 60 F-254) purchased from Merck Inc. Flash column chromatography purification was carried out by gradient elution using *n*-hexane in ethyl acetate (EtOAc) unless otherwise stated. ^{1}H-NMR was recorded at 400, 500, or 600 MHz and ^{13}C-NMR recorded at 100, 125, or 150 MHz, respectively, in DMSO-d_6 as the solvent. The standard abbreviations s, d, t, q, and m refer to the singlet, doublet, triplet, quartet, and multiplet, respectively. Coupling constant (*J*), whenever discernible, have been reported in Hz. Infrared spectra (IR) were recorded as neat solutions or solids; mass spectra were recorded using electron impact or electrospray ionization techniques. The wavenumbers reported are referenced to the polystyrene 1601 cm^{-1} absorption. ESI-MS analyses were performed on an Applied Biosystems API 300 mass spectrometer. High-resolution mass spectra (HRMS) were recorded on a JEOL JMS-HX110 mass spectrometer with an electron ionization (EI) source The UV-visible absorption and emission spectra were performed on a Perkin-Elmer Lambda 265 and Perkin-Elmer LS50B, a fused quartz cuvette (10 mm × 10 mm) at room temperature, respectively.

Quantum yields were obtained by using quinine sulfate (0.60 in 0.05 M H_2SO_4) as a reference. Stock solutions (1 × 10^{-3} M) of luminol (**1**), compounds of **6a–j** and **7a–i** were prepared in dimethyl sulfoxide (DMSO).

*3.2. Standard Procedure for Synthesis of Pyrazolopyridopyridazine Diones **6a–j***

The reliable procedure involved the treatment of 1,3-diarylpyrazolopyrrolopyridine-6,8-diones (**11a–j**), 1,3-diaryl-7-methylpyrazolopyrrolopyridine-6,8-diones (**12a–c**) 1,3-diarylfuropyrazolopyridine-6, 8-diones (**13a–c**, 1.0 equiv.) with hydrazine monohydrate (~40 equiv.) in neat solution at reflux for 5 h. When the reaction was completed, the reaction mixture was added to water (10 mL) for precipitation. The precipitate was filtered, washed with cold water (10 mL), and *n*-hexane/EA (1/2, 15 mL) to give the corresponding crude pyrazolopyridopyridazine diones **6a–j**. The crude desired products **6a–j** were recrystallized in acetone/THF (1/4) solution to obtain the pure pyrazolopyridopyridazine diones **6a–j** in 11–84% yields. The low solubility of the compounds **6a–j** made the ^{13}C-NMR characterization of quaternary and carbonyl carbons of these substrates unclear [25,26].

1,3-Diphenyl-7,8-dihydro-3H-pyrazolo[4',3':5,6]pyrido[3,4-d]pyridazine-6,9-dione (**6a**), Light yellow solid; yield: 84%; mp 292–295 °C. ^{1}H-NMR (DMSO-d_6, 600 MHz) δ 7.43–7.47 (m, 4H, Ar*H*), 7.60–7.64 (m, 4H, Ar*H*), 8.20 (d, *J* = 7.9 Hz, 2H, Ar*H*), 9.43 (s, 1H, Ar*H*), 10.20 (br, 1H, N*H*); ^{13}C{^{1}H} NMR (DMSO-d_6, 150 MHz) δ 109.55, 122.46 (2 × CH), 126.67, 127.16, 127.78, 129.26 (2 × CH + 2 × CH), 130.21 (2 × CH + CH), 134.99, 138.16, 147.54, 149.49, 151.35, 153.28, 155.21; FT-IR (KBr) *v*: 3161, 3033, 2907, 1662, 1584, 1499, 1414, 1356, 1306, 906 cm^{-1}; MS (EI) *m/z* (relative intensity): 356 (24), 355 (M$^+$, 100), 354 (27), 270 (24), 269 (12), 268 (12), 77 (39); HRMS (EI) *m/z*: [M]$^+$ Calcd for $C_{20}H_{13}N_5O_2$: 355.1069; found: 355.1065.

3-(2-Chlorophenyl)-1-phenyl-7,8-dihydro-3H-pyrazolo[4',3':5,6]pyrido[3,4-d]pyridazine-6,9-dione (**6b**), Yellow-brown solid; yield: 74%; mp 332–335 °C; ^{1}H-NMR (DMSO-d_6, 500 MHz) δ 7.40 (br, 3H, Ar*H*), 7.60–7.61 (m, 2H, Ar*H*), 7.63 (d, *J* = 7.5 Hz, 1H, Ar*H*), 7.67 (t, *J* = 7.5 Hz, 1H, Ar*H*), 7.79–7.80 (m, 2H, Ar*H*), 9.35 (s, 1H, Ar*H*); ^{13}C{^{1}H} NMR (DMSO-d_6, 125 MHz) δ 109.33, 124.53, 126.66, 127.77, 128.01, 128.18, 128.45, 130.23 (2 × CH), 130.30, 130.47, 131.33, 131.36, 134.73, 134.94, 147.84, 149.81, 152.65, 155.49, 156.94; FT-IR (KBr) *v*: 3427, 3281, 3060, 2921, 1621, 1561, 1508, 1430, 1351, 905 cm^{-1}; MS (EI) *m/z* (relative intensity): 391 (M$^+$ + 2, 29), 390 (22), 389 (M$^+$, 100), 355 (12), 354 (52), 304 (15), 268 (17), 111 (13), 77 (44); HRMS (EI) *m/z*: [M]$^+$ Calcd for $C_{20}H_{12}ClN_5O_2$: 389.0680; found: 389.0678.

3-(3-Chlorophenyl)-1-phenyl-7,8-dihydro-3H-pyrazolo[4',3':5,6]pyrido[3,4-d]pyridazine-6,9-dione (**6c**), Yellow solid; yield: 71%; mp 228–229 °C; ^{1}H-NMR (DMSO-d_6, 500 MHz) δ 7.42–7.43 (m, 3H, Ar*H*), 7.50 (d, *J* = 6.7 Hz, 1H, Ar*H*), 7.60 (d, *J* = 5.2 Hz, 2H, Ar*H*), 7.64 (t, *J* = 8.0 Hz, 1H, Ar*H*), 8.26 (d, *J* = 8.0 Hz, 1H, Ar*H*), 8.34 (s, 1H, Ar*H*), 9.44 (s, 1H, Ar*H*), 12.06 (br, 1H, N*H*); ^{13}C{^{1}H} NMR (DMSO-d_6, 125 MHz) δ 107.59, 119.56, 120.40, 121.43, 126.72 (3 × CH), 127.93, 130.14 (2 × CH + 1 × C), 130.99, 133.49, 134.67, 139.34, 148.03, 149.66, 151.53, 155.16, 157.57; FT-IR (KBr) *v*: 3453, 3344, 3296, 1651, 1595, 1483 cm^{-1}; MS (EI) *m/z* (relative intensity): 391 (M$^+$ + 2, 33), 390 (27), 389 (M$^+$, 100), 388 (14), 304 (14), 111 (11), 77 (17); HRMS (EI) *m/z*: [M]$^+$ Calcd for $C_{20}H_{12}ClN_5O_2$: 389.0680; found: 389.0686.

3-(4-Chlorophenyl)-1-phenyl-7,8-dihydro-3H-pyrazolo[4',3':5,6]pyrido[3,4-d]pyridazine-6,9-dione (**6d**), Light yellow solid; yield: 81%; mp 339–341 °C; ^{1}H-NMR (DMSO-d_6, 500 MHz) δ 7.40–7.41 (m, 3H, Ar*H*), 7.58–7.60 (m, 2H, Ar*H*), 7.68 (d, *J* = 8.9 Hz, 2H, Ar*H*), 8.29 (d, *J* = 8.9 Hz, 2H, Ar*H*), 9.41 (s, 1H, Ar*H*); ^{13}C{^{1}H} NMR (DMSO-d_6, 125 MHz) δ 107.56, 123.52 (2 × CH), 126.60 (2 × CH), 126.64, 127.77, 127.82, 129.23 (2 × CH), 130.18 (2 × CH), 131.06, 134.89, 137.12, 147.88, 149.80, 151.35, 156.76, 157.30; FT-IR (KBr) *v*: 3345, 3206, 1656, 1494, 1446, 1307, 1094, 902 cm^{-1}; MS (EI) *m/z* (relative intensity): 391 (M$^+$ + 2, 36), 390 (31), 389 (M$^+$, 100), 388 (20), 354 (12), 304 (19), 268 (11), 111 (15), 77 (24); HRMS (EI) *m/z*: [M]$^+$ Calcd for $C_{20}H_{12}ClN_5O_2$: 389.0680; found: 389.0687.

3-(4-Bromophenyl)-1-phenyl-7,8-dihydro-3H-pyrazolo[4',3':5,6]pyrido[3,4-d]pyridazine-6,9-dione (**6e**), Light yellow solid; yield: 77%;mp 337–339 °C; ^{1}H-NMR (DMSO-d_6, 500 MHz) δ 7.40–7.41 (m, 3H, Ar*H*), 7.59 (d, *J* = 5.5 Hz, 2H, Ar*H*), 7.82 (d, *J* = 9.0 Hz, 2H, Ar*H*), 8.25 (d, *J* = 9.0 Hz, 2H, Ar*H*), 9.42 (s, 1H, Ar*H*); ^{13}C{^{1}H} NMR (DMSO-d_6, 125 MHz) δ 107.67, 119.42, 123.81 (2 × CH), 126.64 (2 × CH), 127.82, 130.16 (2 × CH + C), 132.16 (2 × CH + C), 134.88, 137.56, 147.92, 149.78, 151.36, 157.59, 159.25; FT-IR (KBr) *v*: 3435, 3345, 3266, 1655, 1536, 1492, 1443, 1307, 1094, 916, 902 cm^{-1}; MS (EI) *m/z* (relative intensity): 436 (24), 435 (M$^+$ + 2, 98), 434 (39), 433 (M$^+$, 100), 432 (14), 354 (11), 350 (11), 348 (12), 268 (14), 77 (26); HRMS (EI) *m/z*: [M]$^+$ Calcd for $C_{20}H_{12}BrN_5O_2$: 433.0174; found: 433.0171.

1-Phenyl-3-(p-tolyl)-7,8-dihydro-3H-pyrazolo[4',3':5,6]pyrido[3,4-d]pyridazine-6,9-dione (**6f**), Light yellow solid; yield: 84%; mp 346–348 °C; ^{1}H-NMR (DMSO-d_6, 500 MHz) δ 2.40 (s, 3H, C*H*$_3$), 7.39–7.42 (m, 5H, Ar*H*), 7.58–7.60 (m, 2H, Ar*H*), 8.07 (d, *J* = 8.4 Hz, 2H, Ar*H*), 9.39 (s, 1H, Ar*H*); ^{13}C{^{1}H} NMR (DMSO-d_6, 125 MHz) δ 20.61, 106.97, 118.99, 122.32 (2 × CH), 126.60 (2 × CH), 127.68, 128.86, 129.60 (2 × CH), 130.21 (2 × CH), 135.09, 135.83, 136.55, 147.26, 149.35, 151.18, 152.87, 155.65; FT-IR (KBr) *v*: 3436, 3345, 3206, 2919, 1656, 1534, 1514, 1480, 1453, 1310, 1096, 903 cm^{-1}; MS (EI) *m/z* (relative intensity): 370 (25), 369 (M$^+$, 100), 368 (16), 354 (14), 284 (18), 91 (15), 77 (19); HRMS (EI) *m/z*: [M]$^+$ Calcd for $C_{21}H_{15}N_5O_2$: 369.1226; found: 369.1216.

3-(4-Methoxyphenyl)-1-phenyl-7,8-dihydro-3H-pyrazolo[4',3':5,6]pyrido[3,4-d]pyridazine-6,9-dione (**6g**), Deep yellow solid; yield: 81%; mp 311–313 °C; ^{1}H-NMR (DMSO-d_6, 500 MHz) δ 3.85 (s, 3H, OC*H*$_3$), 7.17 (d, *J* = 11.2 Hz, 2H, Ar*H*), 7.43 (s, 3H, Ar*H*), 7.58–7.60 (m, 2H, Ar*H*), 8.03 (d, *J* = 11.2 Hz, 2H, Ar*H*), 9.38 (s, 1H, Ar*H*); ^{13}C{^{1}H} NMR (DMSO-d_6, 125 MHz) δ 55.49, 106.64, 114.38 (2 × CH), 115.59, 124.31 (2 × CH), 124.60, 126.71 (2 × CH), 127.77, 130.23 (2 × CH), 131.18, 135.03, 147.01, 149.17, 151.12, 152.63, 156.80, 158.33; FT-IR (KBr) *v*: 3435, 3226, 3065, 2886, 1650, 1590, 1535, 1516, 1441, 1362, 1252, 1170, 905 cm^{-1}; MS (EI) *m/z* (relative intensity): 386 (24), 385 (M$^+$, 100), 370 (13), 77 (18); HRMS (EI) *m/z*: [M]$^+$ Calcd for $C_{21}H_{15}N_5O_3$: 385.1175; found: 385.1180.

3-(4-Cyanophenyl)-1-phenyl-7,8-dihydro-3H-pyrazolo[4',3':5,6]pyrido[3,4-d]pyridazine-6,9-dione (**6h**), Yellow solid; yield: 73%; mp 344–347 °C; ^{1}H-NMR (DMSO-d_6, 500 MHz) δ 7.43–7.44 (m, 3H, Ar*H*), 7.60–7.61 (m, 2H, Ar*H*), 8.08 (d, *J* = 6.6 Hz, 2H, Ar*H*), 8.56 (d, *J* = 6.6 Hz, 2H, Ar*H*), 9.45 (s, 1H, Ar*H*); ^{13}C{^{1}H} NMR (DMSO-d_6, 125 MHz) δ 108.84, 118.51, 121.39, 121.70 (2 × CH), 126.74 (2 × CH), 128.07, 128.63, 130.08 (2 × CH), 130.15, 133.62 (2 × CH), 134.51, 141.67, 148.75, 149.75, 151.87, 155.15, 156.71; FT-IR (KBr) *v*: 3397, 3284, 3056, 2228, 1606, 1569, 1516, 1430, 1400, 1317, 905 cm^{-1}; MS (EI) *m/z* (relative intensity): 381 (26), 380 (M$^+$, 100), 379 (23), 295 (17), 102 (13), 77(29); HRMS (EI) *m/z*: [M]$^+$ Calcd for $C_{21}H_{12}N_6O_2$: 380.1022; found: 380.1030.

3-(4-Nitrophenyl)-1-phenyl-7,8-dihydro-3H-pyrazolo[4',3':5,6]pyrido[3,4-d]pyridazine-6,9-dione (**6i**), Yellow solid; yield: 69%; mp 340–342 °C; ^{1}H-NMR (DMSO-d_6, 500 MHz) δ 7.41–7.44 (m, 3H, Ar*H*), 7.61 (d, *J* = 6.3 Hz, 2H, Ar*H*), 8.48 (d, *J* = 8.3 Hz, 2H, Ar*H*), 8.67 (d, *J* = 8.3 Hz, 2H, Ar*H*), 9.47 (s, 1H, Ar*H*), 12.11 (br, 1H, N*H*); ^{13}C{^{1}H} NMR (DMSO-d_6, 125 MHz) δ 113.65, 121.61 (2 × CH), 124.52, 125.13 (2 × CH), 126.85 (2 × CH), 128.22, 130.12 (2 × CH), 130.27, 134.47, 143.27, 145.04, 149.17, 149.92, 152.09, 154.07, 158.18; FT-IR (KBr) *v*: 3435, 1637, 1596, 1522, 1341, 1112, 905 cm^{-1}; MS (EI) *m/z*(relative intensity): 401 (23), 400 (M$^+$, 100), 370 (22), 315 (11), 77(19); HRMS (EI) *m/z*: [M]$^+$ Calcd for For $C_{20}H_{12}N_6O_4$: 400.0920; found: 400.0919.

3-(3-Chlorophenyl)-7,8-dihydro-3H-pyrazolo[4',3':5,6]pyrido[3,4-d]pyridazine-6,9-dione (**6j**), Light yellow solid; yield: 71%; mp 351–352 °C; ^{1}H-NMR (DMSO-d_6, 500 MHz) δ 7.51 (d, *J* = 8.03 Hz, 1H, Ar*H*), 7.66 (t, *J* = 8.0 Hz, 1H, Ar*H*), 8.28 (d, *J* = 8.0 Hz, 1H, Ar*H*), 8.39 (s, 1H, Ar*H*), 8.88 (s, 1H, Ar*H*), 9.40 (s, 1H, Ar*H*), 10.20 (br, 1H, N*H*); ^{13}C{^{1}H} NMR (DMSO-d_6, 125 MHz) δ 109.53, 119.80, 120.84, 126.68 (2 × C), 131.14 (CH + C), 133.57, 136.04, 139.62, 149.36, 150.78, 152.47, 155.87; FT-IR (KBr) *v*: 3433, 3294, 3168, 2974, 1639, 1594, 1568, 1487, 1448, 1274, 1218, 1125 cm^{-1}; MS (EI) *m/z* (relative intensity): 315 (M$^+$ + 2, 35),

314 (28), 313 (M$^+$, 100), 278 (12), 255 (13), 227 (21), 111(12), 75 (11); HRMS (EI) *m/z*: [M]$^+$ Calcd for C$_{14}$H$_8$ClN$_5$O$_2$: 313.0367; found: 313.0367.

3.3. Standard Procedure for Synthesis of N-Aminopyrazolopyrrolopyridine Diones (7a–i)

The reliable procedure involved the treatment of 1,3-diarylpyrazolopyrrolopyridine-6,8-diones (**11a–i**, 1.0 equiv.) with hydrazine monohydrate (~5.0 equiv.) in EtOH/H$_2$O (2.0 mL/2.0 mL) in ice-bath to room temperature within 48 h. When the reaction was completed, the reaction mixture was added to water (10 mL) for precipitation. The precipitate was filtered, washed with cold water (10 mL) and *n*-hexane/EA (1/2, 15 mL) to give the corresponding crude *N*-aminophthalimides **7a–i**. The crude desired products **7a–i** were recrystallized in acetone/THF (1/4) solution to obtain the pure *N*-aminophthalimides **7a–i** in 71–87 % yields [31,32].

7-Amino-1,3-diphenylpyrazolo[3,4-b]pyrrolo[3,4-d]pyridine-6,8-(3H,7H)-dione (**7a**), White solid; yield: 83%; mp 217–219 °C; ^{1}H-NMR (DMSO-*d*$_6$, 400 MHz) δ 4.55 (br, 2H, NH$_2$), 7.42 (t, *J* = 7.5 Hz, 1H, Ar*H*), 7.49–7.55 (m, 3H, Ar*H*), 7.60–7.65 (m, 4H, Ar*H*), 8.25 (d, *J* = 8.1 Hz, 2H, Ar*H*), 8.82 (s, 1 H, Ar*H*); ^{13}C{^{1}H} NMR (DMSO-*d*$_6$, 100 MHz) δ 111.65, 121.43 (2 × CH), 123.37, 126.80, 128.27 (2 × CH), 128.58 (2 × CH), 128.96, 129.41 (2 × CH), 132.04, 138.44, 139.23, 146.02, 148.82, 150.61, 164.18, 164.49; FT-IR (KBr) *v*: 3275, 3208, 3172, 3035, 1633.0, 1572, 1518.7, 1501 cm^{-1}; MS (EI) *m/z* (relative intensity): 356 (20), 355 (M$^+$, 100), 354 (18), 270 (13), 77.0(18); HRMS (EI) *m/z*: [M]$^+$ Calcd for C$_{20}$H$_{13}$N$_5$O$_2$: 355.1069; found: 355.1060.

7-Amino-3-(2-chlorophenyl)-1-phenylpyrazolo[3,4-b]pyrrolo[3,4-d]pyridine-6,8-(3H,7H)-dione (**7b**), Yellow solid; yield: 71%; mp 165–166 °C; ^{1}H-NMR (DMSO-*d*$_6$, 500 MHz) δ 4.52 (br, 2H, NH$_2$), 7.48 (d, *J* = 7.0 Hz, 1H, Ar*H*), 7.52 (d, *J* = 7.1 Hz, 2H, Ar*H*), 7.60–7.68 (m, 4H, Ar*H*), 7.74 (d, *J* = 7.7 Hz, 1H, Ar*H*), 7.79 (d, *J* =7.7 Hz, 1H, Ar*H*), 8.71 (s, 1H, Ar*H*); ^{13}C{^{1}H} NMR (DMSO-*d*$_6$, 125 MHz) δ 108.50, 122.16, 122.88, 123.33, 127.72 (2 × CH), 129.27 (3 × CH), 129.72 (2 × CH), 131.26, 131.46, 136.60, 136.98, 143.86, 145.72, 153.20, 166.81, 168.47; FT-IR (KBr) *v*: 3337, 3296, 2952, 2920, 1778, 1740, 1498, 1375, 1315, 1014 cm^{-1}; MS (EI) *m/z* (relative intensity): 391 (M$^+$ + 2, 32), 390 (22), 389 (M$^+$, 100), 355 (18), 354 (90), 304 (12), 268 (11), 77 (18); HRMS (EI) *m/z*: [M]$^+$ Calcd for C$_{20}$H$_{12}$ClN$_5$O$_2$: 389.0680; found: 389.0672.

7-Amino-3-(3-chlorophenyl)-1-phenylpyrazolo[3,4-b]pyrrolo[3,4-d]pyridine-6,8-(3H,7H)-dione (**7c**), Yellow solid; yield: 73%; mp 173–175 °C; ^{1}H-NMR (DMSO-*d*$_6$, 400 MHz) δ 4.55 (br, 2H, NH$_2$), 7.47–7.54 (m, 4H, Ar*H*), 7.64–7.67 (m, 3H, Ar*H*), 8.31 (d, *J* = 8.9 Hz, 1H, Ar*H*), 8.42 (s, 1H, Ar*H*), 8.86 (s, 1H, Ar*H*); ^{13}C{^{1}H} NMR (DMSO-*d*$_6$, 100 MHz) δ 112.05, 119.33, 120.34, 123.72, 126.32, 128.28 (2 × CH), 128.57 (2 × CH), 129.11, 131.21, 131.73, 133.65, 139.41, 139.64, 146.61, 148.98, 150.72, 163.96, 164.33; FT-IR (KBr) *v*: 3264, 3168, 3034, 1649, 1614, 1595, 1488, 1431, 1300, 803 cm^{-1}; MS (EI) *m/z* (relative intensity): 391 (M$^+$ + 2, 33), 390 (24), 389 (M$^+$, 100), 374 (13), 304 (11), 77 (14); HRMS (EI) *m/z*: [M]$^+$ Calcd for C$_{20}$H$_{12}$ClN$_5$O$_2$: 389.0680; found: 389.0688.

7-Amino-3-(4-chlorophenyl)-1-phenylpyrazolo[3,4-b]pyrrolo[3,4-d]pyridine-6,8-(3H,7H)-dione (**7d**), Light yellow solid; yield: 81%; mp 226–227 °C; ^{1}H-NMR (DMSO-*d*$_6$, 500 MHz) δ 4.55 (br, 2H, NH$_2$), 7.48–7.55 (m, 3H, Ar*H*), 7.64 (d, *J* = 7.6 Hz, 2H, Ar*H*), 7.69 (d, *J* = 8.4 Hz, 2H, Ar*H*), 8.35 (d, *J* = 8.4 Hz, 2H, Ar*H*), 8.84 (s, 1 H, Ar*H*); ^{13}C{^{1}H} NMR (DMSO-*d*$_6$, 125 MHz) δ 111.85, 122.54 (2 × CH), 123.55, 128.24 (2 × CH), 128.53 (2 × CH), 129.03, 129.36 (2 × CH), 130.67, 131.81, 137.31, 139.34, 146.35, 148.88, 150.57, 163.99, 164.34; FT-IR (KBr) *v*: 3275, 3207, 3170, 1633, 1499, 828 cm^{-1}; MS (EI) *m/z* (relative intensity): 391 (M$^+$ + 2, 34), 390 (28), 389 (M$^+$, 100), 388 (15), 304 (12), 77 (13); HRMS (EI) *m/z*: [M]$^+$ Calcd for C$_{20}$H$_{12}$ClN$_5$O$_2$: 389.0680; found: 389.0686.

7-Amino-3-(4-bromophenyl)-1-phenylpyrazolo[3,4-b]pyrrolo[3,4-d]pyridine-6,8-(3H,7H)-dione (**7e**), Yellow solid; yield: 79%; mp 235–239 °C; ^{1}H-NMR (DMSO-*d*$_6$, 500 MHz) δ 4.55 (br, 2H, NH$_2$), 7.49–7.55 (m, 3H, Ar*H*), 7.64 (d, *J* = 6.9 Hz, 2H, Ar*H*), 7.82 (d, *J* = 8.7 Hz, 2H, Ar*H*), 8.29 (d, *J* = 8.7 Hz, 2H, Ar*H*), 8.84 (s, 1 H, Ar*H*); ^{13}C{^{1}H} NMR (DMSO-*d*$_6$, 125 MHz) δ 111.91, 118.95, 122.81 (2 × CH), 123.55, 128.25

(2 × CH), 128.54 (2 × CH), 129.04, 131.80, 132.28 (2 × CH), 137.75, 139.33, 146.40, 148.90, 150.59, 164.00, 164.35; FT-IR (KBr) v: 3275, 3208, 3170, 3071, 3037, 1632, 1495, 826 cm^{-1}; MS (EI) m/z (relative intensity): 436 (22), 435 (M$^+$ + 2, 100), 354 (33), 433 (M$^+$, 99), 432 (12), 420 (19), 419 (11), 418 (19), 354 (11), 350 (10), 348 (11), 268 (17), 77 (26); HRMS (EI) m/z: [M]$^+$ Calcd for $C_{20}H_{12}BrN_5O_2$: 433.0174; found: 433.0171.

7-Amino-1-phenyl-3-(p-tolyl)pyrazolo[3,4-b]pyrrolo[3,4-d]pyridine-6,8-(3H,7H)-dione (**7f**), Yellow solid; yield: 86%; mp 232–233 °C; ^{1}H-NMR (DMSO-d_6, 500 MHz) δ 2.40 (s, 3H, CH$_3$), 4.55 (br, 2H, NH$_2$), 7.42 (d, J = 8.0 Hz, 2H, ArH), 7.49 (d, J = 7.3 Hz, 1H, ArH), 7.52 (t, J = 7.3 Hz, 2H, ArH), 7.64 (d, J = 7.3 Hz, 2H, ArH), 8.12 (d, J = 8.0 Hz, 2H ArH), 8.81 (s, 1H, ArH); ^{13}C{^{1}H} NMR (DMSO-d_6, 125 MHz) δ 20.63, 111.44, 121.35(2 × CH), 123.18, 128.21 (2 × CH), 128.54 (2 × CH), 128.85, 129.73 (2 × CH), 132.09, 136.06, 136.18, 139.13, 145.69, 148.72, 150.44, 164.20, 164.50; FT-IR (KBr) v: 3276, 3209, 3171, 3032, 1634, 1517 cm^{-1}; MS (EI) m/z (relative intensity): 370 (25), 369 (M$^+$, 100), 368 (13), 354 (21), 284 (12), 207 (10), 91.1(11), 77.1(13); HRMS (EI) m/z: [M]$^+$ Calcd for $C_{21}H_{15}N_5O_2$: 369.1226; found: 369.1231.

7-Amino-3-(4-methoxyphenyl)-1-phenylpyrazolo[3,4-b]pyrrolo[3,4-d]pyridine-6,8-(3H,7H)-dione (**7g**), Yellow solid; yield: 87%; mp 331–332 °C; ^{1}H-NMR (DMSO-d_6, 500 MHz) δ 3.83 (s, 3H, OCH$_3$), 4.53 (br, 2H, NH$_2$), 7.17 (d, J = 8.7 Hz, 2H, ArH), 7.48–7.53 (m, 3H, ArH), 7.62 (d, J = 7.1 Hz, 2H, ArH), 8.06 (d, J = 8.7 Hz, 2H, ArH), 8.78 (s, 1H, ArH); ^{13}C{^{1}H} NMR (DMSO-d_6, 125 MHz) δ 55.64, 111.24, 114.62 (2 × CH), 123.14, 123.51 (2 × CH), 128.37 (2 × CH), 128.65 (2 × CH), 128.97, 131.60, 132.25, 139.14, 145.54, 148.83, 150.44, 158.16, 164.43, 164.73; FT-IR (KBr) v: 3215, 3066, 3004, 2963, 2935, 2837, 1639, 1577, 1516, 1462, 1443, 1252 cm^{-1}; MS (EI) m/z (relative intensity): 386 (21), 385 (M$^+$, 100), 370 (14), 77.0(10); HRMS (EI) m/z: [M]$^+$ Calcd for $C_{21}H_{15}N_5O_3$: 385.1175; found: 385.1182.

7-Amino-3-(4-cyanophenyl)-1-phenylpyrazolo[3,4-b]pyrrolo[3,4-d]pyridine-6,8-(3H,7H)-dione (**7h**), Yellow solid; yield: 73%; mp 333–335 °C; ^{1}H-NMR (DMSO-d_6, 500 MHz) δ 4.55 (br, 2H, NH$_2$), 7.50–7.55 (m, 3H, ArH), 7.65 (d, J = 7.1 Hz, 2H, ArH), 8.09 (d, J = 8.5 Hz, 2H, ArH), 8.61 (d, J = 8.5 Hz, 2H, ArH), 8.87 (s, 1 H, ArH); ^{13}C{^{1}H} NMR (DMSO-d_6, 125 MHz) δ 108.41, 112.58, 118.68, 120.75 (2 × CH), 124.12, 128.35 (2 × CH), 128.59 (2 × CH), 129.33, 131.55, 133.84 (2 × CH), 139.53, 141.99, 147.47, 149.11, 151.12, 163.83, 164.27; FT-IR (KBr) v: 3330, 3274, 2227, 1665, 1635, 1607, 1518, 1409, 1255, 844 cm^{-1}; MS (EI) m/z (relative intensity): 381 (22), 380 (M$^+$, 100), 379 (18), 295 (13), 77(14); HRMS (EI) m/z: [M]$^+$ Calcd for $C_{21}H_{12}N_6O_2$: 380.1022; found: 380.1023.

7-Amino-3-(4-nitrophenyl)-1-phenylpyrazolo[3,4-b]pyrrolo[3,4-d]pyridine-6,8-(3H,7H)-dione (**7i**), Yellow solid; yield: 74%; mp 281–282 °C; ^{1}H-NMR (DMSO-d_6, 500 MHz) δ 7.53–7.57 (m, 3H, ArH), 7.93 (d, J = 6.5 Hz, 2H, ArH), 8.48 (d, J = 9.2 Hz, 2H, ArH), 8.67 (d, J = 9.2 Hz, 2H, ArH), 9.19 (s, 1H, ArH); ^{13}C{^{1}H} NMR (DMSO-d_6, 125 MHz) δ 109.57, 121.41 (2 × CH), 123.52, 125.21 (2 × CH), 127.91 (2 × CH), 129.73, 129.90 (2 × CH), 131.21, 136.89, 143.24, 144.26, 145.12, 147.11, 153.90, 166.71, 168.43; FT-IR (KBr) v: 3190, 3120, 3064, 1595, 1500, 1341, 1112, 857 cm^{-1}; MS (EI) m/z (relative intensity): 400 (M$^+$, 4), 386 (26), 385 (100), 338 (13), 236 (10), 77 (13); HRMS (EI) m/z: [M]$^+$ Calcd for $C_{20}H_{12}N_6O_4$: 400.0920; found: 400.0925.

3.4. Determination of the Fluorescence Quantum Yield

The fluorescence quantum yield Φ_x was determined through the comparative method. The quinine sulfate (Φ_{st} = 0.60, λ_{ex}= 350 nm) in H$_2$SO$_4$ 0.05 M was used as the standard, and it was calculated by following equation [48]:

$$\Phi_x/\Phi_{st} = [A_{st}/A_x]\,[n_x^2/n_{st}^2]\,[D_x/D_{st}], \tag{1}$$

where st: standard; x: sample; Φ: quantum yield; A: absorbance at the excitation wavelength; D: area under the fluorescence spectra on an energy scale; n: the refractive index of the solution. In the process of detection, the absorbance should be controlled and lower than 0.1.

4. Conclusions

Pyrazolopyridopyridazine diones **6** and N-aminopyrazolopyrrolopyridine diones **7** can be prepared in three synthesis methods from 1,3-diarylpyrazolopyrrolopyridine-6,8-diones, 1,3-diaryl-7-methylpyrazolopyrrolopyridine-6,8-diones, or 1,3-diarylfuropyrazolopyridine-6,8-diones with hydrazine monohydrate. Based on the experimental results, 1,3-diarylpyrazolopyrrolopyridine-6,8-diones were conceived as the best reactive starting materials. Furthermore, compounds **6** and **7** were also selectively synthesized under kinetic and thermodynamic control reactions. For the further photoluminescence, solvatofluorism, and quantum yield (Φ_f) studies, pyrazolopyridopyridazine diones **6** generally exhibited the stronger fluorescence intensity and possessed the significant substituent effect, particularly for **6c** with a m-chloro group. On the other hand, the best Φ_f value of **6j** was obtained ($\Phi_f = 0.140$) and similar to luminol (**1**, $\Phi_f = 0.175$), possibly caused by the planar skeletal conformation. Based on the above photoluminescence studies, we also found that the efficient introduction of the pyrazole and pyridine chromophores led to an increase in the conjugation and aromaticity of compounds **6** and **7** when compared with the standard luminol.

Supplementary Materials: The following are available online, copies of ^{1}H and ^{13}C-NMR spectra of compounds **6a–6j** and **7a–7i**.

Author Contributions: F.F.W. conceived and designed the experiments; C.-C.T., C.-Y.C., and S.-E.T. performed the experiments; H.T., N.U., C.-Y.L., C.C.T., C.-Y.C., and S.-E.T. analyzed the data; F.F.W., H.T., N.U., C.-Y.L., contributed reagents/materials/analysis tools; F.F.W., C.-Y.L., C.C.T., and S.-E.T. wrote the paper. All authors have read and agreed to the published version of the manuscript.

Funding: This research was funded by the Tsuzuki Institute for Traditional Medicine and the China Medical University, Taiwan (CMU108-N-04)

Acknowledgments: We are grateful to the Tsuzuki Institute for Traditional Medicine and the China Medical University, Taiwan (CMU108-N-04) for financial support.

Conflicts of Interest: The authors declare no conflicts of interest.

References

1. Boivin, D.B.; Tremblay, G.M.; James, F.O. Working on Atypical Schedules. *Sleep Med.* **2007**, *8*, 578–589. [CrossRef] [PubMed]
2. Menegatti, R.; Silva, G.M.; Zapata-Sudo, G.; Raimundo, J.M.; Sudo, R.T.; Barreiro, E.J.; Fraga, C.A.M. Design, Synthesis, and Pharmacological Evaluation of New Neuroactive Pyrazolo[3,4-b]pyrrolo[3,4-d]pyridine Derivatives with in vivo Hypnotic and Analgesic Profile. *Bioorg. Med. Chem.* **2006**, *14*, 632–640. [CrossRef] [PubMed]
3. Nascimento-Junior, N.M.; Mendes, T.C.F.; Leal, D.M.; Correa, C.M.N.; Sudo, R.T.; Zapata-Sudo, G.; Barreiro, E.J.; Fraga, C.A.M. Microwave-assisted Synthesis and Structure–activity Relationships of Neuroactive pyrazolo[3,4-b]pyrrolo[3,4-d]pyridine Derivatives. *Bioorg. Med. Chem. Lett.* **2010**, *20*, 74–77. [CrossRef] [PubMed]
4. Yu, G.; Macor, J.E.; Chung, H.-J.; Humora, M.; Katipally, K.; Wang, Y.; Kim, S. Fused Pyridopyridazine Inhibitors of cGMP Phosphodiesterase. U.S. Patent No. 6,316,438, 13 November 2001.
5. Yu, G.; Mason, H.; Wu, X.; Wang, J.; Chong, S.; Beyer, B.; Henwood, A.; Pongrac, R.; Seliger, L.; He, B.; et al. Substituted Pyrazolopyridopyridazines as Orally Bioavailable Potent and Selective PDE5 Inhibitors: Potential Agents for Treatment of Erectile Dysfunction. *J. Med. Chem.* **2003**, *46*, 457–460. [CrossRef] [PubMed]
6. Li, Y.; Zhu, H.; Trush, M.A. Detection of Mitochondria-derived Reactive Oxygen Species Production by the Chemilumigenic Probes Lucigenin and Luminol. *Biochim. Biophys. Acta* **1999**, *1428*, 1–12. [CrossRef]
7. García-Campaña, A.M.; Baeyens, W.R.G. *Chemiluminescence in Chemical Analysis*; Marcel Dekker: New York, NY, USA, 2001.
8. Roda, A. *Chemiluminescence and Bioluminescence: Past, Present, and Future*; RSC: Cambridge, UK, 2011.
9. Barni, F.; Lewis, S.W.; Berti, A.; Miskelly, G.M.; Lago, G. Forensic Application of the Luminol Reaction as a Presumptive Test for Latent Blood Detection. *Talanta* **2007**, *72*, 896–913. [CrossRef]

10. Zhou, H.; Yue, H.; Zhou, Y.; Wang, L.; Fu, Z. A Novel Disposable Immunosensor based on Quenching of Electrochemiluminescence Emission of Ru(bpy)$_3^{2+}$ by Amorphous Carbon Nanoparticles. *Sens. Actuators B* **2015**, *209*, 744–750. [CrossRef]
11. Yoshida, H.; Ureshino, K.; Ishida, J.; Nohta, H.; Yamaguchi, M. Chemiluminescent Properties of some Luminol related Compounds (II). *Dyes Pigm.* **1999**, *41*, 177–182. [CrossRef]
12. Gu, W.; Deng, X.; Gu, X.; Jia, X.; Lou, B.; Zhang, X.; Li, J.; Wang, E. Stabilized, Superparamagnetic Functionalized Graphene/Fe$_3$O$_4$@Au Nanocomposites for a Magnetically-controlled Solid-state Electrochemiluminescence. Biosensing Application. *Anal. Chem.* **2015**, *87*, 1876–1881. [CrossRef]
13. Wang, Y.-Z.; Hao, N.; Feng, Q.-M.; Shi, H.-W.; Xu, J.-J.; Chen, H.-Y. A Ratiometric Electrochemiluminescence Detection for Cancer Cells using g-C$_3$N$_4$ Nanosheets and Ag–PAMAM–luminol nanocomposites. *Biosens. Bioelectron.* **2016**, *77*, 76–82. [CrossRef]
14. Chan, C.M. An improved Synthesis of a Chemiluminescent Cyclic Hydrazide: *N*-(7-Aminobutyl)-*N*-ethyl-naphthalene-1,2-dicarboxylic hydrazide. *Synth. Commun.* **1989**, *19*, 1981–1985. [CrossRef]
15. Ishida, J.; Yamaguchi, M.; Nakahara, T.; Nakamura, M. 4,5-Diaminophthalhy drazide as a highly Sensitive Chemiluminescence Reagent for α-Keto Acids in Liquid Chromatography. *Anal. Chim. Acta* **1990**, *231*, 1–6. [CrossRef]
16. Sasamoto, K.; Ohkura, Y. A New Chemiluminogenic Substrate for *N*-Acetyl-β-D-glucosaminidase, 4′-(6′-Diethylaminobenzofuranyl)phthalylhydrazido-*N*-acetyl-β-D-glucosaminide. *Chem. Pharm. Bull.* **1991**, *39*, 411–416. [CrossRef]
17. Ishida, J.; Takada, M.; Yakabe, T.; Yamaguchi, M. Chemiluminescent Properties of some Luminol related Compounds. *Dyes Pigm.* **1995**, *27*, 1–7. [CrossRef]
18. Ishida, J.; Takada, M.; Hara, S.; Sasamoto, K.; Kina, K.; Yamaguchi, M. Development of a Novel Chemiluminescent Probe, 4-(5′,6′-dimethoxybenzothiazolyl)phthalhydrazide. *Anal. Chim. Acta* **1995**, *309*, 211–219. [CrossRef]
19. Yoshida, H.; Nakao, R.; Nohta, H.; Yamaguchi, M. Chemiluminescent Properties of some Luminol-related compounds—Part 3. *Dyes Pigm.* **2000**, *47*, 239–245. [CrossRef]
20. Smith, M.B.; March, J. *March's Advanced Organic Chemistry: Reactions, Mechanisms, and Structure*, 6th ed.; John Wiley & Sons: Hoboken, NJ, USA, 2007.
21. Antonov, L. *Tautomerism: Methods and Theories*, 1st ed.; Wiley-VCH: Weinheim, Germany, 2013.
22. Ali Ahmed, H.E.; Abdel-Salam, H.A.; Shaker, M.A. Synthesis, Characterization, Molecular Modeling, and Potential Antimicrobial and Anticancer Activities of Novel 2-Aminoisoindoline-1,3-dione derivatives. *Bioorg. Chem.* **2016**, *66*, 1–11. [CrossRef]
23. Conchon, E.; Anizon, F.; Aboab, B.; Golsteyn, R.M.; Léonce, S.; Pfeiffer, B.; Prudhomme, M. Synthesis, in Vitro Antiproliferative Activities, and Chk1 Inhibitory Properties of Pyrrolo[3,4-*a*]carbazole-1,3-diones, Pyrrolo[3,4-*c*]carbazole-1,3-diones, and 2-Aminopyridazino[3,4-*a*]pyrrolo[3,4-*c*]carbazole-1,3,4,7-tetraone. *Eur. J. Med. Chem.* **2008**, *43*, 282–292. [CrossRef]
24. Tseng, C.-C.; Yen, W.-P.; Tsai, A.-E.; Hu, Y.-T.; Takayama, H.; Kuo, Y.-H.; Wong, F.F. ZnCl$_2$-Catalyzed Aza-Diels–Alder Reaction for the Synthesis of 1*H*-Pyrazolo[3,4-*b*]pyridine-4,5-dicarboxylate Derivatives. *Eur. J. Org. Chem.* **2018**, *2018*, 1567–1571.
25. Deshmukh, M.S.; Sekar, N. Chemiluminescence properties of luminol related quinoxaline analogs: Experimental and DFT based approach to photophysical properties. *Dyes Pigm.* **2015**, *117*, 49–60. [CrossRef]
26. Deshmukh, M.S.; Sekar, N. Chemiluminescence Properties of Luminol related *o*-Hydroxybenzimidazole analogues: Experimental and DFT based Approach to Photophysical Properties. *Dyes Pigm.* **2015**, *113*, 189–199. [CrossRef]
27. Peryasami, G.; Martelo, L.; Baleizã, C.; Berberan-Santos, M.N. Strong Green Chemiluminescence from Naphthalene analogues of Luminol. *New J. Chem.* **2014**, *38*, 2258–2261. [CrossRef]
28. Spurlin, S.R.; Cooper, M.M. A Chemiluminescent Precolumn Labelling Reagent for High-Performance Liquid Chromatography of Amino Acids. *Anal. Lett.* **1986**, *19*, 2277–2283. [CrossRef]
29. Yen, W.-P.; Liu, P.-L.; Uramaru, N.; Wong, F.F. Indium(III) chloride/silica gel catalyzed synthesis of pyrazolo[3,4-*b*]pyrrolo[3,4-*d*]pyridines. *Tetrahedron* **2015**, *71*, 8798–8803. [CrossRef]
30. Neumann, H.; Klaus, S.; Klawona, M.; Strúbina, D.; Hűbner, S.; Gördes, D.; von Wangelin, A.J.; Lalk, M.; Beller, M. A New Efficient Synthesis of Substituted Luminols using Multicomponent Reactions. *Z. Naturforsch.* **2004**, *59b*, 431–438. [CrossRef]

31. Flitsch, W.; Krämer, U.; Zimmerman, H. Cyclische Verbindungen mit Heterobrückenatomen, V. Zur Chemie der 1-Amino-pyrrole. *Chem. Ber.* **1969**, *102*, 3268–3276. [CrossRef]
32. Dey, S.K.; Lightner, D.A. 1,1′-Bipyrroles: Synthesis and Stereochemistry. *J. Org. Chem.* **2007**, *72*, 9395–9397. [CrossRef]
33. Gompper, R.; Sobotta, R. Neue Elektronenreiche Butadiene. *Tetrahedron Lett.* **1979**, *11*, 921–924. [CrossRef]
34. Fang, J.M.; Yang, C.C.; Wang, Y.W. Use of α-Anilino Dienenitriles as Nucleophiles in Cycloadditions. *J. Org. Chem.* **1989**, *54*, 477–481. [CrossRef]
35. Tyagi, P.; Venkateswararao, A.; Thomas, K.R.J. Solution Processable Indoloquinoxaline derivatives containing Bulky Polyaromatic Hydrocarbons: Synthesis, Optical Spectra, and Electroluminescence. *J. Org. Chem.* **2011**, *76*, 4571–4581. [CrossRef]
36. Martelo, L.; Periyasami, G.; Fedorov, A.A.; Baleizão, C.; Berberan-Santos, M.N. Chemiluminescence of Naphthalene analogues of Luminol in Solution and Micellar Media. *Dyes Pigm.* **2019**, *168*, 341–346. [CrossRef]
37. Reichardt, C. Solvatochromic Dyes as Solvent Polarity Indicators. *Chem. Rev.* **1994**, *94*, 2319–2358. [CrossRef]
38. Dutkiewlcz, M. Classification of Organic Solvents based on Correlation between Dielectric β Parameter and Empirical Solvent Polarity Parameter E^N_T. *J. Chem. Soc. Faraday Trans.* **1990**, *86*, 2237–2241. [CrossRef]
39. Behera, S.K.; Karak, A.; Krishnamoorthy, G. Photophysics of 2-(4′-Amino-2′-hydroxyphenyl)-1H-imidazo-[4,5-c]pyridine and its analogues: Intramolecular Proton Transfer Versus Intramolecular Charge Transfer. *J. Phys. Chem. B* **2015**, *119*, 2330–2344. [CrossRef] [PubMed]
40. Behera, S.K.; Murkherjee, A.; Sadhuragiri, G.; Elumalai, P.; Sathiyendiran, M.; Kumar, M.; Mandal, B.B.; Krishnamoorthy, G. Aggregation induced enhanced and Exclusively Highly Stokes Shifted Emission from an Excited State Intramolecular Proton Transfer Exhibiting Molecule. *Faraday Discuss.* **2017**, *196*, 71–90. [CrossRef]
41. Chen, Y.-T.; Wu, P.-J.; Peng, C.-Y.; Shen, J.-Y.; Tsai, C.-C.; Hu, W.-P.; Chou, P.-T. A study of the Competitive Multiple Hydrogen Bonding Effect and its associated Excited-state Proton Transfer Tautomerism. *Phys. Chem. Chem. Phys.* **2017**, *19*, 28641–28646. [CrossRef]
42. Wang, Q.; Niu, Y.; Wang, R.; Wu, H.; Zhang, Y. Acid-induced Shift of Intramolecular Hydrogen Bonding Responsible for Excited-state Intramolecular Proton Transfer. *Chem. Asian J.* **2018**, *13*, 1735–1743. [CrossRef]
43. Tang, Z.; Lu, M.; Liu, K.; Zhao, Y.; Qi, Y.; Wang, Y.; Zhang, P.; Zhou, P. Solvation Effect on the ESIPT Mechanism of 2-(4′-Amino-2′-hydroxyphenyl)-1H-imidazo-[4,5-c]pyridine. *J. Photochem. Photobiol. A* **2018**, *367*, 261–269. [CrossRef]
44. Satapathy, A.K.; Behera, S.K.; Yadav, A.; Laxmi Mahour, L.N.; Yelamaggad, C.V.; Sandhya, K.L.; Sahoo, B. Tuning the Fluorescence Behavior of Liquid Crystal Molecules containing Schiff-base: Effect of Solvent Polarity. *J. Lumin.* **2019**, *210*, 371–375. [CrossRef]
45. Skripnikova, T.A.; Lysova, S.S.; Zevatskii, Y.E.; Myznikov, L.V.; Vorona, S.V.; Artamonva, T.V. Physico-chemical Properties of Isomeric Forms of Luminol in Aqueous Solutions. *J. Mol. Struct.* **2018**, *1154*, 59–63. [CrossRef]
46. Deepa, S.; Reddy, S.R.; Rajendrakumar, K. Green Chemiluminescence of Highly Fluorescent Symmetrical Azo-based Luminol derivative. *Orient. J. Chem.* **2018**, *34*, 894–905. [CrossRef]
47. Brouwer, A.M. Standards for Photoluminescence Quantum Yield Measurements in Solution (IUPAC Technical Report). *Pure Appl. Chem.* **2011**, *83*, 2213–2228. [CrossRef]
48. Kiyama, M.; Iwano, S.; Otsuka, S.; Lu, S.W.; Obata, R.; Miyawaki, A.; Hirano, T.; Maki, S.A. Quantum Yield Improvement of Red-light-emitting Firefly Luciferin Analogues for in vivo Bioluminescence Imaging. *Tetrahedron* **2018**, *74*, 652–660. [CrossRef]

Sample Availability: Samples of the compounds are available from the authors.

© 2020 by the authors. Licensee MDPI, Basel, Switzerland. This article is an open access article distributed under the terms and conditions of the Creative Commons Attribution (CC BY) license (http://creativecommons.org/licenses/by/4.0/).

Review

Functional Pyrazolo[1,5-*a*]pyrimidines: Current Approaches in Synthetic Transformations and Uses As an Antitumor Scaffold

Andres Arias-Gómez, Andrés Godoy and Jaime Portilla *

Bioorganic Compounds Research Group, Department of Chemistry, Universidad de los Andes, Carrera 1 No. 18A-10, Bogotá 111711, Colombia; aj.arias@uniandes.edu.co (A.A.-G.); af.godoy@uniandes.edu.co (A.G.)
* Correspondence: jportill@uniandes.edu.co; Tel.: +571-339-4949

Abstract: Pyrazolo[1,5-*a*]pyrimidine (**PP**) derivatives are an enormous family of *N*-heterocyclic compounds that possess a high impact in medicinal chemistry and have attracted a great deal of attention in material science recently due to their significant photophysical properties. Consequently, various researchers have developed different synthesis pathways for the preparation and post-functionalization of this functional scaffold. These transformations improve the structural diversity and allow a synergic effect between new synthetic routes and the possible applications of these compounds. This contribution focuses on an overview of the current advances (2015–2021) in the synthesis and functionalization of diverse pyrazolo[1,5-*a*]pyrimidines. Moreover, the discussion highlights their anticancer potential and enzymatic inhibitory activity, which hopefully could lead to new rational and efficient designs of drugs bearing the pyrazolo[1,5-*a*]pyrimidine core.

Keywords: antitumor scaffold; enzymatic inhibitory; *N*-heterocyclic compounds; organic synthesis; pyrazolo[1,5-*a*]pyrimidine; functionalization

1. Introduction

Pyrazolo [1,5-*a*]pyrimidine (**PP**) structural motif is a fused, rigid, and planar *N*-heterocyclic system that contains both pyrazole and pyrimidine rings [1]. This fused pyrazole is a privileged scaffold for combinatorial library design and drug discovery because its great synthetic versatility permits structural modifications throughout its periphery. The **PP** derivatives synthesis has been widely studied; thus, various reviews related to the obtention and later derivatization steps have been described in the literature [2–5], after the first critical review involving this attractive scaffold [6] (Figure 1).

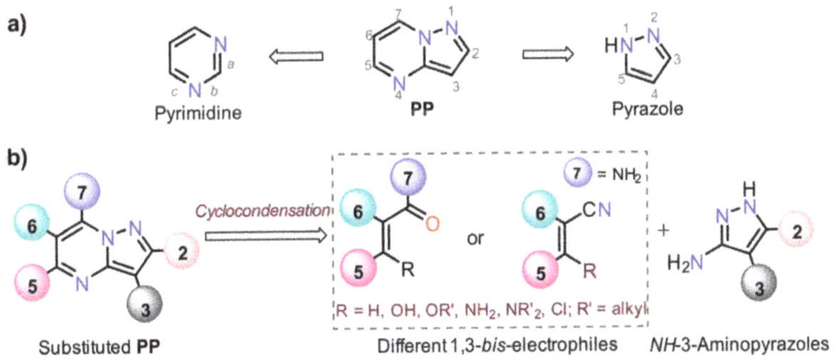

Figure 1. Pyrazolo[1,5-*a*]pyrimidine with (**a**) the constituent rings and (**b**) the modified periphery in accordance with the retrosynthetic analysis.

Despite those reports, the synthetic transformations involving this motif still represent a research priority regarding the process efficiency, environmental impact, and the study of its multiple applications. These reports should address protocols that aim to minimize synthesis pathways, employ cheap reagents and develop processes that prevent or reduce waste production. Usually, **PP** derivative synthesis involves the pyrimidine ring construction via the interaction of *NH*-3-aminopyrazoles with different 1,3-biselectrophilic compounds such as β-dicarbonyls, β-enaminones, β-haloenones, β-ketonitriles, and so on (Figure 1b) [1–6]. Pyrazolo[1,5-*a*]pyrimidine scaffold is part of bioactive compounds with exceptional properties like selective protein inhibitor [7], anticancer [8], psychopharmacological [9], among others [10,11]. Furthermore, the biocompatibility and lower toxicity levels of **PP** derivatives have led them to reach commercial molecules, for instance, Indiplon, Lorediplon, Zaleplon, Dorsomorphin, Reversan, Pyrazophos, Presatovir, and Anagliptin (Figure 2) [1–5]. In recent years, this molecular motif has been a focus of study for promising new applications related to materials sciences [12–21], due to its exceptional photophysical properties as an emergent fluorophore [15–21]. Likewise, the tendency of pyrazol[1,5-*a*]pyrimidine derivatives to form crystals with notable conformational and supramolecular phenomena [17,22,23] could amplify their applications towards the solid-state. Therefore, we aim to cover two main topics related to compounds bearing the pyrazolo[1,5-*a*]pyrimidine core. At the first one, the reader will find relevant synthesis strategies and functionalization reactions. Subsequently, in the second part, we focus on the recent compounds presenting antitumoral and enzymatic inhibitory activity. The examples described and commented herein came from the 2015 to 2021 period.

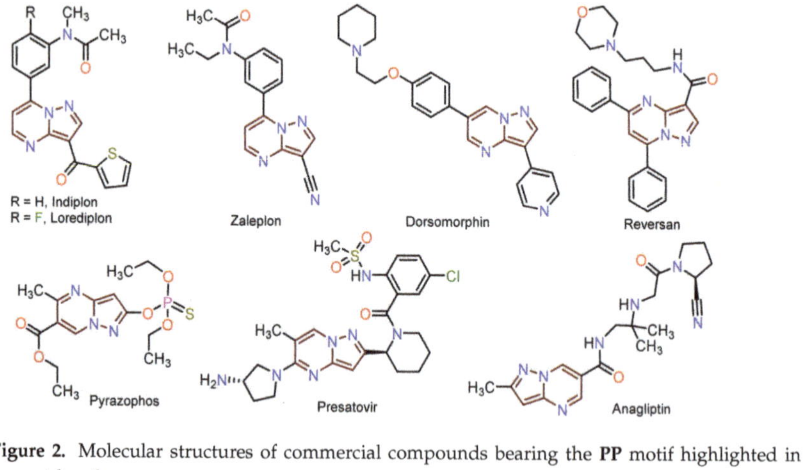

Figure 2. Molecular structures of commercial compounds bearing the **PP** motif highlighted in brownish-red.

2. Synthesis and Functionalization

Firstly, we considered it convenient to carry out a schematic summary depicting the most relevant synthesis and functionalization sections (Table 1).

Table 1. Overview of synthesis and functionalization methods involving pyrazolo[1,5-a]pyrimidine derivatives [a].

Synthetic Methods		Substituted PP	Functionalization Reactions	
Section 2.1.1	1,3-Biselectrophillic systems: β-dicarbonyls, etc.	Heating and/or X: OH, Cl	Metal catalyzed reactions: Suzuki, etc. Ar–B(OH)$_2$ Pd catalyst, solvent R^1 = H, R^2 = Br	Section 2.2.1
Section 2.1.2	Multicomponent reactions	Heating (Solvent and/or catalyst) X: OH, OR, NR$_2$ etc.	Nucleophilic aromatic substitution. R^6-NH$_2$, Solvent R^4 = Cl or Br	Section 2.2.2
Section 2.1.3	Synthesis by a pericyclic reaction	TsN$_3$, Et$_3$N CuCl (10 mol%) MeCN, r.t., 30 min R^1 = R^2 = R^5 = H	Other functionalization reactions. Electrophile = E Solvent/additive R^2 = H	Section 2.2.3

[a] General examples of some relevant synthetic transformations (synthesis and functionalizations) are showed.

2.1. Synthesis

As stated before, the synthesis of **PP** derivatives is the focus of various research; however, more recent studies in this area are focused on improving known reaction protocols. Regardless, innovative synthesis methods still emerge that offer creative ways to modify established ones. Notably, the main synthesis route of pyrazolo[1,5-a]pyrimidines allows versatile structural modifications at positions 2, 3, 5, 6, and 7 via the cyclocondensation reaction of 1,3-biselectrophilic compounds with NH-3-aminopyrazoles as 1,3-bisnucleophilic systems (Figure 1b) [1–6,24,25].

2.1.1. Biselectrophillic Systems

The synthesis starting from 1,3-dicarbonyl compounds is by far the most employed because it provides high functional group tolerance or enables conditions that are easily accessible while employing cheap commercial reagents. Acid and ethanol are commonly used solvents, while common catalysts include sodium ethoxide and amine-based bases [1–5,26–29]. This synthetic strategy is used to obtain large quantities of starting materials (ranging from mg to even kg) under simple conventional heating conditions.

In this context, Yamagami and co-workers developed a method to produce the 5,7-dichloro-2-hetarylpyrazolo[1,5-a]pyrimidine **3** via the cyclocondensation reaction of the 5-amino-3-hetarylpyrazole **1** with malonic acid (**2**) (Scheme 1) [30]. In this approach, the addition of POCl$_3$ in the presence of a catalytic amount of pyridine produces an activated species of malonic acid phosphoric ester. This discovery led to the synthesis of **3** in a higher yield in a reduced time compared to the alternative conditions of dimethyl malonate under basic media as show by the authors. Moreover, this strategy eliminates the need for additional reactions that convert a hydroxyl group to chloride.

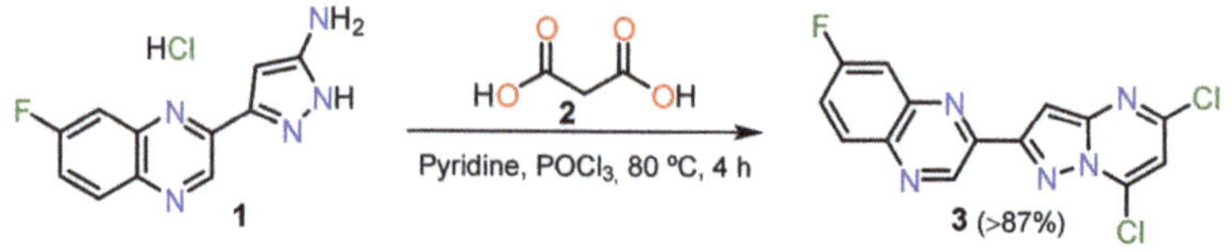

Scheme 1. One-pot protocol to obtain 5,7-dichloro-2-hetarylpyrazolo[1,5-a]pyrimidine **3**.

Similarly, the employment of β-diketones demonstrated to be effective for integrating fluorinated substituents at position 5, improving the electrophilic character of the 1,3-biselectrophile. Petricci and co-workers achieve the synthesis of **6**, where the carbonyl group substituent controls the reaction regioselectivity (Scheme 2a) [31]. Likewise, Lindsley and co-workers presented a protocol to obtain the 7-trifluoromethyl derivative **9** wherein the employment of heating by microwave (MW) irradiation significantly reduces the required reaction time. However, the brominated aminopyrazole **7a** is more reactive than **4** due to the electronic nature of the respective substituents (Scheme 2b) [32].

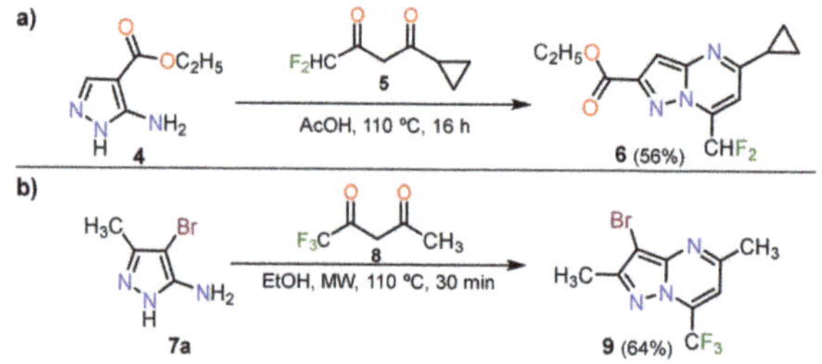

Scheme 2. Synthesis examples of fluorinated pyrazolo[1,5-a]pyrimidines (**a**) **6** and (**b**) **9**.

Considering aryl derivatives, Hylsová and co-workers employed arylketoesters **11a–b** to obtain the 5-arylpyrazolo[1,5-a]pyrimidines **12a–c** [33]. The authors do not report the complete protocol, though it is known that arylketones are a challenging substrate due to the lower electrophilicity of the carbonyl group (Scheme 3a). It would be interesting to evaluate conditions like toluene with tosylic acid [34]. Another report by Lindsley et al. described the synthesis of the 7-fluorophenyl derivative **14** using the brominated aminopyrazole **7b**. The acidic media and the electron-withdrawing effect from the halogen atom could promote the formation of products [35].

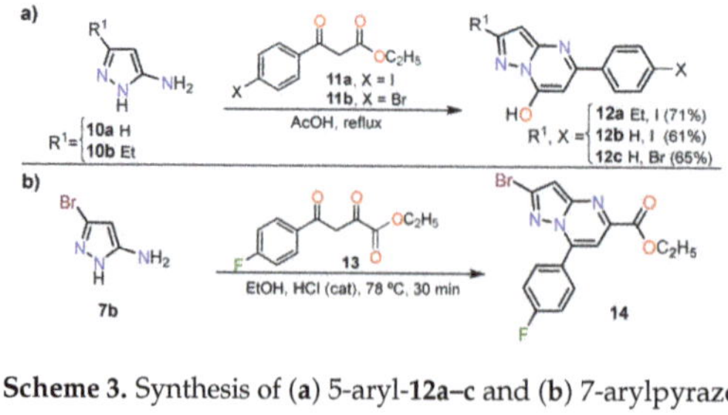

Scheme 3. Synthesis of (**a**) 5-aryl-**12a–c** and (**b**) 7-arylpyrazolo[1,5-a]pyrimidines **14**.

Furthermore, Fu and co-workers reported in 2020 a method to control the regioselectivity of the reaction. The reaction took place as expected with an excess of diethyl malonate (**15a**), and the pyrazolo[1,5-a]pyrimidine **16** is obtained in a 71% yield [36] (Scheme 4). However, adding **15** in stoichiometric quantity and heating the mixture neat produces the exocyclic amine from **10a** as the only available nucleophile. Hence, applying Vilsmeier-

Haack (POCl$_3$/N,N-dimethylaniline) conditions facilitates the annulation and subsequent hydroxyl/chloride exchange for delivering derivative **17**.

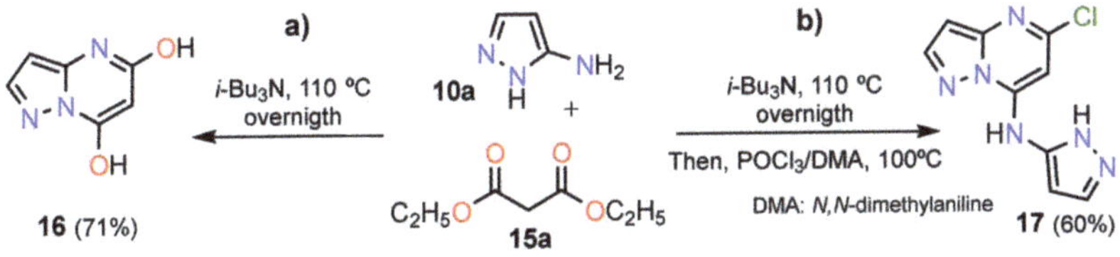

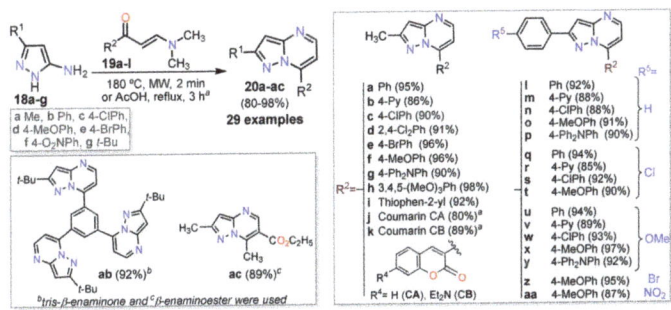

Scheme 4. (a) Obtention of **16** and (b) regioselective synthesis of **17**.

Conversely, if the 1,3-biselectrophilic system is a β-enaminone moiety, it enhances the reactivity or performance compared to 1,3-dicarbonyl compounds [1,37,38]. Portilla and co-workers employed diverse β-enaminone derivatives **19a–l** following a highly efficient methodology under MW irradiation, obtaining various 2,7-disubstituted products **20a–ac** in high yields (Scheme 5) [17,18,21,38]. The regioselectivity of the reaction can be controlled using the dimethylamino leaving group, where the initial condensation proceeds via an addition–elimination mechanism (aza-Michael type), thus bonding the NH$_2$-group of the starting aminopyrazole with the Cβ of **19a–l** [37]. Successively, the cyclocondensation occurs by a nucleophilic attack of the pyrazolic nitrogen to the residual carbonyl group, where the subsequent loss of a water molecule leads to products **20a–ac** [1,37,38].

Scheme 5. Synthesis of pyrazolo[1,5-a]pyrimidines **20a–ac** from β-enaminones **19a–l**.

Following a similar approach, Guo and co-workers in 2019 reported the interesting structure series **23a–h** bearing various types of nitrogenous groups at position 7, which were obtained by using β-enaminone derivatives **22a–h**. Notably, the authors achieved via compound **23d** the discovery and preclinical characterization of Zanubrutinib (BGB-3111): a novel, potent, and selective covalent inhibitor of Bruton's tyrosine kinase. However, the obtained yields for intermediates **23a–h** were not found (Scheme 6) [39]. In other studies, the N-heterocyclic core has been synthetized from β-enaminone derivatives bearing aryl groups substituted with halogen atoms or methoxy groups. Additionally, carboxamides, aryl groups, nitriles, and esters have been employed as substituents on the starting NH-3-aminopyrazole [40–42].

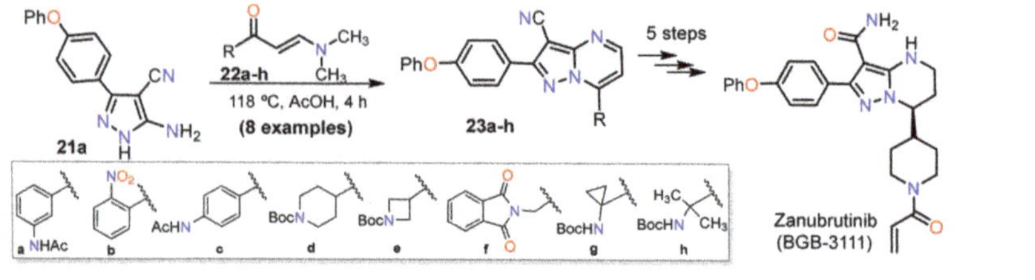

Scheme 6. Examples of pyrazolo[1,5-*a*]pyrimidines with nitrogenous groups at position 7, compounds **23a–h**.

Xu and co-workers reported an interesting method where 1-methyluracil (**25**), a heterocyclic β-enaminone, reacts with the aminopyrazole **24**. Through the uracil ring-opening induced by 24 and a later loss of a methylurea molecule, the 2-aryl-5-hydroxypyrazolo[1,5-*a*]pyrimidine **26** is produced. Employing an excess of **24**, it is possible to avoid the chromatographic separation for the purification of **26** (Scheme 7) [43]. Curiously, β-enaminones or, in general, the enone systems bearing a leaving group at β-position (e.g., Cl, OR, NR_2, etc.) are synthetic analogous of 1,3-biselectrophilic ynones.

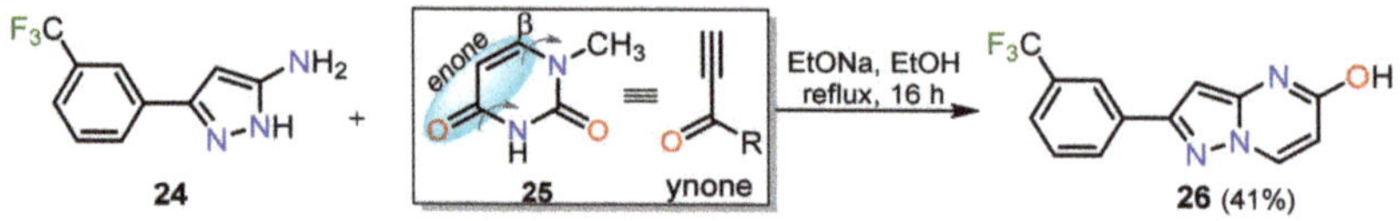

Scheme 7. Synthesis of the PP **26** by using 1-methyluracil (**25**) as a 1,3-biselecrophylic system.

In a similar approach, Pankova and co-workers incorporated an alkyne group at position 7 of the fused ring (compounds **28a–t**) by using the enone **27** having an ethoxy group at β-position [14]. Reaction yields are increased with electron withdrawer groups (EWGs) attached to the 4-aryl moiety in **27**, evidencing an electronic dependence on the nature of substituents. Likewise, the cyclization regiospecificity (enone vs. ynone moiety) emerges from the trimethylsilyl (TMS) group high electron-donating effect in π-electron systems (Scheme 8). This synthetic approach results in an attractive way to obtain the 7-alkynyl-2,6-diarylpyrazolo[1,5-*a*]pyrimidines **28a–t**, which later could undergo Pd-catalyzed carbon–carbon (C–C) cross-coupling reactions.

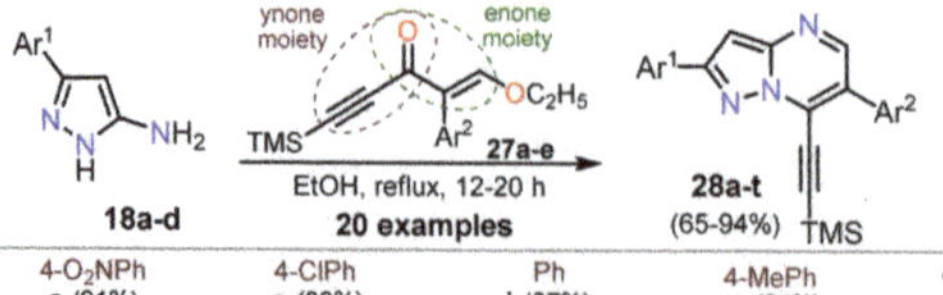

Ar[2]=	4-O_2NPh	4-ClPh	Ph	4-MePh	4-MeOPh
Ar[1]= Ph	**a** (91%)	**e** (89%)	**i** (87%)	**m** (84%)	**q** (61%)
4-ClPh	**b** (94%)	**f** (93%)	**j** (87%)	**n** (91%)	**r** (85%)
4-MePh	**c** (85%)	**g** (88%)	**k** (88%)	**o** (86%)	**s** (61%)
4-MeOPh	**d** (81%)	**h** (85%)	**l** (82%)	**p** (82%)	**t** (65%)

Scheme 8. Synthesis of 7-alkynylpyrazolo[1,5-*a*]pyrimidines **28a–t** under reflux in ethanol.

Moreover, the employment of enones for **PP** derivatives synthesis with two aryl substituents has been a matter of interest with different approaches. Adib and co-workers described using azidochalcones **30**, which yield milder conditions for synthesizing polysubstituted, products **31** in moderate to excellent yields (Scheme 9) [44]. The authors report short times for reactions and also established a recrystallization purification process. For these reactions, catalysis has been carrying out using ionic liquids [13] and nanoparticles [45], resulting in improved yields when the substituents were in the aryl group.

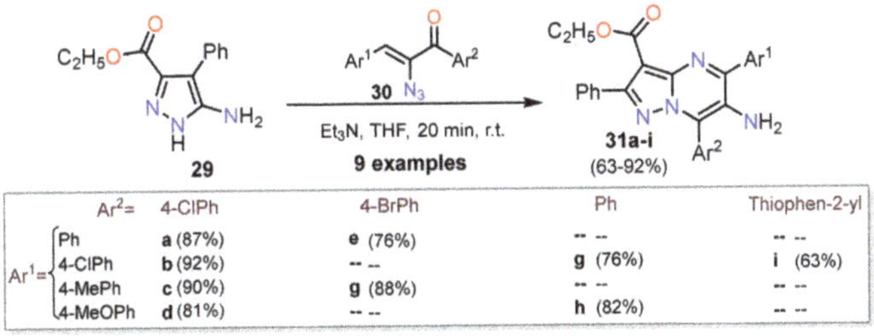

Scheme 9. Use of diarylsubstituted azidochalcones **30** in the synthesis of the 6-amino-2,5,7-triaryl-**PP**.

In the previous examples, enones bearing a leaving group at β-position were used as a 1,3-biselectrophile system. However, the reaction is often tricky when simple enones like chalcone derivatives are used, due to the unsaturation required in products [19,46]. Despite the inconvenience, this methodology provides valuable results for introducing aryl groups at positions 5 and 7 of the aza-heterocyclic core, with the advantage that starting materials are commercial or readily available by simple reactions (e.g., Claisen–Schmidt condensations). In this respect, Portilla and co-workers carried out the synthesis under rigorous (high temperature for a relatively long time) MW conditions of 2,5,7-tris(4-methoxyphenyl)pyrazolo[1,5-*a*]pyrimidine (**33**) starting from the substituted chalcone **32** and aminopyrazole **18d**. This synthesis allowed us to design and implement a probe for cyanide (CN⁻) sensing by a nucleophilic addition reaction on the carbon–carbon double bond of the receptor group and an intramolecular charge transfer (ICT) photophysical phenomenon (Scheme 10) [19].

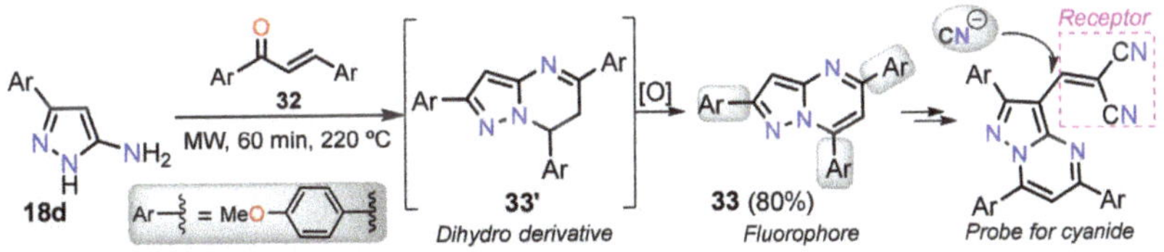

Scheme 10. Synthesis of the cyanide probe **33** from the dimethoxychalcone **32**.

Jismy and co-workers developed protocols for **PP** derivatives synthesis based on 1,3-biselectrophilic ynones [47–49]. Unlike previous approximations, employing ynones delivers to hydroxy/enone substituents at position 5 of the fused pyrazole. Thus, it is possible to obtain an addition acyclic intermediate **37** that, under basic media, delivers derivatives **36** (Scheme 11) [48]. The authors evaluate different conditions such as solvent, temperature, time, Lewis acid catalysis, and MW heating. This synthetic strategy produces compounds substituted at position 7 with inherently electrophilic groups such as CF$_3$ (**36a–d**), which could be added in this manner, avoiding complex post-functionalization steps [49].

Comparatively, Schmitt and co-workers established a methodology that allows the functionalization with aryl or heteroaryl groups at position 7 (compounds **39a–f**) employing substituted alkynes **38a–f** (Scheme 12a). In addition, they took advantage of substituted pyrazoles to functionalize positions 2 and 3 (not shown) [50]. Recently, Akrami and co-workers developed a protocol capable of reduces reaction time employing dimethyl acetylenedicarboxylate (**38g**) and the aminopyrazole **21b** (Scheme 12b) [51]. The reaction was optimized in terms of solvent, temperature, and time, obtaining an ester group at position 7 of the respective product **39g**.

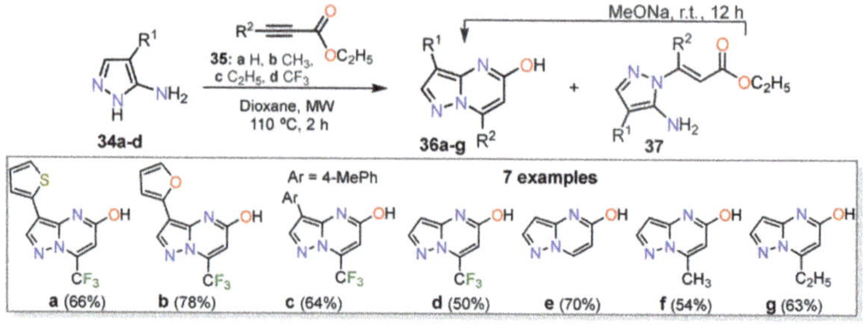

Scheme 11. Examples of regioselective synthesis employing ynones. Highlights the obtention of the intermediate **37**.

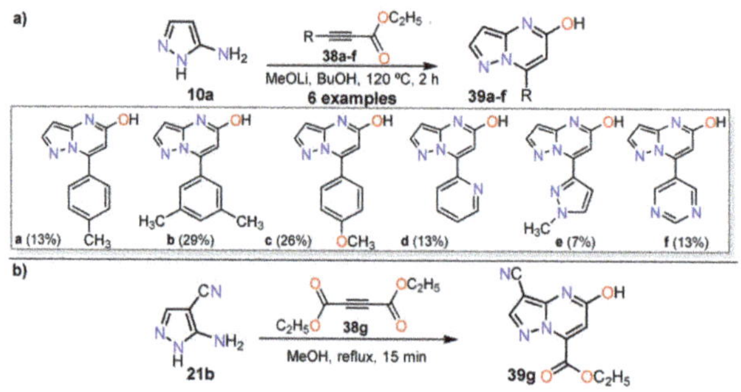

Scheme 12. Synthesis of 5-hydroxypyrazolo[1,5-*a*]pyrimidines (**a**) **39a–f** (poor yields, an opportunity for research) and (**b**) **39g**.

The use of other types of 1,3-biselectrophilic compounds for pyrazolo[1,5-*a*]pyrimidines synthesis have been described in addition to the works mentioned above. In this line, Hebishy and co-workers recently reported the synthesis of highly functionalized derivatives **42** [27]. By employing arylidenemalononitriles **41a–c**, it is possible to introduce an amine and a nitrile group in products **42a–c**, desirable groups for subsequent reactions such as carboxamides synthesis (Scheme 13). Similarly, Fouda and co-workers employ 2-(aryldiazenyl)malononitriles to obtain amines substituents at positions 7 and 5 [52].

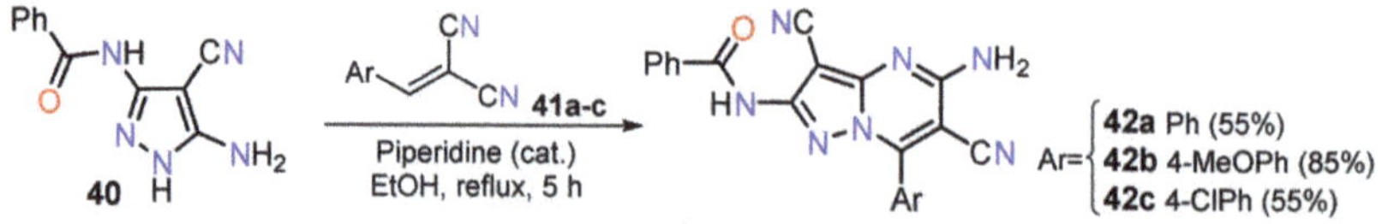

Scheme 13. Synthesis of pyrazolo[1,5-*a*]pyrimidines **42a–c** using arylidenemalononitriles **41a–c**.

Similarly, Portilla and co-workers developed an MW-assisted methodology to obtain 6-(aryldiazenyl)pyrazolo[1,5-*a*]pyrimidines **44a–q** by using 1,3-biselectrophilic derivatives bearing a hydrazone functional group at position 2, that is, reagent **43a–c** (Scheme 14) [53]. Remarkably, the synthesis depicted in Scheme 14 made it possible to introduce an amino group at position 5 of the heterocyclic core via a reductive azo bond cleavage [53]. Additionally, the electronic properties of the starting aminopyrazole **18** from various substituents at position 5 were evaluated. Notably, in this methodology, the authors report no solvent for the reaction, and purification requires little to no effort.

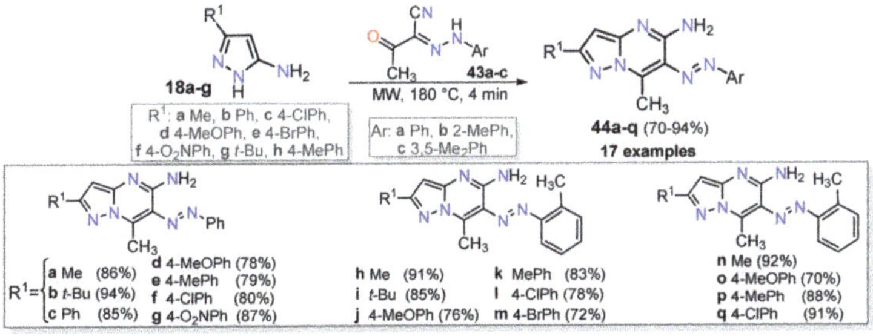

Scheme 14. MW-assisted synthesis of 6-(aryldiazenyl)pyrazolo[1,5-a]pyrimidines **44a–g** using β-ketonitriles **43**.

Furthermore, Zahedifar and co-workers reported the use of freshly prepared ketenes **45** in the preparation of compounds **46** from the corresponding substituted aminopyrazole **18a–e** under reflux in tetrahydrofuran (THF). The authors report high yields for the synthetized pyrazolo[1,5-a]pyrimidines **46a–e** and relatively short reaction times while that purification was carrying out through recrystallization (Scheme 15) [54].

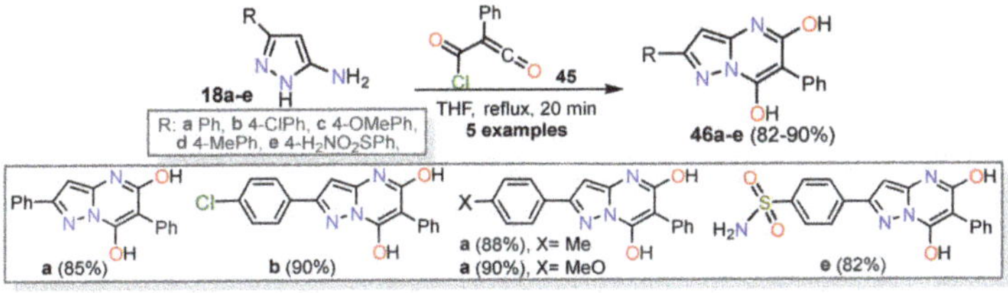

Scheme 15. Synthesis of pyrazolo[1,5-a]pyrimidines **46a–e** using the ketene **45**.

2.1.2. Multicomponent Reactions

Importantly, the pyrazolo[1,5-a]pyrimidines synthesis by multicomponent reactions have also been reported. In this context, the most common approximation resembles a Mannich reaction whose products usually undergo a later oxidation reaction. It is desirable to block any position which could lead to a side reaction, given that the derivatives obtained result highly substituted. Li et al. reported an approach using large quantities of the starting materials **47a** and **18** obtaining a dihydro derivative (not shown). Subsequently, the authors achieved an oxidation with DDQ without an intermediate purification step involved, to produce the unsaturated product **49** in 64% yield (Scheme 16a) [55].

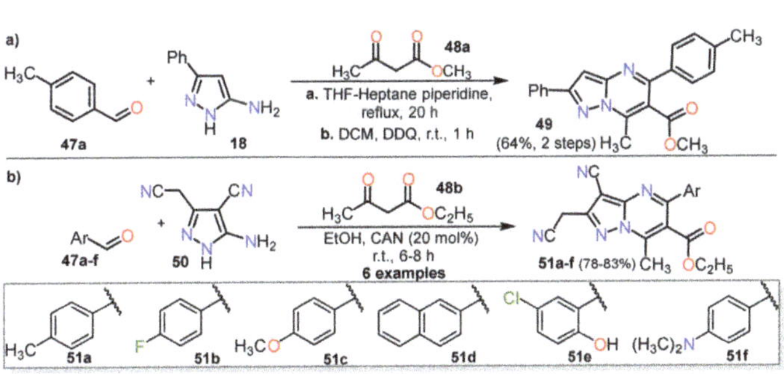

Scheme 16. Examples of multicomponent synthesis of pyrazolo[1,5-a]pyrimidines (a) **49** and (b) **51a–f**.

Notably, the synthetic protocol of compound **49** is suitable for preparative quantities in conditions that are easily reproducible in the laboratory. Similarly, Shastri et al. developed a method focused on obtaining derivatives from different arylaldehydes **47a–f**. In contrast to the previous example, the synthesis of products **51a–f** contemplates mixing all the starting materials in the same vial (Scheme 16b) [56]. In this case, the more extended reaction periods could improve the yields, although aminopyrazole's stereoelectronic properties could improve selectivity, avoiding side reactions like dimerization or aromatic substitutions. Besides, oxidation step relevance could be evaluated from work reported by Ismail, where a similar approach led to products in good yields, but the lack of an oxidation step produces **PP** dihydroderivatives (see Section 3 for detail) [57].

On the other hand, Jiang and co-workers reported an iodine-catalyzed pseudo-multicomponent reaction starting from aroylacetonitriles (β-ketonitriles) **52** and the sulphonyl hydrazine **53** [58]. The authors achieved derivatives **54**, and by involving two molecules of **52**, the positions 2 and 5 become substituted with the same aryl group. The reactions present moderate to good yields in a process that readily gives salts (Scheme 17). Notably, these compounds could have a longer shelf life compared with their analogs **54a–f**.

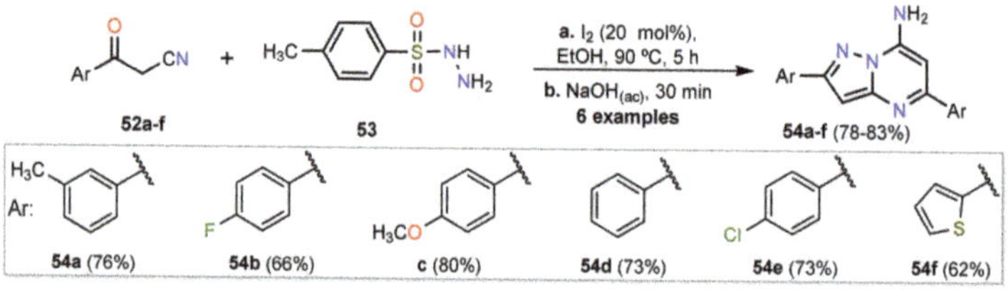

Scheme 17. Pseudo-multicomponent synthesis of 5-amino-2,7-diarylpyrazolo[1,5-*a*]pyrimidines **54**.

Likewise, Ellman et al. [59] designed a method based on catalysis with Rh complexes to obtain the pyrazolo[1,5-*a*]pyrimidines **58a–at** through the multicomponent reaction of aldehydes **55**, aminopyrazoles **56**, and sulfoxonium ylides **57** (Scheme 18). The synthesis of **58** was suitable for aldehydes with electron-donating groups (EDGs), heteroaryl, and haloaryl; however, enolizable aldehydes proved difficult substrates. The authors evaluated diversely substituted sulfoxonium ylides, obtaining good to excellent yields for **58ab–ah**. The reaction yield presents an increased susceptibility to pyrazole modification compared to the aldehydes and sulfoxonium ylides. Notably, the conditions optimized for the reaction gave a setup with benchtop materials, shorts reaction times, and high modulation options. Even though it is intended to be an MW heating protocol, conventional heating was also evaluated, providing good results. The scale-up of the reaction to 1 mmol yield 77% of the expected product using a lower catalyst charge (5 mol%).

Tiwari and co-workers prepared fused heterocycles **60a–c** through a C–C bond formation catalyzed with palladium [60]. The reaction proved efficient in generating the cyclic pyrimidine ring. However, only aryls without halogen or other exchangeable groups should be used by similar approximations reducing the scope of producing side products (Scheme 19).

2.1.3. Synthesis by Pericyclic Reactions

Alternatively, pyrazolo[1,5-*a*]pyrimidines have been obtained by pericyclic reactions and without involving a starting aminopyrazole. In this respect, Ding and co-workers developed a protocol for the fused ring synthesis from acyclic precursors through a [4 + 2] cycloaddition reaction, which the authors report to be scalable and proceed in a one-pot manner [61]. The appropriate *N*-propargylic sulfonylhydrazone **61** is treated with a sulphonyl azide in the presence of catalytic copper (I) chloride since a click reaction drives the substrate **61** to a triazole formation, which discomposes to intermediate **62**. Subsequently, an intramolecular Diels–Alder reaction takes place, forming both rings of **63**.

The dihydro derivative **63** could be treated in basic media to give the desired product **69** by an elimination reaction (Scheme 20).

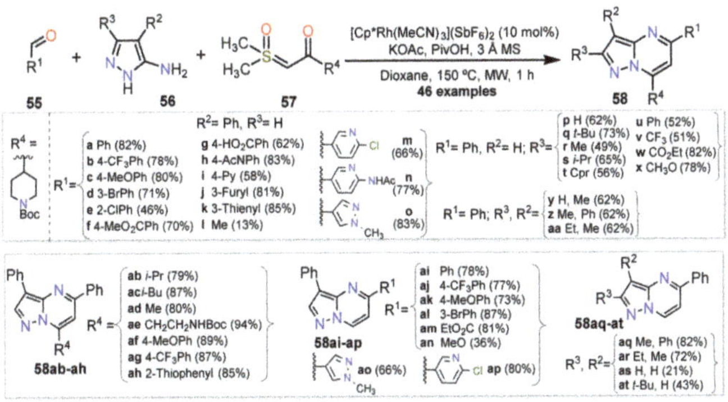

Scheme 18. Rh-catalyzed Multicomponent synthesis of variously substituted pyrazolo[1,5-a]pyrimidines **58a–at**.

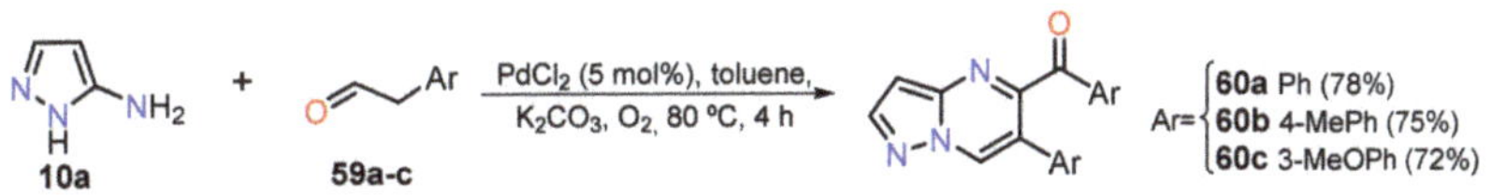

Scheme 19. Synthesis of 6-aryl-5-aroylpyrazolo[1,5-a]pyrimidines **60a–c** by a palladation step.

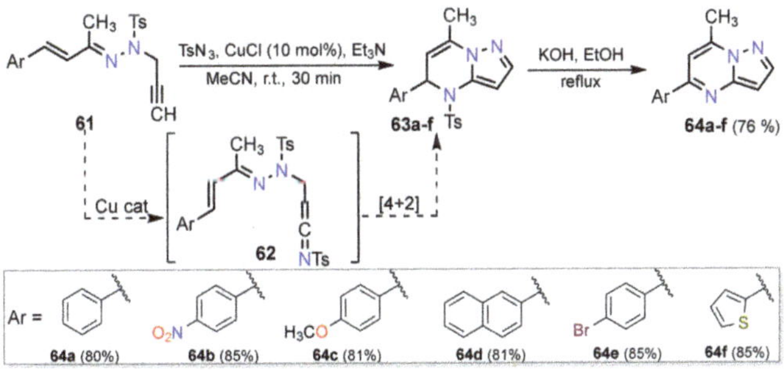

Scheme 20. Synthesis of 5-aryl-7-methylpyrazolo[1,5-a]pyrimidines **64a–f** via intramolecular Diels–Alder reaction.

2.1.4. Synthesis with Fused Cores

Boruah and Nongthombam focused on developing fluorescent probes with biological activity by the androstenol derivatives **66** synthesis, which uses copper (I) iodide as a catalyst [62]. The synthesis proved to be efficient with yields ranging from 78 to 89%, where the pyrazolic nitrogen of **18** attacks the 1,3-biselectrophile **65** probably by a Michael type conjugated addition and the acetamide group as a leaving group (i). Subsequently, the cyclocondensation between the formyl (CHO) and amino (NH_2) groups (ii) of the respective cyclization intermediate leads to the formation of **66a–e** (Scheme 21).

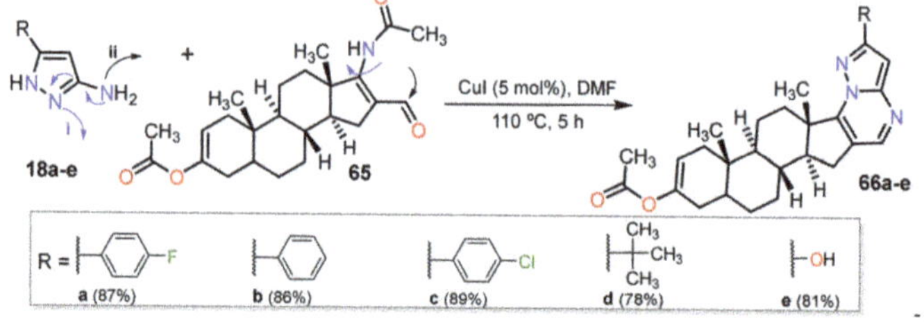

Scheme 21. Development of fluorescent probes with biological activity via the synthesis of fused derivatives **66**.

Comparatively, Mekky and co-workers carried out the synthesis of a bisbenzofuran derivative bearing two pyridopyrazolo[1,5-a]pyrimidine moieties (compound **68**). In addition to β-enaminone **19a**, the authors included β-dicarbonyl compounds or arylidenemalononitriles as 1,3-biselectrophilic systems, which were cyclocondensed with the fused NH-aminopyrazole **67**. The authors also evaluated the repercussion of MW heating achieving better yields and shorter reaction times (Scheme 22) [63].

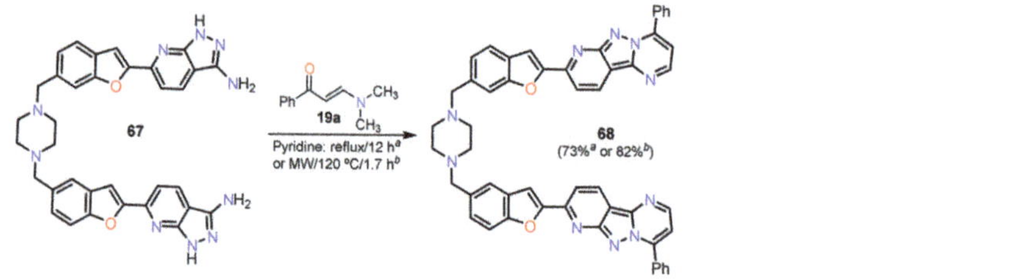

Scheme 22. MW-assisted synthesis of the hybrid system bisbenzofuran—bispyrazolopyrimidine **68**.

Likewise, Jismy and co-workers evaluated a synthesis starting from the indazoles **69** and the appropriate alkyne **35d** to obtain the benzo-condensed derivative **70**, which possesses a CF_3 and hydroxyl group on the pyrimidinic moiety (Scheme 23a) [64]. The authors also evaluated other positions for the bromo substituent in the amino-indazole, obtaining similar yields to that of **70** [64]. Similarly, Song et al. evaluated the obtention of the furan-fused product **72**, which is achieved by a Michael type conjugated addition over the enaminonitrile **71**, conserving the enantiomeric excess from the initial substrate (Scheme 23b) [65].

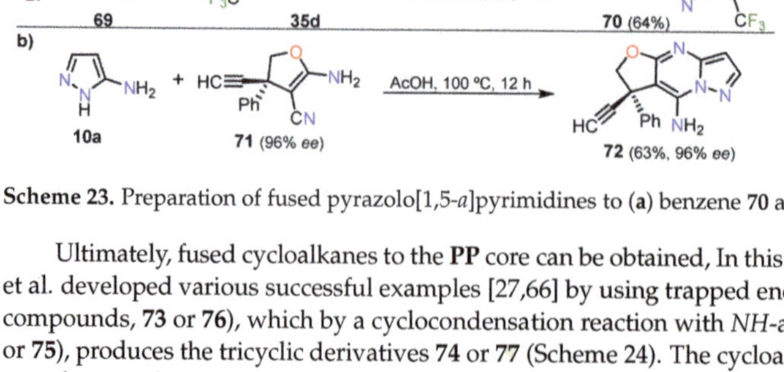

Scheme 23. Preparation of fused pyrazolo[1,5-a]pyrimidines to (**a**) benzene **70** and (**b**) furan **72**.

Ultimately, fused cycloalkanes to the **PP** core can be obtained, In this context, Elgemeie et al. developed various successful examples [27,66] by using trapped enolates (enone type compounds, **73** or **76**), which by a cyclocondensation reaction with NH-aminoprazoles (**40** or **75**), produces the tricyclic derivatives **74** or **77** (Scheme 24). The cycloalkanes to be fused vary from cyclopentane to cyclooctane, though these compounds are not functionalized in the provided examples. Additionally, the electronic effects on the starting pyrazole drive the reaction, showing, for example, better yields from **75** than those obtained from **40** against the same enone.

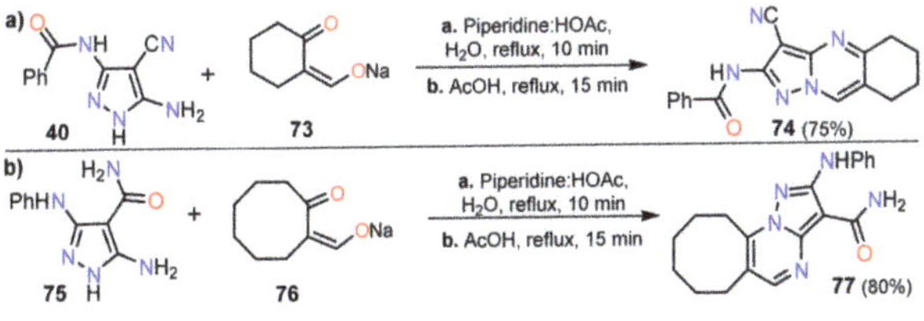

Scheme 24. Synthesis of tricyclic derivatives bearing cycloalkanes such as (**a**) cyclohexane **74** and (**b**) cyclooctane **77**.

2.2. Functionalization
2.2.1. Metal Catalyzed Reactions
Suzuki Couplings

This reaction is one of the most employed to add functionalized aryls on position 3 or 5; several organometallic species have been evaluated with favorable results, such as boronic esters, boronic acids, and fluoroboranes. The reactions have been carried out mainly using solvents like water mixed with some organic solvent (to maximize solubility) or in dioxane, and carbonates appear to be the preferred base, perhaps due to carbon dioxide formation facilitating the workup [67–69].

Jismy and co-workers designed a method for functionalizing **36e** at position 5 using a one-pot synthesis. They produce an exchange of hydroxyl group for a chloride atom by a NAS reaction; thus, employing optimized conditions for the coupling reactions could achieve aryls at position 5 like in product **79** (Scheme 25a) [48]. The reaction was further applied in the amine 3-bromosubstituted **81** and was optimized, finding the highly reproducible conditions to obtain **82** (Scheme 25b) [47]. The reaction shows an improvement in reaction times, and also, regarding the previous work, the authors found that modifying the ligand and catalyst enables the functionalization of the nucleophilic site at the pyrazolic moiety. Recently, they provided a method to add an aryl moiety at position 3 of the fused pyrazole **84**, which has a hydroxyl/enone group at position 5 and a CF_3 group at position 7 (product **85**, Scheme 25c) [64]. The reaction conditions were screened to find an optimal setup, though the reaction conditions with the better performance are those previously reported. Interestingly, the fluorophenyl group and some heterocycles were found to be compatible with this strategy, with yields from 67 to 84% [64].

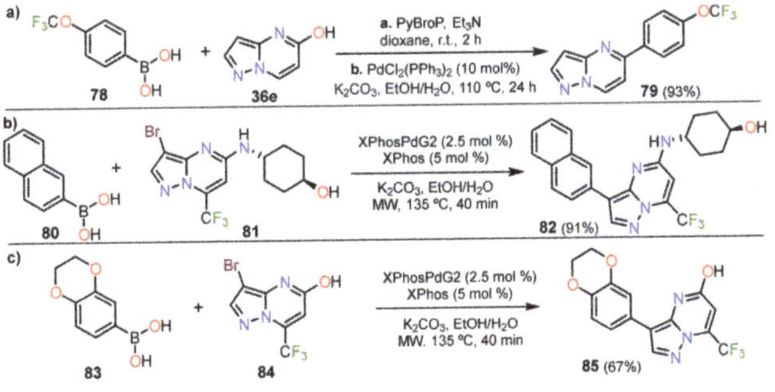

Scheme 25. Functionalization of the position 3 and 5 via Suzuki cross coupling reaction of (**a**) **36**, (**b**) **81** and (**c**) **84**.

Related to this work, the employment of Suzuki coupling results is a common strategy in medicinal chemistry. Liu et al. reported the obtention of 17 examples about **88**, where the amines or ethers present at positions 5 and 7 are modified (Scheme 26a) [67]. Employment of PdCl$_2$(dppf) as a catalyst and heating under MW irradiation appears to produce highly reproducible conditions for short-time reactions, and the authors obtained salt forms of each compound analogs to **88** for reactions with amounts above 200 mg of compounds related to **87** (Scheme 26a). Similarly, Lindsley and co-workers designed a fast reaction to functionalized the CF$_3$-substituted PP **90** with aryls moieties bearing methoxy group (compound **91**) or fluor atoms (not shown) (Scheme 26b) [32]. Related to these advances, Drew et al. employed the same Pd-catalyst bearing dppf as a ligand to perform the coupling of **93** with the isoindolinone **92**, the reaction proceeds by the lability exchange with the halogen added according to the electronic properties of the position 5 against position 3 in the pyrazolo[1,5-*a*]pyrimidine **93** (Scheme 26c) [68].

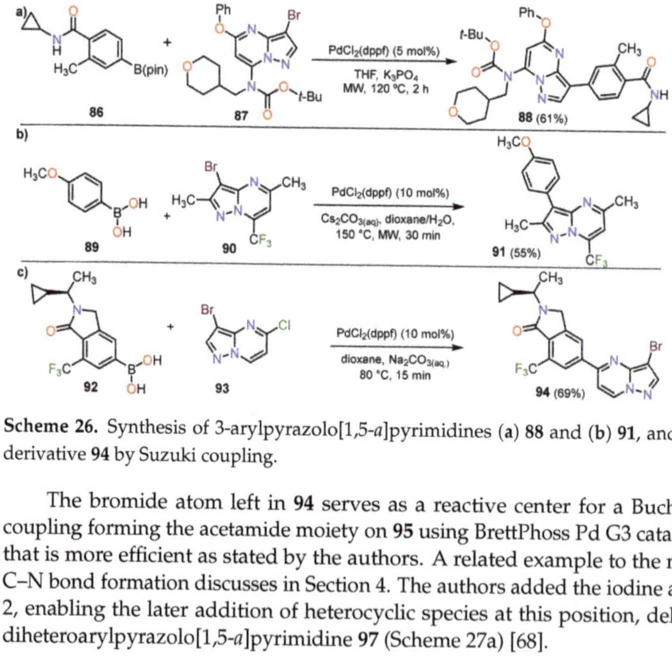

Scheme 26. Synthesis of 3-arylpyrazolo[1,5-*a*]pyrimidines (a) **88** and (b) **91**, and of (c) the 5-aryl derivative **94** by Suzuki coupling.

The bromide atom left in **94** serves as a reactive center for a Buchwald–Hartwig coupling forming the acetamide moiety on **95** using BrettPhoss Pd G3 catalyst, a synthesis that is more efficient as stated by the authors. A related example to the metal-catalyzed C–N bond formation discusses in Section 4. The authors added the iodine atom at position 2, enabling the later addition of heterocyclic species at this position, delivering the 2,5-diheteroarylpyrazolo[1,5-*a*]pyrimidine **97** (Scheme 27a) [68].

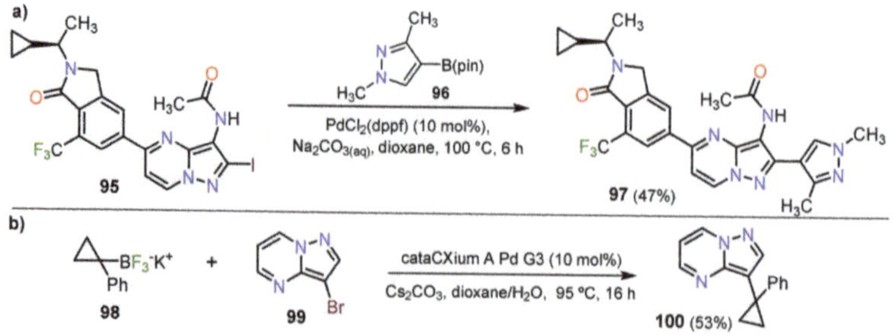

Scheme 27. Examples of functionalization by Suzuki coupling with the addition of (a) pyrazolic and (b) cyclopropyl moieties.

This reaction is probably favored by the labile character on the I–C bond in **95** and the nucleophilic nature of its coupling partner **96**. Likewise, Harris et al. designed a method focused on the generation of disubstituted cyclopropanes, employing the proper borontrifluoride **98** and 3-bromopyrazolo[1,5-*a*]pyrimidine (**99**) which, under optimized reaction conditions, delivers **100** in 53% yield (Scheme 27b) [69]. Similarly, Lindsley and co-workers employed borontrifluoride derivatives as a coupling partner for Suzuki coupling reaction, adding a vinyl over position 3 of **136** (see Section Formylation Reactions for detail).

Sonogashira Couplings

This reaction commonly involves using a terminal alkyne bearing the **PP** core with an aryl/alkenyl halide as the coupling reagent. In almost all scenarios, Pd species are employed as the primary catalyst [38,70]. In this respect, Dong et al. designed a method to functionalize 5-ethynylpyrazolo[1,5-*a*]pyrimidine (**101**) with the alkyl bromide **102**. The reaction employs a copper salt with a ligand quinine derivative (L), achieving the formation of a new C–C (sp)/(sp^3) bond in the absence of any Pd species [70]. The compound **103** is obtained in a high yield and with a high enantiomeric excess (*ee*), providing an approach towards functionalized products readily found in medically relevant molecules (Scheme 28).

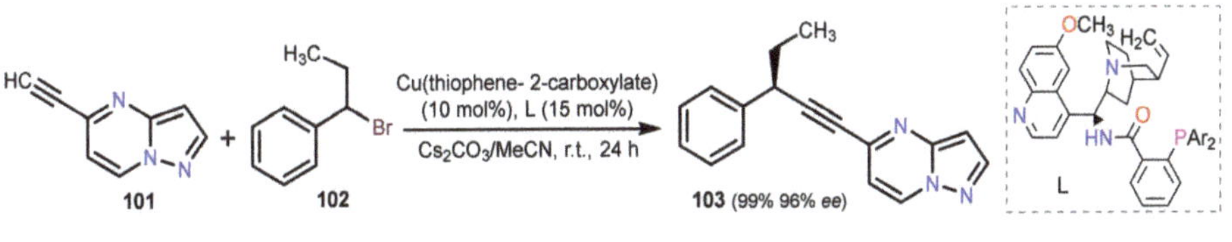

Scheme 28. Generation of a chiral carbon in coupling product **103** by a Cu-catalyzed Sonogashira reaction.

Related to this matter, Childress et al. employed a common Pd catalyst and MW heating to achieve the functionalization of terminal alkyne **105** with an excess of 4-bromo-2-(trifluoromethyl)pyridine (**104**), obtaining the product **106** in 40% after purification by HPLC (Scheme 29a) [35]. The authors also obtain other two pyridine moieties in analogs of **106** employing the same methodology. Similarly, Jismy and co-workers developed a method to functionalized the aza-heterocyclic core at position 5 with wide scope [47–49,64]. Different from other authors, they generate the electrophilic coupling partner substituting the hydroxy/enone group at position 5 of **36d** (and analogs), then using the proper alkyne, the reaction delivers the product **108** (Scheme 29b) [49]. The reaction shows a high scope regarding the alkyne used, with a great influence on the substituent, wherein aromatic or conjugated ones achieve higher yields than alkyl or cycloalkyl substituents.

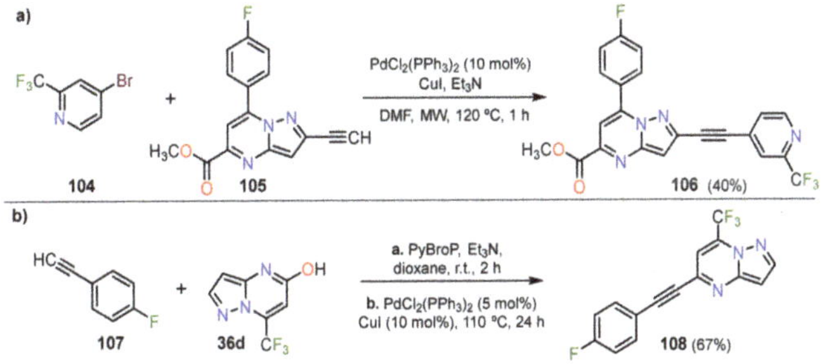

Scheme 29. Pd-Catalyzed synthesis of (a) 2-(pyridin-4-ylethynyl)PP **106** and (b) 5-(4-fluorophenylethynyl)PP **108**.

Other Metal-Catalyzed Reactions

Important reactions related to this matter are the C–H activations employing Pd species where the active species is prepared in situ. Bedford and co-workers developed a protocol that enables functionalizing of the positions 3 or 7 selectively in **PP**. An excess of aryl bromide (Ar–Br) was employed to obtain the 7-pyrazolo[1,5-*a*]pyrimidines **110a–f** a (Scheme 30) [71]. As expected, coupling reactions regioselectivity involving the heterocyclic core and an aryl bromide (Ar–X) depends on the electronic properties of reagents; indeed, the more π-deficient rings (e.g., Ar = pyrimidin-5-yl) behave well as electrophilic partners providing the 3-aryl derivatives **109a–f** in high yields via a coupling at the highly nucleophilic position 3 of **PP**. In contrast, π-excedent rings (e.g., Ar = 4-methoxyphenyl) behaves well when coupled to a more electrophilic place such as position 7, delivering **110a–f**. In order to explain these notable reactivity findings, the authors also report a DFT calculation analysis of the substrate, which are in agreement with those recently reported by Portilla et al. [18].

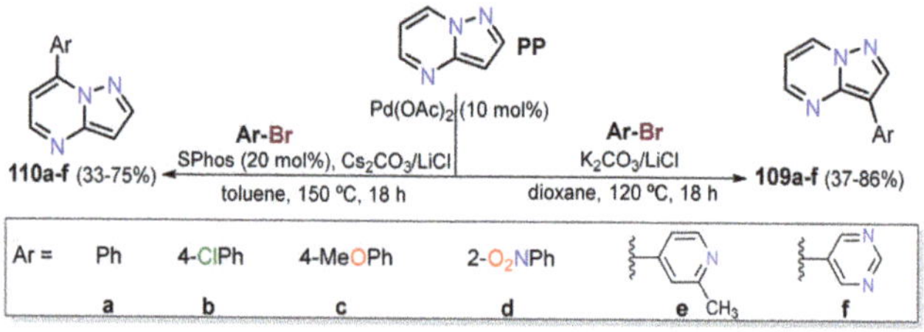

Scheme 30. Pd-catalyzed synthesis of 3-aryl **109** and 7-arylpyrazolo[1,5-*a*]pyrimidines **110** starting from **PP**.

Similarly, Berteina-Raboin and co-workers designed a protocol to functionalize position 3 of the fused pyrazole **111**. The authors optimized the solvent, base, ligand, and palladium source; once the optimal conditions were founded, the authors proved various aryls bromines (Scheme 31) [72]. Furthermore, Gogula et al. provided a method to modulate the C–H activation of (sp^2) or (sp^3) carbon based on the temperature over the fused pyrazole **113** and analogs [73]. In addition, the added palladium generates a stable coordination complex **114** which the authors obtained, this species is responsible for the activation of the methyl C–H leading to a palladation (not shown) and subsequent formation of the C–C bond with the aryl iodine, delivering **115 a–f**. On the other hand, the activation of position 6 in the **PP** ring occurs by a tetramer compound (not shown), which generates a π-aryl palladation at position 6 and later arylation forming products **116a–f**. All reactions were carried out in hexafluoroisopropanol (HFIP) as a solvent, which proved to be efficient in terms of the compound's achieved solubility and reaction temperatures. This solvent is known as a magical solvent for Pd-catalyzed C–H activation [74] (Scheme 32).

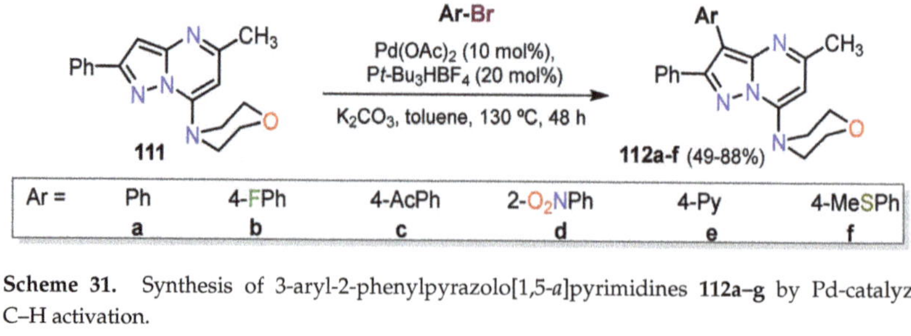

Scheme 31. Synthesis of 3-aryl-2-phenylpyrazolo[1,5-*a*]pyrimidines **112a–g** by Pd-catalyzed C–H activation.

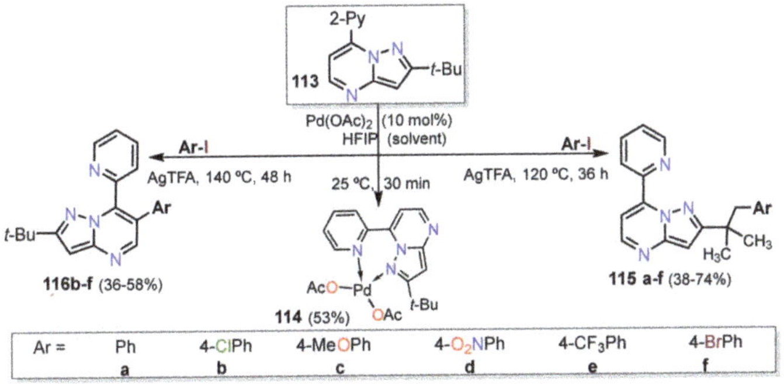

Scheme 32. Pd-catalyzed synthesis of 7-(pyridin-2-yl)pyrazolo[1,5-*a*]pyrimidines **115a–f** and **116a–f**.

Finally, in the context of medicinal chemistry, McCoull and co-workers developed a method that enables the ring closure by olefin metathesis reactions on functionalized pyrazolo[1,5-*a*]pyrimidines **116** (Scheme 33) [75]. The protocol employs considerable catalysts quantities to achieve the reaction, and the authors evaluated various synthesis pathways to maximize the process efficiency. The stereochemistry over **116** correctly in the pyrrolidine fragment controls the final conformation achieve in **117**, which is an atropisomer due to restricted rotation by the aryl fragments of the macrocycle.

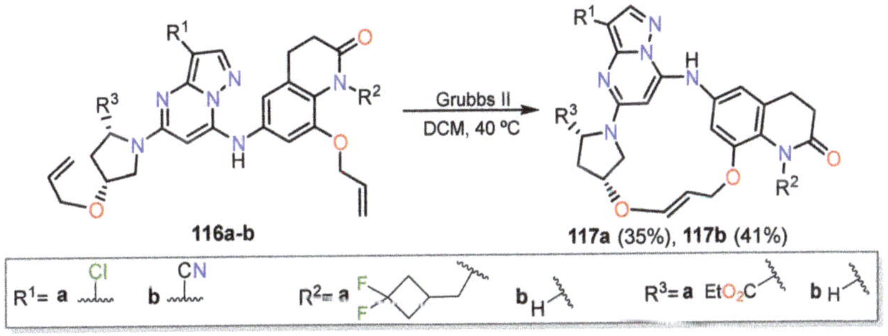

Scheme 33. Use of Grubbs catalyst to obtain macrocycles having **PP** through ring closure metathesis.

2.2.2. Nucleophilic Aromatic Substitution Reactions

The nucleophilic aromatic substitution (NAS) reaction is common to functionalize with nucleophiles the positions 5 and 7 of the fused ring. The reaction results are widely employed in medicinal chemistry because it allows modifications with various structural motifs at electrophilic positions of the pyrimidine ring. Recently, it has been employed with the aim of adding aromatic amines [34,67], alkylamines [48,76], cycloalkylamines [55,77,78], and substituted alkoxides [67]. In this respect, McNally, Paton, and co-workers designed an interesting way of coupling the 5-(diphenylphosphanyl)pyridine **118** to the position 7 of the fused pyrazole **PP**, generating the phosphonium salt **119** according to the authors [79]. This approach opens the door to a new functionalizations with weaker nucleophiles. The authors report the importance of a strong acid in the medium for the reaction mechanism based on pyridines; however, the electronic properties of substituted pyrazolo[1,5-*a*]pyrimidines could enable the avoidance of this requirement. The treatment of **130** with a source of chloride provided **120**, although in a low yield (Scheme 34).

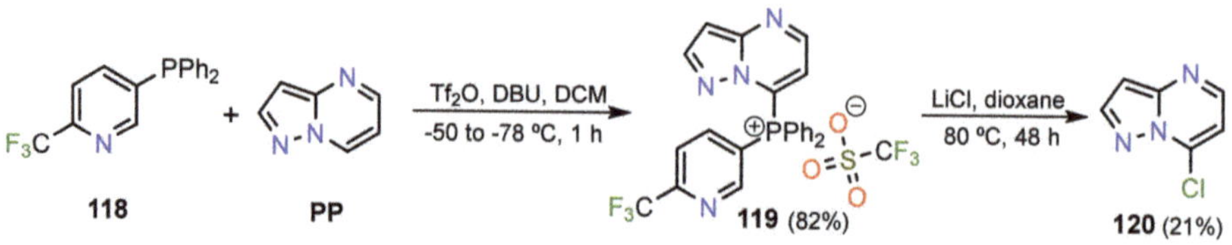

Scheme 34. C–H functionalization of **PP** by using the 5-(diphenylphosphanyl)pyridine **118**.

Similarly, Jismy et al. designed a methodology for the functionalization of position 5 employing PyBroP, probably with the formation of the intermediate species **121**, which facilitates the secondary amine **89** formation by the nucleophilic substitution of **122** over **121** (Scheme 35a) [47,49]. The authors employ this methodology in a one-pot manner achieving high yields. Additionally, the authors evaluated the efficiency of first obtaining **81** and then performing a Suzuki coupling reaction at position 3, obtaining **82** (see Scheme 25b above). As a result, the procedure done in that order delivers the final product in higher yields compared to the Suzuki coupling and then the aromatic nucleophilic substitution reaction [47]. Similarly, Berteina-Raboin developed a multicomponent method to obtain **112a** analogs, controlling the equivalents of added amine (morpholine, Scheme 35b); they were able to decrease the poisoning of the catalyst achieving good yields for various 7-aminoderivatives [72].

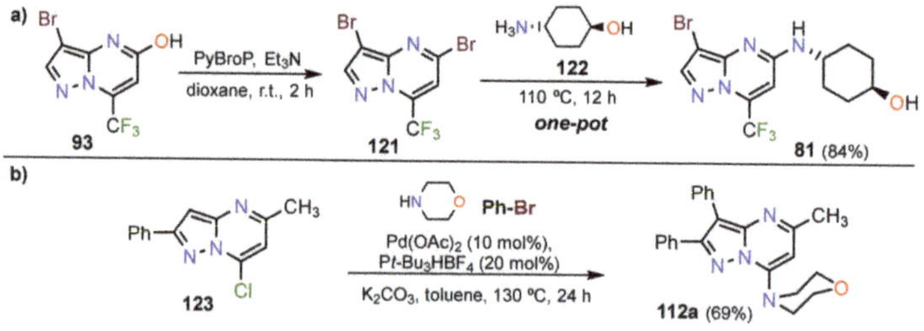

Scheme 35. Synthesis of the amino derivatives (**a**) **81** and (**b**) **112a** via NAS reactions.

Additionally, Jiang and co-workers reported an efficient synthesis of various 7-(*N*-arylamino)pyrazolo[1,5-*a*]pyrimidines **125a–e** due to the electronic properties of the employed aromatic amines and the authors use two synthesis pathways (Scheme 36) [36]. Engaging triethylamine with amines coupled to electron donor groups are conditions that allow to deliver **125a–b** in high yields. In contrast, amines bearing EWGs require strong basic conditions, as is exemplified by products **125c–e**.

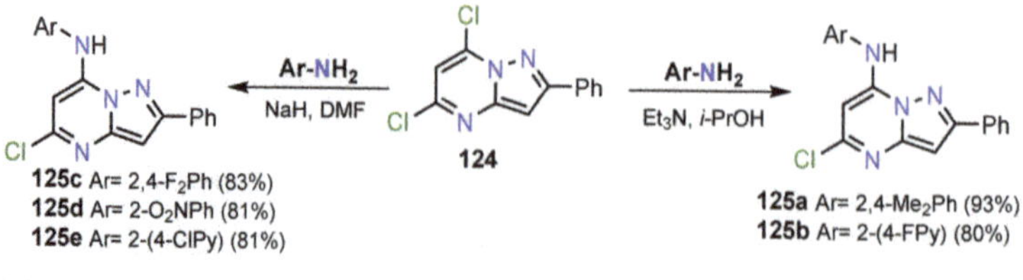

Scheme 36. Synthesis of 7-(*N*-arylamino)pyrazolo[1,5-*a*]pyrimidines **125a–e** starting from the 5,7-dichloroderivative **124**.

2.2.3. Other Functionalization Reactions

Carboxamide Synthesis

Besides the use of amines to expand the structural diversity, a practical secondary way to expand moieties installed over the ring is the formation of amides. Zou et al. developed an important way to functionalize adjacent aryls in **126**, where the added catalyst participates with a C–H activation directed by the pyridine-like nitrogen on the pyrazole side of **126**. The electronic properties of **127** facilitate the addition of the amide fragment by the C–N bond formation with decarboxylation of **127** (Scheme 37). The authors tested the reaction on various substrates where, interestingly, both steric and electronic factors modulate the reaction. However, the halide derivatives could be challenging due to side reactions and electronic properties of the 7-aryl group. From the author's perspective, the bond between the **PP** and the aryl group allows the amide formation and could be a promissory route to generate challenging amides.

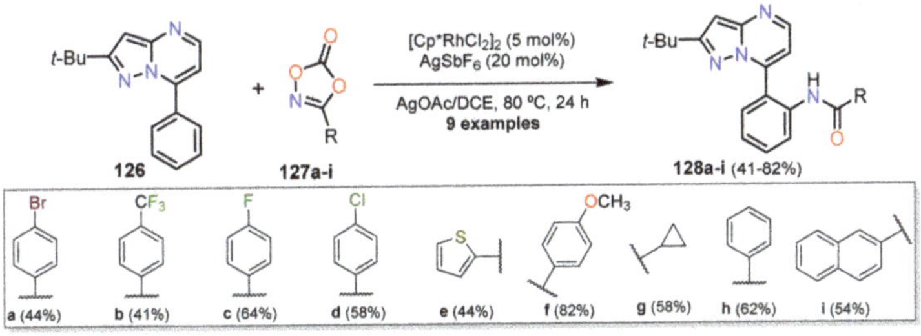

Scheme 37. Rh-catalyzed synthesis of the amides **128a–i** by C–H activation.

A common strategy in the synthesis of amides over the **PP** core consists of using benzotriazole derivatives (HATU) to activate carboxylic acids. Manetti and co-workers reported use of ester **129** which, under basic conditions at room temperature, delivers the corresponding acid (not shown). Subsequently, with an appropriate amine, the weak nucleophilic carboxyl attacks the HATU and generates a species susceptible to the amine **130** at room temperature, obtaining **131**. The authors report the employment of this strategy to achieve diverse amide derivatives (Scheme 38a) [31].

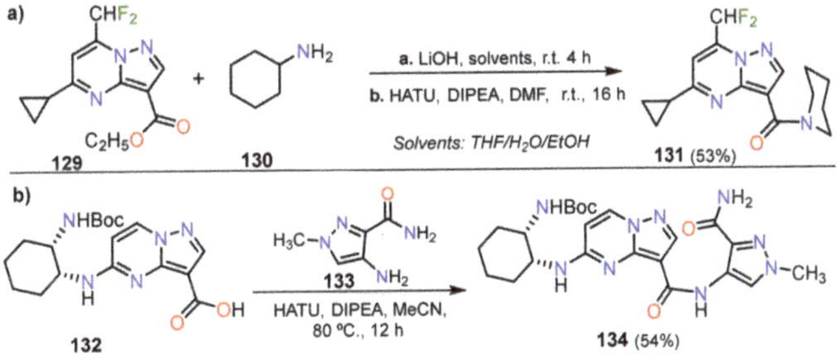

Scheme 38. Synthesis of biologically active amides (**a**) **131** and (**b**) **134** through benzotriazole derivatives (HATU).

Moreover, Lim and co-workers employed HATU, achieving the amide formation between the carboxylic acid **132** and the aminopyrazole **133**. For this reaction, it was necessary to block the pyrrolic nitrogen position on **133** due to its nucleophilic character (Scheme 38b). Other strategies used by the authors to develop amides on position 3 included the formation of acyl chlorides from carboxylic acids [80]. Related to this work, other authors have also employed a similar protocol to access amides [29,39,75,78,81,82].

Formylation Reactions

The procedure focused on the obtention of hetarylaldehydes and represented a way towards various carboxylic acid derivatives such as esters or amides. The reduction or condensation reactions could be proposed as a way for subsequent functionalization to generate the formyl electrophilic group. Thus, the electronic properties of the formed aldehyde will be vastly dependent on the position where the group is added. Portilla and co-workers designed a synthetic approach in which the **PP** ring was functionalized with a formyl group at the highly nucleophilic position 3 in a one-pot manner. By using Vilsmeier-Haack conditions, the 7-arylpyrazolo[1,5-a]pyrimidines **20** were successfully functionalized with a formyl group forming the expected aldehydes **135a–k** [19,21]. The reaction shows a broad substrate scope although with a slight dependence on the electronic properties of the 7-aryl group, that is, when π-excedent rings such as thiophene are in that position, the C3 carbon pyrazolic on **20** is even more nucleophilic, achieving higher yields (**135g**) compared with other π-deficient rings (**135f**) (Scheme 39).

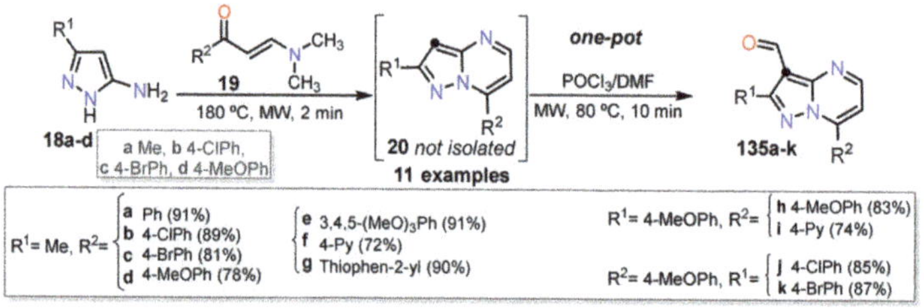

Scheme 39. Synthesis of 7-aryl-3-formylpyrazolo[1,5-a]pyrimidines **135a–k** under Vilsmeier–Haack conditions.

Other formylation protocols have been produced to add the formyl group; for instance, Lindsley et al. employed a Suzuki reaction to add a vinyl group at position 3 of **136**, the reaction proceeds with a commonly used catalyst for this reaction generates **137** in 52% yield [35]. Then, the authors employed Osmium tetroxide in a catalytic amount with a radical oxidant in stoichiometric quantity to produce a diol over the vinyl group at **136**. Lastly, an excess of a strong but selective oxidant (sodium periodate) was added to obtain **138** by a oxidative diol rupture (Scheme 40a) [35]. Similarly, Li and co-workers reported a protocol to obtain formyl group at position 6 of **139** employing an ester moiety. The authors performed a reduction of **139** with a selective hydride donor (DIBAL-H) [55]. However, due to excess reductive reagent, the alcohol **141** was mainly obtained, which after purification suffered an oxidation reaction with PCC, obtaining the desired aldehyde **140** in a good yield (Scheme 40b) [55]. The authors report the reaction performance at room temperature for both protocols; the ester moiety could be less electrophilic than expected because of its electronic participation in the heteroaryl.

Nitration and Halogenation Reactions

Halogen atoms or nitro groups are usually added over the pyrazolic moiety of the **PP** core via electrophilic aromatic substitution reactions. Portilla and co-workers designed an MW-assisted protocol where, by using the suitable electrophile, they could functionalize position 3 of the 7-arylpyrazolo[1,5-a]pyrimidine **20** [38]. The conditions employed to achieve halide derivatives **142a–i** in high yields involve N-halosuccinimides (NXS) as a halogen source at room temperature for 20 min. On the other hand, nitration reactions readily occur at position 3 where, despite the conditions, nitration of the 7-aryl moiety in the substrate was not observed, achieving good yields in a short time for 3-nitroderivatives **143a–d** (Scheme 41).

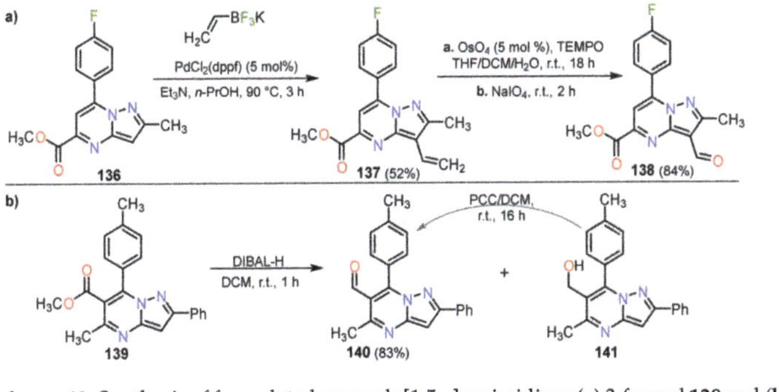

Scheme 40. Synthesis of formylated pyrazolo[1,5-*a*]pyrimidines (**a**) 3-formyl **138** and (**b**) 6-formyl **140**.

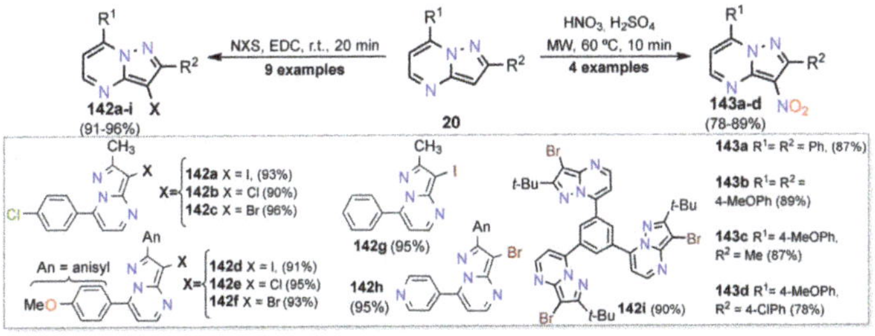

Scheme 41. Synthesis of 3-halo and 3-nitropyrazolo[1,5-*a*]pyrimidines **142a–i** and **143a–d** via aromatic substitution.

Similarly, Gazizov and co-workers employed classic conditions for the formation of the nitronium ion, which delivered products **145a–b** in high yields from **144** [83]. The conditions show interesting tolerance despite the possible amine oxidation reactions or hydrolysis of the nitrile group in **144b** or ester on **144a** (Scheme 42a). Moreover, the authors report the employment of other sources of nitrate ions like potassium nitrate (not shown). Concerning the halogenation reactions, Yamaguhi Sasaki et al. described a reaction that differs from the common Vilsmeier-Haack conditions, developing a methodology that enables the hydroxyl/enone transformation chloride from DMPA (Scheme 42b) [78]. Commonly, this reaction allows for subsequent NAS reactions (as observed in Scheme 4).

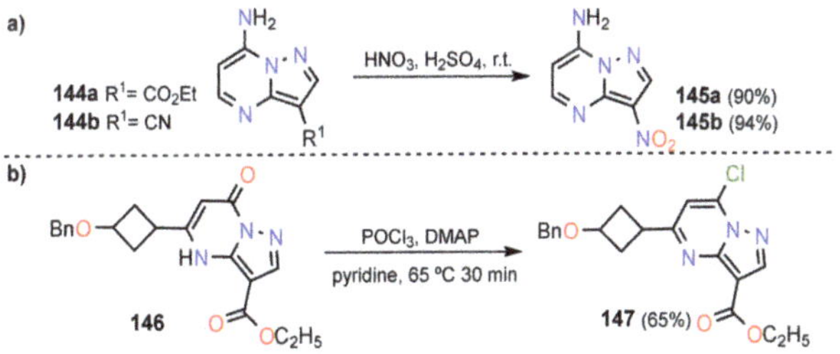

Scheme 42. Examples of (**a**) nitration and (**b**) halogenation reactions of pyrazolo[1,5-*a*]pyrimidines.

Reduction Reactions

Portilla and co-workers designed a strategy that enables the formation of amines at position 6 by using 5-amino-6-(phenydiazenyl)pyrazolo[1,5-*a*]pyrimidines **44a–c**. This strategy delivers the 1,2-diamine system in **148a–c**, which could be used to synthesize other fused rings (Scheme 43a) [53]. Notably, the free amine at position 6 is not easily obtained with a common aromatic substitution. Afterward, the same research group performed the synthesis of the 3-aminoderivatives **149** through catalytic reduction of the appropriate 3-nitroderivative **143** (Scheme 43b), which was described in the previous section (see Scheme 41) [38]. Related to the previous reduction reactions, Wang and co-workers reported the obtention of the tetrahydroderivative **150** from the amide **23a** employing strong reducing conditions (Scheme 43c) [39].

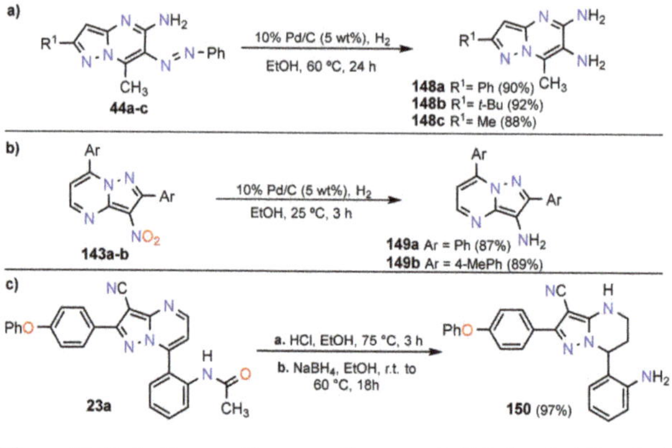

Scheme 43. Reduction reactions over the pyrazolo[1,5-*a*]pyrimidine derivatives (**a**) **44**, (**b**) **143** and (**c**) **23a**.

3. Antitumor Activity

Pyrazole derivatives are involved in many medical applications and are known to be a biologically relevant scaffold [1–6,84,85]. Besides, pyrazole[1,5-*a*]pyrimidines are fused pyrazoles that have attracted particular attention in the cancer treatment field [86–88]. Herein, we analyze the recent publications about new molecules and their respective roles in vitro and in vivo applications, allowing us to identify the principal structure motifs for future novel uses. This section delves into the principal and recent advances of pyrazolo[1,5-*a*]pyrimidines as an antitumor scaffold in bioactive compounds, mainly by inhibiting the reproduction of cancer cells [8,57,89–92] or enzymes directly related to abnormal cell reproduction [82,93–95].

Antiproliferative Activity

McCoull and co-workers [89] built up a novel procedure for macrocyclic motifs development bearing a PP core by obtaining a series of **BCL6** binders from both fragment and virtual screening. Henceforth, dislodging crystallographic water, framing new ligand protein connections, and performing a macrocyclization are actions performed to support the bioactive adaptation of the ligands. The structure-activity relationship (SAR) for **PP** (Scheme 44) indicated that the lactam carbonyl formed a noticeable hydrogen bond. Likewise, the modification in C-3 significantly changes the interaction within the enzyme, where the highest affinity was found with the nitrile group. This approach could indicate a polar interaction with an asparagine residue, which stabilizes the compound inside the enzyme. Despite the good results in SAR terms and its biological action against **BCL6**, its low selectivity decreases the potential activity of these compounds.

In 2020 Lamie et al. [91] published a novel family of pyrazolo[1,5-*a*]pyrimidines that had a great activity against the human breast adenocarcinoma cell line (**MCF-7**) and colon cancer cell line (**HTC-116**). They use conventional heating and a long reaction time (6 h) to obtain the products **158** from aminopyrazoles **156** and β-ketoesters **48** (Scheme 45a). Additionally, the authors heated the compounds **158a–c** under reflux with acetylacetone (**157**) in acetic acid, obtaining the respective 5,7-dimethylpyrazolo[1,5-*a*]pyrimidine-3-carboxamides **159a–c** (Scheme 45b). The authors [91] established that the main interactions between the *N*-heterocycle and the human **PIM-1** enzyme were hydrogen bonds due to high electronegative atoms, like N and O. Furthermore, the **PP** core planar structure and the arylamide group promote π–π interactions with active site residues. They concluded from the experimental results that **158c, 158g, 158h, 159a,** and **159c** showed PIM-1 inhibitory activity in sub-micromolar concentration (Scheme 45). These compounds displayed activity with IC_{50} (μM) of 1.26, 0.95, 0.60, 1.82, and 0.67, respectively; thus, pointing out the relationship between the **PIM-1** hindrance and anticancer action against colon and breast cancer cell lines.

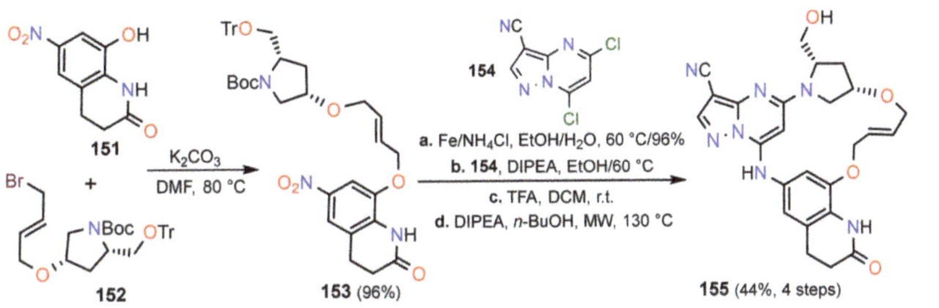

Scheme 44. Synthesis of the macrocyclic pyrazolo[1,5-*a*]pyrimidine **155**.

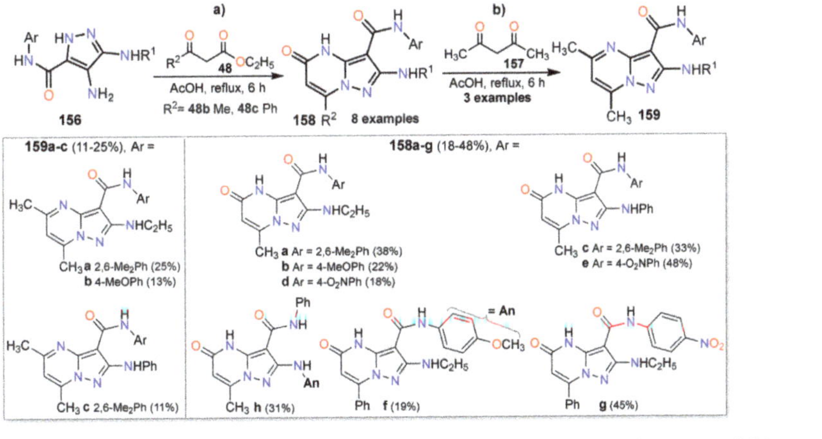

Scheme 45. Lamie's synthesis of (a) 5-oxo-4,5-dihydropyrazolo[1,5-*a*]pyrimidines **158** and (b) pyrazolopyridimines **159**.

Chen et al. [41] published a research article where 24 pyrazolo[1,5-*a*]pyrimidines were synthesized by using the β-enaminone **19h** and 3-bromo-1*H*-pyrazol-5-amine (**7b**) as reagents. In this work, only compounds **162** and **163** exhibited a crucial activity against **B16-F10, HeLa, A549,** and **HCT-116** cancer cell lines when compare against Colchicine (Scheme 46). Two conclusions could been derived from SAR analysis; first, the presence of electro-donating groups (EDGs) as the 4-tolyl group in compound **163** results in higher biological activity in contrast EWGs. Secondly, the substitution of another EDG as the 5-indolyl group (compound **162**) increases the activity against cancer cells. Additionally, the authors proposed that compound **163** was exceptionally viable in restraining melanoma tumor development in vivo with no conspicuous poisonousness.

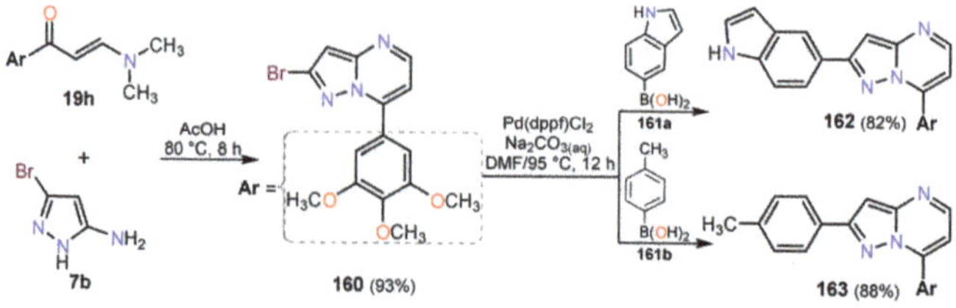

Scheme 46. Synthesis of biologically active 2,7-diarylpyrazolo[1,5-a]pyrimidines **162** and **163**.

Compound	IC$_{50}$ (µM)			
	Hela	A549	HCT-116	B16-F10
162	0.019 ± 0.001	0.015 ± 0.001	0.039 ± 0.003	0.048 ± 0.004
163	0.021 ± 0.001	0.003 ± 0.001	0.048 ± 0.005	0.021 ± 0.002
Colchicine	0.081 ± 0.006	0.107 ± 0.009	0.042 ± 0.004	0.087 ± 0.006

In 2016, Zhang and co-workers published an article in which they carried out a coupling between *N*-mustard residue with **PP** derivatives to afford 43 new compounds [8]. The compound **168** (Scheme 47) possessed antiproliferative activity with IC$_{50}$ (µM) values of 6.023, 0.217, 6.318, 8.317, and 6.82 against the **A549**, **SH-SY5Y**, **HepG-2**, **MCF-7**, and **DU145** cell lines, respectively. For this reason, they chose this compound for further in vitro and in vivo studies. Thus, the authors conclude that **168** is a promising anticancer agent since it showed higher activity and less cytotoxicity than control drugs, Sorafenib and Cyclophosphamide. From the library created by the researchers, they established that for an exhibit potent cytotoxicity in vitro, the *N*-mustard pharmacophore and aniline moieties must be linked at 5 and 7 positions of the **PP** core, respectively (Figure 3).

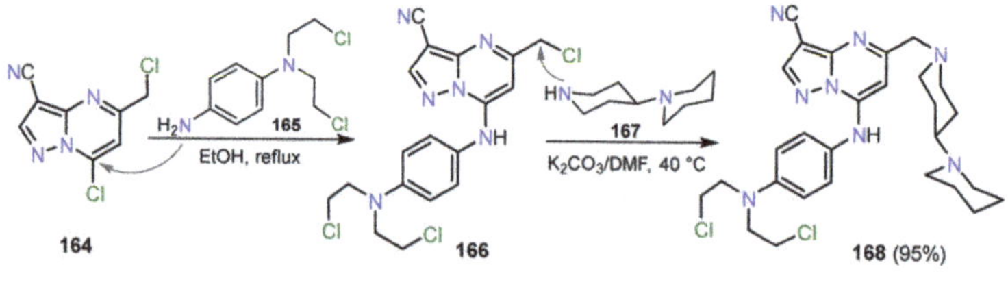

Scheme 47. Synthesis of pyrazolo[1,5-a]pyrimidines with *N*-mustard residue.

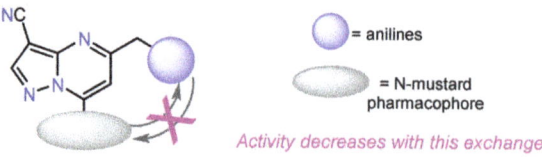

Figure 3. Substituent positions to increase the activity of pyrazolo[1,5-a]pyridimines.

In 2020, Abouzidb et al. [40] published the synthesis of some pyrazolo[1,5-a]pyrimidines as novel larotrectinib analogs using a reaction between β-enaminones **19** and the 4,5-disubstituted 3-aminopyrazole **169** (Scheme 48). The antiproliferative activity of compounds **170e**, **170j**, and **170k** stand out as the most active against three cancer cell lines: hepatocellular carcinoma **Huh-7**, cervical adenocarcinoma **HeLa** and breast adenocarcinoma **MCF-7**. The improvement in the biological activity came from the substitution on the 7-aryl group with methoxy function in **170j** and the higher naphthalene-ring lipophilic character in **170k**. Notably, the activity of compounds **170a–m** is independent of the substituent electronic effect on the 7-aryl group.

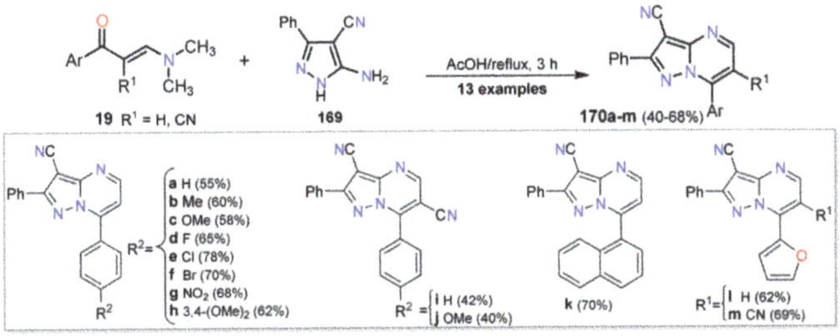

Compound	IC$_{50}$ (µM)			
	Huh-7	HeLa	MCF7	MDA-MB231
170e	6.3	48.8	118.4	74.25
170j	54.1	7.8	106	5.74
170k	95.0	21.6	3.0	4.32
DOX	3.2	8.1	5.9	6.0

Scheme 48. Synthesis of the pyrazolo[1,5-*a*]pyrimidines 170a–m as novel larotrectinib analogs.

In 2016, Narsaiah et al. [92] published a series of pyridine-fused pyrazolo[1,5-*a*]pyrimidines 174 (Scheme 49). The pyrido[2′,3′:3,4]pyrazolo[1,5-*a*]pyrimidines 174 were prepared from the key intermediate pyrazolo[3,4-*b*]pyridine 171 and different 1,3-biselectrophilic reagents (32, 172 and 173). The compounds were screened for relative global growth inhibition against five human cancer cell lines (PC3, MDA-MB-231, HepG2, HeLa, and HUVEC), 5-fluorouracil (5-FU) as a positive control, and DMSO as a negative control. Although the authors mention that all compounds have potential anticancer activity, only compounds 174a and 174f have IC$_{50}$ values lower than the control.

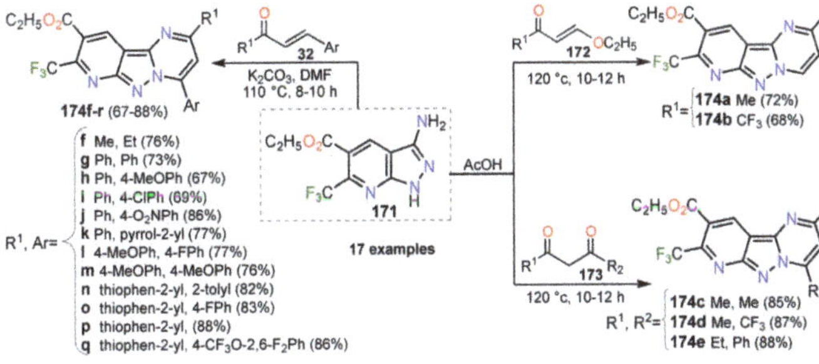

Scheme 49. Synthesis of pyridine-fused pyrazolo[1,5-*a*]pyrimidines 174a–q starting from the pyrazolo[3,4-*b*]pyridine 17.

Through the reaction of β-enaminones 19 with the pyrazolic ester 4, Kumar et al. [42] described the synthesis of 175 as a building block for obtaining different amides 176a–u having the 7-arylpyrazolo[1,5-*a*]pyrimidine fragment, of which some showed great biological activity, inhibiting the reproduction of human cervical cancer cell line (Scheme 50).

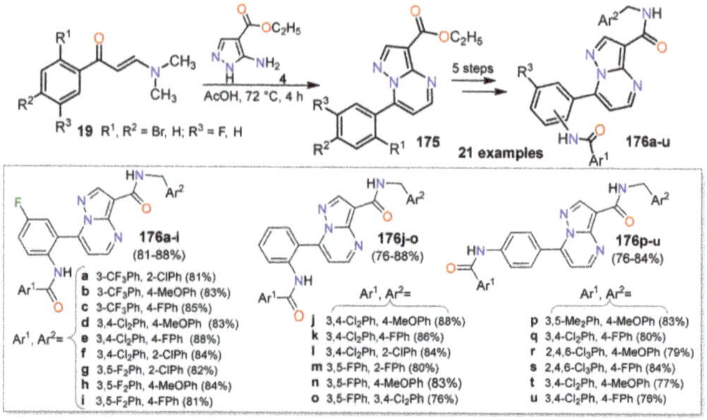

Scheme 50. Synthesis of amides **176a–u** having the 7-arylpyrazolo[1,5-*a*]pyrimidine fragment.

The IC$_{50}$ results showed that compounds with better antiproliferative activity contain at least three halogen atoms. Importantly, there may be a synergy effect between aryl groups of the two amide fragments since when at least one non-halogen substituent is introduced in one of these two rings, the activity decreased substantially. This report is in concordance with the postulated by Narsaiah et al. [92] when introduce the trifluoromethyl substituent to increase the lipophilic activity of pyrazolo[1,5-*a*]pyrimidines and with this have a better global growth inhibition of cancer cell lines (see Scheme 49).

Ismailb and coworkers [57] published in 2019 an investigation focused on the inhibition of CDK2 enzyme; with this purpose, they synthesize ethyl 2-(phenylamino)-4,5-dihydropyrazolo[1,5-*a*]pyrimidine-6-carboxylate derivatives (Scheme 51). Even though the 4,5-dihydropyrazolo[1,5-*a*]pyrimidines derivatives not generated the higher inhibition of the CDK2 enzyme, the compound **178d**, and **178f** revealed the most increased activity against the four tumor cell lines (**HepG2**, **MCF-7**, **A549**, and **Caco2**). This result allows establish the crucial role of fluor atom and nitrile group in the 5-aryl-7-methyl-4,5-dihydropyrazolo[1,5-*a*]pyrimidines **178** for increased the antitumor activity of compounds.

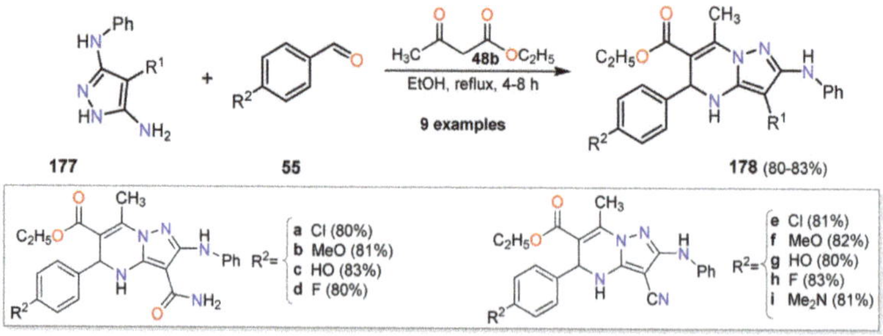

Scheme 51. Synthesis of pyrazolo[1,5-*a*]pyrimidines derivatives **178** with potential activity against CDK2 enzyme.

Based on previous reports [95,96] Husseiny proposed and carried out the synthesis of the 2-(benzothiazol-2-yl)pyrazolo[1,5-*a*]pyrimidine **181** [97] (Scheme 52). The author carried out the condensation reaction between 4-(benzo[*d*]thiazol-2-yl)–N^3-phenyl-1*H*-pyrazole-3,5-diamine (**179**) and diethyl ethoxymethylenemalonate (**180**) under reflux in acetic acid. Compound **181** exhibited re-markable growth inhibition activity, especially against leukemia **CCRF-CEM** and lung cancer **HOP-92** cell lines. The presence of the ethyl carboxylate group at position 6 gave rise to a notable increase in cytotoxic activity against most cancer cell lines. In this work, Husseiny links together successfully two novel antitumoral scaffolds to increase the activity of the final compound against cancer cells;

besides, he shows an easy way to obtain pyrazolo[1,5-a]pyrimidines linked to another aromatic heterocycle with huge pharmacological importance.

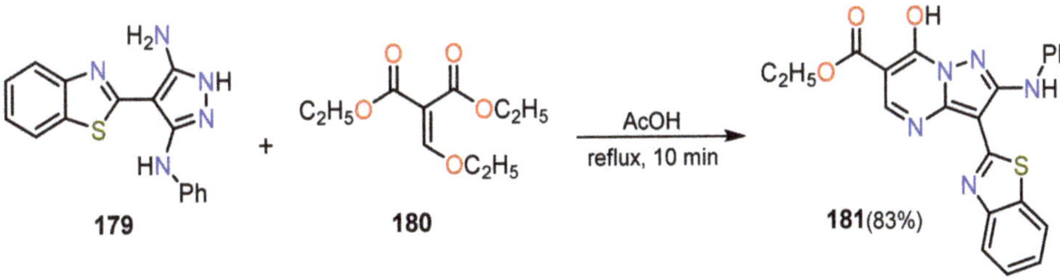

Scheme 52. Synthesis of the 2-(benzothiazol-2-yl)pyrazolo[1,5-a]pyrimidine **181** with great cytotoxic activity.

4. Enzymatic Inhibition

Mikami et al. [82] published in 2017 the synthesis of the pyrazolo[1,5-a]pyrimidine **187a** (Scheme 53), orally bioavailable and selective molecule that possesses CNS drug-like characteristics—including minimal molecular weight, low topological polar surface area, and a limited number of hydrogen bond donors– which translated into excellent brain penetration with no indication of P-glycoprotein (P-gp)-mediated efflux liability.

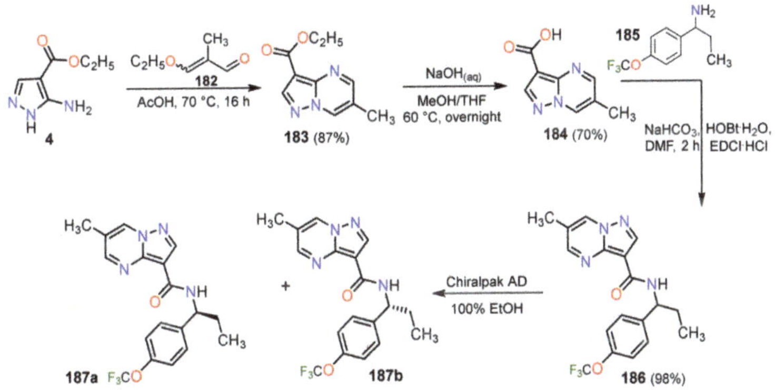

Scheme 53. Synthesis sequence of **PP** derivative **187a** starting from the 3-aminopyrazole **5**.

The high efficiency of fused pyrazoles **187a** must be four different effects (Figure 4):

1. Pazolo[1,5-a]pyrimidine core that makes $\pi - \pi$ and $CH - \pi$ interactions with the residues in the enzyme.
2. The central amide linker generated the hydrogen bond with the more polar residues stabilizing the PDE2A enzyme, while the amide NH plays a crucial role in constraining the binding conformation via intramolecular hydrogen bonding to the pyrimidine nitrogen atom.
3. α-Branched benzylamine portion and the p-CF$_3$O group fits into the sits in the hydrophobic cavity in an energetically favorable orthogonal orientation.
4. The substitution at position 6 combines two effects, the Van der Walls interactions with the residues and the smallest possible size.

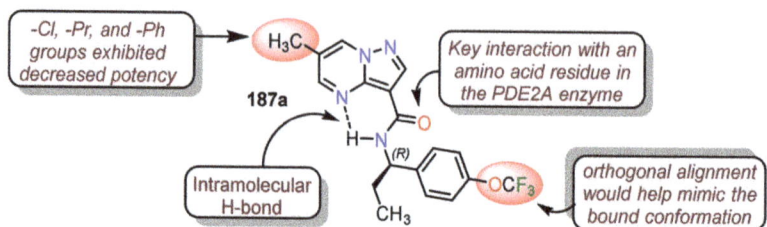

Figure 4. Main interactions that improve the activity of compound **187a** within the PDE2A enzyme.

Finally, the author establishes that a single enantiomer **187a** was more potent than a racemic compound, while **187b** was inactive.

Dowling et al. [94] published in 2017 the synthesis of the **PP** salt **190a** (and other derivatives, Scheme 54a), which exhibits improved cellular activity (Wnt DLD-1 Luciferase, IC_{50} = 50 nM), high solubility, and reduced intrinsic clearance in rat hepatocytes and human microsomes relative to the analogs that have modifications at position 4 of the aniline moiety. Supported by the X-ray crystallographic structures of CK2α with **190a**, the authors identify the importance of the primary amine since directly coordinate and order water molecules and the side-chain carbonyl group the enzyme active site. Moreover, the salt **190a** has physicochemical properties that are ideal for intravenous solution formulation, has shown strong pharmacokinetics in preclinical organisms, and exhibits a high degree of monotherapy activity in xenografts **HCT-116** and **SW-620** (Scheme 54b).

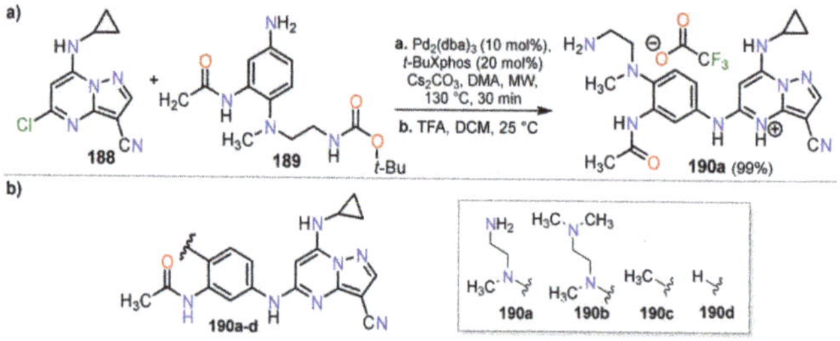

Compound	GI_{50} (μM)	
	HCT-116	SW620
190a	0.7	0.7
190b	0.08	0.1
190c	0.03	0.03
190d	0.01	0.005

Scheme 54. (a) Synthesis of pyrazolo[1,5-*a*]pyrimidine salt **190a**. (b) Analysis of better amino group for the cellular activity.

In 2020 Mathinson and co-workers [77] published the synthesis of a potent RET kinase inhibitor with >500-fold selectivity against KDR (Kinase insert Domain Receptor) in cellular assays, compound **193** (Scheme 55a). The authors identify three different substitutions at position 6 for the synthesized pyrazolo[1,5-*a*]pyrimidines family (i.e., phenyl, *p*-acetamidophenyl, and 2-thiazolyl), structures **193** and **193′** in Scheme 55b. The phenyl group was used due to the synthetic availability of substrate; however, the *p*-acetamidophenyl and 2-thiazolyl groups conferred enhanced potency against a high number of transmembrane receptors tyrosine kinases and cancer cell lines. The selectivity of **193** is because the amino group at position 5 (R^2) favors an H-bonding with a water molecule inside the enzyme active site. The piperidine ring constrained geometry facilitates Van der Waals's interactions with some protein residues. Finally, the researchers concentrated their efforts on the installation of polarity at R^3 to reduce overall lipophilicity; with this objective, they replaced the methoxy group with a variety of amides, the additional hydrophilicity results in a reduction in hERG binding of up to 12-fold without a substantial loss of potency in both KIF5B-RET transfected Ba/F3 cells and the LC-2/ad cell line. Unfortunately, the most promising compound bearing (methylsulfonyl)piperazine moiety, displayed substantially reduced hERG binding and exhibited insufficient oral exposure and bioavailability, precluding advancement to in vivo efficacy studies (Scheme 55b).

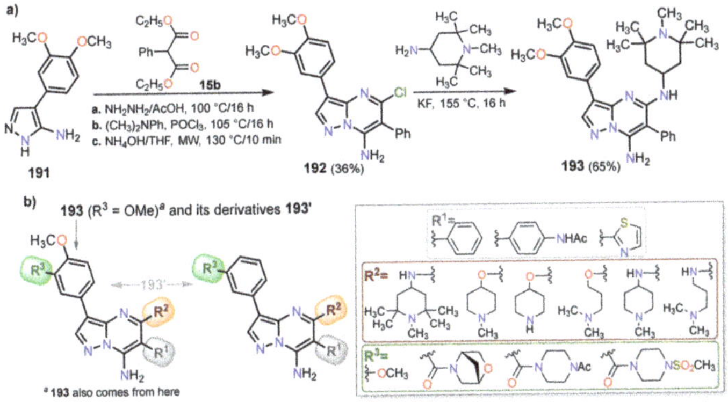

Scheme 55. (a) Synthesis of potent RET kinase inhibitor **193**. (b) Mains substitutions to find the better activity.

In 2018 Hassan et al. [95] synthesized three different *N*-heterocyclic compounds that could inhibit tyrosine kinase in cancer cells. One family of these compounds was the **PP** derivatives **195a–g** (Scheme 56). The authors determined that the compound **195f** had the most potent inhibitory activity against the epidermal growth factor receptor (EGFR) kinase enzyme. According to the docking simulation into EGFR active site is possible to say that the high activity of **195f** is due to: first, the linked by the NH$_2$ backbone to the enzyme residues by a water molecule in the pocket and second, the naphthalene ring, possibly due to π–π interactions, manages to penetrate well into the pocket and out of the cleft.

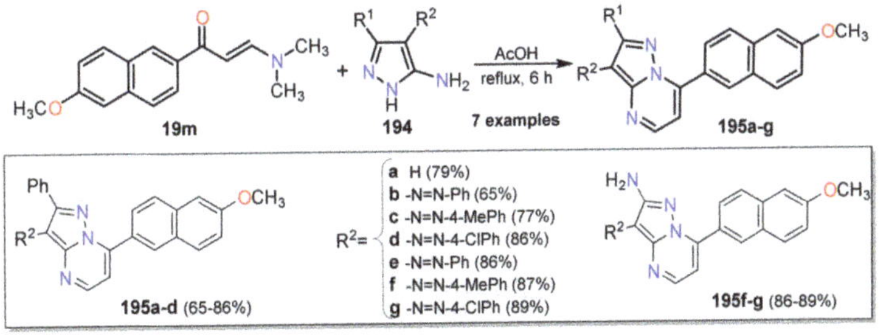

Scheme 56. Synthesis of pyrazolo[1,5-*a*]pyrimidines **195a–g** that were able to inhibit tyrosine kinase in cancer cells.

Metwally et al. [28] published the cyclocondensation reaction of 5-amino-3-cyanomethyl-1*H*-pyrazole-4-carbonitrile (**54**) with acetoacetanilide (**196**) in *N*,*N*-dimethylformamide (DMF), and few drops of acetic acid to obtain the key intermediated **197**, which is used to create new pyrazolo[1,5-*a*]pyrimidines **198** (Scheme 57). Some synthesized compounds were analyzed using MTT assays on two cancer cell lines for their cytotoxic activity (breast and cervical cancer cells). Compounds **197** and **198e–f** showed higher cytotoxicity by using doxorubicin as a reference drug. Likewise, these **PP** derivatives presented inhibitory activity against KDM (histone lysine demethylases). Authors found that heteroarylidene derivatives have the highest cytotoxic activity than arylidene derivatives. Despite this, substituted phenyl group with an EWG such as 4-Cl gave the lowest cytotoxic activity. Moreover, the most active KDM inhibitor **198e** showed that 4-folds of control triggered cell cycle arrest at the G2/M step and induced a total apoptotic effect by 10 folds more than control. These results are due to the π–π interactions between the aromatic ring and the residues of the enzyme active site.

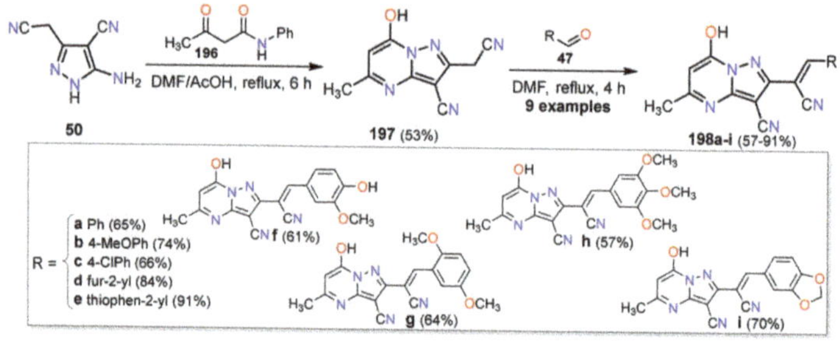

Scheme 57. Synthesis of pyrazolo[1,5-*a*]pyrimidines with potential activity against histone lysine d methylases.

5. Conclusions

Despite ample literature, the innovative discoveries regarding pyrazol[1,5-*a*]pyrimidine scaffold in synthetic and biological fields continue strongly nowadays. The research plenty focused on synthetic transformations shows a strong preference for methods that reliably deliver highly functionalized molecules from strategic reagents to avoid the need for subsequent reaction steps. The most employed synthesis pathway is based on the interaction between 1,3-biselectrophilic substrates and 3-aminopyrazoles, with various examples in diverse applications. Careful substrates selection leads to a modulable substitution pattern in the core, affecting subsequent synthetic steps. As an illustration, malonic acid delivers two hydroxyl groups in products, whereas β-ketonitriles lead to monoamine derivatives. Other significant features of the 3-aminopyrazoles route are the feasibility to scale up the reaction using cheap reagents and that a modular functionalization may also be affordable from multicomponent strategies. Importantly, only one paper of an alternative route, based on [4 + 2] cycloaddition reactions, to access **PP** derivatives was found.

Regarding post-functionalization reactions, in recent years, a strong influence from methods of bond formation supported on metal catalysis has led to the formation of interesting derivatives and structures which could be difficult to achieve before. However, aromatic substitution is still the most common method to carry out functionalization of the pyrazolo[1,5-*a*]pyrimidine core. In this vein, the fused ring prefers the interaction with nucleophiles on the pyrimidine side and with electrophiles on the pyrazole side due to the π-deficient and π-excedent nature of these rings, respectively. The principal functionalization approach resembles aromatic electrophilic substitution on single pyrazole systems (e.g., halogenation, nitration, and formylation reactions). Lastly, given that amides provide a standard industrial method to join residues is quite popular employing this functional group as a linker with **PP** derivatives. Hence, different protocols to insert this functional group have been evaluated in recent years.

Remarkable, the *N*-heterocyclic core allows crucial modifications at C2, C3, C5, C6, and C7 positions during ring-construction or later functionalization steps. These transformations can substantially modify the biological properties of compounds such as antitumoral and enzymatic inhibitory activity. However, bases on the SAR, in most cases, the presence of halogen atoms generates a remarkable effect on the cytotoxicity of the molecule. Additionally, the π–π interactions between pyrazolo[1,5-*a*]pyrimidine ring and enzymatic pocket make possible a significant number of molecules with high anticancer potential. Nevertheless, the insertion of aliphatic motifs generates a greater affinity with some enzymes, like in Kinase insert Domain Receptor. We hope this review is useful in understanding recent advances in the synthesis and functionalization of this privileged scaffold in medicinal chemistry and help the researchers to generate new ideas for rationalistic and efficient designs of pyrazolo[1,5-*a*]pyrimidine-based medical.

Author Contributions: The three individuals listed as authors have contributed substantially to the development of this work, and no other person was involved with its progress. The contribution of the authors is as follows: Literature review, analysis of articles and original draft composition, A.A.-G.; Literature review, analysis of articles and original draft composition, A.G.; Conceptualization, literature review, manuscript review & editing, supervision, and resources, J.P. All authors have read and agreed to the published version of the manuscript.

Funding: This research was funded by the science faculty at the Universidad de los Andes, projects INV-2019-84-1800 and INV-2020-104-2035, and the APC was funded by the science faculty.

Institutional Review Board Statement: Not applicable.

Informed Consent Statement: Not applicable.

Acknowledgments: The authors thank the Chemistry Department and Vicerrectoría de Investigaciones at the Universidad de los Andes for financial support. J.P. wishes to credit current and former Bioorganic Compounds Research Group members whose names appear in the reference section, for their valuable collaboration in the different investigations.

Conflicts of Interest: The authors declare no conflict of interest because of this literature research was conducted in the absence of any commercial or financial relationships.

References

1. Castillo, J.C.; Portilla, J. Recent advances in the synthesis of new pyrazole derivatives. *Targets Heterocycl. Syst.* **2018**, *22*, 194–223. [CrossRef]
2. Salem, M.A.; Helal, M.H.; Gouda, M.A.; Abd EL-Gawad, H.H.; Shehab, M.A.M.; El-Khalafawy, A. Recent synthetic methodologies for pyrazolo[1,5-a]pyrimidine. *Synth. Commun.* **2019**, *49*, 1750–1776. [CrossRef]
3. Al-Azmi, A. Pyrazolo[1,5-a]pyrimidines: A Close Look into their Synthesis and Applications. *Curr. Org. Chem.* **2019**, *23*, 721–743. [CrossRef]
4. Cherukupalli, S.; Karpoormath, R.; Chandrasekaran, B.; Hampannavar, G.A.; Thapliyal, N.; Palakollu, V.N. An insight on synthetic and medicinal aspects of pyrazolo[1,5-a]pyrimidine scaffold. *Eur. J. Med. Chem.* **2017**, *126*, 298–352. [CrossRef]
5. Eftekhari-Sis, B.; Zirak, M. Chemistry of α-oxoesters: A powerful tool for the synthesis of heterocycles. *Chem. Rev.* **2015**, *115*, 151–264. [CrossRef] [PubMed]
6. Abu Elmaati, T.M.; El-Taweel, F.M. New Trends in the Chemistry of 5 Aminopyrazoles. *J. Heterocycl. Chem.* **2004**, *41*, 109–134. [CrossRef]
7. Asano, T.; Yamazaki, H.; Kasahara, C.; Kubota, H.; Kontani, T.; Harayama, Y.; Ohno, K.; Mizuhara, H.; Yokomoto, M.; Misumi, K.; et al. Identification, Synthesis, and Biological Evaluation of 6-[(6R)-2-(4-Fluorophenyl)-6-(hydroxymethyl)-4,5,6,7-tetrahydropyrazolo[1,5-a]pyrimidin-3-yl]-2-(2-methylphenyl)pyridazin-3(2H)-one (AS1940477), a Potent p38 MAP Kinase Inhibitor. *J. Med. Chem.* **2012**, *55*, 7772–7785. [CrossRef] [PubMed]
8. Zhao, M.; Ren, H.; Chang, J.; Zhang, D.; Yang, Y.; He, Y.; Qi, C.; Zhang, H. Design and synthesis of novel pyrazolo[1,5-a]pyrimidine derivatives bearing nitrogen mustard moiety and evaluation of their antitumor activity in vitro and in vivo. *Eur. J. Med. Chem.* **2016**, *119*, 183–196. [CrossRef] [PubMed]
9. Ramsey, S.J.; Attkins, N.J.; Fish, R.; van der Graaf, P.H. Quantitative pharmacological analysis of antagonist binding kinetics at CRF1 receptors in vitro and in vivo. *Br. J. Pharmacol.* **2011**, *164*, 992–1007. [CrossRef]
10. Babaoglu, K.; Boojamra, C.G.; Eisenberg, E.J.; Hui, H.C.; Mackman, R.L.; Parrish, J.P.; Sangi, M.; Saunders, O.L.; Siegel, D.; Sperandio, D.; et al. Pyrazolo[1,5-a]pyrimidines as antiviral agents. Patent WO2011163518A1, 29 December 2011.
11. Naidu, B.N.; Patel, M.; D'andrea, S.; Zheng, Z.B.; Connolly, T.P.; Langley, D.R.; Peese, K.; Wang, Z.; Walker, M.A.; Kadow, J.F. Inhibitors of Human Immunodeficiency Virus Replication. Patent WO2014028384A1, 20 February 2014.
12. Yang, X.Z.; Sun, R.; Guo, X.; Wei, X.R.; Gao, J.; Xu, Y.J.; Ge, J.F. The application of bioactive pyrazolopyrimidine unit for the construction of fluorescent biomarkers. *Dye. Pigment.* **2020**, *173*, 107878. [CrossRef]
13. Singsardar, M.; Sarkar, R.; Majhi, K.; Sinha, S.; Hajra, A. Brønsted Acidic Ionic Liquid-Catalyzed Regioselective Synthesis of Pyrazolopyrimidines and Their Photophysical Properties. *ChemistrySelect* **2018**, *3*, 1404–1410. [CrossRef]
14. Golubev, P.; Karpova, E.A.; Pankova, A.S.; Sorokina, M.; Kuznetsov, M.A. Regioselective Synthesis of 7-(Trimethylsilylethynyl)pyrazolo[1,5-a]pyrimidines via Reaction of Pyrazolamines with Enynones. *J. Org. Chem.* **2016**, *81*, 11268–11275. [CrossRef] [PubMed]
15. Tigreros, A.; Portilla, J. Fluorescent Pyrazole Derivatives: An Attractive Scaffold for Biological Imaging Applications. *Curr. Chinese Sci.* **2021**, *1*, 197–206. [CrossRef]

16. Tigreros, A.; Portilla, J. Recent progress in chemosensors based on pyrazole derivatives. *RSC Adv.* **2020**, *10*, 19693–19712. [CrossRef]
17. Tigreros, A.; Macías, M.; Portilla, J. Photophysical and crystallographic study of three integrated pyrazolo[1,5-a]pyrimidine–triphenylamine systems. *Dye. Pigment.* **2021**, *184*, 108730. [CrossRef]
18. Tigreros, A.; Aranzazu, S.L.; Bravo, N.F.; Zapata-Rivera, J.; Portilla, J. Pyrazolo[1,5-a]pyrimidines-based fluorophores: A comprehensive theoretical-experimental study. *RSC Adv.* **2020**, *10*, 39542–39552. [CrossRef]
19. Tigreros, A.; Castillo, J.C.; Portilla, J. Cyanide chemosensors based on 3-dicyanovinylpyrazolo[1,5-a]pyrimidines: Effects of peripheral 4-anisyl group substitution on the photophysical properties. *Talanta* **2020**, *215*, 120905. [CrossRef]
20. Tigreros, A.; Rosero, H.A.; Castillo, J.C.; Portilla, J. Integrated pyrazolo[1,5-a]pyrimidine–hemicyanine system as a colorimetric and fluorometric chemosensor for cyanide recognition in water. *Talanta* **2019**, *196*, 395–401. [CrossRef]
21. Castillo, J.C.; Tigreros, A.; Portilla, J. 3-Formylpyrazolo[1,5-a]pyrimidines as Key Intermediates for the Preparation of Functional Fluorophores. *J. Org. Chem.* **2018**, *83*, 10887–10897. [CrossRef] [PubMed]
22. Portilla, J.; Quiroga, J.; Cobo, J.; Low, J.N.; Glidewell, C. Hydrogen-bonded chains in isostruo tural 5-methyl-2-(4-methylphenyl)-7,8-dihydro-6H-cyclopenta[g]pyrazolo[1,5-a]pyrimidine, 2-(4-chloro-phenyl)-5-methyl-7,8-dihydro-6H-cyclopenta[g]pyrazolo[1,5-a]pyrimidine and 2-(4-bromophenyl)-5-methyl-7,8-dihydro-6H-. *Acta Crystallogr. Sect. C Cryst. Struct. Commun.* **2005**, *61*, 452–456. [CrossRef]
23. Portilla, J.; Quiroga, J.; Cobo, J.; Low, J.N.; Glidewell, C. 7-Amino-2,5-dimethylpyrazolo[1,5-a]pyrimidine hemihydrate redetermined at 120 K: A three-dimensional hydrogen-bonded framework. *Acta Crystallogr. Sect. C Cryst. Struct. Commun.* **2006**, *62*, 186–189. [CrossRef] [PubMed]
24. Regan, A.C. 11.12- Bicyclic 5-6 Systems with One Bridgehead (Ring Junction) Nitrogen Atom: Two Extra Heteroatoms 1:1. In *Comprehensive Heterocyclic Chemistry III*; Katritzky, A.R., Ramsden, C.A., Scriven, E.F.V., Taylor, R.J.K., Eds.; Elsevier: Oxford, UK, 2008; pp. 551–587. ISBN 978-0-08-044992-0.
25. Candeias, N.R.; Branco, L.C.; Gois, P.M.P.; Afonso, C.A.M.; Trindade, A.F. More sustainable approaches for the synthesis of n-based heterocycles. *Chem. Rev.* **2009**, *109*, 2703–2802. [CrossRef]
26. Bondock, S.; Alqahtani, S.; Fouda, A.M. Synthesis and anticancer evaluation of some new pyrazolo[3,4-d][1,2,3]triazin-4-ones, pyrazolo[1,5-a]pyrimidines, and imidazo[1,2-b]pyrazoles clubbed with carbazole. *J. Heterocycl. Chem.* **2021**, *58*, 56–73. [CrossRef]
27. Hebishy, A.M.S.; Salama, H.T.; Elgemeie, G.H. New Route to the Synthesis of Benzamide-Based 5-Aminopyrazoles and Their Fused Heterocycles Showing Remarkable Antiavian Influenza Virus Activity. *ACS Omega* **2020**, *5*, 25104–25112. [CrossRef]
28. Metwally, N.H.; Mohamed, M.S.; Ragb, E.A. Design, synthesis, anticancer evaluation, molecular docking and cell cycle analysis of 3-methyl-4,7-dihydropyrazolo[1,5-a]pyrimidine derivatives as potent histone lysine demethylases (KDM) inhibitors and apoptosis inducers. *Bioorg. Chem.* **2019**, *88*, 102929. [CrossRef]
29. Yamaguchi-Sasaki, T.; Tokura, S.; Ogata, Y.; Kawaguchi, T.; Sugaya, Y.; Takahashi, R.; Iwakiri, K.; Abe-Kumasaka, T.; Yoshida, I.; Arikawa, K.; et al. Discovery of a Potent Dual Inhibitor of Wild-Type and Mutant Respiratory Syncytial Virus Fusion Proteins. *ACS Med. Chem. Lett.* **2020**, *11*, 1145–1151. [CrossRef]
30. Yamagami, T.; Kobayashi, R.; Moriyama, N.; Horiuchi, H.; Toyofuku, E.; Kadoh, Y.; Kawanishi, E.; Izumoto, S.; Hiramatsu, H.; Nanjo, T.; et al. Scalable Process Design for a PDE10A Inhibitor Consisting of Pyrazolopyrimidine and Quinoxaline as Key Units. *Org. Process Res. Dev.* **2019**, *23*, 578–587. [CrossRef]
31. Manetti, F.; Stecca, B.; Santini, R.; Maresca, L.; Giannini, G.; Taddei, M.; Petricci, E. Pharmacophore-Based Virtual Screening for Identification of Negative Modulators of GLI1 as Potential Anticancer Agents. *ACS Med. Chem. Lett.* **2020**, *11*, 832–838. [CrossRef] [PubMed]
32. Abe, M.; Seto, M.; Gogliotti, R.G.; Loch, M.T.; Bollinger, K.A.; Chang, S.; Engelberg, E.M.; Luscombe, V.B.; Harp, J.M.; Bubser, M.; et al. Discovery of VU6005649, a CNS Penetrant mGlu7/8 Receptor PAM Derived from a Series of Pyrazolo[1,5-a]pyrimidines. *ACS Med. Chem. Lett.* **2017**, *8*, 1110–1115. [CrossRef] [PubMed]
33. Hylsová, M.; Carbain, B.; Fanfrlík, J.; Musilová, L.; Haldar, S.; Köprülüoğlu, C.; Ajani, H.; Brahmkshatriya, P.S.; Jorda, R.; Kryštof, V.; et al. Explicit treatment of active-site waters enhances quantum mechanical/implicit solvent scoring: Inhibition of CDK2 by new pyrazolo[1,5-a]pyrimidines. *Eur. J. Med. Chem.* **2017**, *126*, 1118–1128. [CrossRef] [PubMed]
34. Silveira, F.F.; de Souza, J.O.; Hoelz, L.V.B.; Campos, V.R.; Jabor, V.A.P.; Aguiar, A.C.C.; Nonato, M.C.; Albuquerque, M.G.; Guido, R.V.C.; Boechat, N.; et al. Comparative study between the anti-P. falciparum activity of triazolopyrimidine, pyrazolopyrimidine and quinoline derivatives and the identification of new PfDHODH inhibitors. *Eur. J. Med. Chem.* **2021**, *209*. [CrossRef]
35. Childress, E.S.; Wieting, J.M.; Felts, A.S.; Breiner, M.M.; Long, M.F.; Luscombe, V.B.; Rodriguez, A.L.; Cho, H.P.; Blobaum, A.L.; Niswender, C.M.; et al. Discovery of Novel Central Nervous System Penetrant Metabotropic Glutamate Receptor Subtype 2 (mGlu 2) Negative Allosteric Modulators (NAMs) Based on Functionalized Pyrazolo[1,5-a]pyrimidine-5-carboxamide and Thieno[3,2-b]pyridine-5-carboxamide Cores. *J. Med. Chem.* **2019**, *62*, 378–384. [CrossRef] [PubMed]
36. Shen, J.; Deng, X.; Sun, R.; Tavallaie, M.S.; Wang, J.; Cai, Q.; Lam, C.; Lei, S.; Fu, L.; Jiang, F. Structural optimization of pyrazolo[1,5-a]pyrimidine derivatives as potent and highly selective DPP-4 inhibitors. *Eur. J. Med. Chem.* **2020**, *208*, 112850. [CrossRef] [PubMed]
37. Portilla, J.; Quiroga, J.; Nogueras, M.; Cobo, J. Regioselective synthesis of fused pyrazolo[1,5-a]pyrimidines by reaction of 5-amino-1H-pyrazoles and β-dicarbonyl compounds containing five-membered rings. *Tetrahedron* **2012**, *68*, 988–994. [CrossRef]
38. Castillo, J.C.; Rosero, H.A.; Portilla, J. Simple access toward 3-halo- and 3-nitro-pyrazolo[1,5-a] pyrimidines through a one-pot sequence. *RSC Adv.* **2017**, *7*, 28483–28488. [CrossRef]

39. Guo, Y.; Liu, Y.; Hu, N.; Yu, D.; Zhou, C.; Shi, G.; Zhang, B.; Wei, M.; Liu, J.; Luo, L.; et al. Discovery of Zanubrutinib (BGB-3111), a Novel, Potent, and Selective Covalent Inhibitor of Bruton's Tyrosine Kinase. *J. Med. Chem.* **2019**, *62*, 7923–7940. [CrossRef] [PubMed]
40. Attia, M.H.; Elrazaz, E.Z.; El-Emam, S.Z.; Taher, A.T.; Abdel-Aziz, H.A.; Abouzid, K.A.M. Synthesis and in-vitro anti-proliferative evaluation of some pyrazolo[1,5-a]pyrimidines as novel larotrectinib analogs. *Bioorg. Chem.* **2020**, *94*, 103458. [CrossRef] [PubMed]
41. Li, G.; Wang, Y.; Li, L.; Ren, Y.; Deng, X.; Liu, J.; Wang, W.; Luo, M.; Liu, S.; Chen, J. Design, synthesis, and bioevaluation of pyrazolo[1,5-a]pyrimidine derivatives as tubulin polymerization inhibitors targeting the colchicine binding site with potent anticancer activities. *Eur. J. Med. Chem.* **2020**, *202*, 112519. [CrossRef]
42. Ajeesh Kumar, A.K.; Bodke, Y.D.; Lakra, P.S.; Sambasivam, G.; Bhat, K.G. Design, synthesis and anti-cancer evaluation of a novel series of pyrazolo [1, 5-a] pyrimidine substituted diamide derivatives. *Med. Chem. Res.* **2017**, *26*, 714–744. [CrossRef]
43. Xu, Y.; Brenning, B.G.; Kultgen, S.G.; Foulks, J.M.; Clifford, A.; Lai, S.; Chan, A.; Merx, S.; McCullar, M.V.; Kanner, S.B.; et al. Synthesis and Biological Evaluation of Pyrazolo[1,5-a]pyrimidine Compounds as Potent and Selective Pim-1 Inhibitors. *ACS Med. Chem. Lett.* **2015**, *6*, 63–67. [CrossRef]
44. Peytam, F.; Adib, M.; Shourgeshty, R.; Firoozpour, L.; Rahmanian-Jazi, M.; Jahani, M.; Moghimi, S.; Divsalar, K.; Faramarzi, M.A.; Mojtabavi, S.; et al. An efficient and targeted synthetic approach towards new highly substituted 6-amino-pyrazolo[1,5-a]pyrimidines with α-glucosidase inhibitory activity. *Sci. Rep.* **2020**, *10*, 1–17. [CrossRef] [PubMed]
45. Yu, L.; Xing, S.; Zheng, K.; Sadeghzadeh, S.M. Synthesis of pyrazolopyrimidines in mild conditions by gold nanoparticles supported on magnetic ionic gelation in aqueous solution. *Appl. Organomet. Chem.* **2020**, *34*, 1–9. [CrossRef]
46. Quiroga, J.; Trilleras, J.; Insuasty, B.; Abonía, R.; Nogueras, M.; Cobo, J. Regioselective formylation of pyrazolo[3,4-b]pyridine and pyrazolo[1,5-a]pyrimidine systems using Vilsmeier-Haack conditions. *Tetrahedron Lett.* **2008**, *49*, 2689–2691. [CrossRef]
47. Jismy, B.; Tikad, A.; Akssira, M.; Guillaumet, G. Efficient Access to 3,5-Disubstituted 7-(Trifluoromethyl)pyrazolo[1,5-a]pyrimidines Involving SNAr and Suzuki Cross-Coupling Reactions. *Molecules* **2020**, *25*, 2062. [CrossRef] [PubMed]
48. Jismy, B.; Guillaumet, G.; Allouchi, H.; Akssira, M.; Abarbri, M. Concise and Efficient Access to 5,7-Disubstituted Pyrazolo[1,5-a]pyrimidines by Pd-Catalyzed Sequential Arylation, Alkynylation and SNAr Reaction. *Eur. J. Org. Chem.* **2017**, *2017*, 6168–6178. [CrossRef]
49. Jismy, B.; Allouchi, H.; Guillaumet, G.; Akssira, M.; Abarbri, M. An Efficient Synthesis of New 7-Trifluoromethyl-2,5-disubstituted Pyrazolo[1,5-a]pyrimidines. *Synth.* **2018**, *50*, 1675–1686. [CrossRef]
50. Schmitt, D.C.; Niljianskul, N.; Sach, N.W.; Trujillo, J.I. A Parallel Approach to 7-(Hetero)arylpyrazolo[1,5- a]pyrimidin-5-ones. *ACS Comb. Sci.* **2018**, *20*, 256–260. [CrossRef] [PubMed]
51. Akrami, S.; Karami, B.; Farahi, M. A novel protocol for catalyst-free synthesis of fused six-member rings to triazole and pyrazole. *Mol. Divers.* **2020**, *24*, 225–231. [CrossRef] [PubMed]
52. Fouda, A.M.; Abbas, H.A.S.; Ahmed, E.H.; Shati, A.A.; Alfaifi, M.Y.; Elbehairi, S.E.I. Synthesis, in vitro antimicrobial and cytotoxic activities of some new pyrazolo[1,5-a]pyrimidine derivatives. *Molecules* **2019**, *24*, 1080. [CrossRef]
53. Castillo, J.C.; Estupiñan, D.; Nogueras, M.; Cobo, J.; Portilla, J. 6-(Aryldiazenyl)pyrazolo[1,5-a]pyrimidines as Strategic Intermediates for the Synthesis of Pyrazolo[5,1-b]purines. *J. Org. Chem.* **2016**, *81*, 12364–12373. [CrossRef]
54. Zahedifar, M.; Razavi, R.; Sheibani, H. Reaction of (chloro carbonyl) phenyl ketene with 5-amino pyrazolones: Synthesis, characterization and theoretical studies of 7-hydroxy-6-phenyl-3-(phenyldiazenyl)pyrazolo[1,5-a]pyrimidine-2,5(1H,4H)-dione derivatives. *J. Mol. Struct.* **2016**, *1125*, 730–735. [CrossRef]
55. Li, G.; Meanwell, N.A.; Krystal, M.R.; Langley, D.R.; Naidu, B.N.; Sivaprakasam, P.; Lewis, H.; Kish, K.; Khan, J.A.; Ng, A.; et al. Discovery and Optimization of Novel Pyrazolopyrimidines as Potent and Orally Bioavailable Allosteric HIV-1 Integrase Inhibitors. *J. Med. Chem.* **2020**, *63*, 2620–2637. [CrossRef] [PubMed]
56. Naik, N.S.; Shastri, L.A.; Chougala, B.M.; Samundeeswari, S.; Holiyachi, M.; Joshi, S.D.; Sunagar, V. Synthesis of novel aryl and coumarin substituted pyrazolo[1,5-a]pyrimidine derivatives as potent anti-inflammatory and anticancer agents. *Chem. Data Collect.* **2020**, *30*, 100550. [CrossRef]
57. Ali, G.M.E.; Ibrahim, D.A.; Elmetwali, A.M.; Ismail, N.S.M. Design, synthesis and biological evaluation of certain CDK2 inhibitors based on pyrazole and pyrazolo[1,5-a] pyrimidine scaffold with apoptotic activity. *Bioorg. Chem.* **2019**, *86*, 1–14. [CrossRef] [PubMed]
58. Sun, J.; Qiu, J.K.; Jiang, B.; Hao, W.J.; Guo, C.; Tu, S.J. I2-Catalyzed Multicomponent Reactions for Accessing Densely Functionalized Pyrazolo[1,5-a]pyrimidines and Their Disulphenylated Derivatives. *J. Org. Chem.* **2016**, *81*, 3321–3328. [CrossRef] [PubMed]
59. Hoang, G.L.; Streit, A.D.; Ellman, J.A. Three-Component Coupling of Aldehydes, Aminopyrazoles, and Sulfoxonium Ylides via Rhodium(III)-Catalyzed Imidoyl C–H Activation: Synthesis of Pyrazolo[1,5- a]pyrimidines. *J. Org. Chem.* **2018**, *83*, 15347–15360. [CrossRef] [PubMed]
60. Reddy Kotla, S.K.; Vandavasi, J.K.; Wang, J.J.; Parang, K.; Tiwari, R.K. Palladium-Catalyzed Intramolecular Cross-Dehydrogenative Coupling: Synthesis of Fused Imidazo[1,2-a]pyrimidines and Pyrazolo[1,5-a]pyrimidines. *ACS Omega* **2017**, *2*, 11–19. [CrossRef]
61. Ding, Z.C.; An, X.M.; Zeng, J.H.; Tang, Z.N.; Zhan, Z.P. Copper(I)-Catalyzed Synthesis of 4,5-Dihydropyrazolo[1,5-a]pyrimidines via Cascade Transformation of N-Propargylic Sulfonylhydrazones with Sulfonyl Azides. *Adv. Synth. Catal.* **2017**, *359*, 3319–3324. [CrossRef]

62. Nongthombam, G.S.; Boruah, R.C. Augmentation of steroidal β-formylenamide with pyrazolo and benzimidazo moieties: A tandem approach to highly fluorescent steroidal heterocycles. *Tetrahedron Lett.* **2021**, 152893. [CrossRef]
63. Mekky, A.E.M.; Ahmed, M.S.M.; Sanad, S.M.H.; Abdallah, Z.A. Bis(benzofuran-enaminone) hybrid possessing piperazine linker: Versatile precursor for microwave assisted synthesis of bis(pyrido[2′,3′:3,4]pyrazolo[1,5-a]pyrimidines). *Synth. Commun.* **2021**, *51*, 1–19. [CrossRef]
64. Jismy, B.; Guillaumet, G.; Akssira, M.; Tikad, A.; Abarbri, M. Efficient microwave-assisted Suzuki-Miyaura cross-coupling reaction of 3-bromo pyrazolo[1,5-a] pyrimidin-5(4 H)-one: Towards a new access to 3,5-diarylated 7-(trifluoromethyl)pyrazolo[1,5-a] pyrimidine derivatives. *RSC Adv.* **2021**, *11*, 1287–1302. [CrossRef]
65. Zhang, Y.C.; Zhang, B.W.; Geng, R.L.; Song, J. Enantioselective [3 + 2] Cycloaddition Reaction of Ethynylethylene Carbonates with Malononitrile Enabled by Organo/Metal Cooperative Catalysis. *Org. Lett.* **2018**, *20*, 7907–7911. [CrossRef]
66. Abdallah, A.E.M.; Elgemeie, G.H. Design, synthesis, docking, and antimicrobial evaluation of some novel pyrazolo[1,5-a] pyrimidines and their corresponding cycloalkane ring-fused derivatives as purine analogs. *Drug Des. Devel. Ther.* **2018**, *12*, 1785–1798. [CrossRef] [PubMed]
67. Liu, Y.; Laufer, R.; Patel, N.K.; Ng, G.; Sampson, P.B.; Li, S.-W.; Lang, Y.; Feher, M.; Brokx, R.; Beletskaya, I.; et al. Discovery of Pyrazolo[1,5-a]pyrimidine TTK Inhibitors: CFI-402257 is a Potent, Selective, Bioavailable Anticancer Agent. *ACS Med. Chem. Lett.* **2016**, *7*, 671–675. [CrossRef] [PubMed]
68. Drew, S.L.; Thomas-Tran, R.; Beatty, J.W.; Fournier, J.; Lawson, K.V.; Miles, D.H.; Mata, G.; Sharif, E.U.; Yan, X.; Mailyan, A.K.; et al. Discovery of Potent and Selective PI3Kγ Inhibitors. *J. Med. Chem.* **2020**, *63*, 11235–11257. [CrossRef]
69. Harris, M.R.; Wisniewska, H.M.; Jiao, W.; Wang, X.; Bradow, J.N. A Modular Approach to the Synthesis of gem-Disubstituted Cyclopropanes. *Org. Lett.* **2018**, *20*, 2867–2871. [CrossRef] [PubMed]
70. Dong, X.Y.; Zhang, Y.F.; Ma, C.L.; Gu, Q.S.; Wang, F.L.; Li, Z.L.; Jiang, S.P.; Liu, X.Y. A general asymmetric copper-catalysed Sonogashira C(sp 3)–C(sp) coupling. *Nat. Chem.* **2019**, *11*, 1158–1166. [CrossRef]
71. Bedford, R.B.; Durrant, S.J.; Montgomery, M. Catalyst-Switchable Regiocontrol in the Direct Arylation of Remote C-H Groups in Pyrazolo[1,5-a]pyrimidines. *Angew. Chemie Int. Ed.* **2015**, *54*, 8787–8790. [CrossRef] [PubMed]
72. Loubidi, M.; Manga, C.; Tber, Z.; Bassoude, I.; Essassi, E.M.; Berteina-Raboin, S. One-Pot SNAr/Direct Pd-Catalyzed CH Arylation Functionalization of Pyrazolo[1,5-a]pyrimidine at the C3 and C7 Positions. *Eur. J. Org. Chem.* **2018**, *2018*, 3936–3942. [CrossRef]
73. Gogula, T.; Zhang, J.; Lonka, M.R.; Zhang, S.; Zou, H. Temperature-modulated selective C(sp3)-H or C(sp2)-H arylation through palladium catalysis. *Chem. Sci.* **2020**, *11*, 11461–11467. [CrossRef]
74. Colomer, I.; Chamberlain, A.E.R.; Haughey, M.B.; Donohoe, T.J. Fc12 Hfip. *Nat. Rev. Chem.* **2017**, *1*, 0088. [CrossRef]
75. McCoull, W.; Abrams, R.D.; Anderson, E.; Blades, K.; Barton, P.; Box, M.; Burgess, J.; Byth, K.; Cao, Q.; Chuaqui, C.; et al. Discovery of Pyrazolo[1,5-a]pyrimidine B-Cell Lymphoma 6 (BCL6) Binders and Optimization to High Affinity Macrocyclic Inhibitors. *J. Med. Chem.* **2017**, *60*, 4386–4402. [CrossRef] [PubMed]
76. Mackman, R.L.; Sangi, M.; Sperandio, D.; Parrish, J.P.; Eisenberg, E.; Perron, M.; Hui, H.; Zhang, L.; Siegel, D.; Yang, H.; et al. Discovery of an oral respiratory syncytial virus (RSV) fusion inhibitor (GS-5806) and clinical proof of concept in a human RSV challenge study. *J. Med. Chem.* **2015**, *58*, 1630–1643. [CrossRef]
77. Mathison, C.J.N.; Chianelli, D.; Rucker, P.V.; Nelson, J.; Roland, J.; Huang, Z.; Yang, Y.; Jiang, J.; Xie, Y.F.; Epple, R.; et al. Efficacy and Tolerability of Pyrazolo[1,5-a]pyrimidine RET Kinase Inhibitors for the Treatment of Lung Adenocarcinoma. *ACS Med. Chem. Lett.* **2020**, *11*, 558–565. [CrossRef] [PubMed]
78. Yamaguchi-Sasaki, T.; Kawaguchi, T.; Okada, A.; Tokura, S.; Tanaka-Yamamoto, N.; Takeuchi, T.; Ogata, Y.; Takahashi, R.; Kurimoto-Tsuruta, R.; Tamaoki, T.; et al. Discovery of a potent dual inhibitor of wild-type and mutant respiratory syncytial virus fusion proteins through the modulation of atropisomer interconversion properties. *Bioorg. Med. Chem.* **2020**, *28*, 115818. [CrossRef]
79. Levy, J.N.; Alegre-Requena, J.V.; Liu, R.; Paton, R.S.; McNally, A. Selective Halogenation of Pyridines Using Designed Phosphine Reagents. *J. Am. Chem. Soc.* **2020**, *142*, 11295–11305. [CrossRef] [PubMed]
80. Lim, J.; Altman, M.D.; Baker, J.; Brubaker, J.D.; Chen, H.; Chen, Y.; Fischmann, T.; Gibeau, C.; Kleinschek, M.A.; Leccese, E.; et al. Discovery of 5-Amino- N -(1 H -pyrazol-4-yl)pyrazolo[1,5-a]pyrimidine-3-carboxamide Inhibitors of IRAK4. *ACS Med. Chem. Lett.* **2015**, *6*, 683–688. [CrossRef] [PubMed]
81. Bryan, M.C.; Drobnick, J.; Gobbi, A.; Kolesnikov, A.; Chen, Y.; Rajapaksa, N.; Ndubaku, C.; Feng, J.; Chang, W.; Francis, R.; et al. Development of Potent and Selective Pyrazolopyrimidine IRAK4 Inhibitors. *J. Med. Chem.* **2019**, *62*, 6223–6240. [CrossRef] [PubMed]
82. Mikami, S.; Sasaki, S.; Asano, Y.; Ujikawa, O.; Fukumoto, S.; Nakashima, K.; Oki, H.; Kamiguchi, N.; Imada, H.; Iwashita, H.; et al. Discovery of an Orally Bioavailable, Brain-Penetrating, in Vivo Active Phosphodiesterase 2A Inhibitor Lead Series for the Treatment of Cognitive Disorders. *J. Med. Chem.* **2017**, *60*, 7658–7676. [CrossRef] [PubMed]
83. Gazizov, D.A.; Gorbunov, E.B.; Rusinov, G.L.; Ulomsky, E.N.; Charushin, V.N. A New Family of Fused Azolo[1,5-a]pteridines and Azolo[5,1-b]purines. *ACS Omega* **2020**, *5*, 18226–18233. [CrossRef]
84. Karrouchi, K.; Radi, S.; Ramli, Y.; Taoufik, J.; Mabkhot, Y.N.; Al-aizari, F.A.; Ansar, M. Synthesis and Pharmacological Activities of Pyrazole Derivatives: A Review. *Molecules* **2018**, *23*, 134. [CrossRef] [PubMed]
85. Utecht, G.; Fruziński, A.; Jasiński, M. Polysubstituted 3-trifluoromethylpyrazoles: Regioselective (3 + 2)-cycloaddition of trifluoroacetonitrile imines with enol ethers and functional group transformations. *Org. Biomol. Chem.* **2018**, *16*, 1252–1257. [CrossRef]

86. Li, M.; Zhao, B.X. Progress of the synthesis of condensed pyrazole derivatives (from 2010 to mid-2013). *Eur. J. Med. Chem.* **2014**, *85*, 311–340. [CrossRef]
87. Scott, L.J. Larotrectinib: First Global Approval. *Drugs* **2019**, *79*, 201–206. [CrossRef] [PubMed]
88. Gana, C.C.; Hanssen, K.M.; Yu, D.M.T.; Flemming, C.L.; Wheatley, M.S.; Conseil, G.; Cole, S.P.C.; Norris, M.D.; Haber, M.; Fletcher, J.I. MRP1 modulators synergize with buthionine sulfoximine to exploit collateral sensitivity and selectively kill MRP1-expressing cancer cells. *Biochem. Pharmacol.* **2019**, *168*, 237–248. [CrossRef]
89. McCoull, W.; Abrams, R.D.; Anderson, E.; Blades, K.; Barton, P.; Box, M.; Burgess, J.; Byth, K.; Cao, Q.; Chuaqui, C.; et al. Correction to Discovery of Pyrazolo[1,5-a]pyrimidine B-Cell Lymphoma 6 (BCL6) Binders and Optimization to High Affinity Macrocyclic Inhibitors. *J. Med. Chem.* **2017**, *60*, 6459. [CrossRef] [PubMed]
90. Bou Karroum, N.; Moarbess, G.; Guichou, J.F.; Bonnet, P.A.; Patinote, C.; Bouharoun-Tayoun, H.; Chamat, S.; Cuq, P.; Diab-Assaf, M.; Kassab, I.; et al. Novel and selective TLR7 antagonists among the imidazo[1,2-a]pyrazines, Imidazo[1,5-a]quinoxalines, and Pyrazolo[1,5-a]quinoxalines Series. *J. Med. Chem.* **2019**, *62*, 7015–7031. [CrossRef]
91. Philoppes, J.N.; Khedr, M.A.; Hassan, M.H.A.; Kamel, G.; Lamie, P.F. New pyrazolopyrimidine derivatives with anticancer activity: Design, synthesis, PIM-1 inhibition, molecular docking study and molecular dynamics. *Bioorg. Chem.* **2020**, *100*, 103944. [CrossRef]
92. Ravi Kumar, N.; Poornachandra, Y.; Krishna Swaroop, D.; Jitender Dev, G.; Ganesh Kumar, C.; Narsaiah, B. Synthesis of novel ethyl 2,4-disubstituted 8-(trifluoromethyl)pyrido[2′,3′:3,4]pyrazolo[1,5-a]pyrimidine-9-carboxylate derivatives as promising anticancer agents. *Bioorg. Med. Chem. Lett.* **2016**, *26*, 5203–5206. [CrossRef]
93. Henderson, S.H.; Sorrell, F.; Bennett, J.; Hanley, M.T.; Robinson, S.; Hopkins Navratilova, I.; Elkins, J.M.; Ward, S.E. Mining Public Domain Data to Develop Selective DYRK1A Inhibitors. *ACS Med. Chem. Lett.* **2020**, *11*, 1620–1626. [CrossRef]
94. Dowling, J.E.; Alimzhanov, M.; Bao, L.; Chuaqui, C.; Denz, C.R.; Jenkins, E.; Larsen, N.A.; Lyne, P.D.; Pontz, T.; Ye, Q.; et al. Potent and Selective CK2 Kinase Inhibitors with Effects on Wnt Pathway Signaling in Vivo. *ACS Med. Chem. Lett.* **2016**, *7*, 300–305. [CrossRef] [PubMed]
95. El Sayed, M.T.; Hussein, H.A.R.; Elebiary, N.M.; Hassan, G.S.; Elmessery, S.M.; Elsheakh, A.R.; Nayel, M.; Abdel-Aziz, H.A. Tyrosine kinase inhibition effects of novel Pyrazolo[1,5-a]pyrimidines and Pyrido[2,3-d]pyrimidines ligand: Synthesis, biological screening and molecular modeling studies. *Bioorg. Chem.* **2018**, *78*, 312–323. [CrossRef] [PubMed]
96. Nagaraju, B.; Kovvuri, J.; Kumar, C.G.; Routhu, S.R.; Shareef, M.A.; Kadagathur, M.; Adiyala, P.R.; Alavala, S.; Nagesh, N.; Kamal, A. Synthesis and biological evaluation of pyrazole linked benzothiazole-β-naphthol derivatives as topoisomerase I inhibitors with DNA binding ability. *Bioorg. Med. Chem.* **2019**, *27*, 708–720. [CrossRef] [PubMed]
97. Husseiny, E.M. Synthesis, cytotoxicity of some pyrazoles and pyrazolo[1,5-a]pyrimidines bearing benzothiazole moiety and investigation of their mechanism of action. *Bioorg. Chem.* **2020**, *102*, 104053. [CrossRef] [PubMed]

Review

Pyrazole Scaffold Synthesis, Functionalization, and Applications in Alzheimer's Disease and Parkinson's Disease Treatment (2011–2020)

Xuefei Li, Yanbo Yu and Zhude Tu *

Department of Radiology, Washington University School of Medicine, 510 S. Kingshighway Boulevard St. Louis, MO, 63110, USA; xuefeili@wustl.edu (X.L.); yanboyu@wustl.edu (Y.Y.)
* Correspondence: zhudetu@wustl.edu; Tel.: +1-314-362-8487

Abstract: The remarkable prevalence of pyrazole scaffolds in a versatile array of bioactive molecules ranging from apixaban, an anticoagulant used to treat and prevent blood clots and stroke, to bixafen, a pyrazole-carboxamide fungicide used to control diseases of rapeseed and cereal plants, has encouraged both medicinal and organic chemists to explore new methods in developing pyrazole-containing compounds for different applications. Although numerous synthetic strategies have been developed in the last 10 years, there has not been a comprehensive overview of synthesis and the implication of recent advances for treating neurodegenerative disease. This review first presents the advances in pyrazole scaffold synthesis and their functionalization that have been published during the last decade (2011–2020). We then narrow the focus to the application of these strategies in the development of therapeutics for neurodegenerative diseases, particularly for Alzheimer's disease (AD) and Parkinson's disease (PD).

Keywords: pyrazole; synthesis; functionalization; heterocyclic; neurodegeneration; Alzheimer's disease; Parkinson's disease; inhibitor; antagonist; biological activity

1. Introduction

Pyrazole compounds contain a five-membered aromatic ring composed of three carbon atoms and two adjacent nitrogen atoms. The shifting C-N double bond inside the heterocycle endows pyrazoles with tautomerism. Consequently, regioselectivity can be a challenge in the synthesis and purification of N-substituted and asymmetric pyrazoles. Although a pyrazole core is rare in natural compounds, artificial pyrazole derivatives are prevalent in diverse fields. Applications range from agrochemicals [1] to therapeutics. At least thirty-three pyrazole-containing medicines have been marketed to alleviate or treat diseases ranging from bacterial infections to cancer and neurologic disorder (Figure 1) [2,3]. For example, the anticoagulant apixaban is used to prevent serious blood clots that may cause stroke, heart attack, and is also prescribed to patients with an abnormal heartbeat (atrial fibrillation) or hip/knee joint replacement surgery; in 2018, apixaban was the second most popular blockbuster drug [4]. Investigators in both academic institutions and pharmaceutical industries have put tremendous efforts into the exploration of new pyrazole scaffolds for drug development. This review, building upon previous literature, reviews [3,5] and presents progress in the chemical synthesis of pyrazole scaffold molecules and the functionalization of pyrazole derivatives from 2011 to 2020. We then focus on pyrazole-containing small biomolecules developed as therapeutic candidates for treating neurodegenerative diseases, particularly those that target pathologies found in Alzheimer's disease (AD) and Parkinson's disease (PD). Fluorine-18 radiotracers composed of pyrazole scaffolds have been recently reviewed by Gomes and his colleagues [6] and will not be discussed in this review.

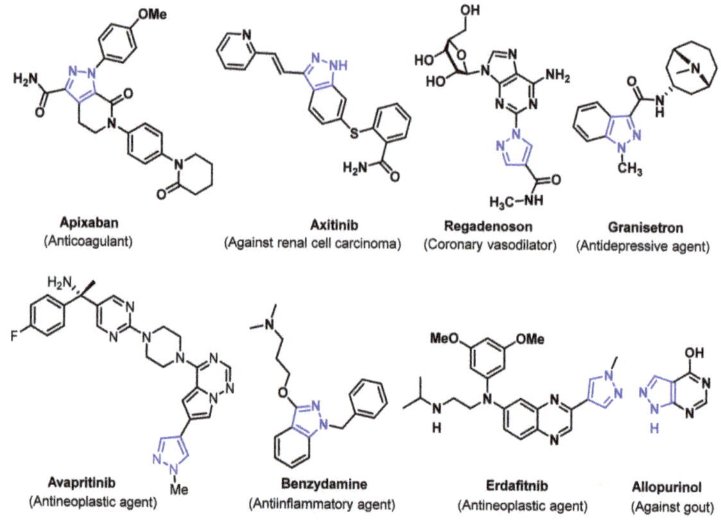

Figure 1. Representative FDA-approved pyrazole-containing medicines. FDA: U.S. Food and Drug Administration.

2. The Synthesis of Pyrazole Scaffold Molecules

The synthesis of substituted pyrazoles has been accomplished by two strategies [3,5]: (1) the cyclocondensation of hydrazines with 1,3-dicarbonyl compounds or their synthetic 1,3-dielectrophilic equivalents (Scheme 1a), and (2) the cycloaddition of 1,3-dipoles to dipolarophiles (Scheme 1b). These two conventional strategies were recently enriched by additional approaches including multicomponent one-pot processes, photoredox reactions, and transition-metal catalyzed reactions. In this section, we will first discuss cyclocondensation, followed with cycloaddition. In addition, new approaches are also discussed.

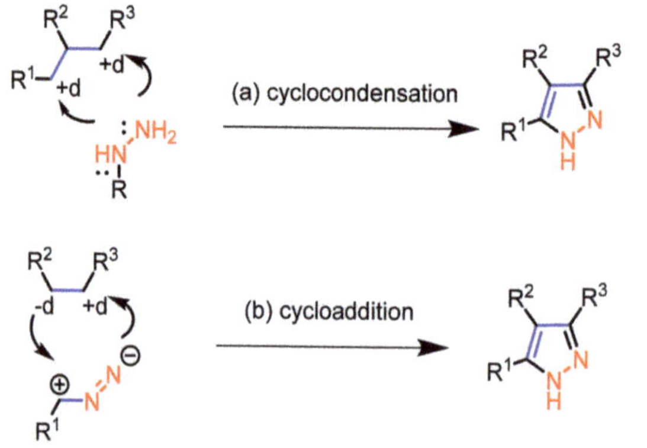

Scheme 1. Conventional pyrazole synthetic strategies. (**a**): cyclocondensation; (**b**): cycloaddition.

2.1. Cyclocondensation of Hydrazines with 1,3-Dielectrophilic Derivatives

Substituted or nonsubstituted hydrazines are readily available as [*NN*] synthons for the synthesis of pyrazole derivatives. Pyrazole molecules can be generated through hydrazines reacting with 1,3-dielectrophilic units such as 1,3-dicarbonyl or with α,β-unsaturated carbonyl compounds; the latter structures include enones, ynones, and vinyl ketones bearing a leaving group (Figure 2) [3].

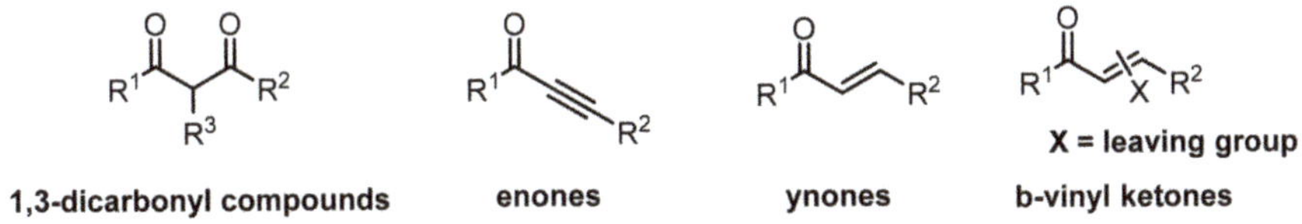

1,3-dicarbonyl compounds   **enones**   **ynones**   **b-vinyl ketones**
X = leaving group

Figure 2. Structures of 1,3-dielectrophilic derivatives.

2.1.1. Cyclocondensation of Hydrazines with 1,3-Dicarbonyl and Related Compounds

1,3-diketones, β-ketoesters, 2,4-diketoesters, and related synthetic equivalents can condense efficiently with hydrazines to generate substituted pyrazoles. Based on this strategy, a series of potent carbonic anhydrase, α-glycosidase, and cholinesterase enzymes inhibitors **1** were synthesized via the cyclocondensation of the 1,3-diketone and appropriate hydrazines (Scheme 2) [7].

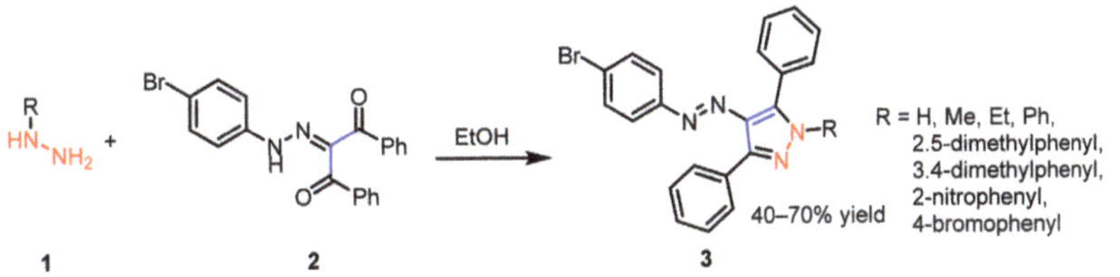

Scheme 2. Bioactive molecules synthesis by cyclocondensation.

However, a mixture of two regioisomers is produced in most cases when unsymmetrical 1,3-dicarbonyl compounds ($R^1 \neq R^2$), and substituted hydrazines are used for cyclocondensation (Scheme 3). Moreover, this kind of cyclocondensation is generally accomplished in a strongly acidic medium (i.e., fluoroboric acid [8]), thus the reaction solution is corrosive and not environmentally-friendly.

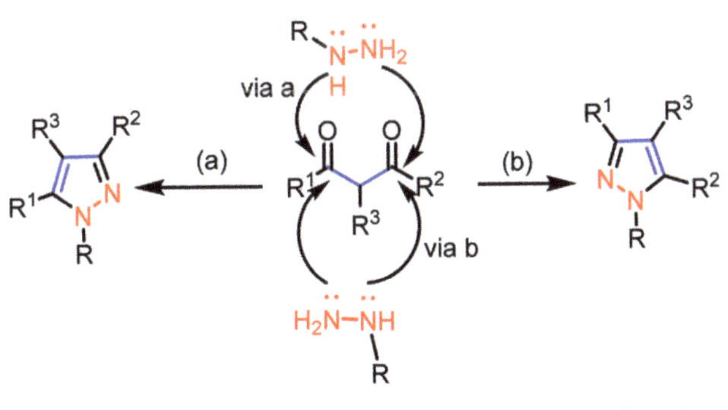

Scheme 3. Regioselectivity of pyrazoles synthesis via cyclocondensation.

A mild and acid-free condensation of 1,3-diketones with substituted hydrazines to generate the 1,3,5-trisubstituted and fully substituted pyrazoles was reported by Wang and co-workers [9]. The optimal conditions were obtained when using copper (II) nitrate as the catalyst, providing the cyclocondensation products at room temperature in less than 60 min. More importantly, this method provided highly regioselective products with good yields (Scheme 4).

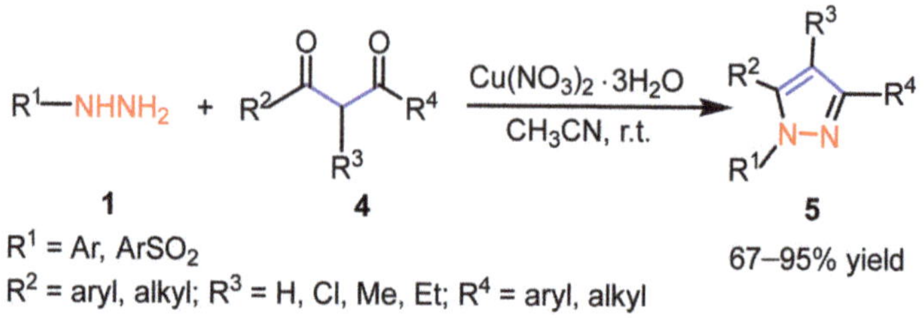

Scheme 4. Cu-catalyzed cyclocondensation for pyrazole synthesis.

As a novel derivative of 1,3-dicarbonyl compounds, 1,3-monothiodiketones are good complements for preparing unsymmetrically substituted 1-aryl-3,5-bis(het)arylpyrazoles with excellent regioselectivity and high yield (Scheme 5) [10].

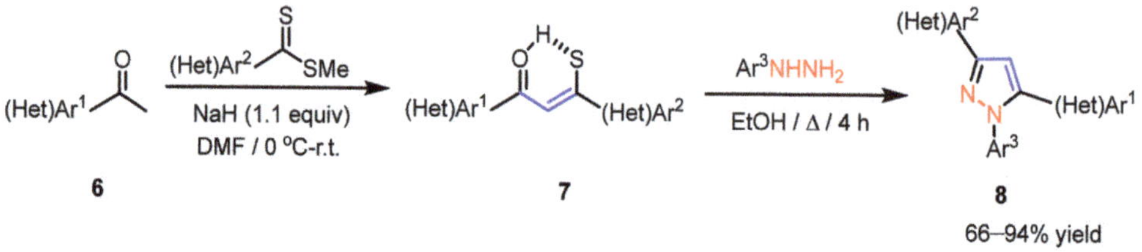

Scheme 5. 1,3-monothiodiketones as substrates for pyrazole synthesis.

Despite the efficiency and flexibility of using 1,3-dicarbonyl compounds as substrates, the synthetic inconvenience and inherent instability of many 1,3-dicarbonyls, particularly dialdehydes, limits the synthesis of some substituted pyrazoles. To overcome this challenge, Schmitt et al. reported the facile generation of 1,4-disubstituted pyrazoles when using ruthenium as a catalyst with 1,3-diols to replace 1,3-dicarbonyl compounds [11]. In this reaction, crotonitrile was used as the reductant to accept hydrogen transfer from ruthenium dihydrides (Scheme 6).

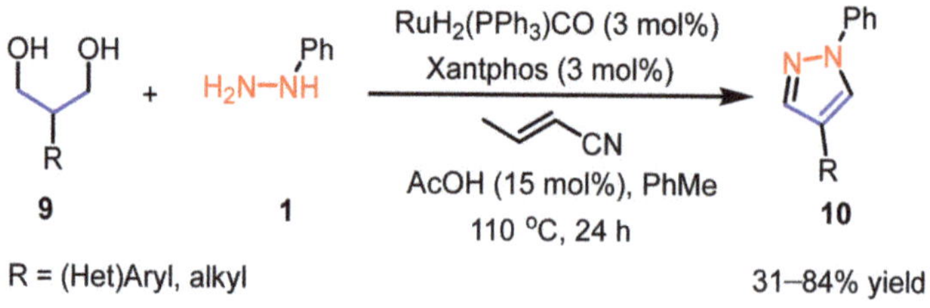

Scheme 6. 1,3-diols as substrates for pyrazole synthesis.

2.1.2. Cyclocondensation of Hydrazines with Enones and Related Compounds

The condensation of hydrazines with α-enones affords pyrazolines, and can be followed by oxidization to generate the corresponding pyrazoles. Zhang et al. reported using iodine to mediate oxidative intramolecular C-N bond formation, and the hydrazone intermediates were cyclized to generate pyrazoles [12]. This one-pot applicable procedure led directly to a variety of di-, tri-, and tetra-substituted (aryl, alkyl, and/or vinyl) pyrazole derivatives with high (64–96%) yields when using β-aryl substituted aldehydes and ketones as substrates. However, the yield decreased sharply when R^1 was a methyl group (23% yield) due to the loss of the extended conjugation to facilitate the nucleophilic attack of the α,β-unsaturated double-bond to iodine (Scheme 7).

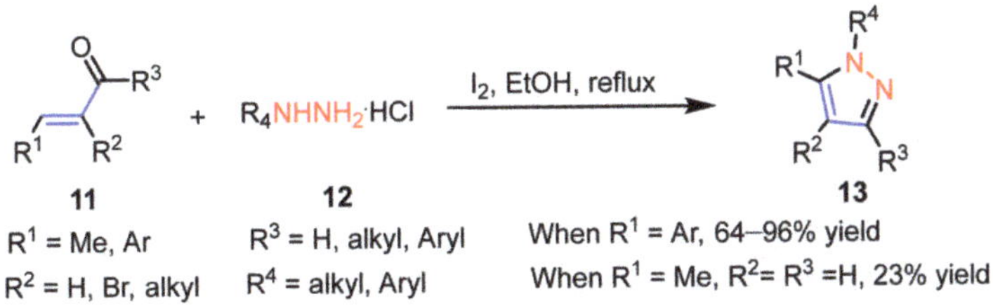

Scheme 7. I_2-mediated cyclocondensation for pyrazole synthesis.

The applicability of this general method has been extended to synthesize an array of novel 3,5-diarylpyrazole derivatives (**15**); some of these are potent acetylcholinesterase inhibitors possessing excellent selectivity, and have been accepted as prospective drug candidates for treating Alzheimer's disease (Scheme 8) [13].

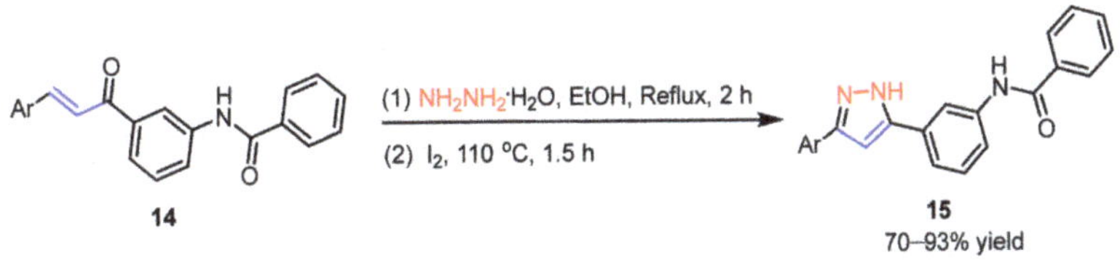

Scheme 8. 3,5-diarylpyrazole synthesis through I_2-mediated cyclocondensation.

In addition, Ding et al. reported an air-promoted photoredox cyclization of substituted hydrazines with activated alkene (Michael addition reaction acceptors) to afford corresponding pyrazoles with good to excellent yields [14]. In this context, hydrazine was oxidized by Ru^{II} to a diazene intermediate that attacks Michael acceptors, followed by an intramolecular cyclization to form pyrazoles. The reduced Ru^{I} was reoxidized by air to recycle this photoredox process. For electron withdrawing groups (EWG) such as cyano or carboxylic acidic groups, this procedure efficiently generated 5-amino and 5-hydroxyl substituted pyrazoles, which are important functionalities for subsequent transformations (Scheme 9).

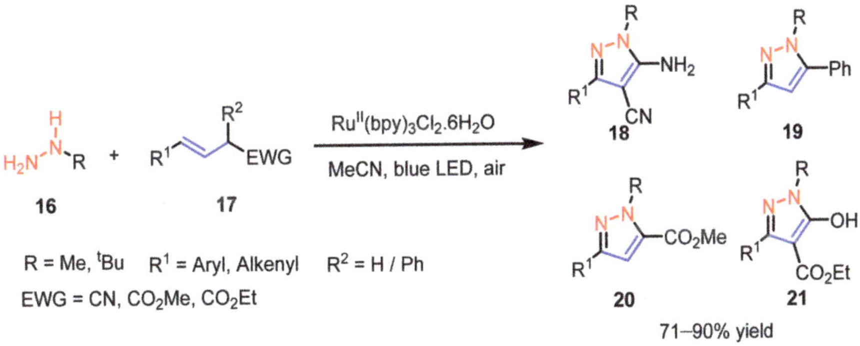

Scheme 9. Ru^{II}-catalyzed photoredox cyclization for pyrazole synthesis.

Pre-installation of enones with hydrazines to form hydrazone intermediates is also accessible. Hu and co-workers [15] reported a Ru^{II}-catalyzed intramolecular oxidative C–N coupling for the facile synthesis of highly diversified tri- and tetrasubstituted pyrazoles. Their method is applicable to a broad scope of substrates with excellent functional group tolerance. More importantly, many of the pyrazoles generated using this strategy were difficult to prepare by conventional methods (Scheme 10).

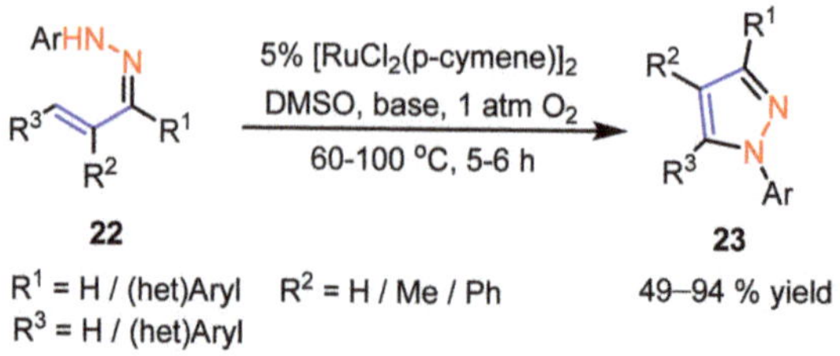

Scheme 10. Ru^{II}-catalyzed oxidative cyclization for pyrazoles synthesis.

2.1.3. Cyclocondensation of Hydrazines with Ynones and Related Compounds

Harigae et al. reported using a one-pot regioselective procedure for the synthesis of 3,5-disubstituted pyrazoles with good yields [16]. In his multicomponent procedure, ynone intermediates were formed through iodine-mediated in situ oxidation of propargylic alcohols that were converted from terminal alkynes and aldehydes (Scheme 11).

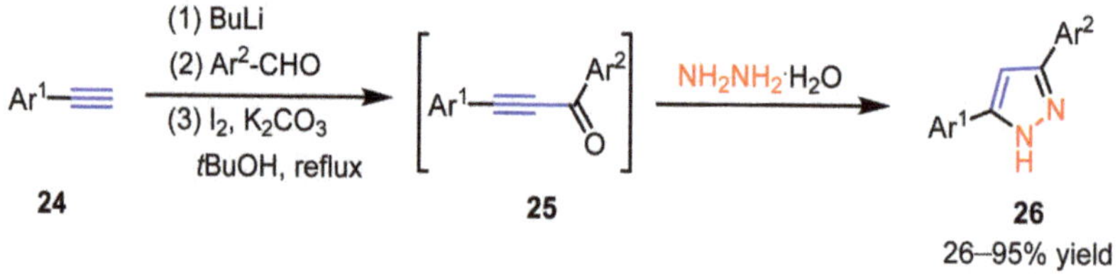

Scheme 11. Multicomponent synthesis of pyrazoles from alkynes.

In addition, propargylic alcohols, the reduced form of ynones, were also used to generate 3,5-disubstituted 1H-pyrazoles [17]. This procedure includes two consecutive steps: (1) Lewis acid catalyzed N-propargylation of propargylic alcohols to propargyl hydrazides with N-acetyl-N-tosylhydrazine, followed by (2) base-mediated intramolecular cyclization (Scheme 12).

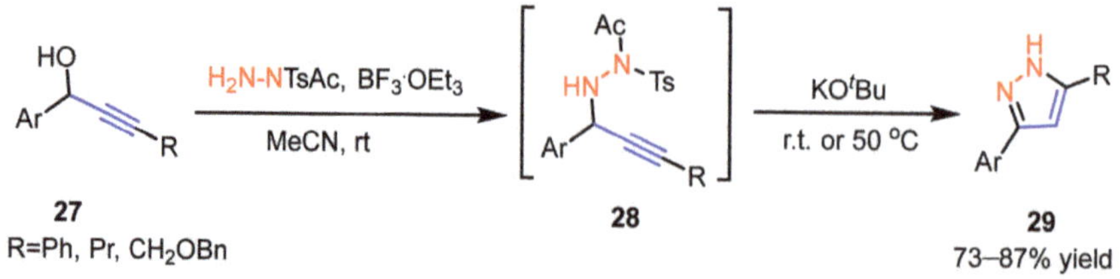

Scheme 12. Propargylic alcohols as substrates for two-step pyrazole synthesis.

2.1.4. Cyclocondensation of Hydrazines with Vinyl Ketones Bearing a Leaving Group

Aminomethylene and (dimethyl)aminomethylene groups are synthetic equivalents of a formyl group. α,β-Vinyl ketones containing these substituents may react with hydrazine derivatives to afford pyrazolines, followed by elimination of the leaving group to generate the desired pyrazoles. The methylthio group has also been reported as an excellent leaving group [10]. Guo et al. reported that in the presence of iodine and *tert*-butyl hydroperoxide (TBHP), β-amino vinyl ketone could cyclize with tosyl hydrazine in water to afford fully substituted pyrazoles [18]. Mechanism studies using ^{15}N-labeled enaminone as a substrate showed that the β-amino group did not participate in the formation of a pyrazole ring, instead, it acted as a leaving group, serving as a hydrogen bond donor to facilitate the reaction in water (Scheme 13).

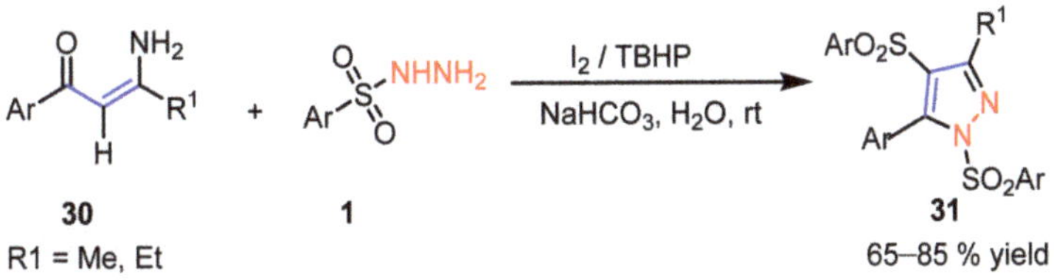

Scheme 13. Pyrazole synthesis in water.

Raghunadh and co-workers reported a copper-catalyzed three-component process for pyrazole synthesis [19]. The authors found that N-substituted pyrazoles could be afforded in high yields and excellent regioselectivity when using β-dimethylamino vinyl ketones as substrates in the presence of aryl halides (Scheme 14).

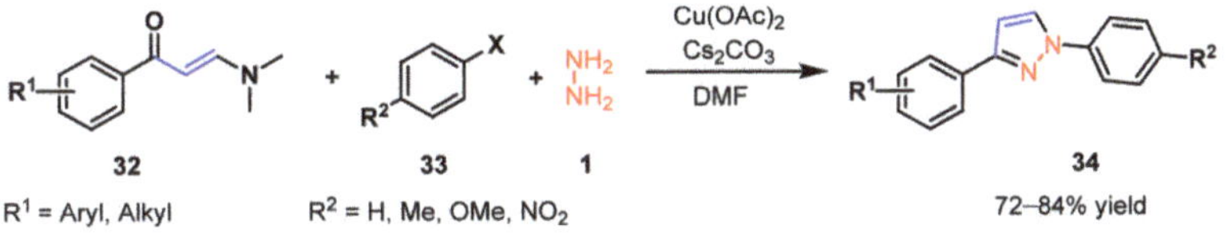

Scheme 14. Cu-catalyzed three-component cyclocondensation for pyrazole synthesis.

In the presence of peroxide (m-CPBA), Chen et al. showed that propargylamines could be converted into enaminones after a one-pot oxidation and rearrangement process. The formed enaminones are key intermediates that would react with hydrazines to afford corresponding pyrazole derivatives [20]. Using this methodology, the researchers realized a one-pot, four-step celecoxib synthesis with a 39% overall yield starting from 4-ethynyltoluene (Scheme 15).

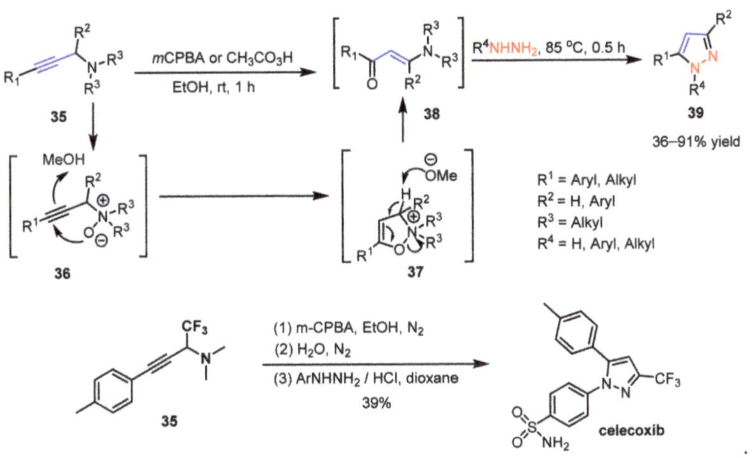

Scheme 15. One-pot oxidative cyclocondensation for pyrazole derivatives synthesis.

Under alkaline conditions, a proton on the β-vinyl substrate can serve as a leaving group to accelerate the conversion of pyrazolines to pyrazoles [21]. In addition, when the phase transfer catalyst Bu$_4$NBr is used, these reactions can be performed smoothly in water [22] (Scheme 16).

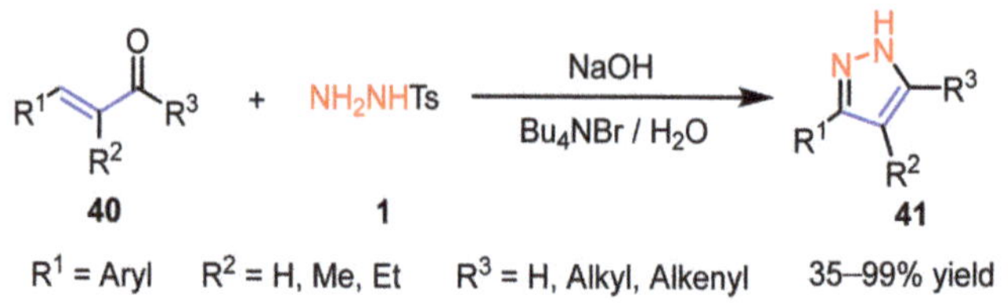

Scheme 16. Proton leaving groups in cyclocondensation for pyrazole synthesis.

2.2. 1,3-Dipolar Cycloadditions

The intrinsically high regioselectivity and efficiency of 1,3-dipolar cycloaddition has led to its prominent role in preparing substituted pyrazoles [5]. Conventional 1,3-dipolar cycloaddition employs a [CNN] fragment and a [CC] fragment. In general, three main classes of 1,3-dipoles have been used as the [CNN] fragment, namely, diazoalkanes, nitrilimines, and azomethine imines, while their [CC] counterparts are alkenes or alkynes (Scheme 17a). Additionally, the combination of a [CCC] fragment plus a [NN] fragment has also been utilized. In these examples, [CCC] comes from alkenes or alkynes, while the [NN] fragment derives from azo compounds (Scheme 17b).

Scheme 17. Strategies for 1,3-dipolar cycloaddition. (**a**): [CNN] + [CC] mode; (**b**): [CCC] + [NN] mode.

In this section, we discuss the synthetic development of 1,3-dipolar cycloaddition reactions according to the source of 1,3-dipoles fragments, namely, from diazoalkanes, nitrilimines, or azomethine imine, respectively. Reactions in which azo compounds act as [NN] fragments are also presented.

2.2.1. Diazoalkanes as 1,3-Dipoles

Electron-rich diazo compounds are toxic and potentially explosive, thus the preparation and handling of these reactants are hazardous. To overcome this problem, an improved method for making aryldiazomethanes from stable tosylhydrazones derivatives was developed to synthesize pyrazoles. Using an operationally simple, multicomponent, one-pot procedure, 3,4,5-trisubstituted 1H-pyrazoles could be prepared from vinyl azide, aldehyde, and tosylhydrazine [23]. In these cases, the azide group serves as a leaving group (Scheme 18).

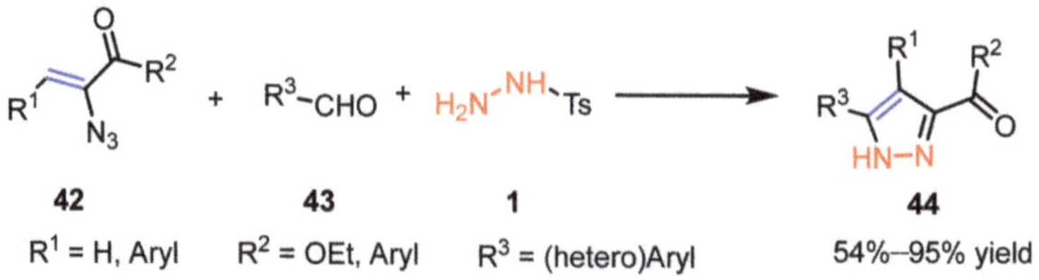

Scheme 18. Azide as leaving groups in diazoalkanes for pyrazole synthesis.

Bromine is also a good leaving group. Sha et al. developed a simple, highly efficient, and regioselective method for the synthesis of 3,5-diaryl-4-bromopyrazoles using *gem*-dibromoalkene as the substrate [24] (Scheme 19).

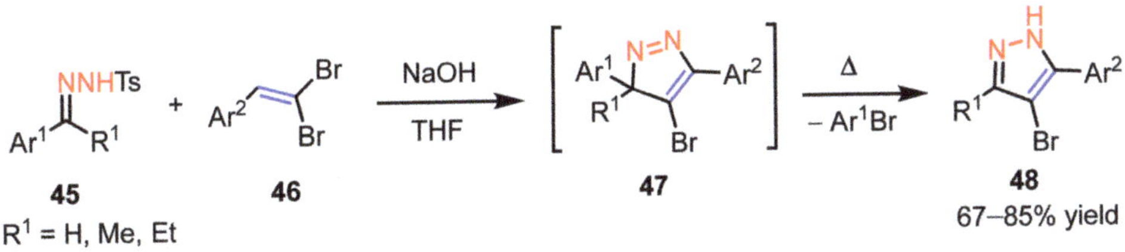

Scheme 19. Bromine as the leaving group in diazoalkanes for pyrazole synthesis.

Using pyrrolidine as a catalyst, 3,4,5-trisubstituted or 3,5-disubstituted pyrazoles could be synthesized from carbonyl compounds through an enamine intermediate under mild conditions [25]. Considering that carbonyl compounds are more readily available than alkenes, this optimization is a remarkable advancement. In addition, bicyclic pyrazoles that are difficult to synthesize by common methods can be prepared straightforwardly using this method (Scheme 20).

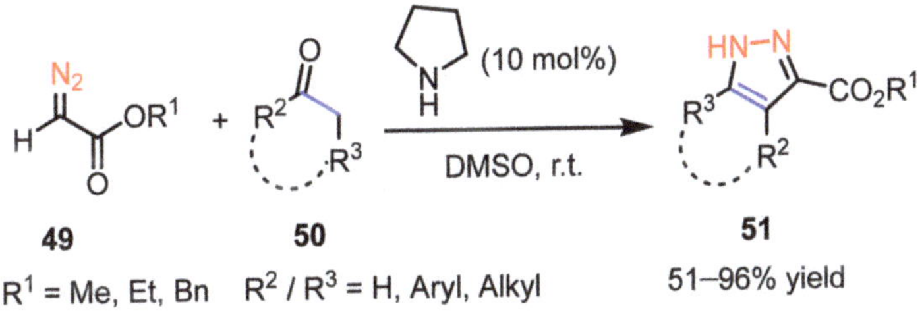

Scheme 20. Ketones as substrates in 1,3-dipolar cycloaddition for pyrazole synthesis.

Jackowski et al. reported a novel method to synthesize organoaluminum heterocyclics via a [3 + 2] cycloaddition route [26]. Using this methodology, 3,4,5-trisubstituted pyrazoles could be prepared from reactive polysubstituted alumino-heteroles intermediates (**53**) after a one-step electrophilic substitution (Scheme 21).

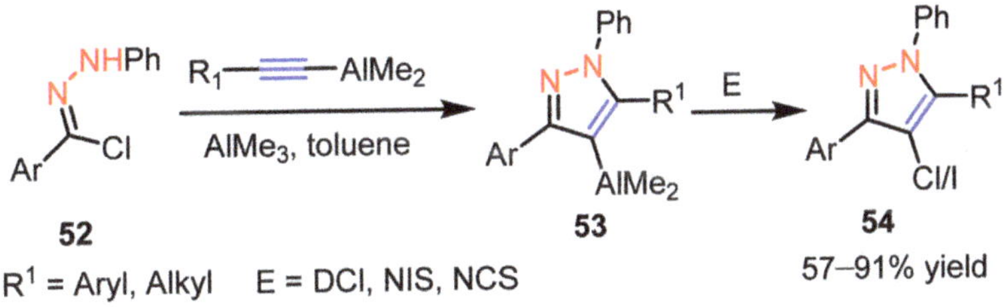

Scheme 21. Trimethylaluminum substituted alkynes as substrates for pyrazoles synthesis.

2.2.2. Nitrilimines as 1,3-Dipoles

Li and co-workers [27] reported a rhodium-catalyzed cycloaddition of hydrazines with dicarboxylic alkynes. This method provides a highly efficient and effective procedure to synthesize 3, 4-dicarboxylic pyrazoles, and offering versatility for subsequent structural modifications (Scheme 22).

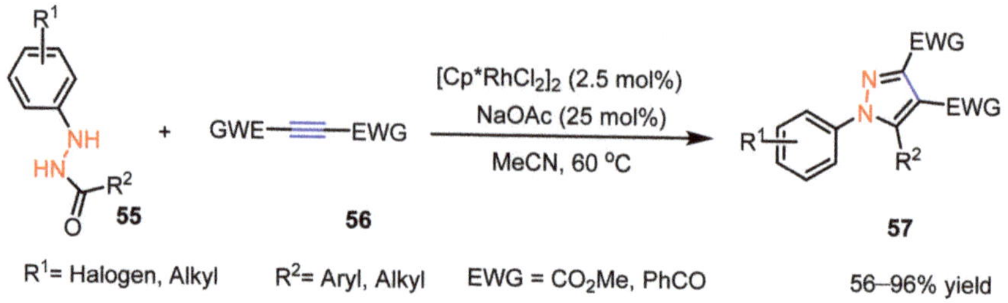

Scheme 22. Ru^{II}-catalyzed 1,3-dipolar cycloaddition for pyrazole synthesis.

Ledovskaya et al. reported that the weak organic base trimethylamine (TEA) could promote the 1,3-dipolar cycloaddition of vinyl ethers and hydrazonoyl chlorides to produce 1,3-disubstituted pyrazoles with absolute regioselectivity [28] (Scheme 23).

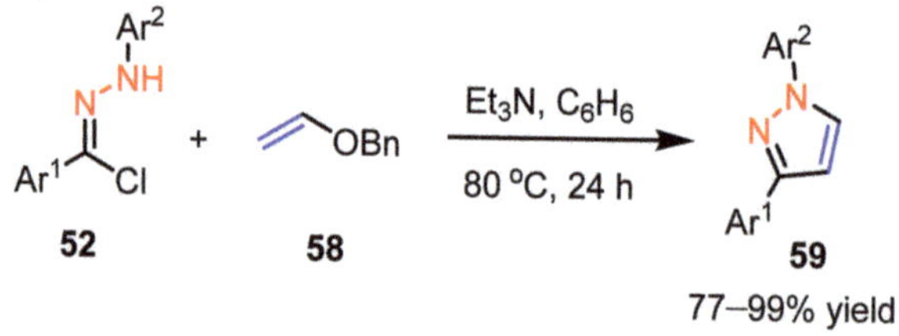

Scheme 23. Trimethylamine (TEA) promoted 1,3-dipolar cycloaddition for pyrazoles synthesis.

Starting from primary alcohols, Kobayashi et al. reported a one-pot, multicomponent procedure for making multisubstituted pyrazoles [29]. According to their description, the primary alcohol was first oxidized by 2,2,6,6-tetramethylpiperidinyloxy (TEMPO) to aldehyde, followed by reacting with hydrazine to form oxime, where the latter was converted into nitrilimine in the presence of *N*-chlorosuccinimide (NCS) and decyl methyl sulfide, followed by a 1,3-dipolar cycloaddition between nitrilimine and diethyl acetylenedicarboxylate (Scheme 24).

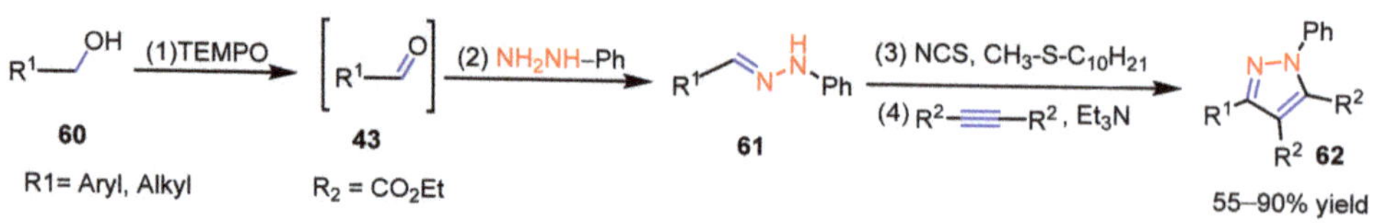

Scheme 24. Primary alcohols as substrates for pyrazole synthesis.

Using alcohols as potential [CC] fragments, Panda et al. reported an iron-catalyzed route [30] for the regioselective synthesis of 1,3- and 1,3,5-substituted pyrazoles by the condensation of diarylhydrazones with α-carbonyl alcohols, which were converted in situ from vicinal diols by ferric chloride in the presence of *tert*-butyl hydroperoxide (TBHP) (Scheme 25).

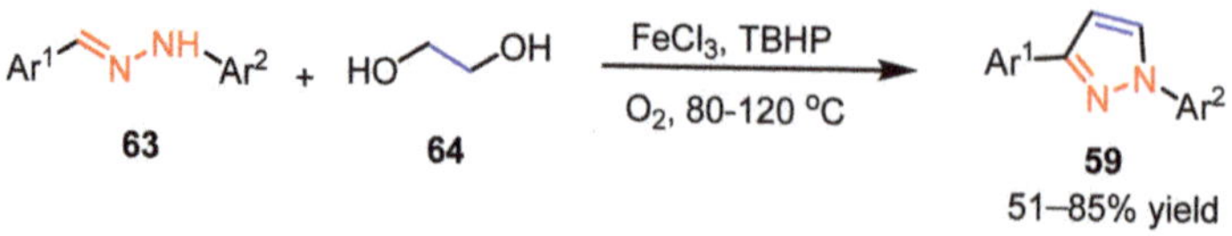

Scheme 25. Fe^{III}-catalyzed 1,3-dipolar cycloaddition for pyrazole synthesis.

A one-pot visible light-promoted single-electron-transfer (SET) process for making 1,3,5-trisubstituted pyrazoles from α-bromoketones was reported by Fan and coworkers [31]. The reaction has good functional group tolerance, even for nitro and nitrile, which are generally not tolerated in SET reactions (Scheme 26).

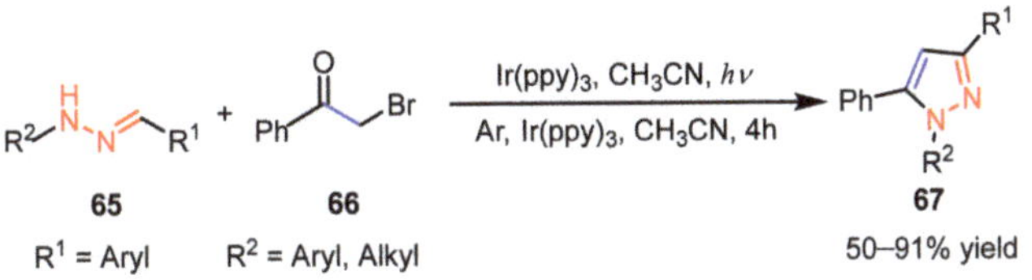

Scheme 26. Ir^{III}-catalyzed photoredox cycloaddition for pyrazole synthesis.

Yi et al. reported a novel silver-mediated [3 + 2] cycloaddition of alkynes and N-isocyanoiminotriphenylphosphorane (NIITP) for the assembly of monosubstituted pyrazoles [32]. The reaction can be accomplished under mild conditions, with broad substrate scope and excellent functional group tolerance. Mechanism studies showed that NIITP was activated by Mo(CO)$_6$, then undergoes a [3 + 2] cycloaddition with a silver acetylide intermediate (Scheme 27).

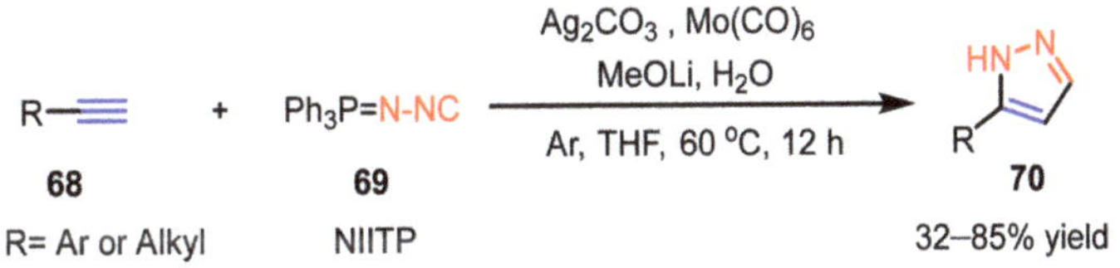

Scheme 27. Ag^{II}-mediated cycloaddition for pyrazole synthesis.

Bioactive molecules with pyrrolopyrazole motifs such as an aurora kinase inhibitor danusertib (**71**), a glycine transporter-1 inhibitor (**72**), an HIV-1 integrase inhibitor (**73**), and an antibacterial agent (**74**) are promising drug candidates [33] (Figure 3). Therefore, general and practical methods for making such compounds are valuable.

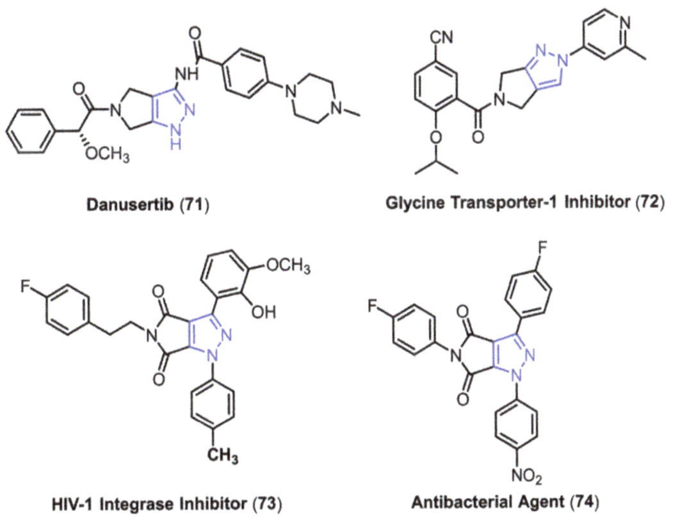

Figure 3. Bioactive molecules containing pyrrolopyrazole motifs.

Zhu et al. reported a CuCl-catalyzed oxidative coupling reaction of aldehyde hydrazones with maleimides to prepare dihydropyrazoles under mild conditions [33]. Using this method, a variety of pyrrolo[3,4-c]pyrazoles could be obtained from dihydropyrazoles by one-step following oxidation (Scheme 28).

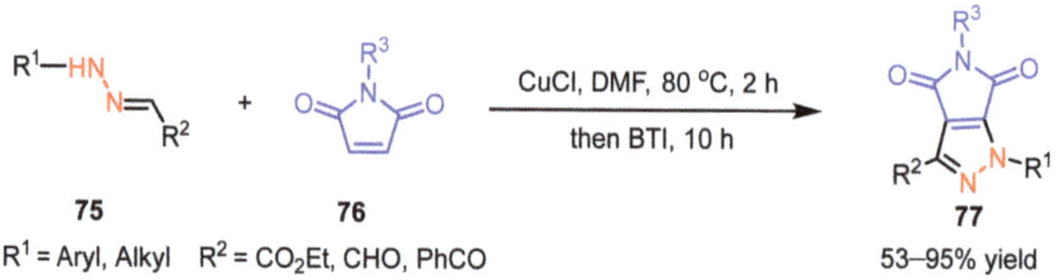

Scheme 28. Cu^I-catalyzed cycloaddition for pyrrolopyrazole synthesis.

2.2.3. Sydnones as 1,3-Dipoles

Specklin et al. reported a one-pot Cu-catalyzed sydnone-alkyne cycloaddition to generate 1,4-disubstituted pyrazoles from readily available arylglycines [34]. This method tolerates various electron-rich or electron-poor N-aryl sydnones, and acetylene components. Using this method, 1,4-pyrazoles were the only products with good to excellent yield; no trace of 1, 3-regioisomers was detected (Scheme 29).

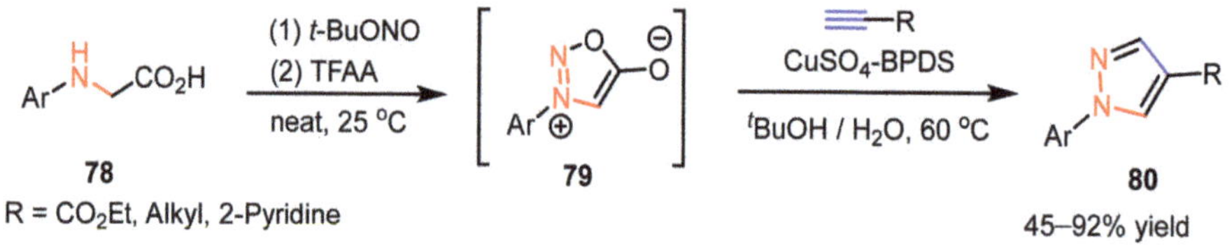

Scheme 29. Sydones as 1,3-dipoles for pyrazole synthesis.

In addition, a visible-light photoredox process using Ru(bpy)$_3$(PF$_6$)$_2$ as the catalyst to accomplish this regioselective reaction was also reported. In this case, a broad scope of 1,4-disubstituted pyrazoles was made in high yields [35] (Scheme 30).

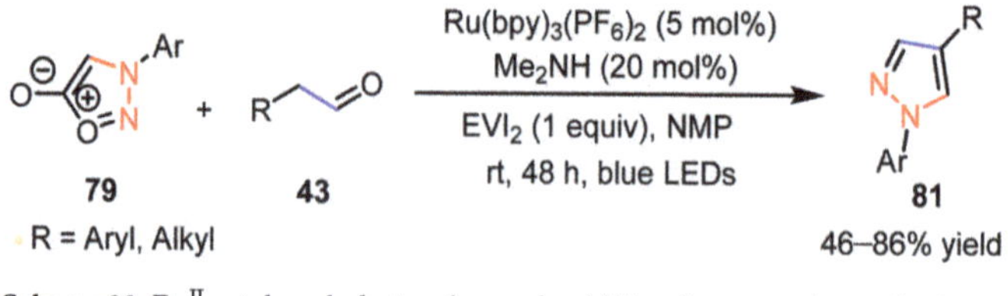

Scheme 30. Ru^{II}-catalyzed photoredox cycloaddition for pyrazole synthesis.

2.2.4. Azo Compounds as [NN] Fragments

Zhang et al. reported a highly efficient nBu$_3$P-catalyzed desulfonylative [3 + 2] cycloadditions of allylic carbonates with arylazosulfones to make 1,4-disubstitued pyrazoles in good to excellent yields under mild conditions [36]. The reaction can be triggered by the Michael-type addition of nBu$_3$P to allylic carbonate, meanwhile, an allylic phosphorus ylide intermediate was formed after decomposition of the Boc group into CO$_2$ and tBuOH. The ylide intermediate subsequently underwent a [3 + 2] cycloaddition with arylazosulfones to afford the substituted pyrazoles (Scheme 31).

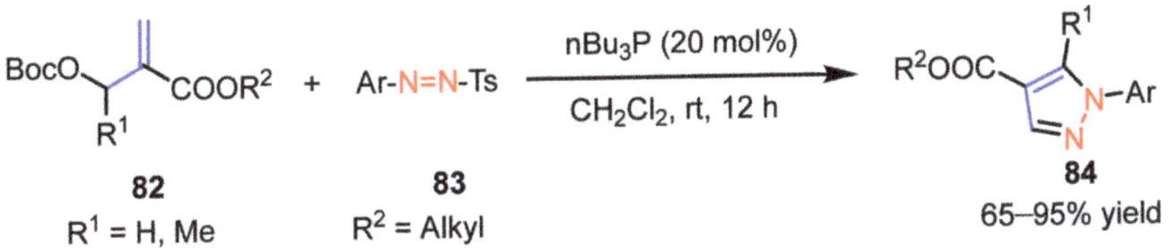

Scheme 31. Azo compounds as [NN] fragment for pyrazole synthesis.

For some reactions, the involvement of organophosphorus in azo-type cycloadditions is useful, but not essential. Zhang and co-workers [37] reported that substituted propargylamines could react with commercially available dialkyl azodicarboxylates (DEAD) in toluene at room temperature without the presence of organophosphorus. This synthetic strategy was also applicable for the synthesis of different pyrazoles in high yields with broad substrate scopes (Scheme 32).

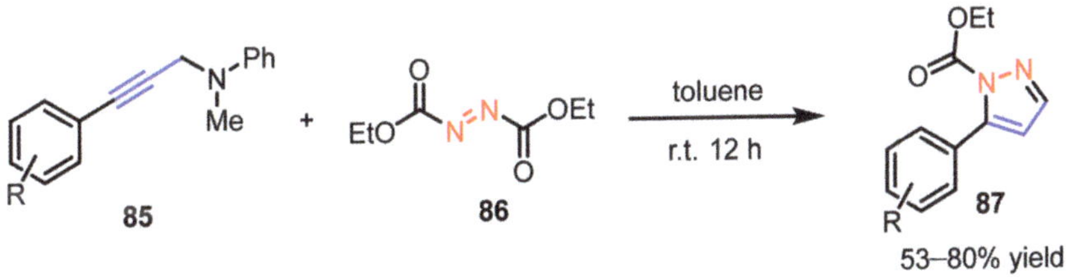

Scheme 32. Dialkyl azodicarboxylates (DEAD) as substrates in cycloaddition for pyrazole synthesis.

2.3. New Approaches to Reactions for Pyrazole Synthesis

In each of the above-mentioned methodologies for pyrazole synthesis, all the [NN] fragments either came from hydrazine derivatives or from azo compounds. Pearce and co-workers recently reported a novel fragment combination mode [NC] + [CC] + [N] using a multicomponent oxidative coupling to make multi-substituted pyrazoles [38]. In these reactions, diazatitana-cyclohexadiene intermediates that were generated from alkynes, nitriles, and titanium imido complexes, could produce pyrazole derivatives after a 2-electron oxidation process triggered by oxidant TEMPO (Scheme 33).

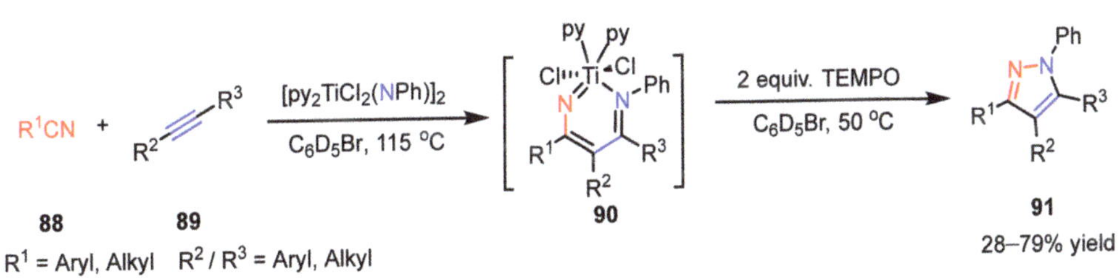

Scheme 33. Organotitanium intermediates for pyrazole synthesis.

3. The Functionalization of Pyrazoles

3.1. The Synthesis of Fluorine-Containing and Fluoroalkyl Substituted Pyrazoles

The incorporation of fluorine or fluoroalkyl groups onto a pyrazole ring [1,39] can positively affect its physicochemical and biological properties. Exploration of these compounds has led to the successful commercial application of numerous fluorine-containing pyrazoles, as shown in Figure 4. Therefore, new methods of synthesizing fluorine containing pyrazoles may facilitate new drug discoveries, both in pharmaceutical and agrochemical industries. In addition, the incorporation of fluorine in pyrazole motifs can enable ^{18}F-labeling for preclinical and clinical position emission tomography (PET) applications. As Gomes et al. have published a recent detailed review on this topic [6], we will not discuss it in this manuscript.

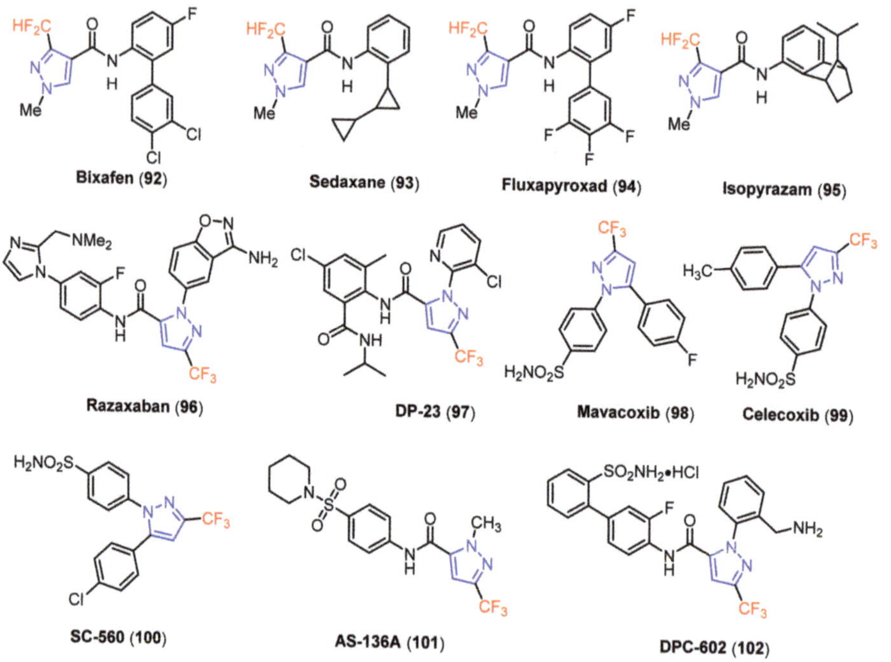

Figure 4. Examples of bioactive molecules with fluorine-containing pyrazole moieties.

3.1.1. The Synthesis of Monofluorine-Substituted Pyrazoles

Prieto and co-workers [40] described an accessible method to synthesize 4-fluoropyrazoles by ruthenium-catalyzed tandem C–H fluoromethylation and cyclization of N-alkylhydrazones. In this context, CBr$_3$F was acting as the fluorine source and a one-carbon unit. Compared to a range of other transition-metal catalysts (Cu, Pd, and Fe), RuCl$_2$(PPh$_3$)$_3$ was the most efficient for this reaction (Scheme 34).

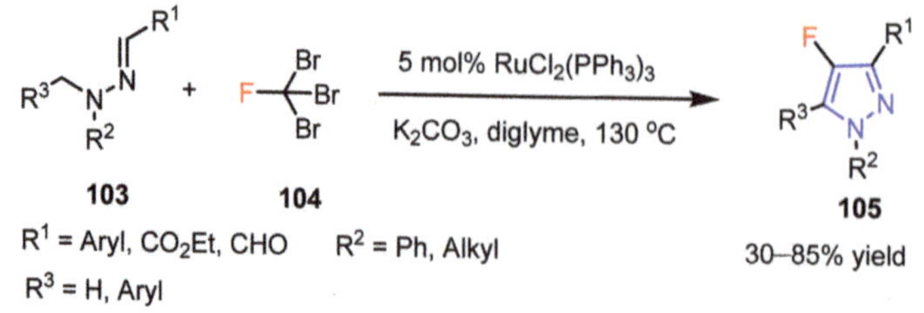

Scheme 34. Ru^{II}-catalyzed synthesis of monofluorine substituted pyrazoles.

3.1.2. The Synthesis of Difluoromethylpyrazoles

Difluoromethyl-substituted pyrazoles are valuable scaffolds for the preparation of fungicides in agriculture (Figure 4) [1]. Recently, industrial materials like ethyl 2,2-difluoroacetate and fluoroalkyl amino reagents (FAR) (Figure 5) have been widely employed as building blocks for making difluoromethyl pyrazoles [41,42].

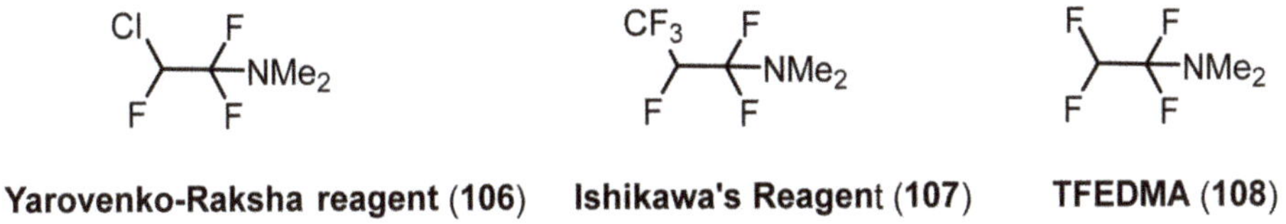

Yarovenko-Raksha reagent (106)   **Ishikawa's Reagent (107)**   **TFEDMA (108)**

Figure 5. Representative fluoroalkyl amino reagents.

Mykhailiuk [43] reported a novel approach to preparing difluoromethyl-substituted pyrazoles by a [3 + 2] cycloaddition between alkynes and CF_2HCHN_2, which was generated in situ from readily available $CF_3CH_2NH_2$. This practical methodology uses a two-step, one-pot procedure, and does not require either the involvement of a catalyst, or the isolation of a potentially toxic and explosive gaseous intermediate. More importantly, this operationally feasible approach supports scale-up synthesis of target pyrazoles (Scheme 35).

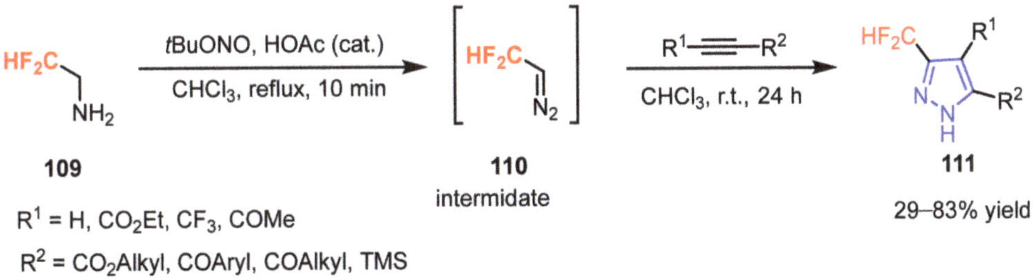

Scheme 35. Synthesis of difluoromethyl pyrazoles.

3.1.3. The Synthesis of Trifluoromethyl Pyrazoles

Among fluorinated pyrazoles, the synthesis of 3-trifluoromethyl pyrazoles has attracted significant attention because trifluoromethyl, a strong electronic withdraw group, may increase the bioactivity of target molecules. The feasibility of this tactic is evidenced by the success of a variety of pharmaceuticals and agrochemicals such as razaxaban (**96**, anticoagulant), and DP-23 (**97**, insecticidal activity), mavacoxi (**98**) and celecoxib (**99**) (both are COX-2 inhibitors), and SC-560 (**100**, human lung cancer inhibitor), AS-136A (**101**, measles virus inhibitor), and DPC-602 (**102** arterial thrombosis) (Figure 4). Conventional synthetic methods often suffer from poor regioselectivity between the 3-isomer and 5-isomer when using trifluoromethyl substituted 1,3-dicarbonyl compounds as starting materials [44].

Li et al. reported a highly regioselective approach to the synthesis of 3-trifluoromethyl pyrazoles using silver-mediated cycloaddition of alkynes with 2,2,2-trifluorodiazoethane, which is generated from readily available $CF_3CH_2NH_2 \cdot HCl$ [45]. This mild procedure is applicable for the synthesis of celecoxib and measles virus inhibitor AS-136. Interestingly, the reaction only proceeds well in the presence of a trace amount of water (0.1–1 equivalent), otherwise, the yield decreased sharply in the presence of either no water or too much water (>10 equivalent). A deuterated substitution experiment suggested that the proton on 4-position is abstracted from the water medium. Further mechanism studies indicated that silver acetylide might be the active species, as the addition of silver oxide was essential to facilitate this cycloaddition reaction (Scheme 36).

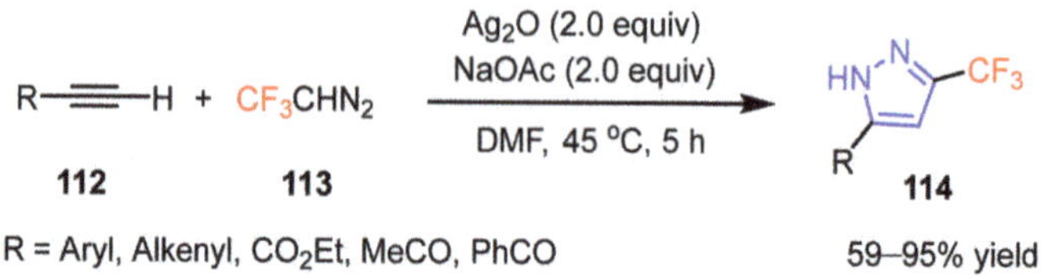

Scheme 36. Ag^I-promoted synthesis of 3-trifluoromethyl pyrazoles.

More importantly, the obtained products through this method are versatile important intermediate chemicals in the pharmaceutical industry. For example, compound **114** is a useful intermediate for the synthesis of measles virus (MV) inhibitor AS-136A (**101**), which displayed nanomolar inhibition against the MV replication in the context of virus infection (Scheme 37) [45,46].

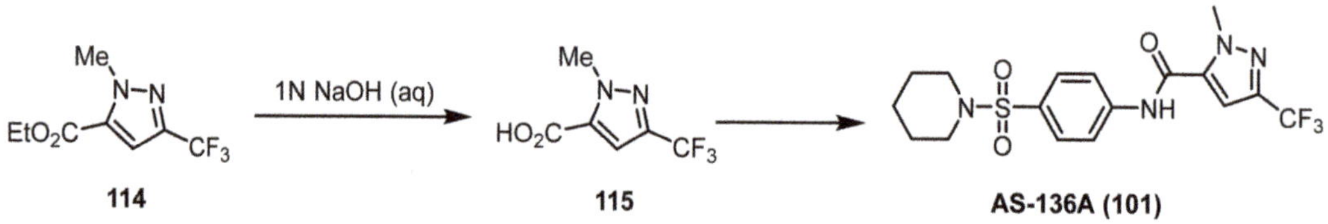

Scheme 37. The synthesis of AS-136A from compound **114**.

Subsequently, in 2014, Ji and co-workers [47] reported on an electrophilic trifluoromethylation to synthesize 3-trifluoromethylpyrazoles. This transition-metal free protocol provided a range of 1,3,5-trisubstituted pyrazoles with moderate yield; preliminary mechanism studies indicated that a trifluoromethyl radical species was involved in this reaction (Scheme 38). For example, celecoxib (**99**), a nonsteroidal anti-inflammatory drug, was made using this method with high efficiency and simplicity (Scheme 39) [47].

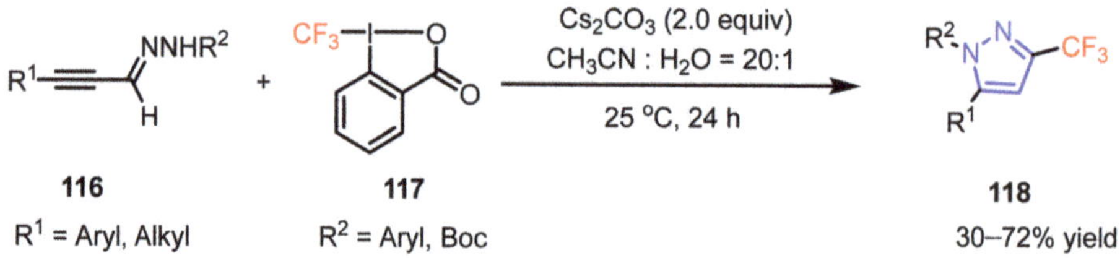

Scheme 38. Base-promoted synthesis of 3-trifluoromethyl pyrazoles.

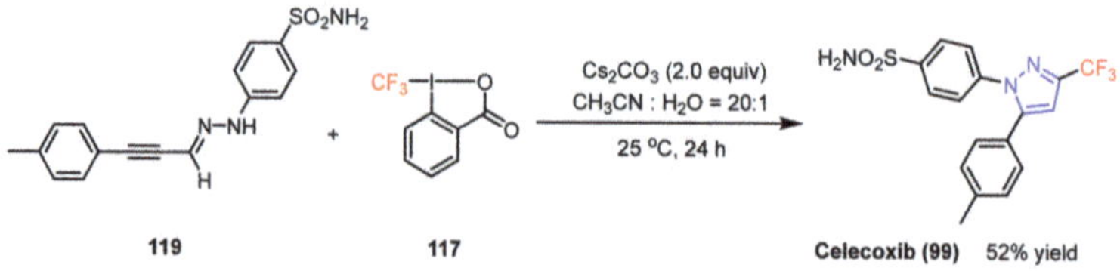

Scheme 39. Celecoxib synthesis using a hypervalent iodine reagent.

Zhu et al. reported a multicomponent method of synthesizing 3-trifluoromethyl pyrazole derivatives through the interactions of readily available 2-bromo-3,3,3-trifluoropropene (BTP), aldehydes, and sulfonyl hydrazides [48]. The protocol features are mild conditions, a broad scope of substrates, high yields, and valuable functional group tolerance. More importantly, this method is highly practical for large-scale production (Scheme 40).

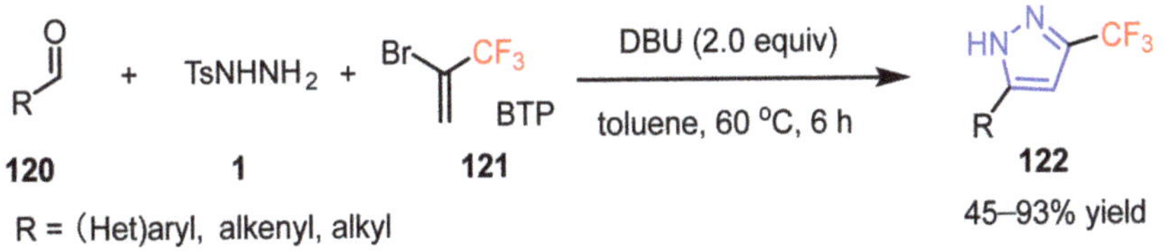

Scheme 40. 2-bromo-3,3,3-trifluoropropene (BTP) as a building block for the synthesis of 3-trifluoromethylpyrazoles.

Wang and co-workers [49] reported a mild method of synthesizing 4-(trifluoromethyl) pyrazoles in high yields; this copper-mediated method, using nucleophilic trifluoromethyltrimethylsilane (TMSCF$_3$) as the CF$_3$ source, is an important supplement to electrophilic trifluoromethylation reactions (Scheme 41).

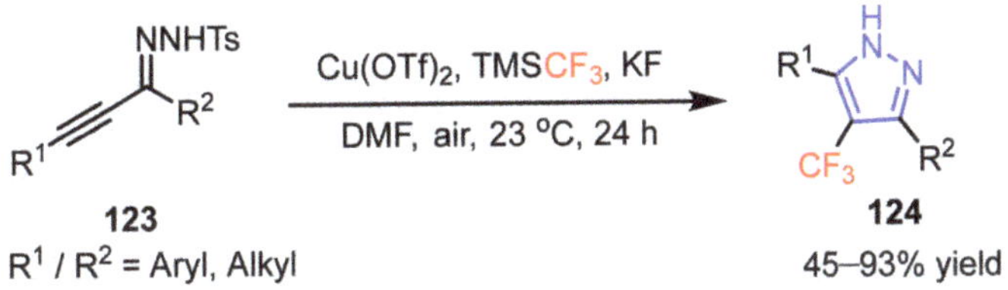

Scheme 41. Cu^{II}-mediated 4-trifluoromethylpyrazoles.

The above-mentioned methodologies offer great convenience in constructing trifluoromethylpyrazoles with various trifluoromethyl-containing building blocks and reagents. However, some of these trifluoromethyl-containing chemicals are expensive or potentially explosive [47,49]. Thus more straightforward methodologies like the direct fluorination of existing functional group into a trifluoromethyl have also been explored. For example, in the presence of HF, sulfur tetrafluoride can directly deoxofluoridize 4-pyrazolecarboxylic acids into corresponding 4-(trifluoromethyl)pyrazoles in high yields [50] (Scheme 42).

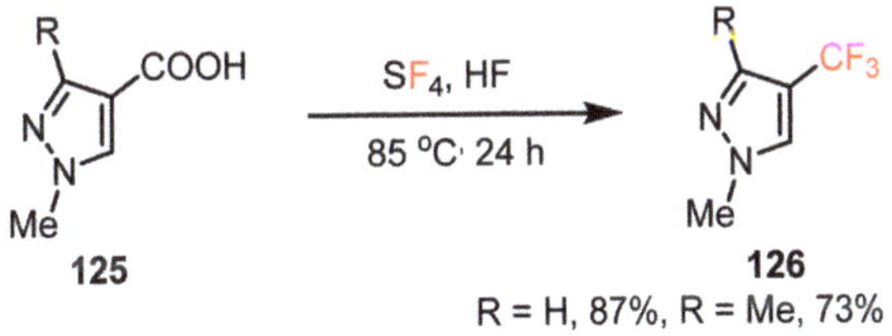

Scheme 42. Direct fluorination to synthesize 4-trifluoromethylpyrazoles.

3.2. N-Alkylation, N-Arylation, and N-Alkenylation of Pyrazoles

The direct N-alkylation of pyrazoles to yield α-pyrazole ketone derivatives is of great importance because of their pharmacological potential in diverse bioactive molecules. Examples including retinoic acid receptor-related orphan receptor gamma (RORγ) modulator 1 (**127**), Janus kinase (JAK) inhibitors (**128**) and (**129**), and insecticides (**130**) are shown in Figure 6 [51]. High regioselectivity in such transformations is readily achieved under optimized reaction conditions. In addition, any minor impurity of regioisomer can be removed easily due to the differences in its physical properties (e.g., polarity, solubility difference) to the main product during purification.

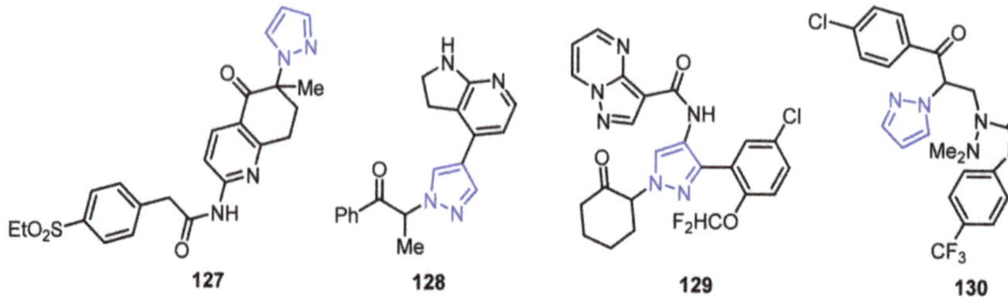

Figure 6. Applications of α-pyrazole ketone scaffolds in bioactive molecules.

3.2.1. The Synthesis of N-Alkylated Pyrazoles

Dhanju et al. reported a general method of using ceric ammonium nitrate to mediate oxidative coupling of enolsilanes and heterocycles for making diverse α-pyrazole ketones [51]. Interestingly, they found that under the same reaction conditions, quaternary α-pyrazole ketone derivatives could be prepared; such pyrazole derivatives are not generally accessible via conventional synthetic methods due to steric hindrance (Scheme 43). Nevertheless, for some asymmetrical pyrazole substrates, this method generated a mixture of regioisomers. For example, **133a** and **133b** were synthesized in a ratio of 1.25:1. This is attributed to the conjugation effect of the phenyl group at the C_2 substitution of the pyrazole ring, the hyperconjugation inside the bicyclic ring system consisting of the pyraozle group, and the phenyl group accelerates the fast shifting of the C–N double bond between the two nitrogen atoms. The two flexible tautomeric structures tend to afford the product as a mixture of regioisomers in almost equal ratio. Conversely, other substitutions (i.e., H, tBu) having no conjugation effect only provided a single regioisomer in high yield [51].

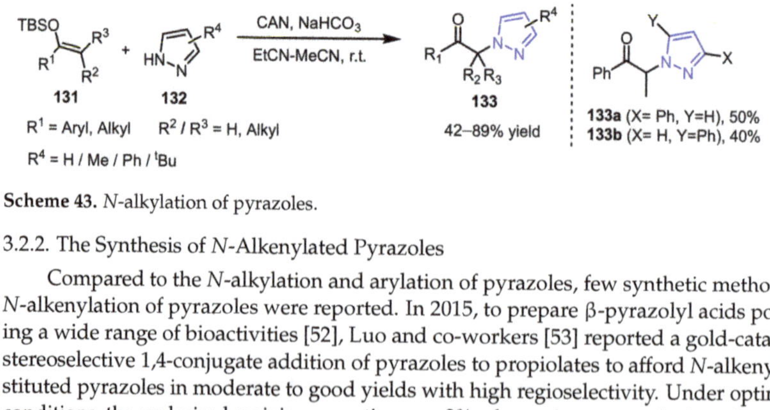

Scheme 43. N-alkylation of pyrazoles.

3.2.2. The Synthesis of N-Alkenylated Pyrazoles

Compared to the N-alkylation and arylation of pyrazoles, few synthetic methods for N-alkenylation of pyrazoles were reported. In 2015, to prepare β-pyrazolyl acids possessing a wide range of bioactivities [52], Luo and co-workers [53] reported a gold-catalyzed stereoselective 1,4-conjugate addition of pyrazoles to propiolates to afford N-alkenyl substituted pyrazoles in moderate to good yields with high regioselectivity. Under optimized conditions, the undesired regioisomer ratio was <3% when using an unsubstituted pyrazole as the substrate. After hydrogenation of the double bond, β-pyrazolyl acid esters, the precursors of β-amino acids, could be obtained effortlessly (Scheme 44).

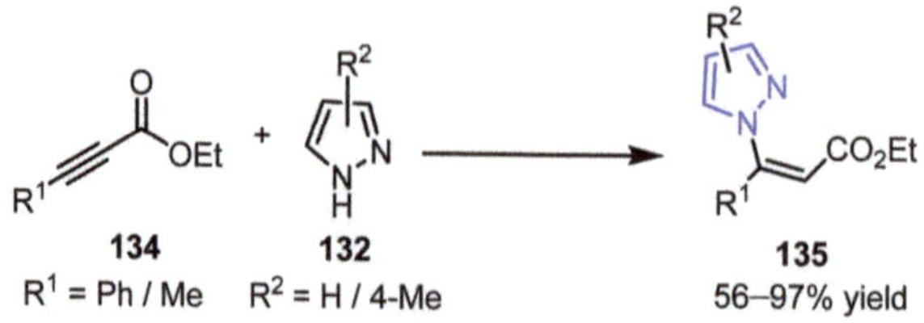

Scheme 44. N-alkenylation of pyrazoles.

3.2.3. The Synthesis of N-Arylated Pyrazoles

N-arylpyrazoles motifs are common in natural products, bioactive molecules, and pharmaceuticals [54]; furthermore, they are also prevalent components of auxiliary ligands for transition-metal catalysis [55]. Conventional methods of synthesizing N-arylpyrazoles rely on using either Buchwald–Hartwig coupling [56] or classical Ullmann reactions [57], requiring noble palladium catalysts or harsh reaction conditions. Onodera et al. reported an efficient method using palladium-catalyzed coupling of aryl halides with 3-trimethylsilylpyrazole to make N-arylpyrazoles in quantitative yields [58]. Interestingly, 3-trimethylsilyl substitution on pyrazole substrates is essential for high yields; in addition, 3-trimethylsilyl substitution also offers considerable synthetic potential and diversity for subsequent structural modifications (Scheme 45).

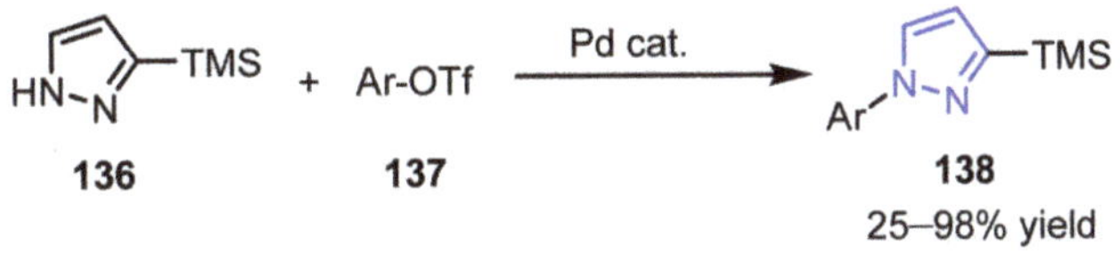

Scheme 45. Pd-catalyzed synthesis of N-arylated pyrazoles.

A general Cu-catalyzed Ullmann-type coupling to synthesize N-arylazoles was reported by Zhou and co-workers [59]; they discovered that using L-(-)-quebrachitol (QCT), a recycled waste in natural rubber industry as the O, O-bidentate ligand, both aryl bromides and aryl chlorides can afford N-arylpyrazoles. In addition, aryl triflates that are readily made from phenolics were also well tolerated to prepare desired products in excellent yields (Scheme 46). Noticeably, using asymmetric imidazole having a tautomerism similar to pyrazole as the substrate, **142a** could be made in a high regioselective ratio of 12:1. This Cu powder involved catalytic system presented a superior regioselectivity and high yield compared to the Cu_2O catalytic system [60]. Copper powder is more reactive in the Ullmann-type reaction than Cu_2O as a catalyst. In addition, the ligand (QCT) can coordinate the in situ formed copper ions to stabilize the reaction intermediates, a more stable tautomeric structure. The collective roles have a net effect of a higher regioselectivity [59].

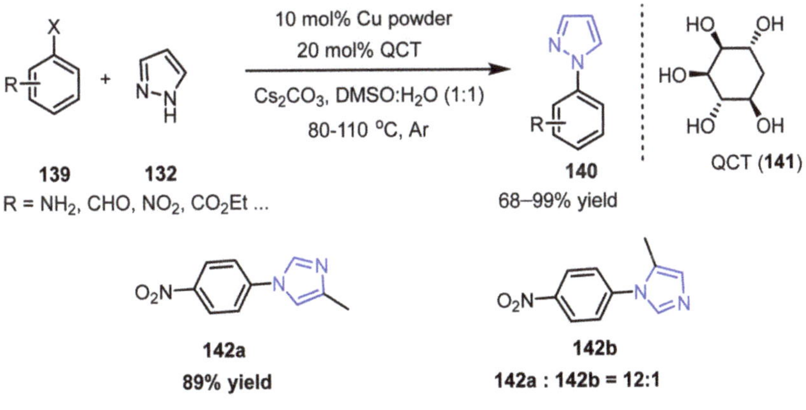

Scheme 46. Cu-catalyzed synthesis of N-arylated pyrazoles.

Wang et al. reported a metal-free oxidative methodology for the coupling of benzoxazoles and pyrazoles to form a bis-heterocyclic system of N_1-benzoxazole substituted pyrazole derivatives [61]; the formation of radical species in the presence of oxidants is the common feature of such protocols (Scheme 47).

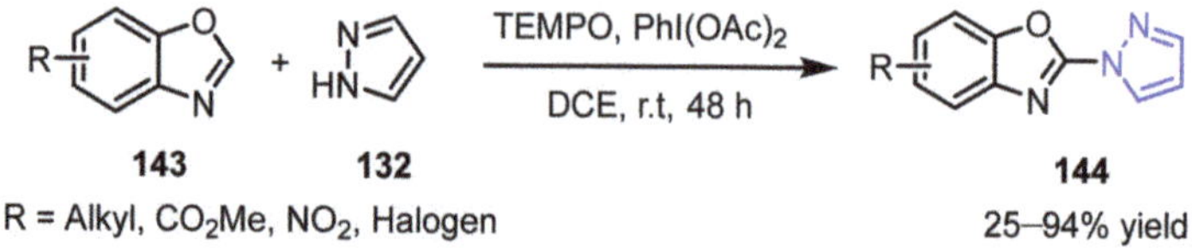

Scheme 47. Oxidative coupling for the synthesis of N-arylated pyrazoles.

3.3. The Amination of Pyrazoles

Aminopyrazoles have broad applications in both pharmaceuticals and agrichemicals, examples such as CDPPB (**145**, positive allosteric modulators), fipronil (**146**, insecticide), sulfaphenazole (**147**, sulfonamide antibacterial), and a Janus kinase inhibitor (**148**) all bear an amino group at the 5-position (Figure 7). In addition, the introduction of an amino group onto pyrazoles is also synthetically beneficial for further transformation [62].

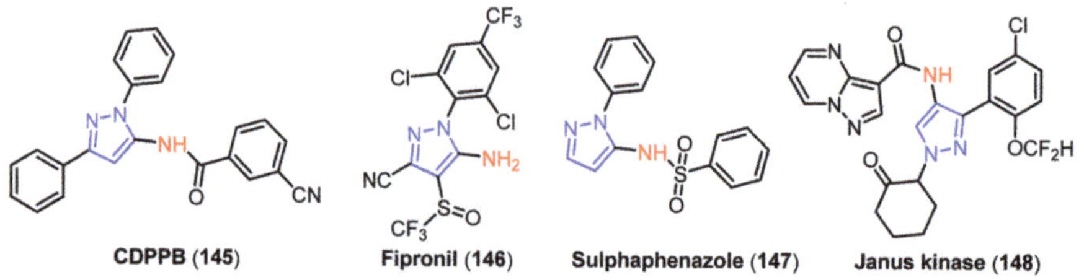

Figure 7. Bioactive aminopyrazoles.

Senadi et al. reported an oxidative cross coupling of N-sulfonyl hydrazones with isocyanides through a formal [4 + 1] annulation process; this metal-free procedure generated 5-aminopyrazoles in good yields with broad substrate scopes [63] (Scheme 48).

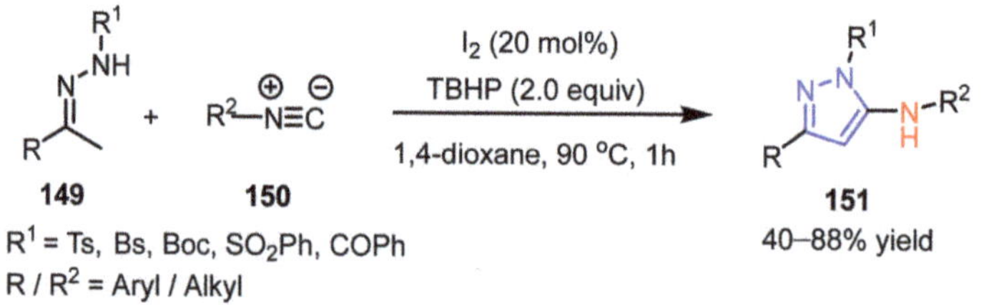

Scheme 48. I_2 catalyzed synthesis of 5-aminopyrazoles.

Kallman et al. reported an efficient method to ammonolyze isoxazoles into aminopyrazoles either in DMSO using one-step, or in an ethanol solution of KOH using two consecutive steps [64]. Both procedures performed well to afford 3-aminopyrazoles selectively. In situ NMR analysis indicated that this reaction underwent a ring-opening process of converting isoxazole into a ketonitrile intermediate, followed by the formation of (Z)-hydrazinylacrylonitrile, then a ring-closure step occurred to afford the target aminopyrazoles (Scheme 49). Unlike other methods providing high regioselectivity, the hydrolyzation of isoxazoles into aminopyrazoles involves the conversion of continuous intermediates. For example, the unstable intermediate hemiaminal can convert into either (Z)-hydrazone or (E)-hydrazone randomly, and this cis-trans isomerism was verified by in situ NMR analysis. The lack of a preponderant Z or E configuration leads to less regioselectivity [64].

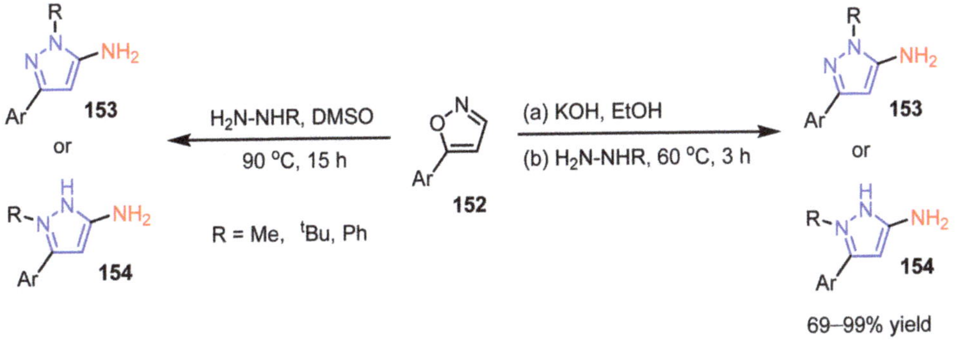

Scheme 49. Isoxazoles as substrates for 3-aminopyrazoles synthesis.

3.4. The Selanylation of Pyrazoles.

Emerging data from pharmacological studies and clinical trials reveal that selenoproteins and organoselenium molecules are participating in a variety of physiological activities [65,66]. For example, a selenium-containing celecoxib derivative was shown to be effective in downregulating the transcription of COX-2 and other pro-inflammatory genes [67]. This highlighted the significance of incorporating selenium into pyrazole motifs. Belladona et al. reported a direct C-H selanylation of pyrazoles with Selectfluor (N-chloromethyl-N-fluorotriethylenediammonium bis(tetrafluoroborate)), a stable and commercially available oxidant, to afford 4-phenylselenol substituted pyrazole derivatives [68]. The reaction was triggered by the interaction of diphenyl selenide with Selectfluor to form electrophilic selanyl species, which were then attacked by nucleophilic pyrazole substrates to give the final product. Mechanism studies suggest that the formation of a Se–F bond between Selectfluor and diselenide is the critical step to restrain the formation of non-fluorinated byproducts (Scheme 50).

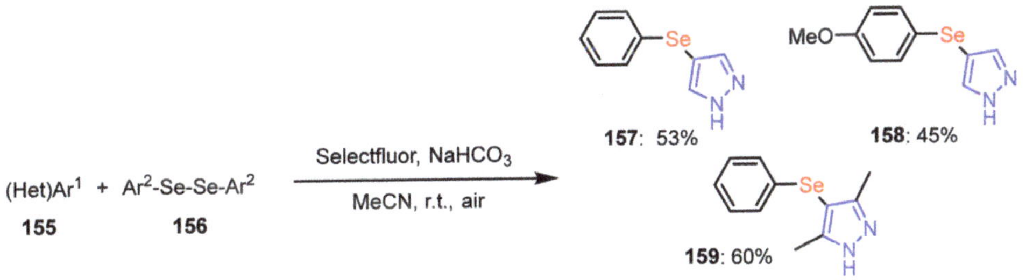

Scheme 50. The synthesis of selanylated pyrazoles.

3.5. The Borylation of Pyrazoles

Borylated pyrazoles are critical building blocks in scalable Suzuki-coupling reactions for preparing pyrazole-containing pharmaceuticals and fine chemicals. Nevertheless, such starting chemicals are commercially scarce due to the unavailability of relevant synthetic methodologies. Recently, Wang and co-workers [69] described that dianthphos, an easily prepared monophosphine ligand, is capable of facilitating the Miyaura borylation of pyrazole substrates with high yields, but only two representatives were presented in their publication (Scheme 51).

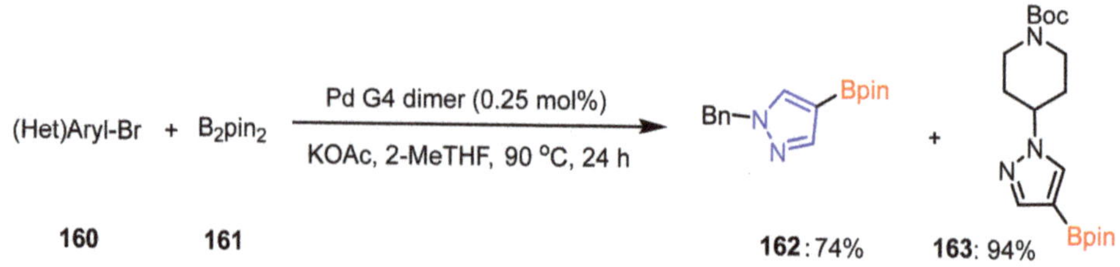

Scheme 51. The synthesis of borylated pyrazoles.

Above, we discussed the synthesis and functionalization of pyrazoles for different purposes. In the next section, we will focus on the implementation of pyrazoles as therapeutic drug candidates, particularly for treating neurological diseases such as Alzheimer's disease (AD) and Parkinson's disease (PD).

4. Applications of Pyrazoles in Neurodegenerative Diseases

Neurodegenerative diseases are characterized by the progressive loss of structure, function, and quantity of neurons, which result in clinical syndromes including motor, cognitive, and behavioral decline. Currently, Alzheimer's disease (AD), and Parkinson's disease (PD) are reported as the top two neurodegenerative diseases, and both are associated with abnormal protein aggregations within the brain. Although academic institutions and pharmaceutical industries have dedicated tremendous efforts to developing drug candidates for treating these two diseases, unfortunately, an effective therapeutic has not been identified yet. Patients and their families continue to bear a huge economic burden as well as enduring physical and emotional pressure. Meanwhile, diverse small-molecule drug candidates have been explored as potential therapeutics for AD and PD. In the development of potential drug candidates for the treatment of PD and AD, pyrazole scaffolds have two intrinsic advantages. First, the pyrazole heterocycle is a bidentate ligand: it can act as both a hydrogen-bond donor and a hydrogen-bond acceptor at the same time, thus allowing it to interact with carbonyl oxygen and amide proton on the peptide backbone complementarily [70]. Second, π–π-stacking interaction between pyrazole and phenylalanine in β-sheet peptides amplifies their high affinity toward peptide backbones [71]. Here, we will present an overview of recently reported potent pyrazole-containing biomolecules that have been evaluated for the treatment of AD and PD. As readers are encouraged to read the recent review by Gomes, Silva, and Silva [6], which summarizes the structures of pyrazole-containing diagnostic radiotracers for neurodegenerative disease, these are not discussed herein.

4.1. Applications of Pyrazoles in Alzheimer's Disease (AD) Treatment

Although a precise understanding of the pathogenesis of AD remains elusive, several iconic hallmarks have been identified. For example, MRI studies on AD patients discovered significant hippocampal and entorhinal cortex atrophy compared to healthy age-matched control subjects [72]. Furthermore, histopathological examination of AD cases has shown that advanced disease is characterized by the deposition of senile plaques (SPs) and neurofibrillary tangles (NFTs), which are composed of aggregates of amyloid-β (Aβ) and hyperphosphorylated tau (ptau), respectively [73]. Based on anatomical studies, several hypotheses including (i) amyloid cascade hypothesis, (ii) tau hypothesis, (iii) metal ion dyshomeostasis, (iv) oxidative stress hypothesis, and (v) cholinergic hypothesis have been proposed in an attempt to guide drug discovery for AD treatment [74].

With respect to their corresponding protein targets, the leading pyrazole-containing therapeutic compounds can be divided into five classes: (1) acetylcholinesterase (AChE) inhibitors; (2) protein aggregation inhibitors; (3) PDE inhibitors; (4) dual leucine zipper kinase inhibitors; and (5) monoamine oxidase B inhibitors.

4.1.1. Acetylcholinesterase (AChE) Inhibitors

Acetylcholine (ACh) is a critical neurotransmitter and neuromodulator in the central nervous system (CNS), its relatively low concentration in the cerebral cortex of AD brains results in the degeneration of cholinergic neurons as well as the deficit of cholinergic neurotransmission [75]. In neurons, ACh is synthesized from choline and acetyl-CoA when catalyzed by choline acetyltransferase (ChAT) and is hydrolyzed into acetate and choline by acetylcholinesterase (AChE) [76]. AChE is primarily found at the synaptic cleft and the neuromuscular junction of the brain. In contrast, outside of the CNS, the cholinergic enzyme subtype butyrylcholinesterase (BuChE) mainly exists in blood plasma. Unlike the substrate specific AChE, BuChE can nonspecifically hydrolyze many different choline or non-choline based esters and amides [77].The colocalization of AChE or BuChE with Aβ aggregates in SPs could accelerate Aβ aggregation, thus increasing neurotoxicity [78–80]. Therefore, cholinesterase inhibitors have been viewed as a promising target for AD treatment.

In 2018, Turkan et al. reported a series of substituted pyrazol-4-yl-diazene derivatives (**3**), and their in vitro enzymatic assay suggested that all were effective AChE and BuChE inhibitors with K_i values at the nanomole level, better than tacrine (**164**), a discontinued acetylcholinesterase inhibitor that was used to treat AD [7]. However, their poor selectivity between AChE and BuChE, and broad activities over α-glycosidase and cytosolic carbonic anhydrase might be problematic. Shaikh et al. reported novel scaffolds of N-substituted pyrazole-derived α-aminophosphonates, two of these compounds (**165**, **166**) exhibited strong potency against AChE and high selectivity over BuChE, and their performances were better than the commercially available drugs galantamine (**167**) and rivastigmine (**168**) [81]. Compound **169** had ~100-fold selectivity for AChE over BuChE, furthermore, it exhibited 45% neuroprotection ratio in rotenone/oligomycin A-induced neuronal death. When compared to tacrine, only a minor activity decrease was observed [82]. Gutti et al. reported a linear pyrazole derivative (**170**) as an AChE inhibitor in 2019 [13]. When tested in MC65 cells at 50 μM, compound **170** improved cell viability by 90% and could reduce $Aβ_{1-42}$ aggregation induced by metals effectively at 20 μM (Figure 8, Table 1).

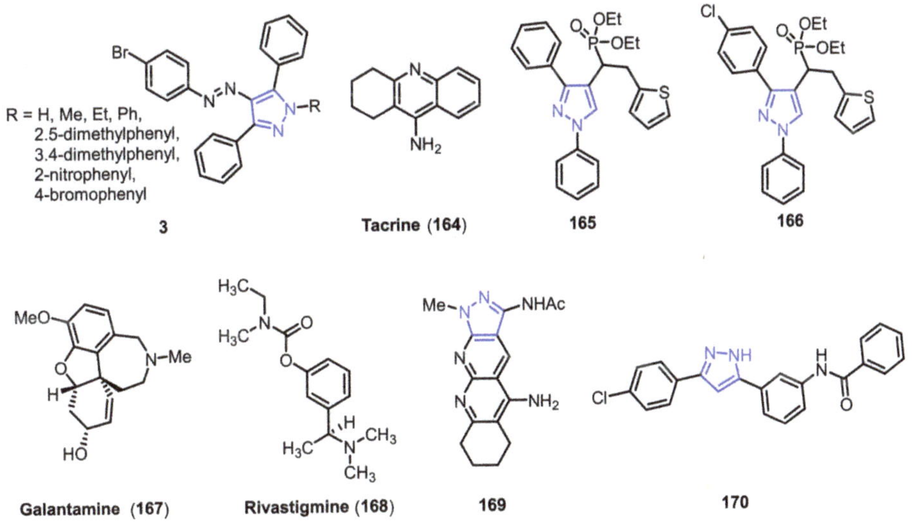

Figure 8. Chemical structures of cholinesterase inhibitors.

Table 1. Affinities of cholinesterase inhibitors.

Compound	R	K_i [1] or IC$_{50}$ [2] Values for Cholinesterase Inhibitors		
		AChE	BuChE	AChE/BuChE
3a [1]	H	58.17 ± 4.84 nM	74.82 ± 18.62 nM	0.77
3b [1]	Me	64.71 ± 13.04 nM	80.36 ± 22.74 nM	0.80
3c [1]	Et	47.93 ± 4.92 nM	58.12 ± 9.27 nM	0.82
3d [1]	Ph	44.66 ± 10.06 nM	50.36 ± 13.88 nM	0.88
3e [1]	2, 5-dimethylphenyl	78.34 ± 17.83 nM	77.62 ± 18.32 nM	1.00
3f [1]	3, 4-dimethylphenyl	48.16 ± 9.63 nM	65.27 ± 12.73 nM	0.73
3g [1]	2-nitrophenyl	56.23 ± 11.74 nM	71.63 ± 8.93 nM	0.78
3h [1]	4-bromophenyl	60.27 ± 15.67 nM	88.36 ± 20.03 nM	0.68
164 [1]	–	126.13 ± 9.37 nM	145.84 ± 13.44 nM	0.86
165 [2]	–	0.055 ± 0.143 µM	8.863 ± 0.22 µM	0.006
166 [2]	–	0.017 ± 0.02 µM	6.331 ± 0.017 µM	0.003
167 [2]	–	3.148 ± 0.139 µM	1.22 ± 0.05 µM	2.58
168 [2]	–	2.632 ± 0.021 µM	6.901 ± 0.01 µM	0.38
169 [2]	–	0.069 ± 0.06 µM	6.3 ± 0.6 µM	0.01
170 [2]	–	1937 ± 66 nM	1166 ± 88 nM	1.60

[1] The values shown are K_i, [2] the values shown are IC$_{50}$

4.1.2. Protein Aggregation Inhibitors

The dysfunction and aggregation of Aβ and tau proteins are the main features of AD pathology. Aβ is a proteolytic product of amyloid precursor protein (APP), which is cleaved by α-secretase and γ-secretase to Aβ monomers with 38 to 43 amino acid residues, and studies have identified Aβ$_{42}$ as a key monomer in pathology.[83,84]. These short Aβ monomers possess a propensity to aggregate through mutual recognition to form Aβ oligomers, which further aggregate into protofibrils and fibrils with a β-sheet conformation [85]. Aβ has also shown a seeding effect in the brain to transmit and expand its influence [86].

Tau is a phosphoprotein, its hyperphosphorylation can destabilize microtubules, thus causing invariably compromised axonal transportation, synaptic dysfunction, and signal transmission failure [87]. Furthermore, hyperphosphorylated tau tends to congregate into paired helical filaments (PHFs) and straight filaments (SFs) in steps [88]. In experimental mouse models of tauopathy, the accumulation of aggregated tau can disrupt anterograde axonal transport and impair long-term potentiation [89,90].

In 2011, Hochdörffer et al. developed a series of Aβ$_{42}$ inhibitors based on a trimeric aminopyrazole carboxylic acid framework. They could convert well-ordered fibrils into less structured aggregates and thin bent filaments through backbone recognition and hydrophobic interactions with Aβ$_{42}$ (Figure 9) [91]. Compound (Trimer-TEG-Lys-OMe, **171**) displayed the most potent inhibition and disaggregation activities against fibril formation, with ~80% of reduced thioflavin fluorescence absorption and 43% of increased PC-12 cell viability in an Aβ existing cytotoxic environment. Notably, the triethyleneglycol (TEG) spacer bridging trimeric aminopyrazole core and lysine appendix are essential to maintain the inhibition activity, as the spacer plays an active role in destabilizing the U-shaped turn of the Aβ protofilament. When the spacer was omitted (**172**), an accelerated aggregation was contrarily observed. Meanwhile, the attached extension (TEG-Lys-OMe, **171**) allows diversity; when it was replaced by a pentapeptide "LPFFD", the corresponding new compound showed a substoichiometric IC$_{50}$ value (3 µM). Considering that Aβ concentration is at nanomolar range in the brain, much lower than the concentration (10 µM) used in in vitro studies, a 3 µM IC$_{50}$ value is remarkable.

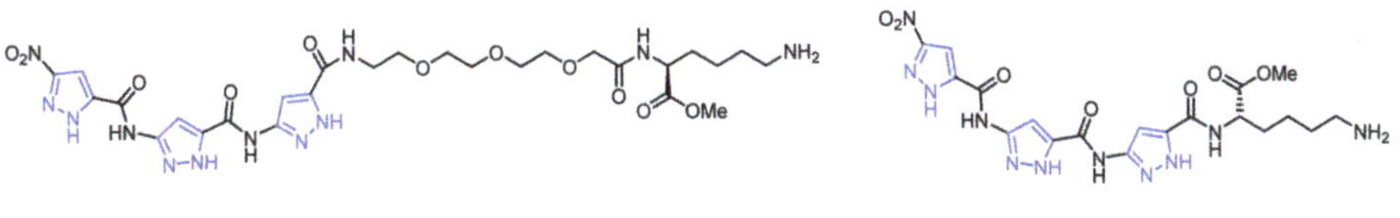

Trimer-TEG-Lys-OMe (171)   Trimer-Lys-OMe (172)

Figure 9. Chemical structures of pyrazole-containing Aβ$_{42}$ inhibitors.

Anle138b (**173**) was first identified as a potent α-synuclein aggregation inhibitor when it was administered in vivo at nanomolar concentrations [92]. Later in vitro and in vivo studies carried in mouse models of AD suggested that it could also inhibit the formation of pathological tau aggregates by specifically binding to pre-aggregated tau species [93]. Molecular dynamic (MD) simulations revealed that anle138b (**173**) could block the conversion of ordered antiparallel β-strands into disordered β-sheet rich conformations by preferentially interacting with pre-aggregated tau fragments to reduce the overall number of intermolecular hydrogen bonds [94]. Structure-activity analysis showed that the pyrazole scaffold plays a crucial role in anle138b's inhibitory effect; when the pyrazole moiety was replaced by imidazole (sery345, **174**) or isoxazole (sery338, **175**), the anti-aggregation effect was weakened due to altered hydrogen bond characteristics. Interestingly, pyrazole anle234b (**176**), a structural isomer of anle138b, is biologically inactive. Anle234b (**176**) is more sterically hindered than anle138b due to placement of the bulky bromine group at the ortho- rather than meta-position, which creates a more stable/less flexible conformation. The torsional inactivation of anle234b (**176**) hampers the interactions of the bromophenyl and pyrazole ring with peptide backbones, resulting in a reduced hydrogen bond strength (Figure 10) [92,94].

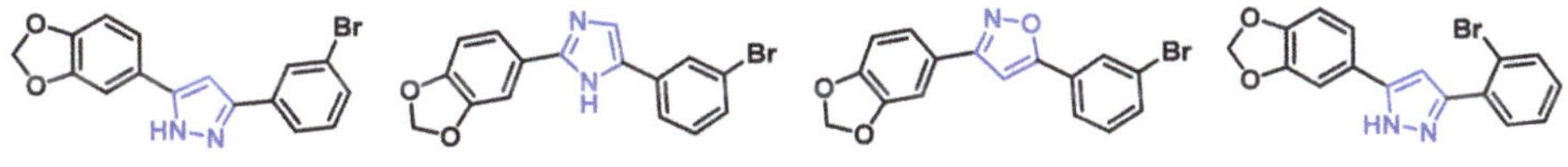

Anle138b (**173**)   Sery345 (**174**)   Sery338 (**175**)   Anle234b (**176**)

Figure 10. Chemical structures of pyrazole-containing inhibitors targeting tauopathy.

Curcumin (**177**), a natural phenolic compound found in the rhizome of Curcuma longa, has been used to treat various ailments in India for centuries. Structural modifications of curcumin to improve its instability and bioavailability have been an attractive strategy in new drug discovery [95,96]; several pyrazole-containing curcumin derivatives have showed potency for AD treatment [97]. CNB-001 (**178**), a pyrazole derivative of curcumin, was initially found to have broad neuroprotective activity [98,99]. Subsequent studies in AD animal models revealed that it could also promote Aβ clearance and improve memory [100]. Kotani et al. further found that CNB-001 (**178**) could stimulate the expression of the endoplasmic reticulum chaperone glucose-regulated protein 78 (GRP78) and increase formation of the amyloid precursor protein APP/GRP78 complex. The subsequent downregulation of intracellular APP trafficking results in a reduced Aβ production without inhibiting β- or γ-secretase activity [101]. Okuda et al. developed the inhibitor PE859 (**179**), this curcumin derivative has IC$_{50}$ values of 1.2 μM and 0.66 μM, respectively, to inhibit Aβ and tau aggregations, and improved in vivo pharmacokinetics and pharmacological efficacy, making it is more competitive as an AD drug candidate than curcumin [102]. Furthermore, studies with PE859 (**179**) in the senescence-accelerated mouse prone 8 (SAMP8) model of accelerated aging showed it also protected cells from Aβ-induced cytotoxicity damage by decreasing Aβ and tau aggregations in the mouse brains (Figure 11) [103].

Figure 11. Chemical structures of pyrazole-containing curcumin derivatives.

4.1.3. Phosphodiesterase (PDE) Inhibitors

Age-related cognitive decline and synaptic dysfunction are closely associated with decreased cAMP (cyclic guanosine monophosphate) and/or cGMP (cyclic guanosine monophosphate) concentration in the brains caused by increased phosphodiesterase (PDE) expression [104,105]. Therefore, PDE inhibitors that can specifically bind with high affinity to specific PDE isomers like PDE1, PDE5, or PDE9 to reduce the hydrolyzing of cAMP and/or cGMP have been evaluated for pharmaceutical potential in clinical trials for the treatment of AD [106].

The PDE1 group has three isoforms: PDE1A, 1B, and 1C, their activities are regulated by intracellular calcium and calmodulin (CaM) [104]. In human and rodent brains, PDE1A is predominantly expressed in the hippocampus, cortex, striatum, thalamus, and cerebellum; PDE1B is prevalent in the hippocampus, cortex, striatum; while PDE1C is more ubiquitously found in the cortex, cerebellum, and amygdala [107,108]. ITI214 (**180**), a potent and highly selective PDE1 inhibitor (K_i = 0.058 nM), which is still in clinical trials, was developed by Intra-Cellular Therapies. This inhibitor was able to significantly enhance memory performance in vivo with a minimum effective dose of 3 mg/kg in various rat models (Figure 12) [109].

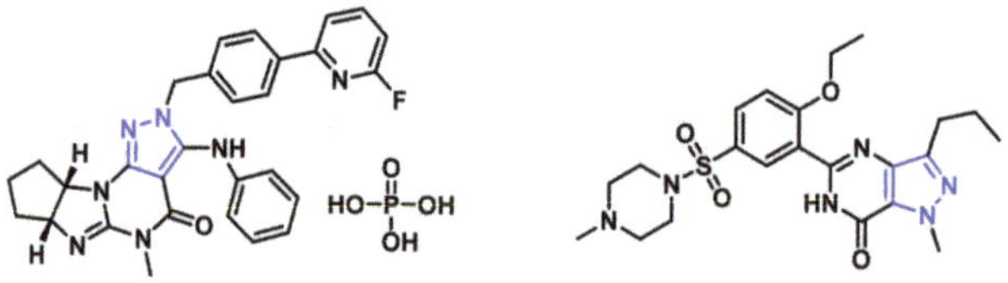

Figure 12. Chemical structures of pyrazole-containing PDE1 and PDE5 inhibitors.

The prominent expression of PDE5 in smooth muscles has led to the extraordinary success of PDE5 inhibitors including sildenafil (**181**), vardenafil, and tadalafil for the treatment of erectile dysfunction [105]. Furthermore, PDE5 is also expressed in human hippocampus and frontal cortex, and particularly abundant in Purkinje neurons [110,111]. Studies in different AD mouse models have indicated that sildenafil (**181**) can improve synaptic function, restore memory loss, and upregulate CREB phosphorylation signaling to reduce Aβ levels over the long-term [112,113]. When administrated to anile Tg2576 transgenic mice (15 mg/kg, intraperitoneally), sildenafil (**181**) could completely reverse cognitive impairment and reduce tau hyperphosphorylation in the hippocampus [107]. More importantly, clinical trials of sildenafil (**181**) in healthy humans showed no central nervous system side effects. Although no overt effect on spatial auditory attention or visual word recognition was observed, sildenafil does help information processing and change specific components of event-related potentials with enhanced attention, and also reduced negativity of electroencephalogram in a memory task [114].

PDE9A is primarily expressed in the brain, with high concentrations in the cerebellum, neocortex, striatum, and hippocampus [115]. Consistent with the specific function of cGMP in controlling neurotransmission and enhancing hippocampal synaptic plasticity, the inhibition of PDE9A results in significant accumulation of cGMP in cerebrospinal fluid of nonhuman primates and humans, thus damaging brain functions like sensory processing, learning, and memory [116]. Compound PF-04447943 (**182**) (K_i = 8.3 nM), a PDE9A inhibitor developed by Pfizer, was found to increase cGMP concentration in cerebrospinal fluid by ~23-fold when dosed in rats, monkeys, and humans [117]. Unfortunately, no cognitive improvement was observed in subsequent phase II clinical trials when administered twice a day for 12 weeks to moderate stage AD patients. Instead, it resulted in a higher incidence of serious adverse events compared to the placebo group [118]. BI 409306 (**183**) is another potent PDE9A inhibitor, and has IC_{50} values of 65 nM and 168 nM in human and rat, respectively [119,120]. Preclinical studies demonstrated that BI 409306 (**183**) could increase cGMP concentrations in rodent prefrontal cortex and cerebrospinal fluid, and promoted long-term potentiation, and improved episodic and working memory performance [120]. Although it was well tolerated in prodromal to mild stage AD patients, its phase II clinical trial was discontinued because no visible efficacy in improving cognitive function was observed (Figure 13) [120].

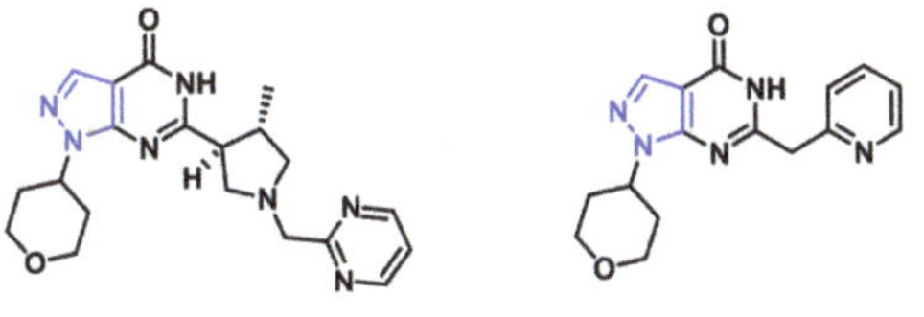

Figure 13. Chemical structures of pyrazole-containing PDE9 inhibitors.

4.1.4. Dual Leucine Zipper Kinase (DLK) Inhibitors

Dual leucine zipper kinase (DLK, MAP3K12) mediates axon degeneration and neuronal apoptosis after activation [121]. DLK-inducible knockout mice displayed increased synaptic transmission and reduced neuronal degeneration when faced with neuronal insult [122]. Ex vivo imaging found that Aβ/plaque-associated synaptic loss is at least partially mediated by DLK signaling [123]. Experiments in mouse models of AD demonstrated that DLK deletion can reverse cognitive deficits and rescue certain phenotypical behaviors. Although the elimination of DLK did not reduce $Aβ_{42}$ production, plaque load, or alter tau pathology, DLK-knockout mice had higher levels of full-length amyloid precursor protein (APP), amyloid plaque, plaque-associated gliosis, and enhanced cell survival in the subiculum of tau [123]. In 2015, Patel et al. reported the first small molecule DLK inhibitor **184** that effectively reduced c-Jun phosphorylation in nerve crush and 1-methyl-4-phenyl-1,2,3,6-tetrahydropyridine (MPTP)-induced neuronal injury [124]. The inhibitor **185** (DLK K_i = 42 nM) was developed using a scaffold-hopping strategy, in which the pyrimidine core was replaced by a pyrazole; it exhibits comparable pharmacological properties and inhibitory activities to compound **184** in a rat model [125]. After further optimization, a more potent compound **186** (DLK K_i = 3 nM) was identified. It had lower plasma clearance (total and unbound), a moderate volume of distribution, a long biological half-life, and good bioavailability when dosed in cynomolgus monkeys. A single dose of **186** (50 mg·kg^{-1}) resulted in significantly decreased phosphorylation of c-Jun, leading to a near-complete inhibition of JNK when tested in a mouse model of AD (Figure 14) [126].

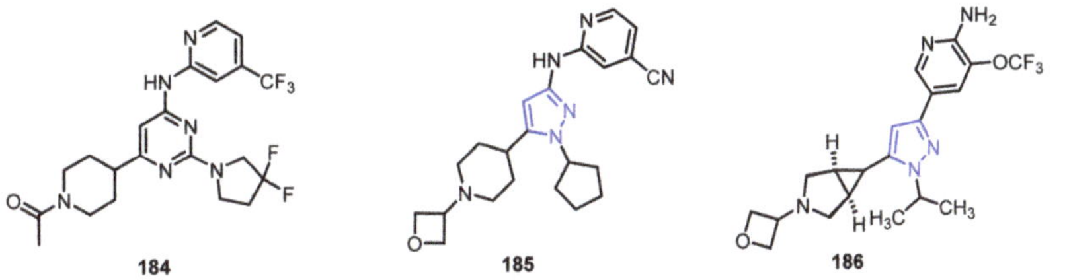

Figure 14. Chemical structures of pyrazole-containing dual leucine zipper kinase (DLK) inhibitors.

4.1.5. Monoamine Oxidases B (MAO-B) Inhibitors

Monoamine oxidases (MAOs) are mitochondrial outer membrane-bound enzymes, existing as two distinct enzymatic isoforms, MAO-A and MAO-B, and are responsible for the metabolism of neurotransmitters such as dopamine, serotonin, adrenaline, and noradrenaline [127]. MAO-B inhibitors selegiline and rasagiline have been used to treat Parkinson's disease clinically. In addition, MAO-B inhibitors are also potential therapies for AD. A post-mortem brain study of AD cases indicated that MAO-B activity increased dramatically in three cortical regions (frontal, parietal, and occipital cortices), thalamus, and white matter [128], the observation was verified by another independent study [129]. MAO-B activity also increases in association with gliosis, resulting in elevated reactive oxygen species (ROS) levels, which in turn enhances Aβ production by decreasing the activity of α-secretase while enhancing the activity of β- and γ-secretases [130,131]. Studies suggest that MAO-B inhibitors like selegiline could slow AD progression in patients suffering from moderately severe impairment [132].

A new strategy for tackling the complexity of AD pathology is incorporating several pharmacological scaffolds into a single molecular entity to produce a synergistic effect [133]. This tactic turned out to be useful when fusing pharmacological scaffolds from a MAO-B inhibitor and an AChE inhibitor together to afford multi-target-directed lead compounds as potential therapeutics against AD [134,135]. Tzvvetkov et al. reported three potent, reversible, and competitive MAO-B inhibitors (**187–189**) with high selectivity against the MAO-A isoform [136]. Enzyme studies suggest that all of them have the ability to bind to Fe(II) and Fe (III) via UV–Vis. These water-soluble, highly blood–brain barrier permeable MAO-B inhibitors are promising drug and radioligand candidates as diagnostic and therapeutic agents for AD (Figure 15, Table 2).

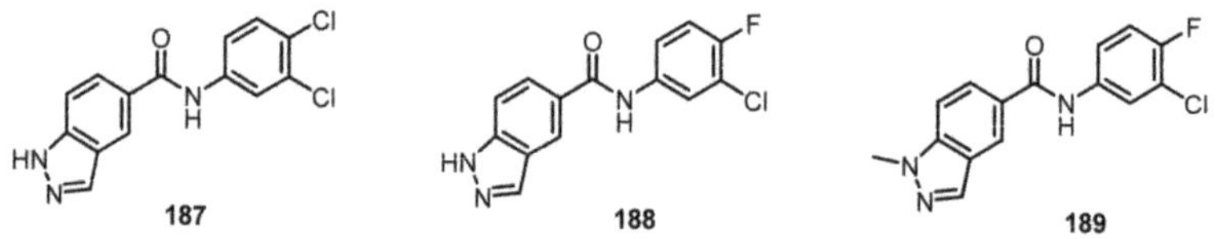

Figure 15. Chemical structures of monoamine oxidases (MAO) inhibitors.

Table 2. Affinities of monoamine oxidases B (MAO-B) inhibitors.

Compound	IC$_{50}$ Value (nM)			K_i (nM)
	hMAO-A	hMAO-B	hMAO-A/hMAO-B	
187	>10,000	0.59 ± 0.09	>16,959	0.26 ± 0.04
188	>10,000	0.68 ± 0.04	>14,706	0.30 ± 0.02
189	≥10,000	0.66 ± 0.06	≥15,151	0.29 ± 0.03

4.2. Applications of Pyrazoles in Parkinson's Disease (PD) Treatment

Parkinson's disease (PD) is the second most prevalent chronic neurodegenerative disease; it affects 1–2% of individuals over 60 years in the U.S. Initially, the most evident motor symptoms of PD are bradykinesia, muscle rigidity, and tremor; as the disease progresses, neurological symptoms including depression, pain, and sleep disturbance become more prevalent. Pyrazole derivatives have wide applications for PD treatment, and their roles can be divided into four classes: (1) antioxidants; (2) protein aggregation inhibitor; (3) adenosine A$_{2A}$ receptor antagonists; and (4) PDE10A inhibitors.

4.2.1. Antioxidants

Mitochondrial dysfunction and subsequent oxidative stress are the main culprits leading to dopaminergic neuronal death in PD. Thus, antioxidants are potentially beneficial therapeutics in PD. Jayaraji and co-workers [137] reported that 2 μM of CNB-001 (**178**) could enhance cell viability insulted by reactive oxygen species (ROS). When applied to adult mice in a model of PD [137], CNB-001 (**178**) could significantly attenuate motor impairment, increase dopamine levels, and reduce inflammation. Two CNB-001 derivatives (**190**, **191**) [138] are also promising therapeutic inhibitors for the treatment of PD caused by α-synuclein amyloidosis, both of them have potent activity in preventing wide-type α-synuclein aggregation and fibrilization, with two-fold higher activity than their isoxazole analogues (Figure 16) [138].

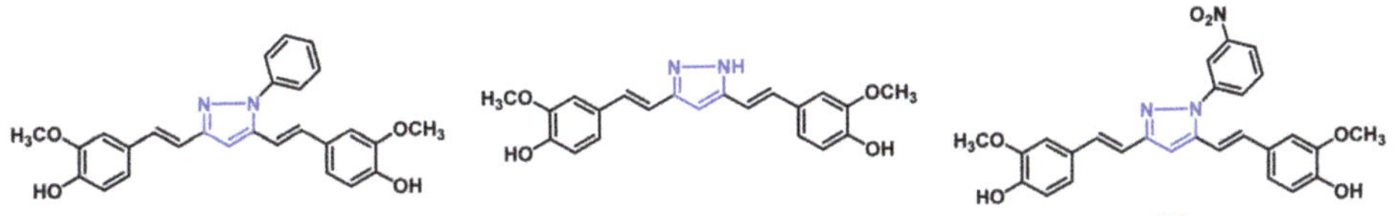

Figure 16. Chemical structures of pyrazole-containing antioxidants.

4.2.2. Protein Aggregation Inhibitors

Anle138b (**173**) was first identified as a pharmaceutical candidate in inhibiting the aggregations of prion protein and of α-synuclein [92]. In different mouse models of PD, anle138b (**173**) strongly inhibited α-synuclein accumulation and neuronal degeneration and had an excellent oral bioavailability, while no detectable toxicity at therapeutic doses was observed [92]. When co-incubated with α-synuclein fibrils, fluorescence spectrum indicated that it binds to the hydrophobic pockets of the fibril with good affinity (K_d = 190 ±120 nM) (Figure 17) [139].

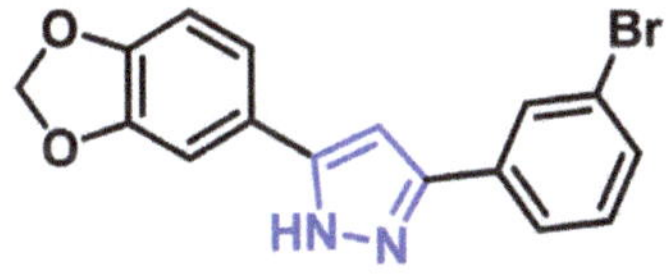

Figure 17. Chemical structure of a protein aggregation inhibitor.

4.2.3. Adenosine Receptor A_{2A} Receptor Antagonists

Adenosine receptor A_{2A} is abundantly expressed and co-localized with dopamine D_2 in the striatum. The blockade of the A_{2A} receptor might antagonize dopaminergic neurotransmission in aspects relevant to motor control in PD patients [140], and the effect of A_{2A} antagonists has been supported by studies on rodent and primate PD models as well as preliminary clinical observations [141,142]. Pyrazole scaffolds have been widely employed in A_{2A} antagonists [143], for example, preladenant is discontinued due to lack of efficacy. Compound **192** [144] exhibited exceptional receptor binding affinity and ligand efficiency against adenosine A_{2A} receptor (K_i = 2.3 nM), however, the 4-acetamide on the pyrimidine ring is unstable and tends to hydrolyze in acidic environments. To overcome this issue, a structurally stable compound **193** (K_i = 0.66 nM) was optimized, unfortunately, it was subject to moderate cytochrome P450 (CYP) inhibition (80% inhibition of CYP3A4 at 10 µM) [145]. Docking experiments based on crystal structure models of the adenosine receptor revealed that the 5-position of the pyrimidine ring was surrounded by multiple residues, which could serve as potential hydrogen bond donors or acceptors in the binding pocket. Analog **194**, which incorporated a cyano group, showed excellent potency and ligand efficiency with low cytochrome P450 inhibition [146]; when dosed orally as low as 3 mg/kg in rats, it could significantly reverse haloperidol-induced catalepsy (Figure 18).

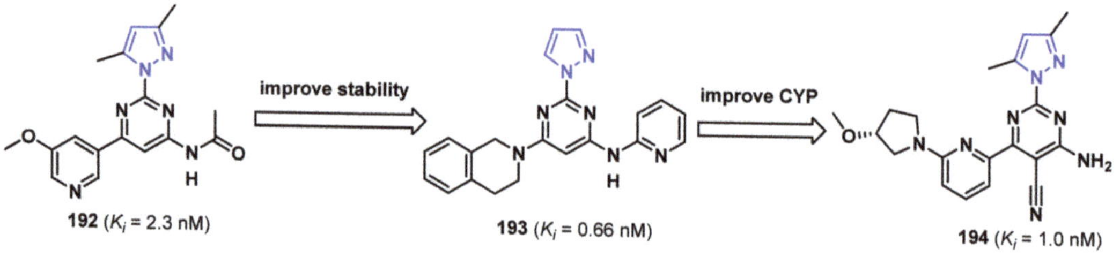

Figure 18. Chemical structures of pyrazole-containing A$_{2A}$ receptor antagonists.

4.2.4. Phosphodiesterase-10A (PDE10A) Enzyme Inhibitors

Phosphodiesterase 10A (PDE10A) is a dual substrate phosphodiesterase enzyme that can hydrolyze both cAMP and cGMP [139,147]. The PDE10A/cAMP interaction is essential for dopamine neurotransmission, and has been implicated in the pathophysiology of Parkinson's disease [148]. Investigators have reported that alteration of PDE10A expression is associated with progression and severity of patients with parkinsonism [149].

Tremendous effort has been dedicated to developing PDE10A inhibitors for the treatment of Parkinson's disease and other neuropsychiatric disorders characterized by regulating medial striatal neuron (MSN) activity. Meanwhile, selectivity of an inhibitor for PDE10A over the other subtypes is also a critical issue. Numerous studies indicate that inhibition of PDE3A/B can lead to arrhythmia and increased mortality [150]; PDE4 plays an important role as a regulator of central nervous system function while inhibition of PDE4A can increase heart and respiratory rates [151].

Among the multiple PDE10A inhibitors, 2-[4-(1-methyl-4-pyridin-4-yl-1*H*-pyrazol-3-yl)-phenoxymethyl]-quinoline (MP-10) is highly potent (IC$_{50}$ = 0.37 nM), selective (>100-fold), and has been extensively evaluated as a therapeutic inhibitor. While MP-10 has progressed to clinical trials without success [152], other PDE10A analogues are under evaluation, though there is no currently FDA-approved PDE10A inhibitor for treating PD.

Tu et al. reported a series of structurally related analogues of MP-10 in which a methoxy group (-OCH$_3$) was added into the 3-, 4-, or 6-position of the 2-methylquinoline moiety of the MP-10 pharmacophore (Figure 19) [153]. The most potent molecule was the 3- and 4-methoxy substituted quinolines (**195**) with an O atom bridge linkage that showed potency similar to MP-10. Interestingly, little difference was observed in the in vitro biological activity of regioisomer **196** (Table 3). More importantly, the six compounds **195a–c** and **196a–c** not only had high potency for PDE10A, but also high selectivity for PDE10A versus PDE3A/3B and 4A/4B (Table 3).

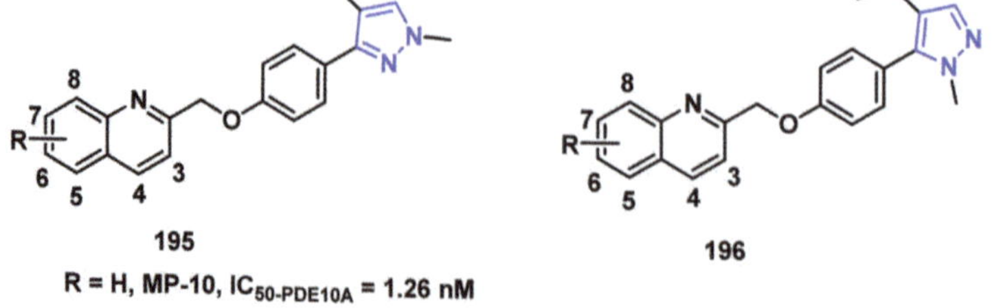

Figure 19. Chemical structures of pyrazole-containing PDE10A inhibitors.

Table 3. PDE affinities of new MP-10 analogues.

Compound	R	PDE10A (nM)	IC$_{50}$ Value PDE3A ($10^3 \times$ nM)	PDE3B ($10^3 \times$ nM)	PDE4A ($10^3 \times$ nM)	PDE4B ($10^3 \times$ nM)
195a	3-OMe	0.40 ± 0.02	123 ± 21	82.7 ± 10	3.85 ± 0.23	3.43 ± 0.21
195b	4-OMe	0.28 ± 0.06	27.5 ± 2.5	3.85 ± 0.95	2.56 ± 0.11	1.79 ± 0.10
195c	6-OMe	1.82 ± 0.25	78.0 ± 4.0	9.75 ± 1.25	3.37 ± 0.18	2.56 ± 0.09
196a	3-OMe	0.24 ± 0.05	18.7 ± 3.1	16.9 ± 2.1	199 ± 16.0	31.5 ± 3.60
196b	4-OMe	0.36 ± 0.03	29.0 ± 4.0	4.80 ± 1.20	4.18 ± 0.33	5.06 ± 0.40
196c	6-OMe	1.78 ± 0.03	1.5 ± 0.5	3.00 ± 0.14	5.60 ± 0.52	7.10 ± 0.58

In 2015, another series of potent and selective PDE10A inhibitors were reported by Tu et al. [154] (Figure 20). Fluorine-containing groups were introduced into the 2-methylquinoline moiety of MP-10. These new potent compounds (**197–202**) (IC$_{50}$ range 0.24–1.80 nM) had >210-fold selectivity.

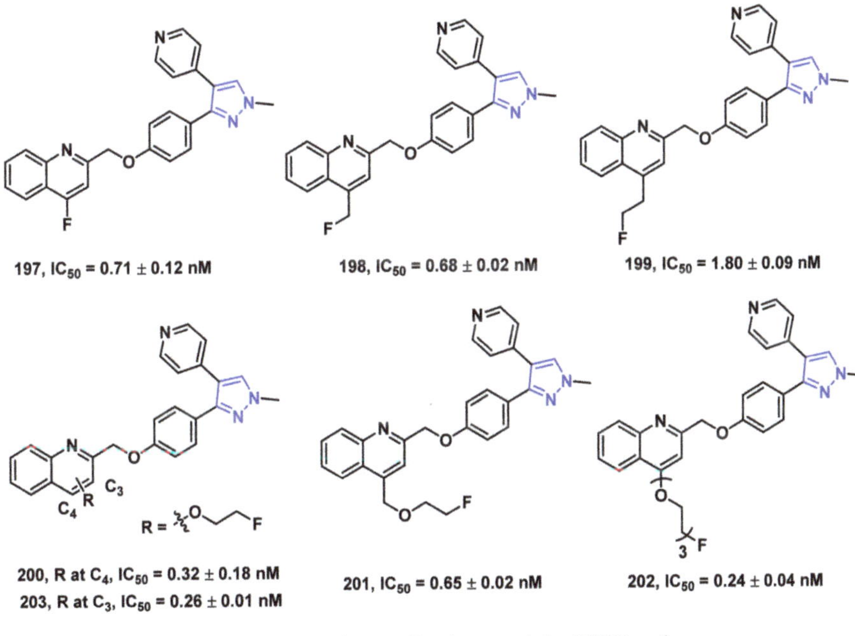

197, IC$_{50}$ = 0.71 ± 0.12 nM

198, IC$_{50}$ = 0.68 ± 0.02 nM

199, IC$_{50}$ = 1.80 ± 0.09 nM

200, R at C$_4$, IC$_{50}$ = 0.32 ± 0.18 nM
203, R at C$_3$, IC$_{50}$ = 0.26 ± 0.01 nM

201, IC$_{50}$ = 0.65 ± 0.02 nM

202, IC$_{50}$ = 0.24 ± 0.04 nM

Figure 20. Structures of potent fluorine-containing MP-10 analogues.

5. Conclusions

The wide applications of pyrazoles in pharmaceuticals have stimulated the rapid methodological development of pyrazole synthesis. During the past decade, many general and practical approaches including the involvement of transition-metal catalysts, photoredox reactions, one-pot multicomponent process, new reactants, and novel reaction type have led to fruitful advances in the fields of the synthesis and functionalization of pyrazole derivatives. This review briefly covers these updates, and highlights the potential of pyrazole scaffolds in pharmaceutical development for AD and PD treatment, two of the most serious chronic neurodegenerative diseases. The varied pyrazole compounds described herein can be regarded as candidate agents for the development of novel neurodegenerative drugs. Although in vitro and in vivo studies with promising therapeutics have shown potential to alleviate neurodegenerative symptoms, they are still far from providing a cure or effective treatment, largely due to a lack of clear understanding of disease pathologies. Therefore, studies based on rational drug design and that incorporate a deep understanding of pathological changes in neurodegenerative disease are urgently needed to explore the potential utility of pyrazoles as therapeutics, and this effort relies on multidisciplinary cooperation ranging from medicinal chemistry to pathophysiology.

Author Contributions: Conceptualization, Z.T. and X.L.; Software, Writing—original draft preparation, X.L.; Writing—review and editing, Z.T., X.L., and Y.Y.; Visualization, X.L. and Y.Y.; Supervision, Z.T.; Funding acquisition, Z.T. All authors have read and agreed to the published version of the manuscript.

Funding: The authors received financial support from the USA National Institutes of Health (NS075527, NS103957, NS103988, U19NS110456).

Acknowledgments: The authors also thank Lynne Jones for assistance with the manuscript.

Conflicts of Interest: The authors declare no conflict of interest.

References

1. Giornal, F.; Pazenok, S.; Rodefeld, L.; Lui, N.; Vors, J.P.; Leroux, F.R. Synthesis of diversely fluorinated pyrazoles as novel active agrochemical ingredients. *J. Fluorine Chem.* **2013**, *152*, 2–11. [CrossRef]
2. Khan, M.F.; Alam, M.M.; Verma, G.; Akhtar, W.; Akhter, M.; Shaquiquzzaman, M. The therapeutic voyage of pyrazole and its analogs: A review. *Eur. J. Med. Chem.* **2016**, *120*, 170–201. [CrossRef] [PubMed]
3. Karrouchi, K.; Radi, S.; Ramli, Y.; Taoufik, J.; Mabkhot, Y.N.; Al-Aizari, F.A.; Ansar, M. Synthesis and pharmacological activities of pyrazole derivatives: A review. *Molecules* **2018**, *23*, 134. [CrossRef] [PubMed]
4. Byon, W.; Garonzik, S.; Boyd, R.A.; Frost, C.E. Apixaban: A clinical pharmacokinetic and pharmacodynamic review. *Clin. Pharmacokinet.* **2019**, *58*, 1265–1279. [CrossRef] [PubMed]
5. Fustero, S.; Sanchez-Rosello, M.; Barrio, P.; Simon-Fuentes, A. From 2000 to mid-2010: A fruitful decade for the synthesis of pyrazoles. *Chem. Rev.* **2011**, *111*, 6984–7034. [CrossRef]
6. Gomes, P.M.O.; Silva, A.M.S.; Silva, V.L.M. Pyrazoles as key scaffolds for the development of fluorine-18-labeled radiotracers for positron emission tomography (PET). *Molecules* **2020**, *25*, 1722. [CrossRef]
7. Turkan, F.; Cetin, A.; Taslimi, P.; Gulcin, I. Some pyrazoles derivatives: Potent carbonic anhydrase, α-glycosidase, and cholinesterase enzymes inhibitors. *Arch. Pharm.* **2018**, *351*, e1800200. [CrossRef]
8. Hazarika, R.; Konwar, M.; Damarla, K.; Kumar, A.; Sarma, D. HBF_4/ACN: A simple and efficient protocol for the synthesis of pyrazoles under ambient reaction conditions. *Synth. Commun.* **2019**, *50*, 329–337. [CrossRef]
9. Wang, H.F.; Sun, X.L.; Zhang, S.L.; Liu, G.L.; Wang, C.J.; Zhu, L.L.; Zhang, H. Efficient copper-catalyzed synthesis of substituted pyrazoles at room temperature. *Synlett* **2018**, *29*, 2689–2692. [CrossRef]
10. Kumar, S.V.; Yadav, S.K.; Raghava, B.; Saraiah, B.; Ila, H.; Rangappa, K.S.; Hazra, A. Cyclocondensation of arylhydrazines with 1,3-bis(het)arylmonothio-1,3-diketones and 1,3-bis(het)aryl-3-(methylthio)-2-propenones: Synthesis of 1-aryl-3,5-bis(het)arylpyrazoles with complementary regioselectivity. *J. Org. Chem.* **2013**, *78*, 4960–4973. [CrossRef] [PubMed]
11. Schmitt, D.C.; Taylor, A.P.; Flick, A.C.; Kyne, R.E., Jr. Synthesis of pyrazoles from 1,3-diols via hydrogen transfer catalysis. *Org. Lett.* **2015**, *17*, 1405–1408. [CrossRef] [PubMed]
12. Zhang, X.; Kang, J.; Niu, P.; Wu, J.; Yu, W.; Chang, J. I_2-mediated oxidative C-N bond formation for metal-free one-pot synthesis of di-, tri-, and tetrasubstituted pyrazoles from α,β-unsaturated aldehydes/ketones and hydrazines. *J. Org. Chem.* **2014**, *79*, 10170–10178. [CrossRef] [PubMed]
13. Gutti, G.; Kumar, D.; Paliwal, P.; Ganeshpurkar, A.; Lahre, K.; Kumar, A.; Krishnamurthy, S.; Singh, S.K. Development of pyrazole and spiropyrazoline analogs as multifunctional agents for treatment of Alzheimer's disease. *Bioorg. Chem.* **2019**, *90*, 103080. [CrossRef] [PubMed]
14. Ding, Y.; Zhang, T.; Chen, Q.Y.; Zhu, C. Visible-light photocatalytic aerobic annulation for the green synthesis of pyrazoles. *Org. Lett.* **2016**, *18*, 4206–4209. [CrossRef]
15. Hu, J.; Chen, S.; Sun, Y.; Yang, J.; Rao, Y. Synthesis of tri- and tetrasubstituted pyrazoles via Ru(II) catalysis: Intramolecular aerobic oxidative C-N coupling. *Org. Lett.* **2012**, *14*, 5030–5033. [CrossRef]
16. Harigae, R.; Moriyama, K.; Togo, H. Preparation of 3,5-disubstituted pyrazoles and isoxazoles from terminal alkynes, aldehydes, hydrazines, and hydroxylamine. *J. Org. Chem.* **2014**, *79*, 2049–2058. [CrossRef] [PubMed]
17. Raji Reddy, C.; Vijaykumar, J.; Grée, R. Facile one-pot synthesis of 3,5-disubstituted 1*H*-pyrazoles from propargylic alcohols via propargyl hydrazides. *Synthesis* **2013**, *45*, 830–836. [CrossRef]
18. Guo, Y.; Wang, G.; Wei, L.; Wan, J.P. Domino C-H sulfonylation and pyrazole annulation for fully substituted pyrazole synthesis in water using hydrophilic enaminones. *J. Org. Chem.* **2019**, *84*, 2984–2990. [CrossRef] [PubMed]
19. Raghunadh, A.; Meruva, S.B.; Mekala, R.; Rao, K.R.; Krishna, T.; Chary, R.G.; Rao, L.V.; Kumar, U.K.S. An efficient regioselective copper catalyzed multi-component synthesis of 1,3-disubstituted pyrazoles. *Tetrahedron Lett.* **2014**, *55*, 2986–2990. [CrossRef]

20. Chen, J.; Properzi, R.; Uccello, D.P.; Young, J.A.; Dushin, R.G.; Starr, J.T. One-pot oxidation and rearrangement of propargylamines and in situ pyrazole synthesis. *Org. Lett.* **2014**, *16*, 4146–4149. [CrossRef] [PubMed]
21. Tang, M.; Zhang, F.M. Efficient one-pot synthesis of substituted pyrazoles. *Tetrahedron* **2013**, *69*, 1427–1433. [CrossRef]
22. Wen, J.; Fu, Y.; Zhang, R.Y.; Zhang, J.; Chen, S.Y.; Yu, X.Q. A simple and efficient synthesis of pyrazoles in water. *Tetrahedron* **2011**, *67*, 9618–9621. [CrossRef]
23. Zhang, G.; Ni, H.; Chen, W.; Shao, J.; Liu, H.; Chen, B.; Yu, Y. One-pot three-component approach to the synthesis of polyfunctional pyrazoles. *Org. Lett.* **2013**, *15*, 5967–5969. [CrossRef]
24. Sha, Q.; Wei, Y.Y. An efficient one-pot synthesis of 3,5-diaryl-4-bromopyrazoles by 1,3-dipolar cycloaddition of in situ generated diazo compounds and 1-bromoalk-1-ynes. *Synthesis* **2013**, *45*, 413–420. [CrossRef]
25. Wang, L.; Huang, J.; Gong, X.; Wang, J. Highly regioselective organocatalyzed synthesis of pyrazoles from diazoacetates and carbonyl compounds. *Chem. Eur. J.* **2013**, *19*, 7555–7560. [CrossRef]
26. Jackowski, O.; Lecourt, T.; Micouin, L. Direct synthesis of polysubstituted aluminoisoxazoles and pyrazoles by a metalative cyclization. *Org. Lett.* **2011**, *13*, 5664–5667. [CrossRef]
27. Li, D.Y.; Mao, X.F.; Chen, H.J.; Chen, G.R.; Liu, P.N. Rhodium-catalyzed addition-cyclization of hydrazines with alkynes: Pyrazole synthesis via unexpected C-N bond cleavage. *Org. Lett.* **2014**, *16*, 3476–3479. [CrossRef]
28. Ledovskaya, M.S.; Voronin, V.V.; Polynski, M.V.; Lebedev, A.N.; Ananikov, V.P. Primary vinyl ethers as acetylene surrogate: A flexible tool for deuterium-labeled pyrazole synthesis. *Eur. J. Org. Chem.* **2020**, *2020*, 4571–4580. [CrossRef]
29. Kobayashi, E.; Togo, H. Facile one-pot transformation of primary alcohols into 3-aryl- and 3-alkyl-isoxazoles and -pyrazoles. *Synthesis* **2019**, *51*, 3723–3735. [CrossRef]
30. Panda, N.; Jena, A.K. Fe-catalyzed one-pot synthesis of 1,3-di- and 1,3,5-trisubstituted pyrazoles from hydrazones and vicinal diols. *J. Org. Chem.* **2012**, *77*, 9401–9406. [CrossRef] [PubMed]
31. Fan, X.W.; Lei, T.; Zhou, C.; Meng, Q.Y.; Chen, B.; Tung, C.H.; Wu, L.Z. Radical addition of hydrazones by α-bromo ketones to prepare 1,3,5-trisubstituted pyrazoles via visible light catalysis. *J. Org. Chem.* **2016**, *81*, 7127–7133. [CrossRef]
32. Yi, F.; Zhao, W.; Wang, Z.; Bi, X. Silver-mediated [3+2] cycloaddition of alkynes and N-isocyanoiminotriphenylphosphorane: Access to monosubstituted pyrazoles. *Org. Lett.* **2019**, *21*, 3158–3161. [CrossRef]
33. Zhu, J.N.; Wang, W.K.; Jin, Z.H.; Wang, Q.K.; Zhao, S.Y. Pyrrolo[3,4-c]pyrazole synthesis via copper(I) chloride-catalyzed oxidative coupling of hydrazones to maleimides. *Org. Lett.* **2019**, *21*, 5046–5050. [CrossRef]
34. Specklin, S.; Decuypere, E.; Plougastel, L.; Aliani, S.; Taran, F. One-pot synthesis of 1,4-disubstituted pyrazoles from arylglycines via copper-catalyzed sydnone-alkyne cycloaddition reaction. *J. Org. Chem.* **2014**, *79*, 7772–7777. [CrossRef]
35. Lakeland, C.P.; Watson, D.W.; Harrity, J.P.A. Exploiting synergistic catalysis for an ambient temperature photocycloaddition to pyrazoles. *Chem. Eur. J.* **2020**, *26*, 155–159. [CrossRef] [PubMed]
36. Zhang, Q.; Meng, L.G.; Wang, K.; Wang, L. nBu3P-catalyzed desulfonylative [3 + 2] cycloadditions of allylic carbonates with arylazosulfones to pyrazole derivatives. *Org. Lett.* **2015**, *17*, 872–875. [CrossRef] [PubMed]
37. Zhang, Y.C.; Liu, J.; Jia, X.S. Phosphine-free [3+2] cycloaddition of propargylamines with dialkyl azodicarboxylates: An efficient access to pyrazole backbone. *Synthesis* **2018**, *50*, 3499–3505.
38. Pearce, A.J.; Harkins, R.P.; Reiner, B.R.; Wotal, A.C.; Dunscomb, R.J.; Tonks, I.A. Multicomponent pyrazole synthesis from alkynes, nitriles, and titanium imido complexes via oxidatively induced N-N bond coupling. *J. Am. Chem. Soc.* **2020**, *142*, 4390–4399. [CrossRef]
39. Sloop, J.C.; Holder, C.; Henary, M. Selective incorporation of fluorine in pyrazoles. *Eur. J. Org. Chem.* **2015**, *2015*, 3405–3422. [CrossRef]
40. Prieto, A.; Bouyssi, D.; Monteiro, N. Ruthenium-catalyzed tandem C-H fluoromethylation/cyclization of N-alkylhydrazones with CBr3F: Access to 4-fluoropyrazoles. *J. Org. Chem.* **2017**, *82*, 3311–3316. [CrossRef]
41. Schmitt, E.; Panossian, A.; Vors, J.P.; Funke, C.; Lui, N.; Pazenok, S.; Leroux, F.R. A major advance in the synthesis of fluoroalkyl pyrazoles: Tuneable regioselectivity and broad substitution patterns. *Chem. Eur. J.* **2016**, *22*, 11239–11244. [CrossRef] [PubMed]
42. Schmitt, E.; Bouvet, S.; Pegot, B.; Panossian, A.; Vors, J.P.; Pazenok, S.; Magnier, E.; Leroux, F.R. Fluoroalkyl amino reagents for the introduction of the fluoro(trifluoromethoxy)methyl group onto arenes and heterocycles. *Org. Lett.* **2017**, *19*, 4960–4963. [CrossRef] [PubMed]
43. Mykhailiuk, P.K. In situ generation of difluoromethyl diazomethane for [3+2] cycloadditions with alkynes. *Angew. Chem. Int. Ed.* **2015**, *54*, 6558–6561. [CrossRef] [PubMed]
44. Montoya, V.; Pons, J.; García-Antón, J.; Solans, X.; Font-Bardia, M.; Ros, J. Reaction of 2-hydroxyethylhydrazine with a trifluoromethyl-β-diketone: Study and structural characterization of a new 5-hydroxy-5-trifluoromethyl-4,5-dihydropyrazole intermediate. *J. Fluorine Chem.* **2007**, *128*, 1007–1011. [CrossRef]
45. Li, F.; Nie, J.; Sun, L.; Zheng, Y.; Ma, J.A. Silver-mediated cycloaddition of alkynes with CF3CHN2: Highly regioselective synthesis of 3-trifluoromethylpyrazoles. *Angew. Chem. Int. Ed.* **2013**, *52*, 6255–6258. [CrossRef] [PubMed]
46. Sun, A.; Chandrakumar, N.; Yoon, J.-J.; Plemper, R.K.; Snyder, J.P. Non-nucleoside inhibitors of the measles virus RNA-dependent RNA polymerase complex activity: Synthesis and in vitro evaluation. *Bioorg. Med. Chem. Lett.* **2007**, *17*, 5199–5203. [CrossRef]
47. Ji, G.; Wang, X.; Zhang, S.; Xu, Y.; Ye, Y.; Li, M.; Zhang, Y.; Wang, J. Synthesis of 3-trifluoromethylpyrazoles via trifluoromethylation/cyclization of α,β-alkynic hydrazones using a hypervalent iodine reagent. *Chem. Commun.* **2014**, *50*, 4361–4363. [CrossRef]
48. 50 Liu, H.; Laforest, R.; Gu, J.; Luo, Z.; Jones, L.A.; Gropler, R.J.; Benzinger, T.L.S.; Tu, Z. Acute rodent tolerability, toxicity, and radiation dosimetry estimates of the S1P1-specific radioligand [^{11}C]CS$_1$P$_1$. *Mol. Imaging. Biol.* **2020**, *22*, 285–292. [CrossRef] [PubMed]

49. Wang, Q.; He, L.; Li, K.K.; Tsui, G.C. Copper-mediated domino cyclization/trifluoromethylation/deprotection with TMSCF$_3$: Synthesis of 4-(trifluoromethyl)pyrazoles. *Org. Lett.* **2017**, *19*, 658–661. [CrossRef]
50. Trofymchuk, S.; Bugera, M.Y.; Klipkov, A.A.; Razhyk, B.; Semenov, S.; Tarasenko, K.; Starova, V.S.; Zaporozhets, O.A.; Tananaiko, O.Y.; Alekseenko, A.N.; et al. Deoxofluorination of (hetero)aromatic acids. *J. Org. Chem.* **2020**, *85*, 3110–3124. [CrossRef] [PubMed]
51. Dhanju, S.; Caravana, A.C.; Thomson, R.J. Access to α-pyrazole and α-triazole derivatives of ketones from oxidative heteroarylation of silyl enolethers. *Org. Lett.* **2020**, *22*, 8055–8058. [CrossRef] [PubMed]
52. Kumar, V.; Kaur, K.; Gupta, G.K.; Sharma, A.K. Pyrazole containing natural products: Synthetic preview and biological significance. *Eur. J. Med. Chem.* **2013**, *69*, 735–753. [CrossRef]
53. Luo, G.; Chen, L. Gold-catalyzed stereoselective 1,4-conjugate addition of pyrazoles to propiolates and their hydrogenation to β-pyrazolyl acid esters. *Tetrahedron Lett.* **2015**, *56*, 6276–6278. [CrossRef]
54. Roughley, S.D.; Jordan, A.M. The medicinal chemist's toolbox: An analysis of reactions used in the pursuit of drug candidates. *J. Med. Chem.* **2011**, *54*, 3451–3479. [CrossRef] [PubMed]
55. Viciano-Chumillas, M.; Tanase, S.; de Jongh, L.J.; Reedijk, J. Coordination versatility of pyrazole-based ligands towards high-nuclearity transition-metal and rare-earth clusters. *Eur. J. Inorg. Chem.* **2010**, *2010*, 3403–3418. [CrossRef]
56. Ruiz-Castillo, P.; Buchwald, S.L. Applications of palladium-catalyzed C-N cross-coupling reactions. *Chem. Rev.* **2016**, *116*, 12564–12649. [CrossRef]
57. Ma, D.; Cai, Q.; Zhang, H. Mild method for Ullmann coupling reaction of amines and aryl halides. *Org. Lett.* **2003**, *5*, 2453–2455. [CrossRef]
58. Onodera, S.; Kochi, T.; Kakiuchi, F. Synthesis of N-arylpyrazoles by palladium-catalyzed coupling of aryl triflates with pyrazole derivatives. *J. Org. Chem.* **2019**, *84*, 6508–6515. [CrossRef]
59. Zhou, Q.; Du, F.; Chen, Y.; Fu, Y.; Sun, W.; Wu, Y.; Chen, G. L-(-)-Quebrachitol as a ligand for selective copper(0)-catalyzed N-arylation of nitrogen-containing heterocycles. *J. Org. Chem.* **2019**, *84*, 8160–8167. [CrossRef] [PubMed]
60. Cristau, H.-J.; Cellier, P.P.; Spindler, J.-F.; Taillefer, M. Mild conditions for copper-catalysed N-arylation of pyrazoles. *Eur. J. Org. Chem.* **2004**, *2004*, 695–709. [CrossRef]
61. Wang, J.; Li, J.H.; Guo, Y.; Dong, H.; Liu, Q.; Yu, X.Q. TEMPO-mediated C-H amination of benzoxazoles with N-heterocycles. *J. Org. Chem.* **2020**, *85*, 12797–12803. [CrossRef]
62. Aggarwal, R.; Kumar, V.; Kumar, R.; Singh, S.P. Approaches towards the synthesis of 5-aminopyrazoles. *Beilstein J. Org. Chem.* **2011**, *7*, 179–197. [CrossRef] [PubMed]
63. Senadi, G.C.; Hu, W.P.; Lu, T.Y.; Garkhedkar, A.M.; Vandavasi, J.K.; Wang, J.J. I$_2$-TBHP-catalyzed oxidative cross-coupling of N-sulfonyl hydrazones and isocyanides to 5-aminopyrazoles. *Org. Lett.* **2015**, *17*, 1521–1524. [CrossRef]
64. Kallman, N.J.; Cole, K.P.; Koenig, T.M.; Buser, J.Y.; McFarland, A.D.; McNulty, L.M.; Mitchell, D. Synthesis of aminopyrazoles from isoxazoles: Comparison of preparative methods by in situ NMR analysis. *Synthesis* **2016**, *48*, 3537–3543.
65. Soriano-Garcia, M. Organoselenium compounds as potential therapeutic and chemopreventive agents: A review. *Curr. Med. Chem.* **2004**, *11*, 1657–1669. [CrossRef] [PubMed]
66. Gandin, V.; Khalkar, P.; Braude, J.; Fernandes, A.P. Organic selenium compounds as potential chemotherapeutic agents for improved cancer treatment. *Free Radic. Biol. Med.* **2018**, *127*, 80–97. [CrossRef]
67. Desai, D.; Kaushal, N.; Gandhi, U.H.; Arner, R.J.; D'Souza, C.; Chen, G.; Vunta, H.; El-Bayoumy, K.; Amin, S.; Prabhu, K.S. Synthesis and evaluation of the anti-inflammatory properties of selenium-derivatives of celecoxib. *Chem. Biol. Interact.* **2010**, *188*, 446–456. [CrossRef] [PubMed]
68. Belladona, A.L.; Cervo, R.; Alves, D.; Barcellos, T.; Cargnelutti, R.; Schumacher, R.F. C-H functionalization of (hetero)arenes: Direct selanylation mediated by Selectfluor. *Tetrahedron Lett.* **2020**, *61*, 152035. [CrossRef]
69. Wang, X.; Liu, W.G.; Tung, C.H.; Wu, L.Z.; Cong, H. A monophosphine ligand derived from anthracene photodimer: Synthetic applications for palladium-catalyzed coupling reactions. *Org. Lett.* **2019**, *21*, 8158–8163. [CrossRef]
70. Kirsten, C.N.; Schrader, T.H. Intermolecular β-sheet stabilization with aminopyrazoles. *J. Am. Chem. Soc.* **1997**, *119*, 12061–12068. [CrossRef]
71. Rzepecki, P.; Schrader, T. β-Sheet ligands in action: KLVFF recognition by aminopyrazole hybrid receptors in water. *J. Am. Chem. Soc.* **2005**, *127*, 3016–3025. [CrossRef] [PubMed]
72. Frisoni, G.B.; Laakso, M.P.; Beltramello, A.; Geroldi, C.; Bianchetti, A.; Soininen, H.; Trabucchi, M. Hippocampal and entorhinal cortex atrophy in frontotemporal dementia and Alzheimer's disease. *Neurology* **1999**, *52*, 91–100. [CrossRef]
73. Hardy, J.A.; Higgins, G.A. Alzheimer's disease: The amyloid cascade hypothesis. *Science* **1992**, *256*, 184–185. [CrossRef] [PubMed]
74. Savelieff, M.G.; Nam, G.; Kang, J.; Lee, H.J.; Lee, M.; Lim, M.H. Development of multifunctional molecules as potential therapeutic candidates for Alzheimer's disease, Parkinson's disease, and amyotrophic lateral sclerosis in the last decade. *Chem. Rev.* **2019**, *119*, 1221–1322. [CrossRef]
75. Francis, P.T.; Palmer, A.M.; Snape, M.; Wilcock, G.K. The cholinergic hypothesis of Alzheimer's disease: A review of progress. *J. Neurol. Neurosurg. Psychiatry* **1999**, *66*, 137–147. [CrossRef] [PubMed]
76. Fonnum, F. Radiochemical micro assays for the determination of choline acetyltransferase and acetylcholinesterase activities. *Biochem. J.* **1969**, *115*, 465–472. [CrossRef] [PubMed]
77. Lockridge, O. Review of human butyrylcholinesterase structure, function, genetic variants, history of use in the clinic, and potential therapeutic uses. *Pharmacol. Ther.* **2015**, *148*, 34–46. [CrossRef]

78. Bartolini, M.; Bertucci, C.; Cavrini, V.; Andrisano, V. β-Amyloid aggregation induced by human acetylcholinesterase: Inhibition studies. *Biochem. Pharmacol.* **2003**, *65*, 407–416. [CrossRef]
79. Greig, N.H.; Utsuki, T.; Ingram, D.K.; Wang, Y.; Pepeu, G.; Scali, C.; Yu, Q.S.; Mamczarz, J.; Holloway, H.W.; Giordano, T.; et al. Selective butyrylcholinesterase inhibition elevates brain acetylcholine, augments learning and lowers Alzheimer β-amyloid peptide in rodent. *Proc. Natl. Acad. Sci. USA* **2005**, *102*, 17213–17218. [CrossRef]
80. Nordberg, A.; Ballard, C.; Bullock, R.; Darreh-Shori, T.; Somogyi, M. A review of butyrylcholinesterase as a therapeutic target in the treatment of Alzheimer's disease. *Prim. Care. Companion CNS Disord.* **2013**, *15*. [CrossRef]
81. Shaikh, S.; Dhavan, P.; Pavale, G.; Ramana, M.M.V.; Jadhav, B.L. Design, synthesis and evaluation of pyrazole bearing α-aminophosphonate derivatives as potential acetylcholinesterase inhibitors against Alzheimer's disease. *Bioorg. Chem.* **2020**, *96*, 103589–103611. [CrossRef] [PubMed]
82. Silva, D.; Chioua, M.; Samadi, A.; Carmo Carreiras, M.; Jimeno, M.L.; Mendes, E.; Rios Cde, L.; Romero, A.; Villarroya, M.; Lopez, M.G.; et al. Synthesis and pharmacological assessment of diversely substituted pyrazolo[3,4-b]quinoline, and benzo[b]pyrazolo[4,3-g][1,8]naphthyridine derivatives. *Eur. J. Med. Chem.* **2011**, *46*, 4676–4681. [CrossRef]
83. Esch, F.S.; Keim, P.S.; Beattie, E.C.; Blacher, R.W.; Culwell, A.R.; Oltersdorf, T.; McClure, D.; Ward, P.J. Cleavage of amyloid beta peptide during constitutive processing of its precursor. *Science* **1990**, *248*, 1122–1124. [CrossRef] [PubMed]
84. Zhang, Y.W.; Thompson, R.; Zhang, H.; Xu, H. APP processing in Alzheimer's disease. *Mol. Brain* **2011**, *4*, 3–15. [CrossRef] [PubMed]
85. Lee, S.J.; Nam, E.; Lee, H.J.; Savelieff, M.G.; Lim, M.H. Towards an understanding of amyloid-β oligomers: Characterization, toxicity mechanisms, and inhibitors. *Chem. Soc. Rev.* **2017**, *46*, 310–323. [CrossRef]
86. Kane, M.D.; Lipinski, W.J.; Callahan, M.J.; Bian, F.; Durham, R.A.; Schwarz, R.D.; Roher, A.E.; Walker, L.C. Evidence for seeding of β-amyloid by intracerebral infusion of Alzheimer brain extracts in β-amyloid precursor protein-transgenic mice. *J. Neurosci.* **2000**, *20*, 3606–3611. [CrossRef]
87. Ballatore, C.; Lee, V.M.; Trojanowski, J.Q. Tau-mediated neurodegeneration in Alzheimer's disease and related disorders. *Nat. Rev. Neurosci.* **2007**, *8*, 663–672. [CrossRef]
88. Grundke-Iqbal, I.; Iqbal, K.; Tung, Y.C.; Quinlan, M.; Wisniewski, H.M.; Binder, L.I. Abnormal phosphorylation of the microtubule-associated protein τ (tau) in Alzheimer cytoskeletal pathology. *Proc. Natl. Acad. Sci. USA* **1986**, *83*, 4913–4917. [CrossRef] [PubMed]
89. Polydoro, M.; Acker, C.M.; Duff, K.; Castillo, P.E.; Davies, P. Age-dependent impairment of cognitive and synaptic function in the htau mouse model of tau pathology. *J. Neurosci.* **2009**, *29*, 10741–10749. [CrossRef]
90. Ward, S.M.; Himmelstein, D.S.; Lancia, J.K.; Binder, L.I. Tau oligomers and tau toxicity in neurodegenerative disease. *Biochem. Soc. Trans.* **2012**, *40*, 667–671. [CrossRef] [PubMed]
91. Hochdorffer, K.; Marz-Berberich, J.; Nagel-Steger, L.; Epple, M.; Meyer-Zaika, W.; Horn, A.H.; Sticht, H.; Sinha, S.; Bitan, G.; Schrader, T. Rational design of β-sheet ligands against Aβ42-induced toxicity. *J. Am. Chem. Soc.* **2011**, *133*, 4348–4358. [CrossRef] [PubMed]
92. Wagner, J.; Ryazanov, S.; Leonov, A.; Levin, J.; Shi, S.; Schmidt, F.; Prix, C.; Pan-Montojo, F.; Bertsch, U.; Mitteregger-Kretzschmar, G.; et al. Anle138b: A novel oligomer modulator for disease-modifying therapy of neurodegenerative diseases such as prion and Parkinson's disease. *Acta. Neuropathol.* **2013**, *125*, 795–813. [CrossRef]
93. Wagner, J.; Krauss, S.; Shi, S.; Ryazanov, S.; Steffen, J.; Miklitz, C.; Leonov, A.; Kleinknecht, A.; Goricke, B.; Weishaupt, J.H.; et al. Reducing tau aggregates with anle138b delays disease progression in a mouse model of tauopathies. *Acta. Neuropathol.* **2015**, *130*, 619–631. [CrossRef] [PubMed]
94. Matthes, D.; Gapsys, V.; Griesinger, C.; de Groot, B.L. Resolving the atomistic modes of anle138b inhibitory action on peptide oligomer formation. *ACS Chem. Neurosci.* **2017**, *8*, 2791–2808. [CrossRef]
95. Belkacemi, A.; Doggui, S.; Dao, L.; Ramassamy, C. Challenges associated with curcumin therapy in Alzheimer disease. *Expert Rev. Mol. Med.* **2011**, *13*, e34. [CrossRef]
96. Mishra, S.; Patel, S.; Halpani, C.G. Recent updates in curcumin pyrazole and isoxazole derivatives: Synthesis and biological application. *Chem. Biodivers.* **2019**, *16*, e1800366. [CrossRef]
97. Chainoglou, E.; Hadjipavlou-Litina, D. Curcumin in health and diseases: Alzheimer's disease and curcumin analogues, derivatives, and hybrids. *Int. J. Mol. Sci.* **2020**, *21*, 1975. [CrossRef]
98. Narlawar, R.; Pickhardt, M.; Leuchtenberger, S.; Baumann, K.; Krause, S.; Dyrks, T.; Weggen, S.; Mandelkow, E.; Schmidt, B. Curcumin-derived pyrazoles and isoxazoles: Swiss army knives or blunt tools for Alzheimer's disease? *ChemMedChem* **2008**, *3*, 165–172. [CrossRef]
99. Liu, Y.; Dargusch, R.; Maher, P.; Schubert, D. A broadly neuroprotective derivative of curcumin. *J. Neurochem.* **2008**, *105*, 1336–1345. [CrossRef]
100. Valera, E.; Dargusch, R.; Maher, P.A.; Schubert, D. Modulation of 5-lipoxygenase in proteotoxicity and Alzheimer's disease. *J. Neurosci.* **2013**, *33*, 10512–10525. [CrossRef]
101. Kotani, R.; Urano, Y.; Sugimoto, H.; Noguchi, N. Decrease of amyloid-β levels by curcumin derivative via modulation of amyloid-β protein precursor trafficking. *J. Alzheimers Dis.* **2017**, *56*, 529–542. [CrossRef]
102. Okuda, M.; Hijikuro, I.; Fujita, Y.; Teruya, T.; Kawakami, H.; Takahashi, T.; Sugimoto, H. Design and synthesis of curcumin derivatives as tau and amyloid β dual aggregation inhibitors. *Bioorg. Med. Chem. Lett.* **2016**, *26*, 5024–5028. [CrossRef]
103. Okuda, M.; Fujita, Y.; Hijikuro, I.; Wada, M.; Uemura, T.; Kobayashi, Y.; Waku, T.; Tanaka, N.; Nishimoto, T.; Izumi, Y.; et al. PE859, a novel curcumin derivative, inhibits amyloid-β and tau aggregation, and ameliorates cognitive dysfunction in senescence-accelerated mouse prone 8. *J. Alzheimers. Dis.* **2017**, *59*, 313–328. [CrossRef] [PubMed]

104. Menniti, F.S.; Faraci, W.S.; Schmidt, C.J. Phosphodiesterases in the CNS: Targets for drug development. *Nat. Rev. Drug. Discov.* **2006**, *5*, 660–670. [CrossRef]
105. Bender, A.T.; Beavo, J.A. Cyclic nucleotide phosphodiesterases: Molecular regulation to clinical use. *Pharmacol. Rev.* **2006**, *58*, 488–520. [CrossRef]
106. Prickaerts, J.; Heckman, P.R.A.; Blokland, A. Investigational phosphodiesterase inhibitors in phase I and phase II clinical trials for Alzheimer's disease. *Expert. Opin. Investig. Drugs* **2017**, *26*, 1033–1048. [CrossRef]
107. Cuadrado-Tejedor, M.; Hervias, I.; Ricobaraza, A.; Puerta, E.; Perez-Roldan, J.M.; Garcia-Barroso, C.; Franco, R.; Aguirre, N.; Garcia-Osta, A. Sildenafil restores cognitive function without affecting β-amyloid burden in a mouse model of Alzheimer's disease. *Br. J. Pharmacol.* **2011**, *164*, 2029–2041. [CrossRef]
108. Heckman, P.R.; Wouters, C.; Prickaerts, J. Phosphodiesterase inhibitors as a target for cognition enhancement in aging and Alzheimer's disease: A translational overview. *Curr. Pharm. Des.* **2015**, *21*, 317–331. [CrossRef]
109. Li, P.; Zheng, H.; Zhao, J.; Zhang, L.; Yao, W.; Zhu, H.; Beard, J.D.; Ida, K.; Lane, W.; Snell, G.; et al. Discovery of potent and selective inhibitors of phosphodiesterase 1 for the treatment of cognitive impairment associated with neurodegenerative and neuropsychiatric diseases. *J. Med. Chem.* **2016**, *59*, 1149–1164. [CrossRef] [PubMed]
110. Lakics, V.; Karran, E.H.; Boess, F.G. Quantitative comparison of phosphodiesterase mRNA distribution in human brain and peripheral tissues. *Neuropharmacology* **2010**, *59*, 367–374. [CrossRef]
111. Shimizu-Albergine, M.; Rybalkin, S.D.; Rybalkina, I.G.; Feil, R.; Wolfsgruber, W.; Hofmann, F.; Beavo, J.A. Individual cerebellar Purkinje cells express different cGMP phosphodiesterases (PDEs): In vivo phosphorylation of cGMP-specific PDE (PDE5) as an indicator of cGMP-dependent protein kinase (PKG) activation. *J. Neurosci.* **2003**, *23*, 6452–6459. [CrossRef]
112. Puzzo, D.; Staniszewski, A.; Deng, S.X.; Privitera, L.; Leznik, E.; Liu, S.; Zhang, H.; Feng, Y.; Palmeri, A.; Landry, D.W.; et al. Phosphodiesterase 5 inhibition improves synaptic function, memory, and amyloid-β load in an Alzheimer's disease mouse model. *J. Neurosci.* **2009**, *29*, 8075–8086. [CrossRef] [PubMed]
113. Zhang, J.; Guo, J.; Zhao, X.; Chen, Z.; Wang, G.; Liu, A.; Wang, Q.; Zhou, W.; Xu, Y.; Wang, C. Phosphodiesterase-5 inhibitor sildenafil prevents neuroinflammation, lowers beta-amyloid levels and improves cognitive performance in APP/PS1 transgenic mice. *Behav. Brain Res.* **2013**, *250*, 230–237. [CrossRef] [PubMed]
114. Schultheiss, D.; Muller, S.V.; Nager, W.; Stief, C.G.; Schlote, N.; Jonas, U.; Asvestis, C.; Johannes, S.; Munte, T.F. Central effects of sildenafil (Viagra) on auditory selective attention and verbal recognition memory in humans: A study with event-related brain potentials. *World J. Urol.* **2001**, *19*, 46–50. [CrossRef]
115. Fisher, D.A.; Smith, J.F.; Pillar, J.S.; St Denis, S.H.; Cheng, J.B. Isolation and characterization of PDE9A, a novel human cGMP-specific phosphodiesterase. *J. Biol. Chem.* **1998**, *273*, 15559–15564. [CrossRef]
116. Kleiman, R.J.; Chapin, D.S.; Christoffersen, C.; Freeman, J.; Fonseca, K.R.; Geoghegan, K.F.; Grimwood, S.; Guanowsky, V.; Hajos, M.; Harms, J.F.; et al. Phosphodiesterase 9A regulates central cGMP and modulates responses to cholinergic and monoaminergic perturbation in vivo. *J. Pharmacol. Exp. Ther.* **2012**, *341*, 396–409. [CrossRef]
117. Verhoest, P.R.; Fonseca, K.R.; Hou, X.; Proulx-Lafrance, C.; Corman, M.; Helal, C.J.; Claffey, M.M.; Tuttle, J.B.; Coffman, K.J.; Liu, S.; et al. Design and discovery of 6-[(3S,4S)-4-methyl-1-(pyrimidin-2-ylmethyl)pyrrolidin-3-yl]-1-(tetrahydro-2H-pyran-4-yl)-1,5-dihydro-4H-pyrazolo[3,4-d]pyrimidin-4-one (PF-04447943), a selective brain penetrant PDE9A inhibitor for the treatment of cognitive disorders. *J. Med. Chem.* **2012**, *55*, 9045–9054. [CrossRef]
118. Schwam, E.M.; Nicholas, T.; Chew, R.; Billing, C.B.; Davidson, W.; Ambrose, D.; Altstiel, L.D. A multicenter, double-blind, placebo-controlled trial of the PDE9A inhibitor, PF-04447943, in Alzheimer's disease. *Curr. Alzheimer Res.* **2014**, *11*, 413–421. [CrossRef]
119. Moschetti, V.; Boland, K.; Feifel, U.; Hoch, A.; Zimdahl-Gelling, H.; Sand, M. First-in-human study assessing safety, tolerability and pharmacokinetics of BI 409306, a selective phosphodiesterase 9A inhibitor, in healthy males. *Br. J. Clin. Pharmacol.* **2016**, *82*, 1315–1324. [CrossRef]
120. Rosenbrock, H.; Giovannini, R.; Schanzle, G.; Koros, E.; Runge, F.; Fuchs, H.; Marti, A.; Reymann, K.G.; Schroder, U.H.; Fedele, E.; et al. The novel phosphodiesterase 9A inhibitor BI 409306 increases cyclic guanosine monophosphate levels in the brain, promotes synaptic plasticity, and enhances memory function in rodents. *J. Pharmacol. Exp. Ther.* **2019**, *371*, 633–641. [CrossRef]
121. Ghosh, A.S.; Wang, B.; Pozniak, C.D.; Chen, M.; Watts, R.J.; Lewcock, J.W. DLK induces developmental neuronal degeneration via selective regulation of proapoptotic JNK activity. *J. Cell. Biol.* **2011**, *194*, 751–764. [CrossRef]
122. Pozniak, C.D.; Sengupta Ghosh, A.; Gogineni, A.; Hanson, J.E.; Lee, S.H.; Larson, J.L.; Solanoy, H.; Bustos, D.; Li, H.; Ngu, H.; et al. Dual leucine zipper kinase is required for excitotoxicity-induced neuronal degeneration. *J. Exp. Med.* **2013**, *210*, 2553–2567. [CrossRef]
123. Le Pichon, C.E.; Meilandt, W.J.; Dominguez, S.; Solanoy, H.; Lin, H.; Ngu, H.; Gogineni, A.; Sengupta Ghosh, A.; Jiang, Z.; Lee, S.H.; et al. Loss of dual leucine zipper kinase signaling is protective in animal models of neurodegenerative disease. *Sci. Transl. Med.* **2017**, *9*, eaag0394. [CrossRef]
124. Patel, S.; Cohen, F.; Dean, B.J.; De La Torre, K.; Deshmukh, G.; Estrada, A.A.; Ghosh, A.S.; Gibbons, P.; Gustafson, A.; Huestis, M.P.; et al. Discovery of dual leucine zipper kinase (DLK, MAP3K12) inhibitors with activity in neurodegeneration models. *J. Med. Chem.* **2015**, *58*, 401–418. [CrossRef]
125. Patel, S.; Harris, S.F.; Gibbons, P.; Deshmukh, G.; Gustafson, A.; Kellar, T.; Lin, H.; Liu, X.; Liu, Y.; Liu, Y.; et al. Scaffold-hopping and structure-based discovery of potent, selective, and brain penetrant N-(1H-pyrazol-3-yl)pyridin-2-amine inhibitors of dual leucine zipper kinase (DLK, MAP3K12). *J. Med. Chem.* **2015**, *58*, 8182–8199. [CrossRef]

126. Patel, S.; Meilandt, W.J.; Erickson, R.I.; Chen, J.; Deshmukh, G.; Estrada, A.A.; Fuji, R.N.; Gibbons, P.; Gustafson, A.; Harris, S.F.; et al. Selective inhibitors of dual leucine zipper kinase (DLK, MAP3K12) with activity in a model of Alzheimer's disease. *J. Med. Chem.* **2017**, *60*, 8083–8102. [CrossRef]
127. Shih, J.C.; Chen, K.; Ridd, M.J. Monoamine oxidase: From genes to behavior. *Annu. Rev. Neurosci.* **1999**, *22*, 197–217. [CrossRef]
128. Sherif, F.; Gottfries, C.G.; Alafuzoff, I.; Oreland, L. Brain gamma-aminobutyrate aminotransferase (GABA-T) and monoamine oxidase (MAO) in patients with Alzheimer's disease. *J. Neural. Transm.* **1992**, *4*, 227–240. [CrossRef]
129. Kennedy, B.P.; Ziegler, M.G.; Alford, M.; Hansen, L.A.; Thal, L.J.; Masliah, E. Early and persistent alterations in prefrontal cortex MAO A and B in Alzheimer's disease. *J. Neural. Transm.* **2003**, *110*, 789–801. [CrossRef] [PubMed]
130. Saura, J.; Luque, J.M.; Cesura, A.M.; Prada, M.D.; Chan-Palay, V.; Huber, G.; Löffler, J.; Richards, J.G. Increased monoamine oxidase b activity in plaque-associated astrocytes of Alzheimer brains revealed by quantitative enzyme radioautography. *Neuroscience* **1994**, *62*, 15–30. [CrossRef]
131. Zhao, Y.; Zhao, B. Oxidative stress and the pathogenesis of alzheimer's disease. *Oxid. Med. Cell.Longev.* **2013**, *2013*, 316523. [CrossRef]
132. Sano, M.; Ernesto, C.; Thomas, R.G.; Klauber, M.R.; Schafer, K.; Grundman, M.; Woodbury, P.; Growdon, J.; Cotman, C.W.; Pfeiffer, E.; et al. A controlled trial of selegiline, alpha-tocopherol, or both as treatment for Alzheimer's disease. *N. Engl. J. Med.* **1997**, *336*, 1216–1222. [CrossRef]
133. Pisani, L.; Catto, M.; Leonetti, F.; Nicolotti, O.; Stefanachi, A.; Campagna, F.; Carotti, A. Targeting monoamine oxidases with multipotent ligands: An emerging strategy in the search of new drugs against neurodegenerative diseases. *Cur. Med. Chem.* **2011**, *18*, 4568–4587. [CrossRef] [PubMed]
134. Sterling, J.; Herzig, Y.; Goren, T.; Finkelstein, N.; Lerner, D.; Goldenberg, W.; Miskolczi, I.; Molnar, S.; Rantal, F.; Tamas, T.; et al. Novel Dual Inhibitors of AChE and MAO Derived from hydroxy aminoindan and phenethylamine as potential treatment for Alzheimer's disease. *J. Med. Chem.* **2002**, *45*, 5260–5279. [CrossRef]
135. Huang, L.; Lu, C.; Sun, Y.; Mao, F.; Luo, Z.; Su, T.; Jiang, H.; Shan, W.; Li, X. Multitarget-directed benzylideneindanone derivatives: Anti-β-amyloid (Aβ) aggregation, antioxidant, metal chelation, and monoamine oxidase B (MAO-B) inhibition properties against Alzheimer's disease. *J. Med. Chem.* **2012**, *55*, 8483–8492. [CrossRef]
136. Tzvetkov, N.T.; Antonov, L. Subnanomolar indazole-5-carboxamide inhibitors of monoamine oxidase B (MAO-B) continued: Indications of iron binding, experimental evidence for optimised solubility and brain penetration. *J. Enzyme Inhib. Med. Chem.* **2017**, *32*, 960–967. [CrossRef]
137. Jayaraj, R.L.; Elangovan, N.; Dhanalakshmi, C.; Manivasagam, T.; Essa, M.M. CNB-001, a novel pyrazole derivative mitigates motor impairments associated with neurodegeneration via suppression of neuroinflammatory and apoptotic response in experimental Parkinson's disease mice. *Chem. Biol. Interact.* **2014**, *220*, 149–157. [CrossRef] [PubMed]
138. Ahsan, N.; Mishra, S.; Jain, M.K.; Surolia, A.; Gupta, S. Curcumin pyrazole and its derivative (N-(3-nitrophenylpyrazole) curcumin inhibit aggregation, disrupt fibrils and modulate toxicity of wild type and mutant α-synuclein. *Sci. Rep.* **2015**, *5*, 9862. [CrossRef] [PubMed]
139. Deeg, A.A.; Reiner, A.M.; Schmidt, F.; Schueder, F.; Ryazanov, S.; Ruf, V.C.; Giller, K.; Becker, S.; Leonov, A.; Griesinger, C.; et al. Anle138b and related compounds are aggregation specific fluorescence markers and reveal high affinity binding to α-synuclein aggregates. *Biochim. Biophys. Acta* **2015**, *1850*, 1884–1890. [CrossRef]
140. Svenningsson, P.; Le Moine, C.; Fisone, G.; Fredholm, B.B. Distribution, biochemistry and function of striatal adenosine A_{2A} receptors. *Prog. Neurobiol.* **1999**, *59*, 355–396. [CrossRef]
141. Schwarzschild, M.A.; Xu, K.; Oztas, E.; Petzer, J.P.; Castagnoli, K.; Castagnoli, N., Jr.; Chen, J.F. Neuroprotection by caffeine and more specific A_{2A} receptor antagonists in animal models of Parkinson's disease. *Neurology* **2003**, *61*, S55–S61. [CrossRef]
142. Bara-Jimenez, W.; Sherzai, A.; Dimitrova, T.; Favit, A.; Bibbiani, F.; Gillespie, M.; Morris, M.J.; Mouradian, M.M.; Chase, T.N. Adenosine A_{2A} receptor antagonist treatment of Parkinson's disease. *Neurology* **2003**, *61*, 293–296. [CrossRef]
143. Armentero, M.T.; Pinna, A.; Ferre, S.; Lanciego, J.L.; Muller, C.E.; Franco, R. Past, present and future of A_{2A} adenosine receptor antagonists in the therapy of Parkinson's disease. *Pharmacol. Ther.* **2011**, *132*, 280–299. [CrossRef]
144. Yang, Z.; Li, X.; Ma, H.; Zheng, J.; Zhen, X.; Zhang, X. Replacement of amide with bioisosteres led to a new series of potent adenosine A_{2A} receptor antagonists. *Bioorg. Med. Chem. Lett.* **2014**, *24*, 152–155. [CrossRef]
145. Zheng, J.; Yang, Z.; Li, X.; Li, L.; Ma, H.; Wang, M.; Zhang, H.; Zhen, X.; Zhang, X. Optimization of 6-heterocyclic-2-(1H-pyrazol-1-yl)-N-(pyridin-2-yl)pyrimidin-4-amine as potent adenosine A_{2A} receptor antagonists for the treatment of Parkinson's disease. *ACS Chem. Neurosci.* **2014**, *5*, 674–682. [CrossRef]
146. Yang, Z.; Li, L.; Zheng, J.; Ma, H.; Tian, S.; Li, J.; Zhang, H.; Zhen, X.; Zhang, X. Identification of a new series of potent adenosine A_{2A} receptor antagonists based on 4-amino-5-carbonitrile pyrimidine template for the treatment of Parkinson's disease. *ACS Chem. Neurosci.* **2016**, *7*, 1575–1584. [CrossRef]
147. Soderling, S.H.; Bayuga, S.J.; Beavo, J.A. Isolation and characterization of a dual-substrate phosphodiesterase gene family: PDE10A. *Proc. Natl. Acad. Sci. USA* **1999**, *96*, 7071–7076. [CrossRef] [PubMed]
148. Sancesario, G.; Morrone, L.A.; D'Angelo, V.; Castelli, V.; Ferrazzoli, D.; Sica, F.; Martorana, A.; Sorge, R.; Cavaliere, F.; Bernardi, G.; et al. Levodopa-induced dyskinesias are associated with transient down-regulation of cAMP and cGMP in the caudate-putamen of hemiparkinsonian rats: Reduced synthesis or increased catabolism? *Neurochem. Int.* **2014**, *79*, 44–56. [CrossRef]

149. Niccolini, F.; Foltynie, T.; Reis Marques, T.; Muhlert, N.; Tziortzi, A.C.; Searle, G.E.; Natesan, S.; Kapur, S.; Rabiner, E.A.; Gunn, R.N.; et al. Loss of phosphodiesterase 10A expression is associated with progression and severity in Parkinson's disease. *Brain* **2015**, *138*, 3003–3015. [CrossRef] [PubMed]
150. Movsesian, M.; Stehlik, J.; Vandeput, F.; Bristow, M.R. Phosphodiesterase inhibition in heart failure. *Heart Fail. Rev.* **2009**, *14*, 255–263. [CrossRef] [PubMed]
151. Heaslip, R.J.; Evans, D.Y. Emetic, central nervous system, and pulmonary activities of rolipram in the dog. *Eur. J. Pharmacol.* **1995**, *286*, 281–290. [CrossRef]
152. Verhoest, P.R.; Chapin, D.S.; Corman, M.; Fonseca, K.; Harms, J.F.; Hou, X.; Marr, E.S.; Menniti, F.S.; Nelson, F.; O'Connor, R.; et al. Discovery of a novel class of phosphodiesterase 10A inhibitors and identification of clinical candidate 2-[4-(1-methyl-4-pyridin-4-yl-1*H*-pyrazol-3-yl)-phenoxymethyl]-quinoline (PF-2545920) for the treatment of schizophrenia. *J. Med. Chem.* **2009**, *52*, 5188–5196. [CrossRef] [PubMed]
153. Li, J.; Jin, H.; Zhou, H.; Rothfuss, J.; Tu, Z. Synthesis and in vitro biological evaluation of pyrazole group-containing analogues for PDE10A. *MedChemComm* **2013**, *4*, 443–449. [CrossRef] [PubMed]
154. Li, J.; Zhang, X.; Jin, H.; Fan, J.; Flores, H.; Perlmutter, J.S.; Tu, Z. Synthesis of fluorine-containing phosphodiesterase 10a (PDE10A) inhibitors and the in vivo evaluation of F-18 labeled PDE10A PET tracers in rodent and nonhuman primate. *J. Med. Chem.* **2015**, *58*, 8584–8600. [CrossRef] [PubMed]

Review

Pyrazolyl-Ureas as Interesting Scaffold in Medicinal Chemistry

Chiara Brullo *[ ], Federica Rapetti[ ] and Olga Bruno

Department of Pharmacy, Section of Medicinal Chemistry, University of Genoa, Viale Benedetto XV 3, I-16132 Genova, Italy; federica.rapetti@edu.unige.it (F.R.); obruno@unige.it (O.B.)
* Correspondence: brullo@difar.unige.it; Tel.: +39-010-353-8368

Academic Editor: Vera L. M. Silva
Received: 10 July 2020; Accepted: 28 July 2020; Published: 29 July 2020

Abstract: The pyrazole nucleus has long been known as a privileged scaffold in the synthesis of biologically active compounds. Within the numerous pyrazole derivatives developed as potential drugs, this review is focused on molecules characterized by a urea function directly linked to the pyrazole nucleus in a different position. In the last 20 years, the interest of numerous researchers has been especially attracted by pyrazolyl-ureas showing a wide spectrum of biological activities, ranging from the antipathogenic activities (bacteria, plasmodium, toxoplasma, and others) to the anticarcinogenic activities. In particular, in the anticancer field, pyrazolyl-ureas have been shown to interact at the intracellular level on many pathways, in particular on different kinases such as Src, p38-MAPK, TrKa, and others. In addition, some of them evidenced an antiangiogenic potential that deserves to be explored. This review therefore summarizes all these biological data (from 2000 to date), including patented compounds.

Keywords: pyrazolyl-ureas; pyrazole nucleus; protein kinase inhibitors; anti-inflammatory agents; anticancer agents; anti-pathogens agents

1. Introduction

Pyrazole consists of a doubly unsaturated five-membered ring containing two nitrogen atoms (named N1 and N2) and, among heterocyclic compounds, represents one of the most important chemical scaffolds in medicinal chemistry and advanced organic materials [1].

The biological and pharmacological properties of pyrazole can be due to its particular chemical characteristics: in detail, pyrazole presents a nitrogen atom 1 (N1), also named "pyrrole-like" because its unshared electrons are conjugated with the aromatic system; and a nitrogen atom 2 (N2), named as "pyridine-like" since the unshared electrons are not compromised with resonance, similarly to pyridine systems. Due to the differences between these two nitrogen atoms, pyrazole can react with both acids and bases [2]. Another important structural characteristic of pyrazole is the prototrophic tautomerism: in fact, three tautomers are possible in unsubstituted pyrazole, while five tautomers can exist in mono-substituted pyrazoles [3].

Pyrazole and its derivatives exhibit a broad spectrum of pharmacological activities from antimicrobial and antitubercular to anticonvulsant, anticancer, analgesic, anti-inflammatory, antidepressant, cardiovascular, and many others [4].

Examples of the most recent pyrazole drugs are Celecoxib, Lonazolac, Tepoxalin, Deracoxib, Mepirizole, Crizotinib, Pyrazomycin, Surinabant, Rimonabant, Difenamizole, Fezolamine, Betazole, and Fomepizole, among many others. All these molecules are uncondensed variously substituted pyrazoles, but condensed pyrazole rings are also an important source of bioactive molecules [5,6].

In recent years, the urea function has led to the development of many drugs that are especially useful in anticancer therapy such as Sorafenib, Cabozatenib, and Regorafenib [7]. In the last ten years, various urea derivatives have been studied as biological modulators of different intracellular targets, confirming the importance of the urea scaffold in medicinal chemistry and in drug development. In fact, the urea NH moiety is a favorable hydrogen bond donor, while the urea oxygen atom is regarded as an excellent acceptor, giving to the function the possibility for interacting with several protein targets in different ways. In addition, due to strong intermolecular hydrogen bonding with different solvents, insertion of the urea moiety can enhance aqueous solubility.

When a urea function was inserted on the pyrazole nucleus in position 3, 4, or 5, different biologically active compounds with a broad spectrum of pharmacological properties were obtained, from antibacterial or antiparasitic to anti-inflammatory and, above all, anticancer ones [8].

This review focuses on the study and classification of new pyrazolyl-ureas (appeared in the literature over the past 15 years) and their interesting pharmacological properties. The numerous articles and patents have been classified firstly on the base of the molecules structure (in particular, the position of the urea function in the pyrazole scaffold has been used as discriminant factor), and then on the base of the most relevant bio-pharmacological activities reported by the authors.

2. 3-Pyrazolyl-Ureas Derivatives

2.1. 3-Pyrazolyl-Ureas as Human Carbonic Anhydrase Inhibitors

Metabolic transformations such as the reversible hydration of CO_2 to HCO_3^-, aldehyde hydration, hydrolysis of alkyl and aryl esters, urine formation, and other similar physiological and physiopathological reactions are catalyzed by the human carbonic anhydrase II (hCA II), a metalloenzyme that uses zinc to catalyze these transformations. In detail, a tetrahedral zinc ion (Zn^{2+}) coordinated with three histidine residues and one water molecule represents the catalytic domain, which is the primary target of human carbonic anhydrase inhibitors (hCAIs). Sulfonamide or sulfamate group linked to an aromatic or heteroaromatic ring seems to play an important role in binding to the Zn^{2+}, as proven by the most powerful inhibitors Celecoxib, Acetazolamide, and ureido-benzenesulfonamides (UBSAs, Figure 1).

Supuran and coworkers recently prepared, as inhibitors of the hCA II, several UBSA derivatives which, shifting the hydroxide molecule due to the sulfonamide nitrogen, bind directly to the active side, masking it [9]. They also reported very important structure–activity relationship (SAR) information about the R group orientation in the different subpockets of the active site and the capacity of linking the Zn^{2+}, probably due to the flexibility of the ureido linker (Figure 1).

On this basis, Sahu and coworkers designed a new set of UBSA ligands with different substituents on the ureido group, including the 3-(1-*p*-Tolyl-4-trifluoromethyl-1*H*-pyrazol-3-yl) moiety (compound **1**, Figure 1). The crystallographic structure of hCA was used to perform a theoretical investigation of electronic structure parameters of hCA II complexed with the new molecules. In particular, the authors used their N-layered integrated molecular orbital and molecular mechanics (ONIOM) method [10].

The analyzed compounds were effectively able to interact with hCA II and showed calculated inhibition constants consistent with the experimental data. The study evidenced that the nature of the R moiety, substituting the second ureido nitrogen in UBSAs, controls the inhibitor potency. This is probably due to the flexibility of the ureido linker and the possibility of the orientation of the R group in different subpockets of the active site cavity. In addition, it was evident that the metal–ligand bond distances and the ligand–metal–ligand angles for **1** were in reasonable agreement with the other sulfonamide ligand, therefore indicating that **1** has a similar binding mode with other sulfonamide inhibitors. The best hCA II inhibitor was the 2-isopropyl-phenyl-substituted compound, but **1** was also revealed as a potent inhibitor of hCA II like other UBSA inhibitors, having a calculated inhibition constant of 220.93 or 224.69 nM, following the optimized geometry at the M1 or M2 level, respectively, in the docking analyses.

2.2. 3-Pyrazolyl-Ureas as Cannabinoid Receptor Antagonist

The cannabinoid (CB) receptors are G protein-coupled receptors which respond to cannabinoids, such as Δ-9-tetrahydrocannabinol^9-THC), the major active constituent extracted from *Cannabis sativa*. The most important and studied are two cannabinoid receptor subtypes named CB1 and CB2. Of these, CB1 receptors are also known as central cannabinoid receptors because of their primary expression in the central nervous system (CNS).

In detail, CB1 receptors are mainly located in the brain (in the basal ganglia, hippocampal cell layer, cerebellum, and cerebral cortex), spinal cord, and peripheral nervous system. CB1 is also the most widely expressed G protein-coupled receptor in the brain. On the other hand, CB2 receptors are minimally expressed there; they are found mostly in the peripheral neurons, the reason for which they are named peripheral cannabinoid receptors. They block the release of neurotransmitters and were found principally in cells related to the immune system. Besides the different organ distribution, the two subtypes differ also for the aminoacidic sequence, signal transduction mechanism, and sensitivity to certain agonists and inhibitors.

Molecules acting as unselective agonists against CB1 and CB2 receptors have undesirable side effects such as drowsiness and impairment of monoamine oxidase function and of non-receptor mediated brain function. The addictive and psychotropic properties of some cannabinoids also limit their therapeutic value. On the contrary, compounds having antagonist activity could provide a pharmacological response useful to treat cannabis abuse, lipid and glucose metabolic disorders, cardiovascular disorders, psychotic disorders (stress, anxiety schizophrenia), epilepsy, migraine, vomiting, memory and cognitive disorders related to neurodegenerative diseases, and more.

Different studies on endogenous ligands have also shown that CB1 regulated food intake and energy consumption for weight management purposes. These studies led to the development, by Sanofi-Aventis, of the first CB1 receptor antagonist, Rimonabant, approved in the UK in July 2006 for the treatment of obesity and as a smoking cessation drug. Because of its serious central side effects, the research is now more interested in the discovery of new peripherally selective CB1 receptor antagonists.

With this intention, Li and researchers [11] tried to find new small molecules as CB1 antagonists with a 3-pyrazolyl-ureas substituted scaffold and also performed the activity test on EGFP-CB1_U20S cells expressing the CB1 receptor. From these tests, the IC$_{50}$ values of the compounds ranged from 5 μM to 10 μM, of which compounds **2** and **3** (Figure 1) were the best CB1 receptor antagonists with IC$_{50}$ values of 5.19 and 3.61 nM, respectively.

Another patent was filled by Makriyannis and coworkers in 2006. It related to biologically active pyrazole analogs having Markus structure **4** (Figure 1), acting as antagonists for CB1 and/or CB2 receptors and having selectivity for one of them [12]. Some of the inventive compounds showed high affinity for at least one of the cannabinoid receptors. The CB1 receptor binding affinities (K_i) for the synthesized analogs ranged between 6 and 1844 nM, while the CB2 K_i ranged between 36.5 and 13,585 nM. Interestingly, some analogs were able to interact with the CB1 receptor without affecting the peripheral (CB2) receptor to the same degree. On the contrary, other analogs were able to interact with the CB2 receptor without affecting the CB1 receptor. The most interesting were the pyrazolyl-ureas **4a–d** (Figure 1) as shown by affinity constant (K_i) and selectivity reported in Table 1.

Table 1. Affinity constant and selectivity of CB1/CB2 antagonists **4a–d**.

	Affinity K_i (nM)		Selectivity
Comp. 4	CB1	CB2	CB1
a	9	4920	546.7
b	29	10863	374.6
c	6.8	4319	635.2
d	26	21791	838.1

2.3. 3-Pyrazolyl-Ureas as Antibacterial Agents

Among 3-ureido-substituted pyrazoles, several examples were functionalized at the ring position 4 with a methyl, ethoxycarbonyl, or carboxy group and were obtained by adding alkyl- or arylisocyanates to the respective 4-aminopyrazoles. Aiming to insert a hydroxymethyl function in position 4, Bratenko and coworkers considered this approach less appropriate, because 3-amino-4-hydroxymethylpyrazoles were unknown in the literature, and even if such compounds were available, selectivity problems would arise when adding isocyanates to the amino group. On this basis, they developed an effective method for the preparation of 1-alkyl(aryl)-3-[4-(hydroxymethyl)-1H-pyrazol-3-yl]urea derivatives (5a–h, Figure 1) based on the interaction of 4-(hydroxymethyl)pyrazole-3-carbonyl azides with primary aliphatic or aromatic amines under Curtius reaction condition. Then, compounds 5 were evaluated as antibacterial and antifungal agents against *Staphylococcus aureus* (ATCC 25923), *Escherichia coli* (ATCC 25922), *Bacillus subtilis* (ATCC 8236F800), and *Candida albicans* (ATCC 885-653), showing a moderate antimicrobial activity. In detail, all the compounds tested present minimum bacteriostatic/fungistatic (MIC), bactericidal (MBC), and fungicidal (MFC) concentrations of 250 µg/mL [13].

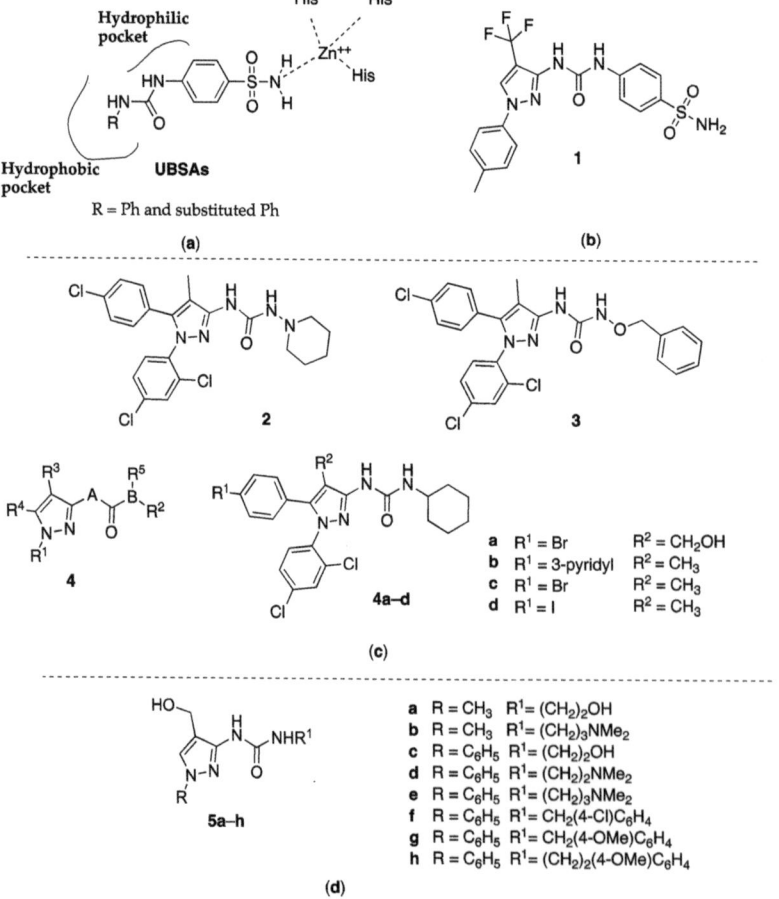

Figure 1. (a) General structure of UBSAs and their localization in the binding pocket of hCA (adapted from Sahu, 2013 [10]). 3-pyrazolyl-urea reported as: (b) hCA inhibitors; (c) CB1/CB2 receptor antagonists; (d) antibacterial/antifungal agents.

3. 4-Pyrazolyl-Ureas Derivatives

3.1. 4-Pyrazolyl-Ureas as Epoxide Hydrolase (sEH) Inhibitors

In mammals, epoxide hydrolases (EH) catalyze the transformation of several carcinogenic epoxides into their corresponding nontoxic diols. Soluble EH (sEH) is also involved in the metabolism of arachidonic and linoleic acid epoxides, and the diols derived from epoxy-linoleates (leukotoxins) have been associated with several inflammatory and vascular pathologies. Therefore, sEH has been suggested as a promising pharmacological target for potential anticancer and anti-inflammatory agents. sEH consists in two globular proteins in which C-terminal domains can bind the different substrates to exerts the hydrolyzing action. 1,3-dicyclohexylurea was reported in 1999 by Morisseau and co-workers as the first stable inhibitor of sEH [14] and numerous symmetrical and asymmetrical diphenyl ureas have been developed [15], giving three new generations of molecules characterized by improved potency and pharmacokinetic properties. Among them, asymmetrical ureas with a flexible side chain, such as 12-(1-adamantan-1-ylureido)-dodecanoic acid (AUDA, Figure 2), have been proposed to increase solubility [16]. In 2019, in pursuing the synthesis and the study of pharmacological properties of adamantyl-containing 1,3-disubstituted ureas, D'yachenko and coworkers [17] reported the synthesis of a series of 1,3-disubstituted ureas containing the 1,3,5-disubstituted pyrazole fragment as possible sEH inhibitors (compounds **6**, Figure 2). Authors declare an inhibitory activity ranging from 16.2 to 50.2 nmol/L and increased bioavailability, while major detailed information about structure–activity relationships is lacking.

Starting from the pyrazole structure of the well-known COX-2 selective inhibitor Celecoxib and the adamantyl urea function of the sEH inhibitor *t*-AUCB (Figure 2), Hwang and coworkers [18] designed new interesting chimera compounds in which an aryl/adamantyl-substituted urea moiety is linked to the pyrazole nucleus in position 3 directly or through an alkyl chain, as reported in general structure **7** (Figure 2). Compounds having the biarylpyrazole ring and the urea group directly connected showed poor sEH inhibitory activity. On the contrary, the addition of methylene groups (total of three) between the COX-2 and the sEH pharmacophore yielded an overall twentyfold improvement in potency against sEH, without changing the efficacy against COX-2. The most active resulted compound had $R^1 = NH_2$, R^2 = phenyl, R^3 = *p*-(trifluoromethyl)phenyl, and three carbon atoms in the chain, which showed IC_{50} values of 0.9 nM, 1.26 µM and >100µM against sEH, COX-2, and COX-1, respectively. These results suggested that the 1,5-diarylpyrazole group is sterically hindered and, by increasing the length of the linker between the 1,5-diarylpyrazole group and the urea group, it is possible to avoid steric repulsion. These compounds have been classified in this subchapter to simplify the reading, despite the presence of some 3-pyrazolyl-ureas in the series.

3.2. 4-Pyrazolyl-Ureas as Antiepileptic Drugs

The incidence of epilepsy in the worldwide population is estimated to be about 0.5–1%, 70% of whom achieve satisfactory seizure control thanks to the large number of new drugs recently approved (e.g., Lamotrigine, Gabapentin, Tiagabine, and Milacemide). However, the need for more effective and less toxic antiepileptic drugs still exists since a large number of patients are drug-resistant. Urea and thioureas derivatives [19,20] have emerged as structurally novel anticonvulsant agents. It was hypothesized that ureas and thioureas display anticonvulsant activity by interacting on the putative aryl binding site, the hydrogen bonding domain, and an auxiliary aryl or other hydrophobic binding site [21]. On these bases, urea and thiourea moieties were attached to a pyrazole ring and to an aryl or alkyl group obtaining compounds **8** (Figure 2) [22]. The anticonvulsant activity of the new compounds was determined using pentylenetetrazol-induced seizure (PTZ) and maximal electroshock seizure (MES) tests, two rodent models widely used to predict protection against generalized tonic-clonic and generalized absence seizures in humans. Thiourea derivative **8a** displayed noteworthy activity in both PTZ and MES screens, while the urea derivatives **8b** and **8c** were much more potent than the thiourea derivatives (82 and 80% protection at 25 mg/kg, respectively). In addition, **8b** was found more effective than **8c** in terms of MES, and both urea derivatives gave good protection when given intraperitoneally at 25 and 50 mg/kg.

3.3. 4-Pyrazolyl-Ureas as Anti-Inflammatory Agents

A large number of compounds were patented in 2005 by Pharmacia Corporation [23] as potential drugs for inflammatory diseases, in particular as inhibitors of IKK-2, an IkB kinase involved in the inflammatory response that is the main cause of rheumatoid arthritis (RA) onset.

The transcription factor NF-κB, which plays a prominent role in immune and inflammatory responses, is normally sequestered in an inactive form in the cytoplasm by a member of the IkB inhibitory proteins, and this prevents the transcription of responsive genes in the nucleus. Cell stimulation leads to the phosphorylation and degradation of IkB, thereby releasing NF-κB to the nucleus for activation of gene transcription. The IkB kinases (IKK-1 and IKK-2), which phosphorylate IkB and thereby initiate its degradation, represent the critical, common denominator in the activation of NF-κB and have been considered as useful targets for new anti-inflammatory agents.

On the basis of literature evidence that active IKK-2 inhibitors could be useful for the treatment of inflammatory diseases [24] and that pyrazolecarboxamide derivatives and other heteroaromatic carboxamide derivatives are useful IKK-2 inhibitors [25], Clare and coworkers patented compounds having the Markus structure **9** (Figure 2) [23] in which a carboxamide and ureas/thioureas function were inserted alternatively in position 3 or 4 of the pyrazole nucleus. In addition, the N1 in the pyrazole was generally aryl substituted. Among the 260 examples reported in the patent compounds, **9a**, **9b**, and **9c** (Figure 2) resulted the most active, having IC_{50} values of 0.013, 0.044, and 0.067 μM, respectively, against human IKK-2, and IC_{50} values of 0.023, 0.033, 0.089 μM, respectively, against rat IKK-2.

In 2014, Sharma and co-inventors [26] patented the preparation of compounds having different heterocyclic moieties (triazole, pyrazole, imidazole, oxazole, isoxazole, thiazole, and pyrrole) linked by a urea function to five- or six-membered hetero or heteroaryl systems. The invention claimed compounds as interleukin 17 (IL-17) inhibitors and tumor necrosis factor alpha (TNFα) inhibitors and further related to their use in the treatment of inflammatory or autoimmune diseases and other related pathologies.

Among the numerous claimed compounds, the 1-(4-trifluoromethylphenyl)-5-methyl-4-pyrazolyl-ureas **10** (Figure 2) exhibited the best IL-17 and TNFα inhibitory activity in the cytokine release assay. In particular, compounds **10a** and **10b** showed IC_{50} values ranging from 0.1 to 1 μM against both IL-17 and TNFα production.

3.4. 4-Pyrazolyl-Ureas as Protein Kinases Inhibitors

3.4.1. CDK8 Inhibitors

Cyclin-dependent kinase 8 and cyclin C (CDK8/CycC) represent a potent kinase oncogene system involved in transcriptional activity and regarded as an attractive drug target. In particular, selective inhibitors could decrease mitogenic signals in cancer cells with reduced toxicity toward normal cells. Type I ligands are able to bind the kinase in the active conformation (DMG-in) and occupy the ATP-binding site, while type II ligands bind to the inactive conformation (DMG-out) and occupy the ATP-binding site and the allosteric site (deep pocket). By rearrangement of the DMG motif, the CDK8 changes from the inactive to the active state. The deep pocket (inaccessible in the DMG-in conformation) is the binding site for many well-known kinase ligands, such as Sorafenib and Imatinib, belonging to the type II category.

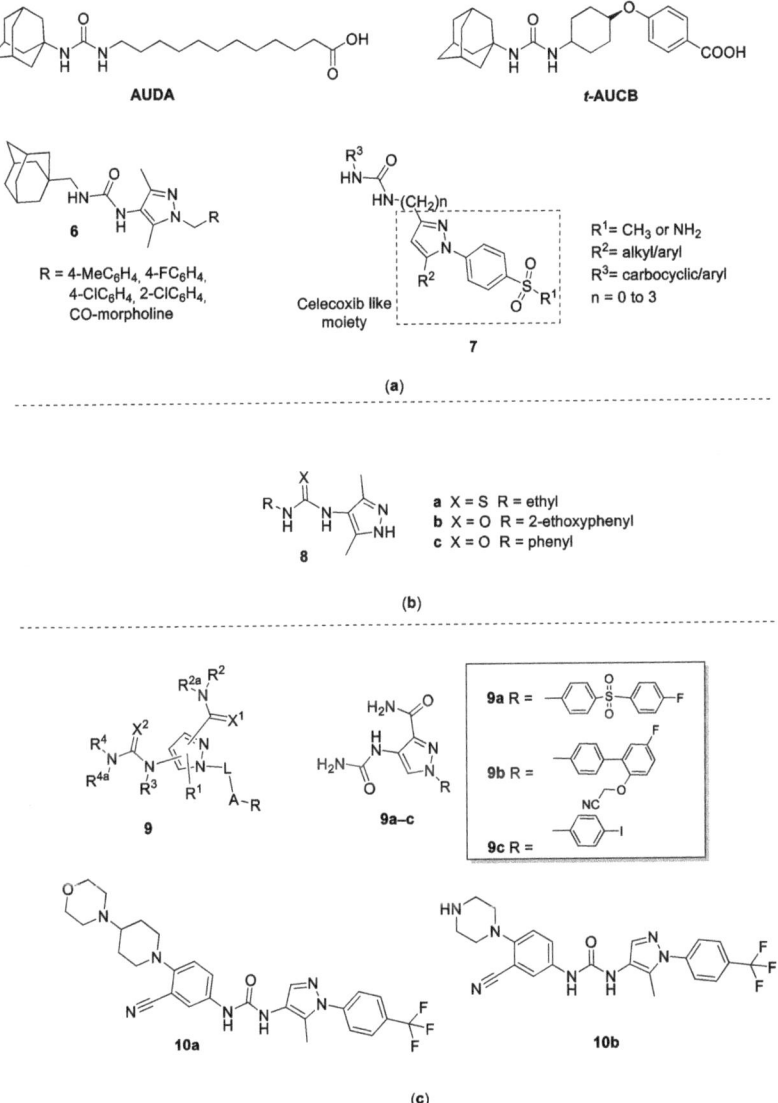

Figure 2. 4-Pyrazolyl-ureas reported as: (**a**) sEH inhibitors; (**b**) anticonvulsant; (**c**) anti-inflammatory agents.

The first pyrazolyl-urea series of type II CDK8 ligands was published in 2013, being actually 5-pyrazolyl-ureas variously substituted in position 4 (**11**, Figure 3) [27]. The ligands identified by Schneider et al., were found to anchor in the CDK8 deep pocket and to extend with diverse functional groups toward the hinge region and the front pocket. These variations can cause the ligands to change from fast to slow binding kinetics, resulting in an improved residence time. Major details about these inhibitors are reported in this review in Section 4 (5-pyrazolyl-ureas).

Recently, Chen and coworkers [28] implemented a novel virtual screening protocol for drug development and applied it to the discovery of new CDK8/CycC type II ligands, starting from the pyrazolyl-ureas previously reported by Schneider et al. [27]. The authors first analyzed the thermodynamic binding of the ligands using molecular dynamics simulations and a free-energy calculation method, then extended the key binding information to assist virtual screening. Starting with the urea moiety, they carried on substructure-based searches on 187,000 compounds and, by the newly developed superposition method and single-point energy evaluation, they isolated 19 possible hits, including the 4-pyrazolyl-urea **12** (Figure 3). Finally, the free-energy calculations provided only three new compounds in which the urea moiety had the expected contacts with CDK8. Among them, the most potent drug-like compound is not a pyrazole, but a thiadiazol-urea derivative that has a K_d value of 42.5 nM, which is similar to those of the most potent pyrazolyl-urea used as a reference ligand. Interestingly, the top two compounds are different by only one atom (the phenyl urea moiety has a 4-methoxy substituent instead of a 4-ethyl one) but have a nearly threefold difference in binding affinity (K_i = 114 nM). This was accurately predicted by both energy evaluation methods. Therefore, the novel virtual screening method is accurate and useful in drug design projects.

3.4.2. RAS/RAF/MEK/ERK-MAPK Inhibitors

Other interesting 4-pyrazoly-ureas, together with a large library of 5-pyrazolyl-ureas, are reported in some patents developed by Deciphera Pharmaceuticals (compounds **13** [29,30] and **14** [31,32], Figure 3). They act on different kinases especially involved in the RAS-RAF/MEK/ERK-MAPK pathways. In these compounds, the pyrazole moiety is linked by the urea function in position 4 to a phenyl, in turn substituted with a dihydropyrimido-pyrimidine scaffold (**13**, Figure 3) or the isosteric dihydropyrido-[2,3-*d*]pyrimidine (**14**, Figure 3). Biological data for these 4-pyrazolyl-urea derivatives were not given in the cited patents. Major considerations are reported in Section 4.4 focused on 5-pyrazolyl analogues.

3.4.3. JNK Inhibitors

The mitogen-activated protein (MAP) kinase family member c-Jun-N-terminal kinase (JNK) has been shown to be a therapeutic target for a variety of diseases including neurodegeneration, metabolic disorders, inflammation, cardiovascular disease, and cancer. Three isoforms, namely as JNK1, JNK2, and JNK3, were identified in the JNK family. The synthesis of new JNK inhibitors has been pursued to develop efficacious therapeutics for Parkinson's disease (PD) and other neurodegenerative diseases, such as Alzheimer's disease (AD), Huntington's disease (HD), amyotrophic lateral sclerosis (ALS), and multiple sclerosis (MS), but also for treatment of myocardial infarction (MI), metabolic disorders, cancer, and inflammatory diseases [33,34].

In 2015, Feng and coworkers [35] developed a large series of pyrazole derivatives aiming to obtain JNK2/JNK3 inhibitors with a high degree of selectivity in respect to the JNK1 isoform. These inhibitors were expected to afford lower toxicity risk for treatment of the conditions associated with JNK activation. Particularly, by targeting the substrate site in JNK, and potentially blocking JNK mitochondrial translocation, they hypothesized to provide an inhibition mechanism that prevents mitochondrial dysfunction and cardiomyocyte cell death.

The authors patented numerous molecules in which a substituted phenyl is always present in N1, while the pyrazole positions 3 and 5 were variously substituted. Among them, compounds of general structure **15** (Figure 3) showed the best results having, in the inhibition test, EC_{50} values <200 nM against JNK3/2, while >1000 nM against JNK1.

3.5. 4-Pyrazolyl-Ureas as Anticancer Agents

In 2008, Aventis Pharma patented a large series of benzimidazole-pyrazole derivatives bearing in position 4 of the pyrazole nucleus the NHCO-R moiety in which R were differently substituted alkyl- or cycloalkyl-amines [36]. Compounds were claimed as anticancer agents and were active in the cancer cells line HeLa with IC_{50} values ranging from 5 to 50 µM. In particular, the IC_{50} of the compound **16** (Figure 3), here reported as an example, is between 5 and 30 µM.

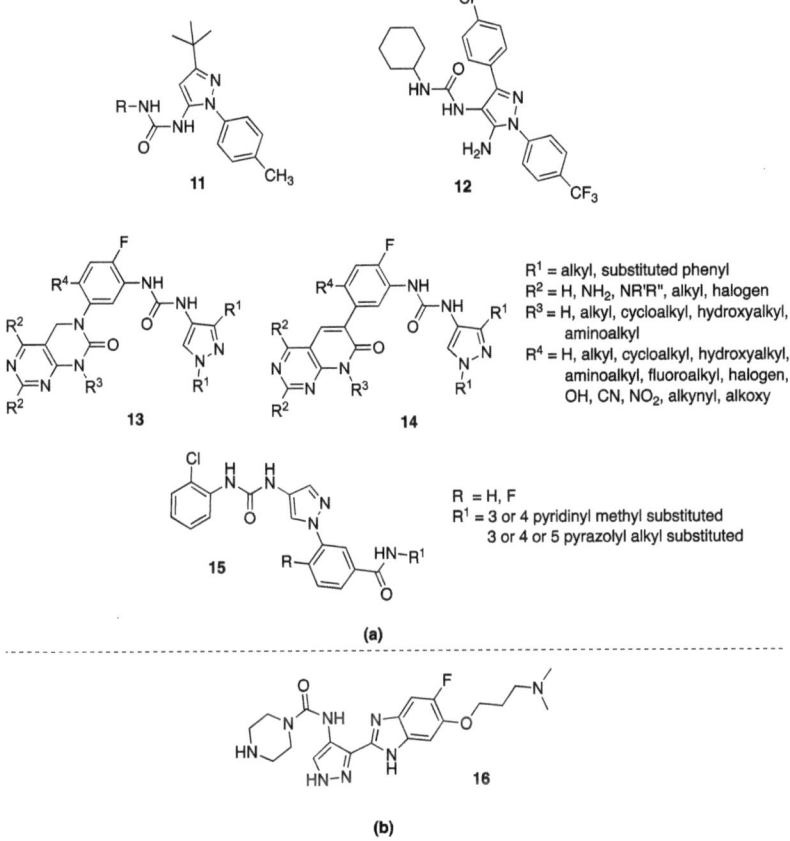

Figure 3. 4-Pyrazolyl-ureas reported as: (a) protein kinases inhibitors; (b) anticancer agents.

4. 5-Pyrazolyl-Ureas

4.1. 5-Pyrazolyl-Ureas as p38MAPK Inhibitors

The p38 mitogen-activated protein kinase (MAPK) is a serine–threonine kinase that includes JNK and extracellular-regulated protein kinase (ERK). It plays an important role in the regulation of a wide range of immunological responses, in particular in production and activation of inflammatory mediators (cytokines such as TNFα and IL-1β), which initiate leukocyte recruitment and activation. Consequently, it shows a crucial role in inflammation (especially in auto-immune diseases such as RA, Crohn's disease, inflammatory bowel disease, and psoriasis), but also in cancer progression (cell proliferation, differentiation, and apoptosis). In detail, p38 is a super family divided into four different isoforms: p38α, p38β, p38γ, and p38δ.

In the past, many p38 inhibitors with different chemical structures have been designed on the basis of their ability to compete with p38 ATP binding or allosteric site [37]. Specifically, p38MAPKα inhibitors, like many other kinase inhibitors, can be classified into two types based on their mode of action: ATP-competitive inhibitors and non-ATP-competitive, or allosteric, inhibitors, which utilize the ATP binding cleft and a hydrophobic allosteric pocket created when the activation loop adopts the inactive "Asp-Phe-Gly (DFG)-out" conformation. Because allosteric inhibitors do not compete directly with ATP or substrate, they can offer a significant kinetic advantage over ATP competitive inhibitors. In addition, because the allosteric pocket is less conserved than the ATP binding region, allosteric inhibitors usually have better kinase selectivity profiles than ATP competitive inhibitors [38].

The first compound discovered in 2000 by Dumas and coworkers at Bayer Pharmaceutical has been a little 5-pyrazolyl-urea which presented a *tert*-butyl substituent on C-3 and two chlorine atoms on the phenyl ring of urea moiety (compound **17**, Figure 4). It showed IC_{50} value of 53 nM on p38 enzyme and inhibited TNFα and IL-1 induced IL-6 production in SW1353 cells (human chondro-sarcoma cell line) with an IC_{50} value of 820 nM [39].

Aiming to obtain more active compounds in the cellular functional assays, the same authors synthesized a large series of 5-pyrazolyl-ureas exploring different substitution on previous lead compound. Firstly, they investigated the substitution on N1 pyrazole, leaving unchanged the urea moiety. In detail, the methyl substituent on N1 has been replaced with a more bulky and lipophilic benzene ring differently decorated. From this investigation, the authors obtained compound **18** (Figure 4), having on the N1 pyrazole a phenyl ring, carrying in turn a free amino group in meta position, that showed IC_{50} = 13 nM on p38 and 42 nM in inhibition of TNFα and IL-1 induced IL-6 production in SW1353 cells [40].

At the same time, structural changes were made on the benzene of the urea segment, removing the halogens and introducing longer and more bulky substituents, such as the phenoxy group (compound **19**, Figure 4) or the 4-methylene-pyridine one (compound **20**, Figure 4). With these structural modifications, the authors obtained p38 inhibition with IC_{50} in the nanomolar range, 270 and 42 nM, respectively, and improved in vivo activity (in acute models of cytokine release and a chronic model of arthritis), confirming that 5-pyrazolyl-urea is the best scaffold in respect to thiophene and pyrrole ones [41,42].

Starting from interesting results, the big pharma Boeheringer developed different 5-pyrazolyl-urea derivatives and patented 1-(3-(*tert*-butyl)-1-(*p*-tolyl)-1*H*-pyrazol-5-yl)-3-(4-(2-morpholinoethoxy)naphthalen-1-yl)urea (BIRB 796, Dorapaminod, Figure 4), one among the most active p38 MAPK inhibitors, now used as pharmacological tools in medicinal chemistry research [43,44].

BIRB 796 resulted the most active compound with picomolar affinity for p38 in enzymatic assay and low nanomolar inhibitory activity in cell culture [45]. In deep investigations, it showed a K_d = 0.046 nM on p38, inhibited TNFα production in THP-1 cell lipopolysaccharide (LPS)-stimulated with EC_{50} value of 18 nM and in human whole blood with EC_{50} value of 780 nM; moreover, it evidenced a good selectivity profile, inhibiting only JNK2α2 kinase with an IC_{50} value of 0.1 µM. In addition, in the LPS-stimulated TNFα synthesis mouse model, a 65% inhibition of TNFα production was observed when BIRB 796 was dosed orally at 10 mg/kg. On a model of established collagen-induced arthritis using B10.RIII mice, BIRB 796 (at 30 mg/kg os) showed a 63% inhibition of arthritis severity [45,46].

In a recent study, the authors evidenced that BIRB 796 can reduce LPS-mediated IL-8 production in THP-1 cells, but not in Raw 264.7 cells. Further analysis of signal molecules revealed that BIRB 796 sufficiently suppressed LPS-mediated phosphorylation of p38MAPK in both cell types, whereas it failed to block inhibition of kappa B (I-κB) degradation in Raw 264.7 cells. Taken together, these results suggested that the anti-inflammatory function of BIRB 796 depends on cell types [47].

In 2001, Regan and coworkers reported a new allosteric binding site for the diaryl urea class of highly potent and selective inhibitors against human p38MAP kinase. The formation of this binding site requires a large conformational change, not previously observed for any serine/threonine kinases, in the highly conserved DGF motif within the active site of the kinase. Authors demonstrated that by improving interactions in this allosteric pocket, as well as establishing binding interactions in the ATP pocket, the affinity of the inhibitors is enhanced [45].

The same authors, using X-ray crystallographic data, deeply investigated the binding kinetics of BIRB 796 to human p38α: interestingly, compound showed slow k_{on} and long k_{off} rates. In the presence of the compound, the DGF must undergo a substantial conformational change, fundamental for slow binding mode. In the binding conformation, the side chain phenyl ring of Phe169 moves about 10 Å ("DFG-out" conformation) and participates in a lipophilic interaction with the aryl portion of the inhibitor, in particular with *tert*-butyl group at C-3 of the pyrazole nucleus. This group occupies the same lipophilic pocket (Phe169 pocket) between the two lobes of the kinase which is exposed upon reorganization of the activation loop; this interaction possibly contributes to the long k_{off} rates.

As evidenced in the X-ray structure of BIRB 796 in p38αMAPK [45], the carbonyl of the urea moiety of BIRB796 accepts hydrogen bonds from the backbone NH of Asp168, while one of the NH forms hydrogen bonds with the Glu71 side chain. These interactions appear to be the most crucial for this class of compounds, since replacement of either NH of the urea by a methylene or an N-methyl group results in a complete loss of p38 inhibition. In addition, other important interactions are lipophilic interaction (π–CH$_2$ interactions) of the pyrazole-tolyl group with the alkyl side portion of Glu71 and of the naphthyl ring system with Lys53, being that the naphthyl system is preferred as compared to simple phenyl ring; additional hydrogen bonding of the morpholine oxygen with the Met109 residue and of the ethoxy morpholine tail with ATP binding region are crucial for activity; in detail, the gauche conformation of the ethoxy linker of BIRB 796 effectively orients the morpholine group so that the morpholine oxygen can achieve a strong hydrogen bond with the NH of Met109. This modification afforded significant improvements in binding affinity, cellular activity, and in vivo reduction of TNFα production and arthritis severity.

For all these reasons, BIRB 796 became an interesting clinical candidate for the treatment of inflammatory diseases, being subjected to 11 clinical trials (some of them in phase II investigation) now completed or terminated [48]; all these trials were focused on autoimmune diseases (RA, psoriasis, and Crohn's disease, in particular) and also on toxicity evaluation on healthy volunteers. Unfortunately, no clinical efficacy was evidenced, but only a significant and transiently dose-dependent decrease of C-reactive protein level after one week of BIRB 796 therapy.

To obtain this compound in a more rapid and simply manner, in 2006, Bagley and co-workers reported a new synthetic approach using microwave irradiation, providing BIRB 796 rapidly in high purity [49].

In 2007, Kulkarni and coworkers reported a combined study of three-dimensional quantitative structure–activity relationship (3D-QSAR) and molecular docking to explore the structural insights of p38 kinase inhibitors with 5-pyrazolyl urea structure, including BIRB 796 and the major part of compounds previously reported. These 3D-QSAR studies involved comparative molecular field analysis (CoMFA) and comparative molecular similarity indices (CoMSIA). Interestingly, the authors confirmed that the hydrophobic properties are the most important for enzyme inhibition and essential structural features to design new potent analogues with enhanced activity [50].

In 2013, Schneider and co-workers, using as reference compounds BIRB 796 and other 5-pyrazolyl-ureas with very similar structure and slow binding kinetics (see compounds 11, Figure 3), reported a structure-kinetic relationship study of CDK8/CycC [27]. Authors evidenced that the scaffold of the 5-pyrazolyl urea compounds is anchored in the kinase deep pocket and extended with diverse functional groups toward the hinge region, causing in this way the change from fast to slow binding kinetics and resulting in an improved residence time. Differently to previous studies, the flip of the DFG motif (named "DMG" in CDK8 enzyme) to the inactive DFG-out conformation appears to have relatively little influence on the binding velocity, whereas hydrogen bonding with the kinase hinge region seems to contribute to the residence time with less impact than hydrophobic complementarities within the kinase front pocket.

More recently, according to the crystal structure of the p38MAPKα/BIRB796 complex (PDB 1VK2) Zhong and coworker synthesized a new series of compounds with the aim to preserve the activity and selectivity, and to eliminate the hepatotoxicity caused by the naphthyl ring [51,52].

They replaced the naphthalene ring of BIRB 796 with two other aromatic hydrophobic scaffolds: 2*H*-chromene and chromane. The best results have been obtained with 1-aryl-3-(2*H*-chromen-5-yl) urea derivatives, in particular with compound 21 (Figure 4), bearing on N1 pyrazole a *p*-nitro phenyl ring. Interestingly, the authors have not tested their derivatives on p38 enzyme, but only towards TNFα production in LPS-stimulated THP-1 cells. The most active compound inhibited TNFα release with an IC$_{50}$ value of 0.033 µM, showing similar potency of BIRB 796 (IC$_{50}$ = 0.032 µM) [53]. The same authors have previously published a short and efficient synthesis of this class of compounds, using Claisen thermal rearrangement of arylpropargyl ethers as a key step to synthesize the chromene core and obtained yield ranging from 27 to 37% [54].

Meanwhile, other compounds with very similar structure to BIRB 796 were patented by Deciphera Pharmaceuticals as p38 inhibitors, compound **22** (Figure 4) being the most representative. In detail, it bears a more hydrophilic substituent in the phenyl on N1 pyrazole and an unsubstituted naphthalene on the urea moiety, and showed an IC$_{50}$ value of 45 nM on p38 enzyme [55–57].

In 2012 Arai and co-workers, using X-ray cocrystal studies, replaced the naphthalene group of BIRB-796 with a benzyl group to increase synthetic flexibility and to obtain more selective p38α inhibitors [58]. They reported the synthesis and pharmacological and toxicological properties of different pyrazole-benzyl urea derivatives as novel allosteric p38α inhibitors, compound **23** (Figure 4) being the most interesting. According to the docking study, the benzyl group appears to occupy the kinase specificity pocket, while the 2-morpholinopyrimidine moiety of **23** can form a hydrogen bond with Met109 in two different modes: (1) between the morpholine oxygen and the backbone NH group; and (2) between N1 of the pyrimidine ring and the same NH group. **23** exhibited highly potent p38a inhibitory activity (IC$_{50}$ = 0.004 µM) and inhibited the production of TNFα in LPS-treated mouse in a dose-dependent manner after a single oral administration (ED$_{50}$ = 16 mg/kg). Furthermore, a five-day repeated oral dose administration suggested that **23** has low hepatotoxicity.

Other patents by RespiVert and TopiVert Pharma in the UK were related to inhibitors of p38 MAP kinase (particularly α and γ isoforms) and Src kinase and to their use in therapy, especially in the treatment of lung inflammatory disease (such as asthma and COPD) or gastrointestinal tract pathologies (such as ulcerative colitis, irritable bowel disease, IBD, and Crohn's disease) or uveitis.

In detail, compounds **24a** and **24b** (Figure 4) showed a long tail inserted on urea moiety and characterized by an amino-pyridinepyrazine substituent; **24a** exhibited inhibitory activities against p38α and p38γ with IC$_{50}$ values of <0.48 nM and 16 nM, respectively [59], whereas **24b** showed IC$_{50}$ values on p38α and p38γ of 26 and 152 nM, respectively, of 55 nM on HcK and of 199 nM on c-Src; in addition, it resulted weak inhibitor of Syk and GSK3α. Therefore, results reported in all these patents confirm the therapeutic potential of kinases inhibitors in inflammatory diseases associated with viral infections. Indeed, it was previously disclosed that compounds inhibiting both c-Src and Syk kinases were effective agents against rhinovirus replication [60] and compounds inhibiting p-Hck were effective against influenza virus replication [61] (see also Section 4.1 in this review). In 2015 Longshaw and coworkers reported different composition formulation and pharmacodynamic evaluation of **24b** [62]. In this second patent, Longoshaw and coworkers patented different formulations (for oral, topic, and aereosol administrations) of compound **24b**, which resulted in the most interesting from all previous patented compounds. Results suggested that the new formulations have similar anti-inflammatory properties to previous patented compounds but, interestingly, they showed a superior therapeutic index (data not reported).

A large library of compounds having general structure **25** (Figure 4), very similar to previous **24**, were patented as p38 inhibitors and claimed as anti-inflammatory/anti-viral agents, for the treatment of several infectious conditions, including influenza. In particular, these derivatives may be suitable for reducing viral load and/or ameliorating symptoms after infection and they may also be administered in combination with one or more other active ingredients (e.g., steroids, β-agonists and/or xanthine derivatives). Compounds **25**, very structurally related to BIRB 796, presented a naphthyl urea moiety characterized by an aryloxy embedded substituents. In detail, **25a** [63] **25b** [64] and **25c** [65] are the most representative of this library and showed an IC$_{50}$ <5 µM against p38α enzyme.

Among this new series of derivatives, **25d** 1-(3-*tert*-butyl-1-*p*-tolyl-1*H*-pyrazol-5-yl)-3-(4-(2-(phenylamino)pyrimidin-4-yloxy)naphthalen-1-yl)urea with a more simple chemical structure was subjected to an additional patent, displaying a very different phenotype and demonstrating great inhibition versus p38MAPK (in particular α isoform, IC$_{50}$ = 26nM) Hck (IC$_{50}$ = 55 nM), c-Src (IC$_{50}$ = 199 nM) without significant effect against GSK3α and Syk [66].

Finally, in another patent, the same company introduced different more embedded compounds; among them **25e** (Figure 4), which presents an additional indole group, was the most representative. All compounds demonstrated a very similar inhibitory profile in respect with the previous **25a–c** in kinase enzyme assays [67].

Other compounds, like **26** (Figure 4), characterized by an isopropyl substituent on C-3 and a morpholino group at the end of urea moiety, were tested for enzyme inhibition against p38 MAPKα, c-Src, and Syk, having IC$_{50}$ values of 2 nM, <1 nM, and 3 nM, respectively. In vitro and in vivo studies, as well as pharmacokinetic analysis were then performed [68].

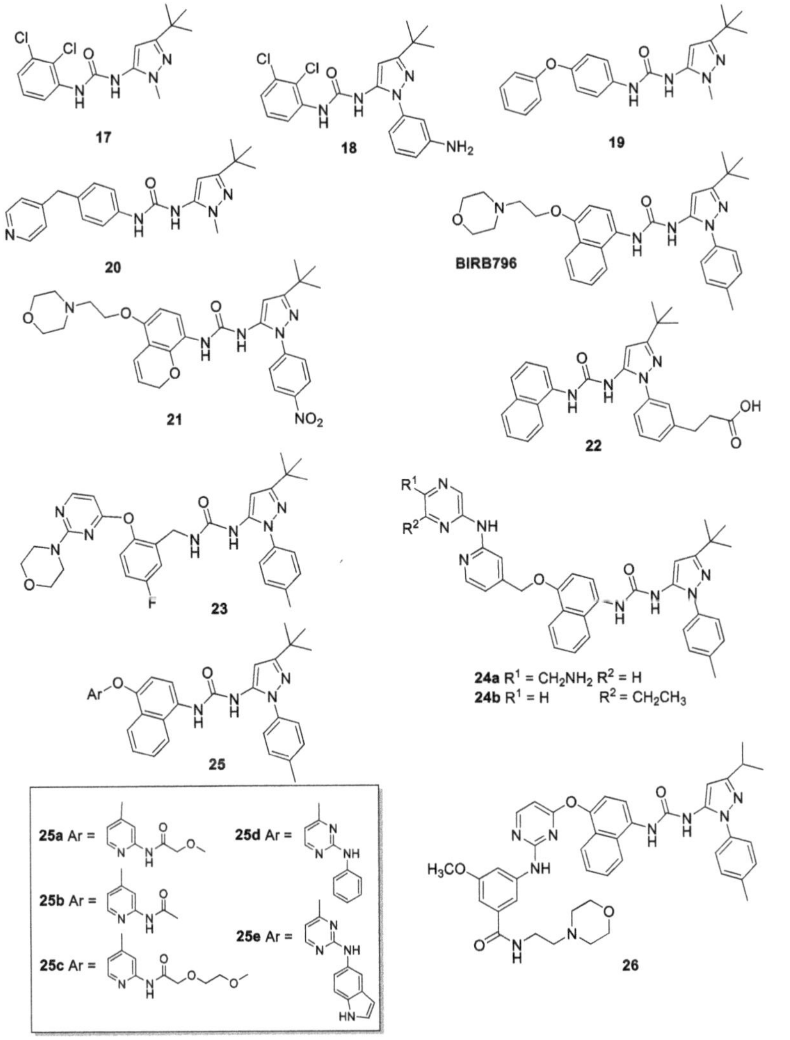

Figure 4. 5-Pyrazolyl-ureas reported as p38 inhibitors.

4.2. 5-Pyrazolyl-Ureas Interfering with Neutrophil Migration

Neutrophil migration (chemotaxis) is a very complex process, which is often the origin of many autoimmune inflammatory diseases, such as RA, MS, and asthma. CXC chemokines (including IL-8, also called CXCL8) and fMLP (N-formyl-methionyl-leucyl-phenylalanine) represent the most common chemoattracting agents.

Bruno et al. in 2007 synthesized a large series of 5-pyrazolyl-ureas (compounds **27**, Figure 5) starting from 5-amino-1-(2-hydroxy-2-phenylethyl)-1*H*-pyrazol-4-ethyl carboxylate and 5-amino-1-(2-hydroxy-2-phenylethyl)-1*H*-pyrazol-4-carbonitrile, subjected to a one-pot reaction with phosgene and proper amines or by reaction with the appropriate phenylisocyanates [69]. New *N*-pyrazolyl-*N'*-alkyl/benzyl/phenyl ureas **27** resulted very active in inhibiting IL-8 induced neutrophils chemotaxis, being the most active 3-benzyl-, 3-(4-benzylpiperazinyl)-, 3-phenyl- and 3-isopropylureido derivatives, with IC_{50} values of 10, 14, 45, and 55 nM, respectively.

Compounds **27** also resulted very selectively, as they do not inhibit fMLP-induced neutrophil migration. Regarding their mechanism of action, different biological tests were performed, being that the compounds are not able to block chemokine receptors (in detail, CXCR1 and CXCR2). The most active derivatives showed phosphorylation inhibition of protein kinase with 50–70 KD, a molecular weight, while specific binding test on different kinases having this molecular weight (Src, Fgr, Hck, Zap-70) gave negative results. Interestingly, derivatives significantly reduced the polymerized actin content and pseudopods formation, mechanisms necessary for migration. In addition, compound carrying in position 4 the carboxyethyl group and in 5 the *m*-fluoroaniline urea moiety showed a good anti-inflammatory action in vivo in a model of zymosane-induced peritonitis in mice and was able to significantly decrease the percentage of cells and granulocytes in the peritoneal cavity at a dose of 100 mg/kg os.

Subsequently, using the same synthetic method, a small library of pyrazolyl-ureas was synthesized (compounds **28**, Figure 5) carrying in position 4 the carboxyethyl group and in position 5 the urea functions more active in the previous series; hydroxyalkyl chains (less bulky than the hydroxyphenyl ethyl chain of the previous ones) were introduced in N1 position.

Many of the new compounds showed an excellent ability to inhibit both the fMLP-OMe-induced chemotaxis (a synthetic derivative of fMLP having the same chemoattracting capacities) with IC_{50} ranging from 0.19 nM to 2 µM, and the IL-8 induced chemotaxis (with IC_{50} in the picomolar range). Two of the most active compounds (bearing both a hydroxypropyl chain on N1 and an *orto*-fluoro- or *meta*-fluorophenyl on the urea moiety) have therefore been tested in vivo, as has the previous, in the zymosane-induced peritonitis inflammation model in mice. At a dose of 50 mg/kg os, they reduced the absolute and relative infiltration of granulocytes into the peritoneal cavity, confirming a possible interference in different neutrophil mechanisms at the inflammation site [70].

During subsequent deep investigations, the authors demonstrated that compounds (in particular some bearing a *meta*-fluoroanilino substituent on the urea scaffold) interfered with ERK1/2, Akt and p38MAPK phosphorylation signaling pathways [71]. Given the interesting results obtained, pyrazolyl-urea molecules by Bruno and coworkers have been patented [72].

In pursuing the investigations in this topic, the same authors synthesized a molecules series in which the urea function was inserted in an imidazo[1,2-*b*]pyrazole cycle (**29**, Figure 5), a constrained and tense form of previous 5-pyrazolyl-ureas, being this cycle not yet investigated for its biological properties [73]. These compounds also were tested in vitro for their ability to interfere with human neutrophils functions. All derivatives were inactive in inhibiting the superoxide anion production and the lysozyme release, while were very active in blocking the fMLP-OMe chemotaxis in a dose dependent manner (IC_{50} between 10 µM and 0.48 nM). In particular, compounds more active, as for previous analogues pyrazolyl-ureas, have a *meta*-fluorophenyl substituent in the urea moiety (**29**, Figure 5). Deep pharmacological investigations evidenced strong p38MAPK inhibition after activation by both fMLP-OMe or IL-8 chemoattractants, while they show different actions on PKC isoforms, suggesting that a fine tuning of the neutrophil activation occurs through differences in the activation of signaling pathways [74].

4.3. 5-Pyrazolyl-Ureas as Antiangiogenic Agents

Taking in account the SAR information given by previous studies, Bruno and coworkers designed and synthesized another small library of 5-pyrazolyl-ureas, combining old and new decorations, their type and position, both on the pyrazole and the urea moieties to increase the molecular diversity.

In detail, the carboxyethyl substituent was moved from position 4 to the uninvestigated position 3 (**30**, Figure 5); in another series, in position 3 was added the *t*-butyl substituent, in analogy with other well-known p38-inhibitors (**31**, Figure 5); in both series the same 2-hydroxy-2-phenylethyl chain was maintained on N1 of the pyrazole core; whereas, in order to modulate the lipophilicity of the molecules, the hydroxy-phenylethyl group was oxidized in a third series (**32**, Figure 5). As concerns the phenyl-urea moiety, different fluorinated substituents in all positions were inserted [75].

All compounds were preliminary screened to evaluate their activity on MAPK and PI3K pathways by monitoring ERK1/2, p38 and Akt phosphorylation. SAR consideration showed that specific substituents and their position on pyrazole nucleus, as well as the type of substituent on the phenyl-urea moiety, play a pivotal role in determining increase or decrease of kinases phosphorylation. Then, by a wound healing assay they were analyzed to assess their capacity of inhibiting endothelial cell migration, a function required for angiogenesis, by using human umbilical vein endothelial cells (HUVEC) stimulated by vascular endothelial growth factor (VEGF). The screening showed significant activity for compound named GeGe-3 (Figure 5). In particular, the studies confirmed that this compound might interfere with cell migration by modulating the activity of different upstream target kinases and could represent a potential angiogenesis inhibitor. Furthermore, it may be used as a tool to identify unknown mediators of endothelial migration and thereby unveiling new therapeutic targets for controlling pathological angiogenesis in diseases such as cancers.

Subsequent investigations showed that GeGe-3 inhibited HUVEC proliferation and endothelial tube formation and, interestingly, impaired inter-segmental angiogenesis during development of zebrafish embryos. In mice, GeGe-3 blocked angiogenesis and tumor growth in transplanted subcutaneous Lewis lung carcinomas. Additional screening revealed that compound could target Aurora B, Aurora C, NEK10, polo-like kinase (PLK2, PLK3), dystrophia myotonica protein kinases-1 (DMPK) and CAMK1, and the authors then speculated that GeGe-3 alters angiogenesis by targeting DMPK in tumor endothelial cells and pericytes. In conclusion, this 5-pyrazolyl-urea, blocking MAPK and PI3K pathways, strongly inhibits physiological and tumor angiogenesis [76].

In 2016, other authors reported synthesis and biological evaluation of different benzothiazoles linked to 5-pyrazolyl-urea nucleus on N1 of the pyrazole. The main objective in this study was the synthesis of new urea, indole, and triazole derivatives of 4-(1,3-benzothiazol-2-yl)-benzoyl-1*H*-pyrazole as anticancer agents able to inhibit VEGF–VEGFR2 complex formation and to suppress proliferation and survival of endothelial cells, angiogenesis and consequently cancer progression.

Some compounds (comprising 5-pyrazolyl derivatives **33**, Figure 5) showed promising cytotoxic activity against breast cancer cell line MCF-7 and inhibited human VEGF in MCF-7 cancer cell lines with comparable activity of reference compound tamoxifen [77].

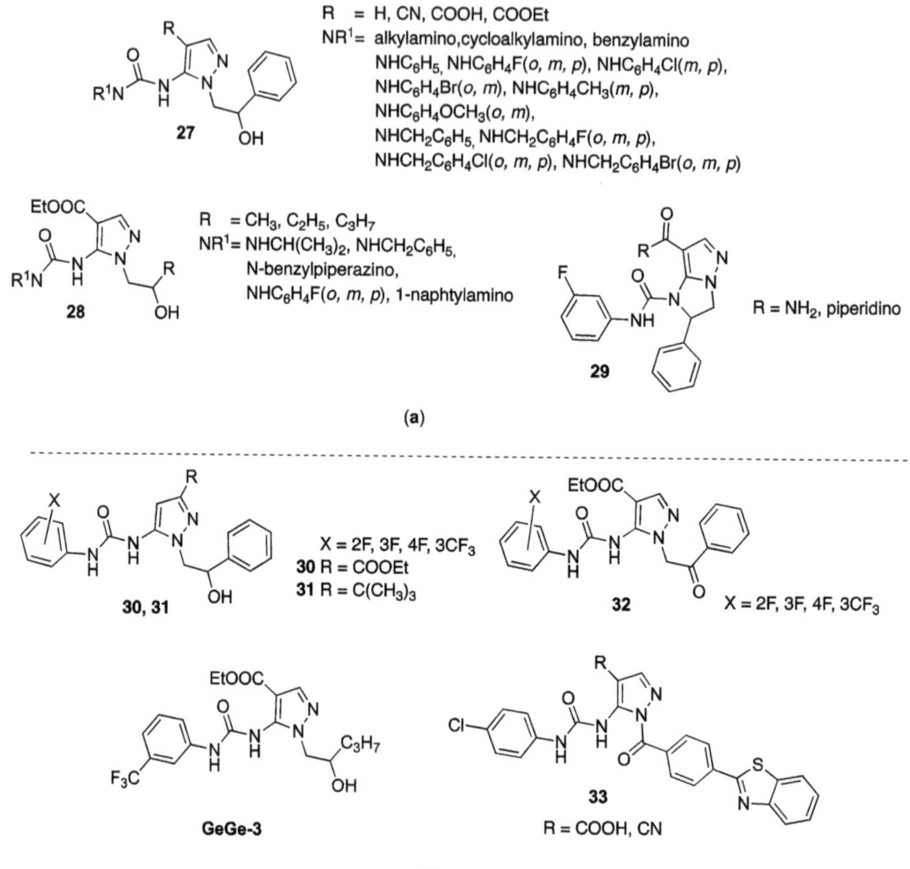

Figure 5. 5-Pyrazolyl-ureas reported as: (**a**) chemotaxis; (**b**) antiangiogenic inhibitors.

4.4. 5-Pyrazolyl-Ureas as Anticancer Agents

Researchers at Deciphera Pharmaceuticals patented and synthesized other 5-pyrazolyl-ureas differently decorated on N1, C-3, and urea moiety as enzyme modulators; in detail, the major part of compounds were characterized by the 2,3-dichlorophenyl-urea moiety (**34**, Figure 6), as the molecules reported by Dumas [39–42]. They evaluated compounds not only on p38 kinase protein, but also on c-Abl, Bcr-Abl, B-Raf, VEGFR, and PDGFR, protein kinases all involved in cancer development. For some derivatives, the authors obtained good results, with IC_{50} values in the low micromolar range [78,79].

In 2007, Smith and coworkers at Bayer Pharmaceutical patented a large series of 5-pyrazolyl-ureas strictly related to BIRB 796 (compounds **35**, Figure 6) [80]. In detail, the authors introduced in position 3, in addition to the *t*-butyl group present in the BIRB 796, bulkier (benzyl, cyclobutyl, cyclopentyl), branched (cyclopropyl, *sec*-butyl) or lipophilic (CF_3) substituents and also varied the aromatic substituent on N1. The greater difference with the lead BIRB 796 was represented by the urea function, which was elongated and characterized by two aromatic rings linked together by oxygen or sulfur atom. In particular, the second aromatic ring is a pyridine bearing in the *orto* position an amide group. The aim of this invention was to evaluate all synthesized compounds on different kinases pathways signaling (p38, but also Raf, VEGFR-2, VEGFR-3, PDGFR-beta, Flt-3), all involved in ongoing cancer, in cancer progression, and in the angiogenesis process. Furthermore, inhibition of Bcr-Abl wild type, an aberrant tyrosine kinase responsible for the onset of chronic myeloid leukemia (CML), and of Bcr-Abl T315I kinase, responsible of imatinib-resistant CML, were evaluated. Some derivatives evidenced antiproliferative properties (IC_{50} < 10 μM) in one or more cell lines such as CAKI-1, MKN45, HCC2998, K562, H441, K812, MEG01, SUP15, HCT116, Ba/F3-Abl (wt) and Ba/F3-Abl(T315l mutant). In detail, a compound (molecular formula not showed) exhibited interesting IC_{50} values on: CAKI-1 (5.2 μM), MKN45 (2.6 μM), HCC2998 (4.7 μM), K562 (<1 μM), H441 (4.4 μM), Ba/F3-Abl(wt) (<1 μM), and Ba/F3-Abl(T315l) (<1 μM).

In other patents, the same authors at Deciphera Pharmaceuticals focused on compounds with possible action on c-Abl, c-Kit, B-Raf, c-Met, Flt-3, VEGF, PDGF and HER family and potentially useful in the treatment of myeloproliferative and other similar diseases. In general, in this new large library (very similar to them patented by Smith [80]) on the N1 pyrazole was introduced a hetero-aromatic bicyclic substituent; the urea moiety was characterized by a phenyl group further substituted with a pyridine nucleus, thus giving a very long and bulky structure (36, Figure 6). The best biological results have been obtained for compound 37 (Figure 6) which showed an IC_{50} value of 0.8 nM for Abl and of 6 nM for T315I mutant Abl kinase. In cellular assay 37 was active on BaF3 cells expressing oncogenic Bcr-Abl kinase (IC_{50} = 6 nM), T315I BaF3 cells (IC_{50} = 8 nM), Y253F BaF3 cells (IC_{50} = 26 nM), E255K BaF3 cells (IC_{50} = 83 nM) and M351T BaF3 cells (IC_{50} = 11 nM). These five forms of Bcr-Abl kinase are oncogenic, causative of human chronic myelogenous leukemia and resistant to Gleevec® [81,82].

Other patents focused on similar compounds bearing on the urea moiety a dihydropyrimido[4,5-*d*]pyrimidine scaffold. Compounds were claimed as potential drugs for cancer, in particular breast cancer, but also for anti-inflammatory pathologies as RA, retinopathies, psoriasis and others. In these patents, the authors reported the synthesis of new urea derivatives, inserting not only pyrazole, but also imidazole, thiadiazol and isoxazole nucleus. Interesting results have been obtained for 5-pyrazolyl-ureas (as compounds 38, Figure 6), but also 4-pyrazolyl-urea derivatives were reported (see compounds 13, Figure 3 [29,30], Section 3.4.2 of this review). Compounds were tested as B-Raf and C-Raf inhibitors and also on A-375 cancer cells to verify their antiproliferative activity. In general, all compounds exhibited >50% inhibition of proliferation at 1–10 μM concentration [29,30].

The same authors patented very similar compounds, isosters of previous, presenting dihydropyrido[2,3-*d*]pyrimidine nucleus instead of dihydropyrimido-pyrimidino one (39, Figure 6). In this larger library, the authors newly introduced the urea function on different heterocycles, such as pyrazole, imidazole, oxazole and others. As in the previous patents, also 4-pyrazolyl-ureas were reported (see compounds 14, Figure 3 [31,32], Section 3.4.2 of this review). Compounds were claimed as useful for chronic myelogenous leukemia, acute lymphocytic leukemia, gastrointestinal stromal tumors, hyper-eosinophilic syndrome, glioblastoma and other solid tumors caused by a mutation in the RAS/RAF/MEK/ERK-MAP kinase pathway, but also for several autoimmune diseases. The invention included the compounds evaluation on Raf (isoforms A, B and C) and other kinases involved in the RAS/RAF/MEK/ERK-MAP kinase pathway. In general, compounds had >50% inhibition activity at 0.2–2 μM concentration against B-Raf and C-Raf and >50% inhibition activity at 1–10 μM concentration against A375 cancer cell line [31,32].

In 2009, Getlik and Rauh [83] designed and synthesized new molecules able to stabilize an enzymatically inactive conformation of the c-Src, a well-known kinase overexpressed or up-regulated in several solid tumors. Compounds were also active on a particular mutation (T338M) at the gatekeeper residue resistant to Dasatinib. Moreover, they seem to be more selective than ATP-competitive inhibitors, as for Scr allosteric inhibitors (type III inhibitors) that bind to sites lying outside the highly conserved ATP pocket, and may circumvent some mechanisms of drug resistance. The authors designed new type II hybrid inhibitors, starting from the analysis of crystal structures of c-Src in complex with different type III pyrazolyl-ureas (previously reported as allosteric inhibitors of p38 and resulted weakly Src inhibitors) and with type I quinazoline-based inhibitors. SAR of the new chimera derivatives confirmed that the geometry is fundamental to direct the binding to c-Src or p38 and to permit the circumvention of larger gatekeeper residue T338M. The most active compound **40** (Figure 6) inhibited wild type c-Src and T338M Src mutant with IC$_{50}$ values of 0.014 µM and 0.023 µM respectively and in prostate carcinoma cells disrupted Fak (focal adhesions Kinase), a non-receptor tyrosine kinase involved in cell cycle progression, cell survival, and cell migration [83].

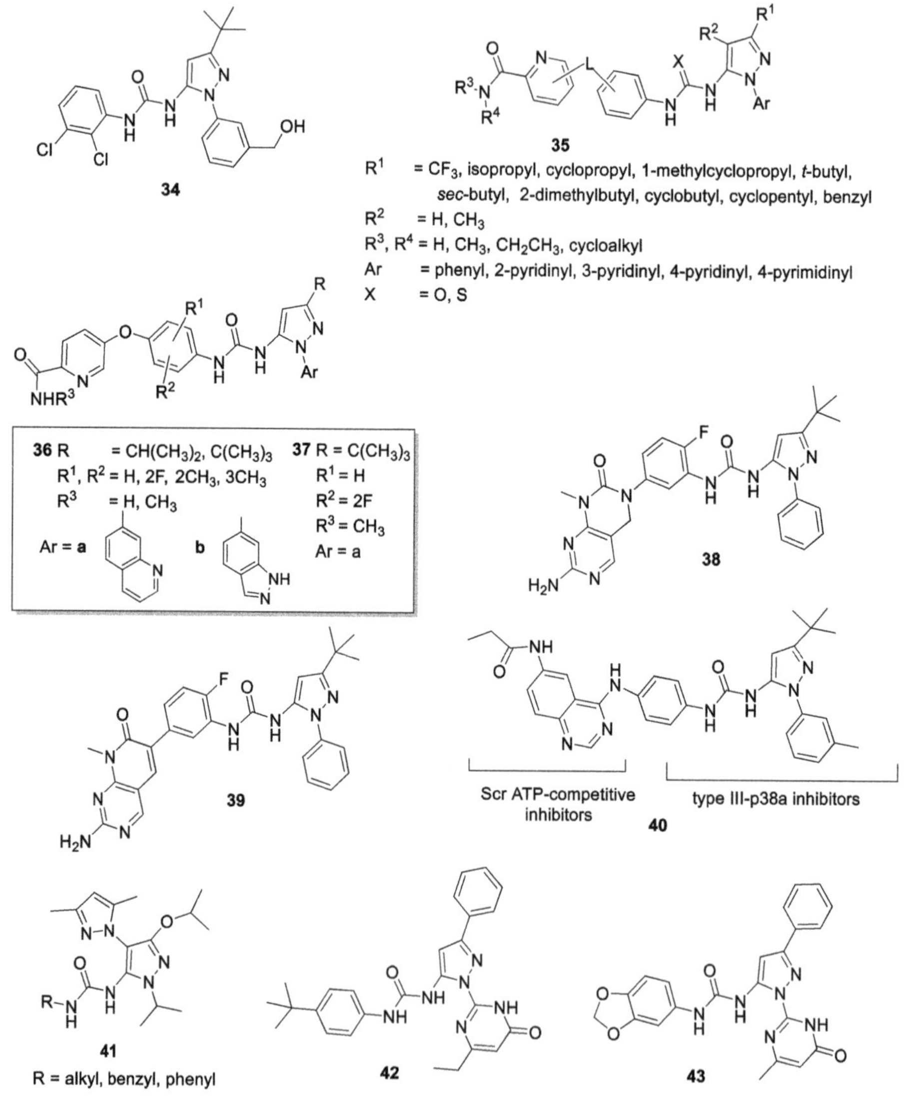

Figure 6. 5-Pyrazolyl-ureas reported as anticancer agents.

More recently, other authors synthesized a series of novel bis-pyrazoles with amide, sulfonamide and urea functions. In particular, compounds presented two different pyrazoles, linked together by N1 bond, one of which is 3,5-dimethyl substituted, whereas the other one has in 5 position amide, sulfonamide or urea moieties [84]. Compounds were evaluated for their in vitro cytotoxicity against Panc-1 (human pancreatic adenocarcinoma), H460 cell lines (human non-small cell lung carcinoma) by a WST-1 cytotoxicity assay. Among them the 5-pyrazolyl-ureas (compounds **41**, Figure 6) were less active than the analogue amide and sulfonamide derivatives.

Other 5-pyrazolyl-ureas were more recently synthesized as SUMO E1 inhibitors. Sumoylation is a cellular mechanism involved in the reversible modification of target proteins with a small ubiquitin-like modifier (SUMO) protein and it showed fundamental role in important cellular functions as DNA replication and repair, nuclear transport, signal transduction and proliferation. Among the sumoylation proteins, higher expression of SUMO E1 has been observed in several cancers [85].

In 2016, Kumar and Zhao studied some thiazole and pyrazole compounds, found to have moderate SUMO E1 inhibitory activity, as starting points for the development of highly potent lead compounds against cancer [86]. In detail, by using molecular docking approach, 3D shape and electrostatic potential investigation, the authors identified some 5-pyrazolyl-ureas able to block SUMO E1 with IC_{50} values in the low micromolar range, being compounds **42** and **43** (Figure 6) the most active (IC_{50} values of 13.8 and 30.4 μM, respectively). These compounds, bearing on N1 pyrazole an oxo-pyrimidinyl ring substituent, resulted more active than derivatives with methyl, cyclopentyl, aryl and aryl halide substitutions. Molecular docking evaluation revealed that **42** and **43** both occupy the SUMO E1 ATP binding pocket with the pyrazole moiety nicely overlapping with the ATP sugar ring, whereas the oxo-pyrimidine moieties overlap with the ATP adenine ring and make hydrophobic contacts with Leu49, Ile96 and Leu116 of the enzyme. However, the aryl-urea group was predicted to extend towards the opening of ATP binding pocket near Asn118, Ile343, Trp344 and Asp345. This interesting class of inhibitors presented broad scope for chemical optimization and could be used as starting points for development of more potent SUMO E1 inhibitors with therapeutic potential.

4.5. 5-Pyrazolyl-Ureas as TrKA Inhibitors

Tropomyosin receptor kinase (Trk) belongs to the receptor tyrosine kinase (RTK) family, which is involved in the development and maintenance of the nervous system. TrkA is a receptor for nerve growth factor (NGF) that mediates neuronal differentiation and programmed cell death prevention [87]. NGF/TrkA pathway plays a central role in the physiological function of chronic pain. TrkA activation triggers intracellular signaling cascades that increase the sensitivity of nociceptors, thus leading to chronic sensitization and pain. Consequently, NGF/TrkA pathway selective inhibition is a potential target for chronic pain treatment.

In 2014, researchers at Array Biopharma Inc. developed and patented different 5-pyrazolyl-ureas able to inhibit TrkA and claimed it as useful for the treatment of pain, cancer, inflammation, Sjogren's syndrome, endometriosis, diabetic peripheral neuropathy, prostatitis, pelvic pain syndrome, neurodegenerative and infectious diseases, and others. Patented compounds have two pyrazole rings linked together by urea, thiourea or guanidine function, always present in position 5 of the pyrazole ring. More interesting compounds are reported in Figure 7 (compounds **44**); they showed IC_{50} values <100 nM in binding test on TrkA and evidenced also a good selectivity profile, being 1000-fold more potent against TrkA than p38α and less active versus a large panel of different kinases [88].

The same company in 2015 patented other pyrazole derivatives carrying on N1 a phenyl, on C-3 bulky aromatic groups and, in position 5, a urea, thiourea or guanidine moiety. From this large library the most interesting compound was 1-((3*S*)-4-(3-fluorophenyl)-1-(2-methoxyethyl)pyrrolidin-3-yl)-3-(4-methyl-3-(2-methylpyrimidin-5-yl)-1-phenyl-1*H*-pyrazol-5-yl)urea (**45**, Figure 7), which showed IC_{50} value of 1.1 nM in binding assay and of 1.9 nM in CHO-Kl cells transfected with human wild type TrkA. One particularly advantageous property of **45** was its predicted human clearance (which is substantially lower than other patented compounds) and its very good PK properties, such as microsomal clearance, determined in different mammalians (rat, dog and monkey). Another important aspect of **45** was its selectivity, being inactive versus hERG potassium channels. In the KinaseProfiler™ at 10 μM concentration **45** showed remarkable and unexpected selectivity and, interestingly, it exerted significantly different potency against TrkA versus TrkB/C. The ability of **45** to selectively inhibit the Trk pathway could indicate that it is essentially free of side-effects, which are generally related to off-target kinases inhibition, therefore representing a safer approach to treat pain, inflammation, cancer and skin diseases [89].

Recently, **45** has been subjected to a deep study by Subramanian and coworkers. The aim of this investigation was, through an initial in silico study, to decipher the allosteric binding mechanism of TrKA inhibitors, including **45**. In detail, the authors provide the basic framework for considering and selecting the appropriate in vitro screening assays or platforms to interrogate selective conformational states and explore unconventional binding pockets and interactions to achieve the desired target specificity. Extensive and exhaustive array of in vitro biochemical and biophysical tools and techniques to characterize and confirm the cryptic allosteric binding pocket and docking mode of the small hTrkA inhibitors were reported. Specifically, assays were designed and implemented to lock the kinase in a predominantly active or inactive conformation and the effect of the kinase inhibitor were probed to understand the hTrkA binding and hTrkB selectivity. The current outcome suggests that inhibitors with a fast association rate take advantage of the inactive protein conformation and lock the kinase state by also exhibiting a slow off-rate. This in turn shifts the inactive/active state protein conformational equilibrium cycle, affecting the subsequent downstream signaling. On the basis of this investigations, the authors suggested for **45** a preferential binding to the inactive enzyme conformation [90].

Similar conclusions were reported for another compound patented by Allen (**46**, Figure 6) recently investigated by other authors [91]. In detail, **46** differs from **45** for the presence of an additional fluorine atom on phenyl of pyrrolidine ring and for the substituent in position 3 of the pyrazole nucleus. To elucidate their efficiency and selectivity among the Trks, Furuya and coworkers determined the X-ray crystal structure of **46** with TrkA, including the kinase domain and Juxtamembrane (JM) region. The results revealed a novel allosteric binding mode of **46** that interacts with the JM region via hydrogen bond, van der Waals interactions, and CH-π interactions at Leu486, His489, and Ile490. In detail, Leu486 interacts with the methyl and ethoxy groups on the pyrazole moiety via van der Waals interactions; His489 interacts with the pyrazole ring via π–π interactions and with the ethoxy group on the pyrazole ring via CH-π interactions; Ile490 forms a hydrogen bond with the urea moiety and interacts with the difluorobenzene group via CH-π interactions. These three residues are not conserved in TrkB and TrkC and, consequently, should play key role for high binding affinity and selectivity. In addition, the role of the JM region in the interaction with **46** has been evaluated by a peptide-based tyrosine kinase assay using two types of TrkA kinase domains (with or without the JM region, JM+ or JM−, respectively). In conclusion, the authors classified **46** as a type III inhibitor (allosteric inhibitor) and clearly demonstrated that one of the most effective strategies for producing highly selective Trk inhibitors is the introduction on the pyrazole of a substituent able to interact with the JM region.

In 2019, Shionogi in Japan patented two very similar classes of 5-pyrazolyl-ureas as TrKA inhibitors claimed as useful for treat osteoarthritis and RA, interstitial cystitis, chronic pancreatitis, prostatitis, and different type of pain such as chronic back pain, nociceptive, neuropathic, postoperative, pelvic, cancer pain, etc. In Figure 7 are reported **47** and **48** as representative compounds. In detail, in the first patent the most active derivatives presented a more embedded pyrrolidine nucleus on the urea moiety, resulting in a more complex structure with respect to compounds patented by Allen (**45**, Figure 7) [89]. The compounds of the present invention have not only a great TrkA inhibitory effect (IC_{50} values < 10 nM), but also a good pharmacokinetic profile, including high bioavailability and moderate clearance, high metabolic stability, high solubility, weak CYP enzyme inhibition, no mutagenic and cardiovascular action. In addition, most of synthesized compounds showed high selectivity of the TrkA receptor [92].

In compounds **48** (Figure 7), the same authors introduced in some cases a bulkier heterocyclic substituent on C-3 and a more simply pyrrolidine or cyclopentane nucleus on the urea moiety. For these derivatives also, interesting inhibitory activity on TrKA (comparable to that of compounds **47**) were obtained [93].

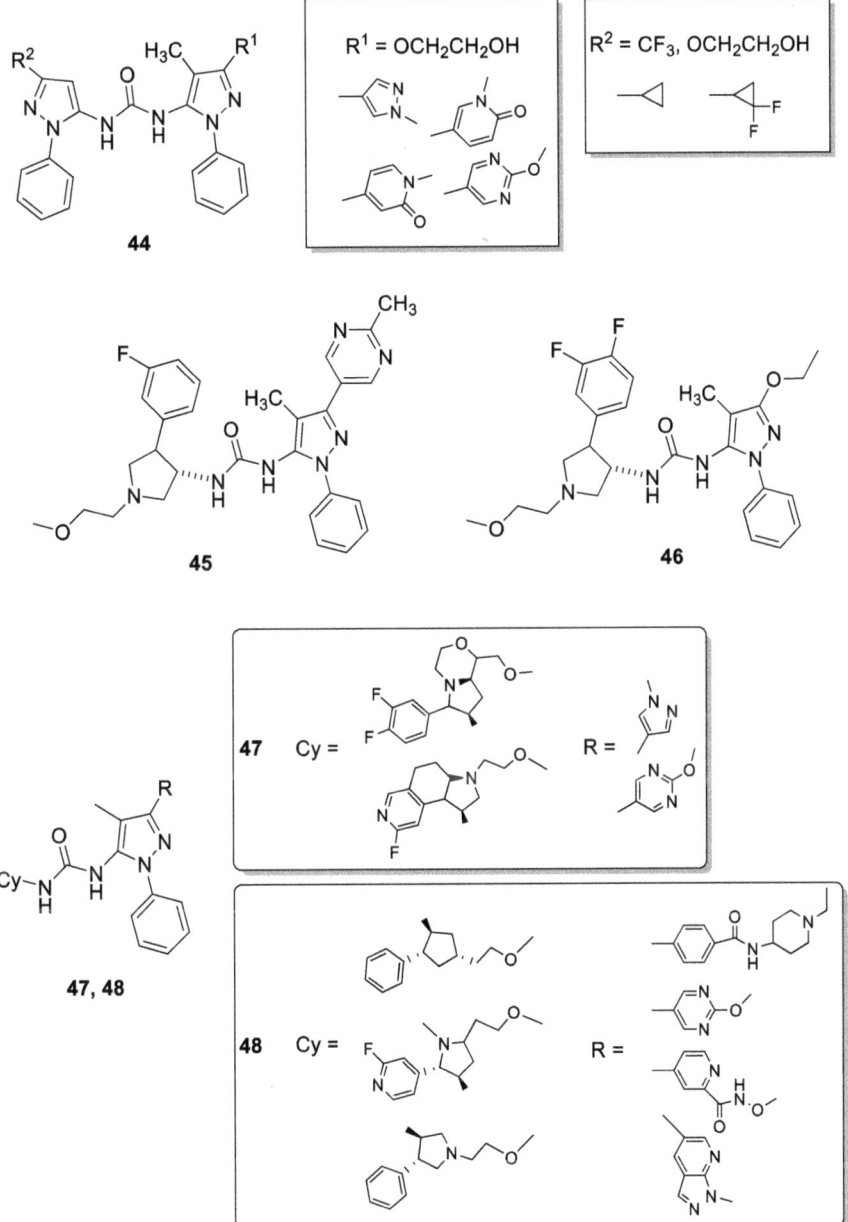

Figure 7. 5-Pyrazolyl-ureas reported as TrKA inhibitors.

4.6. 5-Pyrazolyl-Ureas as Antimalarial Agents

4.6.1. Antimalarial Agents acting as Protein-Protein Interaction Inhibitors

A critical feature of host cell invasion by apicomplexan parasites is the interaction between the carboxy terminal tail of myosin A (MyoA) and the myosin tail interacting protein (MTIP). Parasites can invade the host by a process dependent on actin and myosin machinery that is localized at the parasite's inner membrane complex. This actin-myosin machinery is not present in human host; therefore, it is considered as an attractive drug target for parasite infectious treatment.

Among the apicomplexan parasites that use MTIP-MyoA machinery there is *Plasmodium (P.) falciparum*, the agent responsible for malaria, a devastating parasitic disease that causes more than 3 million deaths every year worldwide. The efforts to control the disease is greatly impeded by the widespread resistance that quickly emerges against the majority of drugs developed in the past and also against the most recent alternative ones.

A fruitful strategy for overcoming microbial drug resistance is to devise new molecular scaffolds that are unrelated to current drugs and able to inhibit novel pathways exclusive for the parasite over the human host. The protein-protein interaction (PPI) between myosin motor components in apicomplexan parasites is the key to the survival of the parasite.

On this basis, Kortagere and co-workers [94,95], have designed small molecule inhibitors against this PPI by using the cocrystal structure of the *Plasmodium knowlesi* MTIP and the MyoA tail peptide. MTIP is composed of three identical subunits while the MyoA protein consists of 13 amino acids forming a hydrophobic alpha helix that occupy the binding site in the subunit C of MTIP. Using this structural information small molecules were designed aiming to competitively inhibit MyoA and disrupt the MTIP-MyoA complex.

The hybrid structure-based virtual screening approach (HSB) [96] that integrates ligand-based and structure-based drug design was then applied to screen two series of small molecules. They successfully isolated eight compounds, having general structure **49** (Figure 8), belonging to the unique pyrazole-urea class, that inhibited the growth of the malarial parasite and altered the gliding motility of rodent malaria sporozoites with EC_{50} values <400 nM. Interestingly, the compounds appeared to act, at several stages of the parasite's life cycle, blocking growth and development. The pyrazole-ureas identified in this study could be effective antimalarial agents because they competitively inhibit the key PPI between MTIP and MyoA responsible for the gliding motility and the invasive features of the malarial parasite [97].

SAR of those compounds showed that hydrophilic groups (such hydroxyl) at the *para* position on the phenyl ring in position 4 of the pyrazole ring reduced the affinity by about 30-fold. Similarly, at position R^1, removing fluorine atoms from the *ortho* and *para* positions led to a minor improvement in the EC_{50} values. Major modification to the structure by shifting the urea linker to the C-4 carbon on the pyrazole ring yielded a compound with an EC_{50} value of about 13 µM.

4.6.2. Antimalarial Agents Acting as PfATP4 Inhibitors

One of the first novel targets for antimalarial agents is the protein *Pf*ATP4, a *P. falciparum* p-type cation ATPase. P-type ATPases are important druggable targets in humans with several clinically relevant inhibitors [98,99]. *Pf*ATP4 belongs to a subfamily of these proteins (family IID) that extrude monovalent cations (sodium, lithium, and potassium). All cells must maintain low cytosolic Na+ concentration to survive and *Pf*ATP4 actively extrudes Na+ from the parasite to maintain low-[Na+]/high-[K+] in the host cell cytosol upon infection. Their presence is limited to lower eukaryotes (fungi, protozoan, and bryophytes) making them an attractive drug target for antiparasitic drugs [100,101]. *Pf*ATP4 is found in the *Plasmodium* plasma membrane and has been suggested to be the target for multiple potential antimalarial candidates, firstly for the spiroindolones, a novel antimalarial chemical class. Compound, KAE609, which is now commercially called Cipargamin, was shown to induce faster parasite clearance times than Artemisinin and was shown to be active against artemisinin resistant parasites [102].

In order to discover new molecular scaffolds acting as antimalarial, a high throughput cellular screen against *P. falciparum* asexual blood stages was performed by Flannery and coworkers on library compounds. Among them, compound **50** (Figure 8) (also known as GNF-Pf4492) showed an IC_{50} value of 184.1 nM against asexual stages of the multidrug resistant *P. falciparum* strain and demonstrated no cytotoxicity against the human hepatoma cell line Huh7 (>30 µM).

Compound **50**, wrongly defined as "aminopyrazole", belong to the pyrazolyl-ureas already patented by Kortagere's team [94,95]. GNF-Pf4492 blocked parasite transmission to mosquitoes and disrupted intracellular sodium homeostasis, similarly to the spiroindolones.

Flannery and co-workers applied a chemical genomic approach [103] to deeply investigate the mechanism of action of these novel antimalarial agents. Whole genome sequencing of three resistant lines showed that each had acquired independent mutations in a P-type cation-transporter ATPase, *Pf*ATP4. The authors showed that the phenotypes of parasites treated with a spiroindolone and the pyrazolyl-urea **50** are similar. They further identified a third chemotype that interacts with *Pf*ATP4. Convergence on this target by multiple chemophores highlights the critical function of this protein in the parasite [104].

As reported before, those pyrazole series was reported as disrupting the PPI between the *P. falciparum* MyoA and MTIP [97]. Anyway, Flannery and co-workers contested the Kortagere hypothesis about the mechanism of action, because GNF-Pf4492 was shown by Flannery as active throughout the asexual life cycle and not just during cell invasion or gliding motility, which requires the MTIP-myosin A interaction.

In our opinion, the mechanism of action of this interesting new class of antimalarial agents deserve further investigation, also to clarify whether they are not, by chance, multitargeted compounds, which would further increase their application interest.

4.6.3. Antimalarial Agents Acting as *P. falciparum* Prolyl-tRNA-synthetase Inhibitors

P. falciparum prolyl-tRNA-synthetase (*Pf*ProRS) is a promising new target that is chemically and genetically validated by halofuginone, a synthetic derivative of the natural product febrifugine which has been known for more than 2000 years to possess antimalarial activity (IC_{50} = 275 nM). However, halofuginone is also a potent inhibitor of *Homo sapiens* prolyl-tRNA-synthetase (*Hs*ProRS) (IC_{50} = 2 µM) which prevents its use in human treatment. The overlapping of the three-dimensional structure of *Pf*ProRS and *Hs*ProRS showed that the active sites, normally occupied by proline, are identical in both enzymes. Halofuginone competes with proline in binding to the active site, making difficult to achieve specificity.

Therefore, Hewitt and coworkers [105] applied a high throughput screen (HTS) method to about 40,000 different compounds belonging to five different libraries and identified compound **51** (TCMDC-124506) (Figure 8) and Glyburide, which are very different compounds but sharing the urea function, as *Pf*ProRS inhibitors. They showed IC_{50} values against *Pf*ProRS at 34 µM and 74 µM, respectively, but <40% inhibitory activity for *Hs*ProRS up to the limit of solubility (1 mM). X-ray crystallographic structures demonstrated that these inhibitors bind outside the active site in an area of the enzyme not previously displayed as a binding pocket.

Moreover, the binding of both glyburide and TCMDC-124506 into the novel binding site significantly distorts the ATP binding site. Further analyses clearly indicated that they act as competitive allosteric inhibitors showing more than 100-fold specificity for *Pf*ProRS compared to *Hs*ProRS, demonstrating these class of compounds could overcome the toxicity of halofuginone and its analogs related to *Hs*ProRS inhibition. A large series of new variants to **51** (Figure 8) was then synthesized, guided by the cocrystal structure with *Pf*ProRS, to find more potent inhibitors while retaining selectivity against *Plasmodium* [105]. Despite the potency of those inhibitors was not greatly increased in respect with the hit (IC_{50} values ranging from 3 to 20 µM against *Pf*ProRS), interesting SAR information have been provided and opened the way for the development of new ligands as possible antimalarial agents.

In detail:

- in N1 is needed the methyl group; NH, or polar substituents are not tolerated;
- the trifluoromethyl substituent in position 3 of the pyrazole acts as vector to the solvent and its removal is not tolerated, while more polar groups (such an example 2-ethyl-morpholine) causes a potency decrease;
- removal of the *p*-fluorophenyl moiety in position 4 of the pyrazole led to inactive compounds, but substitution in the phenyl ring could be changed without affecting the potency.

Finally, the crystal structure showed that the urea moiety forms multiple H-bonds with the protein, and the aromatic substituents fit into a hydrophobic pocket. Removal of the urea led to inactive compound. The *o,p*-difluoro-substituted phenyl urea derivatives retained activity and the *m*-fluoro substitution was tolerated, whereas any other modification around this moiety is detrimental for *Pf*ProRS activity.

4.7. 5-Pyrazolyl-Ureas as Anti-Toxoplasma Agents

Toxoplasma (*T.*) *gondii* is a ubiquitous opportunistic pathogen that infects individuals worldwide, causing severe congenital neurological and ocular disease in humans. Toxoplasmosis is a great risk for patients who are undergoing treatment for cancer and auto-immune diseases, or recipients of tissue or organ transplants. Toxoplasmosis can also lead to morbidity and death in HIV/AIDS patients. No vaccine to protect humans is available, and hypersensitivity and toxicity limit the use of the few available medicines. Therefore, safer and more effective drugs to treat toxoplasmosis are urgently needed.

Starting from their study on MTIP-MyoA that provided new antimalarial agents (see Section 4.6.1) and supported by the hypothesis that the MTIP-MyoA complex is essential also in the in *T. gondii* survival, Kortagere and co-workers constructed a hypothetical structure model of this complex and used the HSB method to screen for small molecule drugs against toxoplasmosis [106]. In the absence of the X-ray crystal structure for the *T. gondii* MTIP-MyoA complex, homology modeling techniques were employed to build a structural model of the *T. gondii* MTIP-MyoA_tail complex using the published X-ray crystal structure of *Plasmodium* MTIP-MyoA tail as the template. The MTIP-MyoA tail binding region was explored to identify the specific points of interaction between corresponding chemical groups in the MTIP pocket and the MyoA helix. The in silico screen of a chemical library containing over three million commercially available compounds made the identification of 150 hits that were submitted to the HBS analyses. Further screening through ADMET filter gave the identification of the 25 best ranking drug-like compounds and led to the discovery of the pyrazolyl-urea **52** (Figure 8) as a lead compound, with an IC_{50} value of 500 nM. In vitro assays have revealed that **52** markedly limits intracellular growth of *T. gondii* tachyzoites but has no effect on host cell human foreskin fibroblasts (HFF) at concentration more than one log greater than the concentration that inhibits the parasites. SAR analyses made information for substituents in the pyrazolyl-urea core to identify a second generation of **52** like compounds. Finally, six new molecules were synthesized and tested on *T. gondii* tachyzoites. Compound **53** (Figure 8), already reported as active against *P. falciparum* [97], performed significantly better than the parent compound in the *T. gondii* inhibition assay. Interestingly, **52** and **53** have nearly similar IC_{50} values against *T. gondii* and *P. falciparum* and may be working through similar mechanism that warrant consideration as broad-spectrum inhibitors of apicomplexan parasites.

4.8. 5-Pyrazolyl-Ureas as Antibacterial Agents

The emergence of resistant strains of pathogenic bacteria make mandatory the development of new antibacterial compounds. In an effort to discover new leads, Kane and coworkers [107] screened mixtures of compounds belonging to their chemical libraries versus several clinically relevant Gram-positive and Gram-negative organisms. Active components of these mixtures were determined through re-synthesis and subsequent retesting. The screening data revealed a variety of heterocyclic urea derivatives, in particular ureas of 5-aminopyrazole (compounds **54**, Figure 8) and 2-aminothiazole, as inhibiting growth of gram-positive bacteria. In all cases, a disubstituted urea is necessary for activity and active compounds require, in addition to the pyrazole scaffold, an N-aryl substituent preferably 3,4 or 3,5 -disubstituted with halogen atoms and/or CF_3 groups. However, despite the high in vitro activity (particularly against S. aureus compounds showed IC_{50} ranging from 0.054 to 4.24 µM), they were only marginally bioavailable under an in vivo test, probably due to their poor solubility.

4.9. 5-Pyrazolyl-Ureas as Anti-Trypanosome Agents

Improved therapies for the treatment of *Trypanosoma brucei*, the etiological agent of the neglected tropical disease human African trypanosomiasis, are urgently needed. *T. brucei* methionyl-tRNA synthetase (MetRS) is an aminoacyl-tRNA synthase (aaRS) that is considered an important drug target due to its role in protein synthesis, cell survival, and its significant differences in structure from its mammalian ortholog.

In 2015, Pedro-Rosa and coworker [108] developed and applied to a large library of more than 300,000 compounds two orthogonal HTS assays to identify inhibitors of *T. brucei* MetRS. First, a chemiluminescence assay was implemented in a 1536-well plate format and used to monitor ATP depletion during the aminoacylation reaction. Hit confirmation then used a counter screen in which adenosine monophosphate production was assessed using fluorescence polarization technology. In addition, a miniaturized cell viability assay was used to triage cytotoxic compounds. Finally, lower throughput assays involving whole parasite growth inhibition of both human and parasite MetRS were used to analyze compound selectivity and efficacy. The outcome of this HTS campaign has led to the discovery of 19 potent and selective *T. brucei* MetRS inhibitors (having IC_{50} as low as 44 nM), including the 5-pyrazolyl-urea derivative **55** (Figure 8) (analogue of **51**, that belong to the series developed by Kortagere in 2010 [97]).

4.10. 5-Pyrazolyl-Ureas as Antiviral Agents

Respivert Limited UK in 2011 patented new compounds very similar to previous derivatives **24** and **25** (Figure 6), but bearing a more embedded imidazo[4,5-*b*]pyridine tail on the urea moiety. The most representative of this series is the 1-(3-(*tert*-butyl)-1-(4-methoxyphenyl)-1*H*-pyrazol-5-yl)-3-(4-((2-oxo-2,3-dihydro-1*H*-imidazo[4,5-*b*]pyridin-7yl)oxy)naphthalen-1-yl)urea **56** (Figure 8) claimed for the treatment of inflammatory diseases of respiratory system (asthma, COPD), and also for the treatment or prevention of inflammation mediated by viral infectious diseases (influenza virus, rhinovirus or RSV) [109]. All synthesized compounds were tested in anti-inflammatory, antiviral and cell viability assays. Moreover, compounds were tested in enzyme inhibition assays (p38MAPK, Src, Hck) in analogy to previous **24** and resulted active against p38MAPK with IC_{50}s in the nanomolar range. An additional test on Rhinovirus-titration assay was performed, while results were not reported.

4.11. 5-Pyrazolyl-Ureas as Potassium Channel Activators

The G protein-gated inwardly-rectifying potassium channels (GIRK, $K_{ir}3$) is a family of various homo- and hetero-tetrameric combinations of four different subunits expressed in different combinations in a variety of regions, throughout the central nervous system and in the periphery. A number of studies, using subunit-specific GIRK knockout mice, suggested the roles for GIRKs in a variety of important physiological processes and pointed to GIRK as a target for therapeutic intervention [110]. However, at the moment very few pharmacological tools have been reported that might allow a better understanding of the roles of GIRKs in normal and pathophysiological conditions.

An HTS-compatible thallium flux assay for Gi/o-coupled GPCRs using GIRK as a readout [111] was used by Kaufmann and coworkers to screen the molecular libraries small molecule repository (MLSMR) as part of the Molecular Libraries Screening Center Network (MLSCN) [112]. The first efforts gave to the identification of the asymmetrical urea CID736191 (**57**, Figure 7) as activator of GIRK1/2 with an EC_{50} of ~1 µM. This molecule was used as starting point for chemical optimization via iterative parallel synthesis. Several libraries were prepared surveying alternative substituents on the aryl ring, alternative heterocycles to replace the N-phenyl pyrazole, as well as alternative linkers for the urea moiety.

Testing of these compounds resulted in clear SAR. The N-phenyl pyrazole was essential for GIRK activity, as both NH moieties of the urea linker. In addition, 2-substituents to the aryl ring generally led to inactive compounds, whereas 3-substituted and 3,4-disubstitiuted analogues proved to be optimal. Chemical optimization afforded ML297 (VU0456810, CID 56642816, **58**, Figure 7), as the first potent GIRK activator. Since previous development of GIRK2 knockout animals has revealed an epilepsy phenotype, suggesting a role for GIRK in regulating excitability, and because the GIRK1/2 subunit combination is most prevalent in the brain, the authors reasoned that a GIRK1/2 activator might produce effects in an epilepsy model in vivo. Therefore, they evaluated ML297 in two model of epilepsy in mice (with PTZ or via electroshock). Despite the fact that ML297's DMPK properties were suboptimal, robust activity was observed in both models. ML297 showed EC_{50} of 160 nM, equal or greater efficacy compared to a clinically active antiseizure medication, sodium valproate. These data support a role for selective GIRK activation in controlling excitability and provide the first evidence for the exciting possibility that GIRK may represent an attractive new target for antiepileptic drugs.

Starting from this study, in 2017 the Wieting's team reported a novel series of 1*H*-pyrazol-5-yl-2-phenylacetamides that showed nanomolar potency as GIRK1/2 activators with improved brain distribution in respect with the analogue urea derivatives [113].

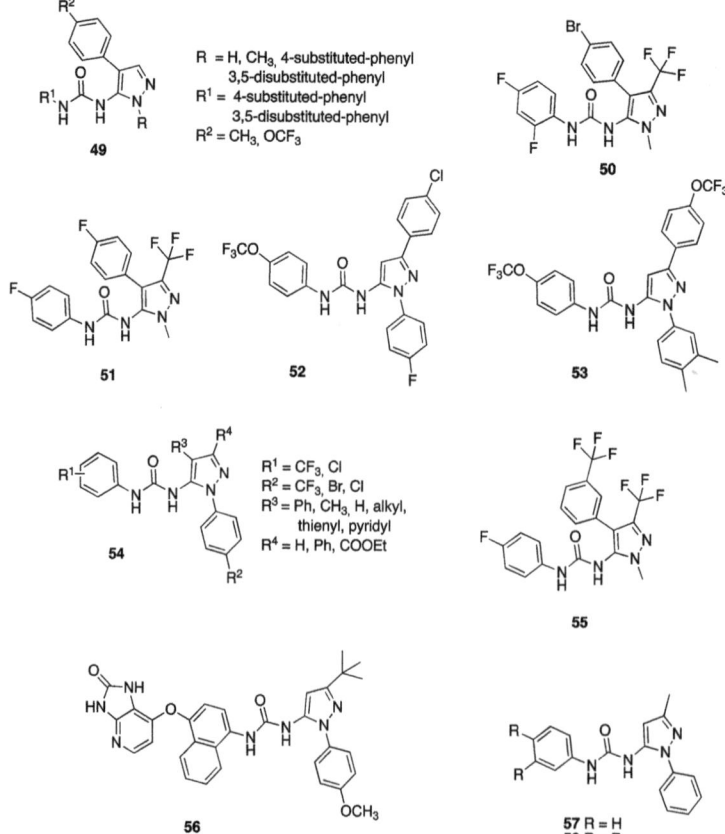

Figure 8. 5-Pyrazolyl-ureas reported as antiparasitic agents and as potassium channel activators.

5. Conclusions

In the last 20 years, linking of the pyrazole nucleus with a urea moiety has led to the design and synthesis of a lot of compounds with interesting biological activities. Many novel compounds have been synthesized and evaluated for their anti-pathogens (bacteria, plasmodium, toxoplasma and others), anti-inflammatory and, in particular, antitumor activity. The urea moiety has been identified as versatile scaffold for kinase inhibitors development, due to its unique binding mode and kinase inhibition profile. On the basis of X-ray crystallography studies, two distinct binding modes of urea binding to kinase have been elucidated: the urea function can bind with a conserved DFG residue (present in the activation loop of protein kinase) or with the hinge-binding site of protein kinase. In particular, urea moiety is used in the type II kinase inhibitors forming one or two hydrogen bonds with a conserved glutamic acid and another one with the backbone amide of the aspartic acid in the DFG motif. Moreover, this structure is used to link pharmacophores with high affinity DNA binder.

The most interesting results have been obtained when urea function was inserted in position 5 of the pyrazole scaffold, probably due to a simpler synthetic procedure. 5-Pyrazolyl-ureas have been largely investigated, being BIRB 796 the most studied compound as p38MAPK inhibitor, useful in autoimmune diseases treatment.

As demonstrated by a multitude of patents reported in the literature from 2000, other important results have been obtained in anticancer field, where 5-pyrazolyl-ureas have shown to interact at intracellular level on many pathways, in detail on different kinases such as Src, Bcr-Abl, Flt3, TrkA and others.

In addition, different 5-pyrazolyl ureas evidenced antiangiogenic and antiproliferative potential that deserves to be explored.

Other important pharmacological results have been obtained in countering malaria, toxoplasma, trypanosome and bacterial infections, that, even today, owing the onset of multi-drug resistance, are not completely eradicated.

4-Pyrazolyl and 3-pyrazolyl ureas, although less investigated, showed varied pharmacological properties, from the inhibition of carbon anhydrase and epoxide hydrolase, to the action on potassium channels and on cannabinoid receptors.

All these data confirm the importance of the urea function which is able to interact with different enzyme substrates depending on its different position on the pyrazole nucleus. Only in some cases, as widely reported above, both 4-pyrazolyl ureas and 5-pyrazolyl ureas have exhibited similar biological activity and have therefore been the subject of common patents.

For all these reasons, the design and synthesis of pyrazolyl-ureas are still ongoing and continue to give interesting insights into medicinal chemistry research.

This review summarizes the large number of pyrazolyl-ureas recently reported (from 2000 to date), including patented compounds and their related biological data.

Author Contributions: Conceptualization, C.B. and O.B.; methodology, C.B.; bibliographic research C.B., O.B., and F.R.; writing—original draft preparation, C.B., O.B., and F.R.; writing—review and editing, C.B., O.B., and F.R. All authors have read and agreed to the published version of the manuscript.

Funding: This research received no external funding.

Acknowledgments: The authors thank the University of Genoa, which provided the technological support for bibliographic research.

Conflicts of Interest: The authors declare IP interest in 5-pyrazoly-ureas derivatives patented as chemotaxis inhibitors.

Abbreviations

3D-QSAR	Three-dimensional quantitative SAR
aaRS	Aminoacyl-tRNA synthase
Abl	Abelson kinase
AD	Alzheimer's disease

ALS	Amyotrophic lateral sclerosis
AUDA	12-(1-Adamantan-1-ylureido)- dodecanoic acid
CAMK	Ca2+/calmodulin- dependent protein kinase
CB1/2	Cannabinoid receptors 1/2
CDK8/CycC	Cyclin-dependent kinase 8 and cyclin C
COMFA	Comparative molecular field analysis
CoMSIA	Comparative molecular similarity
COPD	Chronic obstructive pulmonary disease
COX-2	Cyclooxygenase-type 2
CXCL8	Interleukin 8
CXCR1/R2	CXC Chemokine receptors 1/2
DFG (DMG)	Asp-Phe-Gly motif
DMPK1	Dystrophia myotonica protein kinase 1
ERK	extracellular-regulated protein kinase
Flt3	Fms related tyrosine kinase 3
fMLP	N-formyl-methionyl-leucyl-phenylalanine
GIRK	G protein-gated inwardly-rectifying potassium channels
GSK3α	Glycogen synthase kinases type 3α
hCA II	Human carbonic anhydrase II
hCAIs	Human carbonic anhydrase inhibitors
Hck	Hemopoietic Cell Kinase
HD	Huntington's disease
HeLa	Henrietta Lacks immortal cells
hERG	Human Ether-à-go-go-Related Gene
HFF	Human foreskin fibroblasts
HSB	Hybrid structure based (virtual screening approach)
HTS	High-throughput screening
HUVEC	Human umbilical vein endothelial cells
I-κB	Inhibitor of kappa B kinase
IKK (1/2)	IkB kinases type 1/2
IL-1-/6/-17/-18	Interleukin type 1/6/17/18
JM	Juxta-membrane
JNK	c-Jun NH_2-terminal kinase
LPS	Lipopolysaccharide
MAPK	Mitogen-activated protein kinase
MBC	Minimum Bactericidal concentration
MES	Maximal electroshock seizure
MetRS	Methionyl-t-RNA synthetase
MFC	Minimum fungicidal concentration
MI	Myocardial infarction
MIC	Minimum inhibiting concentration
MLSCN	Molecular libraries screening center network
MLSMR	Molecular libraries small molecule repository
MS	Multiple sclerosis
MTIP	Myosin tail interacting protein
MyoA	Myosin A
NEK10	NIMA (never in mitosis gene a)-related kinase 10
NF-κB	Nuclear factor kappa-light-chain-enhancer of activated B cells
NGF	Nerve growth factor
ONIOM	N-Layered integrated molecular orbital & molecular mechanics
Panc-1	Human pancreatic adenocarcinoma
PD	Parkinson's disease
PDB	Protein data bank
PDGFR	Platelet derived growth factor

PfATP4	Plasmodium falciparum p-type cation ATPase
PfProRS	Plasmodium falciparum prolyl-t-RNA-synthetase
PI3K	Protein inositol kinase 3
PKC	Protein kinase C
PLK2, PLK3	Polo-like kinase 2/3
PPI	Protein-protein interaction
PTZ	Pentylenetetrazol-induced seizure
RAF/RAS	Rapidly accelerated Fibrosarcoma
RSV	Respiratory syncytial virus
SAR	Structure Activity Relationship
sEH	Soluble epoxide hydrolases
SUMO	Small ubiquitin-like modifier
Syk	Spleen tyrosine kinase
TNF-α	Tumor necrosis factor alpha
Trk	Tropomyosin receptor kinase
UBSAs	Ureido benzenesulfonamides
VEGF	Vascular endothelial growth factor
VEGFR1/2	Vascular endothelial growth factor receptor
Zap-70	Zeta Chain of T Cell Receptor Associated Protein Kinase 70

References

1. Kucukguzel, S.G.; Senkardes, S. Recent advances in bioactive pyrazoles. *Eur. J. Med. Chem.* **2015**, *97*, 786–815. [CrossRef] [PubMed]
2. Behr, L.C.; Fusco, R.; Jarboe, C.H. *The Chemistry of Heterocyclic Compounds Pyrazoles, Pyrazoline, Pyrazolidines, Indazoles and Condensed Rings*; Interscience Publisher: New York, NY, USA, 1967.
3. Gilchrist, T.L. *Heterocyclic Chemistry*; Longman Scientific & Technical Publisher: New York, NY, USA, 1992.
4. Bennania, F.E.; Doudachc, L.; Cherraha, Y.; Ramlib, Y.; Karrouchib, K.; Ansarb, M.; Abbes Faouzia, M.E. Overview of recent developments of pyrazole derivatives as an anticancer agent in different cell line. *Bioorg. Chem.* **2020**, *97*, 103470–103532. [CrossRef] [PubMed]
5. Meng, L.; Bao-Xiang, Z. Progress of the synthesis of condensed pyrazole derivatives (from 2010 to mid- 2013). *Eur. J. Med. Chem.* **2014**, *85*, 311–340. [CrossRef]
6. Raffa, D.; Maggio, B.; Raimondi, M.V.; Cascioferro, S.; Plescia, F.; Cancemi, G.; Daidone, G. Recent advanced in bioactive systems containing pyrazole fused with a five membered heterocycle. *Eur. J. Med. Chem.* **2015**, *97*, 732–746. [CrossRef] [PubMed]
7. Jagtap, A.D.; Kondekar, N.B.; Sadani, A.A.; Chern, J.W. Ureas: Applications in Drug Design. *Curr. Med. Chem.* **2017**, *24*, 622–651. [CrossRef]
8. Garuti, L.; Roberti, M.; Bottegoni, G.; Ferraro, M. Diaryl Urea: A Privileged Structure in Anticancer Agents. *Curr. Med. Chem.* **2016**, *23*, 1528–1548. [CrossRef]
9. Supuran, C.T.; Pacchiano, F.; Carta, F.; McDonald, P.C.; Lou, Y.; Vullo, D.; Scozzafava, A.; Dedhar, S. Ureido-Substituted Benzenesulfonamides Potently Inhibit Carbonic Anhydrase IX and Show Antimetastatic Activity in a Model of Breast Cancer Metastasis. *J. Med. Chem.* **2011**, *54*, 1896–1902. [CrossRef]
10. Sahu, C.; Sen, K.; Pakhira, S.; Mondal, B.; Das, A.K. Binding affinity of substituted ureido-benzenesulfonamide ligands to the carbonic anhydrase receptor: A theoretical study of enzyme inhibition. *J. Comput. Chem.* **2013**, *34*, 1907–1916. [CrossRef]
11. Li, S.; Zheng, Z.; Tao, X.; Wang, L.; Zhou, X.; Chen, W.; Zhong, W.; Xiao, J.; Xie, Y.; Li, X.; et al. 4-Methyl-1H-diaryl Pyrazole Derivative as Cannabine Type I Receptor Inhibitor and Its Preparation. WO Patent Application No. 2016184310 A1 20161124, 24 November 2016.
12. Makriyannis, A.; Liu, Q.; Thotapally, R. Preparation of Arylpyrazolecarboxamides as CB1 Cannabinoid Receptor Antagonists. U.S. Patent Application No. 20060100208 A1 20060511, 11 May 2006.
13. Bratenko, M.K.; Barus, M.M.; Rotar, D.V.; Vovk, M.V. Polyfunctional Pyrazoles. 9*. Synthesis of 1-Alkyl(Aryl)-3-[4-(Hydroxymethyl)-1H-Pyrazol-3-Yl]ureas. *Chem. Heterocyc. Compd.* **2014**, *50*, 1252–1258. [CrossRef]

14. Morisseau, C.; Goodrow, M.H.; Dowdy, D.; Zheng, J.; Greene, J.F.; Sanborn, J.R.; Hammock, B.D. Potent urea and carbamate inhibitors of soluble epoxide hydrolases. *Proc. Natl. Acad. Sci. USA* **1999**, *96*, 8849–8854. [CrossRef]
15. Kodani, S.D.; Hammock, B.D. The 2014 Bernard B. Brodie award lecture-epoxide hydrolases: Drug metabolism to therapeutics for chronic pain. *Drug Metab. Dispos.* **2015**, *43*, 788–802. [CrossRef] [PubMed]
16. Hammock, B.D.; Kim, I.-H.; Morisseau, C.; Watanabe, T.; Newmann, J.W. Improved Inhibitors for the Soluble Epoxide Hydrolase. WO Patent Application No. 2004/089296 A2, 21 October 2004.
17. D'yachenko, V.S.; Danilov, D.V.; Shkineva, T.K.; Vatsadze, I.A.; Burmistrov, V.V.; Butov, G.M. Synthesis and properties of 1-[(adamantan-1-yl)methyl]-3-pyrazolyl ureas. *Chem. Heterocycl. Compd.* **2019**, *55*, 129–134. [CrossRef]
18. Hwang, S.H.; Wagner, K.M.; Morisseau, C.; Hua Dong, J.-Y.L.; Wecksler, A.T.; Hammock, B.D. Synthesis and Structure Activity Relationship Studies of Urea-Containing Pyrazoles as Dual Inhibitors of Cyclooxygenase-2 and Soluble Epoxide Hydrolase. *J. Med. Chem.* **2011**, *54*, 3037–3050. [CrossRef] [PubMed]
19. Heinisch, G.; Matuszczak, B.; Rakowitz, D.; Tantisina, B. Synthesis of N-aryl-N'-heteroaryl-substituted urea and thiourea derivatives and evaluation of their anticonvulsant activity. *Arch. Pharm. (Weinheim)* **1997**, *330*, 207–210. [CrossRef]
20. Pandeya, S.N.; Mishra, V.; Sign, P.N.; Rupainwar, D.C. Anticonvulsant activity of thioureido derivatives of acetophenone semicarbazone. *Pharmacol. Res.* **1998**, *37*, 17–22. [CrossRef]
21. Dimmock, J.R.; Vashishtha, S.C.; Stables, J.P. Ureylene anticonvulsants and related compounds. *Pharmazie* **2000**, *55*, 490–494.
22. Kaymakcioglu, B.K.; Rollas, S.; Koercegez, E.; Aricioglu, F. Synthesis and biological evaluation of new N-substituted-N'-(3,5-di/1,3,5-trimethylpyrazole-4-yl)thiourea/urea derivatives. *Eur. J. Pharmac. Sci.* **2005**, *26*, 97–103. [CrossRef]
23. Clare, M.; Fletcher, T.R.; Hamper, B.C.; Hanson, G.A.; Heier, R.F.; Huang, H.; Lennon, P.J.; Oburn, D.S.; Reding, M.T.; Stealey, M.A.; et al. Preparation of Substituted Pyrazole Ureas for the Treatment of Inflammation. WO Patent Application No. 2005037797 A1 20050428, 28 April 2005.
24. Karin, M.; Yamamoto, Y.; Wang, Q.M. The IKK NF-kappa B system: A treasure trove for drug development. *Nat. Rev. Drug Discov.* **2004**, *3*, 17–26. [CrossRef]
25. Baxter, A.; Brough, S.; Faull, A.; Johnstone, C.; Mcinally, T. Heteroaromatic Carboxamide Derivatives and their Use as Inhibitors of the Enzyme IKK-2. WO Patent Application No. 01/58890 Al PCTISE01/00248, 7 February 2001.
26. Sharma, R.; Jain, A.; Sahu, B.; Singh, D.; Mali, S. Preparation of Heterocyclyl Compounds as Inhibitors of IL-17 and TNF-α for the Treatment of Inflammatory Diseases and Autoimmune Disorders. WO Patent Application No. 2014181287 A1 20141113, 8 May 2014.
27. Schneider, E.V.; Bottcher, J.; Huber, R.; Maskos, K.; Neumann, L. Structure–kinetic relationship study of CDK8/CycC specific compounds. *Proc. Natl. Acad. Sci. USA* **2013**, *110*, 8081–8086. [CrossRef]
28. Chen, W.; Ren, X.; Chang, C.A. Discovery of CDK8/CycC Ligands with a New Virtual Screening Tool. *ChemMedChem* **2019**, *14*, 107–118. [CrossRef]
29. Flynn, D.L.; Petillo, P.A.; Kaufman, M.D.; Patt, W.C. Preparation of Heterocyclic Ureas as Kinase Inhibitors Useful for the Treatment of Proliferative Diseases. WO Patent Application No. 2008033999 A2 20080320, 20 March 2008.
30. Flynn, D.L.; Petillo, P.A.; Kaufman, M.D.; Patt, W.C. Preparation of Heterocyclic Ureas as Kinase Inhibitors Useful for the Treatment of Proliferative Diseases. U.S. Patent Application No. 20090099190 A1 20090416, 16 April 2009.
31. Flynn, D.L.; Kaufman, M.D.; Patt, W.C.; Petillo, P.A. Preparation of Heterocyclic Ureas as Kinase Inhibitors Useful for the Treatment of Proliferative and Inflammatory Diseases. WO Patent Application No. 2008034008 A2 20080320, 20 March 2008.
32. Flynn, D.L.; Petillo, P.A.; Kaufman, M.D.; Patt, W.C. Preparation of Dihydropyridopyrimidinyl, Dihydronaphthyidinyl and Related Compounds Useful as Kinase Inhibitors for the Treatment of Proliferative Diseases. U.S. Patent Application No. 8188113 B2 20120529, 29 May 2012.
33. Eshraghi, A.A.; Wang, J.; Adil, E.; He, J.; Zine, A.; Bublik, M.; Bonny, C.; Puel, J.; Balkany, T.J.; Van De Water, T.R. Blocking c-Jun-N-terminal kinase signaling can prevent hearing loss induced by both electrode insertion trauma and neomycin ototoxicity. *Hear Res.* **2007**, *226*, 168–177. [CrossRef] [PubMed]

34. Wang, J.; Van De Water, T.R.; Bonny, C.; de Ribaupierre, F.; Puel, J.L.; Zine, A. A Peptide Inhibitor of c-Jun N-Terminal Kinase Protects against Both Aminoglycoside and Acoustic Trauma-Induced Auditory Hair Cell Death and Hearing Loss. *J. Neurosci.* **2003**, *23*, 8596–8607. [CrossRef] [PubMed]
35. Feng, Y.; LoGrasso, P.; Zheng, K.; Park, C.M. Preparation of Substituted Pyrazolylbenzamides as JNK Kinase Inhibitors. WO Patent Application No. 2015084936 A1 20150611, 11 June 2015.
36. Cherrier, M.P.; Parmantier, E.; Minoux, H.; Clerc, F.; Angouillant-Boniface, O.; Brollo, M.; Schio, L. Pyrazolylbenzimidazole Derivative, their Preparation, Compositions Containing them and their Use for Treating Diseases, Especially Cancer. WO Patent Application No. 2008003857 A1 20080110, 10 January 2008.
37. Swapna, D.; Pooja, V.P.; Sucharitha, D.; Sunitha, K.; Swetha, M. A review of p38 kinase inhibitors as anti-inflammatory drugs targets. *Int. J. Pharm. Technol. (IJPT)* **2010**, *2*, 86–101.
38. Barun, O.; Advait, N.; Francisco, J.A. A general strategy for creating "inactive-conformation" abl inhibitors. *Chem. Biol.* **2006**, *13*, 779–786. [CrossRef]
39. Dumas, J.; Sibley, R.; Riedl, B.; Monahan, M.K.; Lee, W.; Lowinger, T.B.; Redman, A.M.; Johnson, J.S.; Kingery-Wood, J.; Scott, W.J.; et al. Discovery of a new class of p38 kinase inhibitors. *Bioorg. Med. Chem. Lett.* **2000**, *10*, 2047–2050. [CrossRef]
40. Dumas, J.; Hatoum-Mokdad, H.; Sibley, R.; Riedl, B.; Scott, W.J.; Monahan, M.K.; Lowinger, T.B.; Brennan, C.; Natero, R.; Turner, T.; et al. 1-Phenyl-5-pyrazolyl ureas: Potent and selective p38 kinase inhibitors. *Bioorg. Med. Chem. Lett.* **2000**, *10*, 2051–2054. [CrossRef]
41. Redman, A.M.; Johnson, J.S.; Dally, R.; Swartz, S.; Wild, H.; Paulsen, H.; Caringal, Y.; Gunn, D.; Renick, J.; Osterhout, M.; et al. p38 Kinase Inhibitors for the Treatment of Arthritis and Osteoporosis: Thienyl, Furyl, and Pyrrolyl Ureas. *Bioorg. Med. Chem. Lett.* **2001**, *11*, 9–12. [CrossRef]
42. Dumas, J.; Hatoum-Mokdad, H.; Sibley, R.N.; Smith, R.A.; Scott, W.J.; Khire, U.; Lee, W.; Wood, J.; Wolanin, D.; Cooley, J.; et al. Synthesis and pharmacological characterization of a potent, orally active p38 kinase inhibitor. *Bioorg. Med. Chem. Lett.* **2002**, *12*, 1559–1562. [CrossRef]
43. Cirillo, P.F.; Gilmore, T.A.; Hickey, E.R.; Regan, J.R.; Zhang, L.-H. Preparation of Aromatic Heterocyclic Ureas as Antiinflammatory Agents. WO Patent Application No. 2000043384 A1 20000727, 27 July 2000.
44. Zhang, L.-H.; Zhu, L. Novel Process for Synthesis of Heteroaryl-Substituted Ureas. WO Patent Application No. 2001004115 A2 20010118, 18 January 2001.
45. Pargellis, C.; Tong, L.; Churchill, L.; Cirillo, P.F.; Gilmore, T.; Graham, A.G.; Grob, P.M.; Hickey, E.R.; Moss, N.; Pav, S.; et al. Inhibition of p38 MAP kinase by utilizing a novel allosteric binding site. *Nat. Struct. Biol.* **2002**, *9*, 268–272. [CrossRef]
46. Regan, J.; Breitfelder, S.; Cirillo, P.; Gilmore, T.; Graham, A.G.; Hickey, E.; Klaus, B.; Madwed, J.; Moriak, M.; Moss, N.; et al. Pyrazole urea-based inhibitors of p38 MAP kinase: From lead compound to clinical candidate. *J. Med. Chem.* **2002**, *45*, 2994–3008. [CrossRef]
47. Soyoon, R.; Jida, C.; Jaemyung, K.; Suyoung, B.; Jaewoo, H.; Seunghyun, J.; Soohyun, K.; Youngmin, L. BIRB 796 has Distinctive Anti-inflammatory Effects on Different Cell Types. *Immune Netw.* **2013**, *13*, 283–288. [CrossRef]
48. Available online: https://www.clinicaltrials.gov/ (accessed on 5 February 2020).
49. Bagley, M.C.; Davis, T.; Dix, M.C.; Widdowson, C.S.; Kipling, D. Microwave-assisted Synthesis of N-pyrazole Ureas and the p38alpha Inhibitor BIRB 796 for Study into Accelerated Cell Ageing. *Org. Biomol. Chem.* **2006**, *4*, 4158–4164. [CrossRef] [PubMed]
50. Kulkarni, R.G.; Srivani, P.; Achaiah, G.; Sastry, G.N. Strategies to design pyrazolyl urea derivatives for p38 kinase inhibition: A molecular modeling study. *J. Comput.-Aided Mol. Des.* **2007**, *21*, 155–166. [CrossRef] [PubMed]
51. Toru, A.; Hitoshi, Y.; Chiyoshi, K. Identification, synthesis, and biological evaluation of 6-[(6R)-2-(4-fluorophenyl)-6-(hydroxymethyl)-4,5,6,7-tetrahydropyrazolo [1,5-a]pyrimidin-3-yl]-2-(2-methylphenyl)pyridazin-3(2H)-one (AS1940477), a potent *p*38 MAP kinase inhibitor. *J. Med. Chem.* **2012**, *55*, 7772–7785. [CrossRef]
52. Shunsuke, I.; Yoshiji, A.; Hideo, A. A possible mechanism for hepatotoxicity induced by BIRB-796, an orally active p38 mitogen-activated protein kinase inhibitor. *J. Appl. Toxicol.* **2011**, *31*, 671–677. [CrossRef]
53. Li, X.; Zhou, X.; Zhang, J.; Wang, L.; Long, L.; Zheng, Z.; Li, S.; Zhong, W. Synthesis and biological evaluation of chromenylurea and chromanylurea derivatives as anti-TNF-α agents that target the p38 MAPK pathway. *Molecules* **2014**, *19*, 2004–2028. [CrossRef]

54. Li, X.; Zhou, X.; Zheng, Z.; Zhong, W.; Xiao, J.; Li, S. Short Synthesis of 1-(3-tert-Butyl-1-phenyl-1H-pyrazol-5-yl)-3-(5-(2-morpholinoethoxy)-2H-chromen-8-yl) Urea Derivatives. *Synth. Commun.* **2009**, *39*, 3999–4009. [CrossRef]
55. Flynn, D.L.; Petillo, P.A. Preparation of 1-pyrazolyl-3-phenylurea p38 MAP Kinase Inhibitors as Antiinflammatory Medicaments. U.S. Patent Application No. 20050288286 A1 20051229, 29 December 2005.
56. Flynn, D.L.; Petillo, P.A. Preparation of Pyrazolyl aryl Ureas as Modulators of the Protein Kinase Activation State for Treatment of Inflammation and Hyperproliferative Diseases. WO Patent Application No. 2006081034 A2 20060803, 3 August 2006.
57. Flynn, D.L.; Petillo, P.A. Preparation of 1-pyrazolyl-3-phenylurea p38 MAP Kinase Inhibitors as Antiinflammatory Medicaments. U.S. Patent Application No. 20070191336 A1 20070816, 16 August 2007.
58. Arai, T.; Ohno, M.; Inoue, H.; Hayashi, S.; Aoki, T.; Hirokawa, H.; Meguro, H.; Koga, Y.; Oshida, K.; Kainoh, M.; et al. Design and synthesis of novel p38α MAP kinase inhibitors: Discovery of pyrazole-benzyl ureas bearing 2-molpholinopyrimidine moiety. *Bioorg. Med. Chem. Lett.* **2012**, *22*, 5118–5122. [CrossRef]
59. Longshaw, A.I.; Fordyce, E.A.F.; Onions, S.T.; King-Underwood, J.; Venable, J.D.; Walters, I. Preparation of Pyrazolyl-Urea Compounds as p38 MAP Kinase Inhibitors. WO Patent Application No. 2015121660, 20 August 2015.
60. Charron, C.E.; Ito, K.; Rapeport, W.G. Ureido-pyrazole Derivatives for Use in the Treatment of Respiratory Disorders. WO Patent Application No. 2011158042, 22 December 2011.
61. Charron, C.E.; Fenton, R.; Crowe, S.; Ito, K.; Strong, P.; Rapeport, G.; Ray, K. Preparation of Pyrazolylurea Compounds as Hemopoietic Cell Kinase (p59-HCK) Inhibitors for Treatment of Influenza Infection. WO Patent Application No. 2011/070369, 16 June 2011.
62. Longshaw, A.I.; Fordyce, E.A.F.; Onions, S.T.; King-Underwood, J.; Venable, J.D. Preparation of Aromatic Heterocyclic Compounds as p38 MAP Kinase Inhibitors with Antiinflammatory Activity. WO Patent Application No. 2015121444, 20 August 2015.
63. King-Underwood, J.; Murray, P.J.; Williams, J.G.; Buck, I.; Onions, S.T. Preparation of Pyrazolyl Naphthyl Ureas as p38 MAP Kinase Inhibitors. WO Patent Application No. 2011124930, 13 October 2011.
64. King-Underwood, J.; Ito, K.; Strong, P.; Rapeport, W.G.; Charron, C.E.; Murray, P.J.; Williams, J.G.; Onions, S.T. 1-(5-Pyrazolyl)-3-(1-naphthyl)ureas as Enzyme Inhibitors and their Preparation, Pharmaceutical Compositions and Use in the Treatment of Inflammatory Disorders. WO Patent Application No. 2011124923, 13 October 2011.
65. Ito, K.; Strong, P.; Rapeport, W.G.; Murray, P.J.; King-Underwood, J.; Williams, J.G.; Onions, S.T.; Joly, K.; Charron, C.E. Preparation of Pyrazolyl Ureas as p38 MAP Kinase Inhibitors. WO Patent Application No. 2010067130, 21 April 2011.
66. Ito, K.; Charron, C.E.; King-Underwood, J.; Onions, S.T.; Longshaw, A.I.; Broeckx, R.; Filliers, W.; Copmans, A. Preparation of 1-pyrazolyl-3-{4-[(2-anilinopyrimidin-4-yl)oxy]naphththalen-1-yl}ureas as p38 MAP Kinase Inhibitors. WO Patent Application No. 2013050757, 11 April 2013.
67. Duffy, L.A.; King-Underwood, J.; Longshaw, A.I.; Murray, P.J.; Onions, S.T.; Taddei, D.M.A.; Williams, J.G.; Ito, K.; Charron, C.E. Preparation of Pyrazole Ureas as p38 MAP Inhibitors. WO Patent Application No. 2014033448 A1 20140306, 6 March 2014.
68. Cariou, C.A.M.; Charron, C.E.; Fordyce, E.A.F.; Hamza, D.; Fyfe, M.C.T.; Ito, K.; King-Underwood, J.; Murray, P.J.; Onions, S.T.; Thom, S.M.; et al. Preparation of Pyrazolyl-Urea Compounds as Kinase Inhibitors for Treatment of Inflammatory Diseases. WO Patent Application No. 2014027209, 20 February 2014.
69. Bruno, O.; Brullo, C.; Bondavalli, F.; Schenone, S.; Ranise, A.; Arduino, N.; Bertolotto, M.B.; Montecucco, F.; Ottonello, L.; Dallegri, F.; et al. Synthesis and biological evaluation of N-pyrazolyl-N'-alkyl/benzyl/phenylureas: A new class of potent inhibitors of interleukin 8-induced neutrophil chemotaxis. *J. Med. Chem.* **2007**, *50*, 3618–3626. [CrossRef]
70. Bruno, O.; Brullo, C.; Bondavalli, F.; Schenone, S.; Spisani, S.; Falzarano, M.S.; Varani, K.; Barocelli, E.; Ballabeni, V.; Giorgio, C.; et al. 1-Methyl and 1-(2-hydroxyalkyl)-5-(3-alkyl/cycloalkyl/phenyl/naphthylureido)-1H-pyrazole-4-carboxylic acid ethyl esters as potent human neutrophil chemotaxis inhibitors. *Bioorg. Med. Chem.* **2009**, *17*, 3379–3387. [CrossRef]

71. Bertolotto, M.; Brullo, C.; Ottonello, L.; Montecucco, F.; Dallegri, F.; Bruno, O. Treatment with 1-(2-hydroxyalkyl)-5-(3-alkyl/phenyl/naphthylureido)-1Hpyrazole-4-carboxylic acid ethyl esters abrogates neutrophil migration towards fMLP and CXCL8 via the inhibition of defined signalling pathways. *Eur. J. Clin. Invest.* **2014**, *44*, 39.
72. Bruno, O.; Bondavalli, F.; Brullo, C.; Schenone, S. Ureas Derivatives of 1H-pyrazol-4-carboxylic Acid with Neutrophil Chemotaxis Inhibiting Activity. EP Patent Application No. 2033955 A1, 11 March 2009.
73. Brullo, C.; Spisani, S.; Selvatici, R.; Bruno, O. N-Aryl-2-phenyl-2,3-dihydro-imidazo[1,2-b]pyrazole-1-carboxamides 7-substituted strongly inhibiting both fMLP-Ome- and IL-8-induced human neutrophil chemotaxis. *Eur. J. Med. Chem.* **2012**, *47*, 573–579. [CrossRef] [PubMed]
74. Selvatici, R.; Brullo, C.; Bruno, O.; Spisani, S. Differential inhibition of signaling pathways by two new imidazo-pyrazoles molecules in fMLF-Ome- and IL8-stimulated human neutrophil. *Eur. J. Pharmacol.* **2013**, *718*, 428–434. [CrossRef] [PubMed]
75. Meta, E.; Brullo, C.; Sidibé, A.; Imhof, B.A.; Bruno, O. Design, synthesis and biological evaluation of new pyrazolyl-ureas and imidazopyrazolecarboxamides able to interfere with MAPK and PI3K upstream signaling involved in the angiogenesis. *Eur. J. Med. Chem.* **2017**, *133*, 24–35. [CrossRef] [PubMed]
76. Meta, E.; Imhof, B.A.; Roprazb, P.; Fish, R.J.; Brullo, C.; Bruno, O.; Sidibé, A. The pyrazolyl-urea GeGe3 inhibits tumor angiogenesis and reveals dystrophia myotonica protein kinase (DMPK)1 as a novel angiogenesis target. *Oncotarget* **2017**, *8*, 108195–108212. [CrossRef]
77. Abd El-Meguid, E.A.; Ali, M.M. Synthesis of some novel 4-benzothiazol-2-ylbenzoyl-1H-pyrazoles, and evaluation as antiangiogenic agents. *Res. Chem. Intermediat.* **2016**, *42*, 1521–1536. [CrossRef]
78. Flynn, D.L.; Petillo, P.A. Preparation of Pyrazolyl Phenyl Ureas as Enzyme Modulators. WO Patent Apllication No. 2006071940 A2 20060706, 6 July 2006.
79. Flynn, D.L.; Petillo, P.A. Preparation of Pyrazolyl Phenyl Ureas as Enzyme Modulators. U.S. Patent Application No. 20080113967 A1 20080515, 15 May 2008.
80. Smith, R.; Hatoum-Mokdad, H.N.; Cantin, L.D.; Bierer, D.E.; Fu, W.; Nagarathnam, D.; Ladouceur, G.; Wang, Y.; Ogutu, H.; Wilhelm, S.; et al. Pyrazole Urea Compounds Useful in the Treatment of Cancer and their Preparation. WO Patent Application No. 2007064872, 7 June 2007.
81. Flynn, D.L.; Petillo, P.A.; Kaufman, M.D. Preparation of Urea Derivatives as Kinase Inhibitors Useful for the Treatment of Myeloproliferative Diseases and Other Proliferative Diseases. U.S. Patent Application No. 7790756 B2 20100907, 7 September 2010.
82. Flynn, D.L.; Petillo, P.A.; Kaufman, M.D. Preparation of Heterocyclic Urea Derivatives as Kinase Inhibitors Useful for the Treatment of Myeloproliferative Diseases and Other Proliferative Diseases. WO Patent Application No. 2013036232 A2 20130314, 14 March 2013.
83. Getlik, M.; Gruetter, C.; Simard, J.R.; Kluter, S.; Rabiller, M.; Rode, H.B.; Robubi, A.; Rauh, D. Hybrid Compound Design to Overcome the Gatekeeper T338M Mutation in cSrc. *J. Med. Chem.* **2009**, *52*, 3915–3926. [CrossRef]
84. Indrasena, R.K.; Aruna, C.; Manisha, M.; Srihari, K.; Sudhakar, B.K.; Vijayakumar, V.; Sarveswari, S.; Priya, R.; Amrita, A.; Siva, R. Synthesis, DNA binding and in-vitro cytotoxicity studies on novel bis-pyrazoles. *J. Photochem. Photobiol. B Biol.* **2017**, *168*, 89–97. [CrossRef]
85. Kessler, J.D.; Kahle, K.T.; Sun, T.; Meerbrey, K.L.; Schlabach, M.R.; Schmitt, E.M.; Skinner, S.O.; Xu, Q.; Li, M.Z.; Hartman, Z.C.; et al. A SUMOylation-dependent Transcriptional Subprogram Is Required for Myc-driven Tumorigenesis. *Science* **2012**, *335*, 348–353. [CrossRef]
86. Kumar, A.; Ito, A.; Hirohama, M.; Yoshida, M.; Zhang, K.Y.J. Identification of new SUMO activating enzyme 1 inhibitors using virtual screening and scaffold hopping. *Bioorg. Med. Chem. Lett.* **2016**, *26*, 1218–1223. [CrossRef]
87. Benito-Gutiérrez, È.; Garcia-Fernàndez, J.; Comella, J.X. Origin and Evolution of the Trk Family of Neurotrophic Receptors. *Mol. Cell. Neurosci.* **2006**, *31*, 179–192. [CrossRef] [PubMed]
88. Brandhuber, B.J.; Jiang, Y.; Kolakowski, G.R.; Winski, S.L. Pyrazolyl Urea, Thiourea, Guanidine and Cyanoguanidine Compounds as TrkA Kinase Inhibitors and their Preparation. WO Patent Application No. 2014078417 A1 20140522, 22 May 2014.

89. Allen, S.; Andrews, S.W.; Baer, B.; Crane, Z.; Liu, W.; Watson, D.J. Preparation of 1-((3S,4R)-4-(3-fluorophenyl)-1-(2-methoxyethyl)pyrrolidin-3-yl)-3-(4-methyl-3-(2-methylpyrimidin-5-yl)-1-phenyl-1H-pyrazol-5-yl)urea as a TRKA Kinase Inhibitor. WO Patent Application No. 2015175788 A1 20151119, 19 November 2015.
90. Subramanian, G.; Johnson, P.D.; Zachary, T.; Roush, N.; Zhu, Y.; Bowen, S.J.; Janssen, A.; Duclos, B.A.; Williams, T.; Javens, C.; et al. Deciphering the Allosteric Binding Mechanism of the Human Tropomyosin Receptor Kinase A (hTrkA) Inhibitors. *ACS Chem. Biol.* **2019**, *14*, 1205–1216. [CrossRef] [PubMed]
91. Furuya, N.; Momose, T.; Katsuno, K.; Fushimi, N.; Muranaka, H.; Handa, C.; Ozawa, T.; Kinoshita, T. The juxtamembrane region of TrkA kinase is critical for inhibitor selectivity. *Bioorg. Med. Chem. Lett.* **2017**, *27*, 1233–1236. [CrossRef] [PubMed]
92. Shikano, K.; Yamaguchi, H.; Horiguchi, T.; Yasuo, K.; Hata, K. Pain Therapeutic or Preventive Containing Fused Heterocycles and Heterocycle-Fused Carbocycles Having TrkA Inhibitory Activity. JP Patent Application No. 2019189573, 31 October 2019.
93. Yukimasa, A.; Shikano, K.; Horiguchi, T.; Nakamura, K.; Inoue, T.; Fujo, M.; Yamaguchi, H. Pain Treatment and/or Preventive Agent Containing Nitrogen-Containing Heterocycles Having Tropomyosin Receptor Kinase A (TrkA)-Inhibitory Activity. JP Patent Application No. 2019026646, 21 February 2019.
94. Vaidya, A.B.; Welsh, W.J.; Kortagere, S.; Bergman, L.W. Mechanism-Based Small Heterocyclic Molecule Parasite Inhibitors. WO Patent Application No. 2009065096 A1 20090522, 22 May 2009.
95. Welsh, W.J.; Kortagere, S.; Bergman, L.W.; Vaidya, A.B. Mechanism-Based Small-Molecule Parasite Inhibitors. U.S. Patent Application No. 8486987 B2 20130716, 16 July 2013.
96. Kortagere, S.; Welsh, W.J. Development and application of hybrid structure-based method for efficient screening of ligands binding to G-protein coupled receptors. *J. Comput.-Aided. Mol. Des.* **2006**, *20*, 789–802. [CrossRef] [PubMed]
97. Kortagere, S.; Welsh, W.J.; Morrisey, J.M.; Daly, T.; Ejigiri, I.; Sinnis, P.; Vaidya, A.B.; Bergman, L.W. Structure-based design of novel small-molecule inhibitors of Plasmodium falciparum. *J. Chem. Inf. Model.* **2010**, *50*, 840–849. [CrossRef]
98. Kuhlbrandt, W. Biology, structure, and mechanism of P-type ATPases. *Nat. Rev. Mol. Cell Biol.* **2004**, *5*, 282–295. [CrossRef]
99. Yatime, L.; Buch-Pedersen, M.J.; Musgaard, M.; Morth, J.P.; Lund Winther, A.M.; Pedersen, B.P.; Olesen, C.; Andersen, J.P.; Vilsen, B.; Schiott, B.; et al. P-type ATPases as drug targets: Tools for medicine and science. *Biochim. Biophys. Acta* **2009**, *1787*, 207–220. [CrossRef]
100. Rodriguez-Navarro, A.; Benito, B. Sodium or potassium efflux ATPase a fungal, bryophyte, and protozoal ATPase. *Biochim. Biophys. Acta* **2010**, *1798*, 1841–1853. [CrossRef]
101. Spillman, N.J.; Allen, R.J.; McNamara, C.W.; Yeung, B.K.; Winzeler, E.A.; Diagana, T.T.; Kirk, K. Na(+) regulation in the malaria parasite Plasmodium falciparum involves the cation ATPase PfATP4 and is a target of the spiroindolone antimalarials. *Cell Host Microbe* **2013**, *13*, 227–237. [CrossRef]
102. White, N.J.; Pukrittayakamee, S.; Phyo, A.P.; Rueangweerayut, R.; Nosten, F.; Jittamala, P.; Jeeyapant, A.; Jain, J.P.; Lefevre, G.; Li, R.; et al. Spiroindolone KAE609 for falciparum and vivax malaria. *N. Engl. J. Med.* **2014**, *371*, 403–410. [CrossRef]
103. Flannery, E.L.; Fidock, D.A.; Winzeler, E.A. Using genetic methods to define the targets of compounds with antimalarial activity. *J. Med. Chem.* **2013**, *56*, 7761–7771. [CrossRef]
104. Flannery, E.L.; McNamara, C.W.; Kim, S.W.; Kato, T.S.; Li, F.; Teng, C.H.; Gagaring, K.; Manary, M.J.; Barboa, R.; Meister, S.; et al. Mutations in the P-Type Cation-Transporter ATPase 4, PfATP4, Mediate Resistance to Both Aminopyrazole and Spiroindolone Antimalarials. *ACS Chem. Biol.* **2015**, *10*, 413–420. [CrossRef] [PubMed]
105. Hewitt, S.N.; Dranow, D.M.; Horst, B.G.; Abendroth, J.A.; Forte, B.; Hallyburton, I.; Jansen, C.; Baragana, B.; Choi, R.; Rivas, K.L.; et al. Biochemical and Structural Characterization of Selective Allosteric Inhibitors of the Plasmodium falciparum Drug Target, Prolyl-tRNA-synthetase. *ACS Infect. Dis.* **2017**, *3*, 34–44. [CrossRef] [PubMed]
106. Kortagere, S.; Mui, E.; McLeod, R.; Welsh, W.J. Rapid discovery of inhibitors of Toxoplasma gondii using hybrid structure-based computational approach. *J. Comput.-Aided Mol. Des.* **2011**, *25*, 403–411. [CrossRef] [PubMed]

107. Kane, J.L.; Hirth, B.H.; Liang, B.; Gourlie, B.B.; Nahill, S.; Barsomian, G. Ureas of 5-aminopyrazole and 2-aminothiazole inhibit growth of gram-positive bacteria. *Bioorg. Med. Chem. Lett.* **2003**, *13*, 4463–4466. [CrossRef]
108. Pedro-Rosa, L.; Buckner, F.S.; Ranade, R.M.; Eberhart, C.; Madoux, F.; Gillespie, J.R.; Koh, C.Y.; Brown, S.; Lohse, J.; Verlinde, C.L.M.; et al. Identification of potent inhibitors of the Trypanosoma brucei methionyl-tRNA synthetase via high-throughput orthogonal screening. *J. Biomol. Screen.* **2015**, *20*, 122–130. [CrossRef]
109. Murray, P.J.; Onions, S.T.; Williams, J.G.; Joly, K. Preparation of Pyrido[2,3-b]pyrazine Compounds for Use in Drug Formulations for Treating Inflammation, Respiratory Disorders, and Viral Infections. WO Patent Application No. 2011158044 A2 20111222, 22 December 2011.
110. Luscher, C.; Slesinger, P. Emerging roles for G protein-gated inwardly rectifying potassium (GIRK) channels in health and disease. *Nat. Rev. Neurosci.* **2010**, *11*, 301–315. [CrossRef]
111. Niswender, C.M.; Johnson, K.A.; Luo, Q.; Ayala, J.E.; Kim, C.; Conn, P.J.; Weaver, C.D. A novel assay of Gi/o-linked G protein-coupled receptor coupling to potassium channels provides new insights into the pharmacology of the group III metabotropic glutamate receptors. *Mol. Pharmacol.* **2008**, *73*, 1213–1224. [CrossRef]
112. Kaufmann, K.; Romaine, I.; Days, E.; Pascual, C.; Malik, A.; Yang, L.; Zou, B.; Du, Y.; Sliwoski, G.; Morrison, R.D.; et al. ML297 (VU0456810), the first potent and selective activator of the GIRK potassium channel, displays antiepileptic properties in mice. *ACS Chem. Neurosci.* **2013**, *4*, 1278–1286. [CrossRef]
113. Wieting, J.M.; Vadukoot, A.K.; Sharma, S.; Abney, K.K.; Bridges, T.M.; Daniels, J.S.; Morrison, R.D.; Wickman, K.; Weaver, C.D.; Hopkins, C.R. Discovery and Characterization of 1H-Pyrazol-5-yl-2-phenylacetamides as Novel, Non-Urea-Containing GIRK1/2 Potassium Channel Activators. *ACS Chem. Neurosci.* **2017**, *8*, 1873–1879. [CrossRef]

© 2020 by the authors. Licensee MDPI, Basel, Switzerland. This article is an open access article distributed under the terms and conditions of the Creative Commons Attribution (CC BY) license (http://creativecommons.org/licenses/by/4.0/).

Review

Styrylpyrazoles: Properties, Synthesis and Transformations

Pedro M. O. Gomes, Pedro M. S. Ouro, Artur M. S. Silva * and Vera L. M. Silva *

LAQV-REQUIMTE, Department of Chemistry, University of Aveiro, 3810-193 Aveiro, Portugal;
pm.gomes@ua.pt (P.M.O.G.); pedroouro@ua.pt (P.M.S.O.)
* Correspondence: artur.silva@ua.pt (A.M.S.S.); verasilva@ua.pt (V.L.M.S.); Tel.: +351-234-370714 (A.M.S.S.); +351-234-370704 (V.L.M.S.)

Academic Editor: Derek J. McPhee
Received: 21 November 2020; Accepted: 9 December 2020; Published: 12 December 2020

Abstract: The pyrazole nucleus and its reduced forms, pyrazolines and pyrazolidine, are privileged scaffolds in medicinal chemistry due to their remarkable biological activities. A huge number of pyrazole derivatives have been studied and reported over time. This review article gives an overview of pyrazole derivatives that contain a styryl (2-arylvinyl) group linked in different positions of the pyrazole backbone. Although there are studies on the synthesis of styrylpyrazoles dating back to the 1970s and even earlier, this type of compound has rarely been studied. This timely review intends to summarize the properties, biological activity, methods of synthesis and transformation of styrylpyrazoles; thus, highlighting the interest and huge potential for application of this kind of compound.

Keywords: pyrazoles; styrylpyrazoles; biological activity; organic synthesis; reactivity

1. Introduction

The pyrazole (1H-pyrazole, **1**) (Figure 1) is an aromatic five-membered heterocyclic ring constituted by three carbons and two adjacent nitrogen atoms, located at 1- and 2-positions [1]. N-unsubstituted pyrazoles may present three identical and non-separable tautomers, due to rapid interconversion in solution, and it is usually impossible to unequivocally assign the resonances of its proton-nuclear magnetic resonance (^{1}H-NMR) spectrum. Three partially reduced forms may also exist: 1-pyrazolines **2**, 2-pyrazolines **3** and 3-pyrazolines **4**, and a fully reduced form known as pyrazolidine **5** (Figure 1) [1,2].

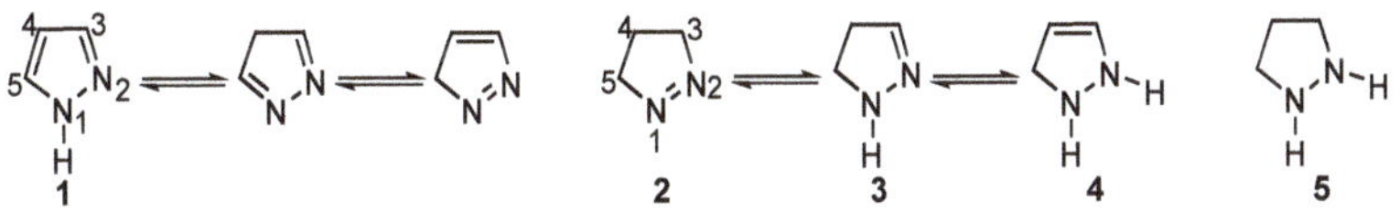

Figure 1. Chemical structures and numbering of pyrazole **1**, dihydropyrazole (pyrazoline) tautomers **2–4** and pyrazolidine **5**.

Pyrazole and its derivatives occupy a prime place in medicinal chemistry because they are present in several drugs with real medicinal applications, such as celecoxib (Celebrex®), sildenafil (Viagra®), rimonabant, lonazolac, fomepizole, penthiopyrad, doramapimod, sulfaphenazole and in several remarkable compounds with a wide range of pharmacological activities, namely anticancer [3], analgesic, anti-inflammatory and antioxidant [4–7], antibacterial and antifungal [8,9] antipyretic, antidepressant and anticonvulsant [10,11], antidiabetic [12] and cannabinoid activities [13,14], among others [15–21]. Along with the medicinal applications, pyrazoles are also useful as agrochemicals, such as herbicides, fungicides, and insecticides [22] and possess other applications, such as dyestuffs, sunscreen materials [23,24], analytical reagents and pluripotent ligands in coordination chemistry [25,26]. For instance, polyaromatic pyrazoles possess important biological, photophysical, optical and electronic properties [10,27–29].

This review is focused on a particular type of pyrazoles, the styrylpyrazoles (or styryl-1H-pyrazoles), which present a styryl (2-arylvinyl) group at one or more positions of the pyrazole nucleus. It is intended to give an overview of the styrylpyrazoles chemistry reported in the literature, including their properties and biological activity, since the 1970s to the present date. Despite the biological activities of styrylpyrazoles [12,14], their interesting physicochemical properties, such as tautomerism, isomerism [30,31] and conjugation extension, due to the presence of the 2-arylvinyl group, as well as the rich chemistry related to their synthesis and transformation, this type of pyrazole has rarely been studied, as can be seen from the low number of publications in a very large time span covered by this review. In our group, we have been working with styrylpyrazoles for a long time. Throughout this review, we intend to describe the properties and methods of synthesis of these compounds, and to show their great potential for application in different fields, from medicine to materials chemistry, including their use as key templates for transformation into other valuable compounds. The articles will be presented based on the molecules' structure, in particular, the position of the styryl group in the pyrazole scaffold.

2. Properties of Styrylpyrazoles

The pharmacological activity of styrylpyrazoles and related compounds has been known for a long time. A study published in 1970 described the anti-inflammatory activity of 1-phenyl-2-styryl-3,5-dioxopirazolidines (**6**). Among these compounds, derivatives **6a** (R^1 = Me, R^2 = R^3 = H, R^4 = n-C_4H_9) and **6b** (R^1 = Cl, R^2 = 4-OMe, R^3 = H, R^4 = n-C_4H_9) (Figure 2) were more active than phenylbutazone or oxyphenbutazone in the carrageenin-induced foot edema test in rats [32]. It was found that a p-methylphenyl ring increased the intrinsic activity in general, while a p-chlorophenyl substituent decreases the activity, except for compound **6b**. The p-methoxy or 3,4-methylenedioxy substituents of the styryl group tend to increase both toxicity and inhibitory activity, while p-alkyl substituents (Me or i-Pr) have the opposite effect.

Styrylpyrazoles were also found to act as dual inhibitors of 5-lipoxygenase (5-LO) and cyclooxygenase (CO) in rat basophilic leukemia cells. In particular, compound **7** (Figure 2) exhibited oral activity in various models of inflammation and, most importantly, is devoid of ulcerogenic potential [33].

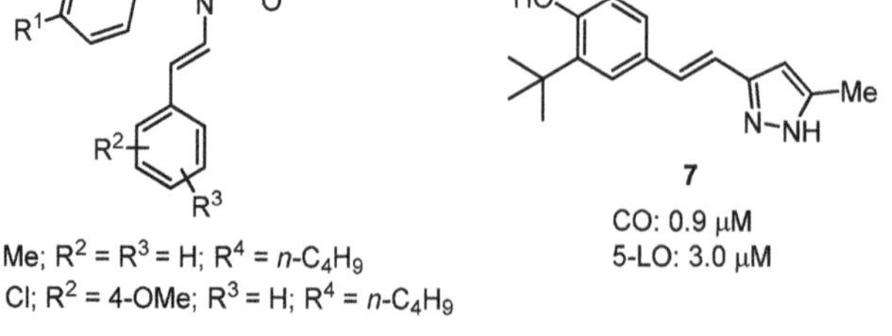

6a: R^1 = Me; R^2 = R^3 = H; R^4 = n-C_4H_9
6b: R^1 = Cl; R^2 = 4-OMe; R^3 = H; R^4 = n-C_4H_9

7
CO: 0.9 µM
5-LO: 3.0 µM

Figure 2. Styrylpyrazole derivatives **6** and **7** presenting anti-inflammatory activity [32,33].

Nauduri et al. studied the antibacterial activity of 1,5-diphenyl-3-styryl-2-pyrazolines **8**, with different substituents (OMe, NMe_2, NO_2, OH and isopropyl) (Figure 3), towards Gram-positive and Gram-negative bacteria and found little variation in the geometric means of minimum inhibitory concentrations (MIC) [34]. The derivatives containing NMe_2, NO_2 and OMe in the p-position of

A and B aromatic rings showed the highest activity (MIC = 3.55–6.56 µg/mL and 3.30–6.27 µg/mL against Gram-positive and Gram-negative bacteria, respectively). This fact was attributed to the strong electronic-driving nature of the substituents (either as electron-donating or electron-withdrawing substituents) and due to a more lipophilic nature (for the NMe$_2$). Low substitution in the aromatic ring also favored the entry of the compound in the bacteria through passive diffusion. They have compared the activity of these compounds with that of 1,3-diphenyl-5-styryl-2-pyrazolines, which showed lower activities than the compounds of the 3-styryl series. Moreover, the 1,5-diphenyl-3-styryl-2-pyrazolines **8** bearing *o*- or *p*-hydroxyphenyl rings showed superior antimycotic (antifungal) activity relative to the other derivatives (MIC = 11.3–24.8 µg/mL and 12.16–13.6 µg/mL against fungi and yeast, respectively). The extended electronic structure of the 3-styryl compared with the 5-styryl derivatives may explain the lower MIC values for the compounds of the 3-styryl series [34].

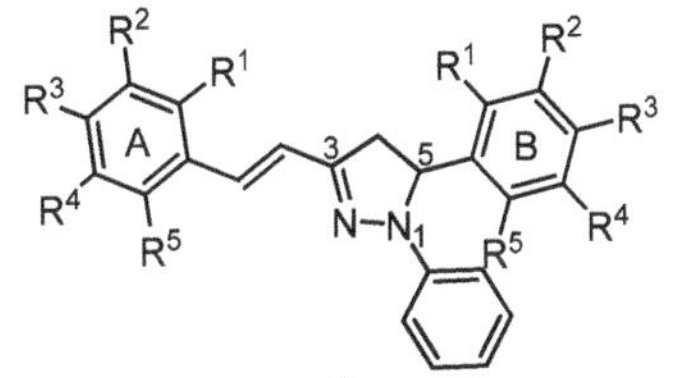

8
R^1 = H, OMe, OH; R^2 = H, OMe
R^3 = H, OMe, NMe$_2$, NO$_2$, OH, isopropyl
R^4 = H, OMe; R^5 = H, OMe

Figure 3. 3-Styryl-2-pyrazolines **8** presenting antibacterial and antifungal activity [34].

In the 2,2-diphenyl-1-picrylhydrazyl (DPPH) and horseradish peroxidase (HRP)/luminol/H$_2$O$_2$ chemiluminescence assays, 5-aryl-3-styryl-1-carboxamidino-1*H*-pyrazole derivative **9** showed higher antioxidant activity (half maximal inhibitory concentration (IC$_{50}$) 45.17–217.22 µg/mL, in DPPH assay) than 2-(5-aryl-3-styryl-1*H*-pyrazol-1-yl)-4-trifluoromethylpyrimidines **10** (Figure 4). The antioxidant activity is dependent on the concentration and type of substituents in each compound, and resonance stabilization. For instance, the presence of SH at Y position strongly enhances antioxidant activity. Some derivatives of compound **9** showed antimicrobial activity against bacteria *Salmonella typhi*, *Staphylococcus aureus* and *Streptococcus pneumoniae*, with inhibition between 1.95 to 15.625 mg/mL. The antimicrobial activity is also affected by the substituent [35].

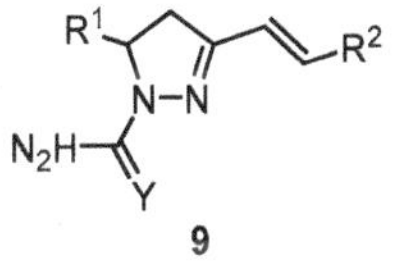

9
R^1, R^2 = 2-MeC$_6$H$_4$; 3,4,5-(MeO)$_3$C$_6$H$_2$; 4-ClC$_6$H$_4$
3,4-(MeO)$_2$C$_6$H$_3$; 4-MeOC$_6$H$_4$
Y = NH; SH

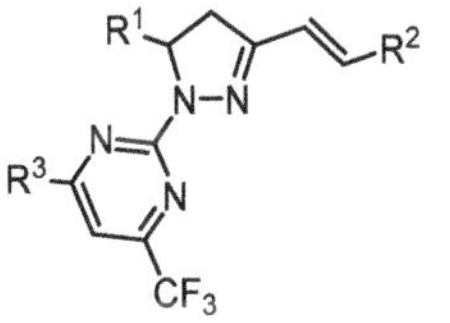

10
R^1 = R^2 = Ph, 2-MeC$_6$H$_4$
R$_3$ = H, CH$_3$, 2-furyl, 2-thienyl

Figure 4. 5-Aryl-3-styryl-1-carboxamidino-1*H*-pyrazole derivatives presenting antioxidant and antimicrobial activity **9** and **10** [35].

Furthermore, 3,5-bis(styryl)pyrazoles **12**, curcumin (**11**) analogues (Figure 5), are well-known for their antioxidant activity. Some derivatives showed superior activity to that of curcumin in DPPH, ferric reducing antioxidant power (FRAP), and β-carotene bleaching assays [36]. For instance, curcumin (**11**) showed lower inhibition of DPPH$^{\bullet}$ (102 mmol), (half maximal effective concentration (EC$_{50}$) = 40 ± 0.06 µmol)) compared with 3,5-bis(4-hydroxy-3-methoxystyryl)pyrazole (**12a**)

(EC_{50} = 14 ± 0.18 µmol). The presence of hydroxy and methoxy groups on the terminal phenyl rings is considered benefic for the antioxidant capacity. Moreover, the 3,5-bis(styryl)pyrazole **CNB-001** has shown interesting results in studies related to neuroprotection and Alzheimer's disease, and was able to restore membrane homeostasis disrupted after brain trauma, being a promising compound for therapeutic use in the treatment of Parkinson's disease [37]. In combination with tissue plasminogen activator, **CNB-001** is efficient in the treatment of stroke [38]. Compound **12a** also acted as inhibitor of β-amyloid secretion and of protein kinases involved in neuronal excitotoxicity [39]. Some pyrazoles that are analogues of **12a** displayed good inhibition of γ-secretase activity, tau aggregation, and/or affinity to fibrillar Aβ42 aggregates [40].

Other properties have been attributed to 3,5-bis(styryl)pyrazoles such as antimalarial, antimycobacterial, antiangiogenic, cytotoxic and antiproliferative activities [39,41,42]. As an example, 3,5-bis(styryl)pyrazoles **12a,b** showed more effective inhibition of chloroquine-sensitive *Plasmodium falciparum* (IC_{50} = 0.48 ± 0.04 µM and 0.87 ± 0.07 µM) and chloroquine-resistant *Plasmodium falciparum* (IC_{50} = 0.45 ± 0.07 µM and 0.89 ± 0.10 µM), respectively, than curcumin (IC_{50} = 3.25 ± 0.6 µM and 4.21 ± 0.8 µM) [41].

Several studies have highlighted the antiproliferative activity of **12a** in PC3 cells, with growth inhibitory concentration (GI_{50}) values of 5.6 µM [43] and in breast cancer cell lines, MCF-7 and SKBR3, with GI_{50} values of 4.19 µM and 0.25 µM, respectively [42]. Liao and coworkers have also shown the antiproliferative activity of 3,5-bis(styryl)pyrazoles **12a** and **13** in the androgen-independent PC3 prostate cancer cell line with GI_{50} values in the low micromolar range. In particular, compound **13** caused significant effects on the PC3 cell cycle, inducing cell death and binding to tubulin (Kd 0.4 ± 0.1 µM), inhibiting tubulin polymerization in vitro, being competitive with paclitaxel for binding to tubulin, and leading to microtubule depolymerization in PC3 cells. Therefore, **13** was considered as a lead compound for the treatment of castration-resistant prostate cancer (CRPC) [44].

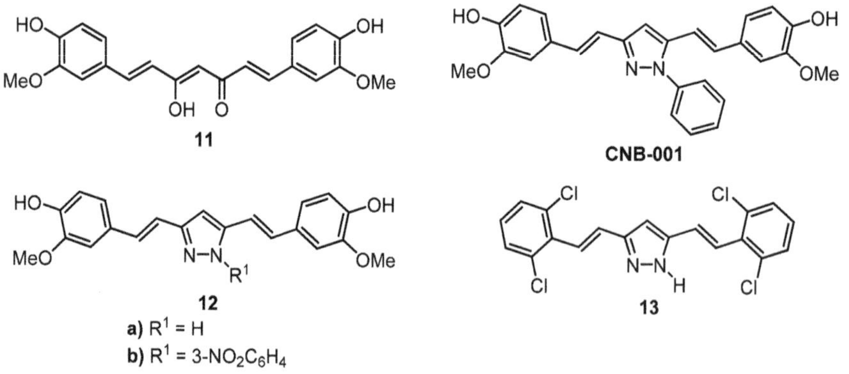

Figure 5. Structures of curcumin (**11**) and 3,5-bis(styryl)pyrazoles **CNB-001**, **12** and **13**, presenting antioxidant, antimalarial, cytotoxic and antiproliferative activities [36–44].

In addition to pharmacological activities, 4-styrylpyrazoles have shown interesting photophysical properties. In particular, 1-(2-pyridyl)-4-styrylpyrazoles (PSPs) substituted at position 3 with donor or acceptor aryl groups have shown strong blue-light emissions with high quantum yields (up to 66%), due to intramolecular charge transfer (ICT) phenomena [45]. The 3-phenyl-1-(2-pyridyl)-4-styrylpyrazole **14** (Figure 6) was studied as a turn-off fluorescent probe in metal ion sensing, showing a high selectivity to Hg^{2+} (Limit of detection (LOD) = 3.1 × 10^{-7} M) in a process that could be reversed with ethylenediamine.

Although pyrazoles are rare in nature, due to the difficulty of living organisms to construct the N–N bond [46,47], a styrylpyrazole natural derivative, (*E*)-1,5-diphenyl-3-styryl-2-pyrazoline (DSDP) **15** (also known as (*E*)-1,5-diphenyl-3-styryl-4,5-dihydro-1*H*-pyrazole) (Figure 6), was isolated from the aerial parts of *Euphorbia guyoniana* [48]. In 2017, Kundu et al. unveiled the binding interaction of DSDP **15** with the calf thymus DNA (ctDNA) [49]. By performing steady state and time resolved spectroscopic measurements, competitive binding studies, circular dichroism and a DNA helix melting study they unequivocally demonstrated that DSDP binds with the ctDNA through groove binding mode with no conformational change. These findings were also supported by molecular docking simulation.

Later, the same authors reported the strong binding interactions of DSDP with two serum transport proteins, human serum albumin (HSA) and bovine serum albumin (BSA) [50]. Exploiting multi-spectroscopic techniques together with in silico molecular docking simulation, they have demonstrated that DSDP is a potent fluorophore. It binds with both proteins with the formation of a 1:2 protein–probe complex at a lower protein concentration range and a 1:1 complex at a higher protein concentration range. The calculated binding constants for the 2:1 DSDP-protein complexes were found to be 1.37×10^{10} M^{-2} and 1.47×10^{10} M^{-2} for HSA and BSA, respectively, while for the 1:1 complexation process, the constants were 1.85×10^5 M^{-1} and 1.73×10^5 M^{-1} for DSDP-HSA and DSDP-BSA systems, respectively. Moreover, they have demonstrated that hydrogen bonding and hydrophobic interactions are the forces primarily responsible for both types of binding. The information gathered from these studies may be useful for the rational design of drugs looking at a greater clinical efficacy.

1,3,5-Trisubstituted pyrazolines, analogues of DSDP, exhibited large fluorescence quantum yields (Φ_f = 0.6–0.8), suited for the design of energy-transfer-based fluorescent probes [51]. In DSDP's structure, two of the aryl rings (phenyl and styryl) are interconnected electronically through the pyrazoline π-system while the 5-phenyl ring is electronically decoupled and can behave as an electron donating receptor for cations or electron deficient centers. The lone pair of electrons on the N1 atom of the pyrazoline ring also takes part in the conjugation, facilitating the ICT process. Thus, DSDP acts as a D-π-A system and the introduction of an electron withdrawing group (such as -CN, -NO$_2$, -COOEt) on the styryl moiety can enhance the push–pull D-π-A capacity of DSDP-like molecular systems [52]. However, a drastic modification of the photophysical properties of DSDP is observed upon dehydrogenation of the pyrazoline ring with subsequent formation of the corresponding pyrazole (DSP) **16** (Figure 6). While DSDP gives dual absorption and dual emission bands corresponding to the locally excited (LE) and ICT species, DSP yields single absorption and emission bands for the LE species only. Comparative steady state and time resolved fluorometric studies have revealed that aromatization of the pyrazoline ring completely inhibits the ICT process. These results were also corroborated by quantum chemical calculations [52].

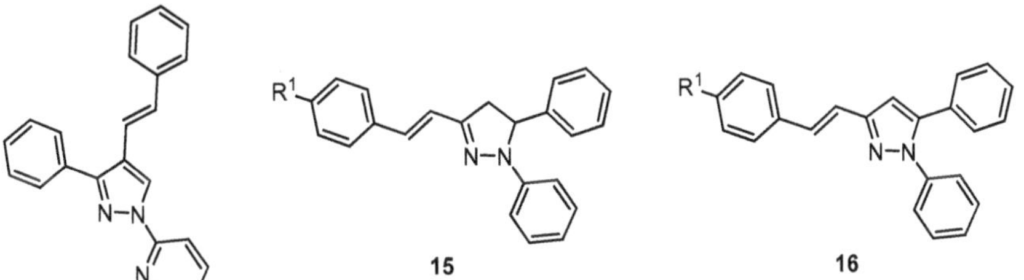

Figure 6. Styrylpyrazoles **14–16** presenting relevant photophysical properties [45,48–52].

3. Synthesis of Styrylpyrazoles

In this section, the methods for the synthesis of styrylpyrazoles are described in a systematic way, based on the compounds' structure, considering the position of the styryl group in the pyrazole scaffold, starting from C-1 to C-5. Bis(styryl)pyrazoles, substituted at C-3 and C-5, are also considered. Moreover, in each subheading, the methods will be presented by alphabetic order. Single examples of a certain reaction will be considered at the end of each section in the miscellaneous subheading.

3.1. *Synthesis of 1-Styrylpyrazoles*

3.1.1. *N*-Cross-Coupling Reaction of Pyrazoles with Styrylboronic Acid

In 2010, Kantam et al. synthesized (*E*)-1-styrylpyrazoles in a simple and efficient way, by the cross-coupling reaction of pyrazoles **17** with styrylboronic acid **18** using a recyclable heterogeneous Cu-exchanged fluorapatite (CuFAP) catalyst, under base-free conditions (Scheme 1) [53]. The lower yield (70%) for the coupling with 3,5-dimethyl-1*H*-pyrazole **17** ($R^1 = R^2$ = Me) was attributed to the steric hindrance caused by the methyl groups. When the reaction was performed in the absence of air, no coupled product was obtained, since O_2 is involved in the oxidation of Cu(I) to Cu(II), which is the active species in the reaction mechanism [53].

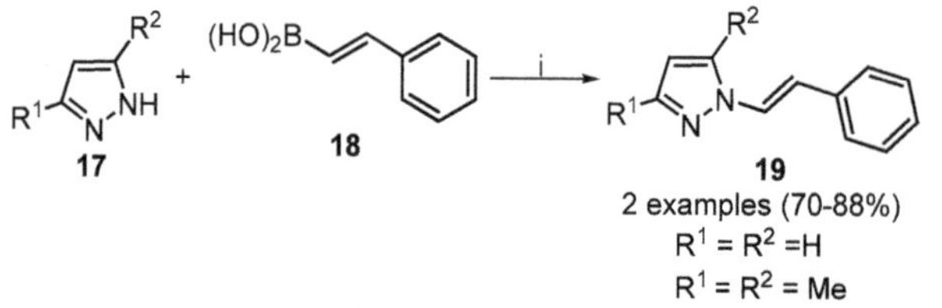
17
18
19
2 examples (70-88%)
R¹ = R² =H
R¹ = R² = Me

(i) **17** (1.2 equiv), CuFAP (100 mg), MeOH, under air, r.t., 13-20 h.

Scheme 1. Cross-coupling reactions of pyrazoles **17** with styrylboronic acid **18** to produce (*E*)-1-styrylpyrazoles **19** [53].

3.1.2. *N*-Styrylation of Pyrazoles with Activated and Non-Activated Alkynes

In 1978, Burgeois et al. described a lithium-catalyzed stereospecific and stereoselective addition of the N–H bond of pyrazole **20** to phenylacetylene, as a method to prepare 1-styrylpyrazoles **21** (Scheme 2) [54]. Under such conditions, (*Z*)-1-styryl-1*H*-pyrazole (**21a**) and (*Z*)-5-amino-1-styryl-1*H*-pyrazole (**21b**) were obtained in 60% and 55% yield, respectively.

Tsuchimoto et al. reported a similar single addition of the N–H bond of pyrazoles to phenylacetylene and prop-1-yn-1-ylbenzene in the presence of a silver catalyst ($AgNO_3$ or AgOTf) to produce a regioisomeric mixture of 1-substituted pyrazoles **22** and **23** (Scheme 2) [55]. Complete stereoselectivity was achieved, since only the (*Z*)-isomer of 1-styrylpyrazole was formed, indicating that *anti*-addition, via a concerted mechanism, in which pyrazole nitrogen attacks the alkyne from the side opposite to a coordinated Lewis acid (LA), is involved as a key step in the reaction process [55].

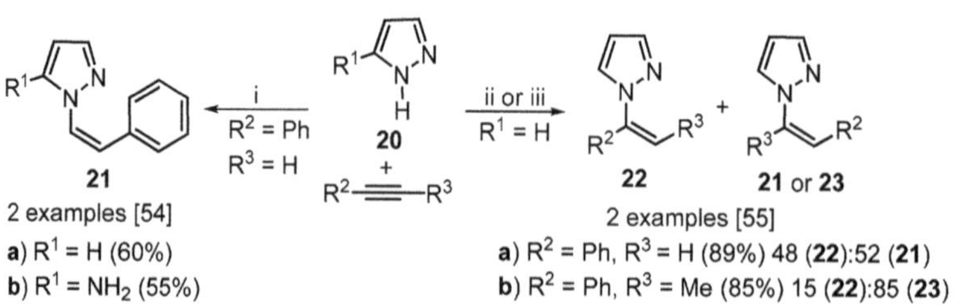
21
2 examples [54]
a) R¹ = H (60%)
b) R¹ = NH₂ (55%)
R² = Ph
R³ = H
20
ii or iii
R¹ = H
22
21 or 23
2 examples [55]
a) R² = Ph, R³ = H (89%) 48 (22):52 (21)
b) R² = Ph, R³ = Me (85%) 15 (22):85 (23)

(i) **20** (1.0 equiv), Li (0.05 g/0.05 mol of alkyne), 130 °C, 10 h.
(ii) $AgNO_3$ (0.05 equiv), 80 °C, 15 min., then **20** (2.0 equiv), PhCl, alkyne (1.0 equiv), argon atmosphere, 130-140 °C, 20 h.
(iii) **20** (2.0 equiv), AgOTf (0.2 equiv), $(iPr)_2NEt$ (0.2 equiv), PhCl, alkyne (1.0 equiv), argon atmosphere, 130-140 °C, 50 h (for R^3 = Me).

Scheme 2. Addition of pyrazole **20** to alkynes to produce (*Z*)-1-styrylpyrazoles **21–23** [54,55].

The role of two ruthenium complexes [Ru(dppe)(PPh3)(CH3CN)2Cl][BPh4] (**24**) (dppe = diphenylphosphinoethane) and [Ru(dppp)2(NCCH3)Cl][BPh4] (**25**) (dppp = diphenylphosphinopropane) as efficient catalysts in the regio- and stereoselective addition of pyrazoles **26** to alkynes was described by Das Kumar et al. (Scheme 3) [56]. In general, the addition catalyzed by complex **24** afforded (*E*)-1-styrylpyrazoles **27** while complex **25** produces (*Z*)-1-styrylpyrazoles **28**. Based on density functional theory calculations, the authors have shown that stereoselectivity is dependent upon the ligand environment around the ruthenium center.

Recently, Garg et al. reported a transition-metal-free chemo-, regio-, and stereoselective synthesis of (*Z*)- and (*E*)-1-styrylpyrazoles by the addition of pyrazoles **26** onto functionalized terminal alkynes using a super basic solution of KOH/dimethyl sulfoxide (DMSO) [31]. The nature of the base seems to be crucial for the reaction, which does not occur in the presence of an organic base, such as triethylamine (Et$_3$N). The reaction stereoselectivity is governed by time and amount of the base (Scheme 3). Deuterium labeling and control experiments highlighted the role of the KOH/DMSO catalytic system in the *cis*→*trans* isomerization, which was further supported by comparative ^{1}H-NMR spectrum studies in DMSO/DMSO-d$_6$. This method is of wide scope and several functionalities, such as OH or Me, NH$_2$, present both in the alkyne and pyrazole are well-tolerated [31].

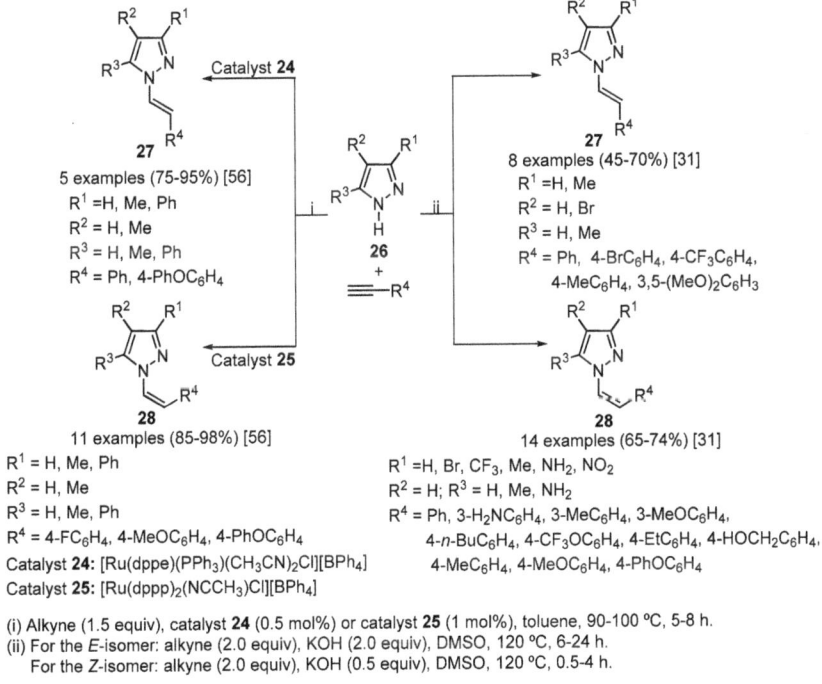

Scheme 3. Stereoselective additions of pyrazoles **26** to alkynes to produce (*E*)- and (*Z*)-1-styrylpyrazoles **27** and **28** [31,56].

3.2. Synthesis of 3(5)-Styrylpyrazoles

3.2.1. Cyclocondensation Reactions

The cyclocondensation of (1*E*,4*E*)-1,5-diarylpenta-1,4-dien-3-one **29** with hydrazine or phenyl hydrazine in glacial acetic acid at reflux [57,58], in the presence of sulfuric acid [59], or cellulose sulfonic acid [60], afforded 5-aryl-3-styryl-4,5-dihydro-1*H*-pyrazoles (pyrazolines) **15** or **30** (Scheme 4). Following a similar approach, Nauduri et al. synthesized a series of 5-aryl-1-phenyl-3-styryl-4,5

-dihydro-1H-pyrazoles by condensation of **29** analogues with phenyl hydrazine hydrochloride in a mixture of ethanol and chloroform, in the presence of a catalytic amount of concentrated HCl (70–77%) [34]. Pathak et al. studied the reaction of **29** with hydrazine hydrate in glacial acetic acid under conventional reflux, ultrasound, microwave irradiation conditions and using mechanochemical mixing, and isolated 1-acetyl-5-aryl-3-(substituted-styryl)pyrazolines **30**. In turn, pyrazolines **15** were converted into the corresponding pyrazoles by oxidation (dehydrogenation) (Scheme 4). Several oxidant agents can be employed for this transformation; some common examples include lead tetraacetate [58], DMSO in open air [61], MnO_2 [62], p-chloranil [63], 2,3-dichloro-5,6-dicyano-1,4-benzoquinone (DDQ) [64], Pd/C [65]. In 2014, Ananthnag et al. developed a simple and high yielding method for the conversion of (E)-3- and 5-styrylpyrazolines into the corresponding pyrazoles via iron(III) catalyzed aerobic oxidative aromatization [65]. The use of $FeCl_3$ as catalyst makes the reaction greener and more economical.

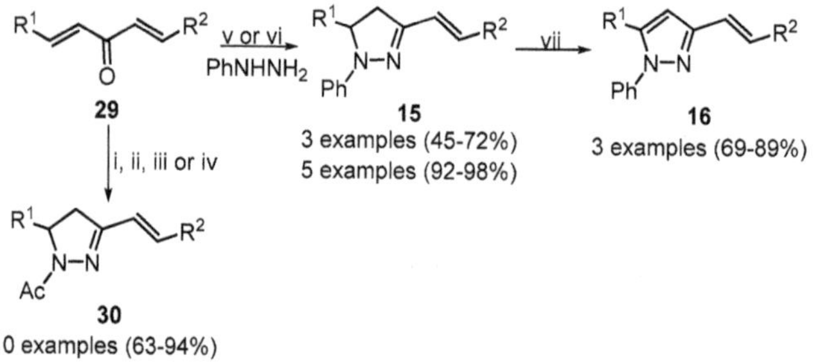

$R^1 = R^2 =$ Ph, 2-ClC_6H_4, 2-FC_6H_4, 2-$MeOC_6H_4$, 3-FC_6H_4, 4-BrC_6H_4, 4-ClC_6H_4, 4-FC_6H_4, 4-$MeOC_6H_4$, 2,4-$(Cl)_2C_6H_3$, 3,4,5-$(MeO)_3C_6H_2$

(i) $NH_2NH_2 \cdot H_2O$ (16%, 1 mL/2.0 mmol of **29**), glacial AcOH in ethanol, reflux, 12–30 h (63–76%).
(ii) $NH_2NH_2 \cdot H_2O$ (16%, 1 mL/2.0 mmol of **29**), glacial AcOH in ethanol, under ultrasonic waves (150W, 37 MHz output frequency), 30 °C, 10-25 min. (76-91%).
(iii) $NH_2NH_2 \cdot H_2O$ (16%, 1 mL/2.0 mmol of **29**), glacial AcOH impregnated on K_2CO_3 as a solid support, 800 W, 0.5-3.0 min. (79-94%).
(iv) $NH_2NH_2 \cdot H_2O$ (16%, 1 mL/2.0 mmol of **29**), $Mg(HSO_4)_2$ (0.3g), glacial AcOH, 12-15 °C, 10-30 min. (70-82%).
(v) **29** (0.4 M solution) in AcOH, $PhNHNH_2$ (1.0 equiv), reflux for 5 min and left overnight.
(vi) **29** (1.0 mmol), cellulose sulfonic acid (10 mol %), EtOH, $PhNHNH_2$ (2.0 equiv), 60 °C, 2-3 h.
(vii) $Pb(OAc)_4$ (1.5 equiv), CH_2Cl_2, r.t., 30 min.

Scheme 4. Synthesis of 3-styrylpyrazolines **15** and **30** and 3-styrylpyrazoles **16** [58].

Gressler et al. described the [3 + 2] cyclocondensation of 1,5-diarylpenta-1,4-dien-3-ones **31** with aminoguanidine hydrochloride **32** in ethanol, in the presence of triethylamine, as a method to prepare 5-aryl-1-carboxamidino-3-styryl-4,5-dihydro-1H-pyrazoles **9** (Scheme 5) [35,66,67].

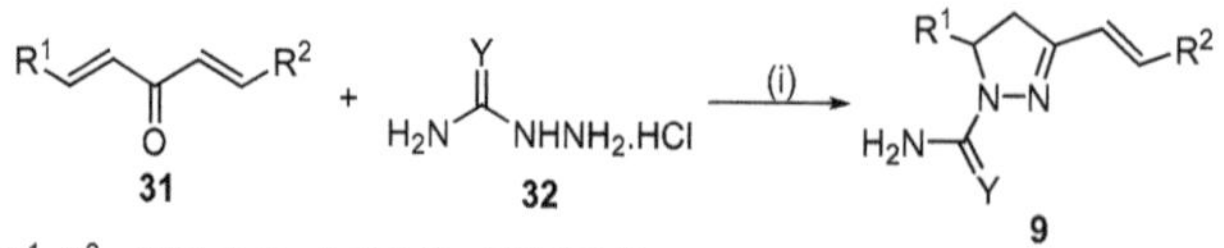

R^1, R^2 = 2-MeC_6H_4; 4-ClC_6H_4; 4-$MeOC_6H_4$; 3,4-$(MeO)_2C_6H_3$; 3,4,5-$(MeO)_3C_6H_2$
Y = NH; SH
(i) $(Et)_3N$, EtOH, reflux, 24 h.

Scheme 5. Synthesis of 5-aryl-1-carboxamidino-3-styryl-4,5-dihydro-1H-pyrazoles **9** [35,66,67].

In 1991, Purkayastha et al. reported the reaction of α-oxoketene dithioacetal **33** with hydrazine hydrate in ethanol, in the presence of acetic acid, as a method to exclusively obtain

(E)-3(5)-styrylpyrazoles (**34⇌35**) (Scheme 6). They have shown that acidic medium is required for the regioselective formation of **34**, otherwise, in neutral conditions, other products are formed [68]. The formation of styrylpyrazoles **34**, in acidic conditions, was explained through the attack of hydrazine on protonated **33** at the β-carbon with a positive charge, which is stabilized by the two methylthio groups. In fact, the (E)-3(5)-styrylpyrazole **34a** (Ar = Ph) was isolated as the sole product (79%) when **33a** (Ar = Ph) reacted with hydrazine hydrate in ethanol:acetic acid (1:1) mixture. Other cinnamoyl oxoketene dithioacetals similarly reacted in these conditions affording a series of (E)-3(5)-styrylpyrazoles. This method is also suitable for the synthesis of 3-(4-arylbuta-1,3-dienyl)pyrazoles and 3-(6-arylhexa-1,3,5-trienyl)pyrazoles starting from the appropriate dienoyl- and trienoyl oxoketene dithioacetals, and has high practical utility, since the introduction of such enyl side-chains in the preformed pyrazole is not possible [68].

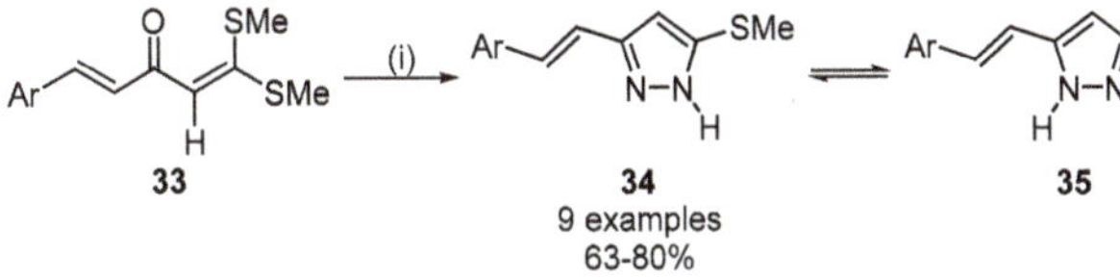

Ar = Ph, 2-ClC$_6$H$_4$, 4-ClC$_6$H$_4$, 4-(Me$_2$N)C$_6$H$_4$, 4-MeOC$_6$H$_4$, 2,6-Cl$_2$C$_6$H$_3$
3,4-methylenedioxyC$_6$H$_3$, 3,4-(MeO)$_2$C$_6$H$_3$, 3,4,5-(MeO)$_3$C$_6$H$_2$
(i) NH$_2$NH$_2$·H$_2$O (4.0 equiv), EtOH:AcOH (1:1), 110 °C, 20 h.

Scheme 6. Reaction of oxoketene dithioacetals **33** with hydrazine hydrate to synthesize (E)-3(5)-styrylpyrazoles **34** [68].

In 2001, Pastine et al. reported the synthesis of (E)-3-styrylpyrazoles starting from benzalacetone **36**, which was converted into the corresponding hydrazones **37** by reaction with hydrazine derivatives. Then, compounds **37** were deprotonated with an excess of lithium diisopropylamide (LDA), and the resulting dilithiated intermediates **38** were condensed at the carbanion center with a variety of substituted benzoate esters, such as methyl benzoate, methyl 4-t-butylbenzoate, (lithiated) methyl 4-hydroxybenzoate, or methyl 3,4,5-trimethoxybenzoate affording **39**. After acid cyclization of **39** with 3 N hydrochloric acid, the 3-styrylpyrazoles **40** were isolated (Scheme 7) [69].

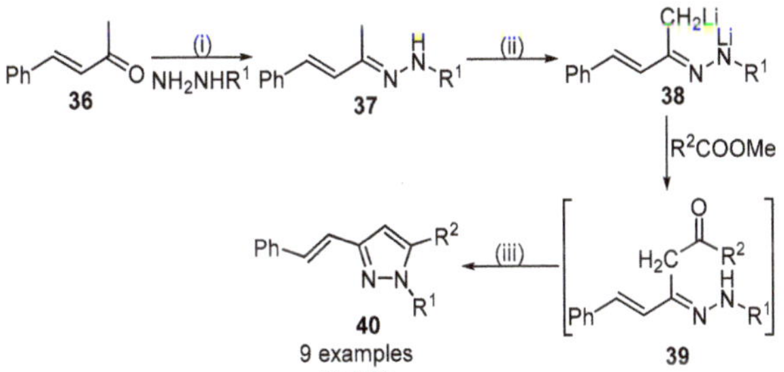

R^1 = COOEt, COOMe, Ph
R^2 = Ph, 4-HOC$_6$H$_4$, 4-(Me)$_3$CC$_6$H$_4$, 4-MeOC$_6$H$_4$, 3,4,5-(MeO)$_3$C$_6$H$_2$

(i) Hydrazine derivative (1.05 equiv), MeOH or EtOH (200 mL), AcOH (1mL), reflux, 45 min.
(ii) 1. 1.6 M n-Butyllithium (1.0 equiv), diisopropylamine (1.0 equiv) in dry THF (dropwise addition, 5 min.), N$_2$ atmosphere, 0 °C, 15-20 min.; 2. Hydrazone dissolved in THF, N$_2$ atmosphere, 0 °C, 45-60 min.; 3. Ester dissolved in THF, (hydrazone:LDA:ester 1:3:1 or 1:4:1), N$_2$ atmosphere, 0 °C, 45-120 min.;
(iii) 3 N HCl, reflux, 45-60 min.

Scheme 7. Reaction of benzalacetone **36** with hydrazines to produce hydrazones **37** and their transformation into (E)-3-styrylpyrazoles **40** [69].

The reaction of (*E*)-3-aryl-1-(3-phenyloxiran-2-yl)-prop-2-en-1-ones (**41**) with tosyl hydrazine under acid catalysis afforded (*E*)-5-hydroxy-5-phenyl-3-styryl-1-tosyl-2-pyrazolines (**42**) (Scheme 8) [70]. Then, (*E*)-5-phenyl-3-styryl-1-tosyl-1*H*-pyrazoles (**43**), which are formed due to dehydration of 5-hydroxypyrazolines **42**, were also isolated, as by-products, by chromatography, after crystallization of the major products.

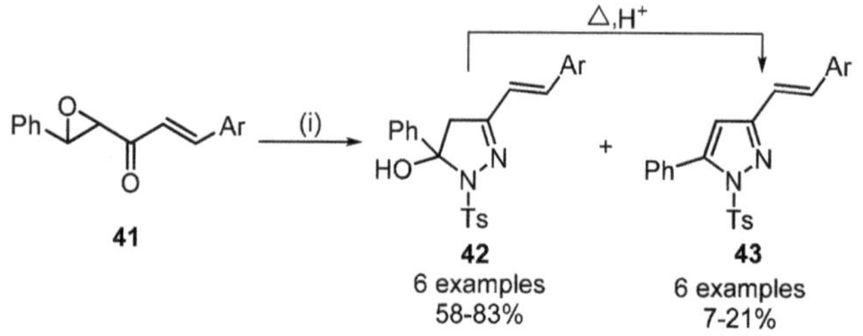

Ar = Ph, 3-ClC_6H_4, 4-BrC_6H_4, 4-ClC_6H_4, 4-FC_6H_4, 4-$NO_2C_6H_4$
(i) $TsNHNH_2$ (1.1 equiv), HCl conc. (cat.), THF, r.t., 12-18 h.

Scheme 8. Synthesis of (*E*)-5-hydroxy-5-phenyl-3-styryl-1-tosyl-2-pyrazolines (**42**) and (*E*)-5-phenyl-3-styryl-1-tosyl-1*H*-pyrazoles (**43**) [70].

The formation of pyrazolines **42** occurred through oxirane ring-opening at the α-carbon, followed by the rearrangement of the azadiene intermediate (**A**) through a hydride [1,5] sigmatropic shift to 1,3-diketone monohydrazone (**B**, Scheme 9). Finally, the intramolecular cyclization led to 5-hydroxypyrazolines **42**. These compounds are only stable if they have an electron-withdrawing group, such as 1-acyl, or if a perfluoroalkyl group is present at the C-5 of the pyrazoline ring [70].

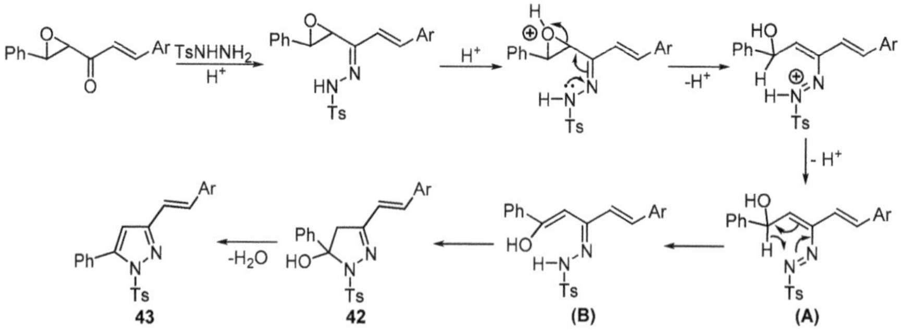

Scheme 9. Mechanism of formation of (*E*)-3-styrylpyrazolines **42** and (*E*)-3-styrylpyrazoles **43** [70].

The condensation of acetylenic ketones **44** with aryl hydrazines **45** produces (*E*)-3(5)-styrylpyrazoles (**46** and/or **47**) in fair to good yield. The substitution pattern in these pyrazoles depends on the nature of the substituents and mainly on the reaction conditions (Scheme 10) [71]. When methanol is used as solvent and the reaction is stirred at room temperature for a period prior to the addition of acid and heating, (*E*)-5-styrylpyrazoles **47** were obtained as the major products. However, if acid is present and heat applied from the onset, a mixture of (*E*)-3- and 5-styrylpyrazoles **46** and **47** is obtained. Regioselectivity, in this case, varies from 39:61% to 83:17% of **46**:**47**, depending on the nature of the substituents. For instance, when R^1 = MeO, R^2 = H and R^3 = $MeSO_2$, the regioisomer **46** is obtained in higher amount. On contrary, when R^1 = R^2 = H and R^3 = NO_2, **46** is the minor regioisomer. In the absence of acid, the initial reaction step involves Michael addition of the more basic terminal nitrogen of the hydrazine derivative to the terminal acetylenic carbon to form enamine, which exists in tautomeric equilibrium with the isomeric hydrazone. The cyclization of the hydrazone in the presence of added acid subsequently afforded only (*E*)-5-styrylpyrazoles **46**. The formation of this isomer seems to be promoted by the strong electron-donating and electron-withdrawing effect of the substituents.

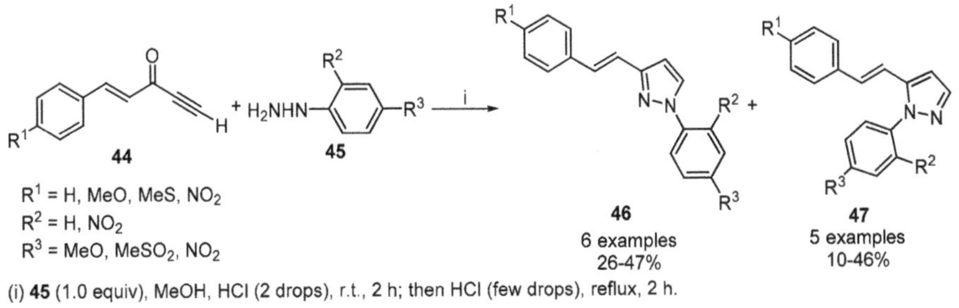
R1
R2
R3
+ H2NHN
i
N–N
R2 +
44
45
R1 = H, MeO, MeS, NO2
R2 = H, NO2
R3 = MeO, MeSO2, NO2
46
6 examples
26-47%
47
5 examples
10-46%
(i) 45 (1.0 equiv), MeOH, HCl (2 drops), r.t., 2 h; then HCl (few drops), reflux, 2 h.

Scheme 10. Condensation of acetylenic ketones **44** with aryl hydrazines **45** to produce (*E*)-3(5)-styrylpyrazoles (**46** and/or **47**) [71].

The microwave-assisted *N*-heterocyclization of metal-diketonic complexes, $Pd(dba)_2$ or $Pd(dba)_3$ (dba = dibenzylideneacetones) with hydrochloride salts of various aryl hydrazines allows the synthesis of 1-aryl-5-phenyl-3-styryl-1*H*-pyrazoles in a single step and with very good yields (Scheme 11) [72]. Metal-diketones act as both catalyst and coupling partner. The hydrazine substrate and the solvent play an important role in the selectivity. Reaction with phenyl hydrazine hydrochloride in water gave both (*E*)-3-styrylpyrazole **48** and (*E*)-3-styrylpyrazoline **49** in approximately equal amounts. If the substrate is phenyl hydrazine (without hydrochloride salt), (*E*)-3-styrylpyrazoline **49** is the major reaction product. The use of DMSO as a solvent instead of water promotes the formation of (*E*)-3-styrylpyrazole **48** in high yields.

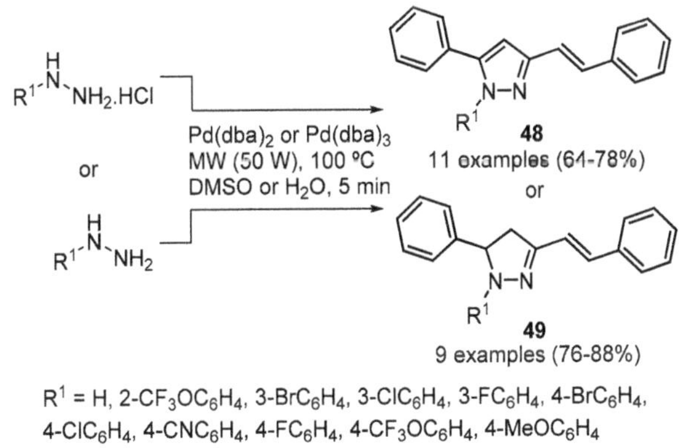
R1-NH-NH2.HCl
or
R1-NH-NH2
Pd(dba)2 or Pd(dba)3
MW (50 W), 100 °C
DMSO or H2O, 5 min
48
11 examples (64-78%)
or
49
9 examples (76-88%)
R1 = H, 2-CF3OC6H4, 3-BrC6H4, 3-ClC6H4, 3-FC6H4, 4-BrC6H4,
4-ClC6H4, 4-CNC6H4, 4-FC6H4, 4-CF3OC6H4, 4-MeOC6H4

Scheme 11. Microwave-assisted synthesis of (*E*)-3-styrylpyrazoles **48** and (*E*)-3-styrylpyrazolines **49** [72].

3.2.2. Intramolecular Oxidative C–N Coupling of Hydrazones

The ruthenium(II)-catalyzed oxidative C–N coupling of 2,4-dinitrophenylhydrazone of (1*E*,4*E*)-1,5-diphenylpenta-1,4-dien-3-one **50** in the presence of oxygen (1 atm) as the oxidant, afforded (*E*)-1-(2,4-dinitrophenyl)-5-phenyl-3-styryl-1*H*-pyrazole (**48**). $[RuCl_2(p\text{-cymene})]_2$ (5 mol%) was found to be the best catalyst for this oxidative $C(sp^2)$–H amination (Scheme 12) [73]. This catalyst allows C–H bond cleavage via an *o*-metalation process that involves chelation with the nitrogen from hydrazone. Consequently, the formation of a C–N bond is possible via reductive elimination to generate the corresponding pyrazole. Finally, the oxygen promotes oxidation of the Ru(0) species to Ru(II) to complete the catalytic cycle.

Other oxidative C(sp^2)–H amination reactions have been reported using copper catalysts. By using $Cu(OAc)_2$ (10 mol%) and 1,4-diazabicyclo[2.2.2]octane (DABCO) (30 mol%) in DMSO at 100 °C in the presence of oxygen (1 atm) as the oxidant, (*E*)-1,5-diphenyl-3-styryl-1*H*-pyrazole (**48**, R^1 = Ph) was obtained (Scheme 12) [74].

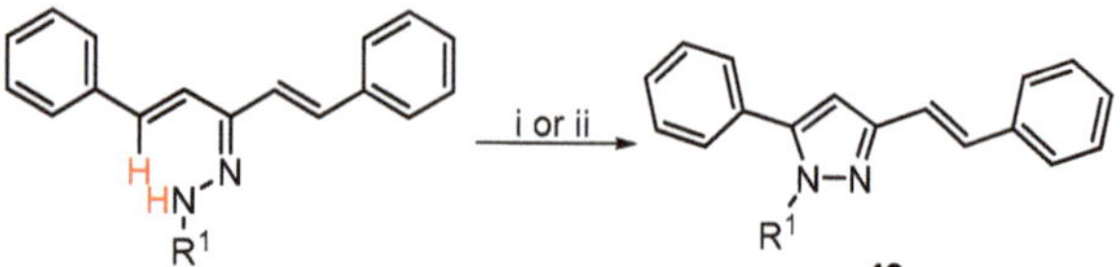

50 **48**

2 examples

R^1 = 2,4-$(NO_2)_2C_6H_3$ (58%)

R^1 = Ph (81%)

(i) $[RuCl_2(p\text{-cymene})]_2$ (5 mol%), $NaHCO_3$ (2.0 equiv), O_2 (1 atm), DMSO, 60-100 °C, 5-6 h.
(ii) $Cu(OAc)_2$ (10 mol%), O_2 (1 atm), DABCO (30 mol%), DMSO, 100 °C, 12 h.

Scheme 12. Synthesis of (*E*)-3-styrylpyrazoles **48** by intramolecular oxidative C–N coupling of hydrazones **50** [73,74].

3.2.3. Miscellaneous

The decarboxylative Knoevenagel reaction of aldehyde **51** with 3-methyl-5-pyrazoleacetic acid **52** directly afforded (*E*)-3-styrylpyrazole **7** (Scheme 13) [33].

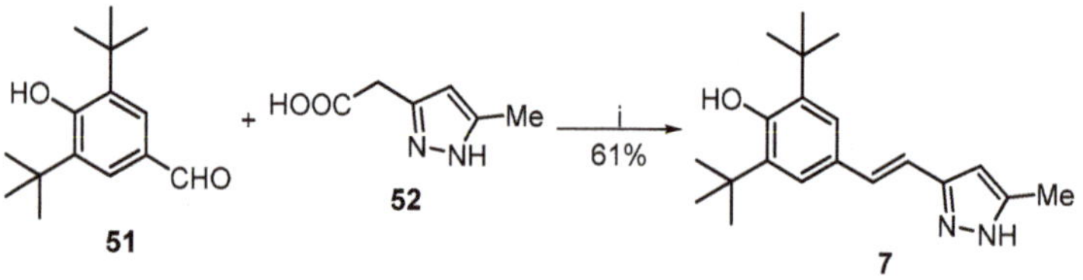

51 **52** **7**

(i) **52** (1.0 equiv), piperidine (0.1 equiv), AcOH (0.1 equiv), toluene, reflux, 24 h (with azeotropic removal of water).

Scheme 13. Synthesis of (*E*)-3-styrylpyrazole **7** by Doebner modification of Knoevenagel condensation [33].

Hydrogenolysis of (*E*)-5-styrylisoxazole **53**, using $Mo(CO)_6$ in the presence of water (1.0 equiv), followed by ring closure with hydrazine also produced (*E*)-3-styrylpyrazole **7** [33]. In a first step, a ketone is obtained (Scheme 14, i), which after reaction with 97% hydrazine in acetic acid gave (*E*)-3-styrylpyrazole **7** (Scheme 14, ii).

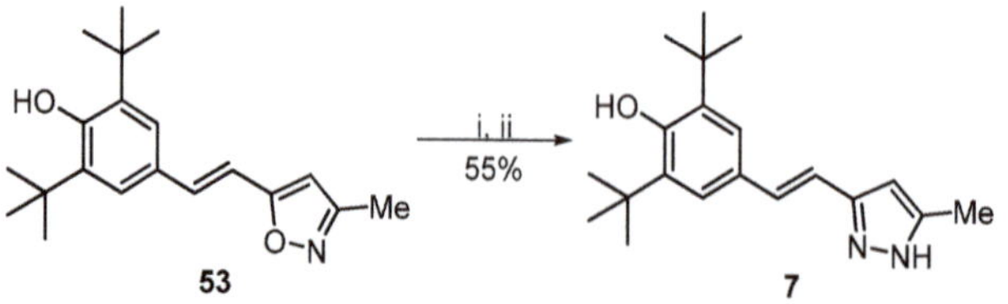

53 **7**

(i) 1.H_2O (1.0 equiv), $Mo(CO)_6$ (7.5 equiv), MeCN, N_2 atmosphere, reflux, 12 h;
2. MeOH, 4N HCl, r.t., 4 h and then 1N NaOH
(ii) Ketone obtained in (i), 97% hydrazine (1.0 mL/1.25 g of ketone), r.t., 12 h.

Scheme 14. Conversion of (*E*)-5-styrylisoxazole **53** into (*E*)-3-styrylpyrazole **7** by hydrogenolysis [33].

Deshayes et al. reported the use of 3-bromomethylpyrazole **54** as template for the synthesis of (E)-3-styrylpyrazole **55**. In a first step, this compound was treated with triethyl phosphite, in the Arbusov reaction, affording the corresponding diethylphosphonomethylpyrazole, which then reacted with benzaldehyde in the presence of sodium hydride to give the corresponding (E)-3-styrylpyrazole **55** (Scheme 15) [75].

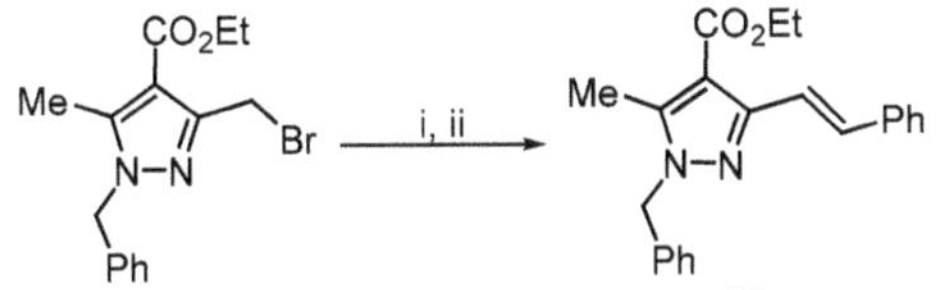

(i) P(OEt)$_3$ (1.0 equiv), 130 °C, 1 h.
(ii) NaH (1.0 equiv, 50% dispersion washed with hexane), DME, benzaldehyde (1.0 equiv, in dimethoxyethane) dropwise, r.t., overnight.

Scheme 15. Synthesis of (E)-3-styrylpyrazole **55** [75].

3.3. Synthesis of 4-Styrylpyrazoles

3.3.1. Cross-Coupling Reactions

(E)-(4-Styryl)aminopyrazoles can be accessed in a straightforward way by the Suzuki–Miyaura cross-coupling reaction of 4-bromo aminopyrazoles **56** and their amides with styryl boronic acids **57** (Scheme 16). Using the pre-catalyst palladacycle XPhos Pd G2 and XPhos, in combination with K$_2$CO$_3$ in a green solvent system (EtOH/H$_2$O), under microwave irradiation, Jedinák et al. obtained the (E)-(4-styryl)aminopyrazoles **58** [76]. Direct comparison of the chloro, bromo, and iodopyrazoles used as substrate in the Suzuki–Miyaura reaction revealed that bromo and chloro derivatives were superior to iodopyrazoles, showing reduced propensity to dehalogenation [76]. Using the same pre-catalyst (XPhos Pd G2), which enabled the coupling with the electron deficient, electron-rich, or sterically demanding boronic acids, Tomanová et al. coupled the 4-bromopyrazoles **59** with styryl boronic acids **57** to prepare the (E)-3,5-dinitro-4-styryl-1H-pyrazoles **60** (Scheme 17) [77]. The introduction of the electron-deficient nitro groups as masked amino functionalities improved the rate of the oxidative addition step and eliminated the Pd-independent side dehalogenation reaction. Both nitro groups were then converted into amine groups by iron-catalyzed reduction with hydrazine hydrate to give (E)-3,5-diamino-4-styryl-1H-pyrazoles (**61**).

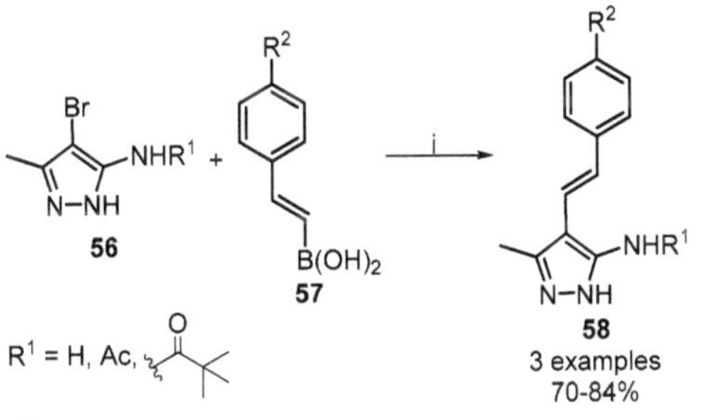

R^1 = H, Ac, pivaloyl
R^2 = H, CF$_3$, OMe

(i) **57** (2.0 equiv), XPhos Pd G2 (1 mol%), XPhos (2 mol%), K$_2$CO$_3$ (2.0 equiv), EtOH/H$_2$O, MW, 110-135 °C, 20 min.

Scheme 16. Suzuki–Miyaura coupling of 4-bromo aminopyrazoles **56** and their amides with styryl boronic acids **57** to prepare (E)-(4-styryl)aminopyrazoles **58** [76].

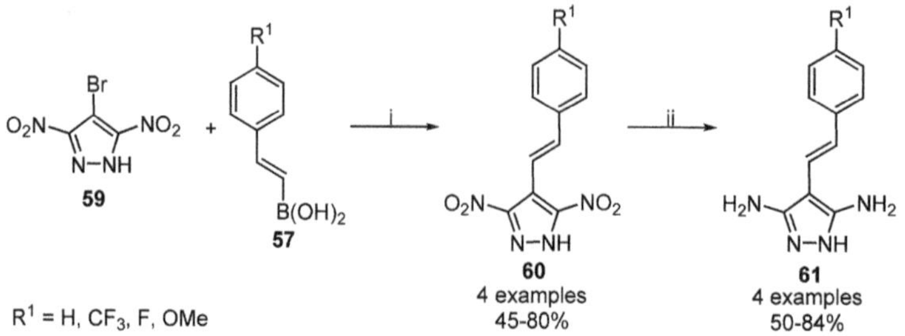

(i) **57** (1.05-4.5 equiv), XPhos Pd G2 (2-4 mol%), K_3PO_4 (4.0 equiv), dioxane/H_2O (4:1), 100 °C, 20 h.
(ii) 5% Fe on activated charcoal (1.0 equiv), $NH_2NH_2.H_2O$ (12.0 equiv), MeOH or dioxane, 55 °C, 1-3 h.

Scheme 17. Synthesis of (*E*)-3,5-dinitro-4-styryl-1*H*-pyrazoles **60** and iron-catalyzed reduction of the nitro groups to produce (*E*)-3,5-diamino-4-styryl-1*H*-pyrazoles **61** [77].

Starting from 4-iodo-1-(4-nitrophenyl)-1*H*-pyrazole **62** and 4-methoxystyrene **63**, Miller et al. synthesized (*E*)-4-(4-methoxystyryl)-1-nitrophenyl-1*H*-pyrazole **64**, using standard palladium-catalyzed coupling conditions (Scheme 18). Furthermore, 4-bromo-1-(4-nitrophenyl)-1*H*-pyrazole did not react under the same conditions [71]. On the other hand, the Mizoroki–Heck coupling of previously prepared 4-vinylpyrazole **65** with halobenzenes **66**, using $Pd(AcO)_2$ (10 mol%) as a catalyst in the presence of a ligand (*o*-MePh)$_3$P (20 mol%), afforded the desired (*E*)-1,3-disubstituted-4-styryl-1*H*-pyrazoles **67** [45] (Scheme 19). This reaction was performed under microwave irradiation to obtain the compound in a shorter reaction time.

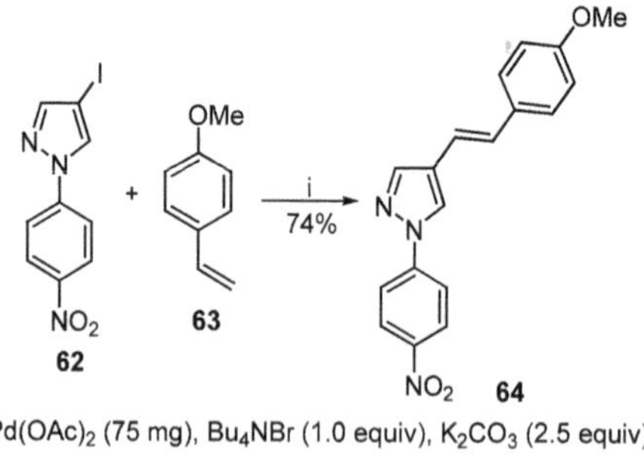

(i) **63** (2.0 equiv), $Pd(OAc)_2$ (75 mg), Bu_4NBr (1.0 equiv), K_2CO_3 (2.5 equiv), DMF, 70 °C, 18 h.

Scheme 18. Synthesis of (*E*)-4-(4-methoxystyryl)-1-nitrophenyl-1*H*-pyrazole **64** [71].

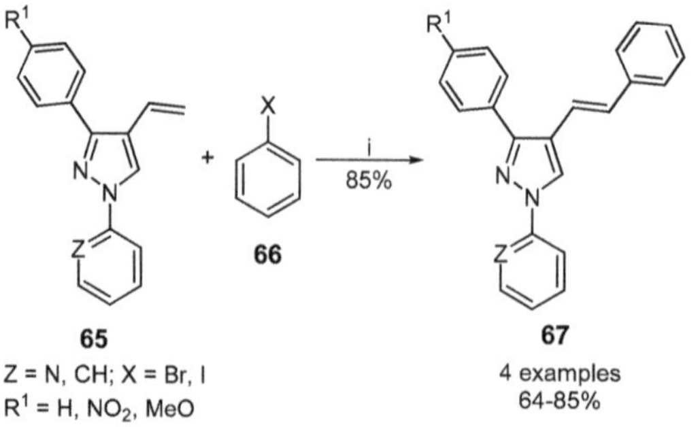

(i) **66** (2.5 equiv), $Pd(AcO)_2$ (10 mol %), (*o*-MePh)$_3$P (20 mol %), Et_3N (1.0 mL/0.4 mmol of **65**), dry DMF, MW, 180 °C for 3 x 30 min.

Scheme 19. Mizoroki–Heck coupling of 4-vinylpyrazole **65** with halobenzenes **66** to prepare (*E*)-1,3-disubstituted-4-styryl-1*H*-pyrazoles **67** [45].

3.3.2. Cyclocondensation Reaction

Kim et al. described a convenient method to prepared (E)-4-styrylpyrazoles starting from α-alkenyl-α,β-enones **68**, readily accessed from the Morita–Baylis–Hillman reaction. The reaction of **68** with aryl hydrazines in ethanol, in the presence of O_2, afforded (E)-4-styrylpyrazoles **69** (Scheme 20), together with trace amounts of the corresponding 4-arylethylpyrazoles and pyridazine derivatives, which were identified but were not isolated [78].

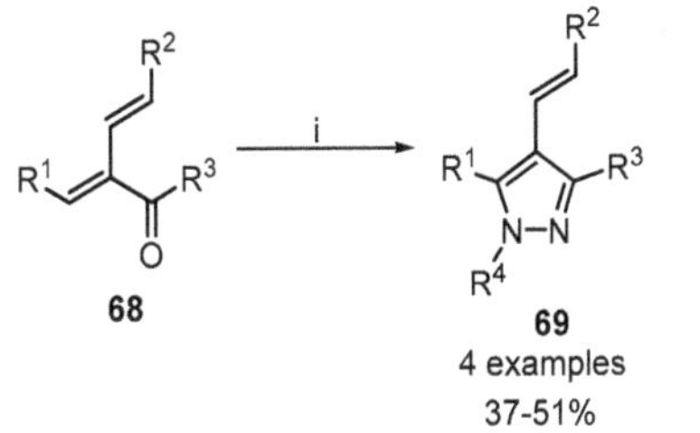

R^1 = Ph, 4-MeOC$_6$H$_4$; R^2 = Ph, 4-MeOC$_6$H$_4$; R^3 = Me;
R^4 = Ph, 4-ClC$_6$H$_4$, 4-MeOC$_6$H$_4$, 4-NH$_2$SO$_2$Ph

(i) R^4NHNH$_2$HCl (1.2 equiv), EtOH, 3 h, under O$_2$ ballon atmosphere.

Scheme 20. Reaction of α-alkenyl-α,β-enones **68** with arylhydrazines to produce (E)-4-styrylpyrazoles **69** [78].

3.3.3. 1,3-Dipolar Cycloaddition Reaction

1,3-Dipolar cycloaddition reaction of (E,E)-cinnamylideneacetophenones **70** and diazomethane, at room temperature or in a refrigerator, lead to the formation of 3-benzoyl-4-styryl-2-pyrazolines **71**. By oxidation with an excess of chloranil, in toluene, pyrazolines **71** were converted into 3(5)-benzoyl-4-styrylpyrazoles **72** (Scheme 21) [79]. The regioselective formation of **71** results from the reaction between the Cα=Cβ double bond of **70** with diazomethane giving rise to 1-pyrazolines **73**, which then isomerize into 2-pyrazoline isomers **71** (Scheme 22). From the reaction between the Cγ=Cδ double bond of **70** with diazomethane, other cycloadducts can be formed; however, their formation was not reported. For some derivatives (R^1 = H, R^2 = H and R^3 = H, Me), small amounts of 3-(2-benzofuranyl)-4-styryl-2-pyrazolines were formed as by-products, as a result of insertion of a methylene group between the carbonyl and the 2-hydroxyphenyl group leading to the formation of compound **74**. Then, the intramolecular reaction of the hydroxy group with the carbonyl carbon and subsequent water elimination from the obtained hemiacetal lead to the formation of the benzofuran ring **75** (Scheme 22) [79].

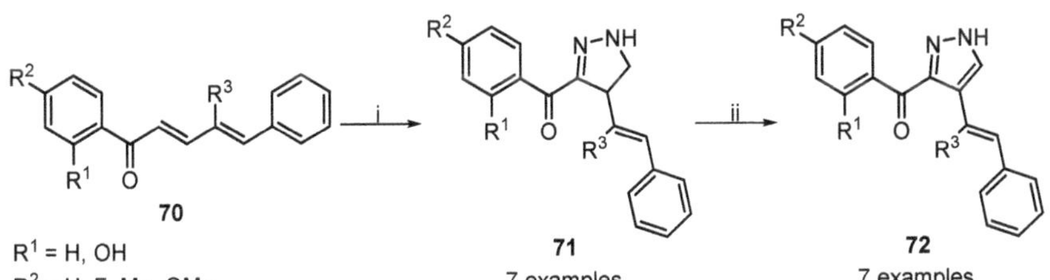

R^1 = H, OH
R^2 = H, F, Me, OMe
R^3 = H, Me

(i) CH$_2$N$_2$, CH$_2$Cl$_2$/diethyl ether, refrigerator, 48 h.
(ii) Chloranil (3.0 equiv), toluene, reflux.

Scheme 21. Synthesis of (E)-3-benzoyl-4-styryl-2-pyrazolines **71** and 3-benzoyl-4-styryl-1H-pyrazoles **72** [79].

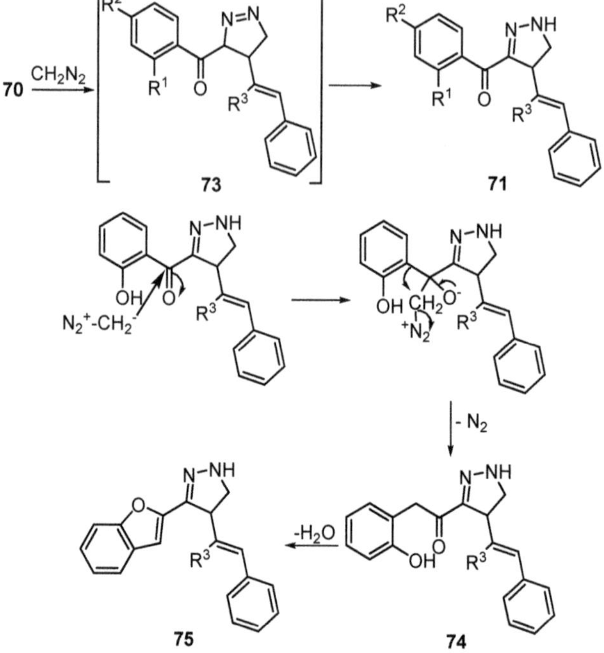

Scheme 22. Mechanism of formation of 3-benzoyl-4-styryl-2-pyrazolines **71** and 3-(2-benzofuranyl)-4-styryl-2-pyrazolines **75** [79].

3.3.4. Reaction of 3-Styrylchromones with Hydrazine Derivatives

The reaction of (Z)- and (E)-3-styryl-4H-chromen-4-ones (**76** and **78**) (also known as 3-styrylchromones) with hydrazine hydrate in methanol, at room temperature, afforded the corresponding (Z)- and (E)-3(5)-(2-hydroxyphenyl)-4-styryl-1H-pyrazoles (**77** and **79**) in 70–94% and 32–98% yield, respectively (Scheme 23) [80]. When the same method was followed to convert (Z)-3-(4-nitrostyryl)-4H-chromen-4-ones (**76**) into the corresponding pyrazoles, the (E)-3(5)-(2-hydroxyphenyl)-4-(4-nitrostyryl)-1H-pyrazoles (**79**) were obtained in great yields (>73%) instead of the expected (Z)-isomer. These results indicate that the strong electron-withdrawing effect of the p-nitro group has an important role in the (Z)→(E) isomerization during the transformation of (Z)-3-(4-nitrostyryl)-4H-chromen-4-ones into the corresponding (E)-4-(4-nitrostyryl)-1H-pyrazoles. Silva et al. have highlighted the role of the nitro group in the mechanism of formation of the (E)-4-styrylpyrazole isomer [81]. After the nucleophilic attack of hydrazine at C-2 of the chromone nucleus, the electronic conjugation moves towards the 4-nitro-3-styryl moiety, allowing the (Z)→(E) isomerization of the vinylic double bond of the styryl group, to adopt the most stable configuration, with consequent ring opening. Finally, an intramolecular reaction of the hydrazine and carbonyl group led to pyrazole ring formation (Scheme 24).

Although the biological activity of 4-styrylpyrazoles has rarely been studied, Silva et al. have shown that some derivatives of 4-styrylpyrazoles **77** and **79** with long alkyl chains of ten or twelve carbons on the N-1 or linked at the oxygen of the 2′-hydroxyphenyl moiety present affinity for CB_1 type cannabinoid receptors in the micromolar range, or even in the nanomolar range, as observed for the (E)-4-(4-chlorostyryl)-3(5)-(2-decyloxyphenyl)-1H-pyrazole (K_i = 53 ± 33 nM) [14].

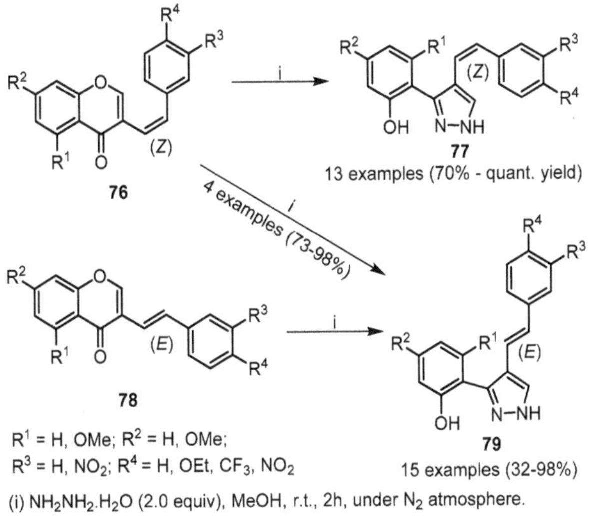
76
77
13 examples (70% - quant. yield)
4 examples (73-98%)
78
79
15 examples (32-98%)
R¹ = H, OMe; R² = H, OMe;
R³ = H, NO₂; R⁴ = H, OEt, CF₃, NO₂
(i) NH₂NH₂.H₂O (2.0 equiv), MeOH, r.t., 2h, under N₂ atmosphere.

Scheme 23. Synthesis of (*Z*)- and (*E*)-3(5)-(2-hydroxyphenyl)-4-styryl-1*H*-pyrazoles **77** and **79** [80,81].

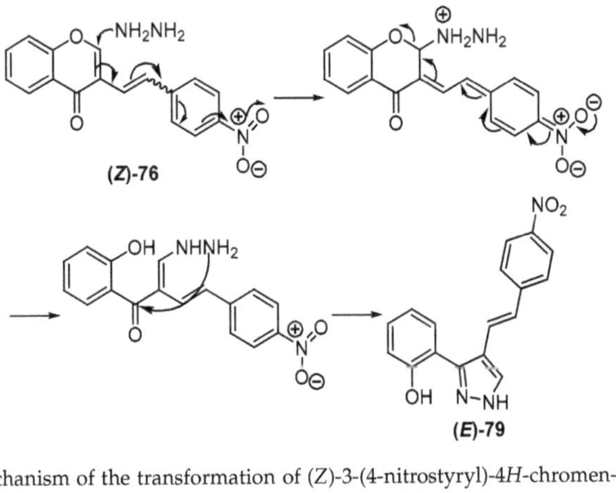
(Z)-76
(E)-79

Scheme 24. Mechanism of the transformation of (*Z*)-3-(4-nitrostyryl)-4*H*-chromen-4-ones (**76**) into (*E*)-3(5)-(2-hydroxyphenyl)-4-(4-nitrostyryl)-1*H*-pyrazoles (**79**) [81].

3.3.5. Miscellaneous

The Wittig reaction of 4-formylpyrazole **80** with benzyltriphenylphosphonium bromide (**81**) in BuLi-THF afforded the corresponding (*E*)- and (*Z*)-4-styrylpyrazoles **82** with an *E*:*Z* ratio of approximately 1:3 (Scheme 25) [45].

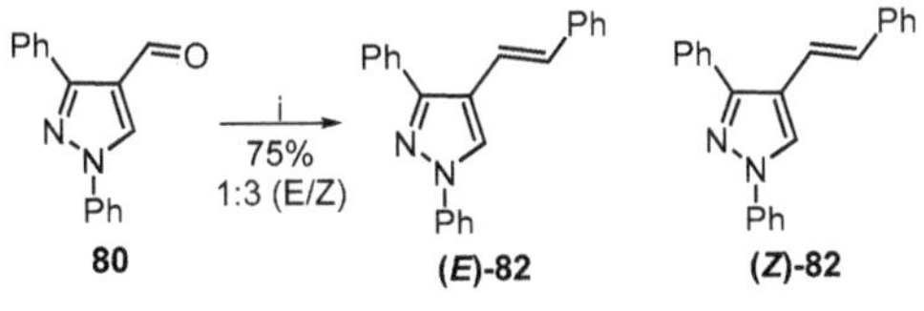
75%
1:3 (E/Z)
80
(E)-82
(Z)-82

(i) $PhCH_2PPh_3Br$ (**81**) (0.5 equiv), BuLi, THF, -50-20 °C, 12 h.

Scheme 25. Wittig reaction of 4-formylpyrazole **80** with benzyltriphenylphosphonium bromide **81** to prepare (*E*)- and (*Z*)-4-styrylpyrazoles **82** [45].

Ahamad et al. described the synthesis of the (E)-4-styryl-5-vinylpyrazole **85** by a domino reaction, which involves the 1,3-dipolar cycloaddition of Bestmann–Ohira reagent (BOR) **84** to (2E,4E)-5-phenylpenta-2,4-dienal **83**, followed by a Horner–Wadsworth–Emmons reaction of the resulting pyrazoline carboxaldehyde and subsequent 1,3-H shift to afford **85**. This methodology has relevant application for the synthesis of 5-vinylpyrazoles (Scheme 26) [82].

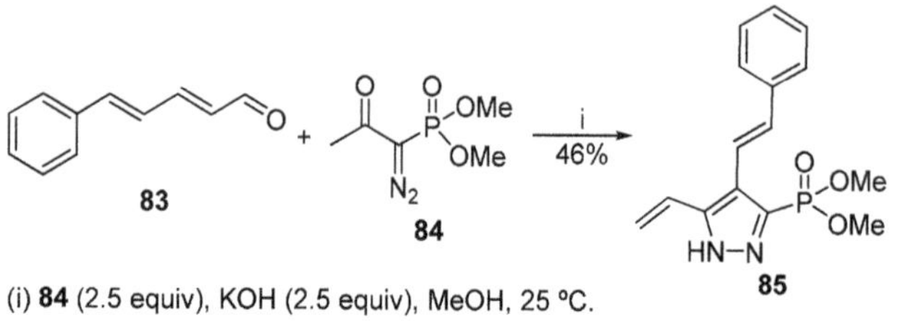

(i) **84** (2.5 equiv), KOH (2.5 equiv), MeOH, 25 °C.

Scheme 26. Synthesis of the (E)-4-styryl-5-vinylpyrazole **85** [82].

3.4. Synthesis of 5(3)-Styrylpyrazoles

3.4.1. Cyclocondensation Reactions

One of the first reports about the synthesis of styrylpyrazoles dates back to 1978 [83]. At that time, Soliman et al. described the reaction of diketoester **86** with aryl/hetaryl hydrazines in ethanol to prepare (E)-5-styrylpyrazoles **87** (Scheme 27).

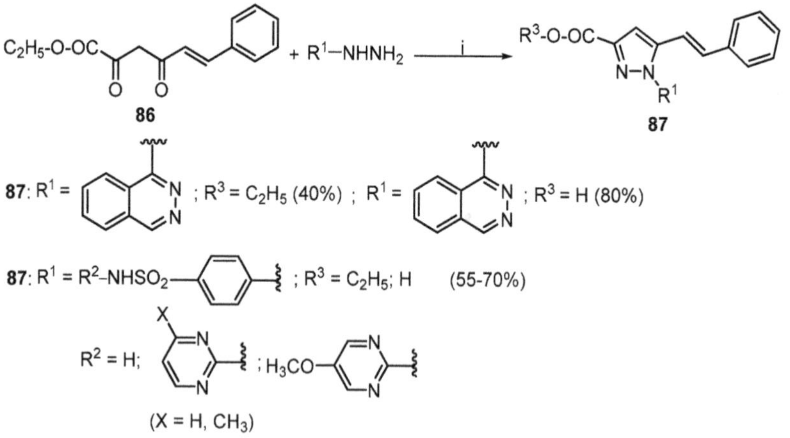

(i) EtOH, 78 °C, 2-3 h.

Scheme 27. Cyclocondensation of diketoester **86** with aryl/hetaryl hydrazines to prepare (E)-5-styrylpyrazoles **87** [83].

The first approach to the synthesis of (E)-3(5)-(2-hydroxyphenyl)-5(3)-styryl-1H-pyrazole (**89a**, R^1 = H) was the treatment of 1-(2-hydroxyphenyl)-5-phenylpent-4-ene-1,3-dione (**88a**, R^1 = H), which exists in equilibrium with the enolic form (**88a′**), with excess hydrazine (formed by treatment of hydrazinium sulfate with potassium carbonate), added gradually (dropwise) and using a 1:1 dichloromethane (DCM)/methanol mixture as solvent (Scheme 28, i) [84]. Using hydrazine hydrate in methanol at room temperature, compounds **88** were converted into pyrazoles **89** (Scheme 28, ii) [84]. Furthermore, (E)-3(5)-(2-hydroxyphenyl)-5(3)-styryl-1H-pyrazoles **89** (R^1 = H, OMe, NO_2; R^2 = H) were also obtained starting from **88** by reaction with hydrazine hydrate, in acetic acid (Scheme 28,

iii) [84]. Treatment of acetic acid solutions of diketones **88** with an excess of phenyl hydrazine afforded a mixture of two 5-styrylpyrazole isomers **90** and **91**, though pyrazoles **91** were obtained in a vestigial amount (Scheme 28, iv). In fact, the reaction of unsymmetrical diketones and monosubstituted hydrazines, such as phenyl hydrazine, although apparently simple, conceal a complex mechanistic problem. Diketones **88** have two tautomeric forms (**88** and **88′**) and phenyl hydrazine can react initially through NH or NH_2. When the reaction is carried out in methanol in neutral conditions, a nucleophilic attack of the primary amine (NH_2) to the more electrophilic position of the diketone (C-1) occurs and only pyrazoles **91** were obtained in low yields. Using acidic conditions (AcOH as solvent), the more basic amine (NH_2) was protonated and subsequently the nucleophilic attack at the more electrophilic position was through the NH affording pyrazoles **90** [84].

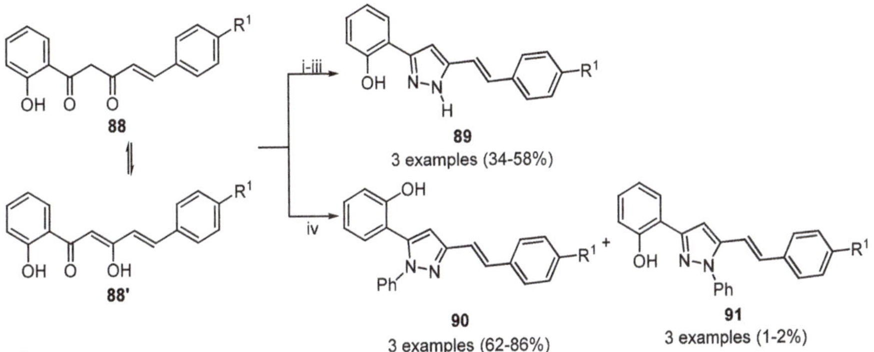

R^1 = H, NO_2, OMe

(i) $NH_2NH_3^+ \cdot HSO_4^-$, K_2CO_3, DCM/MeOH (1:1), r.t., 3 days.
(ii) $NH_2NH_2 \cdot H_2O$ (5-8 equiv), MeOH, r.t., 2h30min.- 24 h.
(iii) $NH_2NH_2 \cdot H_2O$ (10 equiv), AcOH, 50 °C, 24 h.
(iv) $PhNHNH_2$ (10 equiv), AcOH, 50 °C, 24 h.

Scheme 28. Synthesis of (E)-3(5)-(2-hydroxyphenyl)-5(3)-styryl-1H-pyrazoles **89**, **90** and **91** [84].

Deshayes et al. performed the reaction of enamine **92** with phenyl hydrazine in refluxing ethanol, overnight, to produce 4-ethoxycarbonyl-1-phenyl-5-styrylpyrazole **93**. In the same conditions, the reaction with benzyl hydrazine afforded a 3:1 mixture of two pyrazole isomers, the 1-benzyl-4-ethoxycarbonyl-5-styrylpyrazole **94** as the major product together with 1-benzyl-4-ethoxycarbonyl-3-styrylpyrazole **95** (Scheme 29) [85].

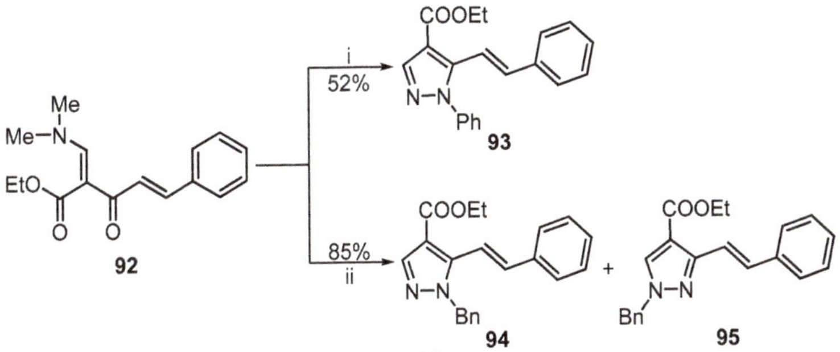

(i) $PhNHNH_2$ (1.0 equiv), EtOH, reflux, overnight.
(ii) $BnNHNH_2$ (1.0 equiv), EtOH, reflux, overnight.

Scheme 29. Synthesis of 4-ethoxycarbonyl-5-/3-styrylpyrazoles **93**, **94** and **95** [85].

3.4.2. Reaction of α,β-Enones with Hydrazines

The reaction of conjugated dienones (or α,β-enones) **96** with phenyl hydrazine hydrochloride in a mixture of ethanol and chloroform, in the presence of a catalytic amount of concentrated HCl, afforded 1,3-diphenyl-5-styryl-4,5-dihydropyrazoles **97** (Ar = Ph) (Scheme 30, i) [34]. No expected dihydropyrazole was obtained in the reaction of 1,5-diphenylpenta-2,4-dien-1-one with phenyl hydrazine due to fast oxidation by atmospheric oxygen affording the corresponding pyrazole [34]. Furthermore, 1-aryl-5-styrylpyrazoles **98** were synthetized by the one-pot reaction of conjugated dienones **96** with aryl hydrazine hydrochlorides in 1,2-dichlorobenzene at 130 °C, under O_2 atmosphere (Scheme 30, ii) [86,87]. Similar conditions for the reaction of Baylis–Hillman adduct **99** with phenyl hydrazine afforded the styrylpyrazole **100** (Scheme 30) [87].

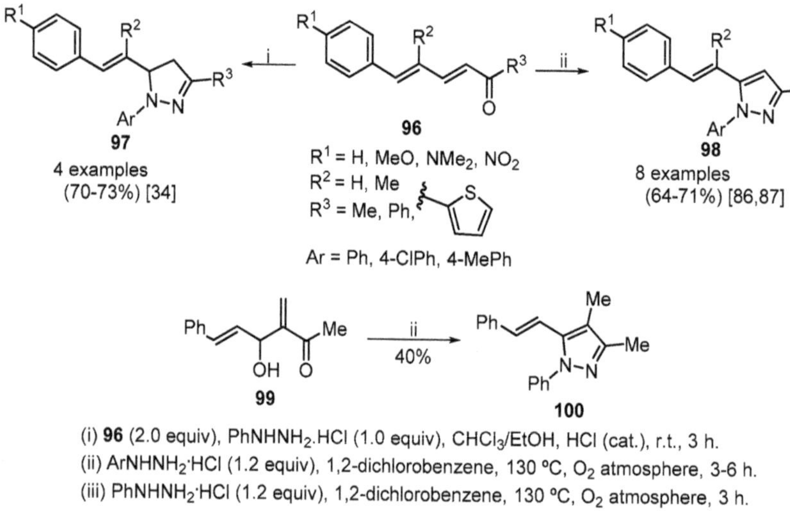

(i) **96** (2.0 equiv), $PhNHNH_2 \cdot HCl$ (1.0 equiv), $CHCl_3$/EtOH, HCl (cat.), r.t., 3 h.
(ii) $ArNHNH_2 \cdot HCl$ (1.2 equiv), 1,2-dichlorobenzene, 130 °C, O_2 atmosphere, 3-6 h.
(iii) $PhNHNH_2 \cdot HCl$ (1.2 equiv), 1,2-dichlorobenzene, 130 °C, O_2 atmosphere, 3 h.

Scheme 30. Conversion of α,β-enones **96** and adduct **99** into (E)-5-styrylpyrazolines **97** and (E)-5-styrylpyrazoles **98** and **100** by reaction with phenyl/aryl hydrazine hydrochlorides [34,86,87].

3.4.3. Reaction of 2-Styrylchromones with Hydrazines

The reaction of 5-benzyloxy-2-styrylchromones **101** with an excess of hydrazine hydrate in methanol at reflux yielded (E)-3-(2-benzyloxy-6-hydroxyphenyl)-5-styryl-1H-pyrazoles **102** (Scheme 31). Small amounts of 3-(2-benzyloxy-6-hydroxyphenyl)-5-(2-phenylethyl)pyrazoles **103a,d,e** and 5-aryl-3-(2-benzyloxy-β,6-dihydroxystyryl)-2-pyrazolines **104a–e** were also formed, in addition to the pyrazoles **102a–e** [88]. Only one isomer was obtained from the reaction of 2-styrylchromones **101** with methylhydrazine. Due to the hydrogen bond between 6′-OH and N-2 in each product, there was the formation of only one tautomer, the 3-(2-benzyloxy-6-hydroxyphenyl)-5-styrylpyrazoles **102f–k** and not the corresponding 5-(2-benzyloxy-6-hydroxyphenyl)-3-styrylpyrazole [89].

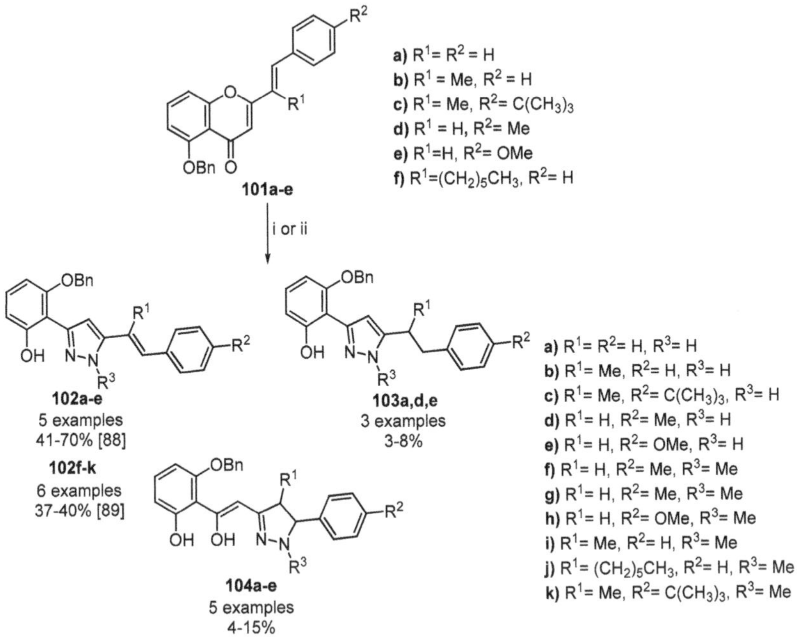

101a-e
a) R¹ = R² = H
b) R¹ = Me, R² = H
c) R¹ = Me, R² = C(CH$_3$)$_3$
d) R¹ = H, R² = Me
e) R¹ = H, R² = OMe
f) R¹ = (CH$_2$)$_5$CH$_3$, R² = H

102a-e 5 examples 41-70% [88]
102f-k 6 examples 37-40% [89]

103a,d,e 3 examples 3-8%

a) R¹ = R² = H, R³ = H
b) R¹ = Me, R² = H, R³ = H
c) R¹ = Me, R² = C(CH$_3$)$_3$, R³ = H
d) R¹ = H, R² = Me, R³ = H
e) R¹ = H, R² = OMe, R³ = H
f) R¹ = H, R² = Me, R³ = Me
g) R¹ = H, R² = Me, R³ = Me
h) R¹ = H, R² = OMe, R³ = Me
i) R¹ = Me, R² = H, R³ = Me
j) R¹ = (CH$_2$)$_5$CH$_3$, R² = H, R³ = Me
k) R¹ = Me, R² = C(CH$_3$)$_3$, R³ = Me

104a-e 5 examples 4-15%

(i) NH$_2$NHR³ (4.1 equiv for R³ = H; 8.2 equiv for R³ = Me), MeOH, reflux, 24 h.

Scheme 31. Synthesis of (*E*)-3-(2-benzyloxy-6-hydroxyphenyl)-5-styryl-1*H*-pyrazoles **102** and some derivatives of pyrazoles **103** and **104** isolated as by-products [88].

3.4.4. Wittig–Horner Reaction

Deshayes et al. reported the reaction of phosponic esters, prepared from 1-substituted 5-bromomethylpyrazoles **105** and triethyl phosphite, with substituted benzaldehydes and furfural, in dimethoxyethane, as a method to prepare (*E*)-4-ethoxycarbonyl-3-methyl-5-styryl-l-substituted pyrazoles **106** (Scheme 32) [75,85].

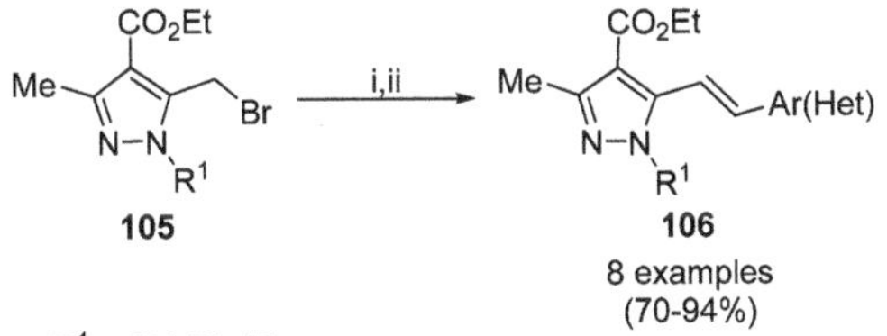

105 → **106** 8 examples (70-94%)

R¹ = CH$_2$Ph, Ph
Ar = Ph, 4-ClC$_6$H$_4$, 4-MeC$_6$H$_4$, 4-MeOC$_6$H$_4$, furyl

(i) P(OEt)$_3$ (1.0 equiv), 130 °C, 1 h
(ii) NaH (1.0 equiv, 50% dispersion washed with hexane), DME, aldehyde (1.0 equiv, in dimethoxyethane) dropwise, r.t., overnight.

Scheme 32. Synthesis of (*E*)-4-ethoxycarbonyl-3-methyl-5-styry-l-substituted pyrazoles **106** [75,85].

3.5. Synthesis of Bis(Styryl)Pyrazoles

Cyclocondensation Reaction

Typically, 3,5-bis(styryl)pyrazole curcumin analogue **12a** and *N*-aryl derivatives have been obtained by treatment of curcumin **11** with hydrazine hydrate or aryl hydrazines in ethanol [42] or toluene [40] at reflux for a long reaction time of 24–40 h. Using glacial acetic acid at reflux, the reaction time can be reduced to 6–8 h [33,39,41]. Room temperature reactions have also been reported but required a longer reaction time [40]. In 2015, Sherin et al. reported a solvent-free, mechanochemical method for the synthesis of curcumin **11** derived 3,5-bis(styryl)pyrazoles **12a–f** [36] (Scheme 33). The heterocyclization of **11** with hydrazine or hydrazine derivatives was performed with vigorous grinding, using an agate mortar and pestle, at room temperature, in the presence of a catalytic amount of acetic acid. A very short reaction time was necessary, in comparison with the previously referred methods that use conventional heating [39–42]. The reaction scope seems to be broad since phenyl hydrazine, *p*-methoxy, *p*-chloro, *p*-nitro, and *p*-carboxyphenyl hydrazines gave bis(styryl)pyrazoles **12b–f** in good yields (79–84%). One year later, the same authors performed the mechanochemical synthesis of 1-phenyl-3,5-bis(styryl)pyrazoles **13a–f**, varying the substituents present in the aromatic ring of both styryl groups [90]. Recently, Liao et al. described a rapid synthesis of similar 3,5-bis(styryl)pyrazole curcumin analogues **13** by using microwave irradiation conditions [44].

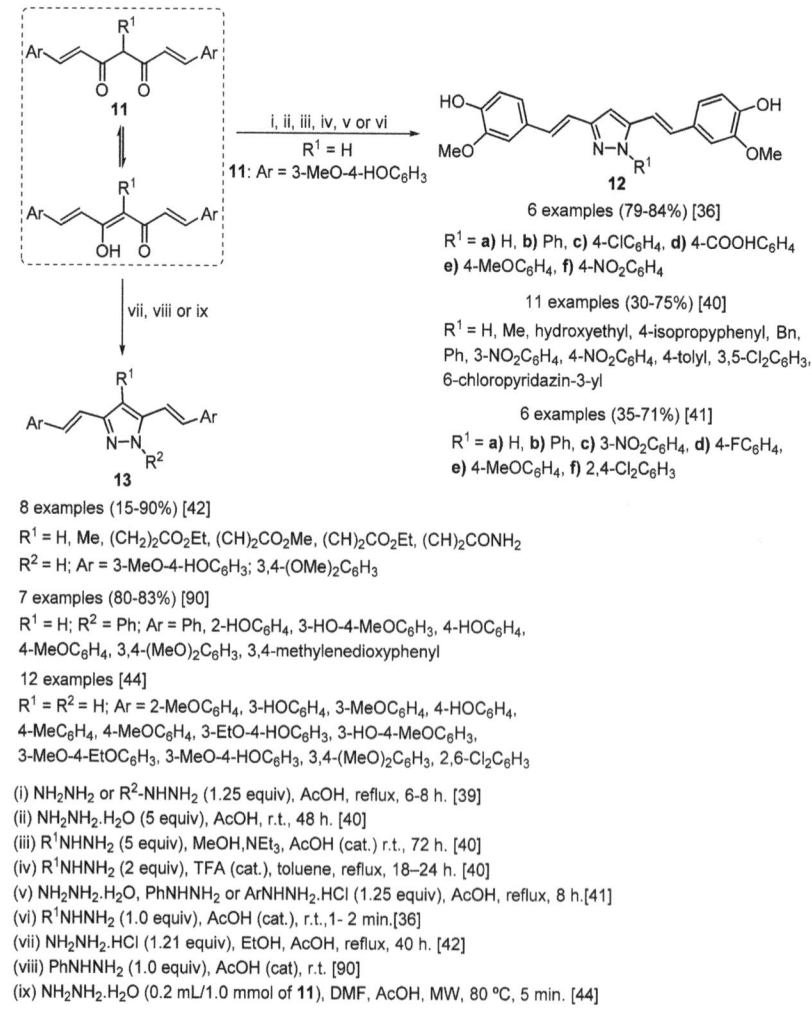

(i) NH$_2$NH$_2$ or R^2-NHNH$_2$ (1.25 equiv), AcOH, reflux, 6-8 h. [39]
(ii) NH$_2$NH$_2$.H$_2$O (5 equiv), AcOH, r.t., 48 h. [40]
(iii) R^{1}NHNH$_2$ (5 equiv), MeOH,NEt$_3$, AcOH (cat.) r.t., 72 h. [40]
(iv) R^{1}NHNH$_2$ (2 equiv), TFA (cat.), toluene, reflux, 18–24 h. [40]
(v) NH$_2$NH$_2$.H$_2$O, PhNHNH$_2$ or ArNHNH$_2$.HCl (1.25 equiv), AcOH, reflux, 8 h.[41]
(vi) R^{1}NHNH$_2$ (1.0 equiv), AcOH (cat.), r.t., 1- 2 min.[36]
(vii) NH$_2$NH$_2$.HCl (1.21 equiv), EtOH, AcOH, reflux, 40 h. [42]
(viii) PhNHNH$_2$ (1.0 equiv), AcOH (cat), r.t. [90]
(ix) NH$_2$NH$_2$.H$_2$O (0.2 mL/1.0 mmol of **11**), DMF, AcOH, MW, 80 °C, 5 min. [44]

Scheme 33. Synthesis of 3,5-bis(styryl)pyrazoles **12** and **13** [36,39–42,44,90].

4. Transformations of Styrylpyrazoles

Styrylpyrazoles are interesting templates for synthetic manipulation towards new heterocycles. Nevertheless, to date, only a small number of transformations involving the 2-arylvinyl moiety of styrylpyrazoles have been reported in the literature. In this section, we describe the most common transformations of the 2-arylvinyl moiety of styrylpyrazoles.

4.1. Transformations of 4-Styrylpyrazoles

Diels–Alder Cycloadditions

4-Styrylpyrazoles (*Z*)-**107** and (*E*)-**108** can participate in Diels–Alder reactions, as dienophiles, involving the exocyclic (Cα = Cβ) double bond, or as dienes when this double bond is conjugated with the C4 = C3(5) double bond of the pyrazole moiety. Silva et al. obtained the tetrahydroindazoles **109** and **110** in good yields and with high selectivities from the Diels–Alder reactions of (*Z*)- and (*E*)-1-acetyl-4-styrylpyrazoles **107** and **108**, which reacted as dienes, with both dienophiles, *N*-methyl- or *N*-phenylmaleimide, using microwave irradiation (800 W) [91,92]. The expected indazoles **111** were obtained by dehydrogenation of the corresponding adducts with DDQ under microwave irradiation or conventional heating conditions (Scheme 34) [91]. They have also studied the cycloadditions of (*E*)-1-acetyl-4-nitrostyrylpyrazole **108** (R^1 = NO_2) with *N*-phenylmaleimide and with dimethyl acetylenedicarboxylate (DMAD). In the reaction with *N*-phenylmaleimide, the cycloadduct **112** was obtained in 49% yield together with the indazole **113**, obtained as a by-product (5% yield). On the other hand, in the reaction with DMAD, a conjugate addition of the pyrazole nitrogen to DMAD occurred with formation of a trace amount of pyrazole **114** (Scheme 35) [91].

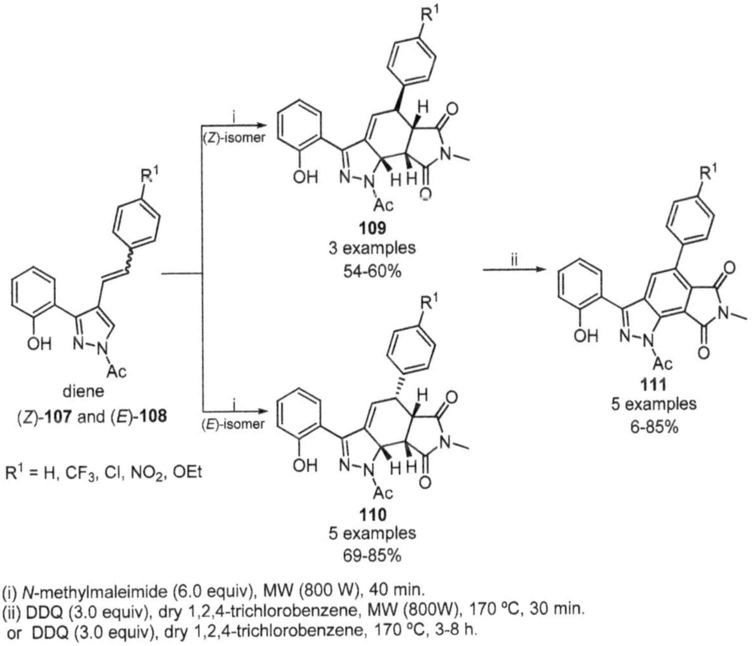

R1
OH N-N Ac
diene
(Z)-107 and (E)-108
R1 = H, CF3, Cl, NO2, OEt
i
(Z)-isomer
(E)-isomer
H O N H H O
109
3 examples
54-60%
110
5 examples
69-85%
ii
111
5 examples
6-85%
(i) N-methylmaleimide (6.0 equiv), MW (800 W), 40 min.
(ii) DDQ (3.0 equiv), dry 1,2,4-trichlorobenzene, MW (800W), 170 °C, 30 min.
or DDQ (3.0 equiv), dry 1,2,4-trichlorobenzene, 170 °C, 3-8 h.

Scheme 34. Transformation of 4-styrylpyrazoles **107** and **108** into indazoles **111** [91,92].

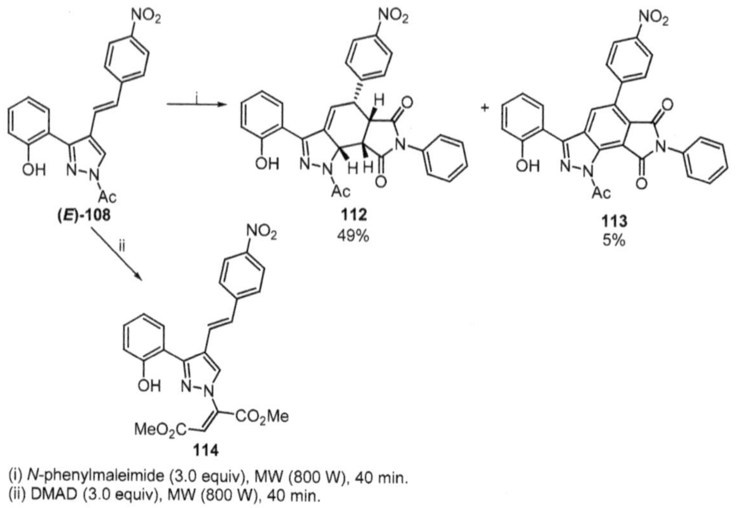

(i) *N*-phenylmaleimide (3.0 equiv), MW (800 W), 40 min.
(ii) DMAD (3.0 equiv), MW (800 W), 40 min.

Scheme 35. Transformation of (*E*)-4-styrylpyrazole **108** into cycloadducts **112**, **113** and pyrazole **114** [92].

4.2. Transformations of 3(5)-Styrylpyrazoles

4.2.1. Diels–Alder Cycloadditions

Starting from 3- and 5-styrylpyrazoles **115** and **116**, Silva et al. developed a synthetic route to prepare naphthylpyrazoles [84]. These styrylpyrazoles, which can behave as dienophiles, undergo a Diels–Alder cycloaddition with *o*-benzoquinodimethane (**117**), the diene, which was formed in situ by the thermal extrusion of sulfur dioxide from 1,3-dihydrobenzo[*c*]thiophene 2,2-dioxide. *N*-Substituted 3-styrylpyrazoles **115** reacted with diene **117** at 250 °C in 1,2,4-trichlorobenzene giving the corresponding 3-[2-(3-aryl-1,2,3,4-tetrahydronaphthyl)]-5-(2-hydroxyphenyl)-1-phenylpyrazoles (**118**) in good yields (69–91%) (Scheme 36, i). The presence of an electron-withdrawing substituent on the *p*-position of the phenyl ring increases the reactivity of the styryl double bond [84]. These authors also performed the Diels–Alder reaction of (*E*)-3(5)-(2-hydroxyphenyl)-5(3)-styrylpyrazoles (**116**) with diene (**117**), but in this case longer reaction times were necessary and the expected cycloadducts were obtained in lower yields (24–48%). The efficiency of this Diels–Alder reaction increased by using a LA, aluminum(III) chloride, which increased the reactivity of the styryl double bond, probably through the formation of an aluminum(III) complex involving the oxygen of the 2′-hydroxy group and the free nitrogen of the pyrazole moiety. Under these conditions, the cycloadduct **119** containing an electron-donating substituent at the *p*-position of styryl moiety was obtained in better yield. On the contrary, the cycloadduct **119** containing the *p*-nitro group as substituent was obtained in better yield without addition of aluminum(III) chloride (Scheme 36, iii) [84]. The formation of naphthylpyrazoles **120** and **121** occurred by dehydrogenation of the corresponding cycloadducts with DDQ (2–6 days). The naphthylpyrazole **120** bearing an electron-donating substituent (R^1 = OMe) at the *p*-position of the phenyl group linked to the hydroaromatic ring was obtained in a shorter time (2 days) and with better yield (59%) than the other derivatives (R^1 = H (25%) and R^1 = NO_2 (17%)). The presence of the *p*-methoxy substituent stabilizes the carbocation formed through a hydride transfer from compound **118** to DDQ. To increase the yield of this oxidation, *p*-toluenesulfonic acid was added and the products were obtained in a shorter time (2–7 h) and with better yields, especially for the derivative containing the nitro group (R^1 = NO_2, 36%) (Scheme 36, ii). The oxidation of cycloadducts **119** was also tried, using both methods (Scheme 36, ii) but only decomposition products were obtained, save the case of 3-(2-hydroxyphenyl)-5-{2-[3-(4-methoxyphenyl)]naphthyl}pyrazole **121** (R^1 = OMe), which was obtained in low yield (13%, method B) [84].

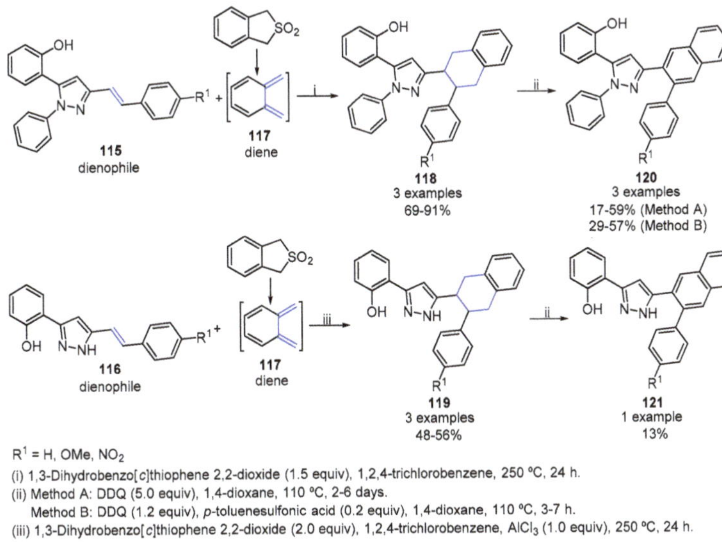

Scheme 36. Transformation of styrylpyrazoles **115** and **116** into tetrahydronaphthylpyrazoles **118** and **119** and their oxidation to naphthylpyrazoles **120** and **121** [84].

4.2.2. Electrophilic Intramolecular Cyclization

C. Deshayes et al. reported the electrophilic intramolecular cyclization of 1-benzyl- and 1-phenyl-5-styrylpyrazoles **122**, in the presence of polyphosphoric acid (PPA), which led to the formation of 5-aryl-3-ethoxycarbonyl-4,5-dihydropyrazolo-type compounds **123** and **124** (Scheme 37) [85]. Both the 5-aryl-3-ethoxycarbonyl-4,5-dihydropyrazolo[1,5-*a*]quinolines (**123**) and 5-aryl-3-ethoxycarbonyl-4,5-dihydro-10*H*-pyrazolo[1,5-*b*][2]benzazepines (**124**) were obtained with very good yields, save for **124** ($R^1 = R^2 = H$).

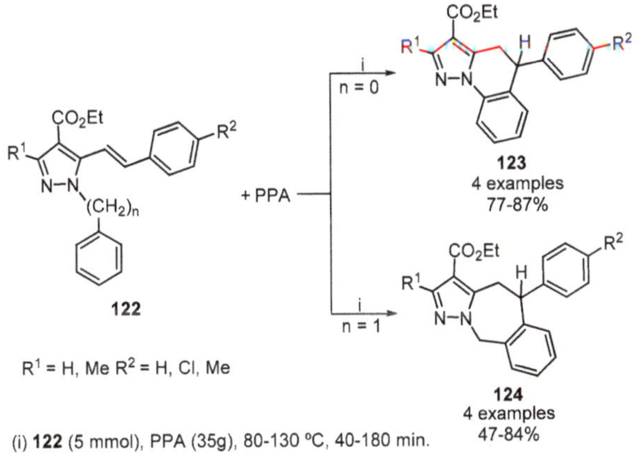

Scheme 37. Transformation of 1-benzyl- and 1-phenyl-5-styrylpyrazoles **122** into 5-aryl-3-ethoxycarbonyl-4,5-dihydropyrazolo-type compounds **123** and **124** [85].

In 2015, Moon et al. reported the acid-catalyzed intramolecular Friedel–Crafts (IMFC) reaction of several 1-aryl-5-styrylpyrazoles **125** in the presence of PPA to produce dihydropyrazolo[1,5-*a*]quinolines **126** (Scheme 38, i) [87]. It is worth mentioning that the corresponding five-membered ring product was not obtained. This fact can be explained by the higher stability of benzylic carbocation when compared to secondary carbocation formed during the reaction. Moreover, the obtained yields show that the efficiency of the reaction is not significantly affected by the nature of the substituents at the 1-, 3-, 4- and 5-positions of pyrazoles **125**. The IMFC reaction of 1-benzyl-5-styrylpyrazole **125** (R^1 = Bn) afforded a 2,9-diphenyl-3,3a-diazabenzo[*f*]azulene derivative [87]. The base-catalyzed aerobic oxidation of dihydropyrazolo[1,5-*a*]quinolines **126** afforded the corresponding pyrazolo[1,5-*a*]quinolines **127** (Scheme 38, ii). However, the reaction only occurred when R^5 = Ph and R^6 = H [87].

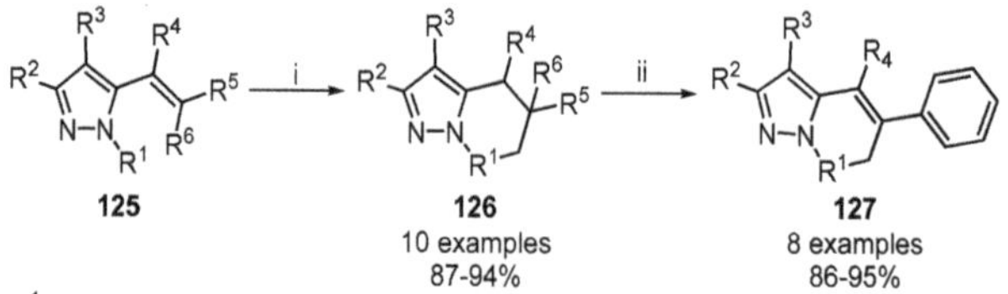

R^1 = Bn, Ph, 4-ClC_6H_4, 4-MeC_6H_4
R^2 = Ph, Me, Thiophene
R^3 = H, Me
R^4 = H, Me
R^5 = Me, Ph
R^6 = H, Ph
(i) PPA, 100-140 °C, 2-10 h.
(ii) Cs_2CO_3 (2.0 equiv), DMF, O_2 atmosphere, 130 °C, 24 h.

Scheme 38. Transformation of 1-aryl-5-styrylpyrazoles **125** into dihydropyrazolo[1,5-*a*]quinolines **126** and their oxidation to pyrazolo[1,5-*a*]quinolines **127** [87].

4.2.3. Oxidative Addition Reactions

In 2004, Ignatenko et al. converted several (*E*)-3- and 5-styrylpyrazoles **128** and **129** into phthalimidoaziridinylpyrazoles **131** and **132** by the oxidative addition of *N*-aminophthalimide (NAPhth) (**130**) to the exocyclic double bond in the presence of lead tetraacetate and potassium carbonate in dichloromethane (Scheme 39). The reaction with 1,5-diphenylpyrazoles **128** afforded the corresponding phthalimidoaziridinylpyrazoles **131** with better yield than with the 1,3-diphenylpyrazoles **129**. Analogous 4,5-dihydropyrazoles were inert in the oxidative addition of NAPhth [57].

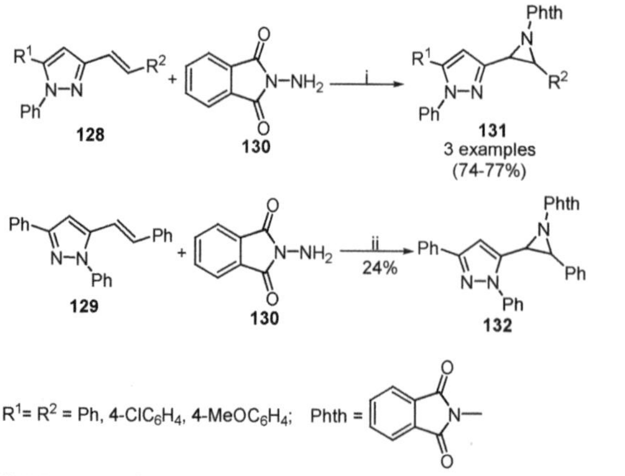

(i) **130** (1.0 equiv), K_2CO_3 (5.0 equiv.), $Pd(OAc)_4$ (1.0 equiv), CH_2Cl_2, -12 °C, 30 min.
(ii) **130** (2.0 equiv), K_2CO_3 (10 equiv.), $Pd(OAc)_4$ (2.0 equiv), CH_2Cl_2, -12 °C, 30 min.

Scheme 39. Transformation of (*E*)-3- and 5-styrylpyrazoles **128** and **129** into phthalimidoaziridinylpyrazoles **131** and **132** [57].

5. Conclusions

Styrylpyrazoles have shown remarkable biological activities, namely as anti-inflammatory, antimicrobial (antibacterial and antifungal) and antioxidant agents. Among these compounds, the 3,5-bis(styryl)pyrazoles, curcumin analogues, should be highlighted herein due to their significant antioxidant, neuroprotective, antimalarial, antimycobacterial, antiangiogenic, cytotoxic and antiproliferative activities. Additionally, styrylpyrazoles present interesting photophysical properties suitable for metal ion sensing, for the design of energy-transfer-based fluorescent probes and are also good DNA groove binders. Although structure–activity relationship studies have been reported for some styrylpyrazole derivatives, there is a gap in the literature regarding a detailed investigation of the specific properties due to the presence of the styryl group, or the different properties that may appear by exchanging the styryl group position. As far as we know, there are no comparative data between styrylpyrazoles and analogous non-styrylpyrazoles, which are very important in order to understand the role of the styryl moiety in the pyrazoles' properties.

The synthesis of 1-styrylpyrazoles has been achieved starting from pyrazoles through N-cross-coupling reactions with styrylboronic acid or alternatively by the addition of N–H to activated and non-activated alkynes. In turn, the cyclocondensation reaction of different substrates, with hydrazine derivatives, is the most common strategy for the synthesis of 3-, 4-, and 5- styrylpyrazoles. Regarding the substrates, penta-1,4-dien-3-ones, α-oxoketene dithioacetals, benzalacetones, acetylenic ketones, 1,3-diketones and diketoesters are good substrates for the synthesis of 3- and 5- styrylpyrazoles while α-alkenyl-α,β-enones and diketones (curcumin analogues) are good synthons for the synthesis of 4-styrylpyrazoles and bis(styryl)pyrazoles, respectively. Moreover, reactions of 2- and 3-styrylchromones with hydrazine derivatives are often used to prepare 3(5)-styrylpyrazoles and 4-styrylpyrazoles, respectively. More recently, the cross-coupling reactions of halopyrazoles with styrylboronic acids gained relevance as a method to access 4-styrylpyrazoles in a straightforward way.

Only a few examples of transformations of styrylpyrazoles involving the 2-arylvinyl moiety have been reported in the literature. These include Diels–Alder reactions, intramolecular cyclisation reactions and oxidative additions to the exocyclic double bond, and they allow the formation of more complex and very interesting heterocycles such as indazoles, naphthylpyrazoles, 4,5-dihydropyrazolo[1,5-a] quinolines, 4,5-dihydro-10H-pyrazolo[1,5-b][2]benzazepines, pyrazolo[1,5-a]quinolines, 2,9-diphenyl-3,3a-diazabenzo [f]azulene and phthalimidoaziridinylpyrazoles. Therefore, further studies towards the investigation of novel transformations of styrylpyrazoles are of high interest.

Author Contributions: Conceptualization, A.M.S.S. and V.L.M.S.; writing—original draft preparation, P.M.O.G. and P.M.S.O.; writing—review and editing, A.M.S.S. and V.L.M.S.; supervision, A.M.S.S. and V.L.M.S. All authors have read and agreed to the published version of the manuscript.

Funding: The authors would like to thank the University of Aveiro and FCT/MEC for the financial support to the LAQV-REQUIMTE (UIDB/50006/2020) research project, financed by national funds and, when appropriate, co-financed by FEDER under the PT2020 Partnership Agreement of the Portuguese NMR network. Vera L. M. Silva thanks the assistant professor position (within CEE-CINST/2018; since 01/09/2019) and the integrated programme of SR&TD "pAGE—Protein Aggregation Across the Lifespan" (reference CENTRO-01-0145FEDER-000003), co-funded by the Centro 2020 program, Portugal 2020 and European Union through the European Regional Development Fund.

Conflicts of Interest: The authors declare no conflict of interest.

References

1. Abrigach, F.; Touzani, R. Pyrazole derivatives with NCN junction and their biological activity: A review. *Med. Chem.* **2016**, *6*, 292–298. [CrossRef]
2. Alam, M.J.; Alam, S.; Alam, P.; Naim, M.J. Green synthesis of pyrazole systems under solvent-free conditions. *Int. J. Pharma Sci. Res.* **2015**, *6*, 1433–1442. [CrossRef]
3. Abdel-Maksoud, M.S.; El-Gamal, M.I.; Gamal El-Din, M.M.; Oh, C.H. Design, synthesis, in vitro anticancer evaluation, kinase inhibitory effects, and pharmacokinetic profile of new 1,3,4-triarylpyrazole derivatives possessing terminal sulfonamide moiety. *J. Enz. Inhib. Med. Chem.* **2019**, *34*, 97–109. [CrossRef]

4. Amir, M.; Kumar, S. Synthesis and anti-inflammatory, analgesic, ulcerogenic and lipid peroxidation activities of 3,5-dimethyl pyrazoles, 3-methyl pyrazol-5-ones and 3,5-disubstituted pyrazolines. *Indian J. Chem.* **2005**, *37*, 2532–2537. [CrossRef]
5. Achutha, D.K.; Kameshwar, V.H.; Ningappa, M.B.; Kariyappa, A.K. Synthesis and in vitro biological evaluation for antioxidant, anti-inflammatory activities and molecular docking studies of novel pyrazole derivatives. *Biointerface Res. Appl. Chem.* **2017**, *7*, 2040–2047.
6. Prabhu, V.V.; Kannan, N.; Guruvayoorappan, C. 1,2-Diazole prevents cisplatin-induced nephrotoxicity in experimental rats. *Pharmacol. Rep.* **2013**, *65*, 980–990. [CrossRef]
7. Silva, V.L.M.; Elguero, J.; Silva, A.M.S. Current progress on antioxidants incorporating the pyrazole core. *Eur. J. Med. Chem.* **2018**, *156*, 394–429. [CrossRef] [PubMed]
8. Chowdary, B.N.; Umashankara, M.; Dinesh, B.; Girish, K.; Baba, A.R. Development of 5-(aryl)-3-phenyl -1*H*-pyrazole derivatives as potent antimicrobial compounds. *Asian J. Chem.* **2019**, *31*, 45–50. [CrossRef]
9. Tanitame, A.; Oyamada, Y.; Ofuji, K.; Fujimoto, M.; Iwai, N.; Hiyama, Y.; Suzuki, K.; Ito, H.; Terauchi, H.; Kawasaki, M.; et al. Synthesis and antibacterial activity of a novel series of potent DNA gyrase inhibitors. Pyrazole derivatives. *J. Med. Chem.* **2004**, *47*, 3693–3696. [CrossRef] [PubMed]
10. Secci, D.; Bolasco, A.; Chimenti, P.; Carradori, S. The state of the art of pyrazole derivatives as monoamine oxidase inhibitors and antidepressant/anticonvulsant agents. *Curr. Med. Chem.* **2011**, *18*, 5114–5144. [CrossRef] [PubMed]
11. Palaska, E.; Aytemir, M.; Uzbay, I.T.; Erol, D. Synthesis and antidepressant activities of some 3,5-diphenyl -2-pyrazolines. *Eur. J. Med. Chem.* **2001**, *36*, 539–543. [CrossRef]
12. Datar, P.A.; Jadhav, S.R. Development of pyrazole compounds as antidiabetic agent: A review. *Lett. Drug Des. Discov.* **2014**, *11*, 686–703. [CrossRef]
13. Silva, V.L.M.; Silva, A.M.S.; Pinto, D.C.G.A.; Jagerovic, N.; Callado, L.F.; Cavaleiro, J.A.S.; Elguero, J. Synthesis and pharmacological evaluation of chlorinated *N*-alkyl-3 and -5-(2-hydroxyphenyl)pyrazoles as CB_1 cannabinoid ligands. *Monatsh. Chem.* **2007**, *138*, 797–811. [CrossRef]
14. Silva, V.L.M.; Silva, A.M.S.; Pinto, D.C.G.A.; Rodríguez, P.; Gomez, M.; Jagerovic, N.; Callado, L.F.; Cavaleiro, J.A.S.; Elguero, J.; Fernandez-Ruiz, J. Synthesis and pharmacological evaluation of new (*E*)- and (*Z*)-3-aryl-4-styryl-1*H*-pyrazoles as potential cannabinoid ligands. *Arkivoc* **2010**, *2010*, 226–247. [CrossRef]
15. Gomes, P.M.O.; Silva, A.M.S.; Silva, V.L.M. Pyrazoles as key scaffolds for the development of fluorine-18-labeled radiotracers for positron emission tomography (PET). *Molecules* **2020**, *25*, 1722. [CrossRef]
16. Ansari, A.; Ali, A.; Asif, M. Biologically active pyrazole derivatives. *New J. Chem.* **2017**, *41*, 16–41. [CrossRef]
17. Küçükgüzel, S.G.; Senkardes, S. Recent advances in bioactive pyrazoles. *Eur. J. Med. Chem.* **2015**, *97*, 786–815. [CrossRef]
18. Santos, C.M.M.; Silva, V.L.M.; Silva, A.M.S. Synthesis of chromone-related pyrazole compounds. *Molecules* **2017**, *22*, 1665. [CrossRef]
19. Faria, J.V.; Vegi, P.F.; Miguita, A.G.C.; dos Santos, M.S.; Boechat, N.; Bernardino, A.M.R. Recently reported biological activities of pyrazole compounds. *Bioorg. Med. Chem.* **2017**, *25*, 5891–5903. [CrossRef]
20. Pérez-Fernández, R.; Goya, P.; Elguero, J. A review of recent progress (2002–2012) on the biological activities of pyrazoles. *Arkivoc* **2014**, *2014*, 233.
21. Marella, A.; Ali, M.R.; Alam, M.T.; Saha, R.; Tanwar, O.; Akhter, M.; Shaquiquzzaman, M.; Alam, M.M. Pyrazolines: A biological review. *Mini Rev. Med. Chem.* **2013**, *13*, 921–931. [CrossRef] [PubMed]
22. Giornal, F.; Pazenok, S.; Rodefeld, L.; Lui, N.; Vors, J.-P.; Leroux, F.R. Synthesis of diversely fluorinated pyrazoles as novel active agrochemical ingredients. *J. Fluorine Chem.* **2013**, *152*, 2–11. [CrossRef]
23. Garcia, H.; Iborra, S.; Miranda, M.A.; Morera, I.M.; Primo, J. Pyrazoles and isoxazoles derived from 2-hydroxyaryl phenylethynyl ketones: Synthesis and spectrophotometric evaluation of their potential applicability as sunscreens. *Heterocycles* **1991**, *32*, 1745–1748. [CrossRef]
24. Catalan, J.; Fabero, F.; Claramunt, R.M.; Santa Maria, M.D.; Foces-Foces, M.C.; Cano, F.H.; Martinez-Ripoll, M.; Elguero, J.; Sastre, R. New ultraviolet stabilizers: 3- and 5-(2′-hydroxyphenyl)pyrazoles. *J. Am. Chem. Soc.* **1992**, *114*, 5039–5048. [CrossRef]
25. Trofimenko, S. Coordination chemistry of pyrazole-derived ligands. *Chem. Rev.* **1972**, *72*, 497–509. [CrossRef]
26. Busev, A.I.; Akimov, V.K.; Gusev, S.I. Pyrazolone Derivatives as Analytical Reagents. *Russ. Chem. Rev.* **1965**, *34*, 237. [CrossRef]

27. Meier, H.; Hormaza, A. Oligo(1,4-phenylenepyrazole-3,5-diyl)s. *Eur. J. Org. Chem.* **2003**, 3372–3377. [CrossRef]
28. Willy, B.; Müller, T.J.J. Rapid one-pot, four-step synthesis of highly fluorescent 1,3,4,5-tetrasubstituted pyrazoles. *Org. Lett.* **2011**, *13*, 2082–2085. [CrossRef]
29. Dorlars, A.; Schellhammer, C.-W.; Schroeder, J. Heterocycles as Structural Units in New Optical Brighteners. *Angew. Chem. Int. Ed.* **1975**, *14*, 665–679. [CrossRef]
30. Perevalov, V.P.; Baryshnenkova, L.I.; Sheban, G.V.; Karim, A.K.K.; Kramarenko, S.S.; Koldaeva, T.Y.; Stepanov, B.I. Spectral peculiarities of isomeric *trans*-styrylpyrazoles. *Chem. Heterocycl. Compd.* **1990**, *26*, 887–890. [CrossRef]
31. Garg, V.; Kumar, P.; Verma, A.K. Chemo-, regio-, and stereoselective *N*-alkenylation of pyrazoles/benzpyrazoles using activated and unactivated alkynes. *J. Org. Chem.* **2017**, *82*, 10247–10262. [CrossRef] [PubMed]
32. Yamamoto, H.; Kaneko, S.-I. Synthesis of 1-phenyl-2-styryl-3,5-dioxopyrazolidines as anti-inflammatory agents. *J. Med. Chem.* **1970**, *13*, 292–295. [CrossRef] [PubMed]
33. Flynn, D.L.; Belliotti, T.R.; Boctor, A.M.; Connor, D.T.; Kostlan, C.R.; Nies, D.E.; Ortwine, D.F.; Schrier, D.J.; Sircar, J.C. Styrylpyrazoles, styrylisoxazoles and styrylisothiazoles. Novel 5-lipoxygenase and cyclooxygenase inhibitors. *J. Med. Chem.* **1991**, *34*, 518–525. [CrossRef] [PubMed]
34. Nauduri, D.; Reddy, G.B.S. Antibacterials and antimycotics: Part 1: Synthesis and activity of 2-pyrazoline derivatives. *Chem. Pharm. Bull.* **1998**, *46*, 1254–1260. [CrossRef] [PubMed]
35. Gressler, V.; Moura, S.; Flores, A.F.C.; Flores, D.C.; Colepicolo, P.; Pinto, E. Antioxidant and antimicrobial properties of 2-(4,5-dihydro-1*H*-pyrazol-1-yl)pyrimidine and 1-carboxamidino-1*H*-pyrazole derivatives. *J. Braz. Chem. Soc.* **2010**, *21*, 1477–1483. [CrossRef]
36. Sherin, D.R.; Rajasekharan, K.N. Mechanochemical synthesis and antioxidant activity of curcumin-templated azoles. *Arch. Pharm. Chem. Life Sci.* **2015**, *348*, 908–914. [CrossRef]
37. Sharma, S.; Ying, Z.; Pinilla, F.G. A pyrazole curcumin derivative restores membrane homeostasis disrupted after brain trauma. *Exp. Neurobiol.* **2010**, *226*, 191–199. [CrossRef]
38. Lapchak, P.A.; Boitano, P.D. Effect of the pleiotropic drug CNB-001 on tissue plasminogen activator (tPA) protease activity in vitro: Support for combination therapy to treat acute ischemic stroke. *J. Neurol. Neurophysiol.* **2014**, *5*, 214–219.
39. Das, J.; Pany, S.; Panchal, S.; Majhi, A.; Rahman, G.M. Binding of isoxazole and pyrazole derivatives of curcumin with the activator binding domain of novel protein kinase C. *Bioorg. Med. Chem.* **2011**, *19*, 6196–6202. [CrossRef]
40. Narlawar, R.; Pickhardt, M.; Leuchtenberger, S.; Baumann, K.; Krause, S.; Dyrks, T.; Weggen, S.; Mandelkow, E.; Schmidt, B. Curcumin-derived pyrazoles and isoxazoles: Swiss army knives or blunt tools for Alzheimer's disease? *ChemMedChem* **2008**, *3*, 165–172. [CrossRef]
41. Mishra, S.; Karmodiya, K.; Surolia, N.; Surolia, A. Synthesis and exploration of novel curcumin analogues as anti-malarial agents. *Bioorg. Med. Chem.* **2008**, *16*, 2894–2902. [CrossRef] [PubMed]
42. Amolins, M.W.; Peterson, L.B.; Blagg, B.S.J. Synthesis and evaluation of electron-rich curcumin analogues. *Bioorg. Med. Chem.* **2009**, *17*, 360–367. [CrossRef] [PubMed]
43. Fuchs, J.R.; Pandit, B.; Bhasin, D.; Etter, J.P.; Regan, N.; Abdelhamid, D.; Li, C.; Lin, J.; Li, P.-K. Structure–activity relationship studies of curcumin analogues. *Bioorg. Med. Chem. Lett.* **2009**, *19*, 2065–2069. [CrossRef] [PubMed]
44. Liao, V.W.Y.; Kumari, A.; Narlawar, R.; Vignarajan, S.; Hibbs, D.E.; Panda, D.; Groundwater, P.W. Tubulin-binding 3,5-bis(styryl)pyrazoles as lead compounds for the treatment of castration-resistant prostate cancer. *Mol. Pharmacol.* **2020**, *97*, 409–422. [CrossRef] [PubMed]
45. Orrego-Hernández, J.; Cobo, J.; Portilla, J. Synthesis, photophysical properties, and metal-ion recognition studies of fluoroionophores based on 1-(2-pyridyl)-4-styrylpyrazoles. *ACS Omega* **2019**, *4*, 16689–16700. [CrossRef]
46. Kumar, V.; Kaur, K.; Gupta, G.K.; Sharma, A.K. Pyrazole containing natural products: Synthetic preview and biological significance. *Eur. J. Med. Chem.* **2013**, *69*, 735–753. [CrossRef]
47. Santos, N.E.; Carreira, A.R.F.; Silva, V.L.M.; Braga, S.S. Natural and biomimetic antitumor pyrazoles, A perspective. *Molecules* **2020**, *25*, 1364. [CrossRef]

48. Boudiara, T.; Hichema, L.; Khalfallaha, A.; Kabouchea, A.; Kabouchea, Z.; Brouardb, I.; Jaime, B.; Bruneauc, C. A new alkaloid and flavonoids from the aerial parts of *Euphorbia guyoniana*. *Nat. Prod. Commun.* **2010**, *5*, 35–37. [CrossRef]
49. Kundu, P.; Chattopadhyay, N. Interaction of a bioactive pyrazole derivative with calf thymus DNA: Deciphering the mode of binding by multi-spectroscopic and molecular docking investigations. *J. Photochem. Photobiol. B Biol.* **2017**, *173*, 485–492. [CrossRef]
50. Kundu, P.; Chattopadhyay, N. Unraveling the binding interaction of a bioactive pyrazole-based probe with serum proteins: Relative concentration dependent 1:1 and 2:1 probe-protein stoichiometries. *Biophys. Chem.* **2018**, *240*, 70–81. [CrossRef]
51. Cody, J.; Mandal, S.; Yang, L.; Fahrni, C.J. Differential tuning of the electron transfer parameters in 1,3,5-triarylpyrazolines: A rational design approach for optimizing the contrast ratio of fluorescent probes. *J. Am. Chem. Soc.* **2008**, *130*, 13023–13032. [CrossRef] [PubMed]
52. Kundu, P.; Banerjee, D.; Maiti, G.; Chattopadhyay, N. Dehydrogenation induced inhibition of intramolecular charge transfer in substituted pyrazoline analogues. *Phys. Chem. Chem. Phys.* **2017**, *19*, 11937–11946. [CrossRef] [PubMed]
53. Kantam, M.L.; Venkanna, G.T.; Kumar, K.B.S.; Subrahmanyam, V.B. An efficient N-arylation of heterocycles with aryl-, heteroaryl-, and vinylboronic acids catalyzed by copper fluorapatite. *Helv. Chim. Acta* **2010**, *93*, 974–979. [CrossRef]
54. Bourgeois, P.; Lucrece, J.; Dunoguès, J. Addition d'hétérocycles azotés aux alcynes arylés; Synthèse de N-styrylazoles. *J. Heterocycl. Chem.* **1978**, *15*, 1543–1545. [CrossRef]
55. Tsuchimoto, T.; Aoki, K.; Wagatsuma, T.; Suzuki, Y. Lewis Acid Catalyzed Addition of Pyrazoles to Alkynes: Selective Synthesis of Double and Single Addition Products. *Eur. J. Org. Chem.* **2008**, *2008*, 4035–4040. [CrossRef]
56. Das, U.K.; Mandal, S.; Anoop, A.; Bhattacharjee, M. Structure–selectivity relationship in ruthenium-catalyzed regio- and stereoselective addition of alkynes to pyrazoles: An experimental and theoretical investigation. *J. Org. Chem.* **2014**, *79*, 9979–9991. [CrossRef]
57. Ignatenko, O.A.; Blandov, A.N.; Kuznetsov, M.A. Oxidative addition of N-Aminophthalimide to alkenyl-4,5-dihydropyrazoles and alkenylpyrazoles. Synthesis of aziridinylpyrazoles. *Russ. J. Org. Chem.* **2005**, *41*, 1793–1801. [CrossRef]
58. Pathak, V.N.; Joshi, R.; Sharma, J.; Gupta, N.; Rao, V.M. Mild and ecofriendly tandem synthesis, and spectral and antimicrobial studies of N_1-acetyl-5-aryl-3-(substituted styryl)pyrazolines. *Phosphorus Sulfur Silicon* **2009**, *184*, 1854–1865. [CrossRef]
59. Lakshmi, K.N.V.C.; Venkatesh, M.; Hareesh, J.; Raju, B.S.; Anupama, B. Synthesis, in silico and in vitro evaluation of some novel 5-[2-phenyl vinyl]-pyrazole and pyrazoline derivatives. *Pharma Chem.* **2016**, *8*, 122–128.
60. Daneshfar, Z.; Rostami, A. Cellulose sulfonic acid as a green, efficient, and reusable catalyst for Nazarov cyclization of unactivated dienones and pyrazoline synthesis. *RSC Adv.* **2015**, *5*, 104695–104707. [CrossRef]
61. Jenkins, R.N.; Hinnant, G.M.; Bean, A.C.; Stephens, C.E. Aerobic aromatization of 1,3,5-triarylpyrazolines in open air and without catalyst. *Arkivoc* **2015**, *7*, 347–353. [CrossRef]
62. Huang, Y.R.; Katzenellenbogen, J.A. Regioselective synthesis of 1,3,5-triaryl-4-alkylpyrazoles: Novel ligands for the estrogen receptor. *Org. Lett.* **2000**, *2*, 2833–2836. [CrossRef] [PubMed]
63. Huisgen, R.; Seidel, M.; Wallbillich, G.; Knupfer, H. Diphenyl-nitrilimin und seine 1,3-dipolaren additionen an alkene und alkine. *Tetrahedron* **1962**, *17*, 3–29. [CrossRef]
64. Walker, D.; Hiebert, J.D. 2,3-Dichloro-5,6-dicyanobenzoquinone and its reactions. *Chem. Rev.* **1967**, *67*, 153–195. [CrossRef]
65. Ananthnag, G.S.; Adhikari, A.; Balakrishna, M.S. Iron-catalyzed aerobic oxidative aromatization of 1,3,5-trisubstituted pyrazolines. *Catal. Commun.* **2014**, *43*, 240–243. [CrossRef]
66. Flores, D.C.; Fiss, G.F.; Wbatuba, L.S.; Martins, M.A.P.; Burrow, R.A.; Flores, A.F.C. Synthesis of new 2-(5-aryl-3-styryl-4,5-dihydro-1H-pyrazol-1-yl)-4-(trifluoromethyl)pyrimidines. *Synthesis* **2006**, *14*, 2349–2356. [CrossRef]
67. Flores, A.F.C.; Flores, D.C.; Vicentia, J.R.M.; Campos, P.T. 2-[5-(2-Methylphenyl)-3-(2-methylstyryl)-4,5-dihydro-1H-pyrazol-1-yl]-6-(thiophen-2-yl)-4-(trifluoromethyl)-pyrimidine chloroform monosolvate. *Acta Cryst.* **2014**, *70*, 789–790. [CrossRef]

68. Purkayastha, M.L.; Patro, B.; Ila, H.; Junjappa, H. Regioselective synthesis of 3 (or 5)-styryl-/(4-aryl-1,3-butadienyl)-/(6-aryl-1,3,5-hexatrienyl)-5 (or 3)-(methylthio)isoxazoles and pyrazoles via α-oxoketene dithioacetals. *J. Heterocycl. Chem.* **1991**, *28*, 1341–1349. [CrossRef]
69. Pastine, S.J.; Kelly, W., Jr.; Bear, J.J.; Templeton III, J.N.; Beam, C.F. The preparation of 3-(2-phenylethenyl)-1*H*-pyrazoles from dilithiated (3*E*)-4-phenyl-3-buten-2-one hydrazones. *Synth. Commun.* **2001**, *31*, 539–548. [CrossRef]
70. Kuz'menok, N.M.; Koval'chuk, T.A.; Zvonok, A.M. Synthesis of 5-hydroxy- and 5-amino-1-tosyl-5-phenyl-3-(2-arylvinyl)-4,5-dihydropyrazoles. *Synlett* **2005**, *2005*, 485–486. [CrossRef]
71. Miller, R.D.; Reiser, O. The synthesis of electron donor-acceptor substituted pyrazoles. *J. Heterocycl. Chem.* **1993**, *30*, 755. [CrossRef]
72. Venkateswarlu, V.; Kour, J.; Kumar, K.A.A.; Verma, P.K.; Reddy, G.L.; Hussain, Y.; Tabassum, A.; Balgotra, S.; Gupta, S.; Hudwekar, A.D.; et al. Direct *N*-heterocyclization of hydrazines to access styrylated pyrazoles: Synthesis of 1,3,5-trisubstituted pyrazoles and dihydropyrazoles. *RSC Adv.* **2018**, *8*, 26523–26527. [CrossRef]
73. Hu, J.; Chen, S.; Sun, Y.; Yang, J.; Rao, Y. Synthesis of tri- and tetrasubstituted pyrazoles via Ru(II) catalysis: Intramolecular aerobic oxidative C-N coupling. *Org. Lett.* **2012**, *14*, 5030–5033. [CrossRef] [PubMed]
74. Li, X.; He, L.; Chen, H.; Wu, W.; Jiang, H. Copper-catalyzed aerobic C(sp^2)–H functionalization for C–N bond formation: Synthesis of pyrazoles and indazoles. *J. Org. Chem.* **2013**, *78*, 3636–3646. [CrossRef]
75. Deshayes, C.; Chabannet, M.; Gelin, S. Synthesis of some 1-substituted-3-or-5-styrylpyrazoles. *J. Heterocycl. Chem.* **1981**, *18*, 1057–1059. [CrossRef]
76. Jedinák, L.; Zátopková, R.; Zemánková, H.; Šustková, A.; Cankař, P. The Suzuki-Miyaura cross-coupling reaction of halogenated aminopyrazoles: Method development, scope, and mechanism of dehalogenation side reaction. *J. Org. Chem.* **2017**, *82*, 157–169. [CrossRef]
77. Tomanová, M.; Jedinák, L.; Košař, J.; Kvapil, L.; Hradil, P.; Cankař, P. Synthesis of 4-substituted pyrazole-3,5-diamines via Suzuki–Miyaura coupling and iron-catalyzed reduction. *Org. Biomol. Chem.* **2017**, *15*, 10200–10211. [CrossRef]
78. Kim, S.H.; Lim, J.W.; Yu, J.; Kim, J.N. Regioselective synthesis of 1,3,4,5-tetrasubstituted pyrazoles from α-alkenyl-α,β-enones derived from Morita-Baylis-Hillman adducts. *Bull. Korean Chem. Soc.* **2013**, *34*, 2915–2920. [CrossRef]
79. Pinto, D.C.G.A.; Silva, A.M.S.; Lévai, A.; Cavaleiro, J.A.S.; Patonay, T.; Elguero, J. Synthesis of 3-benzoyl-4-styryl-2-pyrazolines and their oxidation to the corresponding pyrazoles. *Eur. J. Org. Chem.* **2000**, *2000*, 2593–2599. [CrossRef]
80. Silva, V.L.M.; Silva, A.M.S.; Pinto, D.C.G.A.; Cavaleiro, J.A.S.; Elguero, J. Synthesis of (*E*)- and (*Z*)-3(5)-(2-hydroxyphenyl)-4-styrylpyrazoles. *Monatsh. Chem.* **2009**, *140*, 87–95. [CrossRef]
81. Silva, V.L.M.; Silva, A.M.S.; Pinto, D.C.G.A.; Cavaleiro, J.A.S.; Elguero, J. Novel (*E*)- and (*Z*)-3(5)-(2-hydroxyphenyl)-4-styrylpyrazoles from (*E*)- and (*Z*)-3-styrylchromones: The unexpected case of (*E*)-3(5)-(2-hydroxyphenyl)-4-(4-nitrostyryl)pyrazoles. *Tetrahedron Lett.* **2007**, *48*, 3859–3862. [CrossRef]
82. Ahamad, S.; Gupta, A.K.; Kantc, R.; Mohanan, K. Domino reaction involving the Bestmann–Ohira reagent and α,β-unsaturated aldehydes: Efficient synthesis of functionalized pyrazoles. *Org. Biomol. Chem.* **2015**, *13*, 1492–1499. [CrossRef] [PubMed]
83. Soliman, R.; Mokhtar, H.; El Ashry, E.S.H. The scope of the reactions with hydrazines and hydrazones. Part 4. Trisubstituted pyrazoles of possible hypoglycemic and antibacterial activity. *Pharmazie* **1978**, *33*, 184–185. [CrossRef] [PubMed]
84. Silva, V.L.M.; Silva, A.M.S.; Pinto, D.C.G.A.; Cavaleiro, J.A.S.; Elguero, J. 3(5)-(2-Hydroxyphenyl)-5(3)-styrylpyrazoles: Synthesis and Diels-Alder transformations. *Eur. J. Org. Chem.* **2004**, *2004*, 4348–4356. [CrossRef]
85. Deshayes, C.; Chabannet, M.; Gelin, S. Synthesis of some 5-aryl-4,5-dihydropyrazolo[1,5-*a*]quinolines and 5-aryl-4,5-dihydro-10*H*-pyrazolo[1,5-*b*][2]benzazepines from 1-phenyl- or 1-benzyl-5-styrylpyrazoles. *Synthesis* **1982**, *12*, 1088–1089. [CrossRef]
86. Yu, J.; Kim, K.H.; Moon, H.R.; Kim, J.N. Facile one-pot synthesis of 1,3,5-trisubstituted pyrazoles from α,β-enones. *Bull. Korean Chem. Soc.* **2014**, *35*, 1692–1696. [CrossRef]
87. Moon, H.R.; Yu, J.; Kim, K.H.; Kim, J.N. Synthesis of pyrazolo[1,5-*a*]quinolines from 1-aryl-5-styrylpyrazoles via intramolecular Friedel-Crafts reaction/aerobic oxidation. *Bull. Korean Chem. Soc.* **2015**, *36*, 1189–1195. [CrossRef]

88. Pinto, D.C.G.A.; Silva, A.M.S.; Cavaleiro, J.A.S.; Foces-Foces, C.; Llmas-Saiz, A.L.; Jagerovic, N.; Elguero, J. Synthesis and molecular structure of 3-(2-benzyloxy-6-hydroxyphenyl)-5-styrylpyrazoles. Reaction of 2-styrylchromones with hydrazine hydrate. *Tetrahedron* **1999**, *55*, 10187–10200. [CrossRef]
89. Pinto, D.C.G.A.; Silva, A.M.S.; Cavaleiro, J.A.S. Novel (E)-3-(2′-benzyloxy-6′-hydroxyphenyl)-5-styrylpyrazoles from (E)-2-styrylchromones. *J. Heterocycl. Chem.* **1997**, *3*, 433–436. [CrossRef]
90. Sherin, D.R.; Rajasekharan, K.N. Studies on the antioxidant activities of mechanochemically synthesized 1-aryl-3,5-bis(styryl)pyrazoles as curcuminoids derived CNB 001 analogs. *Indian J. Heterocycl. Chem.* **2016**, *26*, 81–86.
91. Silva, V.L.M.; Silva, A.M.S.; Pinto, D.C.G.A.; Cavaleiro, J.A.S. Efficient microwave-assisted synthesis of tetrahydroindazoles and their oxidation to indazoles. *Synlett* **2006**, *9*, 1369–1373. [CrossRef]
92. Silva, V.L.M.; Silva, A.M.S.; Pinto, D.C.G.A.; Elguero, J.; Cavaleiro, J.A.S. Synthesis of new 1H-indazoles through Diels-Alder transformations of 4-styrylpyrazoles under microwave irradiation conditions. *Eur. J. Org. Chem.* **2009**, 4468–4479. [CrossRef]

Publisher's Note: MDPI stays neutral with regard to jurisdictional claims in published maps and institutional affiliations.

© 2020 by the authors. Licensee MDPI, Basel, Switzerland. This article is an open access article distributed under the terms and conditions of the Creative Commons Attribution (CC BY) license (http://creativecommons.org/licenses/by/4.0/).

Review

Homocoupling Reactions of Azoles and Their Applications in Coordination Chemistry

Steffen B. Mogensen [1], Mercedes K. Taylor [2,*] and Ji-Woong Lee [1,*]

1 Department of Chemistry, University of Copenhagen, Universitetsparken 5, 2100 Copenhagen Ø, Denmark; lng656@alumni.ku.dk
2 Center for Integrated Nanotechnologies, Sandia National Laboratories, Albuquerque, NM 87185, USA
* Correspondence: mercedes.taylor@sandia.gov (M.K.T.); jiwoong.lee@chem.ku.dk (J.-W.L.); Tel.: +45-35333312 (J.-W.L.)

Academic Editors: Vera L. M. Silva and Artur M. S. Silva
Received: 30 November 2020; Accepted: 12 December 2020; Published: 15 December 2020

Abstract: Pyrazole, a member of the structural class of azoles, exhibits molecular properties of interest in pharmaceuticals and materials chemistry, owing to the two adjacent nitrogen atoms in the five-membered ring system. The weakly basic nitrogen atoms of deprotonated pyrazoles have been applied in coordination chemistry, particularly to access coordination polymers and metal-organic frameworks, and homocoupling reactions can in principle provide facile access to bipyrazole ligands. In this context, we summarize recent advances in homocoupling reactions of pyrazoles and other types of azoles (imidazoles, triazoles and tetrazoles) to highlight the utility of homocoupling reactions in synthesizing symmetric bi-heteroaryl systems compared with traditional synthesis. Metal-free reactions and transition-metal catalyzed homocoupling reactions are discussed with reaction mechanisms in detail.

Keywords: homocoupling; bipyrazole; transition-metal catalysts; metal–organic frameworks

1. Introduction

Heterocyclic compounds containing two or more nitrogen atoms display unique molecular properties, which can be extended to macroscale properties in materials chemistry [1]. Furthermore, symmetric, multiply-functionalized organic ligands are required to construct ordered materials, preferably in high crystallinity [2]. Homocoupling reactions of aromatic compounds are often utilized to form carbon–carbon bonds in the presence of (over)stoichiometric or catalytic amounts of transition metal species [3]. For example, Ullmann coupling reactions can enable facile access to symmetric biaryl compounds using copper reagents with or without ligands at high reaction temperatures [4,5]. Based on this method, a plethora of reports was dedicated to the homocoupling of aromatic compounds, accessing symmetric molecules starting from simple aryl halides and derivatives. However, the reported procedures are often limited in terms of substrate scopes, particularly with heterocyclic compounds. Therefore, milder and more practical reaction conditions are required. To this end, this review summarizes recent development in (catalytic)homocoupling reactions of 5-membererd ring heterocycles, particularly focusing on pyrazoles due to their importance in materials chemistry as symmetric bipyrazoles. First, we will discuss traditional syntheses of bipyrazoles without homocoupling reaction methods, to highlight the advantages of (catalytic)homocoupling reactions. The following sub-chapters are organized by metal-catalyzed and metal-free reactions to illustrate the developed synthetic methods. The utility of bipyrazole materials in materials chemistry is discussed in Section 3.3, where we provide ample information on constructing metal–organic frameworks based on bipyrazole ligands.

2. Traditional Syntheses of Bipyrazoles

Condensation reactions of 1,4-dicarbonyl compounds with heteroatom nucleophiles can provide various 5-membered heterocycles: furans, thiophenes, pyrroles and their derivatives (Scheme 1A) [6,7]. To access pyrazoles with the same manner, 1,3-dicarbonyl compounds and hydrazine are required (Scheme 1B). However, the lack of a general method of "dimerization" of pyrazole substrates was a major bottleneck for the synthesis of symmetric bipyrazoles. Alternatively, diamino substrates can be prepared to access pyrazoles via oxidation reactions [8]. In 1964, Trofimenko reported the synthesis of 1,1,2,2-ethanetetracarbaldeyde, which was converted to bipyrazole by a condensation reaction with hydrazine (Scheme 1C) [9]. There are a few synthesis methods to access bipyrazoles via lengthy synthetic steps starting from diethyl succinate via a 3,4-carbonyl furan intermediate (**1**) [9–11].

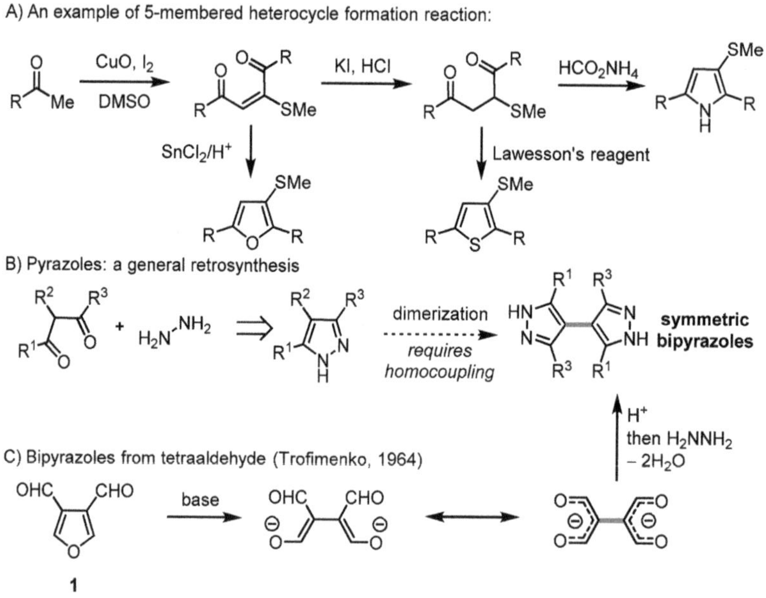

Scheme 1. A comparison of syntheses of 5-membered heterocyclic compounds and pyrazoles.

Starting from succinic ester, a repeated formylation and acetal protection sequence afforded the 3,4-dicarboxylic furan intermediate **2** after treating with concentrated sulfuric acid in 66–68% yield (Scheme 2A). The reaction employed sodium metal and orthoformate for formylation reactions. The author reported that the reaction was amenable to characterize bipyrazole via acetylation, carbamoylation, bromination, dihydrochloride and dinitrate salt. The preparation of 3,4-dicarbonylated furan was further subjected to formylation reactions, and subsequent condensation reactions with hydrazine afforded bipyrazole. For example, the synthesis of furan, thiophene and pyrrole-3,4-dicarboxylic esters was reported, which can allow one to access an intermediate **3** for bipyrazole synthesis [10].

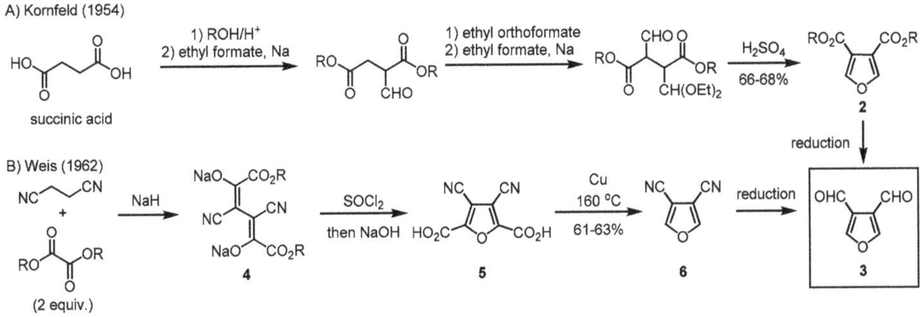

Scheme 2. Syntheses of 3,4-dicarbaldehyde furan **3**.

This procedure started with butyronitril undergoing a Claisen condensation reaction to afford sodium or potassium enolates **4**, which were then treated with SOCl$_2$ to afford 3,4-dicyano-2,5-furandicarboxylate **5**. This compound was further subjected to decarboxylation reaction conditions with copper powder and heated to 160 °C, yielding 61–63% of the product (**6**) [12]. Using this dicyanofuran (**6**), Weis pursued reduction using DIBAL-H to access the dialdehyde compound (**3**). The author found that the compound is stable under acidic conditions; however, it was converted to the ring-opened form under basic conditions, resulting in the formation of the anionic form of tetraaldehyde. This procedure represents a classic synthesis of bipyrazole using a condensation reaction, but Domasevitch and co-workers reported that the procedure was not amenable for the synthesis of bipyrazole [13]. The alternative synthetic approach was started from a dialdehyde, acetylene and formaldehyde to generate an intermediate bis-trimethinium salt (Figure 1) [14]. The synthesis consists of a similar approach as the above-mentioned pathway: a formylation reaction to access tetraaldehyde compounds, which are masked as enamine and iminium functionalities. This sequence requires the use of phosgene and additional steps to finally synthesize bipyrazole.

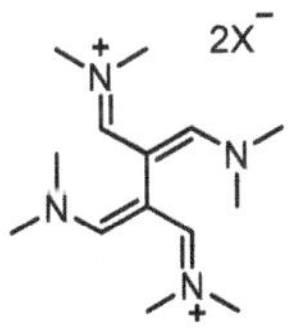

Figure 1. bis-trimethinium salt.

Despite the high application potential of symmetric nitrogen-containing compounds in materials chemistry and coordination chemistry as tunable ligands, the synthetic procedures have been limited to traditional condensation reactions with low atom economy and redox efficiency. Not only are double condensation reactions difficult to control, but the lack of 1,3-dicarbonyl compounds has also hampered the development of facile preparation of pyrazole-based compounds. Therefore, many catalytic systems based on transition-metal catalysts have been investigated owing to the development of cross-coupling reactions, which will be discussed in this review.

3. Recent Advancement of Bipyrazole Syntheses

3.1. Transition Metal-Catalyzed Homocoupling of Azoles

Transition metal catalysts play an important role in the functionalization of C–H bonds of aromatic compounds and heteroaromatics and in the formation of new C–C bonds in the synthesis of homocoupled biaryl compounds [3]. The development in homocoupling reactions of pyrazoles, imidazoles and triazoles showed a broad range of catalytic activities in transition metals, including Cu,

Fe, Ni, Pd, Rh and Ru, which were summarized in a review in 2012 [15]. This review will therefore summarize more recent examples (2010–2020) of homocoupling reactions of pyrazoles and azoles (imidazoles, triazoles, tetrazoles) and related compounds.

3.1.1. Cu-Catalyzed Homocoupling of Various Azoles

The (oxidative) Glaser–Hay coupling reaction utilizes Cu(I) species as catalysts in the presence of a base under air to afford homo-coupled alkyne products [16]. An adaptation of the Glaser–Hay coupling reaction conditions to *N*-substituted 1,2,4-triazoles was realized starting from easily accessible triazole compounds. In 2009, Do and Daugulis developed Cu(II)-catalyzed homocoupling reactions of imidazoles using molecular oxygen as a terminal oxidant (Scheme 3) [17]. The method developed proved useful in the coupling of various heterocycles, including examples of a *N*-butyl-substituted imidazole and triazole, providing the desired products in up to 73% yield. The mechanism of the reaction was suggested to proceed via deprotonation of heteroarenes (pK_a < 35–37 for C2-H), followed by the complexation with metal additives, such as Li^+. The *trans*-metalation step with the copper catalyst and subsequent oxidation yields the intermolecular homocoupling products between two arenes (Scheme 3).

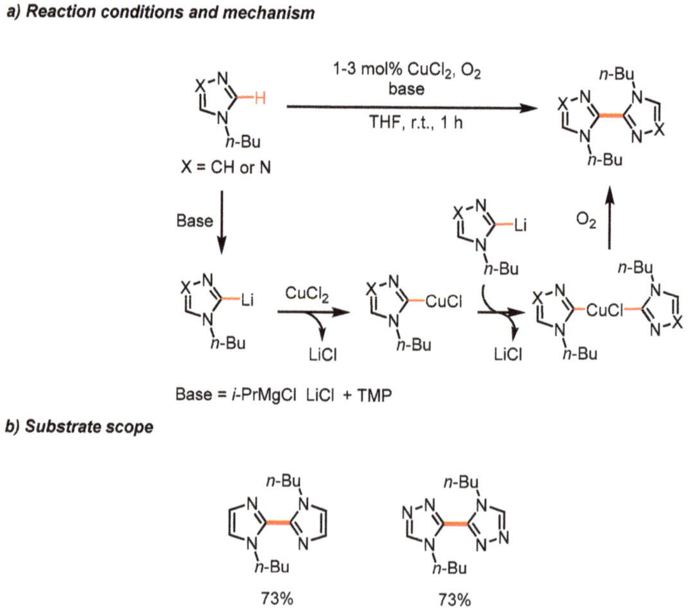

Scheme 3. (a) Glaser–Hay type copper catalyzed homocoupling of various azoles resulting in homocoupling using molecular oxygen as a terminal oxidant. (b) Substrate scope for the copper catalyzed homocoupling.

Li et al. reported Cu(II)-mediated homocoupling reactions using a similar system at elevated temperatures (Scheme 4) [18]. This method utilizes an efficient and convenient approach to form C2-C2 bonds via a Cu(OAc)$_2$ and O$_2$-mediated oxidative intermolecular homocoupling reaction between two azoles. Once again, the oxidative coupling could be used to achieve high yields of homocoupled products.

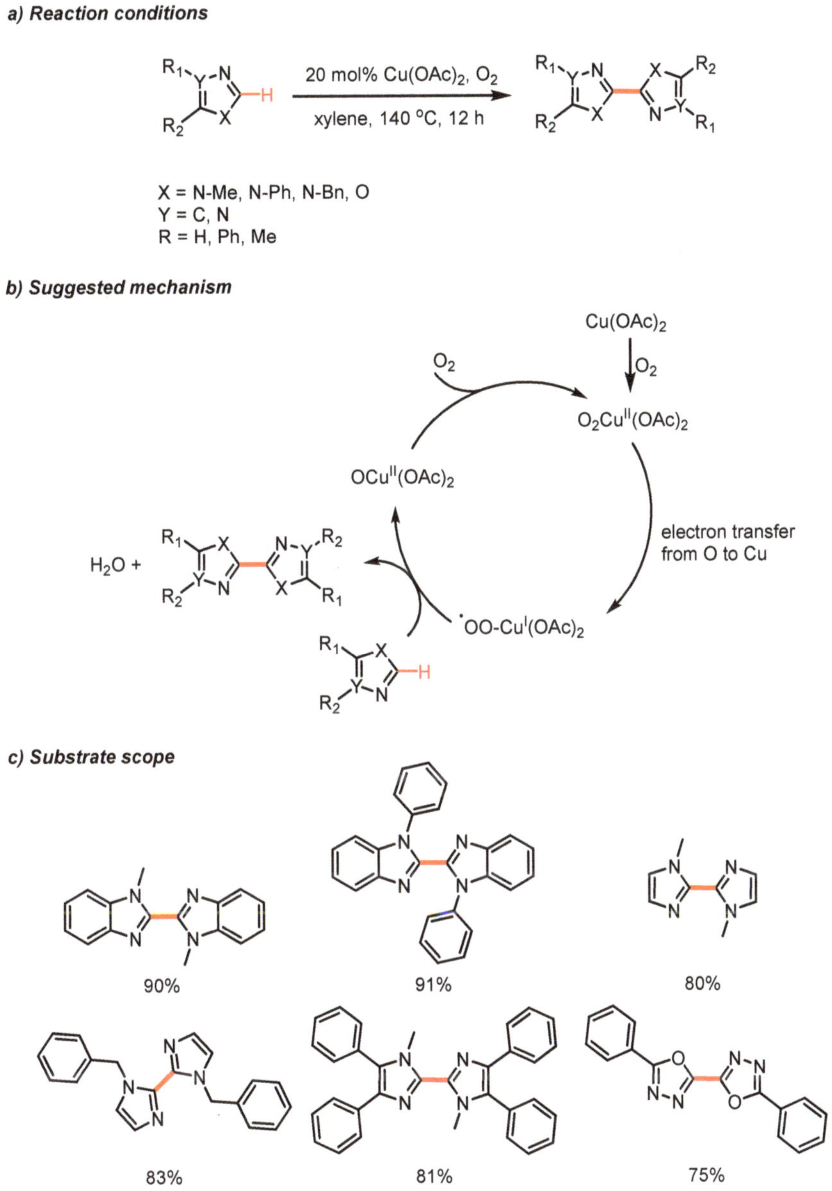

Scheme 4. Copper catalyzed homocoupling of various imidazoles using Cu[II] as the catalyst and molecular oxygen. (**a**) Reaction conditions. (**b**) Suggested mechanism for the homocoupling reaction. (**c**) Substrate scope.

3.1.2. Pd-Catalyzed Homocoupling Reactions of Pyrazoles and Triazoles

Pd-catalyzed cross-coupling reactions are currently classified as some of the most flexible and valuable tools in organic synthesis [19,20]. A common observation is that Pd-catalyzed coupling reactions provide low but significant amounts of homocoupled by-products during cross-coupling reactions [21]. This is likely because the main focus of Pd-catalyzed coupling reactions is to form bonds between two different coupling partners, enabling highly productive homocoupling reactions. A handful of papers [22,23] have used the tactical placement of heteroatoms for the coordination of metals in the activation of specific C–H bonds. Other papers [24,25] utilize the possibility of the oxidative addition of Pd into a carbon halogen bond to form di-azole Pd complexes, which can undergo reductive elimination to form desired intermolecular homocoupled products.

In 2012, the unexpected discovery of the intermolecular homocoupling between two pyrazoles was made by Salanouve et al. (Scheme 5) [22]. In an effort to study the reaction products obtained under different Suzuki–Miyaura conditions, this group discovered the formation of a C3–C3 intermolecular homocoupled product, resulting from C–H activation. The discovered bipyrazole compound was found to originate from a bis(3-ethoxy-1-(pyridin-2-yl)-1H-pyrazol-5-yl) palladium complex. Through oxidation with NCS, this species forms a Pd^{IV} complex, which then allows for reductive elimination of the homocoupled product in a yield of up to 34 % (Scheme 5b).

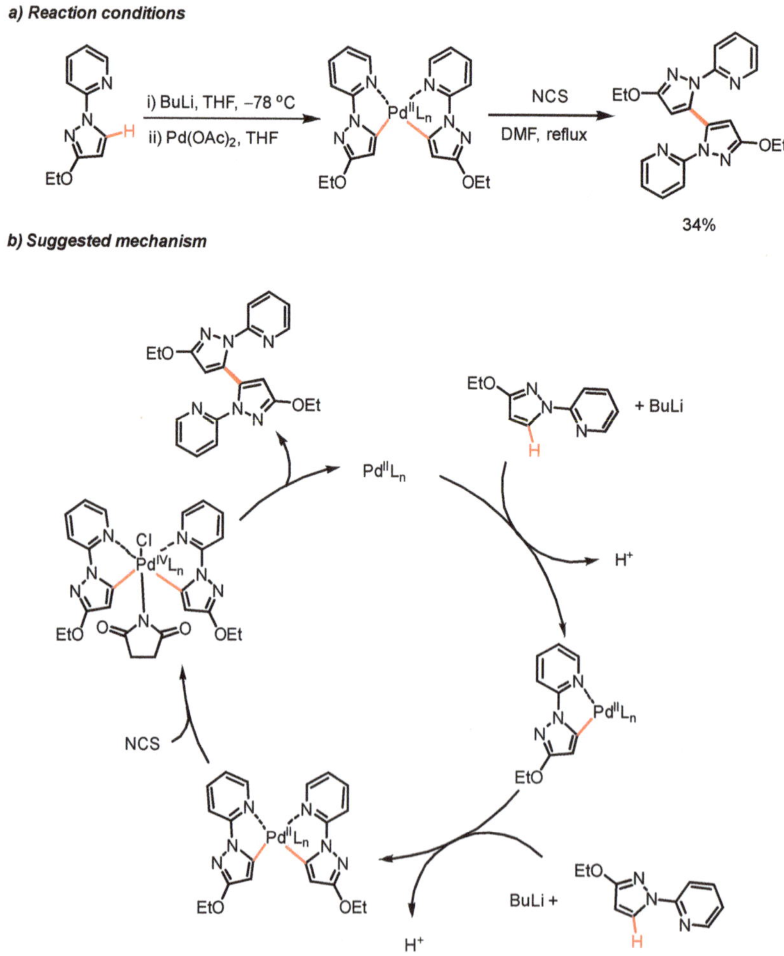

Scheme 5. (**a**) Reaction conditions for the homocoupling of pyrazoles through formation of a bis(3-ethoxy-1-(pyridin-2- yl)-1H-pyrazol-5-yl) palladium complex followed by oxidation by NCS. (**b**) Suggested mechanism for the C–H activation followed by homocoupling.

Afanas'ev et al. reported a method for the homocoupling of 4-bromo-1,2,3-triazoles to access the corresponding 4,4′-bistriazoles in quantitative yields [24]. The intermolecular homocoupling was achieved by mixing 4-brominated triazoles with 1.0 mol% Pd(OAc)$_2$, in the presence of additives (pinB)$_2$, SPhos and KOH. This method is practical due to the use of readily available starting compounds and a simple experimental procedure, without inert atmosphere or solvent. The reaction possibly occurs through the formation of a 4-Bpin-substituted 1,2,3-triazole. Then, in a Suzuki–Miyaura fashion, the homocoupling of the boronic acid can occur (Scheme 6b).

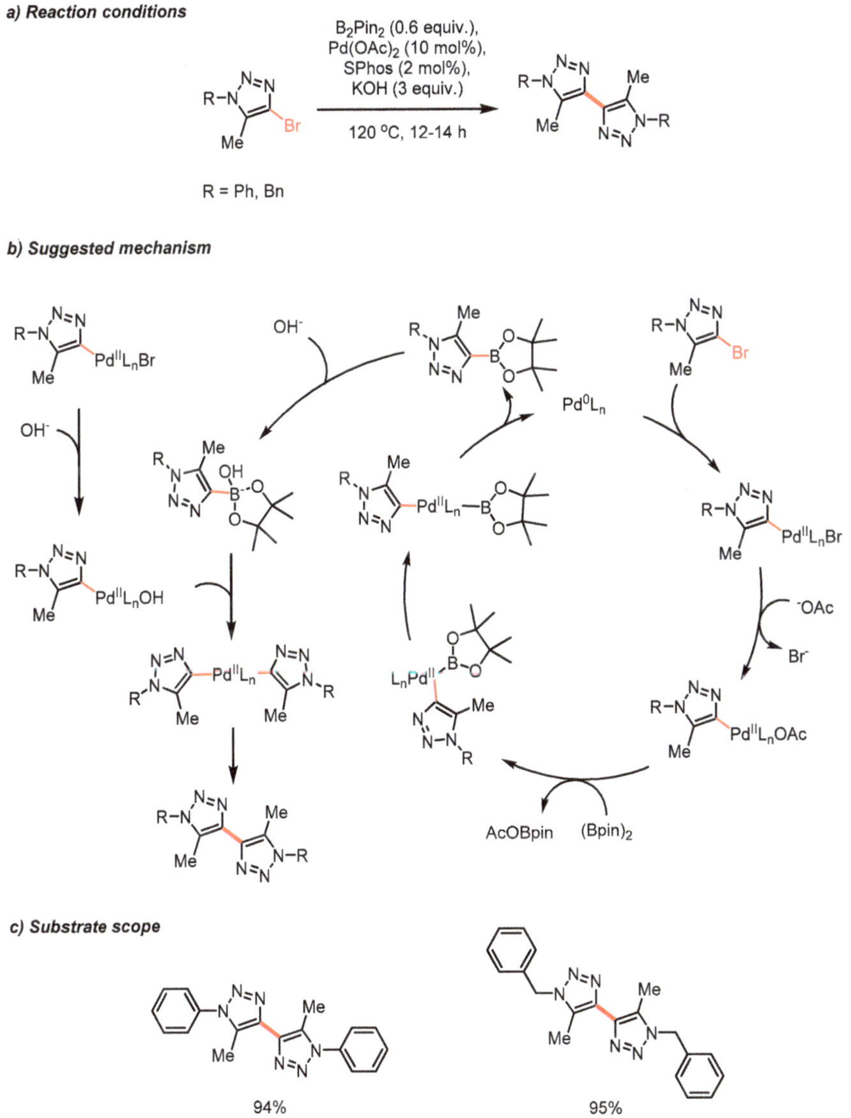

Scheme 6. (a) Conditions for the homocoupling of 4-bromo-1,2,3-triazoles using Suzuki–Miyaura reaction conditions (b) Our suggested mechanism for the homocoupling. (c) Substrate scope for the Suzuki–Miyaura homocoupling reaction.

A palladium-catalyzed, pyrazole-directed C(sp^2)–H functionalization of the *N*-phenyl ring in *N*-phenylpyrazoles to afford a biaryl bis-pyrazole (via dehydrogenative homocoupling) was reported by Batchu et al. [23]. Multiple factors were shown to have significant influences on the outcome of the reaction. For example, it was discovered that if the reaction was performed under ca. 10 times dilution, then no homocoupling product was observed, and *N*-(*o*-hydroxyphenyl)pyrazoles were the major or the sole products. The reaction mechanism is suggested in Scheme 7b. After the first C–H activation, the Pd^{II}-center is oxidized to Pd^{IV}, which undergoes a second C–H activation followed by a reductive elimination to yield the homocoupled bipyrazole.

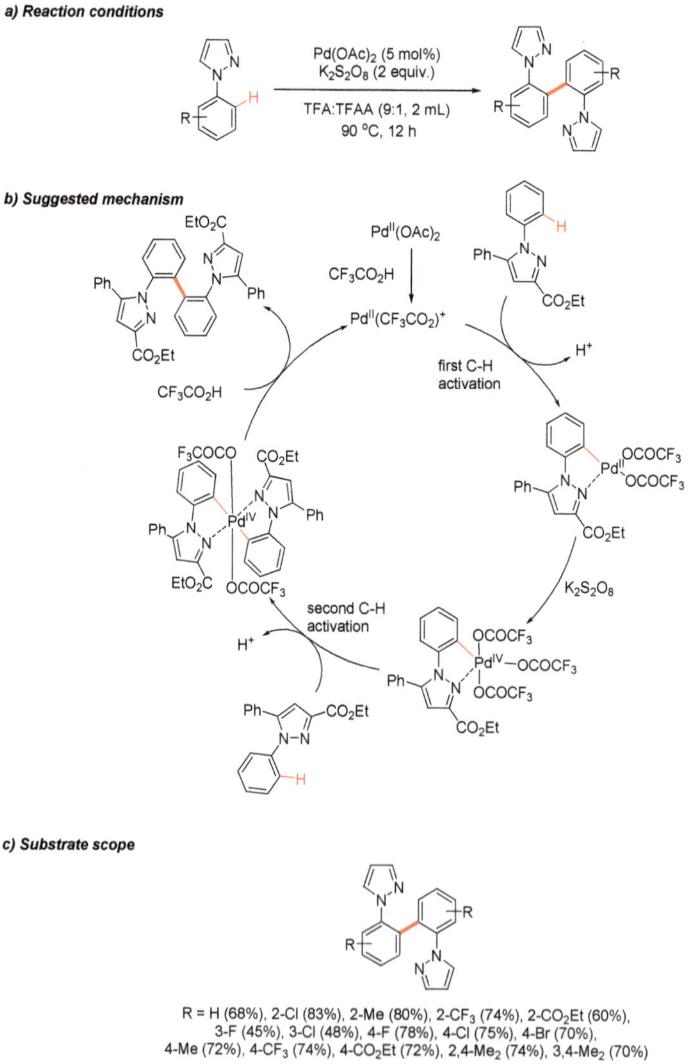

a) Reaction conditions

b) Suggested mechanism

c) Substrate scope

R = H (68%), 2-Cl (83%), 2-Me (80%), 2-CF$_3$ (74%), 2-CO$_2$Et (60%),
3-F (45%), 3-Cl (48%), 4-F (78%), 4-Cl (75%), 4-Br (70%),
4-Me (72%), 4-CF$_3$ (74%), 4-CO$_2$Et (72%), 2,4-Me$_2$ (74%), 3,4-Me$_2$ (70%)

Scheme 7. (a) Homocoupling of *N*-phenylpyrazoles through Pd^{II} catalyzed C–H activation of phenyl protons. (b) The mechanism for the homocoupling is suggested to go through C–H activation on the phenyl ring followed by formation of a Pd^{IV} specie and then subsequent C–H activation which allows for the reductive elimination of the homocoupled product (c) Substrate scope for the homocoupling of various *N*-phenylpyrazole.

The homocoupling reaction between 1,2,3-triazole N-oxides by C–H activation at the C4 position was explored by Zhu et al. (Scheme 8) [26]. The reaction was designed to couple 1,2,3-triazole N-oxides with sodium arenesulfinates via palladium-catalyzed desulfitative cross-coupling, but it showed the possibility of homocoupling between 1,2,3-triazole N-oxides. Under unoptimized reaction conditions, the homocoupled N-oxide bistriazoles were generated with yields of up to 22%. The reaction likely occurs via deprotonation followed by exchange of OAc⁻ for the deprotonated 1,2,3-triazole N-oxides. The di-triazole N-oxide Pd^{II} complex then undergoes reductive elimination to yield the homocoupled bistriazole N-oxide product and Pd^0, which can be oxidized by Ag$_2$CO$_3$.

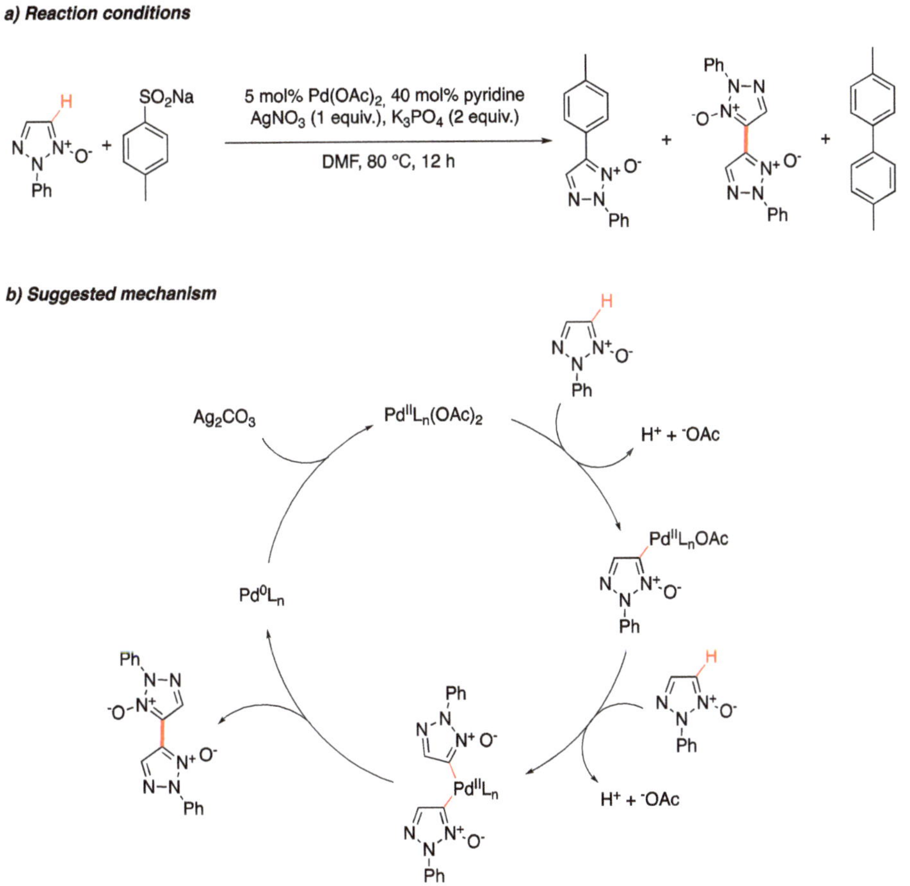

Scheme 8. (a) Reaction conditions for the homocoupling of 1,2,3-triazole N-oxides in the C4 position through C–H activations using Pd^{II} as the catalyst. (b) Suggested mechanism for the homocoupling of 1,2,3-triazole N-oxides in the C4 position.

Later, the full scope of the homocoupling reaction between pyrazole-N-oxides was investigated by Peng et al. (Scheme 9) [27]. The method showed high regioselectivity towards C–H homocoupling of 1,2,3-triazole N-oxides, using a Pd^{II} catalyst (5 mol%), Ag$_2$CO$_3$ and 1,10-phenanthroline under air. Following their previous work, it was found that the addition of the base t-BuOK and a catalytic amount of 1,10-phenanthroline allowed for yields of up to 93%.

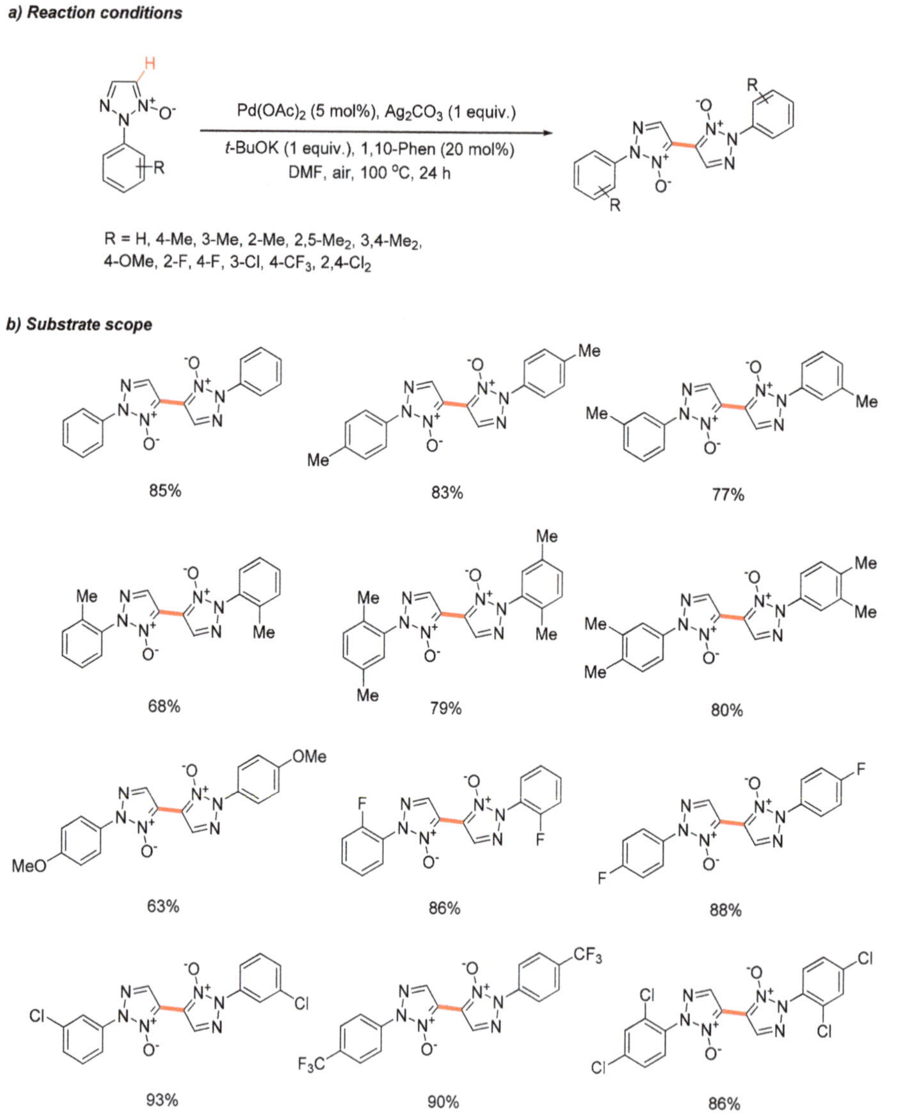
a) Reaction conditions
Pd(OAc)2 (5 mol%), Ag2CO3 (1 equiv.)
t-BuOK (1 equiv.), 1,10-Phen (20 mol%)
DMF, air, 100 °C, 24 h
R = H, 4-Me, 3-Me, 2-Me, 2,5-Me2, 3,4-Me2,
4-OMe, 2-F, 4-F, 3-Cl, 4-CF3, 2,4-Cl2
b) Substrate scope
85%
83%
77%
68%
79%
80%
63%
86%
88%
93%
90%
86%

Scheme 9. (**a**) Reaction conditions for the homocoupling of 1,2,3-triazole *N*-oxides in the C4 position through C–H activations using Pd^{II} as the catalyst. (**b**) Substrate scope of the homocoupling between 1,2,3-triazole *N*-oxides in the C4 position.

The homocoupling reaction between trisubstituted 4-iodo pyrazoles in the C4 position was investigated by Jansa et al. (Scheme 10) [25]. The pyrazoles contain two to three pyridinyl substituents, which are synthesized from the reaction of 1,3-dipyridinyl-1,3-propanediones with 2-hydrazinopyridine or phenylhydrazine, affording the corresponding 1,3,5-trisubstituted pyrazoles. Iodination at the 4-position of the pyrazoles was achieved by treatment with I_2/HIO_3. The coupling of two pyrazole rings was achieved under Negishi cross-coupling reaction conditions by employing organozinc halides in combination with a Pd^{II} catalyst.

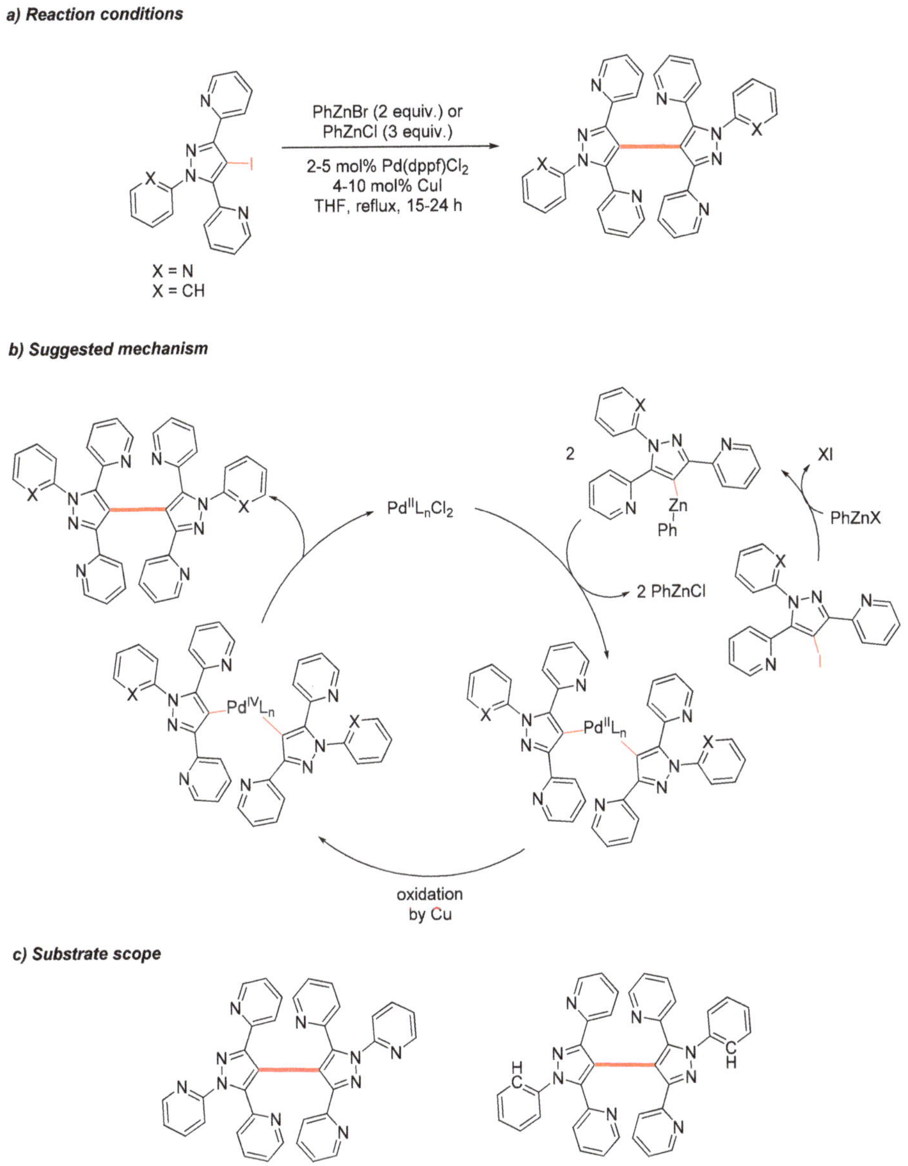

a) Reaction conditions

b) Suggested mechanism

c) Substrate scope

Scheme 10. (a) Negishi reaction conditions allows for the homocoupling of 4-iodopyrazoles in the C4 position. (b) Suggested reaction mechanism for the Negishi cross-coupling. (c) Substrate scope of the homocoupling of 4-iodopyrazoles.

A more recent example of intermolecular homocoupling between two pyrazoles was discovered by Mercedes et al. (Scheme 11): through the palladium-catalyzed homocoupling of pyrazole boronic esters in the presence of air and water, 4,4-bipyrazole (H_2bpz) and other symmetric bipyrazoles could be achieved [28]. The mechanism of the reaction is likely through a Suzuki–Miyaura-type cross-coupling between the boronic acids (Scheme 11b), with O_2 being the oxidant for the Pd^0 formed during the reductive elimination.

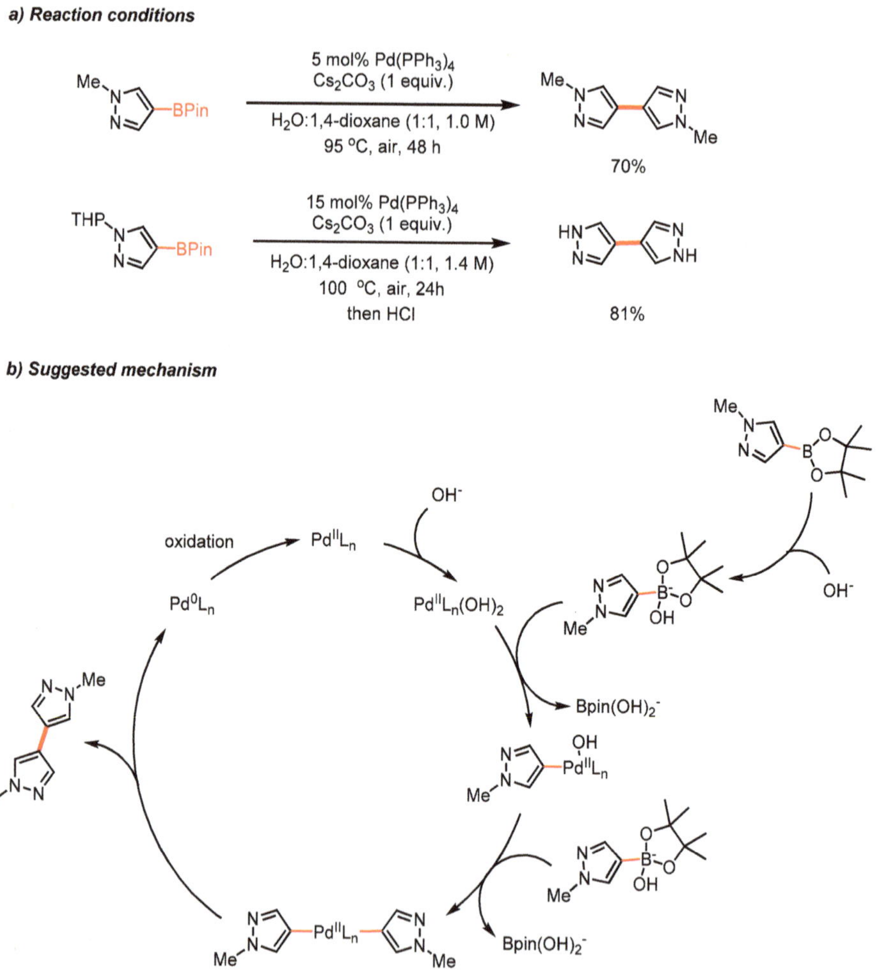

Scheme 11. (**a**) Symmetric bipyrazoles are achieved via the palladium-catalyzed homocoupling of a pyrazole boronic ester in the presence of air and water. (**b**) Suggested mechanism for the homocoupling.

3.1.3. Ru-Catalyzed Homocoupling Reactions

Ackermann et al. demonstrated the chemoselective ruthenium-catalyzed C(sp^2)–H bond arylations on triazol-4-yl and pyrazol-2-yl substituted arenes (Scheme 12) [29]. The authors discovered that *ortho*-substituted arenes favored oxidative intermolecular homocoupling reactions to afford dimerized products in high (or moderate) yields. The suggested reaction mechanism follows a double C–H bond activation with carboxylate assistance, followed by reductive elimination.

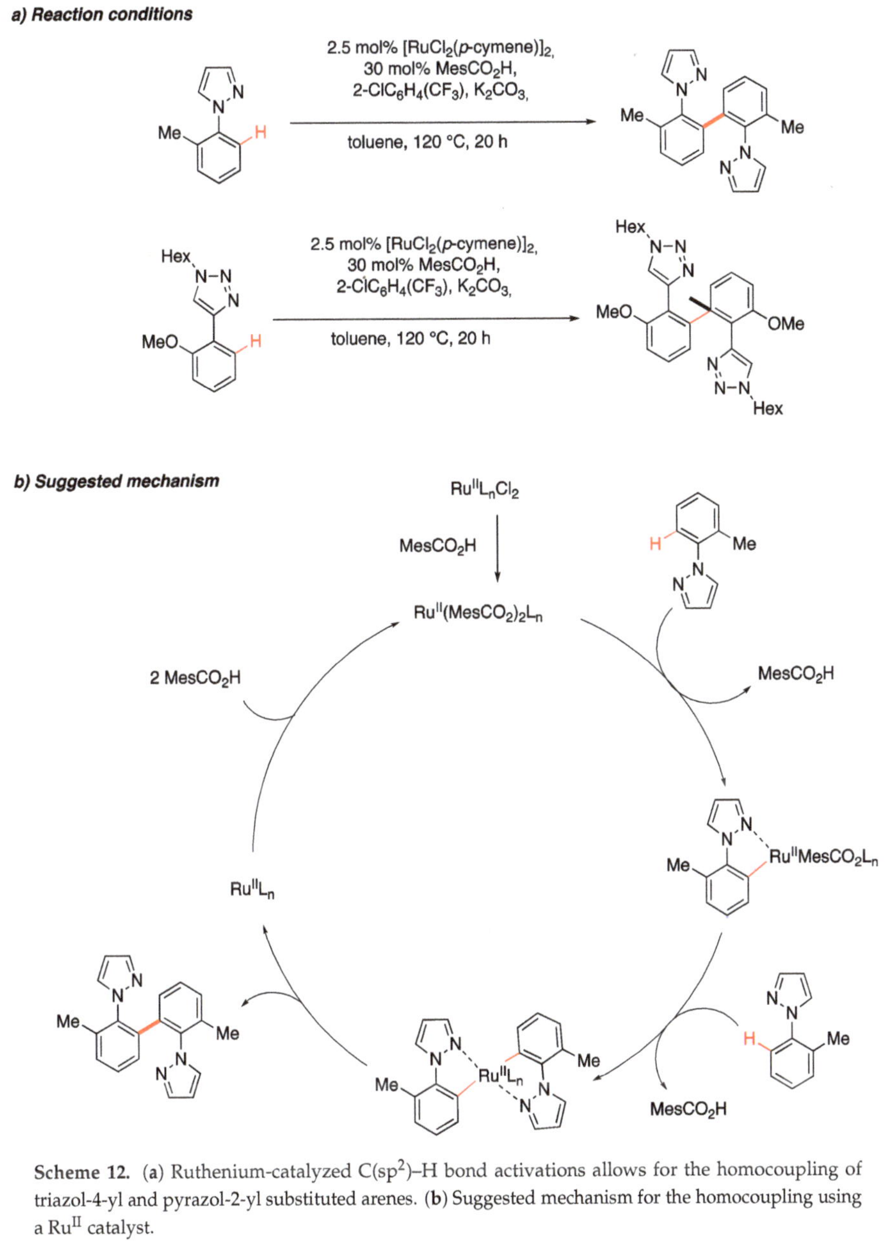

Scheme 12. (**a**) Ruthenium-catalyzed C(sp^2)–H bond activations allows for the homocoupling of triazol-4-yl and pyrazol-2-yl substituted arenes. (**b**) Suggested mechanism for the homocoupling using a Ru^{II} catalyst.

3.1.4. Ni-Catalyzed Reactions

Nickel is an Earth-abundant alternative to Pd and can access various reaction mechanisms, giving a distinct reactivity due to the facile oxidative addition reaction and readily accessible multiple oxidation states [30,31].

In 2000, Fanni et al. reported a new synthetic procedure for the efficient preparation of binuclear Ru(II) polypyridyl complexes, by performing intermolecular homocoupling between two 1,2,4-triazole

bromide in the C3 position (Scheme 13) [32]. The halide functional group is necessary to mediate the Ni-catalyzed reaction in the presence of triphenylphosphine and a zinc reducing reagent.

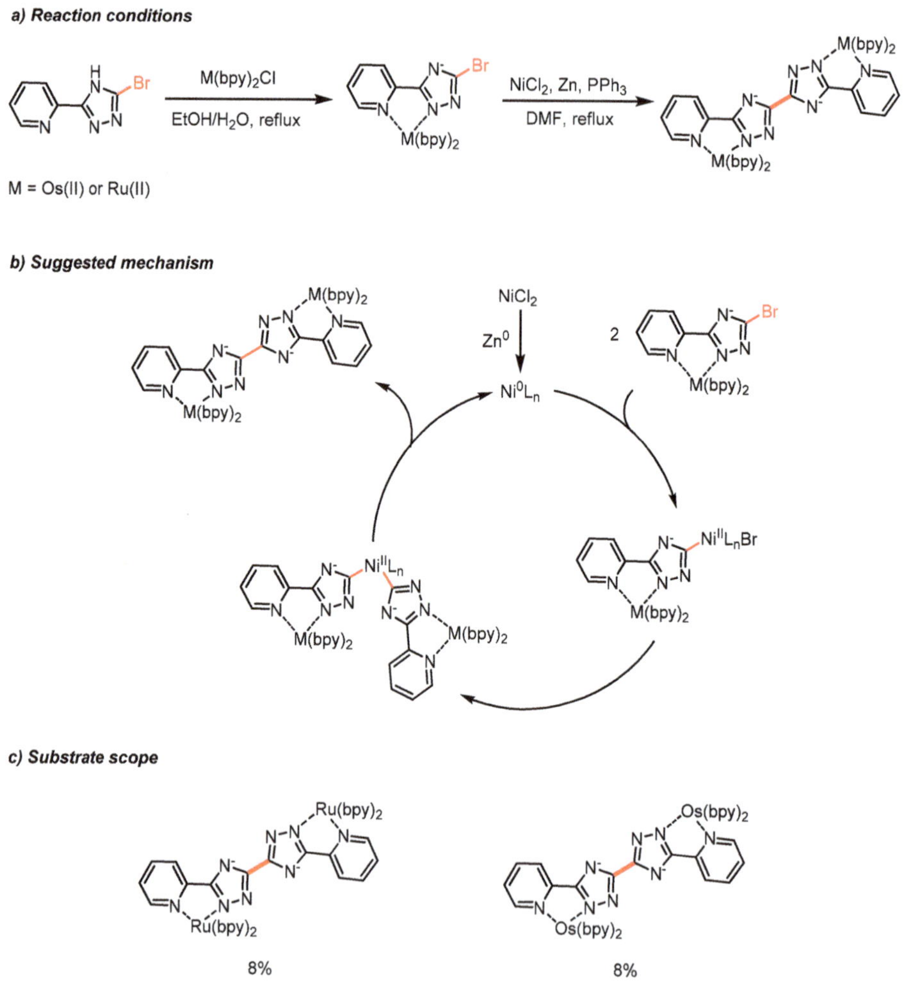

a) Reaction conditions

b) Suggested mechanism

c) Substrate scope

Scheme 13. (**a**) The formation of either ruthenium(II) or osmium(II) complex in water could undergo Ni^0 catalyzed homocoupling from in situ generated Ni^0. (**b**) The mechanism of the homocoupling is suggested to go through double oxidative addition of the 1,2,4-triazole bromides followed by reductive elimination resulting in the homocoupled product. (**c**) Substrate scope for the homocoupling of 1,2,4-triazole bromides in the C3 position.

The homocoupling between 4-bromotriazoles has been reported by Afanas'ev et al. using in situ generated Ni^0 as a catalyst (Scheme 14) [24]. The homocoupling functions via an in situ formation of Ni^0(PPh$_3$)$_4$ from NiCl$_2$ and Zn^0, which undergoes oxidative addition of two equivalents of 4-bromotriazoles, followed by a reductive elimination to form the desired homocoupled products (Scheme 14b).

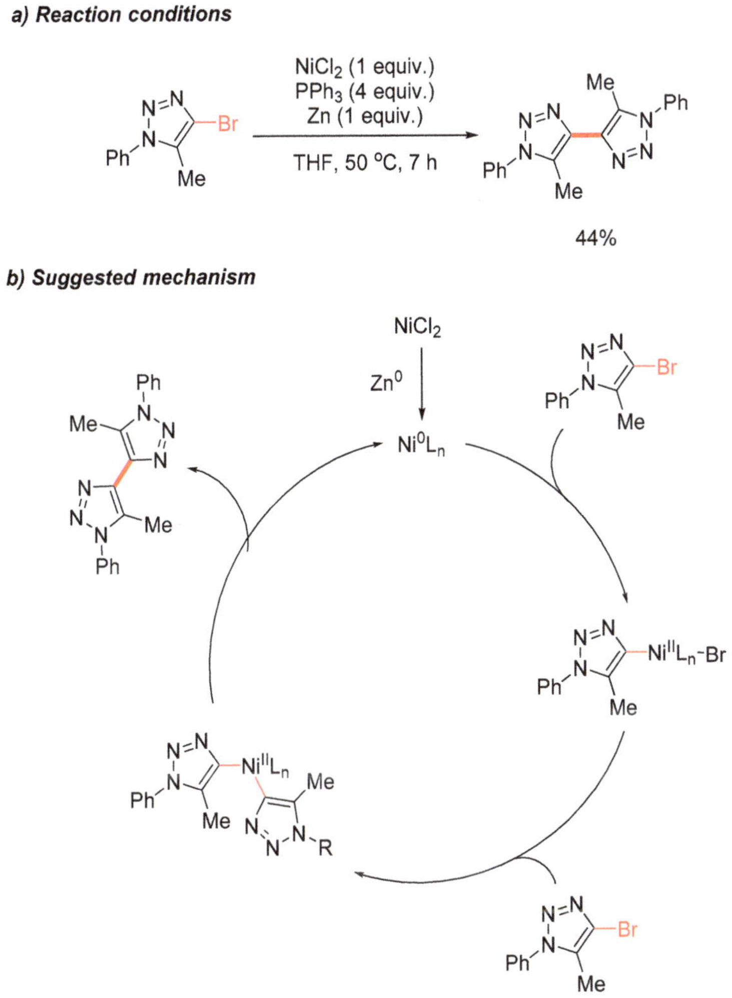

a) Reaction conditions

[Scheme showing 1-Ph-5-Me-4-Br-triazole + NiCl₂ (1 equiv.), PPh₃ (4 equiv.), Zn (1 equiv.), THF, 50 °C, 7 h → homocoupled bistriazole product, 44%]

b) Suggested mechanism

[Catalytic cycle: NiCl₂ + Zn⁰ → Ni⁰L$_n$; oxidative addition of aryl bromide gives Ni^{II}L$_n$–Br aryl intermediate; second oxidative addition gives bis-aryl Ni^{II}L$_n$ species; reductive elimination releases homocoupled product]

Scheme 14. (a) In situ generated Ni⁰ used as the catalyst allows for the homocoupling of 4-bromopyrazoles through oxidative addition. (b) The mechanism is suggested to go through double oxidative addition of the bromine followed by reductive elimination, resulting in the homocoupled products.

3.1.5. Rh-Catalyzed Homocoupling Reaction

Rhodium-catalyzed reactions have shown fruitful successes in organic reactions [33], particularly in C–H activation and C–C bond formation reactions [34,35]. heterocycle formation reactions [36,37], C–X bond formation reactions [38] and asymmetric synthesis [39].

Yue et al. explored the homocoupling of pyrazole silanes using Rh^I as a catalyst [40]. A suggested reaction mechanism includes a double *trans*-metalation reaction between vinylsilane and a rhodium intermediate in a single catalytic cycle (Scheme 15). While investigating an oxidant, the authors found out hexachloroacetone afforded higher yields of the desired product compared to conventional oxidants such as Cu^{II}, TEMPO and chloranil.

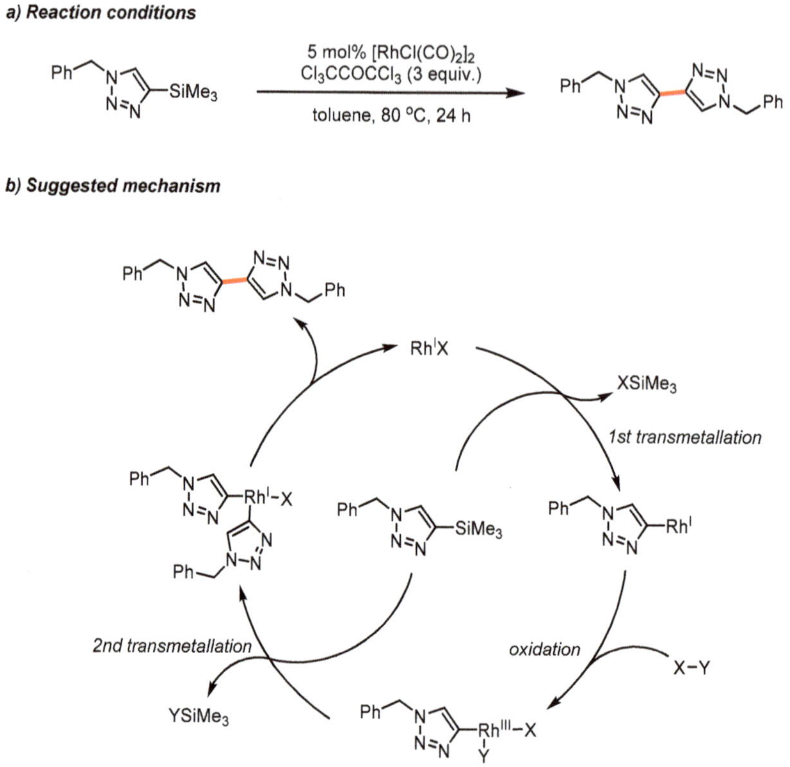

Scheme 15. (**a**) Homocoupling reactions of silane substituted triazoles with catalytic Rh^I. (**b**) Suggested mechanism for the homocoupling of silane substituted triazoles.

3.1.6. Fe-Catalyzed Homocoupling Reactions

Iron is one of the most Earth-abundant metals, and many iron complexes are commercially available. Surprisingly, despite its many advantages, iron has been relatively underrepresented in the catalysis of organic compounds until recently, compared to other transition metals.

In 2010, Luque et al. achieved the homocoupling of two pyrazole boronic acids utilizing nano-ferrite glutathione in a microwave (MW)-assisted aqueous reaction, providing an alternative to the palladium-catalyzed Suzuki coupling reaction (Scheme 16) [41].

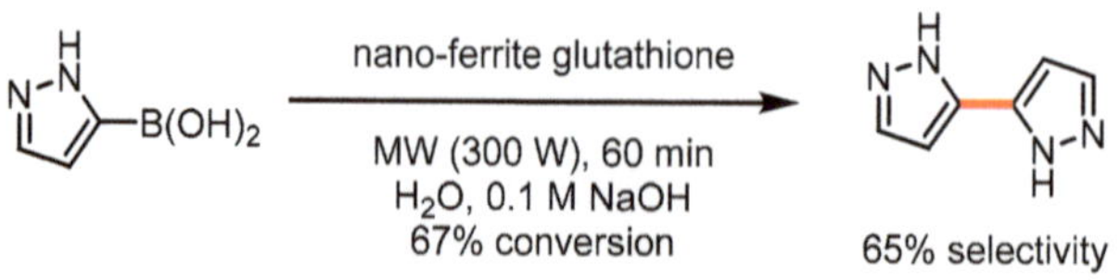

Scheme 16. The homocoupling of boronic acid substituted pyrazoles is possible when nanoparticles of iron are used as the catalyst in water under microwave conditions.

3.2. Metal-Free Homocouplings of Azoles

In addition to the many transition metal-catalyzed routes described above, azole substrates have also been shown to undergo homocoupling reactions in the absence of a metal catalyst. Hypervalent iodine reagents can induce aryl–aryl bond formation in a wide variety of substrates, including pyrroles, pyrazoles, imidazoles and triazoles, as has been recently reviewed [42]. Other approaches to metal-free couplings rely on oxidants or photocatalysis to enact radical-mediated dimerizations. Some recent examples are presented below.

There is precedent for the metal-free homocoupling of pyrazolones through the use of oxidants like phenoxy radicals [43] or O_2 [44]. A recent example extends this approach to a switchable aerobic oxidation, in which a pyrazol-5-one starting material undergoes either hydroxylation or a homocoupling, using O_2 as an oxidant (Scheme 17a) [45]. Simple changes in reaction conditions control the outcome of this reaction; oxidation in dioxane in the presence of base favors hydroxylation, whereas a base-free reaction in acetonitrile favors the homocoupling. The homocoupling reaction was successful for 19 different substrates containing various substituents at positions R^1 and R^2, in yields up to 77% (Scheme 17b). Most reactions led to a mixture of diastereomers, with diastereomeric ratios ranging from 3.1:1 to 5.7:1, although three substrates led to a single diastereomeric product. The authors hypothesize that the homocoupling reaction proceeds through a radical intermediate, although mechanistic studies were not undertaken for this reaction.

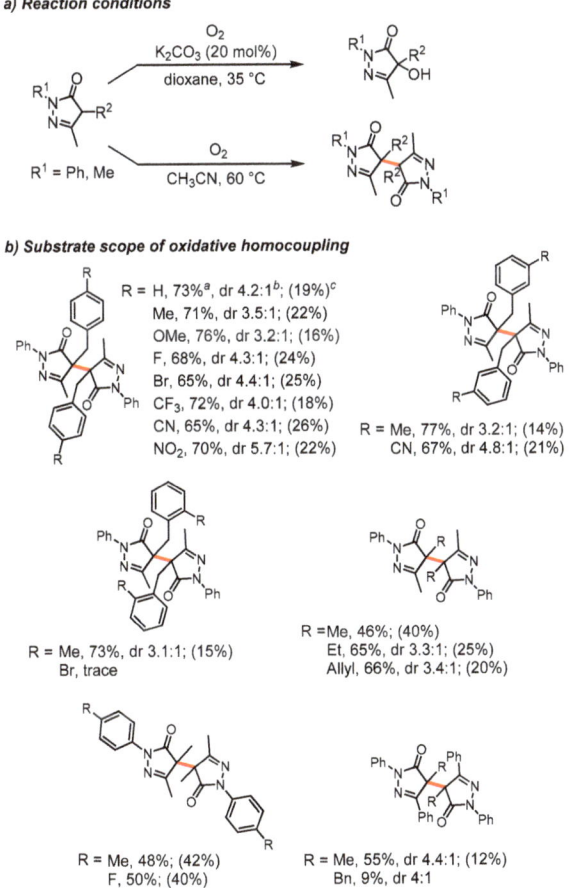

Scheme 17. (a) A switchable aerobic oxidation of a pyrazol-5-one results in either hydroxylation or homocoupling, depending on reaction conditions. (b) Substrate scope of the oxidative homocoupling. [29]. [a] Isolated yields. [b] The dr values were determined by ^{1}H-NMR. [c] Isolated yield of hydroxylated products.

Another homocoupling reaction was recently shown to yield a structurally similar bipyrazolone, but through remarkably different conditions: By refluxing a substituted pyrazole in excess thionyl chloride, the bipyrazolone was produced in high yields (Scheme 18a) [46]. This reaction was successful for nine substrates in yields from 62–89%. (The authors do not provide diastereomeric ratios of meso compounds and racemate for each substrate, but they report that dimeric pyrazolone (R^1 = Ph) appeared to be a racemic mixture by ^{1}H-NMR spectroscopy and X-ray crystallography). The reaction was shown to proceed under an atmosphere of N_2 and under air, ruling out oxidation by air as the mechanism, and the authors propose a hypothetical mechanism via a di(pyrazolyl) sulfite intermediate (Scheme 18b).

a) Reaction conditions and substrate scope

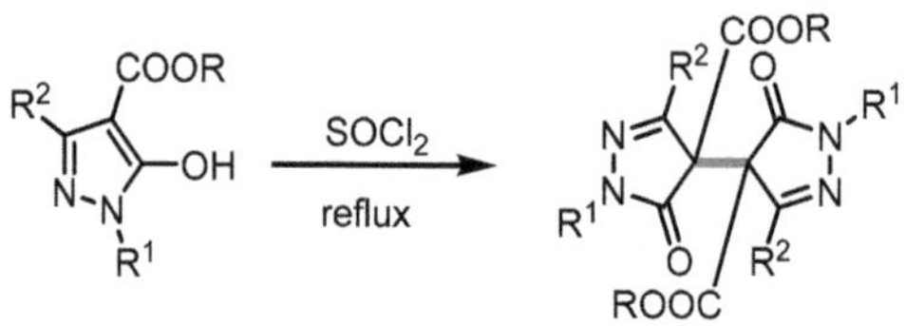

R^1	R^2	R	Yield
Ph	H	Et	74%
Ph	H	Me	89%
4-Br-C$_6$H$_4$	H	Et	72%
3,4-Cl$_2$-C$_6$H$_3$	H	Et	62%
me	H	Et	66%
t-Bu	H	Et	63%
PhCH$_2$	H	Et	84%
Ph	Me	Et	73%
Ph	Ph	Et	75%

b) Proposed mechanism

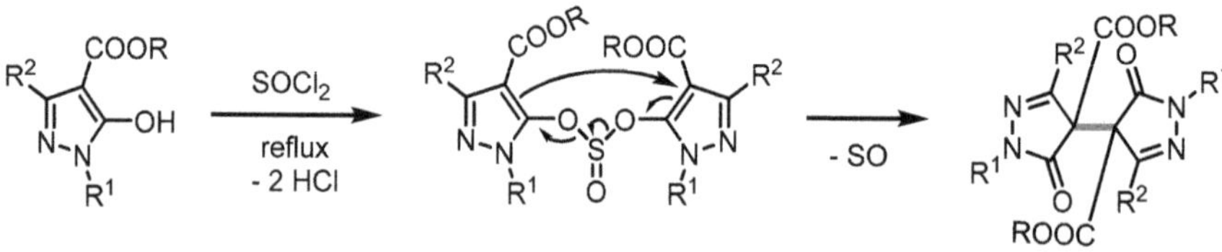

Scheme 18. (a) Reaction conditions and substrate scope for the thionyl chloride-mediated homocoupling of substituted pyrazolones. (b) The reaction is proposed to proceed via a di(pyrazolyl) sulfite intermediate.

Two recent reports describe the oxidative homocoupling of 4-amino-1,2,4-triazoles, yielding the corresponding symmetrical bis(triazoles). Yang and coworkers used the oxidant α-hydroxy-α-methylthioacetophenone to accomplish the homocoupling of five substituted 4-amino-1,2,4-triazoles in yields of 59–73% (Scheme 19a) [47]. Likewise, Zhao and coworkers accomplished the homocoupling of the unsubstituted 4-amino-1,2,4-triazole in 90% yield using sodium dichloroisocyanurate (SDCI) as the oxidant (Scheme 19b) [48].

Photocatalysis represents an alternative metal-free approach to pyrazole homocouplings. An early report of the photochemical oxidation of pyrazolidinones led to a homocoupled product [49]. More recently, Meng and coworkers used photocatalysis to accomplish the homocoupling of a bromopyrazolone [50]. By irradiating a solution of 4-bromo-3-methyl-1-phenyl-4,5-dihydropyrazol-5-one (PyHBr) and naphthalene in acetonitrile with a 300 W high-pressure mercury lamp, the authors obtained the homocoupled product in 64% yield (Scheme 20a). Interestingly, the authors discovered that by substituting phenathrene for naphthalene the dehydrogenated product was favored over the hydrogenated product, and the use of other arenes such as anthracene, benzene, acenaphthylene or indole did not lead to homocoupling at all. As the reaction relies on certain arenes and does not proceed under UV irradiation alone, the authors ruled out the homolytic bond cleavage of the carbon–bromine bond as the mechanism and instead proposed the mechanism shown in Scheme 20b.

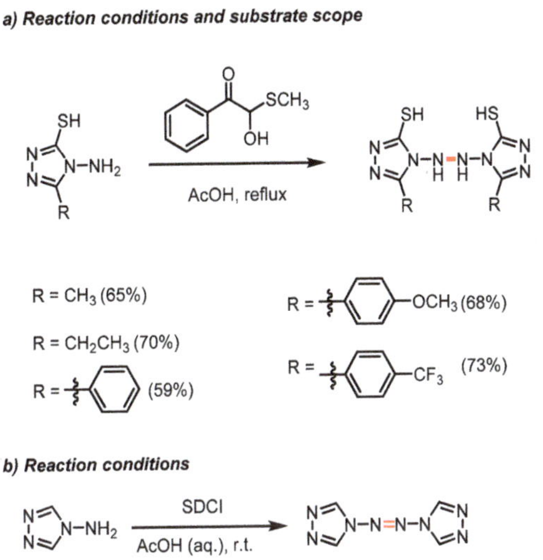

Scheme 19. (a) Reaction conditions and substrate scope for the oxidative homocoupling of substituted 4-amino-1,2,4-triazoles. (b) Reaction conditions for the homocoupling of unsubstituted 4-amino-1,2,4-triazole.

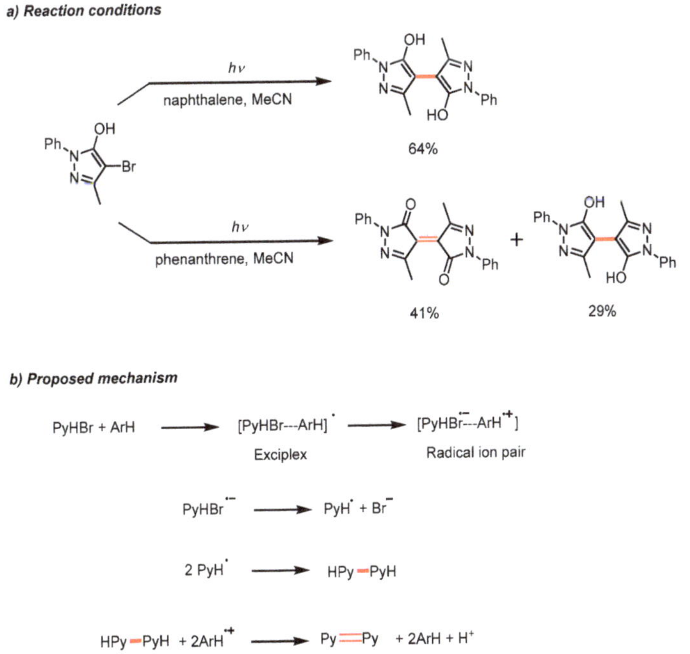

Scheme 20. (a) Photochemical oxidation of pyrazolidinones leads to different products depending on the identity of the auxiliary arene (naphthalene vs. phenanthrene). (b) The proposed reaction mechanism reflects the fact that UV irradiation alone is not sufficient to catalyze the homocoupling [34].

Finally, a recent report of the decarboxylative fluorination of various heterocycles using Selectfluor contained the unexpected observation of homocoupling under certain conditions [51]. Specifically, substituted pyrazole-5-carboxylic acids and indole-2-carboxylic acids underwent homocoupling along with decarboxylative fluorination, yielding the corresponding fluorinated dimers (Scheme 21a). The homocoupling occurred in high yields (70–82%) for the pyrazole-based substrates and in moderate yields (38–58%) for the indole-based substrates (Scheme 21b). In the case of indole-2-carboxylic acid, the biproducts 2-fluoroindole and oxindole were generated along with the dimer (Scheme 21c). Mechanistic investigations indicated that neither of these biproducts is an intermediate in the dimer formation process, but the mechanism of the homocoupling was not elucidated.

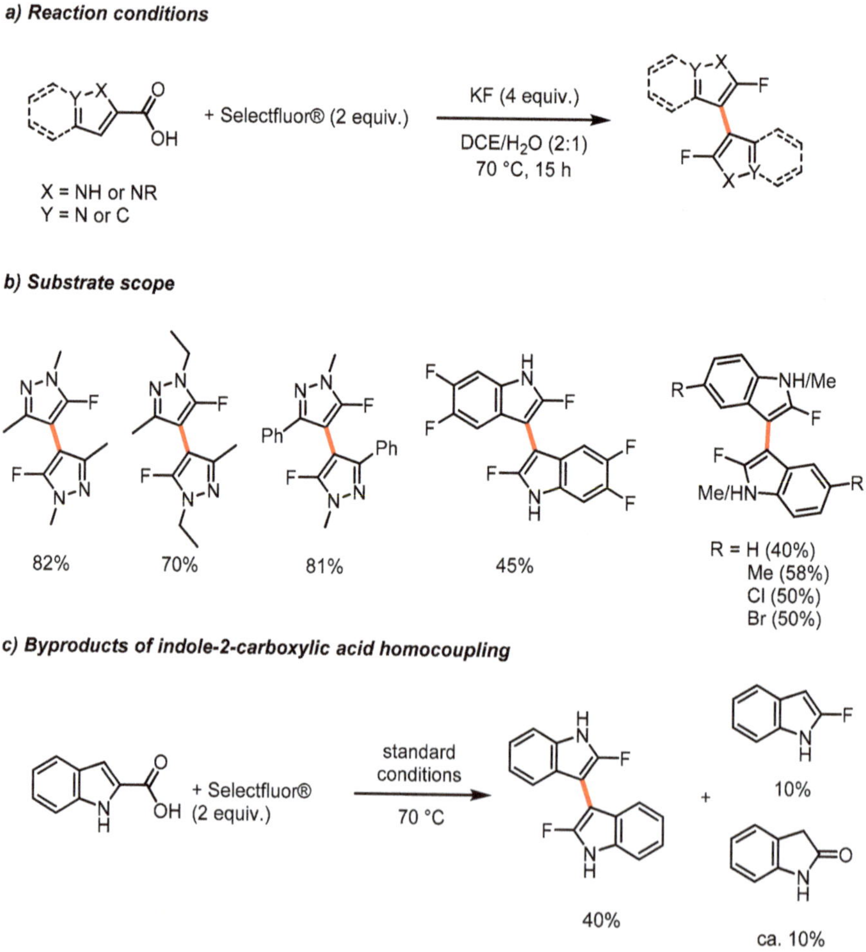

Scheme 21. (a) Reaction conditions for the homocoupling and decarboxylative fluorination of heterocycles in the presence of Selectfluor. (b) Substrate scope for the homocoupling of substituted pyrazoles and indoles using these conditions. (c) The homocoupling of indole-2-carboxylic acid led to the formation of byproducts and the dimer.

3.3. Applications in MOFs

The pyrazole functional group can coordinate with a transition metal as a ligand through the lone pair on the non-protonated nitrogen atom (Figure 2a) [52,53]. When the other nitrogen atom in the ring is deprotonated (to yield the pyrazolate), both nitrogen atoms can coordinate with metals, allowing pyrazolate to bridge metal centers and give rise to clusters, chains or coordination polymers (Figure 2b) [54,55]. Bipyrazolate ligands can coordinate up to four different metal centers (Figure 2c), making possible the construction of extended three-dimensional structures known as metal–organic frameworks (MOFs) [56]. Due to their crystallinity and internal porosity, metal–organic frameworks have shown promise in diverse applications, including gas separations [57], sensing [58], conductivity [59], drug delivery [60] and energy storage [61]. Numerous metal–organic framework structures have been designed and synthesized using poly(pyrazole) ligands as building blocks, as was reviewed by Galli and coworkers in 2016 [62]. Here we give an overview of recent studies on pyrazole-based metal–organic frameworks that have been published since the 2016 review, with an emphasis on 4,4'-bipyrazole (H_2bpz) and its derivatives.

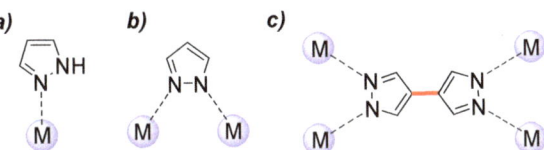

Figure 2. The pyrazole functional group can coordinate with a single metal ion (**a**) or can bridge two metal ions as a deprotonated pyrazolate (**b**). The 4,4'-bipyrazolate ligand can coordinate up to four different metal ions (**c**).

The unsubstituted H_2bpz ligand has recently been used to form two new metal–organic frameworks (Figure 3). Volkmer and coworkers reacted H_2bdp with a mixed-valent iron precursor to yield a nonporous metal–organic framework with Fe(II) ions as the framework nodes (CFA-10-as), and the authors were then able to convert the framework to a porous form upon thermal treatment (CFA-10) [63]. This single-crystal-to-single-crystal transformation was shown to result from the thermal decomposition of formate ions in the pores of CFA-10-as, resulting in a final porous structure that could not be achieved directly through solvothermal synthesis. A subset of these authors also described the synthesis of another H_2bpz-based metal–organic framework, this time with Mn(III) ions as the framework nodes [64]. The structure of this framework, named Mn-CFA-6, was shown to contain two crystallographically and structurally distinct manganese centers, and the framework exhibited structural flexibility in response to the removal of guest molecules. Recently, Lee and coworkers observed a similar flexibility in the H_2bpz-based framework Co(bpz), which undergoes reversible structural changes in response to gas adsorption or temperature change [28].

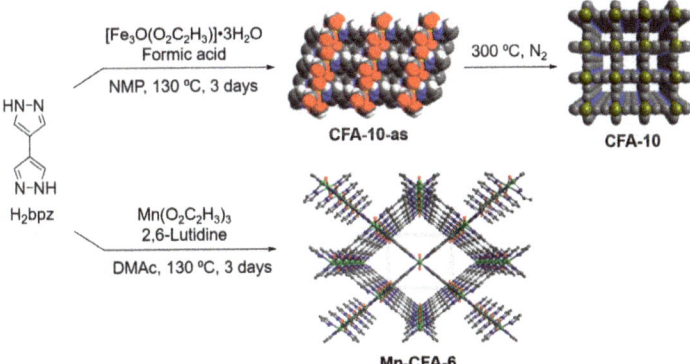

Figure 3. H_2bpz was reacted with an iron source to yield CFA-10-as and CFA-10 [47] and with a manganese source to yield Mn-CFA-6. Adapted with permission from references [63,64]. ©2017 ACS.

The tetramethylated bipyrazole ligand, 3,3′,5,5′-tetramethyl-4,4′-bipyrazole (H_2Me_4bpz), is another popular building block for metal–organic frameworks. In 2015, Sun and coworkers used H_2Me_4bpz to construct a series of Cu(I)-based coordination networks, including two novel metal–organic frameworks (Figure 4) [65]. These two frameworks contain halide and cyanide anions that coordinate with the Cu(I) nodes, along with the Me_4bpz^{2-} ligands. The authors investigated these materials as heterogeneous catalysts for the azide–alkyne click reaction, because of the high density of Cu(I) sites present in the frameworks, by reacting benzyl azide with phenylacetylene in the presence of 1 mol % of a given metal–organic framework. The frameworks proved to be very effective catalysts for the formation of the substituted triazole, with the framework $[Cu_6(Me_4bpz)_6(CH_3CN)_3(CN)_3Br]\bullet 2OH\bullet 14CH_3CN$ leading to a 99% yield within 2 h. The authors ascribed this excellent catalytic performance to the framework's high porosity and large pore diameter, which allow the Cu(I) active sites to be accessible to the reactants. The substrate scope for compound $[Cu_6(Me_4bpz)_6(CH_3CN)_3(CN)_3Br]\bullet 2OH\bullet 14CH_3CN$ was shown to include a variety of terminal azides and alkynes. This work was the first investigation of Cu(I)-based metal–organic frameworks as catalysts for click reactions, and the results indicate that this class of materials holds great promise in heterogeneous azide–alkyne catalysis.

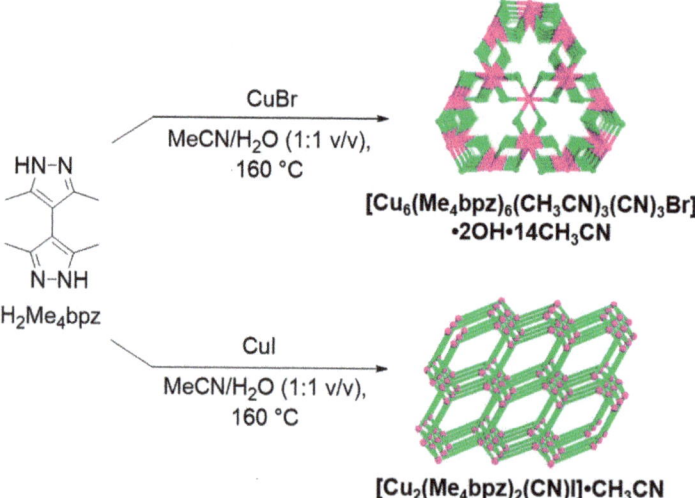

Figure 4. H_2Me_4bpz was reacted with Cu(I) halides to yield two Cu(I)-based metal–organic frameworks, which were shown to be effective catalysts for azide–alkyne click reactions. Adapted with permission from reference [65]. ©2015 ACS.

Chen and coworkers attempted to increase the hydrophobicity of the hallmark metal–organic framework MOF-5 by substituting the traditional 1,4-benzenedicarboxylic acid (H_2bdc) ligand with combinations of H_2Me_4bpz, naphthalene-1,4-dicarboxylic acid (H_2ndc) and biphenyl-4,4′-dicarboxylic acid (H_2bpdc) [66]. The authors synthesized one known framework (MAF-X10) and two new frameworks (MAF-X12 and MAF-X13) from these building blocks (Figure 5). The three metal–organic frameworks were investigated for fluorocarbon adsorption, using $CHClF_2$ as a representative adsorbate, and all three were shown to possess high adsorption capacities for $CHClF_2$. Grand canonical Monte Carlo simulations indicated that the strongest $CHClF_2$ binding site in each framework was located in a hydrophobic pocket formed by two methyl groups from Me_4bpz^{2-} ligands and the aromatic face of an adjacent phenyl ring, bearing out the authors' hypothesis that increased ligand hydrophobicity would be beneficial for fluorocarbon adsorption. The compound MAF-X13 was shown to have the largest saturation uptake and the largest working capacity for $CHClF_2$ adsorption, due to its type-IV isotherm shape.

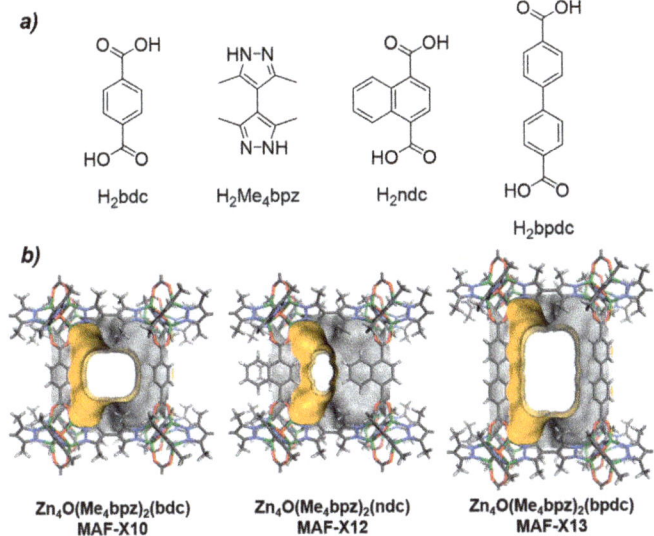

Figure 5. (a) Linkers used to make hydrophobic MOF-5 analogues. (b) Crystal structures of MAF-X10, MAF-X12 and MAF-X13 highlighting the pore surface. Adapted with permission from ref. [66].

Recently, Cao and workers took advantage of the differences in dative bond strength between carboylate ligands and azolate ligands to partially deconstruct mixed-ligand metal–organic frameworks [67]. The authors synthesized a metal–organic framework from a Zn(II) source and a mixture of H_2Me_4bpz and 1,4-benzenedicarboxylate ligands, and then used a pH 13 ammonia solution to dissolve the Zn–carboxylate bonds but not the Zn–pyrazolate bonds. This treatment led to a delamination of the three-dimensional framework, yielding two-dimensional sheets composed of Zn(II) nodes linked by $Me_4bpz_2^-$ ligands. The authors also applied this process to an analogous Zn–triazolate–thiophenedicarboxylate MOF and used the resulting 2D Zn–triazolate sheets as building blocks to construct a library of new 3D frameworks, by combining the 2D sheets with various dicarboxylate linkers (Figure 6). This creative approach to metal–organic framework synthesis highlights the advantages of the strong metal–ligand bonding inherent to bipyrazole ligands like H_2Me_4bpz.

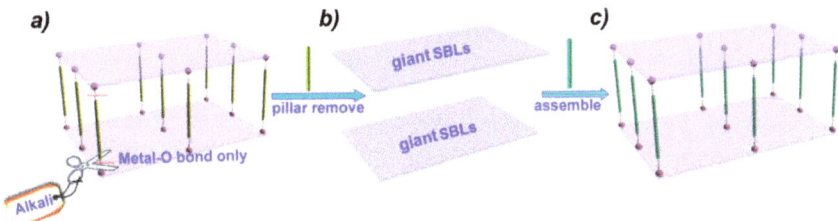

Figure 6. (a) A metal–organic framework composed of sheets of Zn–azolate bonds pillared by Zn–carboxylate bonds undergoes alkaline treatment to break the Zn–carboxylate bonds only. (b) This treatment leaves behind Zn–azolate sheets, which can be thought of as giant supramolecular building layers (SBLs). (c) These layers can then be reassemble using different dicarboylate pillars to yield a variety of metal–organic frameworks. Adapted with permission from reference [67]. ©2015 ACS.

The sheer number of other recent examples of H_2Me_4bpz-based metal–organic frameworks emphasizes the utility of this bipyrazole ligand. These frameworks combine H_2Me_4bpz with various transition metals, including manganese [68], iron [69], cobalt [68,70,71], nickel [68], copper [72], zinc [73], silver [74–76] and cadmium [68,77,78], yielding a diverse array of framework topologies and functionalities. These frameworks were investigated for applications ranging from antibacterial therapy [76] to the photocatalytic degradation of organic pollutants [71], and chemists are sure to push the boundaries of H_2Me_4bpz-based metal–organic frameworks further in the coming years.

The dimethylated analogue of H_2bpz is used in metal–organic framework synthesis less frequently than the tetramethylated analogue, but two recent studies show that H_2Me_2bpz can be used to form a variety of metal–organic frameworks. In 2017, Galli and workers reported two new metal–organic frameworks based on the reaction of H_2bpz with a cobalt source and with a zinc source, respectively (Figure 7a). [79] The authors compared these two H_2Me_2bpz frameworks with the isostructural frameworks M(bpz) (M = Co, Zn) through techniques including gas adsorption and *ab initio* simulation, revealing that the methylated frameworks bound guest CO_2 molecules more strongly than the nonmethylated frameworks but accommodated fewer overall guests because of the increased steric bulk of the H_2Me_2bpz ligand, relative to H_2bpz. Additionally, in 2018 a subset of these authors reported a 3D polymeric network composed of Hg(II) ions linked by $Me_2bpz_2^-$ ligands (Figure 7b). [80]

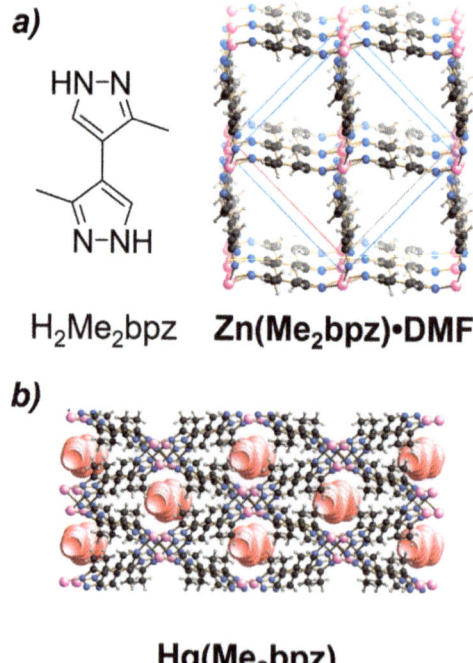

Figure 7. (a) The H_2Me_2bpz ligand and the resulting metal–organic framework $Zn(Me_2bpz)\bullet DMF$ (DMF = N,N-dimethylformamide). DMF molecules in the pores have been removed for clarity. $Co(Me_2bpz)\bullet DMF$ is isostructural to $Zn(Me_2bpz)\bullet DMF$. Adapted with permission from reference [79]. ©2017 ACS. (b) The 3D polymeric network Hg(Me2bpz), with the void spaces highlighted. Adapted with permission from reference [80].

Finally, Galli and coworkers recently demonstrated the power of introducing functional groups onto the H_2bpz scaffold, functionalizing this ligand with either NO_2 or NH_2 groups (Figure 8a) [81]. The authors incorporated these H_2bpz derivatives into a series of Zn(II)-based metal–organic frameworks, each containing a different mixture of ligands. The authors found that frameworks containing a mixture of H_2bpz and H_2bpzNH_2 ligands had the best gas adsorption properties in the context of N_2/CO_2 separations. The following year, Rossin and coworkers used the same three ligands to construct a series of Co(II)-based metal–organic frameworks (Figure 8b) [82]. The authors found that while the unfunctionalized Co(bpz) had higher O_2 gas uptake relative to $Co(bpzNO_2)$ and $Co(bpzNH_2)$, ligand functionalization could tune the catalytic selectivity and performance of these materials when they were used for cumene oxidation and decomposition.

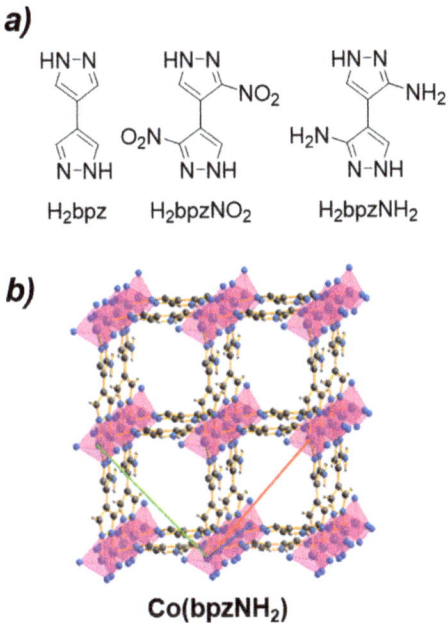

Figure 8. (a) The series of H_2bpz derivatives. (b) Crystal structure of $Co(bpzNH_2)$, which is isostructural to Co(bpz) and $Co(bpzNO_2)$. Adapted with permission from reference [81]. ©2019 ACS.

4. Conclusions

Despite the utility of bipyrazoles and other symmetric heterocycles as building blocks in coordination chemistry, traditional routes to these molecules were laborious, multi-step syntheses with poor atom economy and redox efficiency. In recent years, a number of improved routes to bipyrazoles have been published, employing an efficient homocoupling step that forms an aryl-aryl bond between the two halves of the product. Some of the most promising pyrazole homocoupling reactions rely on transition metal catalysts, especially palladium, as described above. We have also described other unique homocoupling strategies that proceed in the absence of metal catalysts, relying on UV irradiation, aerobic oxidation, or other diverse mechanisms.

We anticipate that the coupling strategies provided in this review will allow metal–organic framework chemists to access a greater variety of bipyrazole-type ligands, beyond the small set of H_2bpz derivatives described herein. As is obvious from the above studies, bipyrazole-based metal–organic frameworks are promising materials for a number of applications, especially gas adsorption and heterogeneous catalysis. The ability to diversify and functionalize bipyrazole ligands will afford chemists greater control over the resultant metal–organic framework properties, leading to even better adsorbents and catalysts.

Author Contributions: Conceptualization, M.K.T. and J.-W.L.; writing—original draft preparation, S.B.M., M.K.T. and J.-W.L.; writing—review and editing, S.B.M., M.K.T. and J.-W.L. All authors have read and agreed to the published version of the manuscript.

Funding: This work was performed, in part, at the Center for Integrated Nanotechnologies, an Office of Science User Facility operated for the U.S. Department of Energy (DOE) Office of Science. Sandia National Laboratories is a multimission laboratory managed and operated by National Technology and Engineering Solutions of Sandia, LLC, a wholly owned subsidiary of Honeywell International, Inc., for the U.S. DOE's National Nuclear Security Administration under contract DE-NA-0003525. The views expressed in the article do not necessarily represent the views of the U.S. DOE or the United States Government. J.-W.L. thanks the Department of Chemistry, University of Copenhagen.

Acknowledgments: We thank reviewers for critical comments.

Conflicts of Interest: The authors declare no conflict of interest.

References

1. Hab, D.; Mlostoń, G. Heterocycles in materials chemistry. *Chem. Heterocycl. Compd.* **2017**, *53*, 1. [CrossRef]
2. Almeida Paz, F.A.; Klinowski, J.; Vilela, S.M.F.; Tomé, J.P.C.; Cavaleiro, J.A.S.; Rocha, J. Ligand design for functional metal–organic frameworks. *Chem. Soc. Rev.* **2012**, *41*, 1088–1110. [CrossRef]
3. Vasconcelos, S.N.S.; Reis, J.S.; de Oliveira, I.M.; Balfour, M.N.; Stefani, H.A. Synthesis of symmetrical biaryl compounds by homocoupling reaction. *Tetrahedron* **2019**, *75*, 1865–1959. [CrossRef]
4. Hassan, J.; Sévignon, M.; Gozzi, C.; Schulz, E.; Lemaire, M. Aryl–Aryl Bond Formation One Century after the Discovery of the Ullmann Reaction. *Chem. Rev.* **2002**, *102*, 1359–1470. [CrossRef]
5. Mondal, S. Recent advancement of Ullmann-type coupling reactions in the formation of C–C bond. *ChemTexts* **2016**, *2*, 17. [CrossRef]
6. Yin, G.; Wang, Z.; Chen, A.; Gao, M.; Wu, A.; Pan, Y. A New Facile Approach to the Synthesis of 3-Methylthio-Substituted Furans, Pyrroles, Thiophenes, and Related Derivatives. *J. Org. Chem.* **2008**, *73*, 3377–3383. [CrossRef]
7. Aponick, A.; Li, C.-Y.; Malinge, J.; Marques, E.F. An Extremely Facile Synthesis of Furans, Pyrroles, and Thiophenes by the Dehydrative Cyclization of Propargyl Alcohols. *Org. Lett.* **2009**, *11*, 4624–4627. [CrossRef]
8. Fustero, S.; Simón-Fuentes, A.; Sanz-Cervera, J.F. Recent Advances in the Synthesis of Pyrazoles. A Review. *Org. Prep. Proced. Int.* **2009**, *41*, 253–290. [CrossRef]
9. Trofimenko, S. 1,1,2,2-Ethanetetracarboxaldehyde and Its Reactions. *J. Org. Chem.* **1964**, *29*, 3046–3049. [CrossRef]
10. Kornfeld, E.C.; Jones, R.G. The Synthesis Oof Furan, Thiophene, and Pyrrole-3,4-Dicarboxylic Esters. *J. Org. Chem.* **1954**, *19*, 1671–1680. [CrossRef]
11. Trofimenko, S. Coordination chemistry of pyrazole-derived ligands. *Chem. Rev.* **1972**, *72*, 497–509. [CrossRef]
12. Weis, C.D. Polycyanofurans. *J. Org. Chem.* **1962**, *27*, 3514–3520. [CrossRef]
13. Boldog, I.; Sieler, J.; Chernega, A.N.; Domasevitch, K.V. 4,4′-Bipyrazolyl: New bitopic connector for construction of coordination networks. *Inorg. Chim. Acta* **2002**, *338*, 69–77. [CrossRef]
14. Arnold, Z. Synthetic reactions of dimethylformamide. XV. Synthesis of symmetrical tetraformylethane. *Collect. Czechoslov. Chem. Commun.* **1962**, *27*, 2993–2995. [CrossRef]
15. Abdel-Wahab, B.F.; Dawood, K.M. Synthesis and applications of bipyrazole systems. *ARKIVOC* **2012**. [CrossRef]
16. Jover, J.; Spuhler, P.; Zhao, L.; McArdle, C.; Maseras, F. Toward a mechanistic understanding of oxidative homocoupling: The Glaser–Hay reaction. *Catal. Sci. Technol.* **2014**, *4*, 4200–4209. [CrossRef]
17. Do, H.-Q.; Daugulis, O. An Aromatic Glaser–Hay Reaction. *J. Am. Chem. Soc.* **2009**, *131*, 17052–17053. [CrossRef]
18. Li, Y.; Jin, J.; Qian, W.; Bao, W. An efficient and convenient Cu(OAc)2/air mediated oxidative coupling of azoles via C–H activation. *Org. Biomol. Chem.* **2010**, *8*, 326–330. [CrossRef]
19. Biffis, A.; Centomo, P.; Del Zotto, A.; Zecca, M. Pd Metal Catalysts for Cross-Couplings and Related Reactions in the 21st Century: A Critical Review. *Chem. Rev.* **2018**, *118*, 2249–2295. [CrossRef]
20. Miyaura, N.; Suzuki, A. Palladium-Catalyzed Cross-Coupling Reactions of Organoboron Compounds. *Chem. Rev.* **1995**, *95*, 2457–2483. [CrossRef]

21. Adamo, C.; Amatore, C.; Ciofini, I.; Jutand, A.; Lakmini, H. Mechanism of the Palladium-Catalyzed Homocoupling of Arylboronic Acids: Key Involvement of a Palladium Peroxo Complex. *J. Am. Chem. Soc.* **2006**, *128*, 6829–6836. [CrossRef] [PubMed]
22. Salanouve, E.; Retailleau, P.; Janin, Y.L. Few unexpected results from a Suzuki–Miyaura reaction. *Tetrahedron* **2012**, *68*, 2135–2140. [CrossRef]
23. Batchu, H.; Bhattacharyya, S.; Kant, R.; Batra, S. Palladium-Catalyzed Chelation-Assisted Regioselective Oxidative Dehydrogenative Homocoupling/Ortho-Hydroxylation in N-Phenylpyrazoles. *J. Org. Chem.* **2015**, *80*, 7360–7374. [CrossRef] [PubMed]
24. Afanas'ev, O.; Tsyplenkova, O.; Seliverstov, M.; Sosonyuk, S.; Proskurnina, M.; Zefirov, N. Homocoupling of bromotriazole derivatives on metal complex catalysts. *Russ. Chem. Bull.* **2015**, *64*, 1470–1472. [CrossRef]
25. Jansa, J.; Schmidt, R.; Mamuye, A.D.; Castoldi, L.; Roller, A.; Pace, V.; Holzer, W. Synthesis of tetrasubstituted pyrazoles containing pyridinyl substituents. *Beilstein J. Org. Chem.* **2017**, *13*, 895–902. [CrossRef]
26. Zhu, J.; Chen, Y.; Lin, F.; Wang, B.; Huang, Q.; Liu, L. Highly Regioselective Arylation of 1,2,3-Triazole N-Oxides with Sodium Arenesulfinates via Palladium-Catalyzed Desulfitative Cross-Coupling Reaction. *Synlett* **2015**, *26*, 1124–1130. [CrossRef]
27. Peng, X.; Huang, P.; Jiang, L.; Zhu, J.; Liu, L. Palladium-catalyzed highly regioselective oxidative homocoupling of 1,2,3-triazole N-oxides. *Tetrahedron Lett.* **2016**, *57*, 5223–5226. [CrossRef]
28. Taylor, M.K.; Juhl, M.; Hadaf, G.B.; Hwang, D.; Velasquez, E.; Oktawiec, J.; Lefton, J.B.; Runčevski, T.; Long, J.R.; Lee, J.-W. Palladium-catalyzed oxidative homocoupling of pyrazole boronic esters to access versatile bipyrazoles and the flexible metal–organic framework Co(4,4′-bipyrazolate). *Chem. Commun.* **2020**, *56*, 1195–1198. [CrossRef]
29. Ackermann, L.; Novák, P.; Vicente, R.; Pirovano, V.; Potukuchi, H.K. Ruthenium-Catalyzed C-H Bond Functionalizations of 1,2,3-Triazol-4-yl-Substituted Arenes: Dehydrogenative Couplings Versus Direct Arylations. *Synthesis* **2010**, *2010*, 2245–2253. [CrossRef]
30. Tasker, S.Z.; Standley, E.A.; Jamison, T.F. Recent advances in homogeneous nickel catalysis. *Nature* **2014**, *509*, 299–309. [CrossRef]
31. Hazari, N.; Melvin, P.R.; Beromi, M.M. Well-defined nickel and palladium precatalysts for cross-coupling. *Nat. Rev. Chem.* **2017**, *1*, 0025. [CrossRef] [PubMed]
32. Fanni, S.; Di Pietro, C.; Serroni, S.; Campagna, S.; Vos, J.G. Ni(0) catalysed homo-coupling reactions: A novel route towards the synthesis of multinuclear ruthenium polypyridine complexes featuring made-to-order properties. *Inorg. Chem. Commun.* **2000**, *3*, 42–44. [CrossRef]
33. Leahy, D.K.; Evans, P.A. Front Matter. In *Modern Rhodium-Catalyzed Organic Reactions*; Evans, P.A., Ed.; Wiley-VCH: Weinheim, Germany, 2005. [CrossRef]
34. Li, S.-S.; Qin, L.; Dong, L. Rhodium-catalyzed C–C coupling reactions via double C–H activation. *Org. Biomol. Chem.* **2016**, *14*, 4554–4570. [CrossRef] [PubMed]
35. Fagnou, K.; Lautens, M. Rhodium-Catalyzed Carbon–Carbon Bond Forming Reactions of Organometallic Compounds. *Chem. Rev.* **2003**, *103*, 169–196. [CrossRef] [PubMed]
36. Albano, G.; Aronica, L.A. From Alkynes to Heterocycles through Metal-Promoted Silylformylation and Silylcarbocyclization Reactions. *Catalysts* **2020**, *10*, 1012. [CrossRef]
37. Kaur, N. Recent developments in the synthesis of nitrogen containing five-membered polyheterocycles using rhodium catalysts. *Synth. Commun.* **2018**, *48*, 2457–2474. [CrossRef]
38. Song, G.; Wang, F.; Li, X. C–C, C–O and C–N bond formation via rhodium(iii)-catalyzed oxidative C–H activation. *Chem. Soc. Rev.* **2012**, *41*, 3651–3678. [CrossRef]
39. Chen, W.-W.; Xu, M.-H. Recent advances in rhodium-catalyzed asymmetric synthesis of heterocycles. *Org. Biomol. Chem.* **2017**, *15*, 1029–1050. [CrossRef]
40. Yue, Y.; Yamamoto, H.; Yamane, M. Rhodium-Catalyzed Homocoupling of (1-Acyloxyvinyl)silanes: Synthesis of 1,3-Diene-2,3-diyl Diesters and Their Derivatives. *Synlett* **2009**, *2009*, 2831–2835. [CrossRef]
41. Luque, R.; Baruwati, B.; Varma, R.S. Magnetically separable nanoferrite-anchored glutathione: Aqueous homocoupling of arylboronic acids under microwave irradiation. *Green Chem.* **2010**, *12*, 1540–1543. [CrossRef]
42. Toshifumi, D.; Yasuyuki, K. Metal-Free Oxidative Biaryl Coupling by Hypervalent Iodine Reagents. *Curr. Org. Chem.* **2016**, *20*, 580–615. [CrossRef]
43. Schulz, M.; Meske, M. Radikalreaktionen an N-Heterocyclen. XI [1] Reaktionen von 3-Methyl-pyrazolin-5-onen mit Phenoxyl-Radikalen. *J. Für Prakt. Chem./Chem.-Ztg.* **1993**, *335*, 607–615. [CrossRef]

44. Veibel, S. Pyrazole Studies. Xiii. Oxidation by Air of 4-Substituted Pyrazole-5-ones and Stereochemistry of the Oxidation Products. *Acta Chem. Scand.* **1972**, *26*, 3685–3990. [CrossRef]
45. Sheng, X.; Zhang, J.; Yang, H.; Jiang, G. Tunable Aerobic Oxidative Hydroxylation/Dehydrogenative Homocoupling of Pyrazol-5-ones under Transition-Metal-Free Conditions. *Org. Lett.* **2017**, *19*, 2618–2621. [CrossRef]
46. Eller, G.A.; Vilkauskaitė, G.; Šačkus, A.; Martynaitis, V.; Mamuye, A.D.; Pace, V.; Holzer, W. An unusual thionyl chloride-promoted C–C bond formation to obtain 4,4′-bipyrazolones. *Beilstein J. Org. Chem.* **2018**, *14*, 1287–1292. [CrossRef]
47. Li, M.; Zhao, G.; Wen, L.; Wang, X.; Zou, X.; Yang, H. A Coupling Reaction of 4-Amino-5-mercapto-3-substituted-1,2,4-triazoles to Generate Symmetrically Substituted Hydrazines. *Mon. Für Chem./Chem. Mon.* **2005**, *136*, 2045–2049. [CrossRef]
48. Li, S.H.; Pang, S.P.; Li, X.T.; Yu, Y.Z.; Zhao, X.Q. Synthesis of new tetrazene(N–NN–N)-linked bi(1,2,4-triazole). *Chin. Chem. Lett.* **2007**, *18*, 1176–1178. [CrossRef]
49. Eğe, S.N.; Tien, C.J.; Dlesk, A.; Potter, B.E.; Eagleson, B.K. Photochemical oxidation of pyrazolidinones. *Chem. Soc. Chem. Commun.* **1972**, 682–683. [CrossRef]
50. Xiao-Liu, L.I.; Song, Z.-Y.; Meng, J.-B. A novel photochemical reaction of 4-bromo-3-methyl-1-phenyl-4,5-dihydro-pyrazol-5-one. *Chin. J. Chem.* **2004**, *22*, 1215–1218. [CrossRef]
51. Yuan, X.; Yao, J.-F.; Tang, Z.-Y. Decarboxylative Fluorination of Electron-Rich Heteroaromatic Carboxylic Acids with Selectfluor. *Org. Lett.* **2017**, *19*, 1410–1413. [CrossRef]
52. Khripun, A.V.; Selivanov, S.I.; Kukushkin, V.Y.; Haukka, M. Hydrogen bonding patterns in pyrazole Pt(II- and IV) chloride complexes. *Inorg. Chim. Acta* **2006**, *359*, 320–326. [CrossRef]
53. Dang, D.; Li, H.; Zheng, G.; Bai, Y. Synthesis, Crystal Structure and Luminescent Properties of One Silver Complex with 3,5-Diphenylpyrazole. *J. Chem. Crystallogr.* **2011**, *41*, 1612–1615. [CrossRef]
54. Brandi-Blanco, P.; Miguel, P.J.S.; Lippert, B. Coordination of two different metal ions as reason for N-chirality in μ-amide complexes. *Dalton Trans.* **2011**, *40*, 10316–10318. [CrossRef] [PubMed]
55. Omary, M.A.; Elbjeirami, O.; Gamage, C.S.P.; Sherman, K.M.; Dias, H.V.R. Sensitization of Naphthalene Monomer Phosphorescence in a Sandwich Adduct with an Electron-Poor Trinuclear Silver(I) Pyrazolate Complex. *Inorg. Chem.* **2009**, *48*, 1784–1786. [CrossRef] [PubMed]
56. Pettinari, C.; Tăbăcaru, A.; Boldog, I.; Domasevitch, K.V.; Galli, S.; Masciocchi, N. Novel Coordination Frameworks Incorporating the 4,4′-Bipyrazolyl Ditopic Ligand. *Inorg. Chem.* **2012**, *51*, 5235–5245. [CrossRef] [PubMed]
57. Cui, W.-G.; Hu, T.-L.; Bu, X.-H. Metal–Organic Framework Materials for the Separation and Purification of Light Hydrocarbons. *Adv. Mater.* **2020**, *32*, 1806445. [CrossRef]
58. Kreno, L.E.; Leong, K.; Farha, O.K.; Allendorf, M.; Van Duyne, R.P.; Hupp, J.T. Metal–Organic Framework Materials as Chemical Sensors. *Chem. Rev.* **2012**, *112*, 1105–1125. [CrossRef]
59. Xie, L.S.; Skorupskii, G.; Dincă, M. Electrically Conductive Metal–Organic Frameworks. *Chem. Rev.* **2020**, *120*, 8536–8580. [CrossRef]
60. Horcajada, P.; Gref, R.; Baati, T.; Allan, P.K.; Maurin, G.; Couvreur, P.; Férey, G.; Morris, R.E.; Serre, C. Metal–Organic Frameworks in Biomedicine. *Chem. Rev.* **2012**, *112*, 1232–1268. [CrossRef]
61. Wang, H.; Zhu, Q.-L.; Zou, R.; Xu, Q. Metal-Organic Frameworks for Energy Applications. *Chem* **2017**, *2*, 52–80. [CrossRef]
62. Pettinari, C.; Tăbăcaru, A.; Galli, S. Coordination polymers and metal–organic frameworks based on poly(pyrazole)-containing ligands. *Coord. Chem. Rev.* **2016**, *307*, 1–31. [CrossRef]
63. Spirkl, S.; Grzywa, M.; Reschke, S.; Fischer, J.K.H.; Sippel, P.; Demeshko, S.; Krug von Nidda, H.-A.; Volkmer, D. Single-Crystal to Single-Crystal Transformation of a Nonporous Fe(II) Metal–Organic Framework into a Porous Metal–Organic Framework via a Solid-State Reaction. *Inorg. Chem.* **2017**, *56*, 12337–12347. [CrossRef] [PubMed]
64. Spirkl, S.; Grzywa, M.; Volkmer, D. Synthesis and characterization of a flexible metal organic framework generated from MnIII and the 4,4′-bipyrazolate-ligand. *Dalton Trans.* **2018**, *47*, 8779–8786. [CrossRef] [PubMed]
65. Xu, Z.; Han, L.L.; Zhuang, G.L.; Bai, J.; Sun, D. In Situ Construction of Three Anion-Dependent Cu(I) Coordination Networks as Promising Heterogeneous Catalysts for Azide–Alkyne "Click" Reactions. *Inorg. Chem.* **2015**, *54*, 4737–4743. [CrossRef]

66. Lin, R.-B.; Li, T.-Y.; Zhou, H.-L.; He, C.-T.; Zhang, J.-P.; Chen, X.-M. Tuning fluorocarbon adsorption in new isoreticular porous coordination frameworks for heat transformation applications. *Chem. Sci.* **2015**, *6*, 2516–2521. [CrossRef]
67. Li, L.; Yi, J.-D.; Fang, Z.-B.; Wang, X.-S.; Liu, N.; Chen, Y.-N.; Liu, T.-F.; Cao, R. Creating Giant Secondary Building Layers via Alkali-Etching Exfoliation for Precise Synthesis of Metal–Organic Frameworks. *Chem. Mater.* **2019**, *31*, 7584–7589. [CrossRef]
68. Tomar, K.; Rajak, R.; Sanda, S.; Konar, S. Synthesis and Characterization of Polyhedral-Based Metal–Organic Frameworks Using a Flexible Bipyrazole Ligand: Topological Analysis and Sorption Property Studies. *Cryst. Growth Des.* **2015**, *15*, 2732–2741. [CrossRef]
69. Kumar, S.; Arora, A.; Kaushal, J.; Oswal, P.; Kumar, A.; Tomar, K. Room temperature synthesis of an Fe(ii)-based porous MOF with multiple open metal sites for high gas adsorption properties. *New J. Chem.* **2019**, *43*, 4338–4341. [CrossRef]
70. Lin, X.-Y.; Zheng, Q.-Q.; Liu, X.-Z.; Jiang, X.-Y.; Lin, C. Synthesis and thermal stability study of a cobalt-organic framework with tetrahedral cages. *Inorg. Chem. Commun.* **2016**, *67*, 51–54. [CrossRef]
71. Ling, X.-Y.; Wang, J.; Gong, C.; Lu, L.; Singh, A.K.; Kumar, A.; Sakiyama, H.; Yang, Q.; Liu, J. Modular construction, magnetism and photocatalytic properties of two new metal-organic frameworks based on a semi-rigid tetracarboxylate ligand. *J. Solid State Chem.* **2019**, *277*, 673–679. [CrossRef]
72. Huang, M.-J.; Deng, X.; Xian, W.-R.; Liao, W.-M.; He, J. Anion-directed structures and luminescences of two Cu(I) coordination polymers based on bipyrazole. *Inorg. Chem. Commun.* **2019**, *101*, 121–124. [CrossRef]
73. Han, L.-L.; Hu, T.-P.; Mei, K.; Guo, Z.-M.; Yin, C.; Wang, Y.-X.; Zheng, J.; Wang, X.-P.; Sun, D. Solvent-controlled three families of Zn(ii) coordination compounds: Synthesis, crystal structure, solvent-induced structural transformation, supramolecular isomerism and photoluminescence. *Dalton Trans.* **2015**, *44*, 6052–6061. [CrossRef] [PubMed]
74. Han, L.-L.; Wang, Y.-X.; Guo, Z.-M.; Yin, C.; Hu, T.-P.; Wang, X.-P.; Sun, D. Synthesis, crystal structure, thermal stability, and photoluminescence of a 3-D silver(I) network with twofold interpenetrated dia-f topology. *J. Coord. Chem.* **2015**, *68*, 1754–1764. [CrossRef]
75. Huang, X.-H.; Xiao, Z.-P.; Wang, F.-L.; Li, T.; Wen, M.; Wu, S.-T. Syntheses and characterizations of two silver(I) coordination polymers constructed from bipyrazole and dicarboxylate ligands. *J. Coord. Chem.* **2015**, *68*, 1743–1753. [CrossRef]
76. Young, R.J.; Begg, S.L.; Coghlan, C.J.; McDevitt, C.A.; Sumby, C.J. Exploring the Use of Structure and Polymer Incorporation to Tune Silver Ion Release and Antibacterial Activity of Silver Coordination Polymers. *Eur. J. Inorg. Chem.* **2018**, *2018*, 3512–3518. [CrossRef]
77. Singh, U.P.; Singh, N.; Chandra, S. Construction and structural diversity of Cd-MOFs with pyrazole based flexible ligands and positional isomer of naphthalenedisulfonate. *Inorg. Chem. Commun.* **2015**, *61*, 35–40. [CrossRef]
78. Tomar, K.; Gupta, A.K.; Gupta, M. Change in synthetic strategy for MOF fabrication: From 2D non-porous to 3D porous architecture and its sorption and emission property studies. *New J. Chem.* **2016**, *40*, 1953–1956. [CrossRef]
79. Mosca, N.; Vismara, R.; Fernandes, J.A.; Casassa, S.; Domasevitch, K.V.; Bailón-García, E.; Maldonado-Hódar, F.J.; Pettinari, C.; Galli, S. CH3-Tagged Bis(pyrazolato)-Based Coordination Polymers and Metal–Organic Frameworks: An Experimental and Theoretical Insight. *Cryst. Growth Des.* **2017**, *17*, 3854–3867. [CrossRef]
80. Mosca, N.; Vismara, R.; Fernandes, J.A.; Pettinari, C.; Galli, S. The Hg(3,3'-dimethyl-1H,1H'-4,4'-bipyrazolate) coordination polymer: Synthesis, crystal structure and thermal behavior. *Inorg. Chim. Acta* **2018**, *470*, 423–427. [CrossRef]
81. Vismara, R.; Tuci, G.; Tombesi, A.; Domasevitch, K.V.; Di Nicola, C.; Giambastiani, G.; Chierotti, M.R.; Bordignon, S.; Gobetto, R.; Pettinari, C.; et al. Tuning Carbon Dioxide Adsorption Affinity of Zinc(II) MOFs by Mixing Bis(pyrazolate) Ligands with N-Containing Tags. *ACS Appl. Mater. Interfaces* **2019**, *11*, 26956–26969. [CrossRef]

82. Nowacka, A.; Vismara, R.; Mercuri, G.; Moroni, M.; Palomino, M.; Domasevitch, K.V.; Di Nicola, C.; Pettinari, C.; Giambastiani, G.; Llabrés i Xamena, F.X.; et al. Cobalt(II) Bipyrazolate Metal–Organic Frameworks as Heterogeneous Catalysts in Cumene Aerobic Oxidation: A Tag-Dependent Selectivity. *Inorg. Chem.* **2020**, *59*, 8161–8172. [CrossRef] [PubMed]

Publisher's Note: MDPI stays neutral with regard to jurisdictional claims in published maps and institutional affiliations.

© 2020 by the authors. Licensee MDPI, Basel, Switzerland. This article is an open access article distributed under the terms and conditions of the Creative Commons Attribution (CC BY) license (http://creativecommons.org/licenses/by/4.0/).

Review

Recent Advances in Synthesis and Properties of Nitrated-Pyrazoles Based Energetic Compounds

Shijie Zhang [1,†], Zhenguo Gao [1,†], Di Lan [1], Qian Jia [1], Ning Liu [2], Jiaoqiang Zhang [1,*] and Kaichang Kou [1,*]

1. MOE Key Laboratory of Material Physics and Chemistry under Extraordinary, School of Chemistry and Chemical Engineering, Northwestern Polytechnical University, Xi'an 710072, China; zsj562389@sina.com (S.Z.); gaozhenguo@mail.nwpu.edu.cn (Z.G.); landi@mail.nwpu.edu.cn (D.L.); qqianjia@163.com (Q.J.)
2. Xi'an Modern Chemistry Institute, Xi'an 710065, China; flackliu@sina.com
* Correspondence: zhangjq@nwpu.edu.cn (J.Z.); koukc@nwpu.edu.cn (K.K.)
† These authors equally contributed to the work.

Academic Editor: Derek J. McPhee
Received: 30 June 2020; Accepted: 23 July 2020; Published: 30 July 2020

Abstract: Nitrated-pyrazole-based energetic compounds have attracted wide publicity in the field of energetic materials (EMs) due to their high heat of formation, high density, tailored thermal stability, and detonation performance. Many nitrated-pyrazole-based energetic compounds have been developed to meet the increasing demands of high power, low sensitivity, and eco-friendly environment, and they have good applications in explosives, propellants, and pyrotechnics. Continuous and growing efforts have been committed to promote the rapid development of nitrated-pyrazole-based EMs in the last decade, especially through large amounts of Chinese research. Some of the ultimate aims of nitrated-pyrazole-based materials are to develop potential candidates of castable explosives, explore novel insensitive high energy materials, search for low cost synthesis strategies, high efficiency, and green environmental protection, and further widen the applications of EMs. This review article aims to present the recent processes in the synthesis and physical and explosive performances of the nitrated-pyrazole-based Ems, including monopyrazoles with nitro, bispyrazoles with nitro, nitropyrazolo[4,3-c]pyrazoles, and their derivatives, and to comb the development trend of these compounds. This review intends to prompt fresh concepts for designing prominent high-performance nitropyrazole-based EMs.

Keywords: nitrated pyrazoles-based; energetic salts; synthesis; high energy density material; insensitivity

1. Introduction

Energetic materials (EMs), including explosives, propellants, and pyrotechnics, are a significant class of compounds containing large amounts of stored chemical energy, which can liberate heat and exert high pressure under some stimuli, like impact, shock, or thermal effect [1–8]. With the development of science and technology, more and more attention has been paid to the high energy density materials (HEDMs) used for energy and as explosives or propellants [9]. Thus, the representatives of traditional HEDMs are 2,4,6-trinitrotoluene (TNT) [10,11], 1,3,5,7-tetranitro-1,3,5,7-tetrazocine (HMX) [12,13], 1,3,5-trinitro-1,3,5-triazine (RDX) [14], triaminotrinitrobenzene (TATB) [15], and 2,4,6,8,10,12-hexanitro-2,4,6,8,10,12-hexaazaisowurtzitane (CL-20) [16]. The key properties for HEDMs include density (ρ), melting point (T_m), decomposition temperature (T_d), heat of formation (HOF), calculated detonation velocity (D, calculated propagation velocity of detonation wave in explosive grain), calculated detonation pressure (P, calculated pressure on the front of detonation wave), oxygen balance (OB, residual amount of oxygen when explosive explodes to produce CO_2 and H_2O, OB = $16[c-(2a + b/2)]/M$ for molecule $C_aH_bO_cN_d$), specific impulse (I_{sp}, impulse produced by the unit quantity of propellant), content of nitrogen (N), impact sensitivity (IS, sensitivity of explosive to impact), friction sensitivity (FS, sensitivity of explosive to friction), electrostatic discharge sensitivity (ESD), and sensitivity of explosive to electrostatic discharge, etc. There are several standards a novel HEDM should meet if it would be applied widely, including insensitivity toward mechanical stimuli (heat, impact, fraction, and electrostatic discharge) to ensure the safety of operation, high performance for various purposes, less toxicity, and producing less hazardous waste after detonation [17–19]. Among them, conflict between the increasing energetic level and decreasing sensitivity has become more and more severe. Therefore, the exploration and development of high energy density compounds with low sensitivity have been a priority. A significant amount of effort has been made to resolve this problem, such as recrystallization of Ems [20], preparing polymer bonded explosives (PBXs) [21,22], forming energetic cocrystals [23–25], and synthesizing novel energetic compounds [26–29]. In contrast with other technologies, synthesizing new HEDMs may be the most direct and effective method.

Nitrogen heterocyclic energetic materials that have large numbers of N–N bonds and C–N bonds with high energy can form the large π bond similar to benzene, which endows this kind of compounds low sensitivity, high positive heat of formation, and good thermal stability. In addition, the low percentage of C and N in these compounds always lead to high density and good oxygen balance. The decomposition of these compounds can result in the N_2, which is environmentally friendly [30]. There is a big difference between nitrogen-rich energetic compounds and traditional explosives, namely the energy of nitrogen heterocyclic compounds is released from the high positive heat of formation rather than the oxidation of carbon backbone like traditional explosive (such as TNT and TATB) [19,31]. Therefore, nitrogen heterocyclic materials have garnered large interest in the research areas of HEDMs.

As an outstanding representative of nitrogen heterocyclic compounds, nitropyrazoles and their derivatives are aromatic stable substances with π electrons in their structures. The system is easy to carry out electrophilic substitution reactions such as nitration, sulfonation and halogenation, etc. [32]. These compounds are characterized by oxidation resistance, heat resistance and hydrolysis resistance [19], and are widely applied in civil fields, such as medicine, pesticide, photosensitive materials, and fine chemicals [33–35]. Due to the compactness, stability, and modifiability of the molecular structure of pyrazoles, nitration and derivatization of pyrazoles are relatively easy. The ring tension in the structures of nitropyrazoles and their derivatives is large. The density and nitrogen content of nitropyrazoles increase with the presence of nitro groups on the ring, and the oxygen balance is closer to the ideal value, which can improve the detonation performance of the target compounds. Many energetic compounds based on nitropyrazoles have been synthesized successively, which have good applications in highly energy insensitive explosives, propellants, pyrotechnic agents, and other fields [2,3,19,36,37].

In the past decade, a lot of papers on the synthesis and properties of nitrated pyrazoles have been published, including many Chinese references which are not accessible for most Western researchers due to language barriers. This review article presents the recent processes in synthesis, physical and explosive performances of the nitropyrazole-based Ems, including monopyrazoles with nitro, bispyrazoles with nitro, nitropyrazolopyrazoles and their derivatives, and to comb the development trend of these compounds. The aim of this review is to provide readers with an overview of the relationship between structures and properties and guide the future design of novel HEDMs. This review also intends to prompt fresh concepts for designing prominent high-performances nitropyrazole-based EMs.

2. Nitrated-Monopyrazole Based Compounds

In this section, the sum of nitro group substituted on carbon position of pyrazole ring in mononitropyrazoles, binitropyrazoles, and trinitropyrazoles are one, two, and three, respectively.

For example, mononitropyrazole represents that only one C position in pyrazole ring is substituted by the nitro group.

2.1. Mononitropyrazoles and Their Derivatives

Mononitropyrazoles and their derivatives due to their energetic property are favored by people in many fields, such as medicine, pesticide, energetic material and so on. Among them, 3-nitropyrazole (3-NP), 4-nitropyrazole (4-NP), 1-methyl-3-nitropyrazole (3-MNP), and 1-methyl-4-nitropyrazole (4MNP) are typical examples, which are commonly used as energetic materials and intermediates for further products of other energetic materials because they contain only one nitro group and have relatively low energy. The syntheses of these compounds is often facile and can meet the development requirements of green chemistry.

As a typical heterocyclic compound, 3-NP is an important intermediate in the synthesis of pyrazole-based compounds such as 3,4-dinitropyrazole (DNP) and other new explosives [36,38]. In 1970, Habraken and co-authors [39] firstly reported synthesis of 3-NP by dissolving N-nitropyrazole in anisole for 10 h at 145 °C. Later, Verbruggen et al. [40] synthesized 3-NP from diazomethane and chloronitroethylene by one-step cyclization, while this reaction was high riskful due to the extremely vivacious raw materials. Nowadays, the main synthesis method of 3-NP was a two-step reaction, that is, nitration of pyrazole to obtain N-nitropyrazole and then rearrangement of N-nitropyrazole in organic solvent to acquire 3-NP (Figure 1, Scheme A). The nitration agents could be HNO_3/H_2SO_4 or $HNO_3/Ac_2O/HAc$, and the organic solvent for rearrangement could be anisole, n-octanol and benzonitrile [41–43]. Among these solvents, benzonitrile was always preferred to be the rearrangement medium since anisole could require an excessively long time and n-octanol would lead to poor-quality product. In 2014, Zhao et al. [44] reported one convenient and green approach to synthesizing the 3-NP. They chose the oxone as the nitration agents of 3-aminopyrazole and water as the solvent (Figure 1, Scheme B). This approach owns some advantages over the previous approach: simple operation, safety, economical reagents, the use of water as solvent, and mild conditions. As shown in Figure 1, 3-MNP is one of the most important derivatives of 3-NP. Its synthesis is mainly accomplished by nitrated 1-methylpyrazole with various nitration agents. Katritzky et al. [45] added 1-methylpyrazole to trifluoroacetic anhydride for 1 h in ice bath, and then concentrated nitric acid was added in the solution. After stirring for 12 h, and evaporation of trifluoroacetic anhydride and nitric acid, the 3-MNP could be obtained (Figure 1, Scheme C). In 2013, Ravi et al. [46] proposed that 1-methylpyrazole could reacted with silicon oxide-bismuth nitrate or silicon dioxide-sulfuric acid-bismuth nitrate in tetrahydrofuran (THF) to produce 3-MNP (Figure 1, Scheme D), this facile route is a synthetic method of low toxicity, high efficiency, and green environmental protection. In addition, metal salts of 3-NP expand its derivatives. Li et al. [42] prepared the metal Cu(II) salt and basic Pb salt of 3-NP, by dissolving 3-NP in NaOH solution and reacting with the $CuSO_4 \cdot 5H_2O$ solution and $Pb(NO_3)_2$ solution, respectively (Figure 1, Scheme E).

4-NP is an isomer of 3-NP with melting point of 163–165 °C, density of 1.52 g/cm^3, detonation velocity of 6.68 km/s and detonation pressure of 18.81 Gpa [47]. Similar to 3-NP, 4-NP can be obtained by nitro group rearrangement. As Rao et al. [48] reported N-nitropyrazole could be rearranged to 4-NP in sulfuric acid at room temperature (Figure 2, Scheme A). Ravi et al. [49] synthesized 4-NP in THF with 4-iodopyrazole as raw material, fuming HNO_3 as nitration agents, octahedral zeolite or silica as solid catalyst (Figure 2, Scheme B). Li et al. [50] reported one-pot two steps route that pyrazole could be nitrated to 4-NP by fuming HNO_3 (90%)/fuming H_2SO_4 (20%) (Figure 2, Scheme C). 4-MNP is another important derivative of nitropyrazole with the similar performance to 3-MNP (Table 1). In 2015, Corte et al. [51] reported that 4-MNP could be synthesized by adding sodium hydride and iodomethane into the THF solution of 4-NP at room temperature for overnight. Ioannidis et al. [52] improved the method by adding sodium hydride and iodomethane to the acetonitrile solution of 4-NP under nitrogen protection for 16 h. However, it is dangerous to handle sodium hydride due to its high chemical reaction activity which can easily cause combustion and explosion, limiting the further application of this method. Han et al. [53] simplified the above method and replaced sodium hydride with potassium carbonate. They added potassium carbonate and iodomethane to the N,N-dimethylformamide (DMF) solution of 4-NP at 25 °C for 14 h. This method not only reduces the risk in the process, but improves the reaction yield (80–98%).

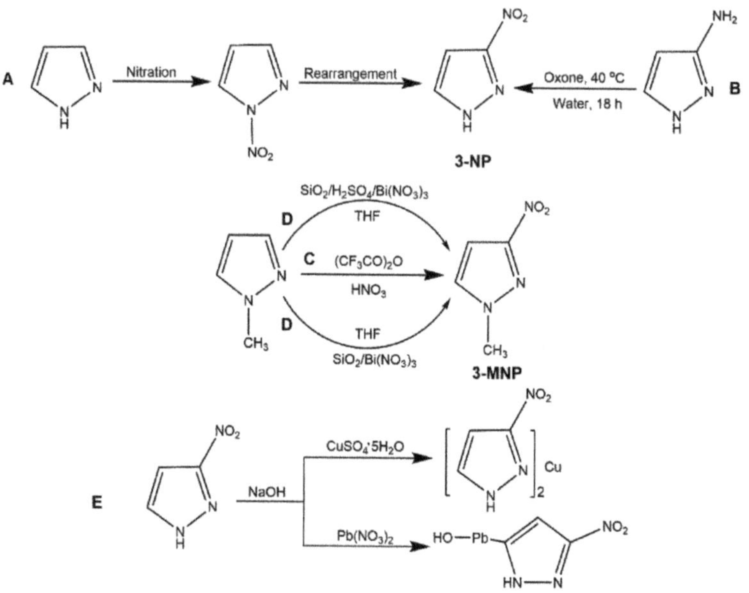

Figure 1. Summary of synthesis of 3-NP, 3-MNP and metal salts of 3-NP.

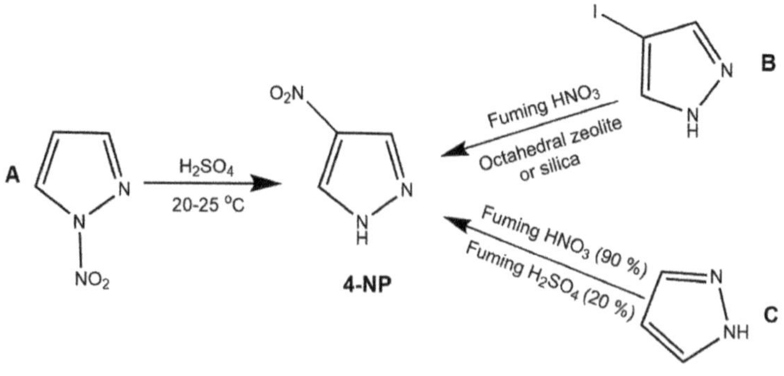

Figure 2. Synthesis of 4-NP.

Table 1 shows the energetic performances of the four typical monopyrazoles. We can see that these energetic performances of pyrazole-based compounds are not satisfying, especially the detonation properties and the nitrogen content. So, these nitropyrazoles are always used as intermediates for the preparation of novel high-performance energetic materials. Furthermore, it is also necessary to explore new high performances energetic materials based on mononitropyrazoles. For example, Deng et al. [54] prepared 5-methyl-4-nitro-1H-pyrazol-3(2H)-one (MNPO) and its energetic salts, showing better performances than these above mononitropyrazoles.

Table 1. Properties of 3-NP, 4-NP, 3-MNP and 4-MNP.

Explosive	ρ/g·cm^{-3}	D/km·s^{-1}	P/GPa	T_m/°C	OB/%	N/%	Ref.
3-NP	1.57	7.02	20.08	174–175	−77.88	37.17	[55]
4-NP	1.52	6.86	18.81	163–165	−77.88	37.17	[55]
3-MNP	1.47	6.62	17.11	80–83	−107.09	26.77	[3]
4-MNP	1.40	6.42	15.52	82	−107.09	26.77	[3]

The introduction of a polynitromethyl group into a heterocyclic compound is interesting for energetic field, because it can increase the oxygen content and improve the energetic properties of energetic material. Generally, the incorporation of a polynitromethyl group (trinitromethyl and dinitromethyl) to nitropyrazoles is essentially equivalent to introducing at least one -NO$_2$ (since one -NO$_2$ is used for the complete oxidation of the C atom in -CH$_3$) [56]. For the trinitromethyl group, it can be incorporated into N position or C position of nitropyrazoles with different energetic properties. The N-H bond of nitropyrazole is relatively active which could provide a reaction site for functionalization easily. In 2014, Yin et al. [57] obtained the carbon and nitrogen functionalization of nitropyrazole with N-trinitroethylamino group (Figure 3, Scheme A). Thereby, 4-NP reacted with NH$_2$OSO$_3$H acid and K$_2$CO$_3$ to accomplish amination, and after functionalization of amino group, the 1-amino-4-nitropyrazole underwent the Mannich reaction with trinitroethanol to get 4-nitro-N-(2,2,2-trinitroethyl)-1H-pyrazol-1-amine (**1**). In 2015, Dalinger et al. [58] prepared and characterized a nitropyrazole bearing a trinitromethyl moiety at N atom, 4-nitro-1-(trinitromethyl)-pyrazoles (**2**). They synthesized the target compound by a destructive nitration of 4-nitro-1-acetonpyrazole with a mixture of concentrated HNO$_3$ and H$_2$SO$_4$ (Figure 3, Scheme B). Although the compound **1** was successfully synthesized, the yield was very low (28%) and this process was comparatively too time-consuming (15 d). To explore new high-performance EM, several C-trinitromethyl-substituted mononitropyrazoles have been reported. In 2018, Zhang and co-authors [56] first synthesized the C-trinitromethyl-substituted nitropyrazole (Figure 4, Scheme A). The reaction of 3-pyrazolecarbaldehyde oxime with N$_2$O$_4$ produced the 3-trinitromethylpyrazole and 1-nitro-3-trinitromethylpyrazole (**3**). They found that the increasing N$_2$O$_4$ concentration could improve the proportion of **3** and 3-trinitromethylpyrazole reacting with N$_2$O$_4$ also form **3**, indicating N$_2$O$_4$ enable nitrate the N position of pyrazole. After the introduction of trinitromethyl group on C position, the 4-nitro-3-trinitromethylpyrazole (**4**) could be obtained with fuming nitric acid and oleum by -NO$_2$ rearrangement of **3** or nitration of 3-trinitromethylpyrazole. In 2019, Xiong et al. [59] further designed 3-Trinitromethyl-4-nitro-5-nitramine-1H-pyrazole (**5**). It was notable that the yield of **5** could improve with the concentration of HNO$_3$ increasing in the last nitration step of Scheme B (Figure 4). For the dinitromethyl group, Semenov et al. [60] prepared the 4-nitro-1-dinitromethylpyrazole by nitrating 4-nitro-1-acetonylpyrazole using H$_2$SO$_4$/H$_2$O mixture, and while the yield was low and it was not investigated as energetic material. In 2019, Pang et al. [61] introduced the dinitromethyl group into nitropyrazole and developed the salt, hydrazinium 5-nitro-3-dinitromethyl-2H-pyrazole (**6**), according to Scheme A in Figure 5. In 2020, Cheng et al. [62] synthesized 3-nitro-4-dinitromethyl-2H-pyrazole (**7**) and its salts, further exploring the application of dinitromethyl group in mononitropyrazolle. Table 2 shows the energetic properties of the polynitromethyl-substituted mononitropyrazoles and salts compared with TNT and RDX. All the density of the derivatives of mononitropyrazole was higher than TNT and close to that of RDX, especially **7a** showed the highest density. **3** and **5** owned the desirable detonation properties, while exhibited poor safety. It was notable that C-trinitromethyl-substituted derivatives owned higher heat of formation than those of N-trinitromethyl-substituted derivatives, and the derivatives with dinitromethyl group owned lower heat of formation than derivatives with trinitromethyl group. Most of the neutral derivatives hold low decomposition temperatures owing to the instability of the polynitromethyl moiety. Compound **4** had the highest decomposition temperature possibly because of the strong intermolecular hydrogen bonding interactions. By comparing **4** and **5**, we can see the nitramino group could further increase the power with low sensitivities. For the salts of compound **7**, **7d** with high detonation properties (comparing with RDX) and low sensitivities could serve as a promising candidate as a new high energy density oxidizer.

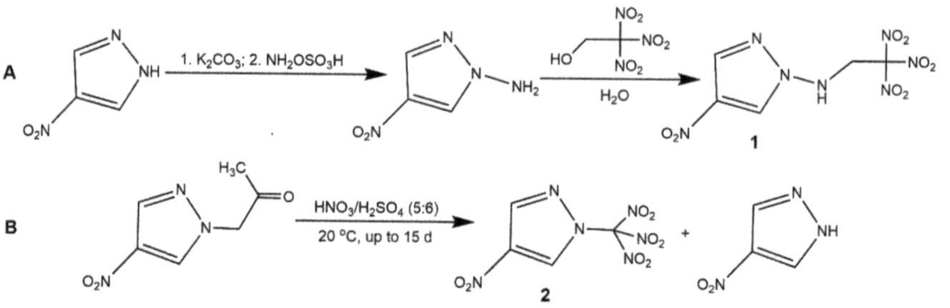

Figure 3. Synthesis of compounds **1** and **2**.

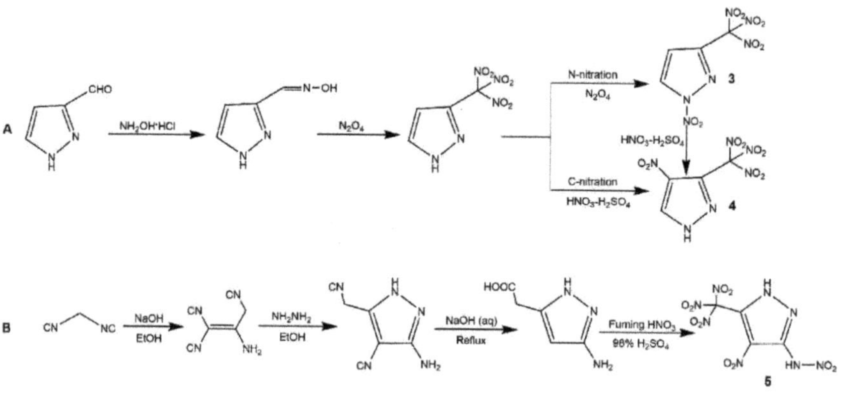

Figure 4. Synthesis of compounds **3–5**.

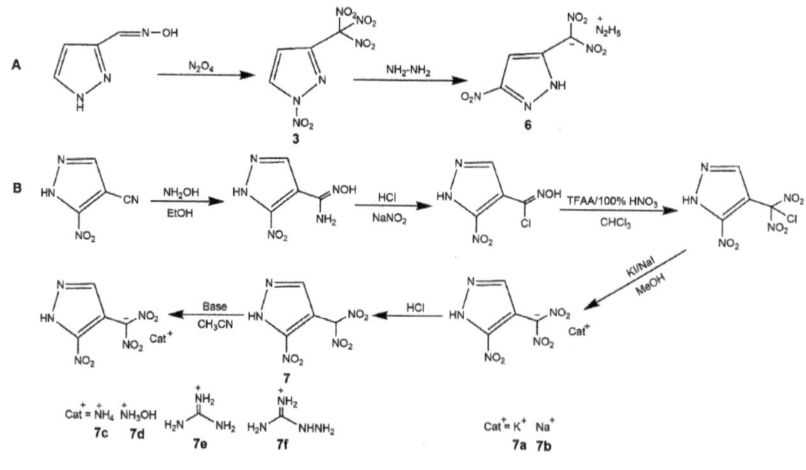

Figure 5. Synthesis of salts of trinitromethyl and dinitromethyl-substituted mononitropyrazoles.

Table 2. Properties of the derivatives of compounds 1–7.

Entry	ρ/g·cm^{-3}	D/km·s^{-1}	P/GPa	T_m/°C	T_d/°C	OB/%	N/%	HOF/kJ·mol^{-1}	IS/J	FS/N	Ref.
1	1.74	8.39	30.8	109.0	112	+2.75	33.68	140.9	15.0	360	[57]
2	1.80	8.65	33.9	39.0	145	−6.10	32.07	181.0	7.5	120	[58]
3	1.85	8.93	35.9	59.9	113	+18.3	32.07	311.4	2.5	36	[56]
4	1.80	8.60	32.2	147.2	154	+18.3	32.07	208.1	3.0	80	[56]
5	1.90	9.12	37.2	74	124	+24.8	34.79	320.2	5.0	80	[59]
6	1.84	8.79	33.8	-	128	−9.64	39.35	194.8	7.0	192	[61]
7	1.76	8.53	30.8	-	117	−25.79	32.26	205.6	17.0	114	[62]
7a	2.01	8.13	29.5	-	171	−21.94	27.44	−55.7	4.0	36	[62]
7b	1.87	8.26	28.6	-	203	−23.42	29.29	28.2	9.0	120	[62]
7c	1.78	8.60	33.0	-	141	−34.16	35.90	85.4	>20.0	192	[62]
7d	1.80	8.70	34.1	-	166	−24.20	33.60	135.5	>20.0	162	[62]
7e	1.72	8.26	28.7	-	161	−46.37	40.58	43.8	>20.0	240	[62]
7f	1.71	8.48	30.1	-	140	−46.74	43.29	231.4	>20.0	252	[62]
TNT	1.65	6.88	19.5	80.5	295	−73.8	18.49	−67.0	15.0	358	[62]
RDX	1.80	8.75	34.9	204.1	210	−21.6	37.82	70.0	7.0	120	[62]

Connecting nitropyrazoles with nitrogen-rich compounds (including tetrazole, triazole, furazan, tetrazine, triazine, and others) has attracted more interest in many fields, it also be an effective approach to increasing the content of nitrogen and getting new high-performance energetic materials. In 2015, Yin et al. [63] synthesized energetic salts based on N-methyl 6-nitropyrazolo[3,4-d][1,2,3]triazol-3(4H)-olate in a similar manner exhibiting good detonation performance with relatively low sensitivities. In 2016, Dalinger et al. [64] synthesized and investigated systematically a series of 1- and 5-(pyrazolyl)tetrazole amino and nitro derivatives which could be components of dyes and luminophores, and high-energy materials. Some of them were always used as intermediates due to their poor energetic properties. In 2017, Zyuzin et al. [65] introduced the 2,2-bis(methoxy-NNO-azoxy)ethyl group to nitropyrazoles to increase the hydrogen content for some special application (gun propellants, solid rocket propellants and others). The derivatives of 3-NP and 4-NP showed high heat of formation, while the oxygen balances and calculated detonation velocity were not ideal. Then, Zyuzin et al. [66] further introduced the trinitromethyl moiety owning the most oxygen-rich block into the combination of tetrazole and pyrazole rings to obtain oxygen-balanced energetic materials with high nitrogen content (8–11) (Figure 6). In 2019, Tang et al. [67] developed several compounds and salts based 3,5-diamino-4-nitropyrazole functionalizing the with tetrazole group and triazine group (12–15) (Figure 7). As shown in Table 3, all the compounds had high density, high nitrogen content and good detonation properties, while the thermal stability of 12–15 was better than that of 8–11. In particular, the derivatives 12–15 showed excellent insensitivities. In addition, most compounds owned positive and high heat of formation, but the presence of water molecules in 13a result in its negative heat of formation. Considering the low sensitivities, good detonation properties, and high thermal stabilities, these derivatives with nitrogen-rich groups may be the candidates of insensitive high energetic materials.

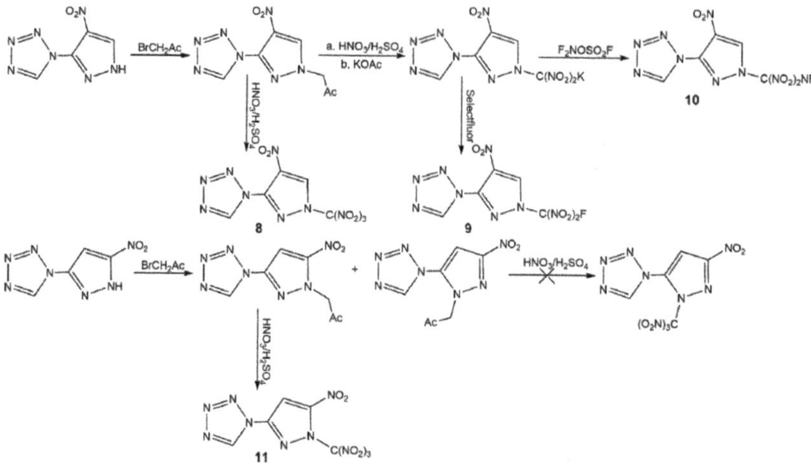

Figure 6. Synthesis of mononitropyrazole derivatives **8–11**.

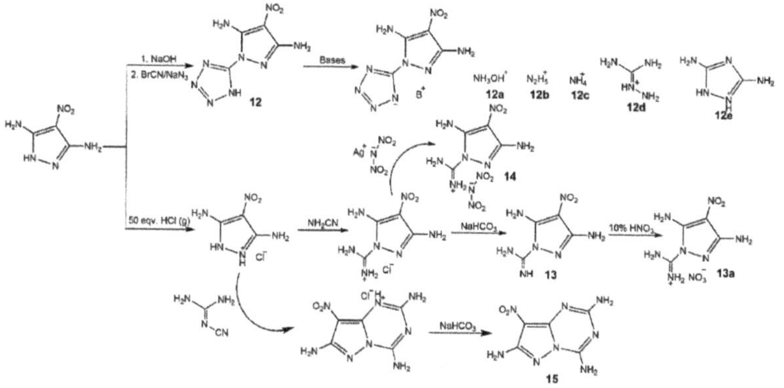

Figure 7. Synthesis of mononitropyrazole derivatives **12–15**.

Table 3. Energetic characteristics of compounds **8–15**. The data of compounds **8–11** are from reference [66], the data of compounds **12–15** are from reference [67].

Entry	ρ/g·cm^{-3}	D/km·s^{-1}	P/GPa	T_d/°C	N/%	HOF/kJ·mol^{-1}	IS/J	FS/N
8	1.79	8.86	34	127	42.43	602	-	-
9	1.81	8.47	31	111	41.59	386	-	-
10	1.91	8.99	36	138	41.67	589	-	-
11	1.76	8.78	32	132	42.43	629	-	-
12	1.76	8.26	25.9	272	59.70	408	30	360
12a	1.78	8.82	29.9	187	56.89	445	32	360
12b	1.75	8.80	28.9	229	63.36	547	35	>360
12c	1.70	8.29	24.9	251	61.39	399	40	>360
12d	1.69	8.24	23.9	224	63.84	486	40	>360
12e	1.72	8.14	23.6	287	63.21	338	40	>360
13	1.72	8.00	22.8	200	52.96	127.6	>40	>360
13a	1.68	8.00	23.8	196	45.15	−483.3	32	360
14	1.75	8.81	32.6	148	47.94	462.8	30	360
15	1.78	7.79	21.8	406	53.52	127.1	>40	>360

Moreover, nitrogen-rich heterocycles with a nitramino moiety could exhibit better performance than the corresponding nitro-substituted analogs as above mentioned [59,68]. In 2019, Shreeve and her group [69] reported a green synthetic route for high-performance nitramino nitropyrazoles. Figure 8 depicted the synthesis of corresponding derivatives, among them the 3,5-dinitramino-4-nitropyrazole (**16**) was quite sensitive to mechanical stimulation. From Table 4, the compound **16b** showed promising properties with a high density (1.87 g·cm^{-3}), good detonation properties (D of 9.58 km·s^{-1} and P of 38.5 GPa), decomposition temperature of 194 °C, and acceptable sensitivities. Xu et al. [70] introduced nitramino and triazole groups into mononitropyrazole to construct multiple hydrogen bonds (**17**), and synthesized the 4-nitro-3,5-bis(1H-1,2,4-triazol-3-nitramino)-1H-pyrazole (**19**) and its ionic derivatives (**19a–i**) as shown in Figure 9. Table 4 also showed their energetic properties. Compound **17** had the highest decomposition temperature (353.6 °C) and excellent low sensitivity (IS > 40, FS > 360), indicating it could be used as heat-resistant insensitive explosive. The compounds (**18–19i**) exhibited moderate detonation properties, high positive heat of formation and ideal insensitivities which had great potential application in green and safe energetic materials. Ma et al. [71] also fused nitropyrazole with triazine and nitramino groups, and prepared a series of salts based on compounds **20** and **21** (Figure 10). These compounds owned high thermal stability and excellent insensitive properties because of the existence of triazine ring.

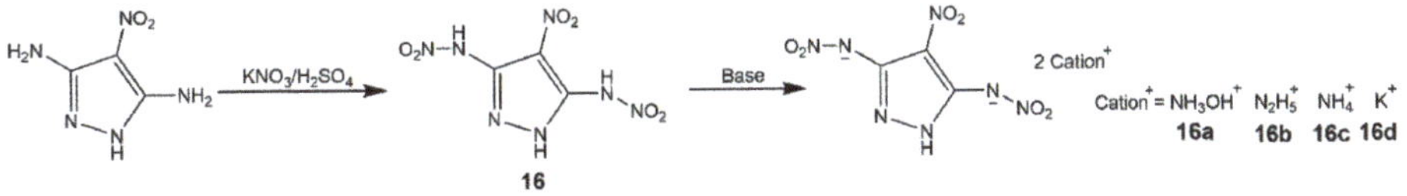

Figure 8. Synthesis of compound **16** and its salts.

Table 4. Energetic characteristics of compounds **16a–21d**. The data of compounds **16a–d** are from reference [69], the data of compounds **17–19i** are from reference [70], the data of compounds **20a·H$_2$O –21d·H$_2$O** are from reference [71].

Entry	ρ/g·cm^{-3}	D/km·s^{-1}	P/GPa	T$_d$/°C	HOF/kJ·mol^{-1}	IS/J	FS/N
16a	1.90	9.39	40.0	155.0	74	6	120
16b	1.87	9.58	38.5	194.0	266	12	160
16c	1.80	8.84	32.6	192.0	−20	15	240
16d	2.12	7.64	26.4	232.0	246	2	120
17	1.77	8.24	23.1	353.6	555.0	>40	>360
18	1.87	8.75	33.0	238.2	737.6	30	360
19	1.92	9.01	35.9	134.4	791.8	20	270
19a	1.76	8.68	30.0	186.6	711.4	>40	>360
19b	1.79	9.08	33.6	171.3	842.0	>40	>360
19c	1.73	8.76	30.2	186.6	1062.4	>40	>360
19d	1.71	8.19	25.1	195.4	677.9	>40	>360
19e	1.71	8.50	27.3	191.3	1068.7	>40	>360
19f	1.75	8.71	29.7	208.2	1014.6	22.4	>360
19g	1.72	8.12	25.7	168.5	1300.5	>40	>360
19h	1.74	8.16	26.0	189.7	1270.5	>40	>360
19i	1.72	8.14	25.9	175.9	1511.2	>40	>360
20a	1.82	8.39	28.2	180.0	60.0	>40	360
20b	1.83	8.10	28.0	279.0	105.0	40	240
21	1.89	8.71	31.9	248.0	314.6	>40	>360
21a·H$_2$O	1.95	8.29	29.1	341.0	260.9	>40	>360
21b·H$_2$O	1.81	8.98	32.1	218.0	386.2	>40	>360
21c·H$_2$O	1.80	9.06	31.7	190.0	557.5	>40	>360
21d·H$_2$O	1.60	8.22	24.6	223.0	690.0	>40	>360

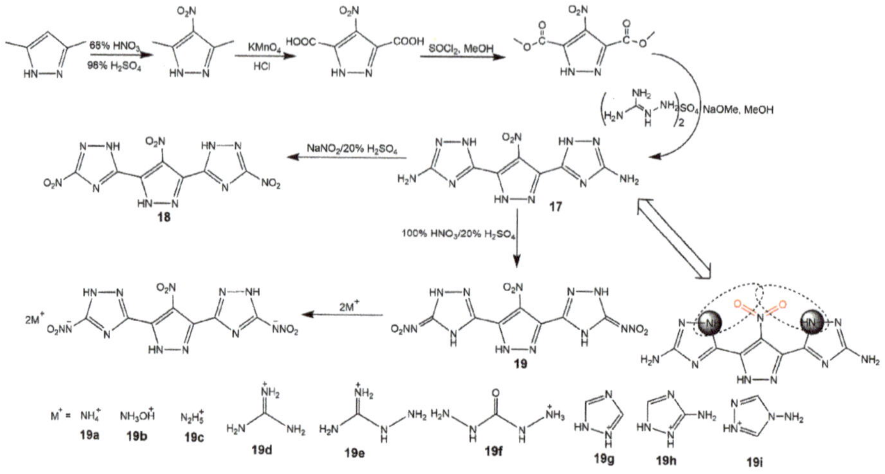

Figure 9. Synthesis of compounds 17–19i.

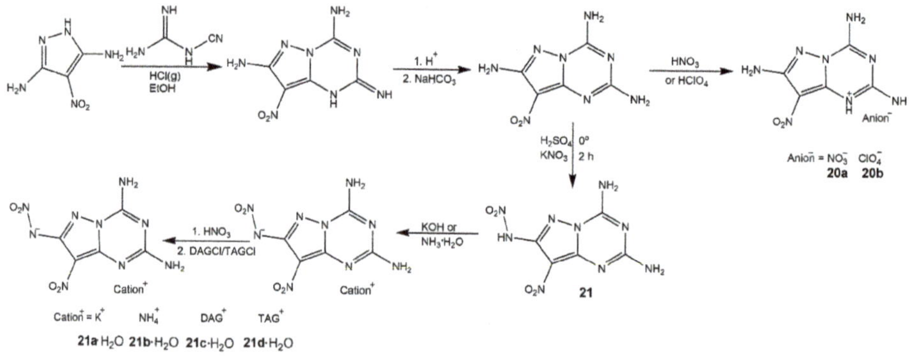

Figure 10. Synthesis of compounds 20a–21d.

In summary, most of mononitropyrazoles and their derivatives owned relatively low thermal properties and detonation properties. They are always used as intermediates for novel complicated energetic materials. The introduction of polynitromethyl group can improve the oxygen balance efficiently, while have a little influence on the heats of formation. The nitramino group and nitrogen-rich heterocyclic can enhance the detonation properties, improve the safety, and increase the heats of formation of mononitropyrazoles. The choice of solvent and nitrification in synthesis routes should be more environmental and facile.

2.2. Dinitropyrazoles and Their Derivatives

Dinitropyrazoles own higher density and better detonation performance than mononitropyrazoles attributing to one more nitro group. The typical dinitropyrazoles include 3,4-dinitropyrazole (3,4-DNP), 3,5-dinitropyrazole (3,5-DNP), 1-methyl-3,4-dinitropyrazole (3,4-MDNP), 1-methyl-3,5-dinitropyrazole (3,5-MDNP), and 4-amino-3,5-dinitropyrazole (LLM-116).

3,4-DNP is a kind of white crystal, possessing higher density (1.87 g·cm^{-3}), lower melting point (86–88 °C), higher decomposition temperature (285 °C), higher detonation velocity (8.1 km·s^{-1}) and detonation pressure (29.4 GPa) than TNT. This compound was first reported by Biffin's team in 1966 [72]. In an earlier study, pyrazole, 4-NP, 3-nitro-4-cyanopyrazole and other raw materials have been investigated to prepare 3,4-DNP, while most of the methods did not satisfied industrialization due to complex process, high production cost or low yield [45,55,73–76]. At present, the three-step synthetic route as shown in Figure 11 (Scheme A), and the two-step route (Scheme B) are the most widely used [77–80]. 3,4-MDNP is a typical thermal stability nitropyrazole, exhibiting stable thermodynamic state at 300 °C. Its melting point and density are lower than those of 3,4-DNP (20–23 °C, 1.67 g·cm^{-3}), and 3,4-DNP shows low detonation velocity (7.76 km s^{-1}) and detonation pressure (25.57 GPa) due to the introduction of methyl group. It has potential application in liquid explosive, which can reduce the melting point of liquid phase carrier in castable explosive [32]. Recently, Ravi et al. [73] had synthesized 3,4-MDNP by nitrating 1-methylpyrazole or 1-methyl-3-nitropyrazole with montmorillonite (K-10) and Bi(NO$_3$)$_3$, while this method was high cost and the products were difficult to separate. Li et al. [81] reacted 3,4-DNP and dimethyl carbonate (DMC) in DMF with K$_2$CO$_3$ as catalyst, then, his group further synthesized 3,4-MDPN with 3-NP as raw material (Figure 11, Scheme C) [82]. In this method, DMC was used as methylation agent and the yield of methylation was high (95.6%), which could meet the requirement of green chemistry. As 3,5-DNP with a melting point of 173–174 °C and density of 1.80 g·cm^{-3}, the decomposition temperature of 316.8 °C owns higher detonation properties than 3,4-DNP (7.76 km·$^{-1}$ and 25.57 GPa). Moreover, 3,5-DNP is relatively stable because of the symmetrical molecular distribution, it can be used as a simple explosive or as a key intermediate in the synthesis of insensitive explosives [55]. Generally, the starting materials for preparing 3,5-DNP could be pyrazole and 3-NP. Wang et al. [83] nitrated 3-NP to get 1,3-dinitropyrazole, then 1,3-dinitropyrazole was reacted with NH$_3$ in PhCN to produce the ammonium salt of 3,5-DNP. After neutralization with hydrochloric acid, the 3,5-DNP could be obtained (Figure 12, Scheme A). Liu et al. [28] also nitrated 3-NP, and rearranged 1,3-dinitropyrazole to get 3,5-DNP (Figure 12, Scheme B). For pyrazole as starting material, 3,5-DNP was always prepared by a four-step route (nitration of pyrazole, rearrangement of N-nitropyrazole, nitration of 3-NP, and rearrangement of 1,3-dinitropyrazole). 3,5-MDNP owns the similar energetic properties with 3,4-MDNP, while it has a higher melting point (about 60 °C). Moreover, 3,5-MDNP could be synthesized by methylation of 3,5-DNP [84]. However, most methylation agents were extremely toxic, thus searching for a green methylation agent would be the key factor.

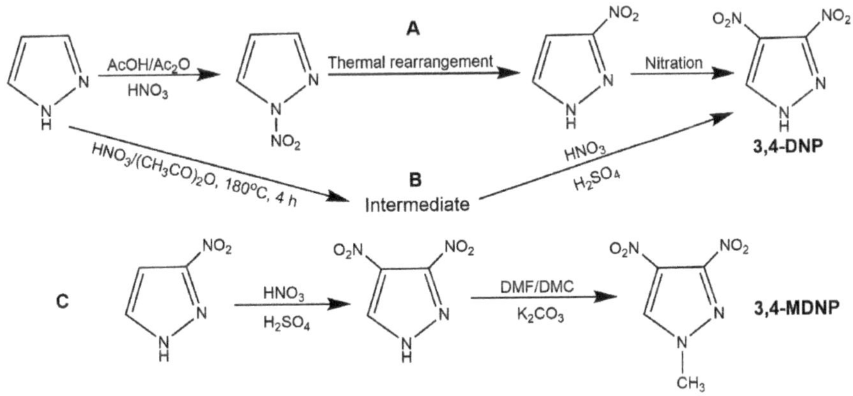

Figure 11. Synthesis of 3,4-DNP and 3,4-MDNP.

LLM-116 is a powerful and insensitive explosive, its energy is 90% of HMX and its impact sensitivity is extremely low [55,85]. It was first synthesized by the Lawrence Livermore National Laboratory (LLNL) in 2001, and many studies were performed to assess its synthesis in the following years. Wang et al. [86] utilized vicarious nucleophilic substitution (VNS) of 3,5-DNP and trimethylhydrazine iodideto (TMHI) to prepare LLM-116 with a yield of 60%, while the toxic TMHI was the main factors restricting wide application of this method. In 2014, Stefan et al. [87] developed four synthetic routes of LLM-116, using 4-NP, 3,5-dimethylpyrazole, 3,5-DNP and 4-chloropyrazole as starting materials, respectively (Figure 13, Scheme A–D). Table 5 shows the comparison of the four routes. The synthesis of

Scheme D was simple and its yield was high, which was suitable for industrialization. Zhang et al. [88] also used 4-chloropyrazole as a starting material to synthesize LLM-116 with an overall yield of 65%.

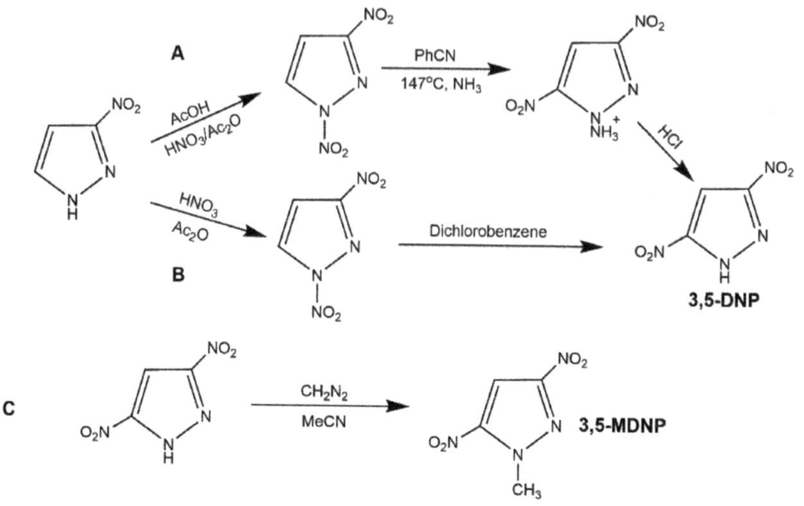

Figure 12. Synthesis of 3,5-DNP and 3,5-MDNP.

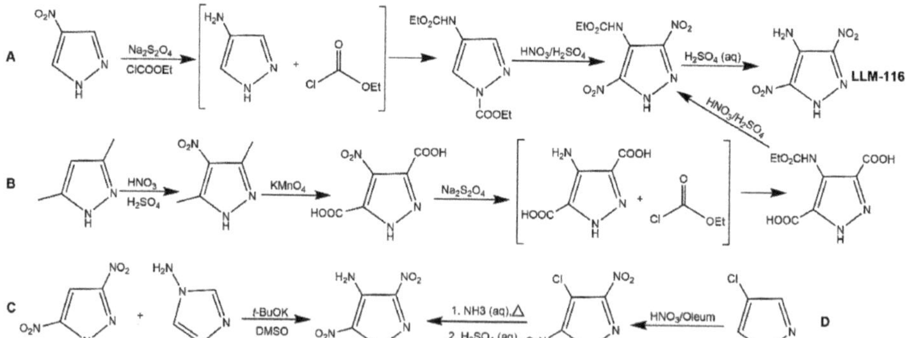

Figure 13. Synthesis of LLM-116.

Table 5. A brief comparison of four routes by Stefan.

Via 4-NP (Method A)	Via 3,5-Dimethylpyrazole (Method B)	Via 3,5-DNP (Method C)	Via 4-Chloropyrazole (Method D)
Four steps Moderate amount of waste No unfavorable solvents required Moderate overall yield, 40% Average yield/step: 80%	Six steps High amount of waste No unfavorable solvents required Moderate overall yield, 37% Average yield/step: 85%	Five steps Moderate amount of waste DMSO used in the last step Low overall yield, 21% Average yield/step: 73%	Two steps Small amount of waste No unfavorable solvents required Moderate overall yield, 61% Average yield/step: 78%

In addition, 4-Chloro-3,5-dinitropyrazole was a useful intermediate in the preparation of various 3,5-DNP [89], owning good reactivity towards nucleophiles. He et al. [90] synthesized a series of 3,5-DNP derivatives based on 4-chloro-3,5-dinitropyrazole and 1-methyl-4-chloro-3,5-dinitropyrazole shown in Figure 14. From Table 6, all compounds exhibited better detonation properties than those of TNT, and these compounds owned better IS than RDX except compound **33**. Compounds **26** and **28** had an especially good balance between physical properties and detonation properties as well as excellent insensitivity, making them potential replacement of RDX.

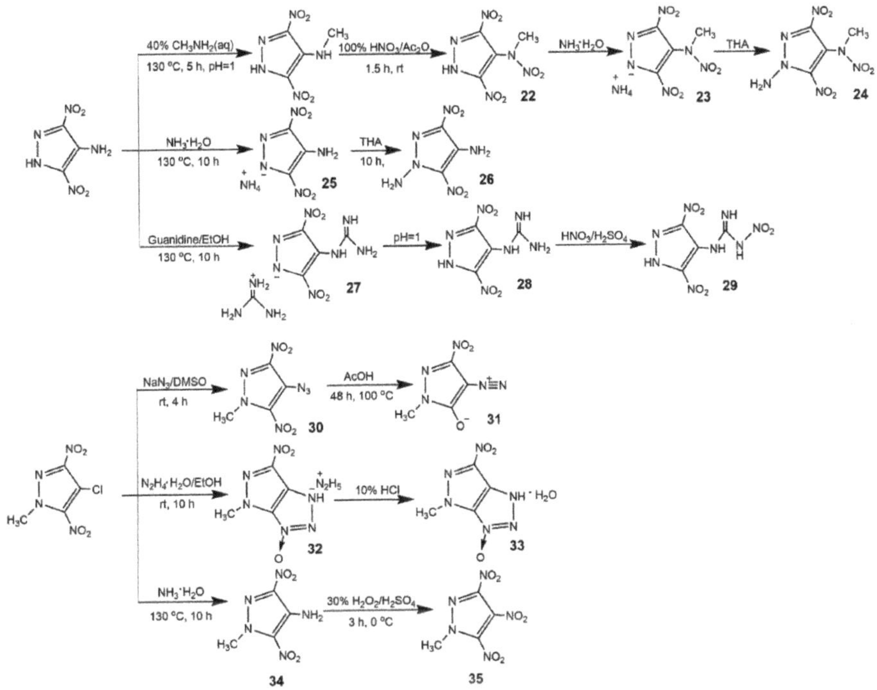

Figure 14. Synthesis of derivatives based on 4-amino-3,5-dinitropyrazole and 1-methyl-4-chloro-3,5-dinitropyrazole.

Table 6. Physical and detonation properties of compounds **22–35**. The data of compounds **22–35** are from reference [90].

Entry	ρ/g·cm^{-3}	D/km·s^{-1}	P/GPa	T_d/°C	HOF/kJ·mol^{-1}	IS/J
22	1.74	8.22	30.1	178	137.0	17
23	1.69	8.25	28.7	176	104.6	35
24	1.72	8.31	30.2	176	220.7	18
25	1.63	8.14	26.3	275	64.8	>60
26	1.88	8.73	35.0	241	166.0	>40
27	1.66	7.82	23.4	245	133.5	>40
28	1.84	8.46	31.0	308	182.6	>40
29	1.78	8.39	31.4	233	236.6	10
30	1.74	8.41	31.0	161	436.0	14
31	1.63	7.42	21.7	228	177.0	22
32	1.71	8.72	30.9	146	549.6	8
33	1.70	8.18	27.6	101	414.4	6
34	1.67	7.80	24.6	270	64.5	>40
35	1.78	8.25	31.2	285	109.1	>40

Energetic salts often possess superior properties comparing with non-ionic species since they always show lower vapor pressures, lower impact and friction sensitivities, and enhanced thermal stabilities [19]. In addition to the derivatives mentioned above, Klapötke group [26] developed the ionic salts of 3,4-DNP and 3,5-DNP shown in Figure 15, and these salts were extremely insensitive in Table 7. Comparing with 3,4-DNP, **36** and **38** owned much lower decomposition temperatures, similar to that of **37**, **39** and 3,5-DNP. Zhang et al. [91] developed the ionic salts of LLM-116 with several nitrogen-rich cations as shown in Figure 16. These compounds showed extraordinary insensitivity to impact (>60 J), as the detonation properties of **40i** and **41k** were comparable to those of TATB (31.15 GPa, 8.11 km·s^{-1}) (Table 7).

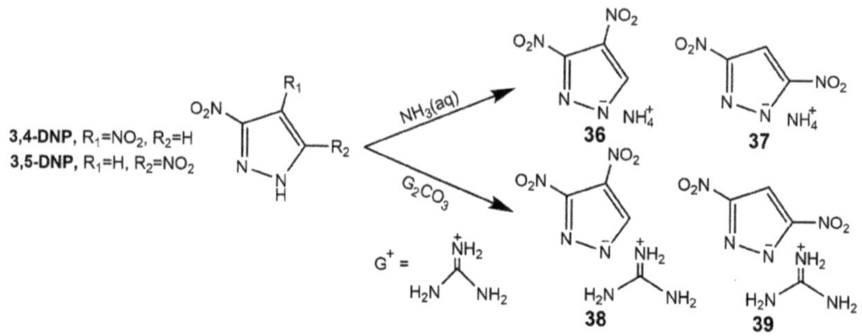

Figure 15. Synthesis of ionic salts of 3,4-DNP and 3,5-DNP.

Table 7. Physical and detonation properties of ionic salts of dinitropyrazoles. The data of compounds **36–39** are from reference [26], the data of compounds **40a–40o** and TATB are from reference [91].

Entry	ρ/g·cm^{-3}	D/km·s^{-1}	P/GPa	T_d/°C	HOF/kJ·mol^{-1}	IS/J	FS/N
36	1.69	-	-	127	-	40	360
37	1.70	8.11	25.9	300	-	40	360
38	1.63	7.59	21.1	156	-	40	360
39	1.59	7.32	19.1	295	-	40	360
40a	1.63	8.14	26.3	275	64.8	>60	-
40b	1.64	8.19	26.4	221	222.6	>60	-
40c	1.63	7.72	21.6	303	36.1	>60	-
40d	1.69	8.24	25.2	223	140.1	>60	-
40e	1.62	7.44	22.7	179	310.4	>60	-
40f	1.67	7.73	22.4	257	283.6	>60	-
40g	1.73	8.12	25.8	223	411.1	>60	-
40h	1.79	8.42	27.2	270	241.6	>60	-
40i	1.84	8.74	32.6	193	211.9	>60	-
41j	1.67	8.35	25.9	201	250.5	>60	-
41k	1.71	8.75	28.9	229	356.9	>60	-
41l	1.72	7.98	24.2	169	100.4	>60	-
41m	1.73	7.94	23.1	243	−166.3	>60	-
41n	1.54	7.71	21.0	206	389.3	>60	-
41o	1.60	7.78	22.4	173	471.8	>60	-
TATB	1.93	8.11	31.2	324	−140.0	50	-

N-oxidation of nitrogen-rich heterocycles including transformation of amino group to nitroso, azoxy, or nitro groups is another approach to designing HEDMs, which opens new avenues for the development of HEDMs [92,93]. The efforts to developing N-oxidation of dinitropyrazoles have been made recently. Bölter et al. [94] introduced -OH on N atom of 3,4-DNP and 3,5-DNP, and obtained several salts (Figure 17, Scheme A). From Table 8, these compounds were less sensitive than RDX, and did not exhibited excellent detonation properties. Yin et al. [95] synthesized a family of 4-amino-3,5-dinitro-1H-pyrazol-1-ol (**44**) and its ionic derivatives (**44a–f**) (Figure 17, Scheme B). Except **44·H$_2$O**, all the compounds (**44a–f**, and **45**) with thermal decomposition temperatures (169–216 °C) shown good balance between detonation properties and insensitive properties as shown in Table 8. Zhang et al. [96] synthesized the 4-nitramino-3,5-dinitropyrazole by nitrating the -NH$_2$ of LLM-116, and prepared several energetic salts which exhibited good insensitivity and moderate detonation properties.

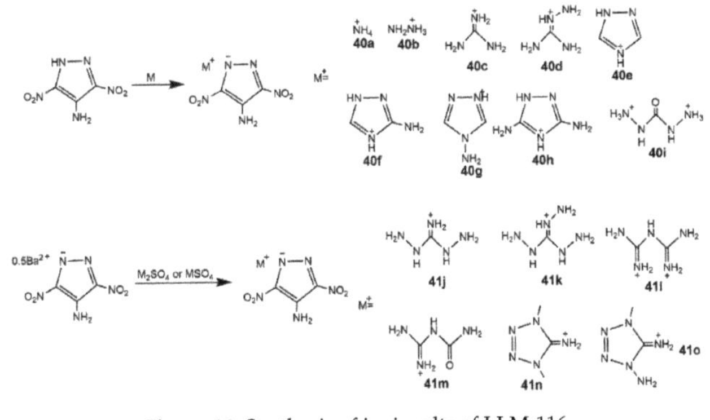

Figure 16. Synthesis of ionic salts of LLM-116.

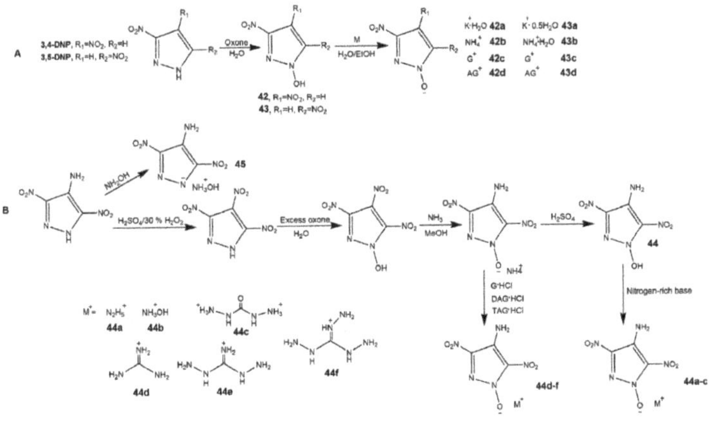

Figure 17. Synthesis of ionic salts of dinitropyrazoles.

Table 8. Physical and computational properties of ionic salts of dinitropyrazoles. The data of compounds 42a–43d are from reference [94], the data of compounds 44·H$_2$O–45 are from reference [95].

Entry	ρ/g·cm^{-3}	D/km·s^{-1}	P/GPa	T_d/°C	IS/J	FS/N
42a	1.96	7.92	26.9	197	5	216
42b	1.68	8.28	28.2	167	10	360
42c	1.70	8.06	25.1	180	30	360
42d	1.64	8.02	24.7	169	10	360
43a	-	-	-	229	6	240
43b	1.62	7.91	24.4	224	30	288
43c	1.68	7.94	24.2	266	40	360
43d	1.68	8.16	25.7	131	10	360
44·H$_2$O	1.86	-	-	93	20	240
44a	1.79	8.94	34.4	216	25	240
44b	1.86	9.00	37.6	182	35	360
44c	1.84	8.80	34.0	175	40	360
44d	1.71	8.20	26.4	204	40	360
44e	1.71	8.54	28.0	169	40	360
44f	175	8.88	30.7	214	40	360
45	1.80	8.81	33.9	212	40	360

As mentioned above, polynitromethyl are considered to be more favorable groups to give remarkable improvements in densities and detonation properties of energetic materials. Especially the N-trinitroethylamination of nitropyrazole is more available since it is stable to be handled safely. The N-trinitroethylamination of dinitropyrazole was firstly proposed by Shreeve team [57]. They obtained several N-amino-dinitropyrazoles firstly, then these compounds underwent Mannich reactions with trinitroethanol to acquire the corresponding derivatives (46–50) (Figure 18, Scheme A). It was noteworthy that 1-amino-3,5-dinitropyrazole and 1-amino-3,4-dinitro-5-cyanopyrazole failed to get the corresponding compounds due to the electron-withdrawing effect of substituent groups bonded to dinitropyrazole ring. In addition, they employed an alternative synthetic method to obtain 1,5-diamino-3,4-dinitropyrazole (51) (Figure 18, Scheme B) because attempted amination of this compound using TsONH$_2$ acid or NH$_2$OSO$_3$H failed. From Table 9, although the azido-functionalized dinitropyrazole (47) decomposed at 121 °C, compound 46 and 51 had high decomposition temperatures, and 47 and 50–52 owned higher density than RDX. These indicated the introduction of an -NH$_2$ could enhance density. In addition, N-trinitroethylamination of dinitropyrazole (48–50 and 52) shown high HOF and good detonation properties. N-trinitromethyl moiety was introduced by Dalinger's team [58], they synthesized 3,4-dinitro-1-(trinitromethyl)-pyrazoles (53) and 3,5-dinitro-1-(trinitromethyl)-pyrazoles (54) with excellent physical and computational properties as shown in Figure 19. They were a little less insensitive than the RDX and PETN, similar to N-trinitroethylamination dinitropyrazoles shown in Table 9. Fluorine and fluorinated functional groups are importantly promising substituents in the field of energetic materials [97]. C(NO$_2$)$_2$F and C(NO$_2$)$_2$NF$_2$ moieties bring high energy, maintaining high density and good thermal property were incorporated into dinitropyrazole by fluorinated compound 55 (Figure 19, Scheme C). The two compounds had high density ($\geq$1.92 g·cm^{-3}), good oxygen balance (+2.55% for 57 and 0% for 56), and high detonation pressure and velocity [98].

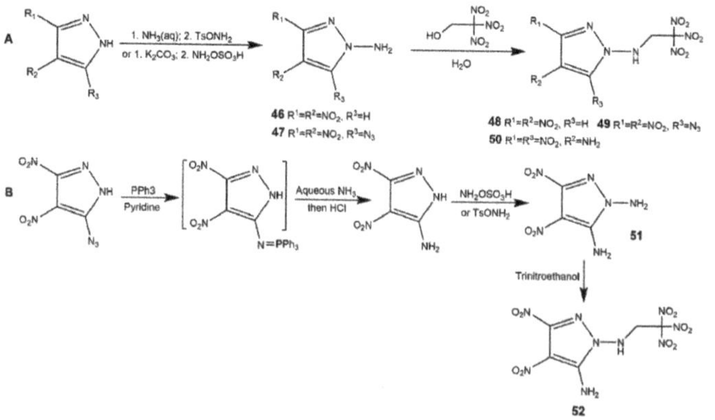

Figure 18. N-Trinitroethylamination of dinitropyrazole.

Table 9. Physical and computational properties of several polynitropyrazoles. The data of compounds 46–52 are from reference [57], the data of compounds 53–54 are from reference [58].

Entry	ρ/g·cm^{-3}	D/km·s^{-1}	P/GPa	T$_m$/°C	T$_d$/°C	HOF/kJ·mol^{-1}	IS/J	FS/N
46	1.71	7.46	20.1	58	241	200.3	>40	360
47	1.82	9.05	35.8	120	121	548.2	1.5	5
48	1.78	8.67	33.1	87	110	142.3	6	80
49	1.82	9.00	35.6	-	117	491.7	2.5	20
50	1.81	8.75	34.3	-	116	124.1	12	120
51	1.82	8.69	32.8	133	238.2	173.0	>40	360
52	1.83	8.80	35.0	-	134.4	112.0	8	80
53	1.91	8.67	35.5	80	157	244.0	8	130
54	1.94	8.73	36.6	81	159	206.0	9	145

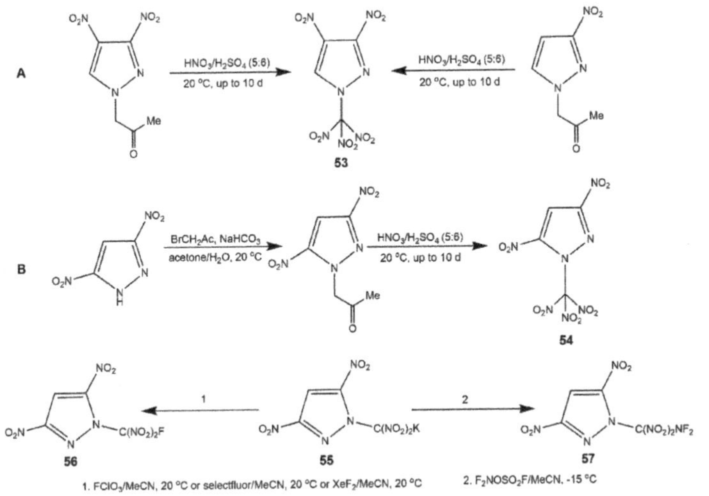

Figure 19. Synthesis of compounds 53–57.

Dinitropyrazoles bearing other heterocycles are also interesting and notable. To obtain the melt-castable explosives with good compatibility, improved oxygen balance and moderate detonation properties, compound **58** incorporating both *N*-trinitromethyl and *C*-methyl substituents in addition to nitro groups was synthesized by Sheremetev's group [99] (Figure 20). This low melting temperature compound has been proved to own higher detonation pressure and velocity values than those of others melt-castable energetic heterocycles bearing methyl group, which provided feasible route to castable energetic materials. In addition, introduction of polynitrogen heterocycle and formation of energetic salts are main methods to improve the thermal stability of explosives [100]. In 2016, a heat-resistant energetic material, compound **59** bearing triazole ring, was synthesized using 5-amino-3-nitro-1*H*-1,2,4-triazole (ANTA) and 3,4,5-trinitrated-1*H*-pyrazole (TNP), and several salts based on it were developed by Zhou et al. [101] (Figure 21, Scheme A). As shown in Table 10, compound **59** had high decomposition temperature (270 °C) and high positive HOF (833 kJ·mol^{-1}). All the salts showed good thermal stability, excellent insensitivity, and good detonation properties. In particular, the guanidinium salt **59d** exhibited the best thermal stability superior than that of most explosives. Considering thermal stability and energetic properties, compounds **59** and **59d** could be used as heat-resistant explosives and it was possible that these compounds can be applied as heat-resistant materials. Afterwards, their group reported a family of unsymmetrical *N*-bridged dinitropyrazoles synthesized by TNP and 5-amino-1*H*-tetrazole (ATZ) and its organic salts (Figure 21, Scheme B). Several compounds (**60**, **60b**, and **60c**) with high N contents exhibited superior detonation velocities but inferior detonation pressures compared to HMX and insensitivities to impact (IS > 40 J) and friction (FS > 360 N) comparable to those of TATB (Table 10), which could be promising insensitive HEDMs for practical application.

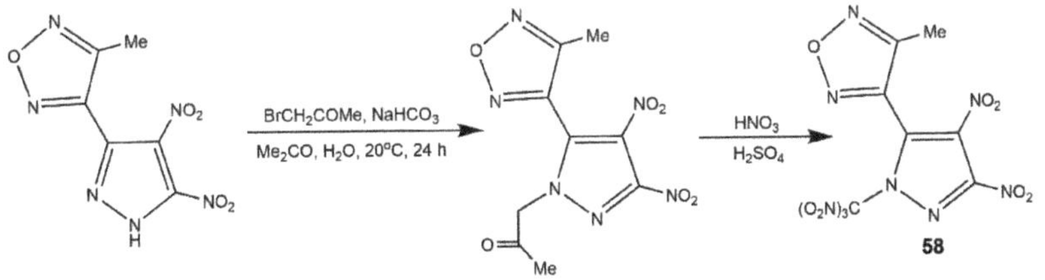

Figure 20. Synthesis of compound 58.

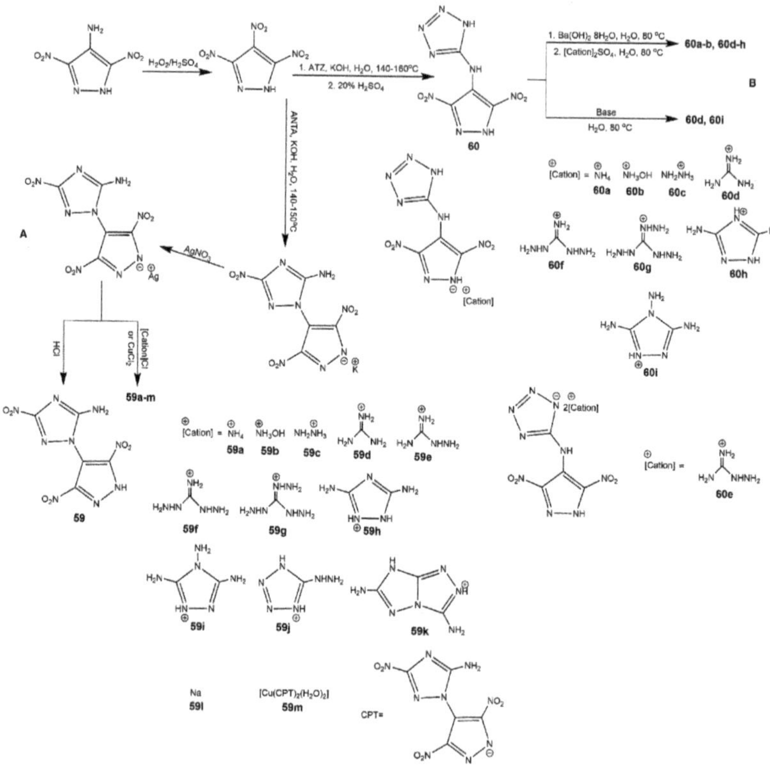

Figure 21. Synthesis of compounds 59–60i.

Table 10. Physical and computational properties of 59–60i. The data of compounds 59–59m are from reference [101], the data of compounds 60–60i and HMX are from reference [88].

Entry	ρ/g·cm^{-3}	D/km·s^{-1}	P/GPa	N/%	T_d/°C	HOF/kJ·mol^{-1}	IS/J	FS/N
59	1.84	9.17	37.8	44.2	270	833.4	9	240
59a	1.73	8.62	31.6	46.4	285	622.8	>40	>360
59b	1.76	8.83	34.4	44.0	215	709.9	33	252
59c	1.74	8.80	32.9	48.6	241	811.4	>40	>360
59d	1.78	8.66	31.1	48.8	340	624.1	>40	>360
59e	1.65	8.24	26.5	50.7	281	728.9	>40	>360
59f	1.70	8.54	28.9	52.4	262	831.7	27	240
59g	1.71	8.69	30.0	54.0	242	941.4	20	216
59h	1.72	8.36	28.8	51.0	279	828.3	>40	252
59i	1.74	8.56	29.5	52.6	292	944.8	>40	>360
59j	1.80	9.03	35.2	54.5	222	1211.7	12	252
59k	1.77	8.65	30.5	54.2	303	1166.1	20	>360
59l	1.75	-	-	41.0	261	-	7.5	252
59m	1.91	-	-	37.7	281	-	5	216
60	1.86	9.29	38.6	52.3	279	856.4	35	240
60a	1.79	8.95	33.3	54.3	299	672.6	>40	168
60b	1.84	9.23	37.4	51.1	296	719.7	>40	216
60c	1.84	9.36	37.0	56.4	290	819.8	>40	360
60d	1.67	8.26	25.9	56.0	256	648.8	>40	360
60e	1.72	8.76	28.6	61.2	216	808.7	>40	32
60f	1.79	9.07	32.5	59.4	285	844.9	>40	288
60g	1.81	9.29	34.2	60.9	287	954.9	>40	84
60h	1.84	8.95	32.2	57.6	286	840.7	>40	360
60i	1.82	9.00	32.3	59.1	261	960.0	>40	360
HMX	1.91	9.19	39.7	37.8	287	104.8	7.4	120

In summary, some dinitropyrazoles and derivatives exhibit low melting points and high decomposition temperatures as well as good detonation, which can make them competitive candidates for a castable explosive. To further improve the performance of dinitropyrazole-based energetic materials, a combination of several functional groups should be better, for example, the combination of nitramine and polynitrogen heterocyclic which can endow them with high thermal stability and good detonation performance.

2.3. Trinitropyrazole and Its Derivatives

TNP is the unique pyrazole compound by total carbon nitrification [102]. This compound owns good thermal stability (260–350 °C of T_d) and chemical stability, and shows high detonation velocity (9.0 km·s^{-1}) and detonation pressure (37.09 GPa). Wu et al. reviewed the synthesis of TNP in recent years in detail [102], including direct nitration methods, amino oxidation method, amino diazotization method, iodo nitrification method and microwave rearrangement method. The typical synthesis of TNP is the oxidation of LLM-116 rather than 5-amino-3,4-dinitropyrazole, and this is partly because the amino group in LLM-116 has higher electron cloud density and steric hindrance than amino group in 5-amino-3,4-dinitropyrazole, which can promote the intermolecular oxidation reaction and avoid the occurrence of intermolecular side reaction effectively, and partly because the "NO$_2$-NH$_2$-NO$_2$" framework in LLM-116 makes it more stable and easier to synthesize. In addition, the nitrification of 3,5-DNP is another typical synthesis route of TNP. Traditional oxidation methods have the following defects: harsh reaction conditions, poor selectivity, by-products, high risk factor, expensive metal catalyst and toxic organic solvent. Although the synthesis of TNP with LLM-116 and 3,5-DNP as starting materials are mature, the synthesis of LLM-116 and 3,5-DNP are complicated. It is necessary to explore novel synthesis method. Zhao et al. [44] used LLM-116 as starting material, water as solvent, and KHSO$_5$ as oxidant to synthesize TNP. Ravi et al. [103] put forward the nitration system of metal nitrate and studied the process of nitration to TNP. These two methods are promising to prepare TNP.

Moreover, 1-methyl-3,4,5-trinitropyrazole (MTNP), a derivative of TNP, is an insensitive energetic material with 91.5 °C of melting point, 248–280 °C of decomposition temperature, 8.65 km·s^{-1} of detonation velocity, and 33.7 GPa of detonation pressure [104]. Ravi et al. [103] added K-10 and TNP to bismuth impregnated in THF to obtain MTNP (Figure 22, Scheme A). There were also many routes to synthesize MTNP. Dalinger et al. [105,106] dissolved TNP in NaHCO$_3$ aqueous solution with Me$_2$SO$_4$ as methylation reagent to acquire MTNP (Figure 22, Scheme B). Guo et al. [107] synthesized MTNP from 1-methyl-pyrazole by one-step method with nitric acid and fuming sulfuric acid (Figure 22, Scheme C). Among these methods, selection of highly efficient catalytic synthesis process and low toxicity methylation reagent are the trend in MTNP synthesis. In addition, 1-amino-3,4,5-trinitropyrazole (ATNP) is also a derivative of TNP with excellent detonation properties (D = 9.17 km·s^{-1} and P = 40.9 GPa) and thermal stability [108]. This was reported by Herve et al. [93], and the synthesis route is shown in Scheme D of Figure 22 (Pic-O-NH$_2$ = 2,4,6-trinitrophenyl-O-hydroxylamine) with a yield of 26%.

The N-H bond in TNP is easy to neutralize with alkali or react with metal salts forming energetic salts due to the stereoscopic structure and spatial effect of pryazole ring. These energetic salts further broaden the application of TNP. Zhang et al. [109] prepared a series of energetic salts of TNP based on nitrogen-rich cations (**61a–m**) (Figure 23, Scheme A), all the salts showed poorer densities and detonation properties than TNP (Table 11), but they owned good thermal stability and excellent insensitivity. Drukenmuller et al. [110] reported the synthesis of alkali and earth alkali trinitropyrazolate (**62a–d**) (Figure 23, Scheme B), compound **62d** exhibited predominantly decomposition temperatures (Table 11). They also prepared pyrotechnic formulations using **62c** and **62d**, which showed good color properties and low sensitivity as well as high T_d. In addition, Shreeve's group [111] synthesized 3,4,5-trinitropyrazole-1-ol (**63**) and its nitrogen-rich salts (**63a–g**) (Figure 24) the corresponding properties are shown in Table 11. Compound **63** with its high oxygen content (51.13%) could be the green replacement of the currently used oxidizer (NH$_4$ClO$_4$), while the high IS

(1 J) restricted its application. Compound **63a–g** with acceptable impact sensitivities and detonation performance could be useful energetic materials.

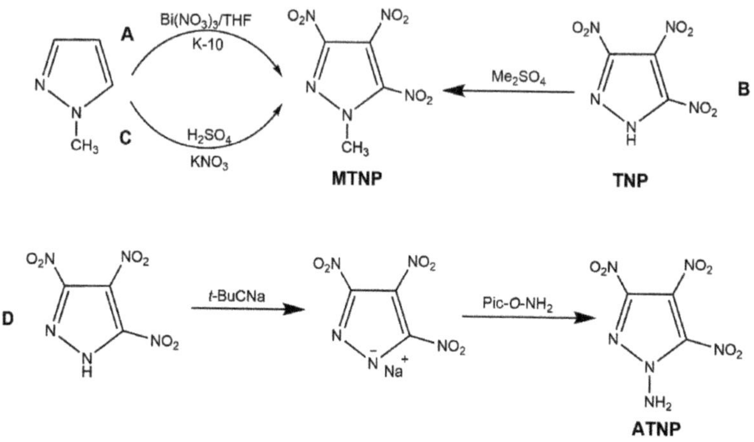

Figure 22. Synthesis of MTNP and ATNP.

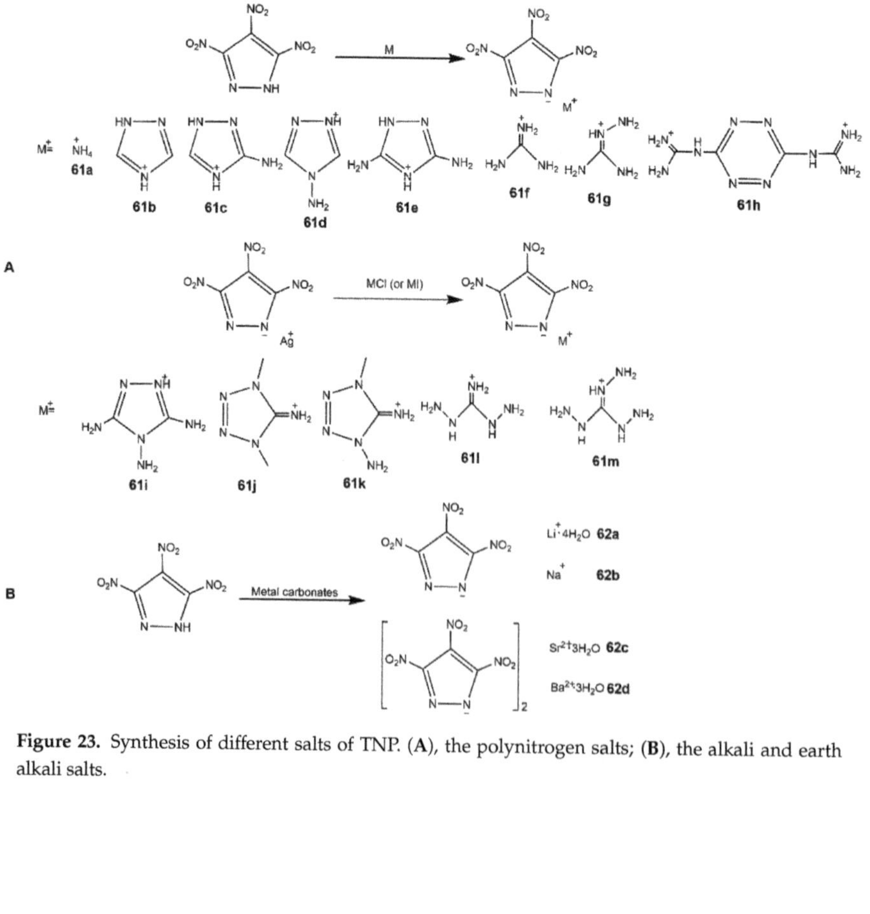

Figure 23. Synthesis of different salts of TNP. (**A**), the polynitrogen salts; (**B**), the alkali and earth alkali salts.

Table 11. Property parameters of salts of TNP. The data of compounds **61a–61m** are from reference [109], the data of compounds **62a–62d** are from reference [110], the data of compounds **63–63g** are from reference [111].

Entry	ρ/g·cm^{-3}	D/km·s^{-1}	P/GPa	T_d/°C	HOF/kJ·mol^{-1}	IS/J	FS/N
61a	1.73	8.46	29.9	224	60.5	40	-
61b	1.69	7.87	25.6	167	299.0	>40	-
61c	1.71	7.97	26.0	171	273.5	>40	-
61d	1.77	8.54	31.9	168	401.2	>40	-
61e	1.76	8.22	27.7	196	235.6	>40	-
61f	1.66	7.87	24.7	235	28.3	>40	-
61g	1.69	8.13	26.9	222	133.6	>40	-
61h	1.68	7.82	24.3	243	452.3	>40	-
61i	1.76	8.36	28.8	206	355.0	>40	-
61j	1.61	7.59	23.7	219	375.0	>40	-
61k	1.64	7.92	25.2	167	459.8	35	-
61l	1.62	7.98	25.3	197	246.5	>40	-
61m	1.65	8.24	27.2	184	352.7	>40	-
62a	-	-	-	274	-	40	96
62b	-	-	-	254	-	25	80
62c	-	-	-	193	-	40	80
62d	-	-	-	302	-	5	144
63	1.90	8.67	36.4	146	118.5	1	-
63a	1.82	8.68	35.1	176	35.1	6	-
63b	1.72	8.18	28.8	171	3.1	>40	-
63c	1.73	8.18	29.5	140	274.9	>40	-
63d	1.73	8.18	29.2	132	250.5	>40	-
63e	1.74	8.15	30.8	118	381.6	>40	-
63f	1.76	8.26	29.7	186	213.7	>40	-
63g	1.77	8.44	31.1	185	331.9	>40	-

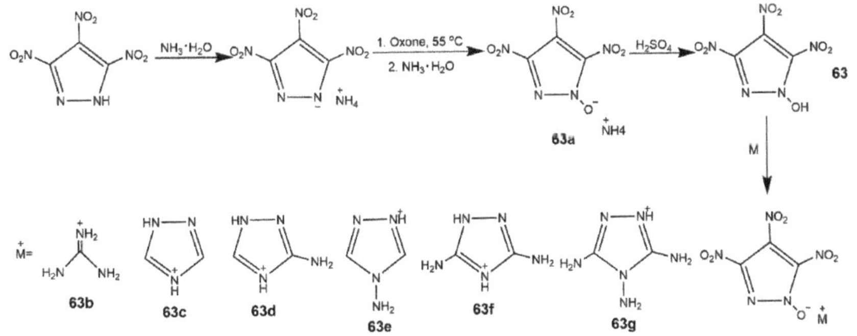

Figure 24. Synthesis of compound 63 and its salts.

Polynitrogen heterocycle linking to TNP is a promising method to reach a balance between the energetic and physical properties of TNP, while there are a few references about it. Shreeve et al. [112] reported the synthesis of asymmetric N,N'-ethylene bridged 5-aminotetrazole and TNP moieties. They prepared 1-(2-(3,4,5-trinitro-1H-pyrazol-1-yl)ethyl)-1H-tetrazol-5-amine and 1-(3-(3,4,5-Trinitro-1H-pyrazol-1-yl)propyl)-1H-tetrazol-5-amine, and the two compounds were excellent insensitive and moderate powerful. In addition, they synthesized 5-((3,4,5-trinitro-1H-pyrazol-1-yl)methyl)-1H-tetrazole by N-methylene-C bridging TNP and tetrazole, which showed outstanding detonation properties and moderate insensitivity [113].

3. Nitrated-Bispyrazoles Based Compounds

Nitropyrazoles can be connected with nitrogen-rich heterocycles to obtain amazing energetic materials. In the previous section, nitropyrazoles bearing some polynitrogen heterocycles have been shown. Generally, these compounds exhibit some special properties, such as high detonation properties, good thermal stability, excellent safety, high density, and heat of formation, etc. Nitrated bispyrazoles also have attracted more and more attention, we will review the nitrated bispyrazole-based energetic materials in this section.

3.1. Directly Bridged Bis(Nitropyrazole)s

In 2014, Li et al. [27] synthesized several polynitro-substituted 1,4'-bridged-bispyrazoles energetic salts (**64–67**) as shown in Figure 25. They found that these compounds showed remarkable and unprecedented comprehensive properties (Table 12), and most of them with low toxicity were not hygroscopic. These compounds exhibited excellent impact sensitivities close to TATB, and the melting points and thermal decomposition temperatures were high, which could be applied as heat-resistant explosive. Compound **64** showed high T_d approximating that of hexanitrostilbene (HNS, 316 °C). The energetic properties of compounds **64, 65, 65a, 66**, and **67** were comparable with or superior to RDX, especially compound **66**. In 2017, Tang et al. [114] prepared 4,4',5,5'-tetranitro-2H,2'H-3,3'-bipyrazole (**69**) and its di-N-amino product (**70**), and the detailed route is described in Figure 26. Compound **70** showed good thermal stability and insensitivities as well as high detonation properties (Table 12). In addition, they synthesized 4,4'-dinitro-5,5'-diamino-2H,2'H-(3,3'-bipyrazole) (consisting of two 3-amino-4-nitropyrazole rings), this compound also show outstanding balance between thermal stability and safety (Table 12) [115]. Afterwards his team reported a variety of energetic materials based on compound **69** shown in Figure 27. Compounds **71, 73b**, and **73h** had high densities and good detonation velocities (Table 12), which were superior to RDX suggesting their use in secondary explosives. The dipotassium salt **73b** had a high density of 2.029 g·cm^{-3} and excellent thermal stability of 323 °C, and could be applied as primary explosives [116]. However, the poor impact sensitivity might restrict their further application. In 2019, Domasevitch and co-authors [117] found an efficient approach towards facile accumulation of nitro functionalities at the pyrazole platform. Compounds **74, 75**, and **76** were synthesized according to Figure 28. From Table 12, the three compounds owned high decomposition temperatures above 290 °C, especially for **75** and **76**. The introduction of three and four -NO$_2$ into the 4,4-bipyrazole scaffold could produce insensitive and thermally stable energetic materials with ideal densities and good detonation properties.

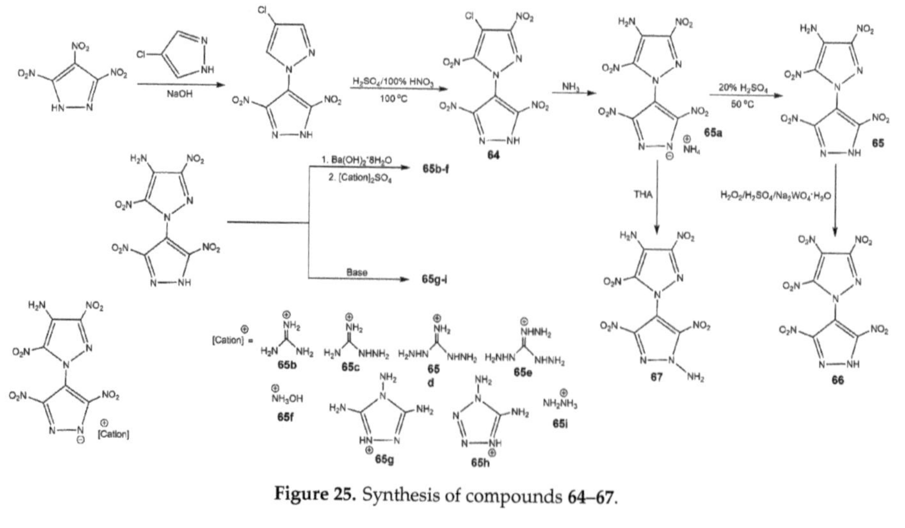

Figure 25. Synthesis of compounds **64–67**.

Table 12. Physicochemical and energetic properties of compounds 64–76. The data of compounds 64–67 are from reference [27], the data of compounds 70 are from reference [114], the data of compounds 71–73i are from reference [116], the data of compounds 74–76 are from reference [117].

Entry	ρ/g·cm^{-3}	D/km·s^{-1}	P/GPa	T_m/°C	T_d/°C	HOF/kJ·mol^{-1}	IS/J	FS/N
64	1.96	8.72	36.0	269	308	185.4	>40	-
65	1.89	8.60	35.0	dec	242	388.1	>40	-
65a	1.88	8.62	34.6	dec	262	274.7	>40	-
65b	1.73	8.04	27.3	dec	228	246.5	>40	-
65c	1.67	8.09	27.1	249	272	506.4	>40	-
65d	1.71	8.20	27.9	210	272	448.8	>40	-
65e	1.72	8.34	29.0	212	266	558.0	>40	-
65f	1.75	8.33	31.1	dec	259	331.2	>40	-
65g	1.82	8.45	31.0	247	297	557.0	>40	-
65h	1.72	8.23	28.9	166	261	700.4	>40	-
65i	1.80	8.54	32.8	dec	260	428.1	>40	-
66	1.82	8.81	37.0	158	297	824.2	28	-
67	1.87	8.65	35.1	260	284	477.9	>40	-
70	1.76	8.50	31.0	-	252	475.7	30	360
71	1.88	8.99	36.0	-	150	347.4	5	240
72	1.92	8.04	28.9	150	228	-50.7	6	120
73a	2.03	7.77	27.3	-	323	-125.2	4	40
73b	1.85	8.85	35.8	-	137	220.6	8	240
73c	1.77	8.67	31.5	94	155	220.9	10	240
73d	1.76	8.34	29.4	-	193	116.2	10	240
73e	1.69	8.14	25.2	-	196	353.3	15	360
73f	1.75	8.31	27.3	185	186	791.9	16	360
73g	1.76	8.22	26.5	-	206	565.4	12	360
73h	1.81	8.95	34.2	187	193	1359.4	10	360
73i	1.80	8.54	28.9	-	250	1269.7	18	360
74	1.79	7.53	22.1	377	382	203.5	30	>360
75	1.81	8.36	28.6	306	314	224.9	20	>360
76	1.86	8.52	31.1	292	298	227.8	4.5	192

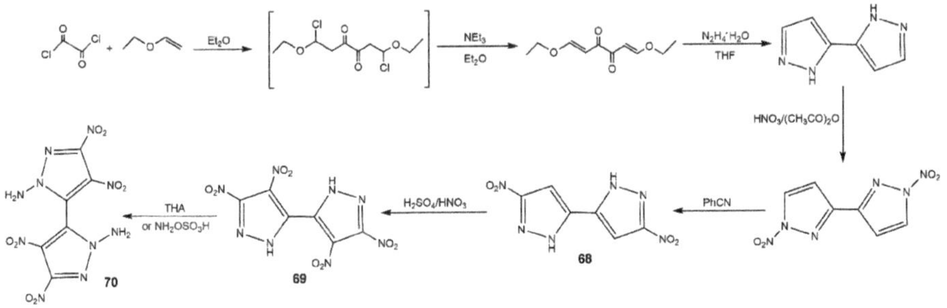

Figure 26. Synthesis of compounds 68–70.

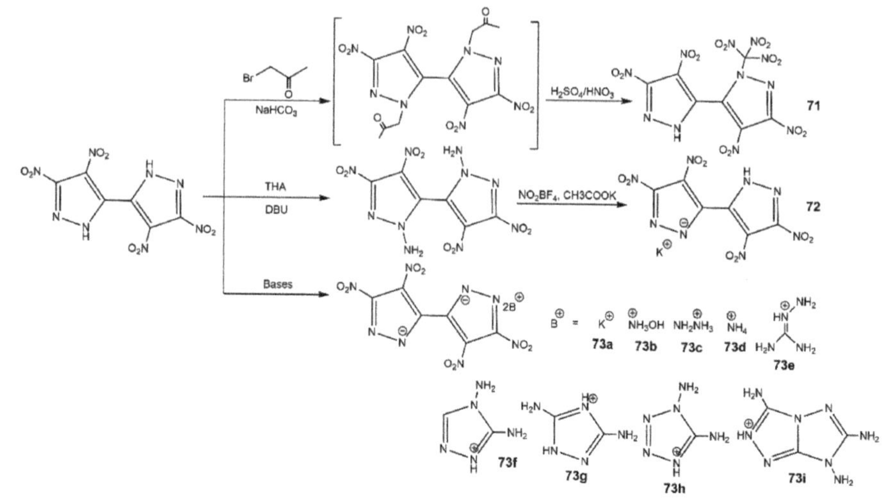

Figure 27. Synthesis of compounds **71–73i**.

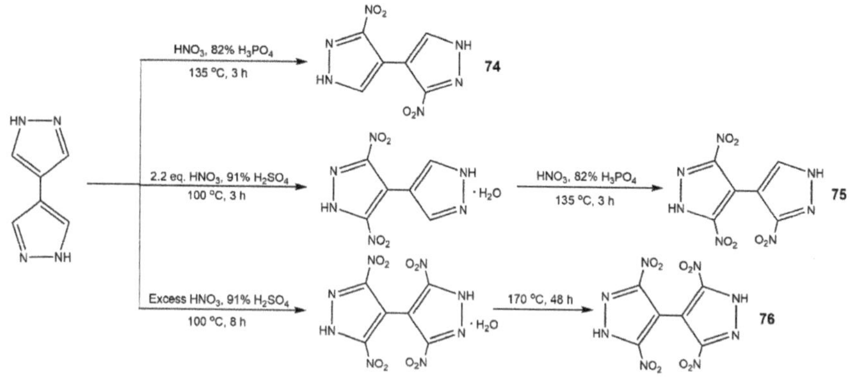

Figure 28. Synthesis of compounds **74–76**.

3.2. Alkyl-Bridged Bis(Nitropyrazoles)

Alkyl is also a good linkage to construct nitrogen-rich moieties, and many *N,N'*-alkyl-bridged energetic materials have been developed [112,118–121]. Yin et al. [122] developed a novel class of *N,N'*-ethylene-bridged bis(nitropyrazoles) with the synthetic route shown in Figure 29. Compounds **77–85** displayed various properties (Table 13) owing to the diversified functionalizations. Diaminobis(pyrazoles) showed good thermal stability, highly insensitivity, and favorable energetic performance; for example, the thermal decomposition temperature (311 °C) and detonation properties (27.9 GPa and 8.19 km·s^{-1}) of **77** were higher than those of TNT, and were comparable to those of TATB. By contrast, *N,N'*-ethylene bridged dinitraminobis(pyrazoles) and diazidobis(pyrazoles) owned better detonation performances, while having higher impact and friction sensitivity. Compound **80** was the most promising energetic material with high density, favorable thermal stability, and good detonation properties, which were comparable to RDX. In addition, the relatively low impact and friction sensitivities of **80** showed good integrated properties, highlighting its potential application as a replacement of RDX. In 2016, Fischer et al. [123] synthesized three different bisnitropyrazole-based energetic materials by *N,N'*-methylene bridge (**86–88**), the detailed synthetic route is displayed in Scheme A of Figure 30. These energetic compounds could be used for different applications according to their properties (Table 13), compound **86** was a secondary explosive with a high T_d (310 °C), enhanced detonation parameters by contrast with HNS, and high sensitivity to external stimuli. Compound **87** exhibited excellent detonation velocity (approximately to CL-20). The higher performance and better thermal stability of **88** was relative to DDNP making it a potential candidate as a green primary explosive. In addition, the synthetic routes are economical. Afterwards, their group used a similar route to prepare bis(3,4-dinitro-1H-pyrazol-1-yl)methane (**89**) and bis(3,5-dinitro-1H-pyrazol-1-yl)methane (**90**) with high decomposition temperature and low sensitivities having capability as future energetic materials (Table 13) [94]. Gozin et al. [124] explored the possible influence factor of the thermostable property of explosives, and under the guidelines they proposed, they synthesized the compounds **91** and **92** with excellent thermal stability and moderate sensitivities shown in Figure 31 and Table 13.

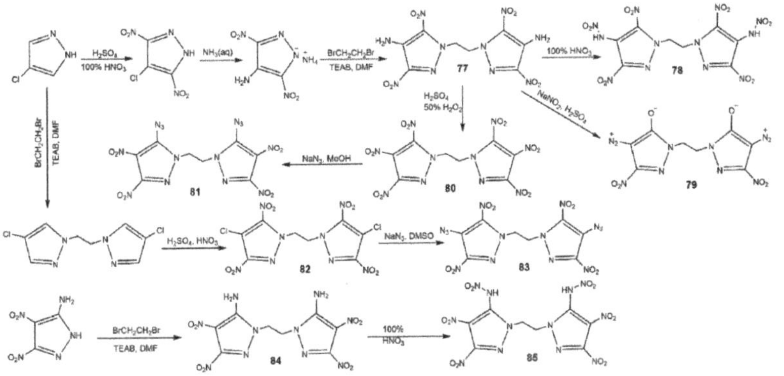

Figure 29. Synthesis of bis(nitropyrazoles) linked by N,N′-ethylene-bridge **77–85**.

Table 13. Physicochemical and energetic properties of **74–87**. The data of compounds **77–85** are from reference [122], the data of compounds **86–88**, CL-20 and DDNP are from reference [123], the data of compounds **89–90** are from reference [94], the data of compounds **91–92** are from reference [124].

Entry	$\rho/\text{g·cm}^{-3}$	$D/\text{km·s}^{-1}$	P/GPa	OB/%	$T_d/°C$	HOF/kJ·mol^{-1}	IS/J	FS/N
77	1.77	8.19	27.9	−17.2	311	218.9	>40	>360
78	1.84	8.75	34.3	3.5	80	380.6	7	80
79	1.72	7.80	24.2	−19.0	247	441.9	20	80
80	1.84	8.76	34.1	7.4	250	306.9	25	160
81	1.78	8.80	33.4	−7.5	112	1233.9	4	60
82	1.88	7.88	27.0	−7.8	319	230.0	>40	>360
83	1.76	8.56	31.0	−7.5	135	1013.9	3	60
84	1.75	8.13	27.3	−17.2	256	237.9	>40	>360
85	1.83	8.71	33.7	3.5	81	368.1	6	60
86	1.80	8.33	29.6	−40.2	310	205	11	>360
87	1.93	9.30	39.1	−11.5	205	379	4	144
88	1.73	8.02	26.0	−44.7	226	497	1.5	40
89	1.76	8.14	28.0	−39.0	319	302	25	360
90	1.72	7.97	26.3	−39.0	330	266	35	360
91	1.81	8.23	28.6	−39.0	262	224.2	14	352
92	1.81	8.36	29.7	−51.4	351	184.3	10	352
CL-20	2.04	9.67	44.9	−11.0	195	365	3	96
DDNP	1.72	76.5	23.8	−60.9	157	139	1	5

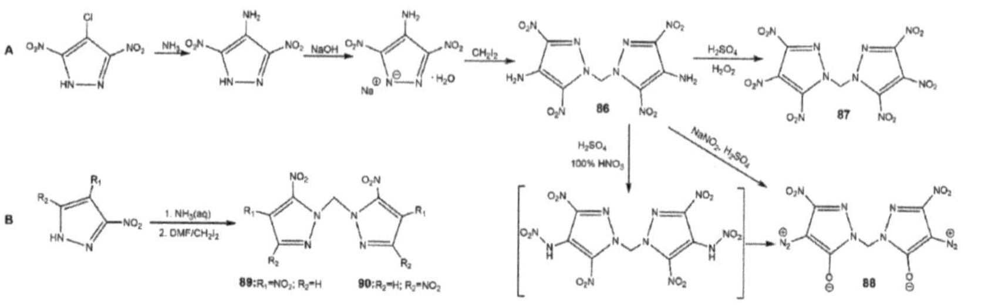

Figure 30. Synthesis of bis(nitropyrazoles) linked by N,N'-methylene-bridge **86–90**.

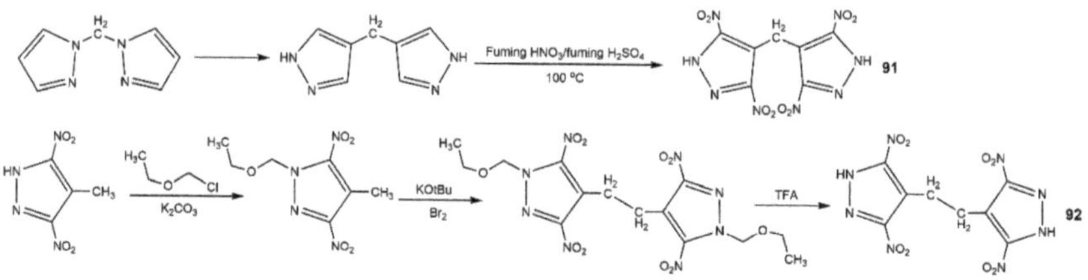

Figure 31. Synthesis of compounds **91** and **92**.

3.3. Ring-Bridged Bis(Nitropyrazoles)

Ring-bridge is an important connector linking bis(nitropyrazoles) to obtain high performance energetic materials. Pagoria et al. [125] reported the trimerization of LLM-116. 4-Diazo-3,5-bis(4-amino-3,5-dinitropyrazol-1-yl) pyrazole (**93**) containing a stable diazo group was synthesized, and the detailed route is shown in Figure 32. Compound **93** was more thermally stable (278 °C of T_d) than LLM-116, attributing to the considerable hydrogen bonding between -NH$_2$ and -NO$_2$, and the short contact between the =N$_2$ and -NO$_2$ through the intermolecular interactions. Moreover, it was insensitive to impact, friction, and spark. Yan et al. [126] designed mono and bi(1,2,4-oxadiazole) rings to bridge polynitropyrazoles (Figure 33). Among compounds **94–99**, **98**, and **99** owned the highest detonation velocity of 8.90 and 8.87 km·s^{-1}, detonation pressure of 35.1 and 34.5 GPa, respectively. **94** and **95** processed good stability (272–274 °C) and good insensitivity (IS > 30 J and FS > 360 N) as well as high detonation properties (8.69–8.74 km·s^{-1} of D and 33.4–34.0 GPa of P). **96** and **97** had the high thermal stability over 310 °C and good sensitivity (IS > 40 J, FS > 360 N). Comparing with the conventional heat resistant explosive HNS, **96** and **97** owned better detonation properties (7.99–8.03 km·s^{-1} of D, 25.2–26.4 GPa of P). Also, their team used the similar routes to synthesize the bis(nitropyrazoles) with 1,3,4-oxadiazole (**100–105**) [127]. The properties of these compounds are showed in Table 14. Moreover, Li et al. [124] synthesized the compound **106** with the procedure shown in Figure 34. This compound exhibited an excellent decomposition temperature (341 °C), high calculated detonation velocity of 8.52 km·s^{-1}, and detonation pressure of 30.6 GPa. It also showed impressive insensitivities (IS = 22 J, FS = 352, and ESD = 1.05 J). These showed building ring bridged bis(nitropyrazoles) can be an effective approach to enhance the properties of energetic materials.

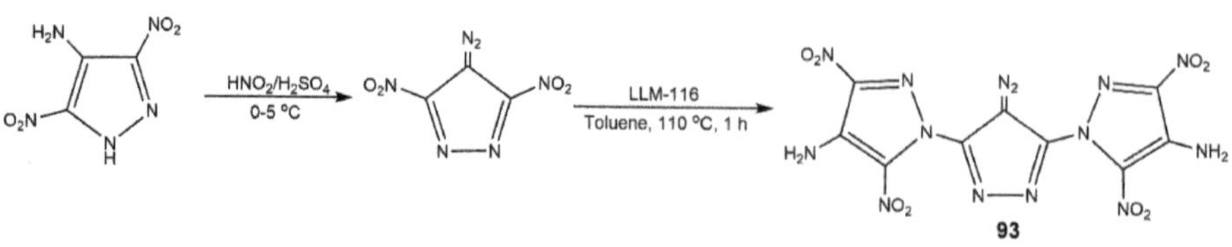

Figure 32. Synthesis of compound **93**.

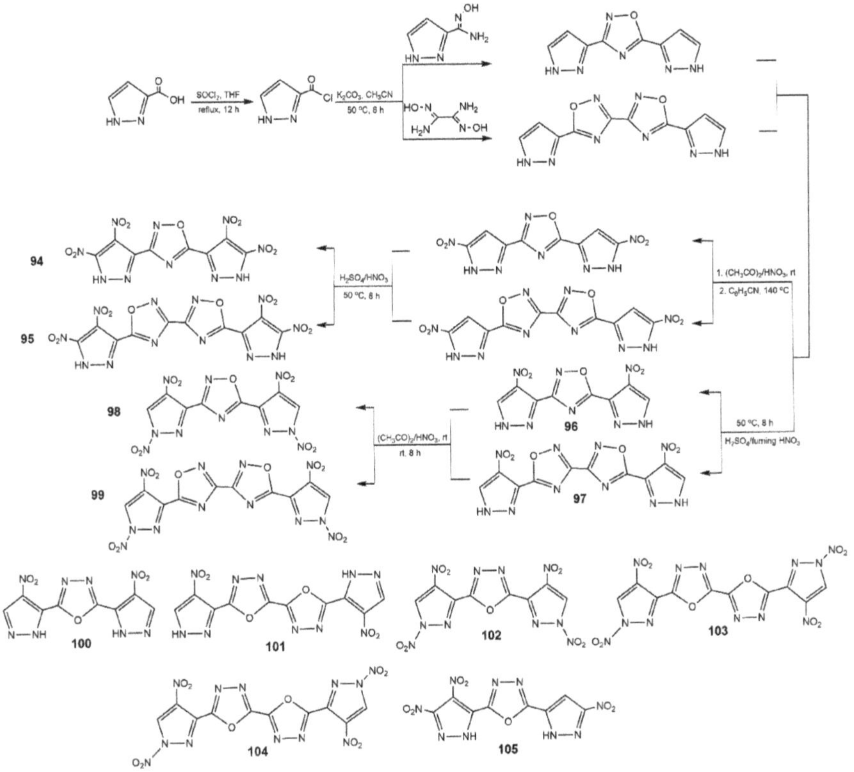

Figure 33. Synthesis of compounds **94–99**, and chemical structures of compounds **100–105**.

Table 14. Physical and energetic properties of energetic compounds **100–105**. The data of compounds **100–105** are from reference [127].

Entry	ρ/g·cm^{-3}	D/km·s^{-1}	P/GPa	T_d/°C	HOF/kJ·mol^{-1}	IS/J	FS/N
100	1.80	8.10	27.1	338	521.6	>40	>360
101	1.81	8.05	26.5	368	639.8	>40	>360
102	1.83	8.86	34.2	159	762.1	8	150
103	1.84	8.77	33.3	186	882.4	13	220
104	1.87	8.71	32.8	265	602.7	30	360
105	1.84	8.54	31.7	254	519.4	35	>360

In addition, there are some other fused ring-bridged bis(nitropyrazoles). In 2017, Yin and co-authors [128] synthesized compound **109** and its derivatives according to the procedure shown in Figure 35, and their physicochemical and energetic properties are shown in Table 15. Among these compounds, **107a** had a high density and decomposition temperature as well as the good safety parameters. The introduction of nitramino group gave **110** and **111** highest detonation velocities and pressures, while they also exhibited sensitive properties to mechanical stimuli. Considering the whole aspect, **108a** was featured with promising integrated energetic performance exceeding those of the benchmark explosive RDX. Shreeve's group prepared (**112**) obtained from compound **69** by N-azo coupling reactions shown in Scheme A of Figure 36 [114]. Compound **112** had a high density of 1.955 g·cm^{-3} and a good thermal stability (233 °C). Its detonation properties (9.63 km·s^{-1} and 44.0 GPa) were comparable to CL-20, much better than those of RDX and HMX. In addition, the IS of 10 J and FS of 240 N showed it was more stable than CL-20. These indicated compound **112** was a superior energetic explosive. In 2018, her team developed an efficient synthetic method of ring closure of polynitropyrazoles with N,N'-ethylene/propylene bridges (Figure 36, Scheme B). Compounds **113** and **114** showed excellent thermal stability (261 °C for **113**, 280 °C for **114**), good detonation properties and moderate insensitivities, making them potential candidates as HEDMs. This ring closure strategy could provide new ideas of designing thermally stable explosives.

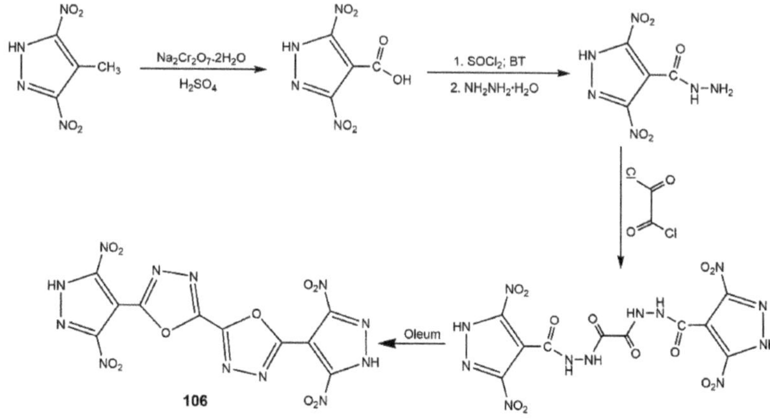

Figure 34. Synthesis of compound **106**.

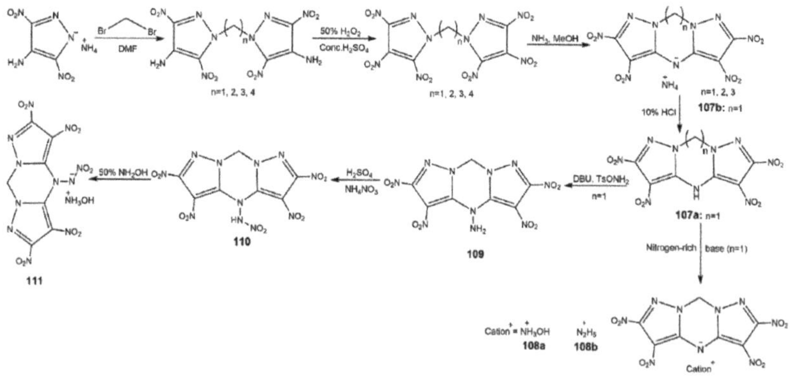

Figure 35. Synthesis of compounds **107a–111**.

Table 15. Physicochemical and energetic properties of compounds **107a–111**. The data of compounds **107a–111** are from reference [128].

Entry	ρ/g·cm^{-3}	D/km·s^{-1}	P/GPa	T_d/°C	HOF/kJ·g^{-1}	IS/J	FS/N
107a	1.90	8.79	34.3	261	1.10	15	240
107b	1.82	8.52	31.7	220	0.97	40	360
108a	1.86	8.89	35.9	221	1.05	35	360
108b	1.83	8.69	33.2	207	1.33	25	360
109	1.79	8.36	29.6	242	0.96	15	160
110	1.94	9.23	38.8	117	1.30	3	20
111	1.87	9.03	37.1	138	1.32	10	80

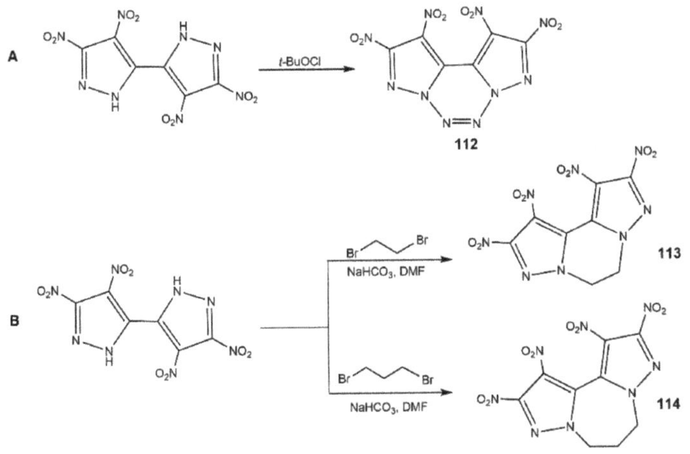

Figure 36. Synthesis of compounds **112–114**.

3.4. DCNP-Bridged Bis(Nitropyrazoles)

It is known that 1,3-Dichloro-2-nitro-2-azapropane (DCNP) is an useful precursor connecting nitropyrazoles via nucleophile substitution [129]. In 2013, Zhang et al. [130] reported a family of functionalized dipyrazolyl *N*-nitromethanamines (compounds **115–122** in Figure 37) using DCNP as the bridge. These compounds exhibited densities between 1.69–1.90 g·cm^{-3} and thermal stabilities range from 166–354 °C. From Table 16, it was easy to see the introduction of the azidodinitropyrazolate group led to the most competitive detonation properties (35.1 GPa and 8.72 km·s^{-1} for **121**, 35.2 GPa and 8.72 km·s^{-1} for **122**). However, they showed high sensitivity (IS = 2 J). Compound **119** exhibited good physical and detonation properties, such as high thermal stability, density, HOF, detonation pressure and velocity, and great impact stability, which could be used a promising HEDM. Klapötke et al. [131] also reported these compounds. They applied a different synthesis method of DNCP by the nitration of hexamethylenetetramine, and the NaBr/acetone system was used to substitution reaction.

Table 16. Properties of nitrated bispyrazoles from 1,3-dichloro-2-nitro-2-azapropane. The data of compounds **115–122** are from reference [130].

Entry	ρ/g·cm^{-3}	D/km·s^{-1}	P/GPa	T_d/°C	HOF/kJ·mol^{-1}	IS/J	OB/%
115	1.69	7.87	25.1	262	377.2	>40	−30.8
116	1.78	8.26	30.9	250	388.0	10	−4.0
117	1.78	8.27	31.0	261	398.0	>40	−4.0
118	1.90	8.06	30.6	252	371.8	11	0
119	1.86	8.64	34.7	232	486.4	>40	−7.4
120	1.89	8.04	30.4	354	381.3	>40	0
121	1.83	8.72	35.1	166	1108.2	2	0
122	1.83	8.72	35.2	169	1118.7	2	0

In general, the physical and energetic properties of bridged bis(nitropyrazole)s can be adjusted by the bridged groups. The design of novel bridged group would be a key factor to synthesize new HEDMs, and forming polycyclic derivatives even cage compounds could be more attractive. In addition, the salts of bridged bis(nitropyrazole)s should be explored in-depth.

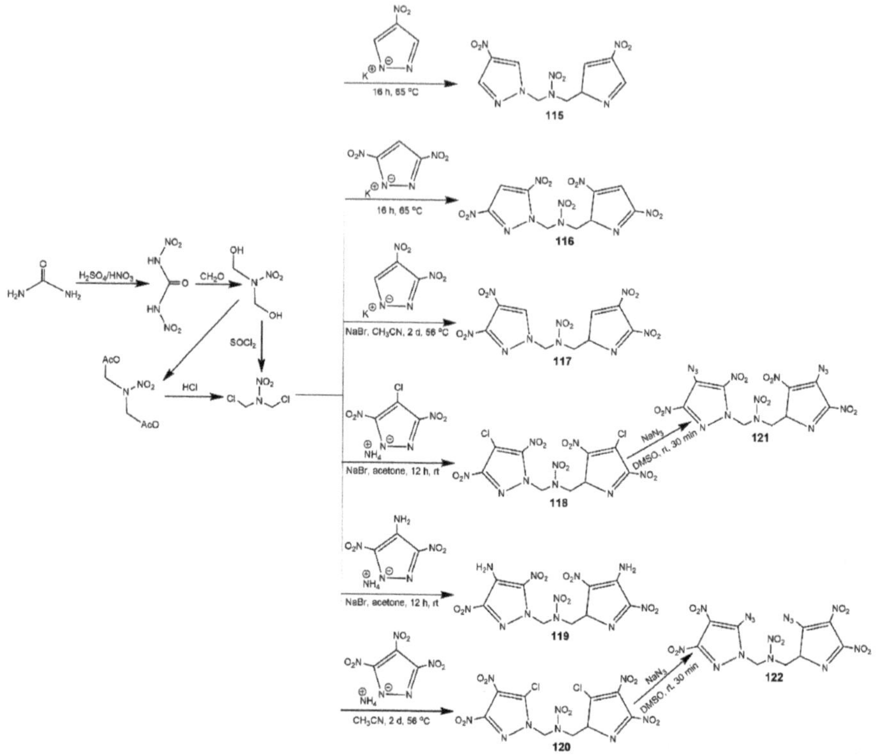

Molecules **2020**, *25*, 3475

Figure 37. Synthesis of nitrated bispyrazoles **115-122** from 1,3-dichloro-2-nitro-2-azapropane.

4. Nitrated Pyrazolo[4,3-c]Pyrazoles and Their Derivatives

Application of molecular design and explosive performance prediction has explored many novel energetic materials based on pyrazolopyrazole ring system [2,132,133]. Heterocycles like pyrazolo-pyrazole always own high density and oxygen balance, good thermal stability, and enhanced energetic performance of an energetic material.

3,6-Dinitropyrazolo[4,3-c]pyrazole (DNPP) is a new type of energetic material with attractive properties (1.865 g·cm^{-1} of ρ, 42.42% of nitrogen content, 273 kJ·mol^{-1}, 330.8 °C of T_d and 68 cm of D_{50}). This compound synthesized from 3,5-dimethylpyrazole was firstly reported by Dalinger and co-workers [134]. Pagoria et al. [135] improved the synthetic route to DNPP as shown in Scheme A of Figure 38. In this procedure, 4-diazo-3,5-dimethylpyrazole salt is an important intermediate. Li et al. [136] improved the process of 4-diazo-3,5-dimethylpyrazole salt using freezing crystallization instead of extraction which avoided large use of organic solvents and improved its yield. This procedure has several advantages, such as ease of synthesis scale-up and better product yield. In addition, Luo et al. [137] proposed that DNPP could be obtained by dehydration condensation, primary nitration, reduction, diazotization, cyclization, secondary nitration, oxidation, and decarboxylation nitration with acetylacetone and hydrazine hydrate as raw materials (Figure 38, Scheme B).

Due to the active N-H bond in molecule of DNPP, it is easy to obtain its energetic salts. In 2014, Zhang et al. [138] reported a series of nitrogen-rich energetic salts based on the anion of DNPP (**123a–m**) shown in Figure 39. Salts **123a–e** could be obtained by reacting DNPP with ammonia, hydrazine, hydroxylamine, 3,5-diamino-1,2,4-triazole, and 3,4,5-triamino-1,2,4-triazole. Salts **123f–j** could be synthesized by reacting Na$_2$DNPP with guanidine nitrate, aminoguanidine, diaminoguanidine, triaminoguanidinium, and 2-iminium-5-nitriminooctahydroimidazo [4,5-d]imidazole hydrochlorides.

Salts **123k–m** were acquired by the reaction of DNPP with NaOH, KOH and AgNO$_3$ respectively. Table 17 displays the properties of these energetic salts. It was notable that the ammonium salt (**123a**), hydroxylammonium salt (**123b**) and guanidinium salt (**123f**) exhibited outstanding decomposition temperatures of >300 °C. Furthermore, the sodium salt (**123k**) and potassium (**123l**) salt of DNPP were thermally stable up to 395 °C and 365 °C, respectively. In addition, most of the salts showed high calculated detonation properties, especially **123b** owned the highest detonation velocity and pressure. Considering the balance of safety and energetic properties as well as physical properties, **123b** could be a competitive candidate in insensitive HEDMs. Luo and co-authors synthesized the basic lead salt of DNPP (Pb-DNPP) and the 3,6-dihydrazine-1,2,4,5-tetrazine salt of DNPP (DHT-DNPP), and studied their thermal decomposition behaviors. Like **123k–m**, the introduction of heavy cations made the salts higher densities and T_d. Combining other organic amines salts of DNPP [139], these salts showed good thermal stabilities.

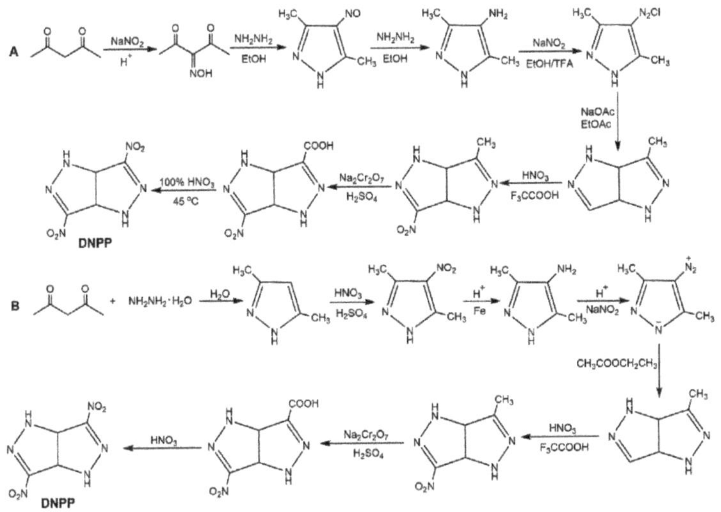

Figure 38. Synthesis of DNPP.

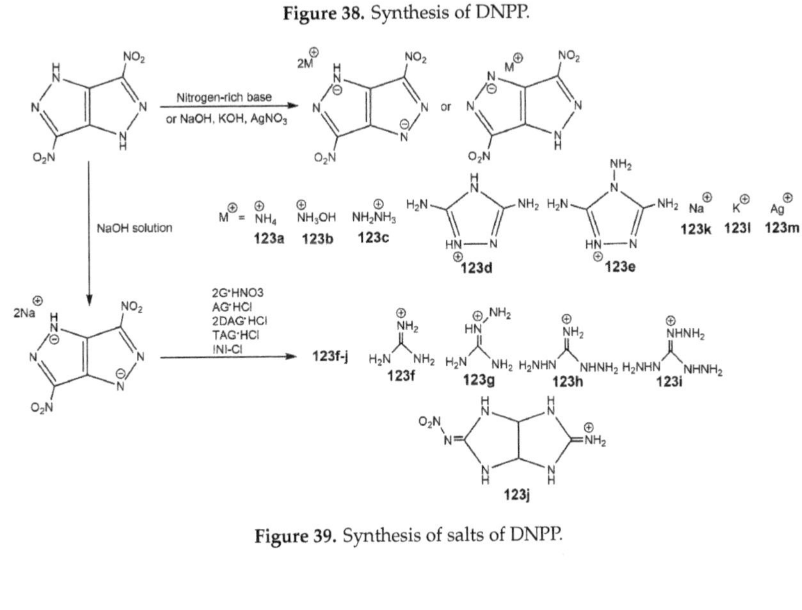

Figure 39. Synthesis of salts of DNPP.

Table 17. Properties of energetic compounds 123a–m. The data of compounds 123a–123m are from reference [138].

Entry	ρ/g·cm^{-3}	D/km·s^{-1}	P/GPa	T_d/°C	HOF/kJ·mol^{-1}	IS/J	FS/N	OB/%
123a	1.69	8.21	25.4	328	158.5	>40	360	−27
123b	1.82	9.01	35.4	327	274.2	29	360	−12
123c	1.72	8.86	30.3	247	501.0	16	160	−30
123d	1.71	8.04	24.5	287	481.9	>40	360	−30
123e	1.67	8.23	24.6	289	963.8	>40	360	−41
123f	1.68	7.95	22.5	324	173.3	>40	360	−40
123g	1.69	8.40	25.6	222	477.0	>40	360	−41
123h	1.71	8.73	28.0	209	679.6	>40	360	−42
123i	1.76	8.81	29.9	215	605.5	12	80	−31
123j	1.79	8.36	27.9	238	505.6	23	160	−27
123k	2.14	-	-	395	-			0
123l	2.20	-	-	365	-			0
123m	3.27	-	-	327	-			0

In addition, 1,4-Diamino-3,6-dinitropyrazolo[4,3-c]pyrazole (LLM-119) is a derivative of DNPP with a predicted energy of 104% HMX and good insensitivity to friction and electric spark stimulation [2]. It is also a very important intermediate of synthesizing novel high-performance energetic materials. Li et al. [140] used NaOH and H$_2$NOSO$_3$H to realize the N-amination of DNPP, while the yield was low (10.4%). Yin reported a modified procedure using 1,8-diazabicycloundec-7-ene (DBU) and O-tosylhydroxylamine (TsONH$_2$) as organic solvents with a good yield [141]. He also developed a series of DNPP derivatives based on N-functionalization strategy including several ionic salts of DNPP, the synthesis route is displayed in Figure 40 (Scheme A). As shown in Table 18, compounds 125, 126 and 126c exhibited high densities and excellent detonation velocities and pressures, which were superior to the current secondary explosive benchmark HMX. These compounds except 126d and 126e were sensitive to stimulation, especially for 126i also showed excellent density and good thermal stability. These could make compound the potassium salt as a green primary explosive. Compounds 126a, 126b, 126c and 126g showed good possibilities for application in bipropellants owing to the high values of (N + O) content and specific impulse. Li and co-author [142] synthesized another four kinds of neutral explosives based on N-functionalization of DNPP shown in Scheme B of Figure 40. Comparing with LLM-119, compound 127 showed slightly lower energetic and physical properties due to the only one -NH$_2$. Compounds 130 owned the relatively high density, good thermal stability, outstanding detonation properties, and reasonable sensitivities, which could be a useful energetic material. Li et al. [143,144] also synthesized several salts of N-nitramino DNPP, which exhibited good energetic properties.

In addition, Zhang et al. [145] introduced the dinitromethyl group and fluorodinitromethyl group into DNPP molecule and synthesized five fused-ring energetic derivatives (131–132) shown in Figure 41. Among these compounds, the dipotassium salt (131a) was formed as an interesting three-dimensional metal-organic framework (MOF) and exhibited outstanding detonation performances (9.02 km·s^{-1} of D and 33.6 GPa of P), which were comparable to that of Pd(N$_3$)$_2$. The compound 132 had a high density of 1.939 g·cm^{-3}, high decomposition temperature of 213 °C and desired mechanical sensitivities (IS: 12 J; FS: 240 N), which could be a competitive candidate of RDX. These energetic compounds containing dinitromethyl or fluorodinitromethyl group enrich the energetic compound library of pyrazolo[4,3-c]pyrazoles. Furthermore, their group incorporated two tetrazole groups into DNPP molecule, and synthesized 3,6-dinitro-1,4-di(1H-tetrazol-5-yl)-pyrazolo[4,3-c]pyrazole (133) and its ionic derivatives (133a–f) shown in Figure 42 [146]. The physicochemical and energetic properties of these compounds are shown in Table 19. These compounds were thermally stable and insensitive to mechanical stimulation. The potassium salt (133a) possessed a high thermal decomposition temperature (329 °C of T_d) and low sensitivities (IS: 25 J; FS: 252 N). In contrast with other derivatives from DNPP, compound 133f owned the best mechanical sensitivities (IS: >60 J; FS: >360 N). Compounds 133, 133a, and **133d** possessed good comprehensive properties, including remarkable thermal decomposition temperatures, excellent insensitivity, and favorable detonation performance.

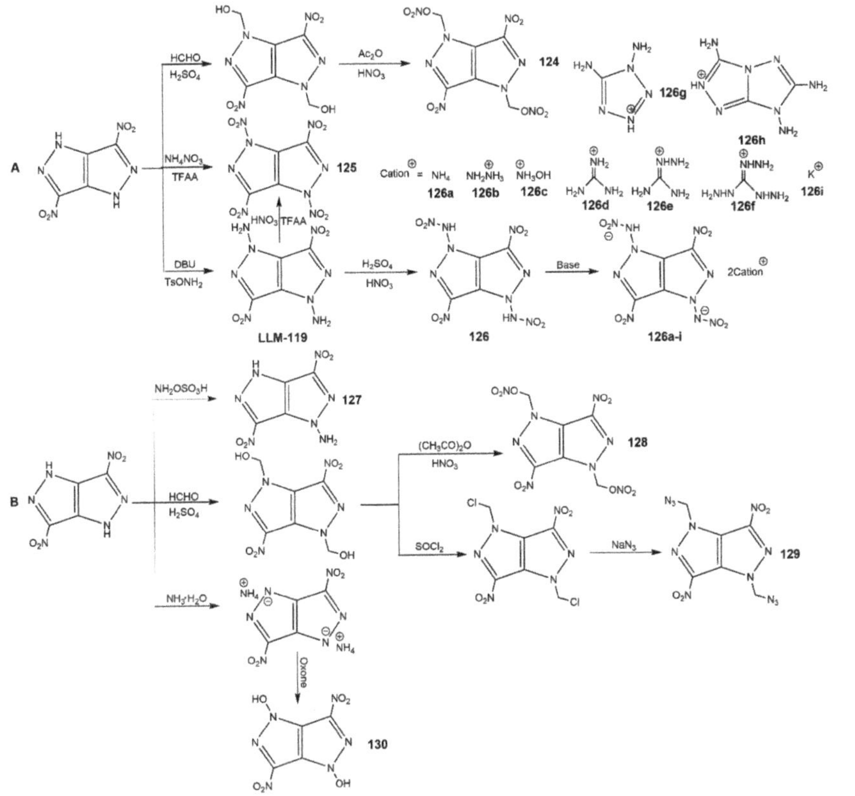

Figure 40. Synthesis of N-functional derivatives of DNPP.

Table 18. Physical and detonation properties of energetic compounds LLM-119 and **124–130**. The data of compounds LLM-119 and **124–126i** are from reference [141], the data of compounds **127–130** are from reference [142].

Entry	ρ/g·cm^{-3}	D/km·s^{-1}	P/GPa	T_d/°C	HOF/kJ·mol^{-1}	IS/J	FS/N	OB/%	N + O/%	I_{sp}/s
LLM-119	1.84	8.86	33.9	230	467.0	15	160	−14.0	77.2	246
124	1.82	8.67	33.1	206	133.7	10	120	9.2	78.2	245
125	1.96	9.46	40.9	145	550.9	3	20	22.2	83.3	269
126	1.93	9.51	41.8	128	595.2	2	20	15.1	84.3	274
126a	1.81	8.98	35.9	181	423.1	10	120	0	84.1	270
126b	1.85	9.40	39.5	174	738.9	5	60	−4.2	84.8	280
126c	1.88	9.50	41.3	170	531.2	7	120	8.3	85.4	282
126d	1.68	8.30	26.9	190	454.7	35	360	−14.7	80.7	239
126e	1.71	8.61	29.3	153	692.9	30	360	−17.2	81.5	247
126f	1.70	8.88	30.8	141	1144.8	10	80	−21.3	82.8	258
126g	1.78	9.17	36.0	163	1683.3	5	60	−9.3	84.2	280
126h	1.83	9.00	33.1	203	1599.5	10	120	−23.0	78.6	244
126i	2.11	8.31	31.2	208	152.9	2	20	16.2	68.0	226
127	1.74	7.93	27.9	178	356.0	14	280	−41.3	76.0	-
128	1.83	8.48	32.8	208	18.8	12	160	−18.4	78.2	-
129	1.74	7.82	27.1	198	863.0	10	240	−51.9	75.3	-
130	1.90	8.84	36.5	296	269.0	16	300	−20.9	78.2	-

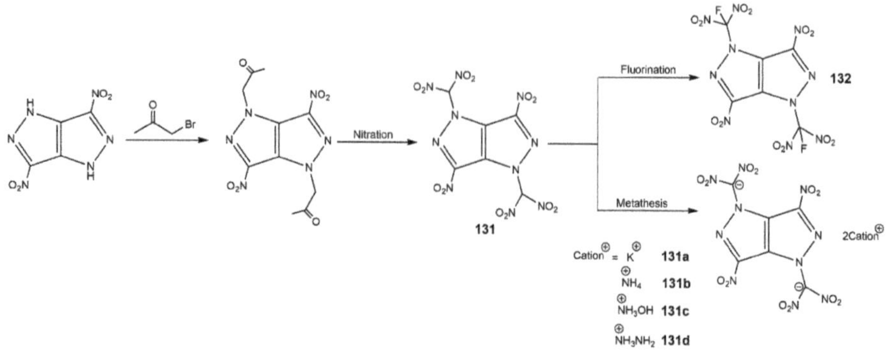

Figure 41. Synthesis of derivatives of DNPP containing dinitromethyl and fluorodinitromethyl group.

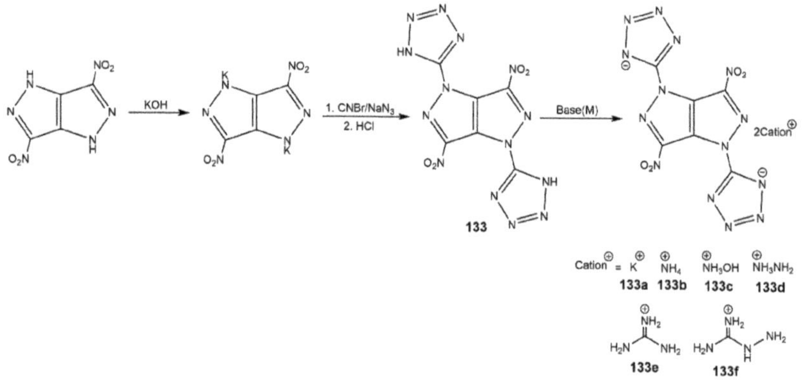

Figure 42. Synthesis of derivatives of DNPPP containing tetrazole groups.

Table 19. Physicochemical and energetic properties of **133** and its ionic salts. The data of compounds **133–133f** are from reference [146].

Entry	ρ/g·cm^{-3}	D/km·s^{-1}	P/GPa	T_d/°C	HOF/kJ·mol^{-1}	IS/J	FS/N
133	1.79	8.72	30.9	281	1111.5	15	192
133a	2.00	8.81	28.5	329	638.9	25	252
133b	1.69	8.40	26.2	280	916.8	19	>360
133c	1.61	8.24	26.0	178	1062.2	27.5	324
133d	1.75	9.08	31.3	221	1223.0	12	144
133e	1.62	8.02	22.4	299	926.9	>60	>360
133f	1.64	8.40	24.9	255	1143.3	35	>360

Nitrated pyrazolo[4,3-c]pyrazoles own acceptable performances both the energetic and physical properties, further functionalization of these compounds could be interesting. However, the synthesis of DNPP are still multistep reactions with unsatisfactory yield. The more efficient and facile synthesis technology should be investigated.

5. Conclusions

In recent years, a lot of scholars over the world have paid much attention to the development of nitrogen-rich heterocyclic energetic materials, due to their high positive heat of formation, low sensitivity, tailored thermal stability, and attractive detonation performance. According to the reference [37], a new energetic compound should be environmentally friendly, easy and economical to synthesize, thermal stable (T_d > 200 °C), insensitive to mechanical stimulation (IS > 7 J; FS > 120 N), good detonation properties ($D > 8.5$ km·s^{-1}), and not insoluble in water. For the nitropyrazoles-based energetic materials, most of them can meet these requirements. Some nitropyrazole-based compounds show good performance as castable explosives, such as compounds **1, 2, 3, 5**, 3,4-DNP, **46, 48, 53, 54**, and MTNP, which are competitive candidates of TNT. Some exhibited excellent thermal stability such as compounds **17, 28, 37, 64, 75, 86, 90, 92, 101, 120**, DNPP, etc. Further, many showed a balance between good safety and high detonation performance. The introduction of high-nitrogen groups (including fused-ring, polynitramino group, polynitromethyl group, etc) to nitropyrazoles can be useful approach for the further development of new-generation HEDMs. In addition, the concept of forming ionic salts, bridged structures and pyrazolo-pyrazoles provides novel insights to synthesize high performance energetic materials. It is better to synthesize new energetic compounds under the direction of theoretical calculation, so it is important to understand the relationship between structures and properties for the design and synthesis of new nitropyrazoles-based energetic materials.

Furthermore, there are some areas requiring improvement for the further synthesis of novel nitropyrazoles-based EMs. First, traditional nitration is generally used in the synthesis of nitropyrazoles-based EMs, which does not meet the requirements of modern green chemistry. It is vital to find out the suitable green nitrating agents and catalysts in the future synthesis process. Second, many syntheses of nitropyrazoles-based EMs entail several steps, leading to a low yield and high cost. Therefore, it is necessary to search for an efficient route when preparing new HEDMs.

Author Contributions: Conceptualization, S.Z.; writing—original draft preparation, S.Z. and Q.J.; writing—review and editing, S.Z. and Z.G.; data curation, N.L. and D.L.; supervision, K.K.; project administration, K.K. and J.Z.; funding acquisition, K.K. All authors have read and agreed to the published version of the manuscript.

Funding: This research was funded by the National Natural Science Foundation of China, grant numbers 21673182 and 21703168.

Caution: Readers are reminded that the information given in this review is intended to cover the progress of recent research on energetic azo materials. Most of the molecules collected in this review are energetic materials that may be explosive under certain conditions. Their syntheses should be carried out by experienced personnel and handled with caution. In any case, carefully planned safety protocols and proper protective equipment, such as Kevlar gloves, ear protection, safety shoes and plastic spatulas, should be utilized at all times, especially when working on a large scale (>1 g). The authors strongly suggest that the original references be consulted for detailed safety information.

Conflicts of Interest: The authors declare no conflict of interest.

References

1. Becuwe, A.; Delclos, A. Low-sensitivity explosive compounds for low vulnerability warheads. *Propell. Explos. Pyrot.* **1993**, *18*, 1–10. [CrossRef]
2. Pagoria, P.F.; Lee, G.S.; Mitchell, A.R.; Schmidt, R.D. A review of energetic materials synthesis. *Thermochim. Acta* **2002**, *384*, 187–204. [CrossRef]
3. Guo, J.; Cao, D.; Wang, J.; Wang, Y.; Qiao, R.; Li, Y. Review on synthesis of nitropyrazoles. *Chin. J. Energ. Mater.* **2014**, *22*, 872–879.
4. Talawar, M.B.; Sivabalan, R.; Mukundan, T.; Muthurajan, H.; Sikder, A.K.; Gandhe, B.R.; Rao, A.S. Environmentally compatible next generation green energetic materials (GEMs). *J. Hazard. Mater.* **2009**, *161*, 589–607. [CrossRef]
5. Tsyshevsky, R.V.; Sharia, O.; Kuklja, M.M. Molecular theory of detonation initiation: Insight from first principles modeling of the decomposition mechanisms of organic nitro energetic materials. *Molecules* **2016**, *21*, 236. [CrossRef]
6. Yan, Q.-L.; Cohen, A.; Petrutik, N.; Shlomovich, A.; Zhang, J.-G.; Gozin, M. Formation of highly thermostable copper-containing energetic coordination polymers based on oxidized triaminoguanidine. *ACS Appl. Mater. Inter.* **2016**, *8*, 21674–21682. [CrossRef]
7. Yan, Q.-L.; Gozin, M.; Zhao, F.-Q.; Cohen, A.; Pang, S.-P. Highly energetic compositions based on functionalized carbon nanomaterials. *Nanoscale* **2016**, *8*, 4799–4851. [CrossRef]

8. Chen, X.; Zhang, C.; Bai, Y.; Guo, Z.; Yao, Y.; Song, J.; Ma, H. Synthesis, crystal structure and thermal properties of an unsymmetrical 1, 2, 4, 5-tetrazine energetic derivative. *Acta Crystallogr. C* **2018**, *74*, 666–672. [CrossRef]
9. Klapötke, T.M. *High Energy Density Materials*; Springer: Berlin, Germany, 2007; Volume 125.
10. An, C.W.; Li, F.S.; Song, X.L.; Wang, Y.; Guo, X.D. Surface Coating of RDX with a Composite of TNT and an Energetic-Polymer and its Safety Investigation. *Propell. Explos. Pyrot.* **2009**, *34*, 400–405. [CrossRef]
11. Jia, Q.; Kou, K.-C.; Zhang, J.-Q.; Zhang, S.-J.; Xu, Y.-L. Study on the dissolution behaviors of CL-20/TNT co-crystal in N, N-dimethylformamide (DMF) and dimethyl sulfoxide (DMSO). *J. Therm. Anal. Calorim.* **2018**, *134*, 2375–2382. [CrossRef]
12. Simpson, R.L.; Urtiew, P.A.; Ornellas, D.L.; Moody, G.L.; Scribner, K.J.; Hoffman, D.M. CL-20 Performance Exceeds that of HMX and its Sensitivity is Moderate. *Propell. Explos. Pyrot.* **1997**, *22*, 249–255. [CrossRef]
13. Wang, Y.; Jiang, W.; Song, X.; Deng, G.; Li, F. Insensitive HMX (octahydro-1, 3, 5, 7-tetranitro-1, 3, 5, 7-tetrazocine) nanocrystals fabricated by high-yield, low-cost mechanical milling. *Cent. Eur. J. Energ. Mater.* **2013**, *10*, 277–287.
14. Qiu, H.; Stepanov, V.; Di Stasio, A.R.; Chou, T.; Lee, W.Y. RDX-based nanocomposite microparticles for significantly reduced shock sensitivity. *J. Hazard. Mater.* **2011**, *185*, 489–493. [CrossRef] [PubMed]
15. Kolb, J.R.; Rizzo, H.F. Growth of 1, 3, 5-Triamino-2, 4, 6-trinitrobenzene (TATB) I. Anisotropic thermal expansion. *Propell. Explos. Pyrot.* **1979**, *4*, 10–16. [CrossRef]
16. Nair, U.R.; Sivabalan, R.; Gore, G.M.; Geetha, M.; Asthana, S.N.; Singh, H. Hexanitrohexaazaisowurtzitane (CL-20) and CL-20-based formulations. *Combust. Explo. Shock Waves* **2005**, *41*, 121–132. [CrossRef]
17. Nair, U.R.; Asthana, S.N.; Rao, A.S.; Gandhe, B.R. Advances in high energy materials. *Def. Sci. J.* **2010**, *60*, 137. [CrossRef]
18. Swain, P.K.; Singh, H.; Tewari, S.P. Energetic ionic salts based on nitrogen-rich heterocycles: A prospective study. *J. Mol. Liq.* **2010**, *151*, 87–96. [CrossRef]
19. Gao, H.; Shreeve, J.M. Azole-based energetic salts. *Chem. Rev.* **2011**, *111*, 7377–7436. [CrossRef]
20. Kröber, H.; Teipel, U. Crystallization of insensitive HMX. *Propell. Explos. Pyrot.* **2008**, *33*, 33–36. [CrossRef]
21. Zhang, S.; Kou, K.; Zhang, J.; Jia, Q.; Xu, Y. Compact energetic crystals@ urea-formaldehyde resin micro-composites with evident insensitivity. *Compos. Commun.* **2019**, *15*, 103–107. [CrossRef]
22. Zhang, S.; Gao, Z.; Jia, Q.; Liu, N.; Zhang, J.; Kou, K. Fabrication and characterization of surface modified HMX@ PANI core-shell composites with enhanced thermal properties and desensitization via in situ polymerization. *Appl. Surf. Sci.* **2020**, *515*, 146042. [CrossRef]
23. Liu, N.; Duan, B.; Lu, X.; Mo, H.; Xu, M.; Zhang, Q.; Wang, B. Preparation of CL-20/DNDAP cocrystals by a rapid and continuous spray drying method: An alternative to cocrystal formation. *CrystEngComm* **2018**, *20*, 2060–2067. [CrossRef]
24. Liu, N.; Duan, B.; Lu, X.; Zhang, Q.; Xu, M.; Mo, H.; Wang, B. Preparation of CL-20/TFAZ cocrystals under aqueous conditions: Balancing high performance and low sensitivity. *CrystEngComm* **2019**, *21*, 7271–7279. [CrossRef]
25. Zhang, S.; Zhang, J.; Kou, K.; Jia, Q.; Xu, Y.; Liu, N.; Hu, R. Standard Enthalpy of Formation, Thermal Behavior, and Specific Heat Capacity of 2HNIW· HMX Co-crystals. *J. Chem. Eng. Data* **2018**, *64*, 42–50. [CrossRef]
26. Bolter, M.F.; Harter, A.; Klapotke, T.M.; Stierstorfer, J. Isomers of Dinitropyrazoles: Synthesis, Comparison and Tuning of their Physicochemical Properties. *Chempluschem* **2018**, *83*, 804–811. [CrossRef]
27. Li, C.; Liang, L.; Wang, K.; Bian, C.; Zhang, J.; Zhou, Z. Polynitro-substituted bispyrazoles: A new family of high-performance energetic materials. *J. Mater. Chem. A* **2014**, *2*, 18097–18105. [CrossRef]
28. Liu, W.; Wang, J.; Cao, D.L.; Chen, L.; Liu, Y.; Zhao, X. Synthesis of 3,5-dinitropyrazole. *Appl. Chem. Ind.* **2018**, *47*, 30–34. [CrossRef]
29. Zhang, P.; Kumar, D.; Zhang, L.; Shem-Tov, D.; Petrutik, N.; Chinnam, A.K.; Yao, C.; Pang, S.; Gozin, M. Energetic Butterfly: Heat-Resistant Diaminodinitro trans-Bimane. *Molecules* **2019**, *24*, 4324. [CrossRef]
30. Göbel, M.; Klapötke, T.M. Development and testing of energetic materials: The concept of high densities based on the trinitroethyl functionality. *Adv. Funct. Mater.* **2009**, *19*, 347–365. [CrossRef]
31. Klapötke, T.M.; Sabaté, C.M. Nitrogen-rich tetrazolium azotetrazolate salts: A new family of insensitive energetic materials. *Chem. Mater.* **2008**, *20*, 1750–1763. [CrossRef]
32. Zhao, T.X.; Li, L.; Dong, Z.; Zhang, Y.; Zhang, G.Z.; Huang, M.; Li, H.B. Research Progress on the Synthesis of Energetic Nitroazoles. *Chin. J. Org. Chem.* **2014**, *34*, 304–315. [CrossRef]

33. Huang, A.; Wo, K.; Lee, S.Y.C.; Kneitschel, N.; Chang, J.; Zhu, K.; Mello, T.; Bancroft, L.; Norman, N.J.; Zheng, S.-L. Regioselective synthesis, NMR, and crystallographic analysis of N1-substituted pyrazoles. *J. Org. Chem.* **2017**, *82*, 8864–8872. [CrossRef] [PubMed]
34. Lahm, G.P.; Cordova, D.; Barry, J.D. New and selective ryanodine receptor activators for insect control. *Bioorg. Med. Chem.* **2009**, *17*, 4127–4133. [CrossRef] [PubMed]
35. Cavero, E.; Uriel, S.; Romero, P.; Serrano, J.L.; Giménez, R. Tetrahedral zinc complexes with liquid crystalline and luminescent properties: Interplay between nonconventional molecular shapes and supramolecular mesomorphic order. *J. Am. Chem. Soc.* **2007**, *129*, 11608–11618. [CrossRef]
36. Dalinger, I.L.; Vatsadse, I.A.; Shkineva, T.K.; Popova, G.P.; Ugrak, B.I.; Shevelev, S.A. Nitropyrazoles. *Russ. Chem. Bull.* **2010**, *59*, 1631–1638. [CrossRef]
37. Qu, Y.; Babailov, S.P. Azo-linked high-nitrogen energetic materials. *J. Mater. Chem. A* **2018**, *6*, 1915–1940. [CrossRef]
38. Janssen, J.; Koeners, H.J.; Kruse, C.G.; Habrakern, C.L. Pyrazoles. XII. Preparation of 3 (5)-nitropyrazoles by thermal rearrangement of N-nitropyrazoles. *J. Org. Chem.* **1973**, *38*, 1777–1782. [CrossRef]
39. Habraken, C.L.; Janssen, J. Pyrazoles. VIII. Rearrangement of N-nitropyrazoles. Formation of 3-nitropyrazoles. *J. Org. Chem.* **1971**, *36*, 3081–3084. [CrossRef]
40. Verbruggen, R. Cycloaddition with 2-chloro-1-nitroethelene. *Chimia* **1975**, *29*, 350–352.
41. Li, C.; Sun, T.; Chen, X. Synthesis of 3-Nitropyrozale. *Dyes. Color.* **2004**, *41*, 168–169.
42. Li, H.; Xiong, B.; Jiang, J.; Zheng, X.; Li, Z.; Ma, Y. Synethsis and Characterization of 3-Nitropyrazole and its Salts. *Chin. J. Energ. Mater.* **2008**, *31*, 102–104.
43. Tian, X.; Li, J.; Wang, J. Synthesis of 3,4-Dinitropyrazole. *Chin. J. Syn. Chem.* **2012**, *20*, 117–118.
44. Zhao, X.X.; Zhang, J.C.; Li, S.H.; Yang, Q.P.; Li, Y.C.; Pang, S.P. A Green and Facile Approach for Synthesis of Nitro Heteroaromatics in Water. *Org. Process Res. Dev.* **2014**, *18*, 886–890. [CrossRef]
45. Katritzky, A.R.; Scriven, E.F.V.; Majumder, S.; Akhmedova, R.G.; Akhmedov, N.G.; Vakulenko, A.V. Direct nitration of five membered heterocycles. *Arkivoc* **2005**, *3*, 179–191.
46. Ravi, P.; Gore, G.M.; Tewari, S.P.; Sikder, A.K. A simple and environmentally benign nitration of pyrazoles by impregnated bismuth nitrate. *J. Heterocyclic Chem.* **2013**, *50*, 1322–1327. [CrossRef]
47. Yi, J.H.; Hu, S.Q.; Liu, S.N.; Cao, D.L.; Ren, J. Theoretical study on structures and detonation performances for nitro derivatives of pyrazole by density functional theory. *Chin. J. Energ. Mater.* **2010**, *18*, 252–256.
48. Rao, E.N.; Ravi, P.; Tewari, S.P.; Rao, S.V. Experimental and theoretical studies on the structure and vibrational properties of nitropyrazoles. *J. Mol. Struct.* **2013**, *1043*, 121–131.
49. Ravi, P. Experimental and DFT studies on the structure, infrared and Raman spectral properties of dinitropyrazoles. *J. Mol. Struct.* **2015**, *1079*, 433–447. [CrossRef]
50. Li, Y.; Dang, X.; Cao, D.; Chai, X. One-pot two steps synthesis of 4-nitropyrazole and its crystal structure. *Chin. J. Energ. Mater.* **2018**, *22*, 872–879.
51. Corte, J.R.; De Lucca, I.; Fang, T.; Yang, W.; Wang, Y.; Dilger, A.K.; Pabbisetty, K.B.; Ewing, W.R.; Zhu, Y.; Wexler, R.R. Macrocycles with aromatic P2' groups as factor xia inhibitors. U.S. Patent No. 9,777,001, 3 October 2017.
52. Ioannidis, S.; Talbot, A.C.; Follows, B.; Buckmelter, A.J.; Wang, M.; Campbell, A.M.; Schmidt, D.R.; Guerin, D.J.; Caravella, J.A.; Diebold, R.B. Preparation of Pyrrolo and Pyrazolopyrimidines as Ubiquitin-Specific Protease 7 Inhibitors. US patent 2,016,018,578,5A, 30 June 2016.
53. Han, S.J.; Kim, H.T.; Joo, J.M. Direct C–H Alkenylation of Functionalized Pyrazoles. *J. Org. Chem.* **2016**, *81*, 689–698. [CrossRef]
54. Deng, M.; Wang, Y.; Zhang, W.; Zhang, Q. Synthesis and Properties of 5-Methyl-4-nitro-1H-pyrazol-3(2H)-one and its Energetic Ion Compounds. *Chin. J. Energ. Mater.* **2017**, *25*, 646–650.
55. Pan, Y.; Wang, Y.; Zhao, B.; Gao, F.; Chen, B.; Liu, Y. Research Progress in Synthesis, Properties and Applications of Nitropyrazoles and Their Derivatives. *Chin. J. Energ. Mater.* **2018**, *26*, 796–812.
56. Zhang, Y.; Li, Y.; Hu, J.; Ge, Z.; Sun, C.; Pang, S. Energetic C-trinitromethyl-substituted pyrazoles: Synthesis and characterization. *Dalton Trans.* **2019**, *48*, 1524–1529. [CrossRef]
57. Yin, P.; Zhang, J.; He, C.; Parrish, D.A.; Shreeve, J.N.M. Polynitro-substituted pyrazoles and triazoles as potential energetic materials and oxidizers. *J. Mater. Chem. A* **2014**, *2*, 3200–3208. [CrossRef]

58. Dalinger, I.L.; Vatsadze, I.A.; Shkineva, T.K.; Kormanov, A.V.; Struchkova, M.I.; Suponitsky, K.Y.; Bragin, A.A.; Monogarov, K.A.; Sinditskii, V.P.; Sheremetev, A.B. Novel Highly Energetic Pyrazoles: N-Trinitromethyl-Substituted Nitropyrazoles. *Chem.-Asian. J.* **2015**, *10*, 1987–1996. [CrossRef]
59. Xiong, H.; Yang, H.; Cheng, G. 3-Trinitromethyl-4-nitro-5-nitramine-1H-pyrazole: A high energy density oxidizer. *New J. Chem.* **2019**, *43*, 13827–13831. [CrossRef]
60. Semenov, V.V.; Shevelev, S.A.; Bruskin, A.B.; Kanishchev, M.I.; Baryshnikov, A.T. Mechanism of nitration of nitrogen-containing heterocyclic N-acetonyl derivatives. General approach to the synthesis of N-dinitromethylazoles. *Russ. Chem. Bull.* **2009**, *58*, 2077–2096. [CrossRef]
61. Zhang, Y.; Li, Y.; Yu, T.; Liu, Y.; Chen, S.; Ge, Z.; Sun, C.; Pang, S. Synthesis and Properties of Energetic Hydrazinium 5-Nitro-3-dinitromethyl-2H-pyrazole by Unexpected Isomerization of N-Nitropyrazole. *ACS Omega* **2019**, *4*, 19011–19017. [CrossRef]
62. Lei, C.; Yang, H.; Cheng, G. New pyrazole energetic materials and their energetic salts: Combining the dinitromethyl group with nitropyrazole. *Dalton Trans.* **2020**, *49*, 1660–1667. [CrossRef]
63. Yin, P.; He, C.; Shreeve, J.N.M. Fused heterocycle-based energetic salts: Alliance of pyrazole and 1,2,3-triazole. *J. Mater. Chem. A* **2016**, *4*, 1514–1519. [CrossRef]
64. Dalinger, I.L.; Kormanov, A.V.; Vatsadze, I.A.; Serushkina, O.V.; Shkineva, T.K.; Suponitsky, K.Y.; Pivkina, A.N.; Sheremetev, A.B. Synthesis of 1-and 5-(pyrazolyl) tetrazole amino and nitro derivatives. *Chem. Heterocycl. Com+.* **2016**, *52*, 1025–1034. [CrossRef]
65. Zyuzin, I.N.; Suponitsky, K.Y.; Dalinger, I.L. N-[2,2-Bis(methoxy-NNO-azoxy)ethyl]pyrazoles. *Chem. Heterocycl. Com+.* **2017**, *53*, 702–709. [CrossRef]
66. Dalinger, I.L.; Kormanov, A.V.; Suponitsky, K.Y.; Muravyev, N.V.; Sheremetev, A.B. Pyrazole-Tetrazole Hybrid with Trinitromethyl, Fluorodinitromethyl, or (Difluoroamino)dinitromethyl Groups: High-Performance Energetic Materials. *Chem.-Asian J.* **2018**, *13*, 1165–1172. [CrossRef] [PubMed]
67. Tang, Y.; Ma, J.; Imler, G.H.; Parrish, D.A.; Jean'ne, M.S. Versatile functionalization of 3, 5-diamino-4-nitropyrazole for promising insensitive energetic compounds. *Dalton Trans.* **2019**, *48*, 14490–14496. [CrossRef]
68. Yin, P.; He, C.; Shreeve, J.N.M. Fully C/N-Polynitro-Functionalized 2, 2'-Biimidazole Derivatives as Nitrogen-and Oxygen-Rich Energetic Salts. *Chem.-Eur. J.* **2016**, *22*, 2108–2113. [CrossRef]
69. Tang, Y.; Liu, Y.; Imler, G.H.; Parrish, D.A.; Shreeve, J.N.M. Green Synthetic Approach for High-Performance Energetic Nitramino Azoles. *Org. Lett.* **2019**, *21*, 2610–2614. [CrossRef]
70. Xu, M.; Cheng, G.; Xiong, H.; Wang, B.; Ju, X.; Yang, H. Synthesis of high-performance insensitive energetic materials based on nitropyrazole and 1,2,4-triazole. *New J. Chem.* **2019**, *43*, 11157–11163. [CrossRef]
71. Ma, J.; Tang, Y.; Cheng, G.; Imler, G.H.; Parrish, D.A.; Shreeve, J.M. Energetic Derivatives of 8-Nitropyrazolo[1,5-a][1,3,5]triazine-2,4,7-triamine: Achieving Balanced Explosives by Fusing Pyrazole with Triazine. *Org. Lett.* **2020**, *22*, 1321–1325. [CrossRef]
72. Biffin, M.E.C.; Brown, D.J.; Porter, Q.N. A novel route from 4-methoxy-5-nitropyrimidine to 3-amino-4-nitropyrazole and pyrazolo [3, 4-b] pyrazine. *Tetrahedron Lett.* **1967**, *8*, 2029–2031. [CrossRef]
73. Ravi, P.; Koti Reddy, C.; Saikia, A.; Mohan Gore, G.; Kanti Sikder, A.; Prakash Tewari, S. Nitrodeiodination of polyiodopyrazoles. *Propell. Explos. Pyrot.* **2012**, *37*, 167–171. [CrossRef]
74. Liu, W.; Li, Y. Prospect on studying 3,4-nitropyrazole synthesis. *Jiangxi Chem. Ind.* **2015**, *5*, 8–12.
75. Latypov, N.V.; Silevich, V.A.; Ivanov, P.A.; Pevzner, M.S. Diazotization of aminonitropyrazoles. *Chem. Heterocycl. Com+.* **1976**, *12*, 1355–1359. [CrossRef]
76. Vinogradov, V.M.; Dalinger, I.L.; Gulevskaya, V.I.; Shevelev, S.A. Nitropyrazoles. *Russ. Chem. Bull.* **1993**, *42*, 1369–1371. [CrossRef]
77. Stefan, E.; Wahlström, L.Y.; Latypov, N. Derivatives of 3 (5), 4-dinitropyrazole as potential energetic plasticisers. *Chem. Chem. Eng.* **2011**, *5*, 929–935.
78. Wang, Y.L.; Ji, Y.P.; Chen, B.; Wang, W.; Gao, F.L. Improved Synthesis of 3, 4-Dinitropyrazole. *Chin. J. Energ. Mater.* **2011**, *19*, 377–379.
79. Tian, X.; Li, J.-S. Thermal decomposition and non-isothermal kinetics of 3, 4-dinitropyrazole. *Chem. Res. Appl.* **2013**, *25*, 18–22.
80. Du, S.; Li, Y.-X.; Wang, J.-L.; Wang, Y.-H.; Cao, D.-L. Review for synthetic method and properties of 3, 4-dinitropyrazole. *Chem. Res.* **2011**, *5*, 106–110.

81. Li, Y.; Guo, J.; Song, L. Synthesis, performance and crystal structure of 1-methyl-3, 4-dinitropyrazole. *Chin. J. Explos. Propell.* **2015**, *38*, 64–68.
82. Zhang, Q.; Li, Y.; Qiao, Q.; Ren, H.; Li, Y. Optimization on synthesis process and characterization of 1-methyl-3,4-dinitropyrazoles. *Chin. J. Energ. Mater.* **2015**, *23*, 222–225.
83. Wang, Y.-L.; Zhang, Z.-Z.; Wang, B.-Z.; Zheng, X.-D.; Zhou, Y.-S. Synthesis of 3, 5-Dinitropyrazole. *Chin. J. Energ. Mater.* **2007**, *15*, 574–576.
84. Zaitsev, A.A.; Dalinger, I.L.; Shevelev, S.A. Dinitropyrazoles. *Russ. Chem. Rev.* **2009**, *78*, 589. [CrossRef]
85. Zhao, X.; Qi, C.; Zhang, L.; Wang, Y.; Li, S.; Zhao, F.; Pang, S. Amination of nitroazoles—A comparative study of structural and energetic properties. *Molecules* **2014**, *19*, 896–910. [CrossRef] [PubMed]
86. Wang, Y.-L.; Zhang, Z.-Z.; Wang, B.-Z.; Luo, Y.-F. Synthesis of LLM-116 by VNS Reaction. *Chin. J. Explos. Propell.* **2007**, *30*, 20–23.
87. Stefan, E.; Latypov, N.V. Four Syntheses of 4-Amino-3, 5-dinitropyrazole. *J. Heterocyclic. Chem.* **2014**, *51*, 1621–1627.
88. Zhang, M.; Gao, H.; Li, C.; Fu, W.; Tang, L.; Zhou, Z. Towards improved explosives with a high performance: N-(3, 5-dinitro-1 H-pyrazol-4-yl)-1 H-tetrazol-5-amine and its salts. *J. Mater. Chem. A* **2017**, *5*, 1769–1777. [CrossRef]
89. Dalinger, I.L.; Vatsadze, I.A.; Shkineva, T.K.; Popova, G.P.; Shevelev, S.A. Efficient procedure for high-yield synthesis of 4-substituted 3, 5-dinitropyrazoles using 4-chloro-3, 5-dinitropyrazole. *Synthesis* **2012**, *44*, 2058–2064. [CrossRef]
90. He, C.; Zhang, J.; Parrish, D.A.; Jean'ne, M.S. 4-Chloro-3, 5-dinitropyrazole: A precursor for promising insensitive energetic compounds. *J. Mater. Chem. A* **2013**, *1*, 2863–2868. [CrossRef]
91. Zhang, Y.; Huang, Y.; Parrish, D.A.; Jean'ne, M.S. 4-Amino-3, 5-dinitropyrazolate salts—highly insensitive energetic materials. *J. Mater. Chem.* **2011**, *21*, 6891–6897. [CrossRef]
92. Chavez, D.E.; Bottaro, J.C.; Petrie, M.; Parrish, D.A. Synthesis and Thermal Behavior of a Fused, Tricyclic 1, 2, 3, 4-Tetrazine Ring System. *Angew. Chem. Int. Ed.* **2015**, *54*, 12973–12975. [CrossRef]
93. Hervé, G.; Roussel, C.; Graindorge, H. Selective Preparation of 3, 4, 5-Trinitro-1H-Pyrazole: A Stable All-Carbon-Nitrated Arene. *Angew. Chem. Int. Ed.* **2010**, *49*, 3177–3181. [CrossRef]
94. Bölter, M.F.; Klapötke, T.M.; Kustermann, T.; Lenz, T.; Stierstorfer, J. Improving the Energetic Properties of Dinitropyrazoles by Utilization of Current Concepts. *Eur. J. Inorg. Chem.* **2018**, *2018*, 4125–4132. [CrossRef]
95. Yin, P.; Mitchell, L.A.; Parrish, D.A.; Shreeve, J.M. Comparative Study of Various Pyrazole-based Anions: A Promising Family of Ionic Derivatives as Insensitive Energetic Materials. *Chem.-Asian J.* **2017**, *12*, 378–384. [CrossRef] [PubMed]
96. Zhang, Y.; Parrish, D.A.; Shreeve, J.N.M. 4-Nitramino-3, 5-dinitropyrazole Based Energetic Salts. *Chem.-Eur. J.* **2012**, *18*, 987–994. [CrossRef] [PubMed]
97. Kumari, D.; Balakshe, R.; Banerjee, S.; Singh, H. Energetic plasticizers for gun & rocket propellants. *Rev. J. Chem.* **2012**, *2*, 240–262.
98. Dalinger, I.L.; Shakhnes, A.K.; Monogarov, K.A.; Suponitsky, K.Y.; Sheremetev, A.B. Novel highly energetic pyrazoles: N-fluorodinitromethyl and N-[(difluoroamino)dinitromethyl] derivatives. *Mendeleev Commun.* **2015**, *25*, 429–431. [CrossRef]
99. Dalinger, I.L.; Suponitsky, K.Y.; Pivkina, A.N.; Sheremetev, A.B. Novel Melt-Castable Energetic Pyrazole: A Pyrazolyl-Furazan Framework Bearing Five Nitro Groups. *Propell. Explos. Pyrot.* **2016**, *41*, 789–792. [CrossRef]
100. Agrawal, J.P.; Past, P. Future of Thermally Stable Explosives. *Cent. Eur. J. Energ. Mater.* **2012**, *9*, 273–290.
101. Li, C.; Zhang, M.; Chen, Q.; Li, Y.; Gao, H.; Fu, W.; Zhou, Z. 1-(3,5-Dinitro-1H-pyrazol-4-yl)-3-nitro-1H-1,2,4-triazol-5-amine (HCPT) and its energetic salts: Highly thermally stable energetic materials with high-performance. *Dalton Trans.* **2016**, *45*, 17956–17965. [CrossRef]
102. Wu, J.; Cao, D.; Wang, J.; Liu, Y.; Li, Y. Progress on 3,4,5-Trinitro-1H-pyrazole and Its derivatives. *Chin. J. Energ. Mater.* **20116**, *24*, 1121–1130.
103. Ravi, P.; Tewari, S.P. Facile and environmentally friendly synthesis of nitropyrazoles using montmorillonite K-10 impregnated with bismuth nitrate. *Catal. Commun.* **2012**, *19*, 37–41. [CrossRef]
104. Ravi, P.; Gore, G.M.; Sikder, A.K.; Tewari, S.P. Thermal decomposition kinetics of 1-methyl-3, 4, 5-trinitropyrazole. *Thermochim. acta* **2012**, *528*, 53–57. [CrossRef]

105. Dalinger, I.L.; Vatsadze, I.A.; Shkineva, T.K.; Popova, G.P.; Shevelev, S.A. The specific reactivity of 3, 4, 5-trinitro-1H-pyrazole. *Mendeleev Commun.* **2010**, *20*, 253–254. [CrossRef]
106. Dalinger, I.L.; Vatsadze, I.A.; Shkineva, T.K.; Popova, G.P.; Shevelev, S.A.; Nelyubina, Y.V. Synthesis and Comparison of the Reactivity of 3, 4, 5-1H-Trinitropyrazole and Its N-Methyl Derivative. *J. Heterocyclic. Chem.* **2013**, *50*, 911–924. [CrossRef]
107. Guo, H.-J.; Dang, X.; Yang, F.; Cao, D.-L.; Li, Y.-X.; Hu, W.-H.; Jiang, Z.-M.; Li, Z.-H.; Li, Z.-Y.; Zhang, L.-W. Solubility and thermodynamic properties of a kind of explosives in four binary solvents. *J. Mol. Liq.* **2017**, *247*, 313–327. [CrossRef]
108. Jun, W.; Mei, J.; Zhang, X.-Y. Empirical calculation of the explosive prameters of nitrodiazole explosives (II). *Chin. J. Energ. Mater.* **2013**, *21*, 609–611.
109. Zhang, Y.; Guo, Y.; Joo, Y.H.; Parrish, D.A.; Shreeve, J.N.M. 3, 4, 5-Trinitropyrazole-Based Energetic Salts. *Chem.-Eur. J.* **2010**, *16*, 10778–10784. [CrossRef]
110. Drukenmüller, I.E.; Klapötke, T.M.; Morgenstern, Y.; Rusan, M.; Stierstorfer, J. Metal Salts of Dinitro-, Trinitropyrazole, and Trinitroimidazole. *Z. Anorg. Allg. Chem.* **2014**, *640*, 2139–2148. [CrossRef]
111. Zhang, Y.; Parrish, D.A.; Jean'ne, M.S. Synthesis and properties of 3, 4, 5-trinitropyrazole-1-ol and its energetic salts. *J. Mater. Chem.* **2012**, *22*, 12659–12665. [CrossRef]
112. Kumar, D.; He, C.; Mitchell, L.A.; Parrish, D.A.; Shreeve, J.N.M. Connecting energetic nitropyrazole and aminotetrazole moieties with N,N'-ethylene bridges: A promising approach for fine tuning energetic properties. *J. Mater. Chem. A* **2016**, *4*, 9220–9228. [CrossRef]
113. Kumar, D.; Imler, G.H.; Parrish, D.A.; Shreeve, J.N.M. N-Acetonitrile Functionalized Nitropyrazoles: Precursors to Insensitive Asymmetric N-Methylene-C Linked Azoles. *Chem.-Eur. J.* **2017**, *23*, 7876–7881. [CrossRef]
114. Tang, Y.; Kumar, D.; Shreeve, J.N.M. Balancing excellent performance and high thermal stability in a dinitropyrazole fused 1, 2, 3, 4-tetrazine. *J. Am. Chem. Soc.* **2017**, *139*, 13684–13687. [CrossRef] [PubMed]
115. Tang, Y.; He, C.; Imler, G.H.; Parrish, D.A.; Jean'ne, M.S. AC–C bonded 5, 6-fused bicyclic energetic molecule: Exploring an advanced energetic compound with improved performance. *Chem. Commun.* **2018**, *54*, 10566–10569. [CrossRef] [PubMed]
116. Tang, Y.; He, C.; Imler, G.H.; Parrish, D.A.; Jean'ne, M.S. Energetic derivatives of 4, 4', 5, 5'-tetranitro-2 H, 2' H-3, 3'-bipyrazole (TNBP): Synthesis, characterization and promising properties. *J. Mater. Chem. A* **2018**, *6*, 5136–5142. [CrossRef]
117. Domasevitch, K.V.; Gospodinov, I.; Krautscheid, H.; Klapötke, T.M.; Stierstorfer, J. Facile and selective polynitrations at the 4-pyrazolyl dual backbone: Straightforward access to a series of high-density energetic materials. *New J. Chem.* **2019**, *43*, 1305–1312. [CrossRef]
118. Kumar, D.; Mitchell, L.A.; Parrish, D.A.; Shreeve, J.N.M. Asymmetric N,N'-ethylene-bridged azole-based compounds: Two way control of the energetic properties of compounds. *J. Mater. Chem. A* **2016**, *4*, 9931–9940. [CrossRef]
119. Yin, P.; Parrish, D.A.; Shreeve, J.M. Bis(nitroamino-1,2,4-triazolates): N-bridging strategy toward insensitive energetic materials. *Angew. Chem. Int. Ed.* **2014**, *53*, 12889–12892. [CrossRef]
120. Zhang, Q.; Zhang, J.; Parrish, D.A.; Shreeve, J.N.M. Energetic N-Trinitroethyl-Substituted Mono-, Di-, and Triaminotetrazoles. *Chem.-Eur. J.* **2013**, *19*, 11000–11006. [CrossRef]
121. Joo, Y.H.; Shreeve, J.N.M. Energetic Mono-, Di-, and Trisubstituted Nitroiminotetrazoles. *Angew. Chem. Int. Ed.* **2009**, *48*, 564–567. [CrossRef]
122. Yin, P.; Zhang, J.; Parrish, D.A.; Shreeve, J.M. Energetic N,N'-ethylene-bridged bis(nitropyrazoles): Diversified functionalities and properties. *Chemistry* **2014**, *20*, 16529–16536. [CrossRef]
123. Fischer, D.; Gottfried, J.L.; Klapotke, T.M.; Karaghiosoff, K.; Stierstorfer, J.; Witkowski, T.G. Synthesis and Investigation of Advanced Energetic Materials Based on Bispyrazolylmethanes. *Angew. Chem. Int. Ed.* **2016**, *55*, 16132–16135. [CrossRef]
124. Li, H.; Zhang, L.; Petrutik, N.; Wang, K.; Ma, Q.; Shem-Tov, D.; Zhao, F.; Gozin, M. Molecular and crystal features of thermostable energetic materials: Guidelines for architecture of "bridged" compounds. *ACS Central Sci.* **2019**, *6*, 54–75. [CrossRef] [PubMed]
125. Zhang, M.X.; Pagoria, P.F.; Imler, G.H.; Parrish, D. Trimerization of 4-Amino-3,5-dinitropyrazole: Formation, Preparation, and Characterization of 4-Diazo-3,5-bis(4-amino-3,5-dinitropyrazol-1-yl) pyrazole (LLM-226). *J. Heterocyclic. Chem.* **2019**, *56*, 781–787. [CrossRef]

126. Yan, T.; Cheng, G.; Yang, H. 1,2,4-Oxadiazole-Bridged Polynitropyrazole Energetic Materials with Enhanced Thermal Stability and Low Sensitivity. *Chempluschem* **2019**, *84*, 1567–1577. [CrossRef] [PubMed]
127. Yan, T.; Cheng, G.; Yang, H. 1, 3, 4-Oxadiazole based thermostable energetic materials: Synthesis and structure–property relationship. *New J. Chem.* **2020**, *44*, 6643–6651. [CrossRef]
128. Yin, P.; Zhang, J.; Imler, G.H.; Parrish, D.A.; Shreeve, J.N.M. Polynitro-Functionalized Dipyrazolo-1, 3, 5-triazinanes: Energetic Polycyclization toward High Density and Excellent Molecular Stability. *Angew. Chem. Int. Ed.* **2017**, *129*, 8960–8964. [CrossRef]
129. Il'yasov, S.G.; Danilova, E.O. Preparation of 1, 3-Diazido-2-Nitro-2-Azapropane from Urea. *Propell. Explos. Pyrot.* **2012**, *37*, 427–431. [CrossRef]
130. Zhang, J.; He, C.; Parrish, D.A.; Shreeve, J.N.M. Nitramines with Varying Sensitivities: Functionalized Dipyrazolyl-N-nitromethanamines as Energetic Materials. *Chem.-Eur. J.* **2013**, *19*, 8929–8936. [CrossRef]
131. Klapötke, T.M.; Penger, A.; Pflüger, C.; Stierstorfer, J.; Sućeska, M. Advanced Open-Chain Nitramines as Energetic Materials: Heterocyclic-Substituted 1, 3-Dichloro-2-nitrazapropane. *Eur. J. Inorg. Chem.* **2013**, *2013*, 4667–4678. [CrossRef]
132. Yin, M.; Shu, Y.; Xiong, Y.; Luo, S.; Long, X.; Zhu, Z.; Du, J. Theoretical study on relationship between structures and properties of pyrazole compounds. *Chin. J. Energ. Mater.* **2008**, *15*, 567–571.
133. Türker, L.; Variş, S. A Review of Polycyclic Aromatic Energetic Materials. *Polycycl. Aromat. Comp.* **2009**, *29*, 228–266. [CrossRef]
134. Dalinger, I.L.; Shkineva, T.K.; Shevelev, S.A.; Kanishchev, M.I.; Kral, V.; Arnold, Z. A method for preparation of C-diformylmethylnitropyrazoles. *Russ. Chem. Bull.* **1993**, *42*, 211–212. [CrossRef]
135. Pagoria, P.F. Synthesis, Scale-up and Experimental Testing of ANPZO. In Proceedings of the 1998 Insensitive Munitions and Energetic Materials Technology Symposium, San Diego, CA, USA, 16–19 November 1998.
136. Li, Y.; Wang, B.; Luo, Y.; Yang, W.; Wang, Y.; Li, H. Synthesis of 3,6-Dinitropyrazolo[4, 3-c]pyrazole (DNPP) in Hectogram scale and Crystal Structure of DNPP·H₂O. *Chin. J. Energ. Mater.* **2013**, *21*, 449–454.
137. Luo, Y.; Ge, Z.; Wang, B.; Zhang, H.; Liu, Q. Synthetic improvent of DNPP. *Chin. J. Energ. Mater.* **2007**, *15*, 205–208.
138. Zhang, J.; Parrish, D.A.; Shreeve, J.M. Thermally sTable 3,6-dinitropyrazolo[4,3-c]pyrazole-based energetic materials. *Chem.-Asian J.* **2014**, *9*, 2953–2960. [CrossRef]
139. Li, Y.; Li, X.; Li, H.; Lian, P.; Yifen, L.; Wang, B. Synthesis and thermal performance of organic diamino salts of 3,6-dinitropyrazolo4,3-cpyrazole (DNPP). *J. Solid Rocket Technol.* **2015**, *38*.
140. Li, Y.; Tang, T.; Lian, P.; Luo, Y.; Yang, W.; Wang, Y.; Li, H.; Zhang, Z.; Wang, B. Synthesis, Thermal Performance and Quantum Chemistry Study on 1,4-Diamino-3,6-dinitropyrazolo[4,3-c]pyrazole. *Chin. J. Energ. Mater.* **2012**, *32*, 580. [CrossRef]
141. Yin, P.; Zhang, J.; Mitchell, L.A.; Parrish, D.A.; Shreeve, J.N.M. 3, 6-Dinitropyrazolo [4, 3-c] pyrazole-Based Multipurpose Energetic Materials through Versatile N-Functionalization Strategies. *Angew. Chem. Int. Ed.* **2016**, *128*, 13087–13089. [CrossRef]
142. Li, Y.; Shu, Y.; Wang, B.; Zhang, S.; Zhai, L. Synthesis, structure and properties of neutral energetic materials based on N-functionalization of 3,6-dinitropyrazolo[4,3-c]pyrazole. *RSC Adv.* **2016**, *6*, 84760–84768. [CrossRef]
143. Li, Y.; Chang, P.; Chen, T.; Hu, J.; Wang, B.; Zhang, H.; Li, P.; Wang, B. Synthesis, Characterization are Properties of 1,4-Dinitramino-3,6-dinitropyrazolo[4,3-c]pyrazole and Its Energetic Salts. *Chin. J. Energ. Mater.* **2018**, *26*, 578–584.
144. Li, Y.; Hu, J.; Chang, P.; Chen, T.; Zhang, H.; Wang, B. Synthesis, Crstal Structure and Properties of 1, 4-Dinitramino-3, 6-dinitropyrazolo[4, 3-c]pyrazole's Aminourea Salt. *Chin. J. Explos. Propell.* **2019**, *42*, 341–345.

145. Zhang, W.; Xia, H.; Yu, R.; Zhang, J.; Wang, K.; Zhang, Q. Synthesis and Properties of 3, 6-Dinitropyrazolo [4, 3-c]-pyrazole (DNPP) Derivatives. *Propell. Explos. Pyrot.* **2020**, *45*, 546–553. [CrossRef]
146. Xia, H.; Zhang, W.; Jin, Y.; Song, S.; Wang, K.; Zhang, Q. Synthesis of Thermally Stable and Insensitive Energetic Materials by Incorporating the Tetrazole Functionality into a Fused-Ring 3,6-Dinitropyrazolo-[4,3-c]Pyrazole Framework. *ACS Appl. Mater. Inter.* **2019**, *11*, 45914–45921. [CrossRef] [PubMed]

© 2020 by the authors. Licensee MDPI, Basel, Switzerland. This article is an open access article distributed under the terms and conditions of the Creative Commons Attribution (CC BY) license (http://creativecommons.org/licenses/by/4.0/).

Review

Electrooxidation Is a Promising Approach to Functionalization of Pyrazole-Type Compounds

Boris V. Lyalin [1], Vera L. Sigacheva [1], Anastasia S. Kudinova [1,2], Sergey V. Neverov [1], Vladimir A. Kokorekin [1,2,3,*] and Vladimir A. Petrosyan [1]

1. N.D. Zelinsky Institute of Organic Chemistry, Russian Academy of Sciences, Leninsky Prosp. 47, 119991 Moscow, Russia; bv.lyalin@yandex.ru (B.V.L.); vsigaveva@yandex.ru (V.L.S.); ana_kudinova@mail.ru (A.S.K.); sergei.neverov@yandex.ru (S.V.N.); vladim.petros@yandex.ru (V.A.P.)
2. Institute of Pharmacy, Sechenov First Moscow State Medical University (Sechenov University), Trubetskaya Str. 8, Bldg. 2, 119991 Moscow, Russia
3. All-Russian Research Institute of Phytopathology, Institute Str. 5, 143050 Bol'shiye Vyazemy, Russia
* Correspondence: kokorekin@yandex.ru; Tel.: +7-499-135-9070

Abstract: The review summarizes for the first time the poorly studied electrooxidative functionalization of pyrazole derivatives leading to the C–Cl, C–Br, C–I, C–S and N–N coupling products with applied properties. The introduction discusses some aspects of aromatic hydrogen substitution. Further, we mainly consider our works on effective synthesis of the corresponding halogeno, thiocyanato and azo compounds using cheap, affordable and environmentally promising electric currents.

Keywords: pyrazoles; pyrazolo[1,5-a]pyrimidines; aromatic substitution; electrooxidation; C–H halogenation; C–H thiocyanation; N–N coupling; cyclic voltammetry

1. Introduction

The functionalization of arenes is the key to their diversity, opening the way to practically useful substances. At the beginning of the 21st century, C-H functionalization of arenes became a popular tool for the implementation of such processes [1], and the center for selective C-H functionalization (mainly based on metal complex catalysis) was organized in the USA [2,3]. At the same time, a more attractive metal-free C-H functionalization of arenes has existed for many years. Its development is discussed below, since the review is related to it.

The first to consider is the electrophilic aromatic hydrogen substitution (S_E^H) [4]. It proceeds via the σ_H^+ adduct formation and proton elimination leading to the target product (Scheme 1a).

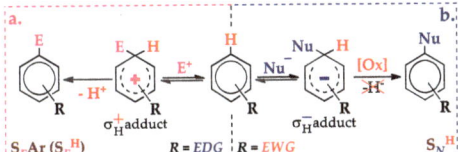

Scheme 1. Electrophilic (**a**) and nucleophilic (**b**) aromatic hydrogen substitution.

The nucleophilic aromatic hydrogen substitution (S_N^H, Scheme 1b) is problematic due to the difficulty of the hydride ion direct elimination from the σ_H^- adduct. Nevertheless, in the mid-1970s, Chupakhin and Postovsky proposed an indirect way of obtaining the target product, the chemical oxidation of the σ_H^- adduct [5]. To date, such processes have been extensively developed by Chupakhin and Charushin [6–9].

This review is devoted to electrooxidative functionalization of pyrazoles. Why is it so attractive? The answer is given by Seebach [10], who showed that the interaction of

two near-polar co-reactants is provided by the polarity inversion of one of them. This is the essence of electrochemistry, where the electrode transformation of the substrate is accompanied by its polarity inversion. For example, 1,4-dimethoxybenzene (DMB) and pyrazole (Scheme 2) are two non-interacting nucleophiles, but polarity inversion of DMB by electrooxidation leads to the electrophilic radical cation that reacts with pyrazole. In the electrochemical literature, this C-H functionalization is called anodic substitution, which has been actively studied since the mid-1950s [11–20].

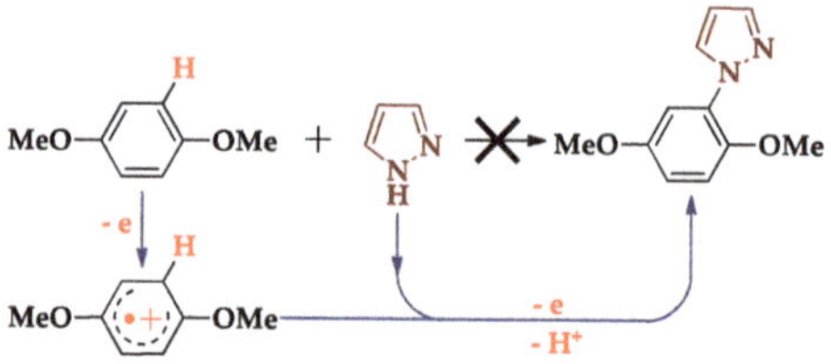

Scheme 2. Electrooxidative N-arylation of pyrazole by 1,4-dimethoxybenzene.

Note that chemical (electrophilic and nucleophilic) substitution and electrochemical (anodic) substitution in arenes have been being developed in parallel and independently of each other for a long time. In addition, the role of anodic substitution among the corresponding chemical processes was unconsidered.

Only recently we have introduced the concept of anodic substitution as an electrochemical aromatic substitution of hydrogen [21,22]. Such reactions were designated as S_N^H An (An is anode), since the key stage is anodic oxidation of the substrate. For electron-rich arenes, the processes proceed along two main routes (Scheme 3), depending on the ease of oxidation of nucleophile vs. arene [21]. Route I (arene oxidizes easier than Nu$^-$) proceeds via the interaction between Nu$^-$ and the arene radical cation. Route II (Nu$^-$ oxidizes easier than arene) proceeds via the formation of Nu$^\bullet$, which either interacts with the arenes homolytically (route II^a) or forms a dimer that interacts with the arene as an electrophile (route II^b).

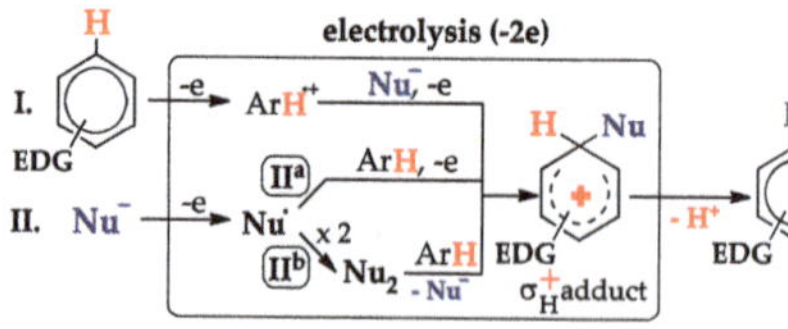

Scheme 3. Main routes of S_N^H An processes: **I** ($E_p^{ox}{}_{Ar} < E_p^{ox}{}_{Nu^-}$) or **II** ($E_p^{ox}{}_{Nu^-} < E_p^{ox}{}_{Ar}$).

In general, the multitude of S_N^H (An) processes, where hydrogen is displaced with a nucleophile, are electrooxidative C-H functionalizations (C-H An) and can be described by Scheme 4. Such strategy opens up the direct method of C-H functionalization of arenes with the C–C and C–Het coupling realization. The latter is especially shown in the examples of electrooxidative C-H halogenation and thiocyanation of pyrazole derivatives (Sections 2 and 3). A special place is occupied by the N–N coupling of amino pyrazoles via the N–H functionalization (Section 4), since its patterns are somehow similar to those described above, but not sufficiently studied.

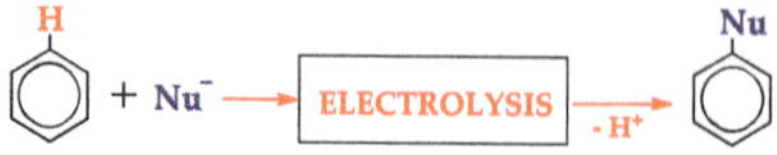

Scheme 4. General scheme of the electrooxidative C-H functionalization (C–H An).

Such processes are attractive for green chemistry [23,24] because they use cheap, affordable and environmentally promising electric current instead of chemical oxidants (often toxic, unrecyclable and used in excess) and complex catalysts (sometimes expensive and toxic). In addition, the varying anode potential eliminates the difficulties of an empirical search for suitable chemical oxidants, while Pt, a frequently used electrode material [25], can be replaced by more attractive ones, e.g., glassy carbon, or ruthenium–titanium oxide (for more information on electrode materials see [25–27]).

Since the review is devoted to the poorly studied electrochemical functionalization of pyrazole derivatives, Sections 2–4 mainly summarize the investigations of the review's authors.

2. Electrooxidative C–H Halogenation of Pyrazole and Its Substituted Derivatives

Halogenated pyrazoles are widely used in organic synthesis; in particular, iodo and bromo pyrazoles are the key reagents in transition metal-catalyzed cross-coupling [28]. Moreover, they are important precursors of drugs, such as antihepatitis, anti-Alzheimer, antiparkinsonian and anti-schizophrenic drugs (chloro-pyrazoles) [29–31], antiglaucoma drugs (bromo-pyrazoles) [32] and antiatherosclerotic, antimalarial, anti-inflammatory and immunocorrective drugs (iodo-pyrazoles) [33–38]. At the same time, chloro- and iodo-pyrazoles are used for preparation of antidiabetic drugs [39,40], bromo- and iodo-pyrazoles—of anticancer [41–44] and antimicrobial [45,46] agents, and chloro- and bromo-pyrazoles—of agrochemicals [47–51].

The active use of halogeno-pyrazoles has stimulated interest in their efficient and ecologically attractive synthesis, including electrosynthesis (see Introduction). At the same time, the electrochemical halogenation of pyrazoles has practically not been studied before us, but it was preceded by chemical halogenation, the aspects of which are briefly given below.

2.1. Chemical Halogenation of Pyrazoles

The processes (Scheme 5) are usually carried out by the interaction of pyrazoles and halogens (or halogenating reagents), and they occur first at position 4 and only then at other positions [28].

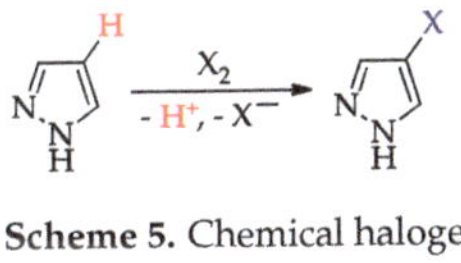

Scheme 5. Chemical halogenation of pyrazoles (X = Cl, Br, I).

2.1.1. Chlorination

The most common is the interaction of pyrazoles and Cl_2. It was used to convert the pyrazole and its alkyl derivatives into the 4-chloro pyrazoles (yields 40–85%, at 0–40 °C, in CH_2Cl_2 or CCl_4). Under severe conditions (at 80–100 °C, in AcOH) dichloropyrazoles and the chloro products of the alkyl groups were also formed [52,53]. At the same time, the corresponding 4-chloro derivatives were obtained by the reaction of 3,5-dimethyl-1H-pyrazole (and its N-substituted derivatives) with N-chlorosuccinimide (yield 95–98%, at 20–25 °C, in CCl_4 or H_2O) [54,55].

2.1.2. Bromination

The reaction of alkyl-substituted pyrazoles and their carboxylic acids with Br_2 (at 20–25 °C) proceeded in yields of 75–96% in non-aqueous media (CH_2Cl_2, $CHCl_3$ or CCl_4) [45,56–58], or 50–75% in water [59,60]. The NaOH additives (which binds with HBr formed) allows bromination of low-reactive pyrazole-3-carboxylic acid in the yield of 90% [61]. A number of 4-bromopyrazoles were also obtained by the reaction of 3,5-dimethyl-1H-pyrazole (and its N-substituted derivatives) with N-bromosuccinimide (yields 90–99%, at 20–25 °C, in CCl_4 or H_2O) [55].

2.1.3. Iodination

Good results in the iodination of pyrazoles with donor substituents were obtained using the I_2–NaI–K_2CO_3 system (yields 75–90% at 20–25 °C in aq. EtOH) [62–64]. A solution of N-iodosuccinimide in acidic media (50% aq. H_2SO_4, CF_3SO_3H, CF_3COOH, AcOH) was also efficient for the iodination [44,62,65]. Finally, the iodination of practically any pyrazoles proceeded efficiently and without toxic waste using the I_2–HIO_3 system in AcOH–CCl_4 [66,67].

In general, the above methods are quite effective, but not ecologically attractive enough due to the frequent use of halogens in their pure form or waste of other halogenating agents (e.g., succinimide). Such problems can be solved using electrochemical methods—C–H An halogenation.

2.2. C-H An Halogenation of Pyrazoles

The halogenation (Scheme 6) usually proceeds [26] via the electrogeneration of a halogen followed by its interaction with pyrazole (cf. Scheme 3, route II^b).

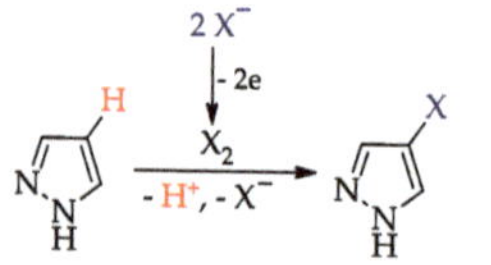

Scheme 6. C–H An halogenation of pyrazoles (X = Cl, Br, I).

Such processes are mainly carried out under mild conditions in an anodic compartment of a divided cell on a Pt-anode under galvanostatic electrolysis with alkali metal halides in H_2O or in H_2O–$CHCl_3$. A series of N–H and N–Alk pyrazoles, including those with donor (acceptor) substituents, were objects of study (Tables 1–3).

2.2.1. Chlorination

C–H An chlorination of pyrazole **1a** (Table 1, entry 1) led to 4-chloropyrazole **1b** (yield 46%) and to by-product **1b'** (yield 8%) in H_2O–$CHCl_3$ at the theoretical amount of electricity passed ($Q/Q_t = 1$). Apparently, the product **1b** undergoes chlorination to 1,4-dichloropyrazole **1b'** (Scheme 7), followed by C–N dehydrogenative cross-coupling to by-product **1b-1b'**.

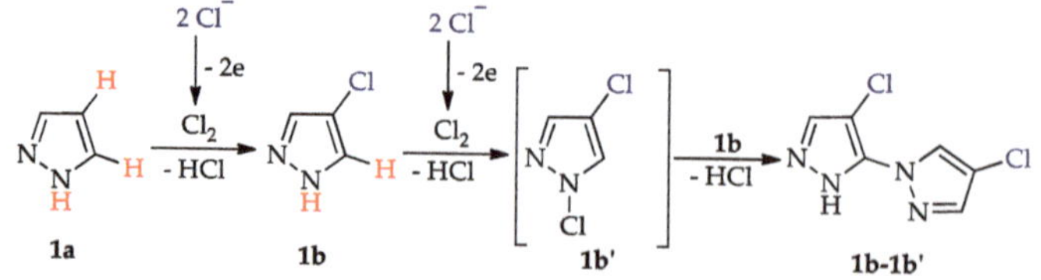

Scheme 7. C-H An chlorination of pyrazole **1a**.

The need for $CHCl_3$ (as an extractant of target product) should be noted, since its absence decreased the yield of product **1b** to 34% and increased the yield of by-product **1b-1b'** to 15%. Pyrazoles **2a–7a** (entries 2–7), gave the monochlorinated products **2b–7b** with different yields (8–71%) depending on the position of Me groups. Di- and trichloroproducts were obtained in entries 5 and 6.

Pyrazoles with acceptor groups (NO_2 or COOH) were chlorinated without $CHCl_3$ additives: the yields of the target products **8b–14b** were 41–93% (entries 8–14). Only pyrazole **14a**, containing both NO_2 and COOH groups, was the least reactive. Therefore, the electrochemical method for the synthesis of 4-chloropyrazolcarboxylic acids [68] is noticeably superior to the corresponding chemical one [69].

Table 1. C-H An chlorination of pyrazoles (Az-H) [1].

Entry	Az-H, Az-Cl and other Products (Yield, %) [70,71]			Entry	Az-H, Az-Cl (Yield, %) [68,70,71]	
1	1a	1b (46 [2])	1b-1b' (8 [2])	8	8a	8b (41 [2])
2	2a	2b (71 [3])		9	9a	9b (64 [3])
3	3a	3b (34 [2])		10	10a	10b (92 [3])
4	4a	4b (70 [3])		11	11a	11b (69 [3])
5	5a	5b (15 [2])	5b' (35 [2])	12	12a	12b (93 [3])
6	6a	6b (47 [2])	6b' (13 [2]), 6b'' (4) [2]	13	13a	13b (84 [3])
7	7a	7b (8 [2])		14	14a	14b (4 [2])

[1] Electrolysis in 100 mL of 4 M solution of NaCl in H_2O–$CHCl_3$ (entries 1–7), H_2O (entries 8–14), 15 °C, pyrazole (12.5–50 mmol), divided cell, Pt anode, Cu cathode, galvanostatic electrolysis (j_{anode} = 100 mA·cm^{-2}), Q_t = 2412–9650 C, Q/Q_t = 1–2; [2] the yield was calculated from the ^{1}H NMR spectroscopic data for the isolated mixture of products with unreacted pyrazoles; [3] the yield was determined for the isolated product.

Table 2. C-H An bromination of pyrazoles (Az-H) [1].

Entry	Az-H, Az-Br and Other Products (Yield, %) [71,72]			Entry	Az-H, Az-Br (Yield, %) [71,72]	
1	1a	1c (70 [2])		8	8a	8c (89 [2])
2	2a	2c (76 [3])	2c' (5 [3])	9	9a	9c (15 [3])

Table 2. Cont.

Entry	Az-H, Az-Br and Other Products (Yield, %) [71,72]		Entry	Az-H, Az-Br (Yield, %) [71,72]	
3	3a	3c (66 [3])	10	10a	10c (68 [3])
4	4a	4c (94 [2])	11	11a	11c (78 [2])
5	5a	5c (55 [3]), 5c′ (26 [3])	12	12a	12c (84 [2])
6	6a	6c (88 [2])	13	13a	13c (84 [2])
7	7a	7c (0 [5])	14	14a	14c (0 [3])

[1] Electrolysis in 100 mL of 1 M solution of NaBr in H_2O–$CHCl_3$ (entries 1–7), H_2O (entries 8–14), 30 °C, pyrazole (12.5–50 mmol), divided cell, Pt anode, Cu cathode, galvanostatic electrolysis (j_{anode} = 30 mA·cm^{-2}), $Q = Q_t$ = 2412–9650 C; [2] the yield was determined for the isolated product; [3] the yield was calculated from the ^{1}H NMR spectroscopic data for the isolated mixture of products with (or) unreacted pyrazoles; [5] unpublished data.

Table 3. C-H (An) iodination of pyrazoles (Az-H).

Entry	Az-H, Az-I (Yield, %) [73,74]		Entry	Az-H, Az-I (Yield, %) [73,74]	
1	1a	1d (57 [1,3], 93 [2,3])	8	8a	8d (2 [1,4], 82 [2,3])
2	2a	2d (5 [1,4], 79 [2,3])	9	9a	9d (0 [1,4])
3	3a	3d (71 [1,3])	10	10a	10d (30 [1,4], 86 [2,3])
4	4a	4d (86 [1,3], 93 [2,3])	11	11a	11d (0 [1,4], 74 [2,3])

Table 3. Cont.

Entry	Az-H, Az-I (Yield, %) [73,74]		Entry	Az-H, Az-I (Yield, %) [73,74]	
5	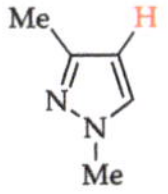 5a (Me, H; N-Me pyrazole)	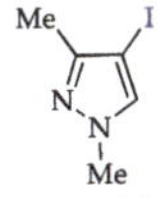 5d (35 [1,4])	12	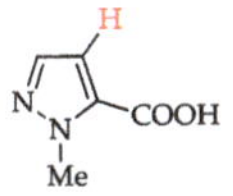 12a (Me, H, COOH)	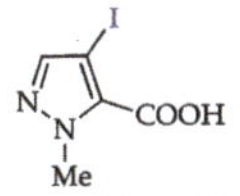 12d (0 [1,4], 78 [2,3])
6	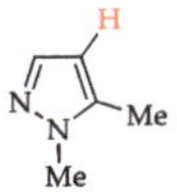 6a	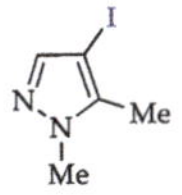 6d (42 [1,5])	13	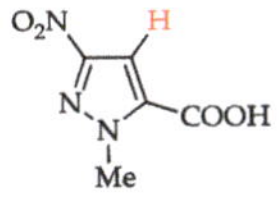 14a (O_2N, Me, COOH)	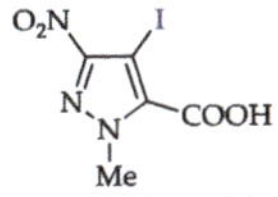 14d (0 [1,5], 79 [2,3])
7	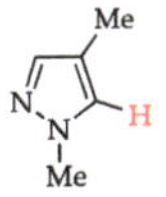 7a	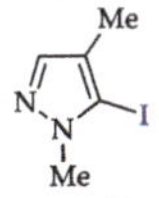 7d (0 [1,5])	14	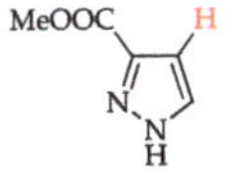 15a (MeOOC, Me)	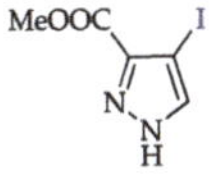 15d (40 [1,4])

[1] Electrolysis in 100 mL of 0.3 M solution of $NaNO_3$ in H_2O–$CHCl_3$, 30 °C, KI (10 mmol), pyrazole (10 mmol), $NaHCO_3$ (15 mmol), divided cell, Pt anode, Cu cathode, galvanostatic electrolysis (j_{anode} = 7.5 mA·cm^{-2}), Q = Q_t = 1930 C; [2] two-step process: 1. electrogeneration of KIO_3 in 1M aq. KOH, 70 °C, KI (30 mmol), $K_2Cr_2O_7$ (0.7 mmol), undivided cell, NiO(OH) anode, Ni cathode, galvanostatic electrolysis (j_{anode} = 200 mA·cm^{-2}), Q_t = 17370 C, Q/Q_t = 0.9–1.1; 2. pyrazole (45–150 mmol), KIO_3 (9–30 mmol), I_2 (18–60 mmol), H_2O–$CHCl_3$ (or H_2O–CCl_4), H_2SO_4 conc., temperature 50–66 °C, 0.5–14 h; [3] the yield was determined for the isolated product; [4] the yields were calculated from the ^1H NMR spectroscopic data for the isolated products with (or) unreacted pyrazoles; [5] unpublished data.

2.2.2. Bromination

Compared with chlorination, the C-H An bromination (Table 2) proceeded more effectively for pyrazole and its methyl derivatives (yields of products 1c–6c 55–94%). In some cases, dibromo by-products (entries 2 and 5), low yield (entry 9), or the absence of any reactions (entries 7 and 14) were observed.

2.2.3. Iodination

C-H An iodination by weakly electrophilic I_2 (Table 3) was generally less effective than bromination [73]. Traces of the target products or no reaction were observed in half of the cases (entries 2, 7–9, 11–13). In other cases, the yields were 35–86% (for pyrazole and its methyl derivatives in entries 1, 3–6) and 30–40% (for nitro- and carboxypyrazoles in entries 10 and 14).

A much more effective iodinating agent was HOI, which can be obtained by the reaction of KIO_3 with KI (or I_2) and H_2SO_4 [75–79]. The original process [74] includes the electrogeneration of KIO_3 (on the NiO(OH) anode [80]), followed by the interaction of HOI generated in situ with the pyrazole (Scheme 8). As a result, the yields of target products increased to 74–93% (entries 1, 2, 4, 8, 10–12, 13).

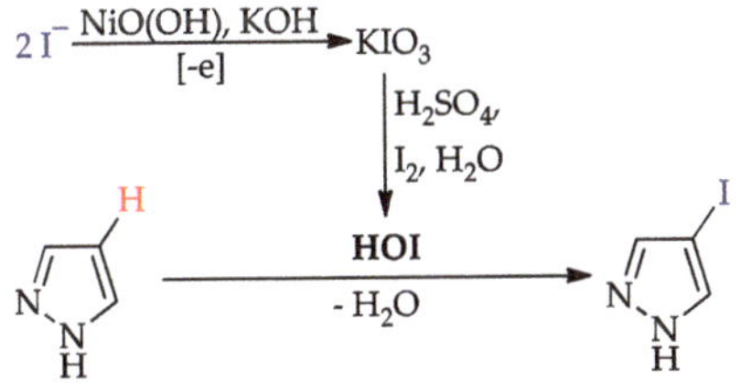

Scheme 8. C–H iodination of pyrazoles via HOI.

2.3. The Mechanistic Aspects of C–H (An) Halogenation of Pyrazoles

Since I^-, Br^-, and Cl^- are commonly oxidized at lower anodic potentials than the studied pyrazoles, the process proceeds via the electrooxidation of Hal^- to Hal_2

followed by interaction of the latter with arenes (see Scheme 3, route II[b] and Scheme 6). The possible mechanism [26,71,79] (Scheme 9) includes the initial attack of the halogen on the N^2 of **Az–H** with the formation of σ_H^+ **adduct 1**. The latter, depending on the R, gives either **N–X** intermediate (R = H) or σ_H^+ **adduct 2** (R = Alk). Therefore, the target **Az–X** is formed either due to N–C rearrangement of **N–X** derivative or due to the deprotonation of σ_H^+ **adduct 2**.

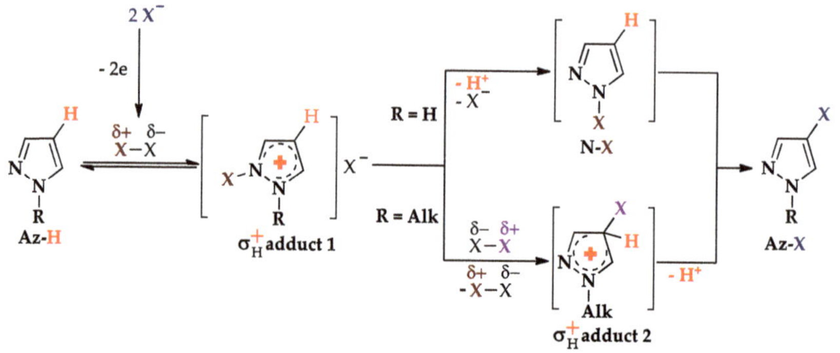

Scheme 9. C–H An halogenation of pyrazoles (possible mechanisms, X = Cl, Br, I).

Iodination by HOI proceeds similarly, but for highly basic N-unsubstituted pyrazole and its alkyl derivatives it most likely occurs (Scheme 10) via **C–I adduct** (the result of protonation of N^2 and HOI attack on C^4) [74,75,77–79].

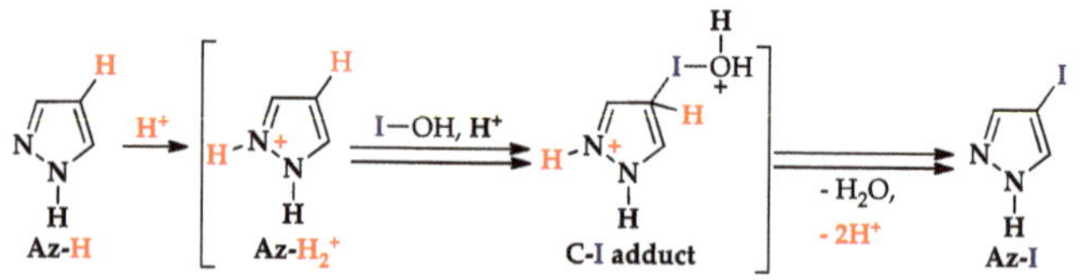

Scheme 10. Iodination of highly basic pyrazoles by HOI.

Additional control experiments showed different properties of N–Cl and N–Br intermediates (Scheme 11). Therefore, the N–C rearrangement of the N–Cl derivative is significantly lower than that for the N–Br (cf. stages **N–X→Az–Cl** and **N–X→Az–Br**). At the same time, for the N–Cl bond, homolytic cleavage is observed, while for N–Br it is heterolytic (cf. stages **N–X→Ar–CH₂-Cl** and **N–X→Ar–Br**).

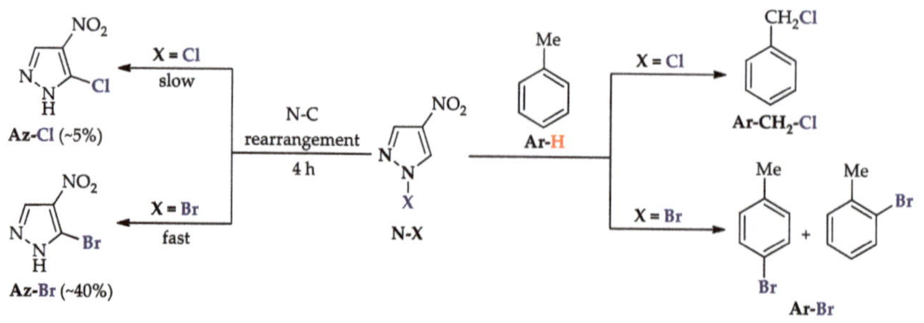

Scheme 11. Different behavior of N–Cl and N–Br intermediates (unpublished control experiments).

The above data not only reveal the essence of pyrazoles C–H An halogenation, but also explain the difference in its efficiency (e.g., the anomalously less efficient chlorination of N-unsubstituted pyrazoles compared to bromination (cf. entries 1, 3 and 4, Tables 1 and 2), and the formation of by-products (e.g., entries 1 and 6, Table 1).

Therefore, this Section describes the basic patterns of C–H An halogenation, and the efficient (up to 94% yield) gram-scale synthesis of a series of chloro-, bromo- and iodo-pyrazoles in aqueous or aqueous-organic media. The following Section reflects the main points on the related C–H An thiocyanation of pyrazole derivatives.

3. Electrooxidative C–H Thiocyanation of 5-Aminopyrazoles and Pyrazolo [1,5-a]pyrimidines

Thiocyanation of the C–H bond of arenes is an effective tool for C–S coupling [81–84]. The resulting aryl thiocyanates are valuable precursors of sulfur and nitrogen-containing compounds (thiols [85], (di)sulfides [86,87], dithiocarbamates [88], thiazoles [89], tetrazoles [90]), and are highly bioactive compounds (antifungal [91], antitumor [92], antiparasitic [93]). Recently synthesized thiocyanates of pyrazole derivatives also have sufficient antifungal [94] and antitumor [95] activity.

One of the key intermediates of C–H thiocyanation of arenes is the well-known [96,97] pseudohalogene thiocyanogen $(SCN)_2$. It is usually obtained in situ by chemical or electrochemical oxidation of the thiocyanate ion (Scheme 12).

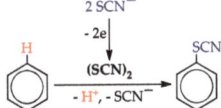

Scheme 12. C–H thiocyanation of arenes via the thiocyanogen.

The chemical approach has been actively developed over the past 10–15 years, but it is often associated with the use of an excess of unrecyclable oxidants, which can sometimes be toxic, scalding or poorly available (e.g., Br_2 [98], I_2 [99], DEAD [100], HIO_3 [101], H_5IO_6 [102], I_2O_5 [103], H_2O_2 [102,104–107], $K_2S_2O_8$ [108,109], CAN [110], $Mn(OAc)_3$ [111], p-TSA [112], NCS [113], NBS [100], NIS [114], NTS [115], DDQ [116,117]). The electrochemical approach (see [22], Scheme 3, route II^b, and Scheme 13) is devoid of such disadvantages, but it is poorly studied in general [118–120]. For pyrazole derivatives, C–H An thiocyanation is studied for the first time in a series of works [22,121–126], which are reflected in this Section.

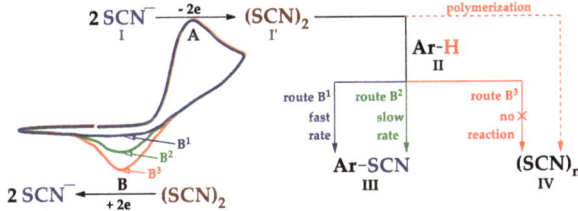

Scheme 13. C–H An thiocyanation of arenes (**II**) via the thiocyanogen (**I′**) and voltammetric test of the process efficiency.

3.1. C–H An Thiocyanation: General Patterns and Approaches

According to the above and developed [22,118–126] concepts, C–H An thiocyanation occurs during the anodic oxidation of the thiocyanate ion in the presence of arene, as a rule, via the thiocyanogen (Scheme 13, step **I→I′**). The latter either interacts with arene (step **I′ + II→III**), or gives polythiocyanogen [127] (step **I′→IV**).

These processes were investigated by cyclic voltammetry (CV) [22,123,125,126]. Scheme 13 shows a typical CV curve of SCN$^-$. Peak **A** corresponds to the oxidation of thiocyanate ion **I** to thiocyanogen **I'**, which is detected on the reverse scan by its reduction peak **B** (**B**^{3}). If after the addition of arene **II**, peak **B** disappears (cf. peaks **B**^{1} and **B**^{3}) or decreases (cf. peaks **B**^{2} and **B**^{3}), then Ar–H **II** interacts with (SCN)$_2$, respectively, via the route **B**^{1} or **B**^{2} to form the target Ar–SCN **III**. If the peak **B** does not change, then Ar–H **II** does not react with (SCN)$_2$ (see peak B^3 and route B^3). In this case the main reaction product is polythiocyanogen **IV**.

Further, we proposed [122,123] the original system of approaches to the C–H An thiocyanation of arenes (Scheme 14) depending on the reactivity of arenes with respect to (SCN)$_2$: via the generation (SCN)$_2$ at the oxidation potential (E_p^{ox}) of thiocyanate ion (approach A, cf. Scheme 3, route II^b, and Scheme 13), via electrogeneration (SCN)$_2$ in the presence of ZnCl$_2$ activating additives (approach B) or via the generation of a highly reactive radical cation at E_p^{ox} of arene (approach C, cf. Scheme 3, route I).

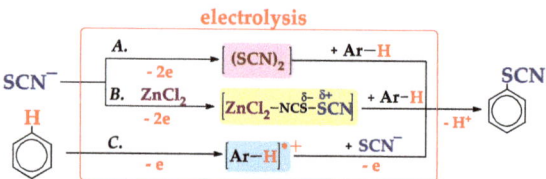

Scheme 14. Approaches and possible mechanisms of C–H An thiocyanation.

Approach A is used for arenes that react with (SCN)$_2$ (Scheme 13, routes B^1 and B^2), whereas approaches B and C are used for arenes that do not interact with non-activated (SCN)$_2$.

These patterns and approaches are considered below on the examples of C–H An thiocyanation of the practically useful [128–130] derivatives of 5-aminopyrazole and pyrazolo[1,5-a]pyrimidine and the original electrosynthesized 1-(hetero)arylpyrazoles [124,131].

3.2. C–H An Thiocyanation of Pyrazole Derivatives

The studies included a preliminary CV test in addition to electrosynthesis. The initial pyrazoles **1e–15e** and their thiocyanation products **1f–15f** are presented in Tables 4 and 5.

Table 4. C–H An thiocyanation of pyrazole derivatives (**Az-H**) via (SCN)$_2$ (approach A) [1].

Entry	Az–H, Az–SCN (Yield, %) [22,122,123,126]		Entry	Az–H, Az–SCN (Yield, %) [22,122,123,126]	
1	1e	1f (83 [2,5,6], 72 [2,4,6], 74 [2,5,7], 69 [3,5,7])	6	6e	6f (86 [2,4,6])
2	2e	2f (87 [2,5,6], 78 [2,5,7], 71 [3,5,7])	7	7e	7f (83 [2,4,6], 80 [2,4,7], 75 [2,5,7], 77 [3,4,7], 71 [3,5,7])
3	3e	3f (65 [2,4,6], 57 [3,4,6])	8	8e	8f (85 [2,4,6], 82 [2,5,6])

Table 4. Cont.

Entry	Az–H, Az–SCN (Yield, %) [22,122,123,126]	Entry	Az–H, Az–SCN (Yield, %) [22,122,123,126]
4	4e, 4f (75 [2,4,6], 68 [3,4,6])	9	9e, 9f (75 [2,4,6], 66 [2,5,6])
5	5e, 5f (89 [2,4,6], 64 [2,5,6])		

[1] Electrolysis in 50–85 mL of 0.1 M solution of NaClO$_4$ in MeCN-H$_2$O (entries 1–2), MeCN (entries 3–5, 7–9), MeCN–MeOH (entry 6), 20–25 °C, NH$_4$SCN (3–20 mmol), pyrazole (entries 1–5, 7–9) or its hydrochloride (entry 6) (1–6 mmol), divided cell (entries 1–2), undivided cell (entries 3–9), Q_t = 193–985 C, Q/Q_t = 1 (entries 1–6), Q/Q_t = 2 (entries 7–9). All yields were determined for the isolated and purified products; [2] CPE—controlled potential electrolysis (E_{anode} = 0.70–1.00 V); [3] GE—galvanostatic electrolysis (j_{anode} = 2.50–12.50 mA·cm^{-2}); [4] Pt electrodes; [5] GC electrodes; [6] milligram scale of electrosynthesis; [7] gram scale of electrosynthesis.

Table 5. C–H An thiocyanation of hardly oxidizable pyrazolo[1,5-a]pyrimidines (**Az-H**) (approaches B and C) [1].

Entry	Az–H (E_p^{ox}, V[2]), Az–SCN (Yield, %) [122,123]	Entry	Az–H (E_p^{ox}, V[2]), Az–SCN (Yield, %) [122,123]
1	10e (1.75), 10f (81 [3,5], 65 [4,5], 60 [4,6])	4	13e (1.85), 13f (80 [3,5], 62 [4,5], 55 [4,6])
2	11e (1.77), 11f (79 [3,5], 60 [4,5])	5	14e (1.85), 14f (73 [3,5], 60 [4,5])
3	12e (1.79), 12f (77 [3,5], 60 [4,5], 52 [4,6])	6	15e (1.88), 15f (69 [3,5], 63 [4,5], 47 [4,6])

[1] Electrolysis in 60 mL 0.1 M solution of NaClO$_4$ in MeCN, 20–25 °C, undivided cell, Q_t = 193 C. All yields was determined for the isolated and purified products; [2] it was determined by CV (working electrode—Pt, reference electrode—SCE, ν = 0.10 V·s^{-1}); [3] Approach B: KSCN (4 mmol), ZnCl$_2$ (2 mmol), pyrazole (1 mmol), CPE (E_{anode} = 1.00 V), Q_t=193 C, Q/Q_t = 3; [4] Approach C: NH$_4$SCN (4 mmol) pyrazole (1 mmol), CPE (E_{anode} = 1.75–1.88 V), Q/Q_t = 3; [5] Pt electrodes; [6] GC electrodes.

3.2.1. CV studies and the Choice of Optimal Approach

Figure 1 shows CV curves of NH$_4$SCN and its mixtures with 3-methyl-1H-pyrazol-5-amine (**1e**), 2-methyl-5-thiophen-2-yl-7-(trifluoromethyl) pyrazolo[1,5-a]pyrimidine (**10e**), as well as curves of individual compounds and their thiocyanato products (**1e,10e,1f,10f**) [125,126]. The CV of NH$_4$SCN (curve 1, Figure 1A,B) has the anodic peak A[1] (E_p^{ox} = 0.70 V) of the thiocyanate ion and a cathodic peak B[1] (E_p^{red} = 0.34 V) of the thiocyanogen. The peak B[1] disappeared after the addition of pyrazole **1e** and did not change after the addition of pyrazole **10e** (cf. corresponding curves 1 and 4). This clearly shows that pyrazole **1e** reacts rapidly with (SCN)$_2$ (see Scheme 13, route B[1]) and approach A (Scheme 14) is suitable for its thiocyanation. From the other side, the pyrazole **10e** does not react with (SCN)$_2$ and approaches B and C may be suitable for its thiocyanation. Note also that on full scans, peaks A^3 of thiocyanates **1f,10f** were observed, in addition to the peaks A^2 of pyrazoles **1e,10e** (see curve 5, Figure 1A,B).

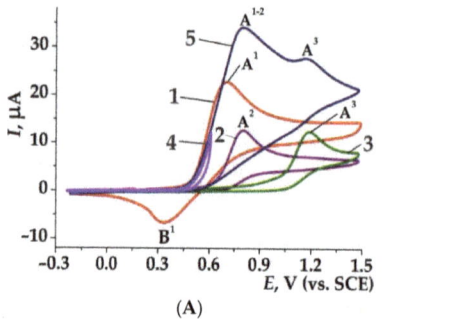 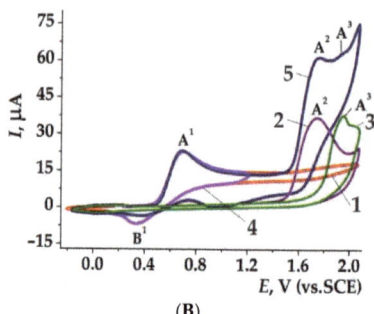

(A) (B)

Figure 1. CV curves on Pt working electrode in 0.1M NaClO$_4$ in MeCN, $\nu = 0.10$ V·s^{-1}. (**A**) NH$_4$SCN (0.002M)—1; 3-methyl-1H-pyrazol-5-amine **1e** (0.002M)—2; 3-methyl-4-thiocyanato-1H-pyrazol-5-amine **1f** (0.002M)—3; mixture NH$_4$SCN/azole **1e** (1:1) with the reverse scan from 0.60 V—4; the same on the reverse scan from 1.45V—5; (**B**) NH$_4$SCN (0.002M)—1; 2-methyl-5-thiophen-2-yl-7-(trifluoromethyl)pyrazolo[1,5-a]pyrimidine **10e**—2; 2-methyl-3-thiocyanato-5-thiophen-2-yl-7-(trifluoromethyl)pyrazolo[1,5-a]pyrimidine **10f**—3; mixture NH$_4$SCN/ azole **10e** (1:1) on the reverse scan from 1.20 V—4; the same on the reverse scan from 2.10 V—5.

3.2.2. Electrosynthesis

Electrolyses were carried out in 0.1M solution of NaClO$_4$ in MeCN (MeCN–H$_2$O) in undivided or divided cells (UC or DC) in controlled-potential or galvanostatic mode (CPE or GE), passing a theoretical or excess amounts of electricity ($Q/Q_t = 1$–3). Pt or glassy carbon (GC) electrodes were used.

The amino compounds **1e–6e** gave thiocyanates **1f–6f** with yields 64–89% (under CPE at $E_p^{ox}{}_{SCN^-}$) and 57–71% (under GE) at $Q/Q_t = 1$ (Table 4, entries 1–6) when implementing approach A (see Scheme 14).

From the less reactive pyrazolo[1,5-a]pyrimidines **7e–9e**, products **7f–9f** were obtained with yields 66–85% at $Q/Q_t = 2$ (entries 7–9). The electrode material affected things differently: the yield of thiocyanate **2a** (entry 1) was 83% (GC) and 72% (Pt), while the yield of thiocyanate **2g** (entry 7) was 64% (GC) and 89% (Pt). Most of the processes (entries 3–9) were successfully carried out in an undivided cell. The possibility of scaling the process was also shown (entries 1, 2 and 7).

It was noted [121–123,125] that approach A is not suitable for the thiocyanation of hardly oxidizable ($E_p^{ox} > 1.70$ V) pyrazoles **10e–15e** with acceptor substituents (Table 5) and leads to trace amounts of target thiocyanates **10f–15f** and polythiocyanogen (see Scheme 13, route B^3). In this case, the process proceeds quite efficiently with an increase in the reactivity of the thiocyanogen (Scheme 14, approach B) or the initial pyrazole (approach C). As a result, CPE in the presence of ZnCl$_2$ activating additives (approach B) allowed us to obtain products **10f–15f** with yields 69–81% [121,123], while metal-free CPE at $E_p^{ox}{}_{AzH}$ with $Q/Q_t = 3$ (approach C) also led to the products **10f–15f** with smaller yields of 47–65% [122,123].

Note that the possibility of approach C realization can also be tested by CV: a decrease in the peak B^1 is observed on the full scan (cf. curves 5 and 1, Figure 1B), which corresponds to the interaction of the thiocyanate ion and the pyrazole cation radical via the ECE mechanism [122,123,132] (see Scheme 14, approach C, and Scheme 3, route I).

In addition to approaches A–C, an equally effective approach was developed [41] based on the HCl-catalyzed condensation of previously obtained 4-thiocyanatopyrazoles **1e, 2e** (see entries 1 and 2, Table 4) with 1,3-dicarbonyl compounds or their derivatives (Scheme 15). As a result, 3-thiocyanatopyrazolo[1,5-a]pyrimidines both without substituents and with donor (acceptor) substituents in the pyrimidine ring were obtained with yields of 77–96% [126].

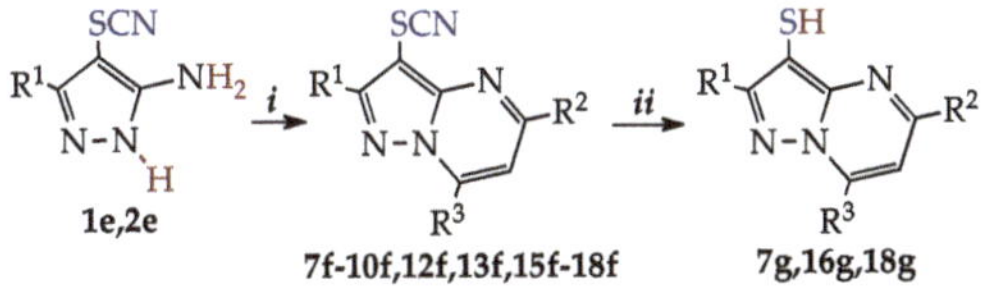

7f (7g): R¹=R²=R³=Me; 96% (69%)	**13f**: R¹=Me, R²=Ph, R³=CF₃; 84%
8f: R¹=cyclo-Pr, R²=R³=Me; 92%	**15f**: R¹=Me, R²=H, R³=CF₃; 78%
9f: R¹=Me, R²=cyclo-Pr, R³=CF₃; 87%	**16f (16g)**: R¹=Me, R²=R³=H; 77% (77%)
10f: R¹=Me, R²=thienyl, R³=CF₃; 91%	**17f**: R¹=cyclo-Pr, R²=R³=H; 84%
12f: R¹=Me, R²=H, R³=CCl₃; 89%	**18f (18g)**: R¹=R²=Me, R³=CF₃; 71% (61%)

i. Azole **1e,2e** (5 mmol), R₂C(O)CHC(O)R₃ or its derivative (6 mmol), H₂O (**7f,8f,16f,17f**), H₂O-EtOH (1:4) (**9f,12f,15f,18f**) or EtOH (**10f,13f**), HCl, 20-25 °C, 24 h. *ii*. H₂O, HCl, argon or nitrogen barbotation, 20-25 °C, 72 h.

Scheme 15. Synthesis of 3-thiocyanatopyrazolo[1,5-a]pyrimidines and pyrazolo[1,5-a]pyrimidine-3-thiols.

Developing this direction, the opportunity of transformation of the SCN group into the SH group [94,123] was shown, which opens the way to thiols as promising nucleophiles for C–H functionalization (e.g., see [133–135]). Hydrolysis with HCl was the most effective (yields of thiols **7g**, **16g**, **18g** were 61–77%), while the use of chemical reductants or strong acids (LiAlH₄, NaBH₄, Zn in AcOH, HClO₄, H₂SO₄) was ineffective.

Thus, a series of thiocyanates of substituted pyrazoles and pyrazolo[1,5-a]pyrimidines were obtained on the basis of electrolysis of the "thiocyanate ion/pyrazole" mixture.

In addition, during the development of research on the electrosynthesis of aryl thiocyanates [22] and N-arylpyrazoles [131,136–138] we showed [124] the possibility of synthesizing new molecules with pyrazole and thiocyanate fragments (Scheme 16) as promising hybrid polyfunctional [139] structures.

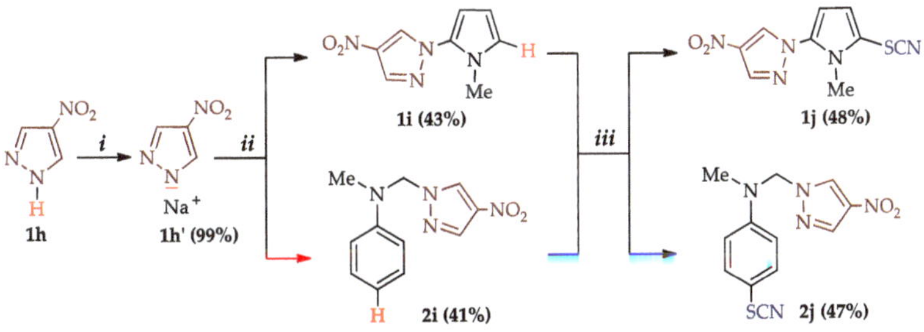

i. Na, dry MeOH, argone, 8h; *ii*. Electrolysis in 60 mL 0.1 M NaClO₄ in MeCN-MeOH, 30 °C, Pt electrodes, divided cell, Q$_t$=386 C, 4-nitropyrazolate **1h'** (4 mmol), N-methylpyrrole or N,N-dimethylaniline (2 mmol), E$_{anode}$=1.28 V or 0.75 V, Q/Q$_t$=1 or 1.5; *iii*. Electrolysis in 60 mL 0.1 M NaClO₄ in MeCN, 25 °C, Pt electrodes, divided cell, Q$_t$=386 C, E$_{anode}$=0.70 V, NH₄SCN (8 mmol), N-arylazole **1i** or **2i** (2 mmol), Q/Q$_t$=2 or 1.5

Scheme 16. N–H An arylation of 4-nitropyrazole followed by C–H An thiocyanation of resulting N-arylazoles.

The N–H arylation of pyrazole **1h** was carried out by activating its N–H bond (i) followed by the introduction of electrolysis with N-methylpyrrole or N,N-dimethylaniline (ii). In the latter case, the reaction proceeded selectively at the Me group without affecting the aromatic ring. Subsequent C–H An thiocyanation (iii) of the isolated N-arylazoles **1i**, **2i** led to the target products **1j**, **2j**.

3.3. Antifungal and Antibacterial Activity of Thiocyanated Pyrazole Derivatives

Tests for antifungal (*C. albicans*, *A. niger*) and antibacterial (*S. aureus*, *E. coli*) activity [91,94,123,124] showed that thiocyanate-pyrazoles are more active against fungi than bacteria. The greatest activity is observed against *A. niger* at thiocyanate **7f** [94] and thiocyanatoazolylaniline **2j** [124], whose minimum inhibitory concentration (MIC) is

0.24–0.48 µg/mL (it is superior to the antifungal drugs amphotericin B and fluconazole and is comparable to itraconazole).

The contribution of thiocyanate and pyrazole fragments to antifungal and antibacterial activity was clearly shown in the individual examples (Figure 2). Thus, the activity of compound **1j** increased more than 2000-fold for *A. niger* and more than 16-fold for *C. albicans* after the introduction of the SCN group. The presence of 4-nitropyrazole in 4-thiocyanatoaniline **2j'** provided a selective increase in antifungal activity by a factor of 16–64, while *N*-arylazole **2i** was inactive in all cases.

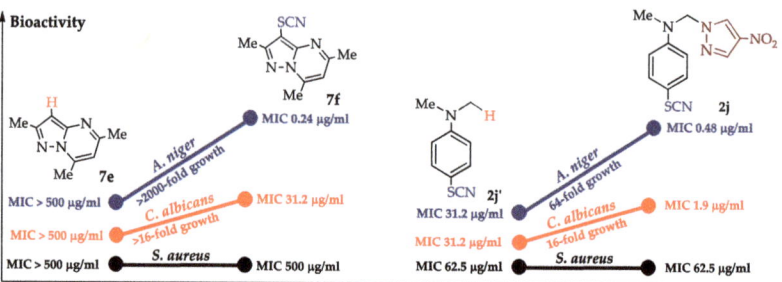

Figure 2. Effect of thiocyanate and pyrazole fragments on antifungal and antibacterial activity.

Therefore, this Section is devoted to the efficient C–H An thiocyanation of various pyrazole derivatives (in some cases, their N–H An arylation), leading to pharmacologically active target mono- and polyfunctional products. The next Section is devoted to the N–H functionalization of amino pyrazoles followed by their N–N coupling and obtaining azopyrazoles.

4. (Electro)oxidative N–N Coupling of Aminopyrazoles

Azoarenes are widely used in practice: from dyes and pharmaceuticals [140–142] to reagents in syntheses [143,144] and energy-rich materials [145,146]. One of the most popular methods for the synthesis of azoarenes is the oxidation of corresponding amines (Scheme 17), predominantly by chemical oxidants (BaMnO$_4$ [147], Pb(OAc)$_4$ [148], HgO [149], K$_2$FeO$_4$ [150], TCICA [151], t-BuOI [152]) or oxidation systems (CuBr-pyridine-O$_2$ [153], I$_2$-t-BuOOH [154], t-BuOCl-NaI [155]). The synthesis of polyfunctional azopyrazoles by silver catalyzed cascade conversion of diazo compounds [156] is also of interest.

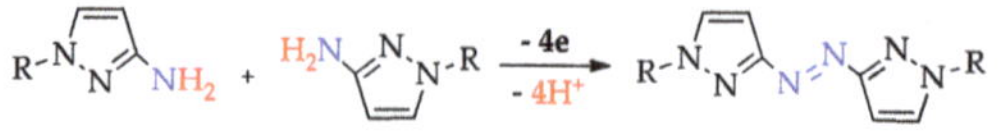

Scheme 17. Synthesis of azopyrazoles by (electro)oxidative N–N coupling of aminopyrazoles.

At the same time, a more promising electrochemical approach is poorly studied. In particular, electrosyntheses of azobenzene on the Pt anode [157,158] or N,N'-bis(morpholino)diazene on the NiO(OH) anode [159] are described. Note that NiO(OH) is one of the popular electrogenerated redox mediators [80,159]. The use of such redox-mediators is a trend in modern electroorganic chemistry [18], since it allows the processes to be carried out under milder conditions, increasing their efficiency and selectivity. This Section describes the original approaches to the synthesis of azopyrazoles using electrogenerated redox mediators NiO(OH) [160–162] and Br$_2$ [163], or electrogenerated hypohalites as oxidants [164,165].

4.1. (Electro)oxidative N–N Coupling of Aminopyrazoles: Approaches and General Patterns

One-stage Approach A (Scheme 18) is carried out in alkaline medium via the anodic dissolution of the Ni and the formation of adsorbed Ni(OH)$_2$, followed by its anodic oxidation to adsorbed NiO(OH). It oxidizes aminopyrazoles (**Az–NH$_2$**) to azopyrazoles (**Az–N = N–Az**) and forms Ni(OH)$_2$, after which the cycle repeats [80,161]. In Approach B, the metal-free oxidant is Br$_2$ [164], which is effectively electro(re)generated on the ruthenium–titanium oxide anode (RTOA).

Scheme 18. One-stage redox-mediated N–N coupling of aminopyrazoles using electrogenerated NiO(OH) (**Approach A**) and Br$_2$ (**Approach B**).

In addition, special voltammetric tests showed that an increase in the Ni(OH)$_2$ peak (Figure 3, A, peak A^1, E_p^{ox} = 0.46 V) or a decrease in the Br$_2$ peak (Figure 3B, peak B^2, E_p^{red} = 0.69 V) after the adding of aminopyrazole is proportional to the process efficiency [162,164].

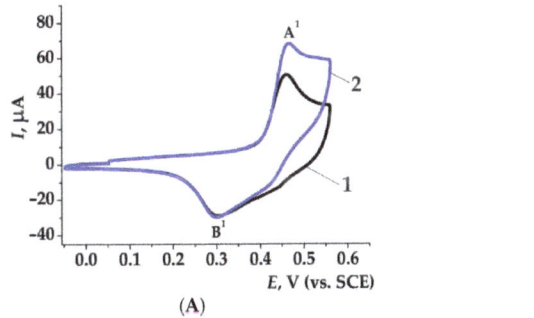 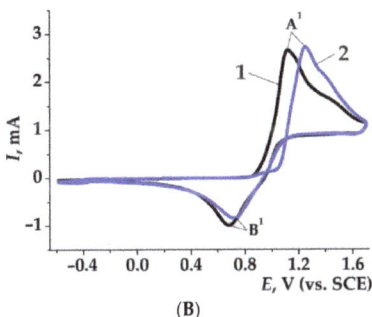

Figure 3. CV curves, ν = 0.10 V·s^{-1}. (**A**) On Ni working electrode in 0.2M aq. NaOH: electrogenerated Ni(OH)$_2$—1; after addition of 1-methyl-1*H*-pyrazol-3-amine **1k** (0.002M)—2; (**B**) on Pt working electrode in 1M aq. NaNO$_3$: NaBr (0.3M)—1; after addition of 1-methyl-1*H*-pyrazol-3-amine **1k** (0.002M)—2.

Two-stage approaches (Scheme 19) include preliminary electrogeneration of hypogalites followed by addition of aminopyrazoles [164,165]. Note, that hypohalites exist in equilibrium forms: predominantly HOCl (HOBr) in a neutral medium, and predominantly NaOCl (NaOBr) after adding NaOH (approaches C, D, C', D', respectively).

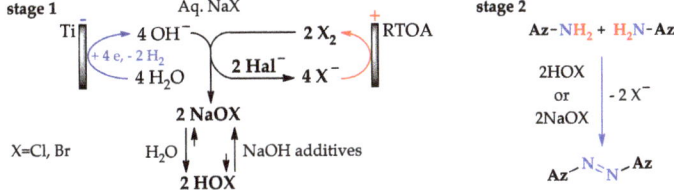

Scheme 19. Two-stage N–N coupling of aminopyrazoles using electrogenerated HOCl (HOBr) or NaOCl (NaOBr) (Approaches C, D, C', D', respectively).

According to the data [80,152,154,155,159,163–165], the possible mechanisms (Scheme 20) involve the oxidation of 1-methyl-1*H*-pyrazol-3-amine **1k** (step **1k**→[**Az–NH$_2$**]$^{•+}$) or its

N–H halogenation (step **1k**→[NH–X]) followed by N–H amination to hydrazopyrazole [NH–NH] and its oxidation to the target azopyrazole **1k–1k** (see also Schemes 3 and 9).

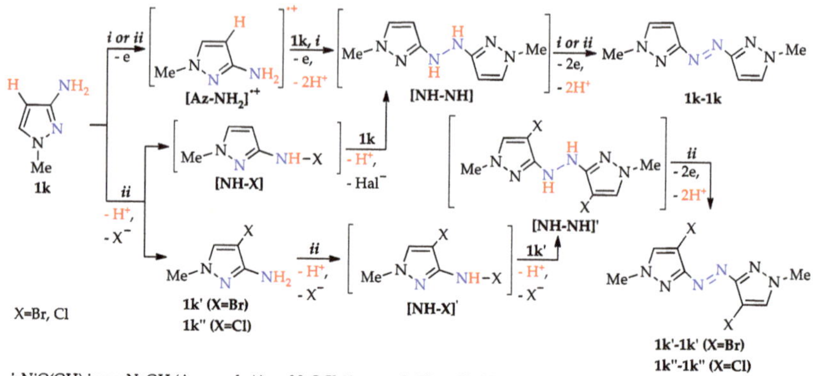

Scheme 20. N–N coupling of aminopyrazoles (possible mechanisms).

i: NiO(OH) in aq. NaOH (Approach A); or NaOCl (Approach C'); or NaOBr (Approach D')
ii: Br$_2$ in H$_2$O (Approach B); or HOCl (Approach C); or HOBr (Approach D)

When Br$_2$ and HOX are used, C–H halogenation of aminopyrazole **1k** also occurs (step **1k**→**1k′**(**1k″**), see Scheme 9), followed by the formation of halogenated azopyrazole (steps **1k′**→[NH–X]′→[NH–NH]′→**1k′–1k′**(**1k″–1k″**)). These patterns are consistent with the experimental results below.

4.2. Synthesis of Azopyrazoles

Approach A is most versatile and allows us to obtain target azopyrazoles with yields of 52–88% (Table 6, entries 1–6, 8–11). On the contrary, approaches B, C and D are more suitable for 4-substituted pyrazoles (entries 2, 6 and 7) or pyrazole with acceptor (CF$_3$) group (entry 12), where the yields of the corresponding azo products were 62–93%.

Table 6. N–N homo-coupling of aminopyrazoles (Az–NH$_2$) using approaches A, B, C (C'), D (D') [1].

Entry	Az–NH$_2$, Az–N=N–Az and Other Products (Yield, %) [161,163,165]	Entry	Az–NH$_2$, Az–N=N–Az and Other Products (Yield, %) [161,163,165]
1	**1k**; **1k-1k** (82[2], 34[4], 72[5], 40[6], 75[7]); **1k′** (59[3], 22[4], 1[5]); **1k′-1k′** (28[3], 48[4]); **1k″** (7[6], 5[7]); **1k″-1k″** (40[6], 14[7])	7	**5k**; **5k-5k** (86[3])
		8	**6k**; **6k-6k** (79[2], 2[6]); **6k′** (3[3]); **6k′-6k′** (80[3])

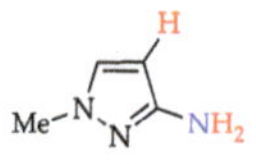

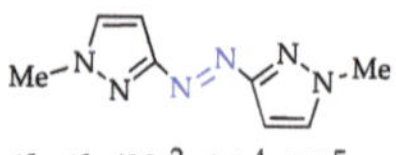

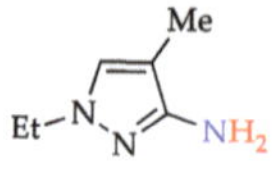

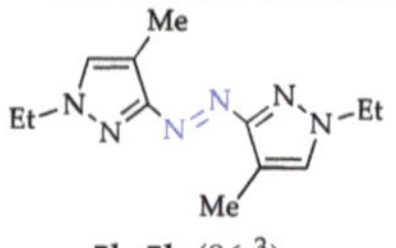

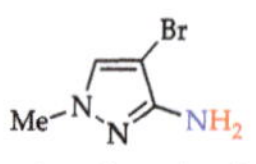

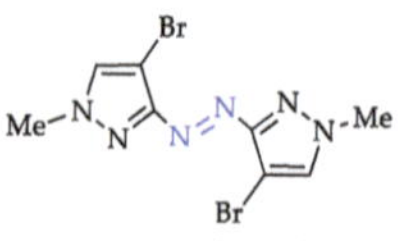

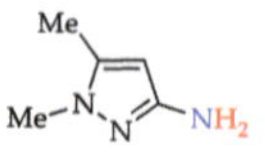

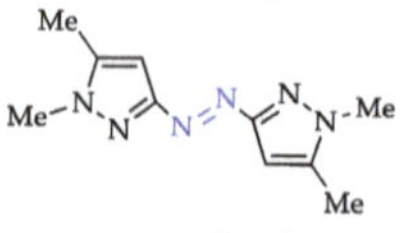

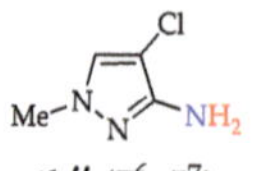

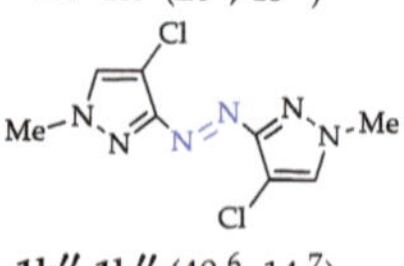

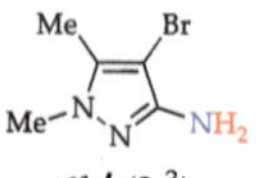

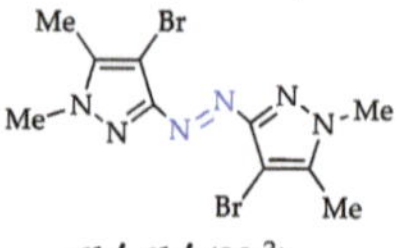

Table 6. Cont.

Entry	Az–NH₂, Az–N=N–Az and Other Products (Yield, %) [161,163,165]		Entry	Az–NH₂, Az–N=N–Az and Other Products (Yield, %) [161,163,165]	
2	1k′	1k′-1k′ (77 [2], 62 [4])	8	6k″ (6)	6k″-6k″ (79 [6])
3	1k″	1k″-1k″ (87 [2])	9	7k	7k-7k (55 [2])
4	2k	2k-2k (71 [2], 73 [5])	10	8k	8k-8k (52 [2])
5	2k′ (2 [5])	2k′-2k′ (6 [5])	11	9k	9k-9k [8] (67 [2])
5	3k	3k-3k (86 [2])	12	10k	10k-10k (93 [4], 86 [6])
6	4k	4k-4k (88 [2], 62 [3], 70 [4])			

[1] Electrolysis in 100 mL of supporting electrolyte, 20–25 °C, undivided cell galvanostatic electrolysis; [2] Approach A: 0.5M aq. NaOH, NiO(OH) anode, pyrazole (3 mmol), j_{anode} = 6 mA·cm^{-2}, Q_t = 579 C, Q/Q_t = 1–4; [3] Approach B: 2M aq. NaBr, RTOA, pyrazole (3 mmol), j_{anode} = 100 mA·cm^{-2}, 48% HBr additives during electrolysis until reaching pH~7 (entries 1, 8), Q_t = 579 C, Q/Q_t = 1–4; [4] Approach C: 1. electrogeneration of NaOBr (HOBr) in 2M aq. NaBr, RTOA, j_{anode} = 100 mA·cm^{-2}, Q = 661–1983 C; 2. pyrazole (2 mmol), NaOBr (HOBr) (2–4 mmol), 5 h; [5] Approach C′: see Approach C, but with NaOH (6 mmol) additives; [6] Approach D: 1. electrogeneration of HOCl (NaOCl) in 4M aq. NaCl, RTOA, j_{anode} = 161.5 mA·cm^{-2}, Q= 588–1764 C; 2. Pyrazole (2 mmol), HOBr (NaOBr) (2–4 mmol), 5 h; [7] Approach D′: see Approach D, but with NaOH (6 mmol) additives; [8] it was identified after preparation and isolation of the corresponding methyl ester.

Nevertheless, approaches C′ and D′ (entries 1 and 4) allow to obtain rather selectively azopyrazoles **1k–1k** and **2k–2k** (yields 72–75%), and approaches C and D (entry 8) open the way to azohalogenopyrazoles **6k″–6k″** and **6k″–6k″** (yields 79–80%).

Moreover, the approach A is useful for previously unexplored chemical and electrochemical N-N cross-coupling of aminopyrazoles [162] (Table 7), and yields of the target azo compounds **1k–2k** and **4k–5k** were 48–50%. Such results create prospects for obtaining useful multifunctional azo compounds [152,156].

Table 7. N–N cross-coupling of aminopyrazoles (Az–NH$_2$) using approach A [1].

Entry	Az^1–NH$_2$	H$_2$N–Az^2	Az^1–N=N–Az^2	Yield, %
1	1k	2k	1k-2k	50 (1k-2k) 36 (1k-1k) 37 (2k-2k)
2	4k	5k	4k-5k	48 (4k-5k) 29 (1k-1k) 23 (2k-2k)

[1] Electrolysis in 100 mL of 0.5 M aq. NaOH, 20–25 °C, undivided cell,; [2] Approach A: NiO(OH) anode, Ti cathode, pyrazole (1.5 mmol), galvanostatic electrolysis (j_{anode} = 6 mA·cm^{-2}), Q = 2Q$_t$ = 579 C.

5. Conclusions

This review is the first step in summarizing the data on promising, but poorly studied electrooxidative functionalization of C-H and N-H bonds in pyrazole derivatives. It paves the way for the efficient synthesis of C-Cl, C-Br, C-I, C-S and N-N coupling products using cheap, affordable and environmentally promising electric currents.

Additional advantages are the predominantly galvanostatic electrolysis mode and the reusability of commercially available electrodes, salts and solvents, as well as the gram-scalability of the processes. In half of the cases, a simple isolation of pure target products without chromatography is also possible. Moreover, the key regularities of the corresponding processes are considered, including the dependence of the efficiency of functionalization of pyrazoles on their structure and oxidation potential. An increasingly important role is played by cyclic voltammetry, which makes it possible both to study mechanisms and to predict the efficiency of synthesis.

All this makes the electrooxidative functionalization of pyrazole-type compounds very viable for further application and development.

Author Contributions: Conceptualization, V.A.P.; writing—original draft preparation, V.A.K. and V.A.P.; writing—review and editing, B.V.L., V.L.S., A.S.K. and S.V.N. All authors have read and agreed to the published version of the manuscript.

Funding: This work was supported by the Russian Science Foundation. Grant 19-73-20259.

Conflicts of Interest: The authors declare no conflict of interest.

Sample Availability: Samples of the compounds 2b, 4b, 9b–13b, 1c–4c, 6c, 8c, 10c–13c, 1d–4d, 8d, 10d–14d, 1f–15f, 1j, 2j, 1k-1k–8k-8k are available from the authors.

References

1. Godula, K.; Sames, D. C-H Bond Functionalization in Complex Organic Synthesis. *Science* **2006**, *312*, 67–72. [CrossRef]
2. Davies, H.M.L.; Morton, D. C-H Functionalization: Collaborative Methods to Redefine Chemical Logic. *Angew. Chem. Int. Ed.* **2014**, *53*, 10256–10258. [CrossRef] [PubMed]
3. Davies, H.M.L.; Morton, D. Recent Advances in C–H Functionalization. *J. Org. Chem.* **2016**, *81*, 343–350. [CrossRef] [PubMed]
4. Smith, M.B. Chapter 11. Aromatic Substitution, Electrophilic. In *March's Advanced Organic Chemistry: Reactions, Mechanisms, and Structure*, 8th ed.; John Wiley & Sons: Hoboken, NJ, USA, 2020; pp. 607–686.
5. Chupakhin, O.N.; Postovskii, I.Y. Nucleophilic Substitution of Hydrogen in Aromatic Systems. *Russ. Chem. Rev.* **1976**, *45*, 454–468. [CrossRef]
6. Chupakhin, O.N.; Charushin, V.N.; van der Plas, H.C. *Nucleophilic Aromatic Substitution of Hydrogen*; Academic Press: New York, NY, USA, 1994.
7. Charushin, V.N.; Chupakhin, O.N. Nucleophilic aromatic substitution of hydrogen and related reactions. *Mendeleev Commun.* **2007**, *17*, 249–254. [CrossRef]
8. Chupakhin, O.N.; Charushin, V.N. Recent advances in the field of nucleophilic aromatic substitution of hydrogen. *Tetrahedron Lett.* **2016**, *57*, 2665–2672. [CrossRef]

9. Charushin, V.N.; Chupakhin, O.N. Nucleophilic C—H functionalization of arenes: A contribution to green chemistry. *Russ. Chem. Bull.* **2019**, *68*, 453–471. [CrossRef]
10. Seebach, D. Methods of Reactivity Umpolung. *Angew. Chem. Int. Ed.* **1979**, *18*, 239–258. [CrossRef]
11. Weinberg, N.L.; Weinberg, H.R. Electrochemical oxidation of organic compounds. *Chem. Rev.* **1968**, *68*, 449–523. [CrossRef]
12. *Organic Electrochemistry*, 1st ed.; Baizer, M.M. (Ed.) M. Dekker: New York, NY, USA, 1973.
13. *Organic Electrochemistry*, 2nd ed.; Baizer, M.M. (Ed.) M. Dekker: New York, NY, USA, 1983.
14. Yoshida, K. *Electrooxidation in Organic Chemistry. The Role of Cation Radicals as Synthetic Intermediates*; Wiley: New York, NY, USA, 1984.
15. *Organic Electrochemistry*, 3rd ed.; Lund, H.; Baizer, M.M. (Eds.) M. Dekker: New York, NY, USA, 1991.
16. Hammerich, O.; Utley, J.H.P.; Eberson, L. *Organic Electrochemistry*, 4th ed.; Lund, H., Hammerich, O., Eds.; M. Dekker: New York, NY, USA, 2001.
17. Shchepochkin, A.V.; Chupakhin, O.N.; Charushin, V.N.; Petrosyan, V.A. Direct nucleophilic functionalization of C(sp^2)–H-bonds in arenes and hetarenes by electrochemical methods. *Russ. Chem. Rev.* **2013**, *82*, 747–771. [CrossRef]
18. Francke, R.; Little, R.D. Redox catalysis in organic electrosynthesis: Basic principles and recent developments. *Chem. Soc. Rev.* **2014**, *43*, 2492–2521. [CrossRef]
19. Karkas, M.D. Electrochemical strategies for C–H functionalization and C–N bond formation. *Chem. Soc. Rev.* **2018**, *47*, 5786–5865. [CrossRef] [PubMed]
20. Yang, Q.-L.; Fang, P.; Mei, T.-S. Recent Advances in Organic Electrochemical C–H Functionalization. *Chin. J. Chem.* **2018**, *36*, 338–352. [CrossRef]
21. Petrosyan, V.A. Reactions of anodic and chemical aromatic substitution. *Mendeleev Commun.* **2011**, *21*, 115–121. [CrossRef]
22. Kokorekin, V.A.; Sigacheva, V.L.; Petrosyan, V.A. New data on heteroarene thiocyanation by anodic oxidation of NH$_4$SCN. The processes of electroinduced nucleophilic aromatic substitution of hydrogen. *Tetrahedron Lett.* **2014**, *55*, 4306–4309. [CrossRef]
23. Sheldon, R.A.; Arends, I.; Hanefeld, U. *Green Chemistry and Catalysis*; Wiley-VCH: Weinheim, Germany, 2007.
24. Frontana-Uribe, B.A.; Little, R.D.; Ibanez, J.G.; Palma, A.; Vasquez-Medrano, R. Organic electrosynthesis: A promising green methodology in organic chemistry. *Green Chem.* **2010**, *12*, 2099–2119. [CrossRef]
25. Heard, D.M.; Lennox, A.J.J. Electrode Materials in Modern Organic Electrochemistry. *Angew. Chem. Int. Ed.* **2020**, *59*, 18866–18884. [CrossRef]
26. Lyalin, B.V.; Petrosyan, V.A. Electrochemical halogenation of organic compounds. *Russ. J. Electrochem.* **2013**, *49*, 497–529. [CrossRef]
27. Beer, H.B. Dimensionally Stable Anodes. In *Electrochemistry in Industry: New Directions*; Landau, U., Yeager, E., Kortan, D., Eds.; Springer: Boston, MA, USA, 1982; pp. 19–28. [CrossRef]
28. Janin, Y.L. Preparation and Chemistry of 3/5-Halogenopyrazoles. *Chem. Rev.* **2012**, *112*, 3924–3958. [CrossRef] [PubMed]
29. Slomczynska, U.; Olivo, P.; Beattie, J.; Starkey, G.; Noueiry, A.; Roth, R. Compounds, Compositions and Methods for Control of Hepatitis C Viral Infections. International Patent 2010096115, 26 August 2010.
30. Tanigchi, T.; Kawada, A.; Kondo, M.; Quinn, J.F.; Kanitomo, J.; Yoshikawa, M.; Fashimi, M. Pyridazinone Compounds. U.S. Patent 20100197651, 5 August 2010.
31. Alberati, D.; Alvares, R.S.; Bleicher, K.; Flohr, A.; Markus, R.; Groeke, K.Z.; Koerner, M.; Kuhn, B.; Peters, J.-U.; Rudolph, M. Novel Imidazopyridines. U.S. Patent 20110071128, 24 March 2011.
32. Linn, D.M.; Fong, E.H. Nicotinic Acetylcholine Agonists in the Treatment of Glaucoma and Retinal Neuropathy. International Patent 2004039366, 13 May 2004.
33. Tomokazu, N. Therapeutic Agent for Hyperlipemia, Arteriosclerosis, and/or Metabolic Syndrome. Japan Patent 2006182668, 13 July 2006.
34. D'orchumont, H.; Fraisse, L.; Zimmerman, A. Use of Indazolecarboxamide Derivatives for the Preparation of a Medicament that is Intended for the Treatment and Prevention of Paludism. International Patent 2005099703, 27 October 2005.
35. Pennell, A.M.K.; Aggen, J.B.; Wright, J.J.; Sen, S.; McMaster, B.E.; Brian, E.; Dairaghi, J.D. 1-Aryl-4-substituted Piperazines Derivatives for Use as CCR1 Antagonists for the Treatment of Inflammation and Immune Disorders. International Patent 03105853, 24 December 2003.
36. Singh, J.; Gurney, M.E.; Burgin, A.; Sandanayaka, V.; Kiselyov, A.; Motta, A.; Schultz, G.; Hategan, G. Biaryl as PDE4 inhibitors for treating inflammation. International Patent 2009067600, 28 May 2009.
37. Talley, J.J.; Sprott, K.; Pearson, J.P.; Game, P.; Milene, G.T.; Schairer, W.; Yang, J.J.; Kim, C.; Barden, T.; Lundigran, R. Indole compounds. International Patent 2008019357, 14 February 2008.
38. Pelcman, B.; Sanin, A.; Nilsson, P.; Groth, T.; Kromann, H. Pyrazoles Useful in the Treatment of Inflammation. International Patent 2007051981, 10 May 2007.
39. Banner, D.; Helpert, H.; Mauser, H.; Mayweg, V.A.; Rogers-Evans, M. Amino oxazine Derivatives. International Patent 2011070029, 16 June 2011.
40. Bamba, M.; Eiki, J.; Mitsuya, M.; Nishimura, T.; Sakai, F.; Sasaki, Y.; Watanabe, H. Novel 2-Pyridinecarboxamide Derivatives. International Patent 2004081001, 23 September 2004.
41. Fix-Stenzel, S.; Judd, A.; Michaelides, M. Pyrrolopyridine and Pyrrolopyrimidine Inhibitors of Kinases. International Patent 2011143459, 17 November 2011.

42. You, H.; Youn, H.-S.; Im, I.; Bae, M.-H.; Lee, S.-K.; Ko, H.; Eom, S.H.; Kim, Y.-C. Design, synthesis and X-ray crystallographic study of NAmPRTase inhibitors as anti-cancer agents. *Eur. J. Med. Chem.* **2011**, *46*, 1153–1164. [CrossRef]
43. Ahearn, S.P.; Altman, M.; Chichetti, S.; Czako, B.; Daniels, M.H.; Falcone, D.; Guerin, D.; Lipford, K.; Martinez, M.; Osimboni, E.; et al. Tyrosine kinase Inhibitors. International Patent 2011084402, 14 July 2011.
44. Basso, A.D. Anti-mitotic Agent and Aurora Kinase Inhibitor Combination as Anti-cancer Treatment. International Patent 2009017701, 5 February 2009.
45. Hattori, K.; Matsuda, K.; Murata, M.; Nakajima, T.; Tsutsumi, H. Substituted 3-Pyrrolidinylthio-carbapenems as Antimicrobial Agents. International Patent 9321186, 28 October 1993.
46. Barrett, D.; Matsuda, H.; Matsuda, K.; Matsuya, T.; Mizuno, H.; Murano, K.; Ogino, T.; Toda, A. New Compound. International Patent 02072621, 19 September 2002.
47. Kudo, N.; Furuta, S.; Taniguchi, M.; Endo, T.; Sato, K. Synthesis and Herbicidal Activity of 1,5-Diarylpyrazole Derivatives. *Chem. Pharm. Bull.* **1999**, *47*, 857–868. [CrossRef]
48. Fustero, S.; Román, R.; Sanz-Cervera, J.F.; Simón-Fuentes, A.; Cuñat, A.C.; Villanova, S.; Murguía, M. Improved Regioselectivity in Pyrazole Formation through the Use of Fluorinated Alcohols as Solvents: Synthesis and Biological Activity of Fluorinated Tebufenpyrad Analogs. *J. Org. Chem.* **2008**, *73*, 3523–3529. [CrossRef]
49. Higashino, Y.; Ikeda, Y.; Koike, S.; Kyomura, N.; Okui, S.; Tomita, H. Pyrazoles and Agricultural Chemicals Containing Them as Active Ingredients. International Patent 9737990, 16 October 1997.
50. Auler, T.; Bojack, G.; Bretschneider, T.; Drewes, M.W.; Feucht, D.; Fischer, R.; Hills, M.; Kehne, H.; Konze, J.; Kuck, K.-H.; et al. N-Heterocyclyl Phenyl-substituted Cyclic Ketoenols. International Patent 2004111042, 23 October 2004.
51. Fischer, R.; Gomibuchi, T.; Nakakura, N.; Otsu, Y.; Shibuya, K.; Wada, K.; Yoneta, Y. N1-((Pyrazol-1-ymethyl)-2-methylphenyl)-phatalamide Derivatives and Related Compounds Insecticides. International Patent 2005095351, 13 October 2005.
52. Hüttel, R.; Schäfer, O.; Welzel, G. Die Chlorierung der Pyrazole. *Justus Liebigs Ann. Chem.* **1956**, *598*, 186–197. [CrossRef]
53. Ohsawa, A.; Kaihoh, T.; Itoh, T.; Okada, M.; Kawabata, C.; Yamaguchi, K.; Igeta, H. Reactions of N-Aminopyrazoles with Halogenating Reagents and Synthesis of 1, 2, 3-Triazines. *Chem. Pharm. Bull.* **1988**, *36*, 3838–3848. [CrossRef]
54. Stefani, H.A.; Pereira, C.M.P.; Almeida, R.B.; Braga, R.C.; Guzen, K.P.; Cella, R. A mild and efficient method for halogenation of 3,5-dimethyl pyrazoles by ultrasound irradiation using N-halosuccinimides. *Tetrahedron Lett.* **2005**, *46*, 6833–6837. [CrossRef]
55. Zhao, Z.G.; Wang, Z.X. Halogenation of Pyrazoles Using N-Halosuccinimides in CCl_4 and in Water. *Synthetic Commun.* **2007**, *37*, 137–147. [CrossRef]
56. Mullens, P.R. An improved synthesis of 1-methyl-1H-pyrazole-4-boronic acid pinacol ester and its corresponding lithium hydroxy ate complex: Application in Suzuki couplings. *Tetrahedron Lett.* **2009**, *50*, 6783–6786. [CrossRef]
57. Khan, K.M.; Maharvi, G.M.; Choudhary, M.I.; Rahman, A.-U.; Perveen, S. A modified, economical and efficient synthesis of variably substituted pyrazolo[4,3-d]pyrimidin-7-ones. *J. Het. Chem.* **2005**, *42*, 1085–1093. [CrossRef]
58. Kerr, E.R.; Mather, W.B. Electrochemical Chlorination of Hydrocarbon in Hydrochloric Acid-Acetic Acid Solytion. U.S. Patent 3692646, 19 September 1972.
59. Ehlert, M.K.; Storr, A.; Thompson, R.C. Metal pyrazolate polymers. Part 3. Synthesis and study of Cu(I) and Cu(II) complexes of 4-Xdmpz (where X = H, Cl, Br, I, and CH_3 for Cu(I) and X = H, Cl, Br, and CH_3 for Cu(II); dmpz = 3,5-dimethylpyrazolate). *Can. J. Chem.* **1992**, *70*, 1121–1128. [CrossRef]
60. Petko, K.; Sokolenko, T.; Yagupolskii, L. Chemical properties of derivatives of N-difluoromethyl-and N-2-H-tetrafluoroethylpyrazoles. *Chem. Heterocycl. Compd.* **2006**, *42*, 1177–1184. [CrossRef]
61. Akopyan, G.A. Bromination of pyrazole-3(5)-carboxylic acid. *Russ. J. Gen. Chem.* **2007**, *77*, 1652–1653. [CrossRef]
62. Guillou, S.; Bonhomme, F.J.; Ermolenko, M.S.; Janin, Y.L. Simple preparations of 4 and 5-iodinated pyrazoles as useful building blocks. *Tetrahedron* **2011**, *67*, 8451–8457. [CrossRef]
63. Salanouve, E.; Guillou, S.; Bizouarne, M.; Bonhomme, F.J.; Janin, Y.L. 3-Methoxypyrazoles from 1,1-dimethoxyethene, few original results. *Tetrahedron* **2012**, *68*, 3165–3171. [CrossRef]
64. Guillou, S.; Janin, Y.L. 5-Iodo-3-Ethoxypyrazoles: An Entry Point to New Chemical Entities. *Chem. Eur. J.* **2010**, *16*, 4669–4677. [CrossRef]
65. Sinyakov, A.; Vasilevskii, S.; Shvartsberg, M. A new rearrangement of chloroethynylpyrazoles. *Russ. Chem. Bull.* **1977**, *26*, 2142–2147. [CrossRef]
66. Vasilevskii, S.F.; Shvartsberg, M.S. Oxidative iodination of substituted N-methylpyrazoles. *Russ. Chem. Bull.* **1980**, *29*, 778–784. [CrossRef]
67. Tret'yakov, E.V.; Vasitevsky, S.F. Nitrodeiodination of 4-iodo-l-methylpyrazoles. *Russ. Chem. Bull.* **1996**, *45*, 2581–2584. [CrossRef]
68. Lyalin, B.V.; Petrosyan, V.A.; Ugrak, B.I. Electrosynthesis of 4-chloropyrazolecarboxylic acids. *Russ. Chem. Bull.* **2009**, *58*, 291–296. [CrossRef]
69. Perevalov, V.P.; Manaev, Y.A.; Bezborodov, B.V.; Stepanov, B.I. Chlorination of 1,5-dimethylpyrazole. *Chem. Heterocycl. Compd.* **1990**, *26*, 301–303. [CrossRef]
70. Lyalin, B.V.; Petrosyan, V.A.; Ugrak, B.I. Electrosynthesis of 4-chloro derivatives of pyrazole and alkylpyrazoles. *Russ. J. Electrochem.* **2008**, *44*, 1320–1326. [CrossRef]
71. Lyalin, B.V.; Petrosyan, V.A. The Reactivity Trends in Electrochemical Chlorination and Bromination of N-Substituted and N-Unsubstituted Pyrazoles. *Chem. Heterocycl. Compd.* **2014**, *49*, 1599–1610. [CrossRef]

72. Lyalin, B.V.; Petrosyan, V.A.; Ugrak, B.I. Electrosynthesis of 4-bromosubstituted pyrazole and its derivatives. *Russ. J. Electrochem.* **2010**, *46*, 123–129. [CrossRef]
73. Lyalin, B.V.; Petrosyan, V.A.; Ugrak, B.I. Electrosynthesis of 4-iodopyrazole and its derivatives. *Russ. Chem. Bull.* **2010**, *59*, 1549–1555. [CrossRef]
74. Lyalin, B.V.; Petrosyan, V.A. New approach to electrochemical iodination of arenes exemplified by the synthesis of 4-iodopyrazoles of different structures. *Russ. Chem. Bull.* **2014**, *63*, 360–367. [CrossRef]
75. Lyalin, B.V.; Petrosyan, V.A. Efficient iodination of structurally varying pyrazoles in heterophase medium. *Russ. Chem. Bull.* **2013**, *62*, 1044–1051. [CrossRef]
76. Noszticzius, Z.; Noszticzius, E.; Schelly, Z.A. Use of ion-selective electrodes for monitoring oscillating reactions. 1. Potential response of the silver halide membrane electrodes to hypohalous acids. *J. Am. Chem. Soc.* **1982**, *104*, 6194–6199. [CrossRef]
77. Kanishchev, M.I.; Korneeva, N.V.; Shevelev, S.A.; Fainzil'berg, A.A. Nitropyrazoles (review). *Chem. Heterocycl. Compd.* **1988**, *24*, 353–370. [CrossRef]
78. Belen'kii, L.I.; Chuvylkin, N.D. Relationships and features of electrophilic substitution reactions in the azole series. *Chem. Heterocycl. Compd.* **1996**, *32*, 1319–1343. [CrossRef]
79. Pevzner, M.S. Aromatic N-Haloazoles. In *Advances in Heterocyclic Chemistry*; Katritzky, A.R., Ed.; Academic Press: Cambridge, MA, USA, 1999; Volume 75, pp. 1–77.
80. Lyalin, B.V.; Petrosyan, V.A. Oxidation of organic compounds on NiOOH electrode. *Russ. J. Electrochem.* **2010**, *46*, 1199–1214. [CrossRef]
81. Nikoofar, K. A Brief on Thiocyanation of N-Activated Arenes and N-Bearing Heteroaromatic Compounds. *Chem. Sci. Trans.* **2013**, *3*, 691–700. [CrossRef]
82. Castanheiro, T.; Suffert, J.; Donnard, M.; Gulea, M. Recent advances in the chemistry of organic thiocyanates. *Chem. Soc. Rev.* **2016**, *45*, 494–505. [CrossRef]
83. Majedi, S.; Sreerama, L.; Vessally, E.; Behmagham, F. Metal-Free Regioselective Thiocyanation of (Hetero) Aromatic C-H Bonds using Ammonium Thiocyanate: An Overview. *J. Chem. Lett.* **2020**, *1*, 25–31. [CrossRef]
84. Rezayati, S.; Ramazani, A. A review on electrophilic thiocyanation of aromatic and heteroaromatic compounds. *Tetrahedron* **2020**, *76*, 131382. [CrossRef]
85. Maurya, C.K.; Mazumder, A.; Gupta, P.K. Phosphorus pentasulfide mediated conversion of organic thiocyanates to thiols. *Beilstein J. Org. Chem.* **2017**, *13*, 1184–1188. [CrossRef]
86. Noland, W.E.; Lanzatella, N.P.; Dickson, R.R.; Messner, M.E.; Nguyen, H.H. Access to Indoles via Diels–Alder Reactions of 5-Methylthio-2-vinylpyrroles with Maleimides. *J. Heterocycl. Chem.* **2013**, *50*, 795–808. [CrossRef]
87. Maurya, C.K.; Mazumder, A.; Kumar, A.; Gupta, P.K. Synthesis of Disulfanes from Organic Thiocyanates Mediated by Sodium in Silica Gel. *Synlett* **2016**, *27*, 409–411. [CrossRef]
88. Biswas, K.; Ghosh, S.; Ghosh, P.; Basu, B. Cyclic ammonium salts of dithiocarbamic acid: Stable alternative reagents for the synthesis of S-alkyl carbodithioates from organyl thiocyanates in water. *J. Sulfur Chem.* **2016**, *37*, 361–376. [CrossRef]
89. Li, W.-T.; Wu, W.-H.; Tang, C.-H.; Tai, R.; Chen, S.-T. One-Pot Tandem Copper-Catalyzed Library Synthesis of 1-Thiazolyl-1,2,3-triazoles as Anticancer Agents. *ACS Comb. Sci.* **2011**, *13*, 72–78. [CrossRef] [PubMed]
90. Baghershiroudi, M.; Safa, K.D.; Adibkia, K.; Lotfipour, F. Synthesis and antibacterial evaluation of new sulfanyltetrazole derivatives bearing piperidine dithiocarbamate moiety. *Synth. Commun.* **2018**, *48*, 323–328. [CrossRef]
91. Kokorekin, V.A.; Terent'ev, A.O.; Ramenskaya, G.V.; Grammatikova, N.É.; Rodionova, G.M.; Ilovaiskii, A.I. Synthesis and Antifungal Activity of Arylthiocyanates. *Pharm. Chem. J.* **2013**, *47*, 422–425. [CrossRef]
92. Fortes, M.P.; da Silva, P.B.N.; da Silva, T.G.; Kaufman, T.S.; Militão, G.C.G.; Silveira, C.C. Synthesis and preliminary evaluation of 3-thiocyanato-1H-indoles as potential anticancer agents. *Eur. J. Med. Chem.* **2016**, *118*, 21–26. [CrossRef]
93. Chao, M.N.; Matiuzzi, C.E.; Storey, M.; Li, C.; Szajnman, S.H.; Docampo, R.; Moreno, S.N.J.; Rodriguez, J.B. Aryloxyethyl Thiocyanates Are Potent Growth Inhibitors of Trypanosoma cruzi and Toxoplasma gondii. *ChemMedChem* **2015**, *10*, 1094–1108. [CrossRef]
94. Kokorekin, V.A.; Khodonov, V.M.; Neverov, S.V.; Grammatikova, N.É.; Petrosyan, V.A. "Metal-free" synthesis and antifungal activity of 3-thiocyanatopyrazolo[1,5-a]pyrimidines. *Russ. Chem. Bull.* **2021**, *70*, 600–604. [CrossRef]
95. Yi, X.-J.; El-Idreesy, T.T.; Eldebss, T.M.A.; Farag, A.M.; Abdulla, M.M.; Hassan, S.A.; Mabkhot, Y.N. Synthesis, biological evaluation, and molecular docking studies of new pyrazol-3-one derivatives with aromatase inhibition activities. *Chem. Biol. Drug Des.* **2016**, *88*, 832–843. [CrossRef] [PubMed]
96. Wood, J.L. Substitution and Addition Reactions of Thiocyanogen. In *Organic Reactions*; John Wiley & Sons, Inc.: New York, NY, USA, 1946; Volume 3, pp. 240–266.
97. Guy, R.G. Syntheses and preparative applications of thiocyanates. In *Cyanates and Their Thio Derivatives*; Patai, S., Ed.; John Wiley & Sons, Ltd.: New York, NY, USA, 1977; Volume 2, pp. 819–886.
98. Jason, A.; Phadke, A.S.; Deshpande, M.; Agarwak, A.; Chen, D.; Gadhachanda, V.R.; Hashimoto, A.; Pais, G.; Wang, Q.; Wang, X. Aryl, Heteroaryl, and Heterocyclic Compounds for Treatment of Medical Disorders. WO2017035353A1, 2 March 2017.
99. Rodríguez, R.; Camargo, P.; Sierra, C.A.; Soto, C.Y.; Cobo, J.; Nogueras, M. Iodine mediated an efficient and greener thiocyanation of aminopyrimidines by a modification of the Kaufmann's reaction. *Tetrahedron Lett.* **2011**, *52*, 2652–2654. [CrossRef]

100. Reddy, B.V.S.; Reddy, S.M.S.; Madan, C. NBS or DEAD as effective reagents in α-thiocyanation of enolizable ketones with ammonium thiocyanate. *Tetr. Lett.* **2011**, *52*, 1432–1435. [CrossRef]
101. Mahajan, U.S.; Akamanchi, K.G. Facile Method for Thiocyanation of Activated Arenes Using Iodic Acid in Combination with Ammonium Thiocyanate. *Synthetic Commun.* **2009**, *39*, 2674–2682. [CrossRef]
102. Khazaei, A.; Zolfigol, M.A.; Mokhlesi, M.; Panah, F.D.; Sajjadifar, S. Simple and Highly Efficient Catalytic Thiocyanation of Aromatic Compounds in Aqueous Media. *Helv. Chim. Acta* **2012**, *95*, 106–114. [CrossRef]
103. Wu, J.; Wu, G.; Wu, L. Thiocyanation of Aromatic and Heteroaromatic Compounds using Ammonium Thiocyanate and I_2O_5. *Synthetic Commun.* **2008**, *38*, 2367–2373. [CrossRef]
104. Ali, D.; Panday, A.K.; Choudhury, L.H. Hydrogen Peroxide-Mediated Rapid Room Temperature Metal-Free C(sp2)-H Thiocyanation of Amino Pyrazoles, Amino Uracils, and Enamines. *J. Org. Chem.* **2020**, *85*, 13610–13620. [CrossRef]
105. Sajjadifar, S.; Louie, O. Regioselective Thiocyanation of Aromatic and Heteroaromatic Compounds by Using Boron Sulfonic Acid as a New, Efficient, and Cheap Catalyst in Water. *J. Chem.* **2013**, *2013*, 6. [CrossRef]
106. Khalili, D. Highly efficient and regioselective thiocyanation of aromatic amines, anisols and activated phenols with H_2O_2/NH_4SCN catalyzed by nanomagnetic Fe_3O_4. *Chin. Chem. Lett.* **2015**, *26*, 547–552. [CrossRef]
107. Khazaei, A.; Zolfigol, M.A.; Mokhlesi, M.; Pirveysian, M. Citric acid as a trifunctional organocatalyst for thiocyanation of aromatic and heteroaromatic compounds in aqueous media. *Can. J. Chem.* **2012**, *90*, 427–432. [CrossRef]
108. Songsichan, T.; Katrun, P.; Khaikate, O.; Soorukram, D.; Pohmakotr, M.; Reutrakul, V.; Kuhakarn, C. Thiocyanation of Pyrazoles Using $KSCN/K_2S_2O_8$ Combination. *SynOpen* **2018**, *2*, 6–16. [CrossRef]
109. Mete, T.B.; Khopade, T.M.; Bhat, R.G. Transition-metal-free regioselective thiocyanation of phenols, anilines and heterocycles. *Tetrahedron Lett.* **2017**, *58*, 415–418. [CrossRef]
110. Karimi Zarchi, M.A.; Banihashemi, R. Thiocyanation of aromatic and heteroaromatic compounds using polymer-supported thiocyanate ion as the versatile reagent and ceric ammonium nitrate as the versatile single-electron oxidant. *J. Sulfur Chem.* **2016**, *37*, 282–295. [CrossRef]
111. Pan, X.-Q.; Lei, M.-Y.; Zou, J.-P.; Zhang, W. $Mn(OAc)_3$-promoted regioselective free radical thiocyanation of indoles and anilines. *Tetrahedron Lett.* **2009**, *50*, 347–349. [CrossRef]
112. Das, B.; Kumar, A.S. Efficient Thiocyanation of Indoles Using Para-Toluene Sulfonic Acid. *Synth. Commun.* **2010**, *40*, 337–341. [CrossRef]
113. Zhang, H.; Wei, Q.; Wei, S.; Qu, J.; Wang, B. Highly Efficient and Practical Thiocyanation of Imidazopyridines Using an N-Chlorosuccinimide/NaSCN Combination. *Eur. J. Org. Chem.* **2016**, *2016*, 3373–3379. [CrossRef]
114. Muniraj, N.; Dhineshkumar, J.; Prabhu, K.R. N-Iodosuccinimide Catalyzed Oxidative Selenocyanation and Thiocyanation of Electron Rich Arenes. *ChemistrySelect* **2016**, *1*, 1033–1038. [CrossRef]
115. Mokhtari, B.; Azadi, R.; Rahmani-Nezhad, S. In situ-generated N-thiocyanatosuccinimide (NTS) as a highly efficient reagent for the one-pot thiocyanation or isothiocyanation of alcohols. *Tetrahedron Lett.* **2009**, *50*, 6588–6589. [CrossRef]
116. Memarian, H.R.; Mohammadpoor-Baltork, I.; Nikoofar, K. DDQ-promoted thiocyanation of aromatic and heteroaromatic compounds. *Can. J. Chem.* **2007**, *85*, 930–937. [CrossRef]
117. Memarian, H.R.; Mohammadpoor-Baltork, I.; Nikoofar, K. Ultrasound-assisted thiocyanation of aromatic and heteroaromatic compounds using ammonium thiocyanate and DDQ. *Ultrason. Sonochem.* **2008**, *15*, 456–462. [CrossRef] [PubMed]
118. Gitkis, A.; Becker, J.Y. Anodic thiocyanation of mono- and disubstituted aromatic compounds. *Electrochim. Acta* **2010**, *55*, 5854–5859. [CrossRef]
119. Fotouhi, L.; Nikoofar, K. Electrochemical thiocyanation of nitrogen-containing aromatic and heteroaromatic compounds. *Tetrahedron Lett.* **2013**, *54*, 2903–2905. [CrossRef]
120. Zhang, X.; Wang, C.; Jiang, H.; Sun, L. A low-cost electrochemical thio- and selenocyanation strategy for electron-rich arenes under catalyst- and oxidant-free conditions. *RSC Adv.* **2018**, *8*, 22042–22045. [CrossRef]
121. Kokorekin, V.A.; Yaubasarova, R.R.; Neverov, S.V.; Petrosyan, V.A. Reactivity of electrogenerated thiocyanogen in the thiocyanation of pyrazolo[1,5-a]pyrimidines. *Mendeleev Commun.* **2016**, *26*, 413–414. [CrossRef]
122. Kokorekin, V.A.; Mel'nikova, E.I.; Yaubasarova, R.R.; Gorpinchenko, N.V.; Petrosyan, V.A. "Metal-free" electrooxidative C—H thiocyanation of arenes. *Russ. Chem. Bull.* **2019**, *68*, 2140–2141. [CrossRef]
123. Kokorekin, V.A.; Yaubasarova, R.R.; Neverov, S.V.; Petrosyan, V.A. Electrooxidative C–H Functionalization of Heteroarenes. Thiocyanation of Pyrazolo[1,5-a]pyrimidines. *Eur. J. Org. Chem.* **2019**, *2019*, 4233–4238. [CrossRef]
124. Yaubasarova, R.R.; Kokorekin, V.A.; Ramenskaya, G.V.; Petrosyan, V.A. Double electrooxidative C–H functionalization of (het)arenes with thiocyanate and 4-nitropyrazolate ions. *Mendeleev Commun.* **2019**, *29*, 334–336. [CrossRef]
125. Kokorekin, V.A.; Melnikova, E.I.; Yaubasarova, R.R.; Petrosyan, V.A. Electrooxidative C–H thiocyanation of hetarenes: Voltammetric assessment of thiocyanogen reactivity. *Mendeleev Commun.* **2020**, *30*, 70–72. [CrossRef]
126. Kokorekin, V.A.; Neverov, S.V.; Kuzina, V.N.; Petrosyan, V.A. A New Method for the Synthesis of 3-Thiocyanatopyrazolo[1,5-a]pyrimidines. *Molecules* **2020**, *25*, 4169. [CrossRef] [PubMed]
127. Cataldo, F. New Developments in the Study of the Structure of Parathiocyanogen: $(SCN)_x$, An Inorganic Polymer. *J. Inorg. Organomet. Polym.* **1997**, *7*, 35–50. [CrossRef]
128. Shaabani, A.; Nazeri, M.T.; Afshari, R. 5-Amino-pyrazoles: Potent reagents in organic and medicinal synthesis. *Mol. Divers.* **2019**, *23*, 751–807. [CrossRef] [PubMed]

129. Cherukupalli, S.; Karpoormath, R.; Chandrasekaran, B.; Hampannavar, G.A.; Thapliyal, N.; Palakollu, V.N. An insight on synthetic and medicinal aspects of pyrazolo[1,5-a]pyrimidine scaffold. *Eur. J. Med. Chem.* **2017**, *126*, 298–352. [CrossRef] [PubMed]
130. Arias-Gómez, A.; Godoy, A.; Portilla, J. Functional Pyrazolo[1,5-a]pyrimidines: Current Approaches in Synthetic Transformations and Uses as an Antitumor Scaffold. *Molecules* **2021**, *26*, 2708. [CrossRef] [PubMed]
131. Sigacheva, V.L.; Kokorekin, V.A.; Strelenko, Y.A.; Neverov, S.V.; Petrosyan, V.A. Electrochemical Azolation of N-substituted Pyrroles: A New Case in $S_N^H(An)$ Reactions. *Mendeleev Commun.* **2012**, *22*, 270–272. [CrossRef]
132. Henton, D.R.; McCreery, R.A.; Swenton, J.S. Anodic oxidation of 1,4-dimethoxy aromatic compounds. A facile route to functionalized quinone bisketals. *J. Org. Chem.* **1980**, *45*, 369–378. [CrossRef]
133. Adibi, H.; Rashidi, A.; Khodaei, M.M.; Alizadeh, A.; Majnooni, M.B.; Pakravan, N.; Abiri, R.; Nematollahi, D. Catecholthioether Derivatives: Preliminary Study of in-Vitro Antimicrobial and Antioxidant Activities. *Chem. Pharm. Bull.* **2011**, *59*, 1149–1152. [CrossRef]
134. Kokorekin, V.A.; Solomatin, Y.A.; Gening, M.L.; Petrosyan, V.A. Acid-catalyzed $S_N^H(An)$ thiolation of p-dihydroxybenzene. *Mendeleev Commun.* **2017**, *27*, 586–588. [CrossRef]
135. Chen, C.; Niu, P.; Shen, Z.; Li, M. Electrochemical Sulfenylation of Indoles with Disulfides Mediated by Potassium Iodide. *J. Electrochem. Soc.* **2018**, *165*, G67–G74. [CrossRef]
136. Petrosyan, V.A.; Burasov, A.V. Anodic azolation of 1,2- and 1,3-dimethoxybenzenes. *Russ. Chem. Bull.* **2010**, *59*, 522–527. [CrossRef]
137. Feng, P.; Ma, G.; Chen, X.; Wu, X.; Lin, L.; Liu, P.; Chen, T. Electrooxidative and Regioselective C−H Azolation of Phenol and Aniline Derivatives. *Angew. Chem. Int. Ed.* **2019**, *58*, 8400–8404. [CrossRef]
138. Wang, J.-H.; Lei, T.; Nan, X.-L.; Wu, H.-L.; Li, X.-B.; Chen, B.; Tung, C.-H.; Wu, L.-Z. Regioselective Ortho Amination of an Aromatic C−H Bond by Trifluoroacetic Acid via Electrochemistry. *Org. Lett.* **2019**, *21*, 5581–5585. [CrossRef] [PubMed]
139. Ananikov, V.P.; Eremin, D.B.; Yakukhnov, S.A.; Dilman, A.D.; Levin, V.V.; Egorov, M.P.; Karlov, S.S.; Kustov, L.M.; Tarasov, A.L.; Greish, A.A.; et al. Organic and hybrid systems: From science to practice. *Mendeleev Commun.* **2017**, *27*, 425–438. [CrossRef]
140. Hunger, K. *Industrial Dyes: Chemistry, Properties, Applications*; Wiley-VCH: Weinheim, Germany, 2003.
141. Sandborn, W.J. Rational selection of oral 5-aminosalicylate formulations and prodrugs for the treatment of ulcerative colitis. *Am. J. Gastroenterol.* **2002**, *97*, 2939–2941. [CrossRef]
142. Ovchinnikov, I.V.; Kulikov, A.S.; Makhova, N.N.; Tosco, P.; Di Stilo, A.; Fruttero, R.; Gasco, A. Synthesis and vasodilating properties of N-alkylamide derivatives of 4-amino-3-furoxancarboxylic acid and related azo derivatives. *Il Farmaco* **2003**, *58*, 677–681. [CrossRef]
143. Iranpoor, N.; Firouzabadi, H.; Khalili, D.; Motevalli, S. Easily Prepared Azopyridines as Potent and Recyclable Reagents for Facile Esterification Reactions. An Efficient Modified Mitsunobu Reaction. *J. Org. Chem.* **2008**, *73*, 4882–4887. [CrossRef]
144. Iranpoor, N.; Firouzabadi, H.; Shahin, R.; Khalili, D. 2,2′-Azobenzothiazole as a New Recyclable Oxidant for Heterogeneous Thiocyanation of Aromatic Compounds with Ammonium Thiocyanate. *Synth. Commun.* **2011**, *42*, 2040–2047. [CrossRef]
145. Ovchinnikov, I.V.; Makhova, N.N.; Khmel'nitsky, L.I.; Kuz'min, V.S.; Akimova, L.N.; Pepekin, V.I. Dinitrodiazenofuroxan as a new energetic explosive. *Dokl. Chem.* **1998**, *359*, 67–70.
146. Veauthier, J.M.; Chavez, D.E.; Tappan, B.C.; Parrish, D.A. Synthesis and Characterization of Furazan Energetics ADAAF and DOATF. *J. Energ. Mater.* **2010**, *28*, 229–249. [CrossRef]
147. Firouzabadi, H.; Mostafavipoor, Z. Barium Manganate. A Versatile Oxidant in Organic Synthesis. *Bull. Chem. Soc. Jpn.* **1983**, *56*, 914–917. [CrossRef]
148. Birchall, J.M.; Haszeldine, R.N.; Kemp, J.E.G. Polyfluoroarenes. Part X. Polyfluoroaromatic azo-compounds. *J. Chem. Soc. C* **1970**, 449–455. [CrossRef]
149. Farhadi, S.; Zaringhadama, P.; Sahamiehb, R.Z. Photo-assisted Oxidation of Anilines and Other Primary Aromatic Amines to Azo Compounds Using Mercury (II) Oxide as a Photo-Oxidant. *Acta Chim. Slov.* **2007**, *54*, 647–653.
150. Huang, H.; Sommerfeld, D.; Dunn, B.C.; Lloyd, C.R.; Eyring, E.M. Ferrate(VI) oxidation of aniline. *J. Chem. Soc. Dalton Trans.* **2001**, 1301–1305. [CrossRef]
151. Chavez, D.E.; Parrish, D.A.; Leonard, P. The Synthesis and Characterization of a New Furazan Heterocyclic System. *Synlett* **2012**, *23*, 2126–2128. [CrossRef]
152. Okumura, S.; Lin, C.-H.; Takeda, Y.; Minakata, S. Oxidative Dimerization of (Hetero)aromatic Amines Utilizing t-BuOI Leading to (Hetero)aromatic Azo Compounds: Scope and Mechanistic Studies. *J. Org. Chem.* **2013**, *78*, 12090–12105. [CrossRef] [PubMed]
153. Zhang, C.; Jiao, N. Copper-Catalyzed Aerobic Oxidative Dehydrogenative Coupling of Anilines Leading to Aromatic Azo Compounds using Dioxygen as an Oxidant. *Angew. Chem.* **2010**, *122*, 6310–6313. [CrossRef]
154. Jiang, B.; Ning, Y.; Fan, W.; Tu, S.-J.; Li, G. Oxidative Dehydrogenative Couplings of Pyrazol-5-amines Selectively Forming Azopyrroles. *J. Org. Chem.* **2014**, *79*, 4018–4024. [CrossRef]
155. Takeda, Y.; Okumura, S.; Minakata, S. Oxidative Dimerization of Aromatic Amines using tBuOI: Entry to Unsymmetric Aromatic Azo Compounds. *Angew. Chem. Int. Ed.* **2012**, *51*, 7804–7808. [CrossRef]
156. Pandit, R.P.; Lee, Y.R. Construction of Multifunctionalized Azopyrazoles by Silver- Catalyzed Cascade Reaction of Diazo Compounds. *Adv. Synth. Catal.* **2015**, *357*, 2657–2664. [CrossRef]
157. Wawzonek, S.; McIntyre, T. Electrolytic oxidation of aromatic amines. *J. Electrochem. Soc.* **1967**, *114*, 1025. [CrossRef]

158. Wawzonek, S.; McIntyre, T. Electrolytic preparation of azobenzenes. *J. Electrochem. Soc.* **1972**, *119*, 1350. [CrossRef]
159. Schäfer, H.-J. Oxidation of organic compounds at the nickel hydroxide electrode. In *Electrochemistry I. Top. Curr. Chem*; Steckhan, E., Ed.; Springer: Berlin, Germany; New York, NY, USA, 1987; Volume 142, pp. 101–129.
160. Lyalin, B.V.; Sigacheva, V.L.; Kokorekin, V.A.; Petrosyan, V.A. A new synthesis of azopyrazoles by oxidation of C-aminopyrazoles on a NiO(OH) electrode. *Mendeleev Commun.* **2015**, *25*, 479–481. [CrossRef]
161. Lyalin, B.V.; Sigacheva, V.L.; Kokorekin, V.A.; Petrosyan, V.A. Electrosynthesis of azopyrazoles via the oxidation of N-alkylaminopyrazoles on a NiO(OH) anode in aqueous alkali—A green method for N-N homocoupling. *Tetrahedron Lett.* **2018**, *59*, 2741–2744. [CrossRef]
162. Lyalin, B.V.; Sigacheva, V.L.; Petrosyan, V.A. Electrocatalytic N—N cross-coupling of N-alkyl-3-aminopyrazoles at a nickel anode. *Russ. Chem. Bull.* **2020**, *69*, 2020–2021. [CrossRef]
163. Lyalin, B.V.; Sigacheva, V.L.; Kokorekin, V.A.; Dutova, T.Y.; Rodionova, G.M.; Petrosyan, V.A. Oxidative transformation of N-substituted 3-aminopyrazoles to azopyrazoles using electrogenerated bromine as a mediator. *Russ. Chem. Bull.* **2018**, *67*, 510–516. [CrossRef]

164. Lyalin, B.V.; Sigacheva, V.L.; Kokorekin, V.A.; Petrosyan, V.A. Oxidative conversion of N-substituted 3-aminopyrazoles to azopyrazoles using electrogenerated NaOCl as the mediator. *Arkivoc* **2017**, *2017*, 55–62. [CrossRef]
165. Lyalin, B.V.; Sigacheva, V.L.; Ugrak, B.I.; Petrosyan, V.A. Oxidative N—N coupling of N-alkyl-3-aminopyrazoles to azopyrazoles in aqueous solutions of NaOCl and NaOBr. *Russ. Chem. Bull.* **2021**, *70*, 164–170. [CrossRef]

MDPI
St. Alban-Anlage 66
4052 Basel
Switzerland
Tel. +41 61 683 77 34
Fax +41 61 302 89 18
www.mdpi.com

Molecules Editorial Office
E-mail: molecules@mdpi.com
www.mdpi.com/journal/molecules

www.ingramcontent.com/pod-product-compliance
Lightning Source LLC
LaVergne TN
LVHW070248100526
838202LV00015B/2194